Frugal Franklin's
Principles of Anatomy and Physiology I
Essential Texts and Review

AUTHORS

Phillip Y.P. Jen, Ph.D.
Associate Professor
Gordon State College

Amanda L.J. Duffus, Ph.D.
Assistant Professor
Gordon State College

CONTRIBUTORS

Donna Jen, M.Ed.
Cathy Lee, Ph.D.
Wyatt T. Cramblet
Chante' Venette Domenech
Abigail Turner
Aliyah Hugh
Amanda Austin
Caitlin Rentz
Caylee Wilson
Cori Harding
Crystal Wynne
Josue Padilla
Katherine Taing
Katie Dixon
Kristin Walker
Marquisha Bussey
Nicole Carroll
Priscilla Brown
Regina D. Schaeffer
Sydney Travers
Sarah Hegedus
Sarah Phillips
Taylor Eubanks
Victoria Grant

Front cover: Image ID: csp8162847. Designed and released by Pixelchaos (Author ID 25R557385G309690X) on December 22, 2011. Reprint license purchased through Can Stock Photo Inc. 6139 Quinpool Rd., Suite 33027 Halifax, NS, B3L 4T6 Canada. Transaction ID: 9494662

Back cover (top): This image is acquired from a unit of currency issued by the United States of America. Public domain graphics

Back cover (bottom): Human cranial muscles. Original art work by Jeremy Ashley at Just the Arts Creative Firm

Principles of Anatomy and Physiology I

Published by Frugal Franklin™, a business unit of Asclepius Research LLC. Copyright© 2015 by Phillip Y.P. Jen PhD. and the Asclepius Research LLC. All rights reserved. No part of this publication may be reproduced or distributed in any form or by any means, or stored in a database or retrieval system, without the prior written consent of the Asclepius Research LLC., Including, but not limited to, in any network or other electronic storage or transmission, or broadcast for distance learning.

ISBN-10: 0692503811
ISBN-13: 978-0692503812 (Phillip Y.P. Jen)

Library of Congress Control Number: 2015912700
CreateSpace Independent Publishing Platform
North Charleston, South Carolina

TABLE OF CONTENTS

Chapter	Subchapter	Section	Page Number
1	LEVELS OF ORGANIZATION		1-2
	HOMEOSTAISIS		2
	ANATOMICAL POSITION AND PLANE		2
	ANATOMICAL AREAS		2-3
	ANATOMICAL DIRECTIONAL TERMS		3-5
		Abdominal Quadrants	5
	QUADRANT DIFFERENTIAL DIAGNOSIS		5-7
		Abdominal Regions	7
	BODY CAVITIES		7-8
	NURING CARE		8
		Auscultating Assessment	8-10
		Neuromuscular and Vascular Assessment	10
		Vital Signs	10-11
		Nursing Process	11
		Complication of Immobility	11
		Integumentary	11
		Elimination	11
		Urinary	11-12
		Respiratory	12
		Cardiovascular	12
	QUESTIONS		12-16
	ANSWERS		16
	REFERENCES		16
	PHOTO AND GRAPHIC BIBLIOGRAPHY		16-17
2	ELEMENTS AND SUBATOMIC PARTICLE		18
	ISOTOPES		18-19
	NUCLEAR MEDICINE		19-20
	OCTET RULE		20-21
	CHEMICAL BONDING		21
		Ionic Bond	21
		Non-Polar Covalent Bond	22
		Polar Covalent Bond	22-23
		Hydrogen Bond	23
	ENERGY		23
	ACID AND BASES		23-24
	pH SCALE		24
	ACIDOSIS AND ALKALOSIS		24-25
	ORGANIC MOLECULES		24-25
		Carbohydrates	26
		Lipids	26-27
		Fats and Health	27-28
		Steroids	28
		Proteins	28-29
		Genetic Material	29
		Deoxyribonucleic Acid	29-30
		Ribonucleic Acid	30-31
		Transcription	31-32
	TRANSCRIPTION AND TRANSLATION		32
	SICKLE CELL ANEMIA		32-33
	ADENOSINE TRIPHOSPHATE		33-34
	QUESTIONS		34-39
	ANSWERS		39
	REFERENCES		39
	PHOTO AND GRAPHIC BIBLIOGRAPHY		39
	APPENDIX I		40
	APPENDIX II		41
	APENDIX III		42
3	EUKARYOTIC CELLS		43
		Plasma Membrane	43-45
		Cellular Junctions	45-47
		Membrane Transport Proteins	47
		Receptor Proteins	47-50
		Membrane Assisted Transport	50-51
	ORGANELLES		51
		Nucleus	51-52
		Ribosome	52-53
		Mitochondrion	53

Chapter	Subchapter	Section	Page Number
		Endoplasmic Reticulum	53-54
		Golgi Apparatus	54
		Intracellular Lipid Transport	54-55
		Lysosomes	55
		Lysosomal Storage Disease	55
		Peroxisomes	55
		Cytoskeleton	55-56
		Centrosomes	56
		Cilia and Flagella	56
		Microvilli	56-57
	QUESTIONS		57-64
	ANSWERS		64
	REFERENCES		64
	PHOTO AND GRAPHIC BIBLIOGRAPHY		64-65
4	EPITHELIAL TISSUE		66-68
		Simple Squamous Epithelium	68
		Simple Cuboidal Epithelium	68
		Simple Columnar Epithelium	68-69
		Pseudostratified Columnar Epithelium	69
		Stratified Squamous Epithelium	69
		Stratified Cuboidal Epithelium	69
		Stratified Columnar Epithelium	69
		Transitional Epithelium	69-70
	MUSCLES		70
		Skeletal Muscle	70
		Cardiac Muscle	70
		Smooth Muscle	70-71
	CONNECTIVE TISSUE		71
		Extracellular Matrix	71-72
		Marfan Syndrome	72
	GLANDS		72-73
	CONNECTIVE TISSUE CELLS		73
		Fixed Cells	73-74
		Wandering Cells	74-75
	TYPES OF CONNECTIVE TISSUE		75
		Embryonic Connective Tissues	75
		General Connective Tissue	75-76
		Specialized Connective Tissue	76-77
	INTEGUMENT SYSTEM		78-80
	PRESSURE ULCERS		80-81
		Staging	81
		Assessment and Intervention of Wounds	81-82
		Wound Complications	82
		Sterile Technique and Standard Precautions	82-83
		Burns	83-86
		Hair	86
		Nails	86-87
		Glands of the Skin	87
	TEMPERATURE REGULATION		87-88
	QUESTIONS		88-95
	ANSWERS		95
	REFERENCES		95-96
	PHOTO AND GRAPHIC BIBLIOGRAPHY		96
5	SKELETAL SYSTEM		97
	HISTOLOGY OF BONES		97
		Bone Cells	97-98
		Bone Matrix	99
	CLASSIFICATION OF BONES		99
		Woven and Lamellar Bone	99
		Cancellous and Compact Bone	99-101
		Shapes of Bones	101-102
	BONE DEVELOPMENT		102
		Intramembranous Ossification	102
		Endochondrial Ossification	102-104
		Bone Growth in Length	104-105
		Bone Growth in Width	105
		Osteoporosis	106
	BONE FRACTURES		106-108

Chapter	Subchapter	Section	Page Number
	BONE REPAIR		108-109
	SKELETAL SYSTEM		109
		Bone injury	109
		Casting	109-110
		External and Internal Fixator	110
		Hip Fracture	110-111
		Hip Replacement	111-112
		Pelvic Fracture	112
		Total Knee Replacement	112
		Amputation	112
		Bone Disorders	112
		RICE	112-113
		Morse Fall Scale	113
		Pediatric Fractures	113
	QUESTIONS		113-120
	ANSWERS		120
	REFERENCES		120
	PHOTO AND GRAPHIC BIBLIOGRAPHY		120
6	MUSCULAR SYSTEM		121
		Muscle Fibers	121
		Myofilaments, Regulatory and Structure Proteins	121-122
		Troponin Test	122-123
		Sarcomere	123-124
		Sliding Filament Theory	124-125
	SKELETAL MUSCLE		125-126
		Neuromuscular Junctions	126-127
		Membrane Potential	127-129
		Skeletal Muscle Excitation-Contraction Coupling	129
		Polarization	129
		Depolarization	129-130
		Repolarization	130-131
		Muscle Relaxation	131
		Length-Tension relationship	131-132
		Muscle Twitch	132
		Strength of Muscle Contraction	132-133
		Treppe	133
		Types of Muscle Contractions	133-134
	AEROBIC RESPIRATION		134-135
	ENERGY MOLECULES		135-136
	TETANY		136-137
	ANAEROBIC RESPIRATION		137
	TYPES OF SKELETAL MUSCLE FIBERS		137
	MYOGLOBIN TEST		137-138
	SMOOTH MUSCLE		138
	QUESTIONS		138-146
	ANSWERS		146-147
	REFERENCES		147
	PHOTO AND GRAPHIC BIBLIOGRAPHY		147
7	NERVOUS SYSTEM		148
		Nerve Cell Structure	148-149
		Axonal Transport	149-151
		Neurotransmitters	151-153
		Neuron Types	153
		Synapses	153-154
	NEUROGLIAL CELLS		154
		Astrocytes	154-156
		Ependymal Cells	156
		Oligodendrocytes	156
		Microglial Cell	156-157
		Schwann Cells	157-158
		Satellite Glial Cells	158
	ELECTRICAL PHYSIOLOGY		158-159
		Local Potential	159-160
		Action Potential	160
	SUMMARY OF ACTION POTENTIAL FORMATION		160
		Polarization	160
		Depolarization: Local Potential	161-162
		Depolarization: Action Potential	162-163

Chapter	Subchapter	Section	Page Number
		Repolarization	163
		nhibiting Action Potential Formation	163-164
	QUESTIONS		164-171
	REFERENCES		171-172
	REFERENCES		172
	PHOTO AND GRAPHIC BIBLIOGRAPHY		172
8	CENTRAL NERVOUS SYSTEM		173
		Cerebrum	173-174
		Cerebral Cortex	174-175
		Basal Ganglia	175-176
	LIMBIC SYSTEM		176-177
	DIENCEPHALON		177-179
	BRAIN STEM		179-181
		Reticular Formation	181-182
	CEREBELLUM		182
	SPINAL CORD		182-184
		Spinal Tracts	184
	MENINGES		184-185
	CEREBRAL SPINAL FLUID		185-186
	PERIPHERAL NERVOUS SYSTEM		186-187
		Cranial Nerves	187-188
	SPINAL NERVES		188-189
		Cervical Plexus	189
		Brachial Plexus	189-190
		Lumbar Plexus	190
		Sacral Plexus	190-191
		Coccygeal Plexus	191
	AUTONOMIC NERVOUS SYSTEM		191
		Sympathetic Division	191
		Parasympathetic Division	191-192
		Reflex	192-193
	QUESTIONS		193-208
	ANSWERS		208
	REFERENCES		208
	PHOTO AND GRAPHIC BIBLIOGRAPHY		208-209
9	SENSES		210
		Sensory Perception	210
		Sensory Receptors	210
		Sensory Receptor Classifications	210-211
	SIMPLE RECEPTORS		211
		Unencapsulated Thermoreceptors	211
		Unencapsulated Nociceptors	211-212
		Unencapsulated Pressure Receptors	212
		Encapsulated Pressure Receptors	212-213
		Somatosensory Projection Pathways	213-215
	SENSE ORGANS		215
		Taste	215-216
		Taste Receptors	216
		Types of Taste Buds	216-217
		Olfactory Sense	217-218
		Odor Receptors	218
	HEARING AND BALANCE		
		Anatomy of the Ear	218-221
		Hair Cells	221
		Physiology of Balance	221-223
		Hearing	223
		Physiology of Hearing	223-225
	SIGHT		225-226
		Accessory Structures of the Eyes	226-227
		Anatomy of the Eyes	227-228
		Other Vision Apparatus	228-229
		Neuronal Apparatus	229
		Retina	229-231
		Stargardt Disease	231-232
		Action Potential Formation	232-233
	QUESTIONS		233-247
	ANSWERS		247
	REFERENCES		247
	PHOTO AND GRAPHIC BIBLIOGRAPHY		248

Anatomy is the study of structure of the body. The term anatomy is derived from the Greek word ἀνατέμνω (pronounce anatemnō) which is defined as "to dissect or cut open." The anatomical discipline is comprised of many subsidiary fields, which include the gross analysis of anatomical parts (gross anatomy), microscopic organization (cytology and histology), and developmental anatomy.

Physiology, on the other hand, is the study of the functions of a living organism and its chemical and physical processes. The term physiology is derived from the Greek word φύσις (pronounce phusis), which is defined as the examination of nature and its origin. The physiological discipline is also comprised of numerous supplementary fields, which include the examination of animal physiology, human physiology, cellular physiology, systemic physiology, and the examination of bodily functions in a diseased state, pathophysiology.

LEVELS OF ORGANIZATION

In order to gain a better understanding of the anatomical and physiological construction of an organism, we must first recognize the levels of organization that form its basic structure. For example, at the most rudimentary level is the atom. The atom is the basic unit of matter, which in turn, is defined as anything that takes up space and has mass. The atom is composed of a dense central core comprised of positive charged particle(s) called proton(s), and neutron(s) which possess no electrical charge, but has a mass slightly greater than the proton. Circling the nucleus in orbit(s) is (are) negatively charged subatomic particle(s) called electron(s). It is understood that electrons are bound in their specific orbits by an electromagnetic force generated by the nucleus. The mass of the electron is considered to be insignificant, as it has calculated mass of approximately 1/1836 of the proton (Figure 1).

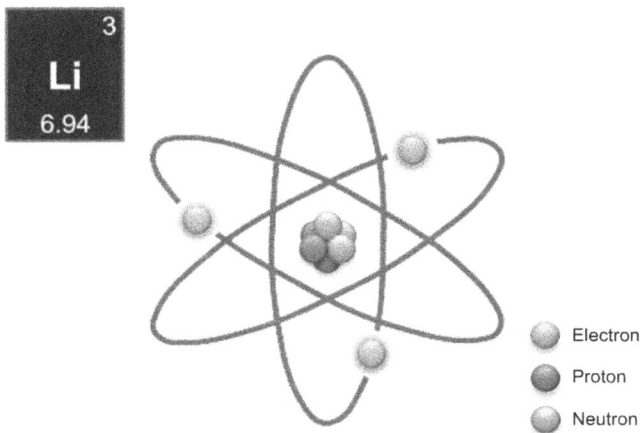

● Electron

● Proton

● Neutron

Figure 1: Illustration of a lithium (Li) atom. Please note that the atomic number for this element is 3 which equates to 3 protons, while the atomic weight is 6.94, which equates to 4 neutrons located within the nucleus

It is the interaction between atoms that create either homonuclear molecules or chemical compounds. A homonuclear molecule is composed of a single element. On the other hand, a chemical compound is a molecule composed of more than one type of element. For example, a homonuclear molecule of hydrogen or oxygen is composed of two hydrogen or two oxygen atoms respectively. Chemical compounds, on the other hand, are much more complex and involve various elements in unique combinations. Molecules such as water, proteins, lipids, carbohydrates, deoxyribose nucleic acids (DNA), etc. are found in all living organisms on earth.

The next level of organization is the cell. The cell is the smallest living organism and is the basic building block of all multicellular organisms. The cell is constructed through the interactions of millions of molecules with DNA serving as the blue-print from which they are derived.

There are two groups of cells that exist on earth. The first group consists of prokaryotic cells and is considered to be the simplest, as well as, being the most ancient (radioactive dating: first appeared on earth ~3.5 billion years ago) of all organisms. Prokaryotic organisms consist of two groups: Bacteria and Achaea. The second group consists of more modern organisms known as eukaryotic cells (radioactive dating: suggests they first appeared on earth ~2.0 billion years ago). Eukaryotic cells can exist in unicellular forms, such as the free living amoeba, or in multicellular form, such as the human body. For example, it is estimated that the human body is composed of approximately 73 trillion eukaryotic cells, working together to respond to the environment, transforming energy, and reproduction, as well as, maintaining homeostasis.

The level of organization above cellular level is the tissues and they are composed of interacting eukaryotic cells with similar functions. Tissues are defined as a collective of interacting cells with a similar embryonic origin. It is understood that the type of cells that makes up the tissue determines its function within a living organism. In the human body, there are four primary tissue types: epithelial tissue, connective tissue, muscular tissue and nervous tissue.

When two or more tissue types interact an organ is formed and it performs specific functions. Organs are typically found in functional groups called organ systems. Organ systems maintain the basic functions of a living organism (homeostasis). For example, the heart is an organ that is composed of simple epithelial tissues, connective tissues, and thick layers of cardiac muscular tissue. The heart is basically an organic pump that provides positive pressure to push a fluid tissue, called blood, through numerous interconnecting pipes and their braches, called blood vessels. All together, the heart, along with blood vessels, and other related tissues form the circulatory system also known as the cardiovascular system.

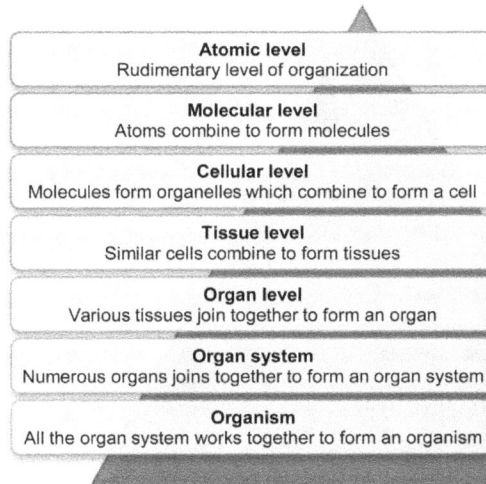

| **Atomic level** |
| Rudimentary level of organization |
| **Molecular level** |
| Atoms combine to form molecules |
| **Cellular level** |
| Molecules form organelles which combine to form a cell |
| **Tissue level** |
| Similar cells combine to form tissues |
| **Organ level** |
| Various tissues join together to form an organ |
| **Organ system** |
| Numerous organs joins together to form an organ system |
| **Organism** |
| All the organ system works together to form an organism |

Figure 2: Levels of organization

Most animals, including humans, possess ten organ systems. For example, the integumentary system, the skin, provides protection from abrasion, regulates temperature, and prevents uncontrollable water gain and water loss. The musculoskeletal system provides protection, support and movement. The respiratory system includes the trachea, bronchi and alveoli which provides the necessary location for gas exchange. The digestive system is involved in the breakdown and absorption of nutrients, minerals, and water, as well as the subsequent excretion of wastes. The circulatory system is involved in the distribution of nutrients, minerals and water, and removes wastes from the tissues. The urinary system is a filtration system for the circulatory system, which is involved in the removal of excess water and waste materials from the blood stream. The immune system mostly operates within the circulatory system and provides the necessary protection against viruses, bacteria, or other foreign entities that attempt to invade the organism. The reproductive system produces sperm or eggs, which are necessary for the propagation of the species. The endocrine system is involved in the production and secretion of hormones, which regulate the functions of internal organs, as well as, maintaining homeostasis. Finally, the nervous system is the control center of the organism and it is involved in the coordination of movements, organ functions, reception of stimuli and memory formation (Figure 2).

HOMEOSTAISIS

Homeostasis is defined as the maintenance of a constant environment for the ~73 trillion cells in the human body. For the cells to operate normally, the variables such as salinity, fluid volume, temperature, and nutrient concentration, to name a few, must be maintained within a narrow range at all times. Any disruption of this homeostasis may result in disease and at times death.

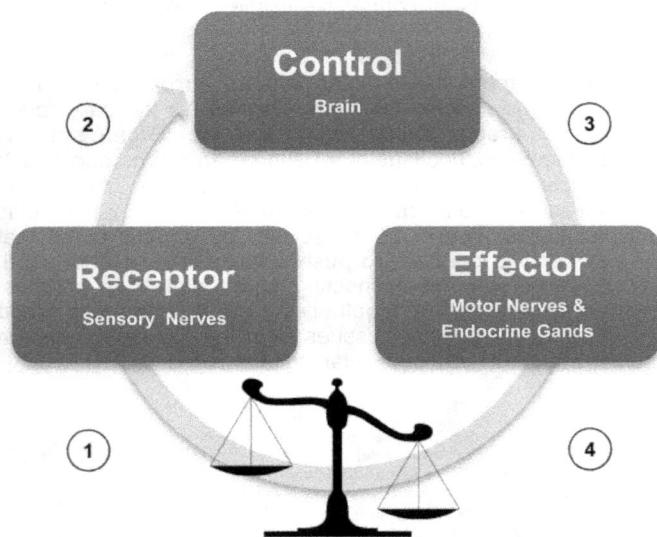

Figure 3: A negative feedback system where any deviation from the norm is counteracted and brought back to homeostasis. ① An imbalance is detected by the sensory receptors and this information is ② sent to the brain via sensory nerves (afferents). ③ The brain analyzes the physiological condition and sends commands via efferent nerves to the effector organs ④ which in turn restores homeostasis

In the body, homeostasis of most systems is maintained by a negative feedback system, coordinated by two major organ systems: nervous and endocrine. Negative feedback is defined as a type of internal regulation mechanism where any deviation from the norm is immediately brought back to the preset homeostatic levels. Many negative feedback systems possess components that are involved in detection, analysis and regulation of internal environmental conditions. The first component is the receptor that monitors the values of the variable. For example, sensory receptors monitor the external and systemic conditions and send signals to the brain, where the information is analyzed and assessed. Subsequently, if a deviation occurs, the regulatory center of the brain uses the effectors to activate the endocrine system to secrete hormones or uses motor nerves to initiate counteraction to bring the body back to homeostasis (Figure 3).

ANATOMICAL POSITION AND PLANE

Anatomical positions are the universally accepted "standard" positions of the human body. Anatomical positions are used by medical professionals and scientists when describing a particular area of the human body and are essential in diagnosis. In a proper anatomical position, the human body is standing erect with the feet slightly (~1 foot) apart. The head and toes of the individual should be facing forward, while their arms should be hanging by the side of the torso. The hands of the individual should be turned with their palms facing forward. Anatomical positions can also be used when a person is supine (lying face upward) and prone (lying face down) (Figure 4).

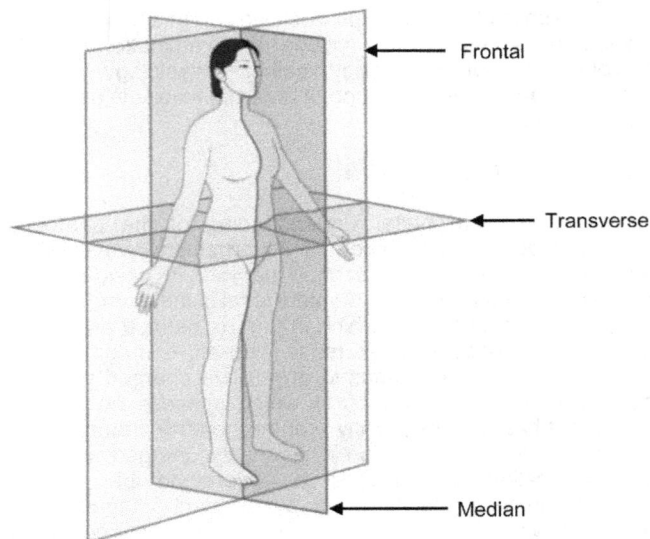

Figure 4: Accepted anatomical position and anatomical planes

Once the subject is in an anatomical position, imagine the vertical and horizontal planes transecting through the torso. This imaginary plane is a two-dimensional surface (length and width). These planes are essential to locate or describe the location of structures in the body. For example, these planes are used in medical imaging, such as computed tomography (CT) scan, positron emission tomography (PET scans) and magnetic resonance imaging (MRI), where the scans take pictures of the body in flat planar slices.

These anatomical planes are the best way to describe any body part of interest or an entire body. Understanding body planes will make it easier to study anatomy. The various planes will help in visualizing spatial and positional locations of anatomical structures and navigate directionally from one area to another.

There are three basic reference planes used in anatomy and the medical sciences. A median plane, also known as the sagittal plane, is a plane perpendicular to the ground and divides the body into right and left portions. A frontal or coronal plane, divides the body into anterior and posterior portions. Finally, the transverse plane, also known as the axial plane or cross-section, divides the body into cranial and caudal portions (Figure 4). The oblique plane (not shown on Figure 4) is an angled (slanted or diagonal) plane that is relative to the transverse and median planes being described.

ANATOMICAL AREAS

The surface anatomy of the human body provides a wealth of visible landmarks. In general, the human body is divided into two major areas. The first is known as the axial region, which relates to the head, neck and trunk (also known as the axis) of the body. The second is known as the appendicular area, and pertains to the limbs and various other attachments to the axis. The two described areas of the human body are further divided into sections. In anatomy, the human body is divided into eight sections. For example, when an individual is standing in the anatomical position, the cephalic section (cranial section), also known as the head, is located at the top of the body and is visible either anteriorly or posteriorly. The cervical section (neck) starts below the head, ends at the thorax, and is visible either anteriorly or posteriorly. The thoracic section (chest) is visible anteriorly and is located immediately below the cervical section, beginning at the clavicles, and ends along the bottom of the ribcage. The abdominal section (stomach region) is found between the distal aspects of the ribcage and extends to the hips, this section is visible anteriorly. The pelvic section is located between the hips and the thighs, while the perineum is located between the thighs and is not readily visible when an individual is standing in an anatomical position. The upper extremity sections include the shoulders, arms, forearms, elbows, wrists, and hands. This section is visible both anteriorly and posteriorly. Finally, the lower extremity sections include the hips, buttocks, thighs, knees, calves, ankles, and feet. The buttocks are visible only posteriorly, while the rest of the lower extremity sections are visible both anteriorly and posteriorly (Tables 1 and 2; Figure 5).

ANATOMICAL DIRECTIONAL TERMS

Anatomical directional terms are used to describe the locations of body parts or locations of the body. Proper terminology provides a method of communication that prevents confusion when identifying various structures. For example, right and left are retained in medical terminology, however, other terms such as up and down are replaced by superior (or cephalic) and inferior (or caudal), respectively. For example, the umbilicus is superior to the genital while inversely the genital is inferior to the umbilicus.

In anatomical directional terms, proximal means nearest, while distal describes away or distant. For example, the brachia is proximal to the torso of the body, while the hand is distal (in comparison) to the torso of the body.

The term front is replaced by anterior (or ventral in animals) and rear is replaced by posterior (or dorsal in animals). While describing a structure that is at the midline of the body, the term median is used. For example, the umbilicus is a median structure.

When describing a structure that is towards the midline or away from the sides, the term used is medial. In contrast, when describing a structure that is towards the sides or away from the midline, the term used is lateral. For example, the torso is medial to the limbs while the limbs are lateral to the torso. Please examine Figure 6 and Table 3 for the illustration and examples of directional terms.

When describing a structure in relation to the surface of the skin, the terms to use are superficial, intermediate, or deep. For example, the arrector pili muscles are superficial to the brachioradialis muscle. The brachioradialis muscle is intermediate between the skin and the radius (bone of the forearm), and finally, the radius

Table 1: Posterior anatomical regions

External Anatomy	Location on the Human Body	External Anatomy	Location on the Human Body
Acromial	Point of the shoulder	Lumbar	Between the ribs and hips
Antebrachial	Forearm	Manus	Hand
Axillary	Armpit	Nuchal	Back of the neck
Brachial	Upper arm	Occipital	Lower back of the skull
Calcaneal	Heel of the foot	Olecranal	Posterior aspect of the elbow
Carpal	Wrist	Otic	Ear
Cephalic	Head	Perineal	Area between anus & genitals
Cervical	Neck	Plantar	Sole of the foot
Cranial	Head	Popliteal	Back of the knee
Coxal	Tail bone	Sacral	Between the hips
Cubital	Elbow	Scapular	Scapula shoulder blade
Dorsum	Back of the hand	Sural	Calf
Femoral	Thigh	Tarsal	Ankle
Gluteal	Buttocks	Thoracic	Chest
Interscapular	On the two sides of the spine	Vertebral	Spinal column

Table 2: Anterior anatomical regions

External Anatomy	Location on the Human Body	External Anatomy	Location on the Human Body
Abdominal	Anterior body trunk inferior to ribs	Hallux	Great toe
Acromial	Point of the shoulder	Inguinal	Groin
Antebrachial	Forearm	Mammary	Breast
Antecubital	Anterior surfaces of the elbow	Manus	Hand
Axillary	Armpit	Mental	Chin
Brachial	Arm	Nasal	Nose
Buccal	Cheek	Oral	Mouth
Carpal	Wrist	Orbital (orbit)	Eye socket
Cervical	Neck region	Otic	Ear
Clavicular	Clavicle	Palmar	Palm of the hand
Coxal	Hip	Patellar	Anterior knee – kneecap
Cranial	Head	Pectoral	Chest
Crural	Between knee and ankle	Pedal	Foot
Cubital	Elbow	Pelvic	Pelvis
Digital	Finger or toe	Pollex	Thumb
Dorsum	Top of the foot	Pubic	Genital
Facial	Face	Sternal	Breast bone
Femoral	Thigh	Tarsal	Ankle
Fibular (peroneal)	Lateral side of the lower leg	Thoracic	Chest
Frontal	Forehead	Umbilical	Navel

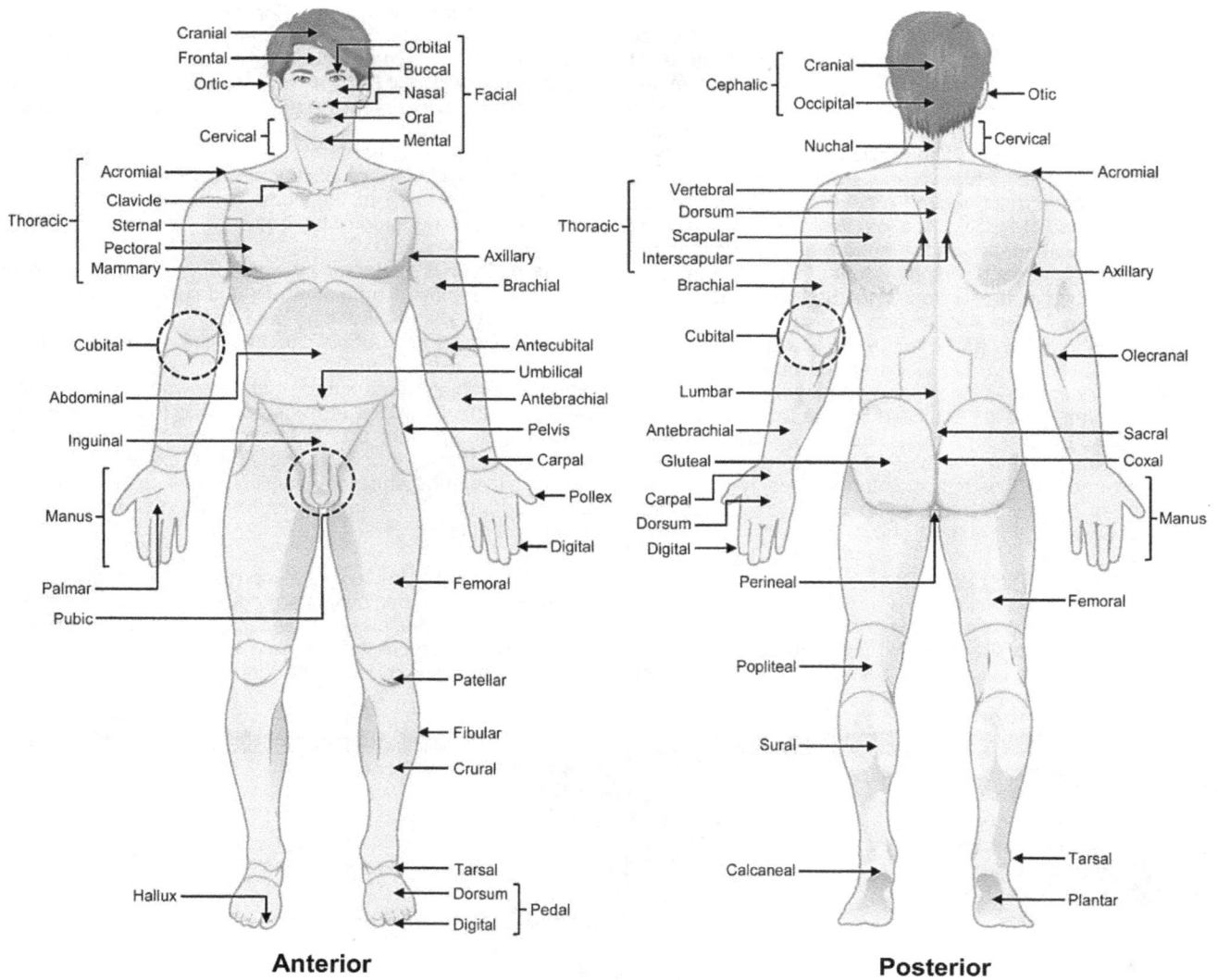

Anterior | **Posterior**

Figure 5: Graphic illustration of anatomical regions of the human body

is deep to the skin (Table 3).

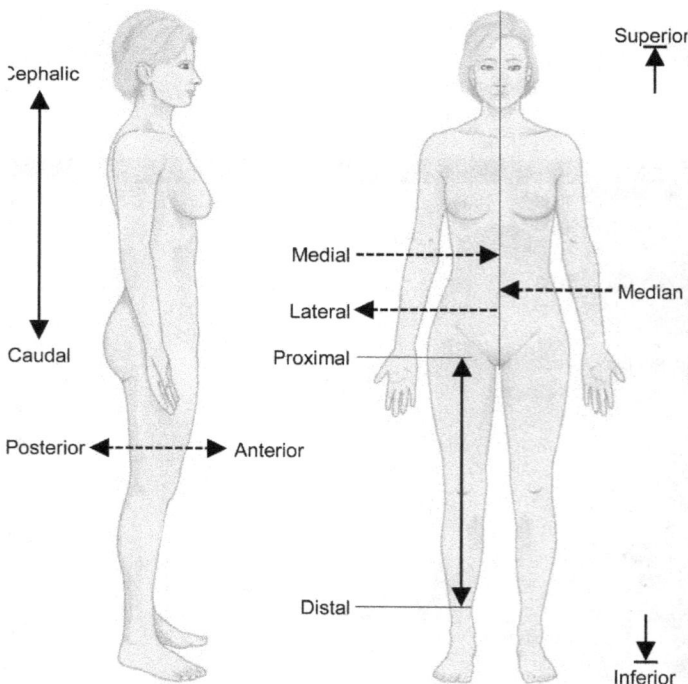

Figure 6: Graphic illustration of directional terminology

Table 3: Body orientation terminology and definitions

Directional Terms	Definition
Superior	Above
Inferior	Below
Anterior	In front of / most forward
Posterior	Back of or backside of the body
Medial	Towards the midline or middle
Lateral	Away from the midline (side)
Cephalad	Towards the head
Cranial	Towards the head
Caudal	Towards the backside of the body
Proximal	Closer to the trunk of the body
Distal	Further away from the trunk of the body
Superficial	Towards the surface of the body
Intermediate	In between
Deep	Away from the surface of the body
Dorsal	Backside of animals
Ventral	Front or belly side of animals
Unilateral	On one side
Bilateral	On both sides
Ipsilateral	On the same side
Contralateral	On opposite sides

When describing structures that are located to the sides of the body, the terms to use are unilateral, bilateral, ipsilateral, and contralateral. For example, the term unilateral is used to describe a structure that is located only on one side of the body, while bilateral is use to describe structures that exist on both sides of the body. The term ipsilateral is used to describe structures that are located on the same side of the body, while contralateral is used to describe structures that are located on the opposite side of the body (Table 3).

Abdominal Quadrants

Out of all the eight bodily regions defined in this text, the abdominal region, being the most widely described during diagnosis, requires further elucidation. For diagnostic purposes, the abdominal region is divided into four quadrants (Right Upper Quadrant, Right Lower Quadrant, Left Upper Quadrant, and Left Lower Quadrant) by drawing an imaginary line vertically and horizontally though the umbilicus (Figure 7). The abdominal quadrants are often used to describe sites of discomfort or pain during diagnosis.

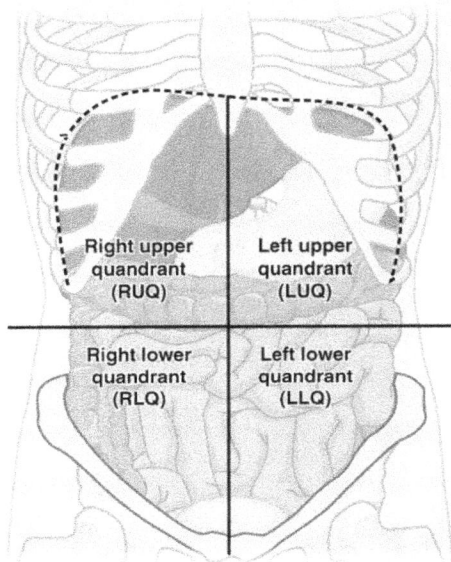

Figure 7: Illustration of quadrants

The right upper quadrant (RUQ) includes right lobe of the liver, gallbladder, head of the pancreas, a segment of the ascending and transverse colon and pylorus (region of the stomach that connects to the duodenum, the first segment of the small intestines). In addition, this quadrant also includes the hepatic flexure (the sharp bend between the ascending and the transverse colon), upper lobe of the right kidney, right adrenal gland and portion of the duodenum.

The right lower quadrant (RLQ) includes the cecum, section of the ascending colon, part of the duodenum, jejunum, and ilium. Additionally, this quadrant also includes the appendix, a portion of the extended bladder, right ureter, right spermatic cord, right ovary, right fallopian tube, and the lower lobe of the right kidney. In pregnant women, the RLQ will also include the enlarged uterus.

The left upper quadrant (LUQ) includes the left lobe of liver, stomach, a section of the transverse and descending colon. In addition, this quadrant also includes the splenic flexure, pancreas, upper lobe of the left kidney, left adrenal gland, and spleen.

The left lower quadrant (LLQ) includes portions of the duodenum, jejunum and ilium, left ureter, sigmoid colon, a segment of the descending colon, and a portion of distended bladder. Additionally, the LLQ includes the lower lobe of left kidney, left ureter, left spermatic cord, left ovary and left fallopian tube. In pregnant women, the LLQ will also include the enlarged uterus.

QUADRANT DIFFERENTIAL DIAGNOSIS

Abdominal pains generally represent a myriad of possible conditions ranging from benign ailments, to surgical emergencies. For physicians, the particular quadrant which the pain emanates can be useful during differential diagnosis. Put simply, the physician utilizes the pain or physical discomfort described by the patient to narrow the prognostic possibilities. Please note that in addition to the information acquired through physical examination of the abdominal quadrants, patient history (including information regarding patient's sex, age and generally status of their health), laboratory analysis and imaging studies (X-rays, CT scan, etc.) would be necessary to formulate a more accurate diagnosis.

Table 4: Probable cause of pains originated from LUQ

Organs/Structure	Causes
Diaphragm & peritoneum	Intra-abdominal bleeding
	Subphrenic abscess (disease caused by an accumulation of fluids between the diaphragm, liver and the spleen
	Pericolic abscess (abscess formation just outside the colon)
	Perforated peptic ulcer
	Pressure formed just following laparoscopy
	Peritonitis
Thoracic region	Angina (chest discomfort due to poor circulation of the heart)
	Empyema (infections of the pleura)
	Myocardial infarction
	Pneumonia of the left lower lobe of the lung
	Inflammation of the pleura (also known as pleurisy)
Gastrointestinal	Diverticular disease (inflammation of the haustra of the colon)
	Diverticulitis (inflammation of small pouches called diverticula in the intestines)
	Gastritis (inflamed stomach)
	Hiatal hernia (stomach sticks upward into the chest, through an opening in the diaphragm)
	Inflammatory bowel disease
	Ischemic colitis (inflammation of the large intestine result from inadequate blood flow)
	Colon tumor/cancer
	Irritable bowel syndrome
	Severe constipation or fecal impaction (hardened stool that can't be eliminated)
Abdominal/pelvic region	Aortic aneurysm
	Ruptured spleen,
	Splenic infarction (spleen necrosis due to lack of blood flow)
	Acute splenic sequestration (blockage of blood vessel due to sickle cell anemia which prevented blood from leaving the spleen and resulting in acute splenic engorgement)
	Splenomegaly
	Pancreatitis
	Pancreatic tumor/cancer
	Gastric ulcer
	Gastritis (inflammation of the stomach lining)
	Gastric carcinoma
	Renal colic (dilation, stretching, and spasm due to acute ureteral obstruction)
	Kidney stones
	Pyelonephritis (infection of the kidneys)
	Renal adenomas (kidney tumors)

Left upper quadrant (LUQ) pain may involve actual and radiating pains from several structures and/or organs. Some of these pains may be benign, while others may possibly be life-threatening (Table 4).

Table 5: Probable cause of pains originated from RUQ

Organs/Structure	Causes
Gastrointestinal	Hepatic fixture lesions (lesions at the sharp bend between the ascending and the transverse colon)
	Hiatal hernia (stomach sticks upward into the chest, through an opening in the diaphragm)
	Appendicitis
	Stomach cancer
	Peptic ulcer
	Gastritis (inflammation of the stomach lining)
	Intestinal obstruction
	Colon tumor/cancer
	Diverticulitis (inflammation of small pouches called diverticula in the intestines)
	Diverticular disease (herniation of mucosa through the thickened colonic muscle)
	Ischemic colitis (blood flow to part of the colon is reduced due to narrowed or blocked)
	Severe constipation or fecal impaction (hardened stool that can't be eliminated)
	Crohn's disease
	Acute appendicitis
	Irritable bowel syndrome
	Pyloric stenosis (thickening of the pylorus, a sphincter connecting stomach with duodenum, which blocks food from entering the small intestines). This condition only occurs in infants
Cardiovascular	Abdominal aortic aneurysm
	Pericarditis (inflammation of the pericardium)
	Congestive cardiac failure
Liver and gallbladder	Hepatosplenomegaly
	Hepatitis
	Primary biliary cirrhosis
	Liver abscess (pus-filled pocket in the liver)
	Liver cancer
	Liver hemangioma (benign mass that occurs in the liver)
	Budd-Chiari syndrome (caused by the obstruction of the hepatic veins)
	Carcinoma of the gallbladder
	Cholangitis (inflammation of the bile duct)
	Cancer of the gallbladder
	Gallbladder stones
	Acute cholecystitis (acute inflammatory disease of the gallbladder)
Pancreas	Pancreatic cancer
	Pancreatitis (inflammation of the pancreas)
Kidneys	Acute pyelonephritis (bacterial invasion of the renal parenchyma)
	Nephrolithiasis (kidney stones)
	Nephritis (inflammation of the kidneys and may involve the glomeruli, tubules, or interstitial tissue)
	Hydronephrosis (swelling of the kidneys)
	Renal carcinoma
	Urinary tract obstruction
Lung (right lower lobe)	Pleurisy (inflammation of the pleura)
	Pneumonia
	Pulmonary infarction (blockage of blood flow to the lungs)
	Lobar pneumonia
	Pulmonary embolism
Endocrine & exocrine conditions	Diabetic ketoacidosis
	Addison crisis
	Adrenal tuberculosis (spread of Mycobacterium tuberculosis, into the adrenal gland)
	Metastatic carcinoma
	Pancreatic carcinoma
Other conditions	Shingles (a viral infection caused by varicella-zoster virus)

Some of the structures and organs involved may include the diaphragm, peritoneum, thoracic region, gastrointestinal tract, and abdominal, and pelvic regions. For a list of possible causes for LUQ pains, please examine Table 4 for more information.

Pains that originated in the right upper quadrant (RUQ) could be the result of a wide variety of medical conditions; some may be benign while others may be life-threatening. RUQ pains usually originate from the gastrointestinal tract or are indicators of cardio-vascular disease. Additionally, the pain can originate from the liver, gallbladder, kidneys, pancreas, and lungs, as well as being a symptom of certain endocrine or exocrine conditions. Please examine Table 5 for a list of possible causes of RUQ pains.

Table 6: Probable cause of pains originated from LLQ

Organs/Structure	Causes
Abdominal wall	Abdominal wall abscess
	Abdominal wall hematoma (collection of blood in the abdominal wall)
	Psoas abscess (collection of pus in the iliopsoas muscle compartment)
	Inguinal hernia (soft tissue protrudes through a weak point in the abdominal muscles)
	Retroperitoneal hemorrhage accumulation of blood found in the behind the peritoneal)
Gastrointestinal tract	Constipation
	Intestinal obstruction
	Incarcerated hernia (a loop of intestine becomes stuck in the hernia)
	Appendicitis
	Infectious colitis (colon inflammation/infection)
	Inflammatory bowel disease
	Ischemic bowel (injury of the large intestine result from inadequate blood)
	Diverticulitis (inflammation of small pouches called diverticula in the intestines)
	Crohn's disease
	Omental infarction (blockage of blood flow to the omentum)
	Sigmoid diverticulitis (faulty contraction of the sigmoid colon)
Genitourinary tract	Prostatitis (inflammation of the prostate)
	Nephrolithiasis (kidney stones)
	Nephritis (inflammation of the kidneys and may involve the glomeruli, tubules, or interstitial tissue)
	Seminal vasculitis (inflammation of the seminal vesicle)
	Urinary tract infection
Gynecologic	Cervical infection
	Inflammation of the cervix
	Growths on the cervix
	Ectopic pregnancy
	Endometriosis
	Hemorrhagic ovarian cyst
	Ruptured ovarian cyst
	Malignancy (tumor found in the gynecological tract)
	Miscarriage
	Mittelschmerz
	Tuboovarian abscess (pus-filled pocket located at the fallopian tube and ovary)
	Ovarian torsion (rotation of the ovary to such a degree as to occlude the ovarian artery)
	Ovarian cysts
	Pelvic congestion syndrome (pooling of blood in the pelvic varicose veins)
	Ruptured corpus luteum
	Uterine fibroids (non-cancerous tumors that develop in the uterus)
Vascular conditions	Aortitis
	Vasculitis
	Thoracic aortic aneurysm

A variety of conditions may cause pains in the left lower quadrant (LLQ). Some of these pains could result from self-limiting conditions while others may be life-threatening illness. LLQ pains could originate from conditions afflicting the abdominal walls, gastro-intestinal and genitourinary tracts. In addition, certain gynecological and vascular conditions can also possess symptomatic LLQ pains. Please examine Table 6 for a list of possible causes of LLQ pains.

Table 7: Probable cause of pains originated from RLQ

Organ/Structure	Causes
Gastrointestinal tract	Appendicitis
	Tumor/cancer
	Cholecystitis (inflammed gallbladder)
	Diverticulitis (inflammation of small pouches called diverticula in the intestines)
	Intestinal obstruction
	Viral gastroenteritis (inflammation of the stomach caused by a viral infection)
Genitourinary tract	Inflammed cervix
	Cervical infection
	Endometriosis
	Kidney infection
	Kidney stones
	Seminal vesiculitis (inflamed seminal vesicles)
Gynecological conditions	Mittelschmerz (dull pains associated with ovulation)
	Ovarian cysts
	Salpingitis (inflammation of the fallopian tubes)
	Tuboovarian abscess (pus-filled pocket involving a fallopian tube and an ovary)
Abdominal region	Inguinal hernia (soft tissue protrudes through a weak point in the abdominal muscles)
Cardiovascular	Thoracic aortic aneurysm

Differential diagnosis of pains originating from the right lower quadrant (RLQ) may include conditions affecting the gastrointestinal tract and genitourinary tract. Additionally, gynecological and vascular conditions also may display symptomatic pains emanating from RLQ. Please examine Table 7 for a list of possible causes of RLQ pains.

Abdominal Regions

In addition to dividing the abdominal area into quadrants, it is further divided into nine regions by drawing three horizontal and two vertical lines. The division of the abdominal area into nine regions grants the physician the ability to more accurately diagnose the origin of pain or discomfort (Figure 8).

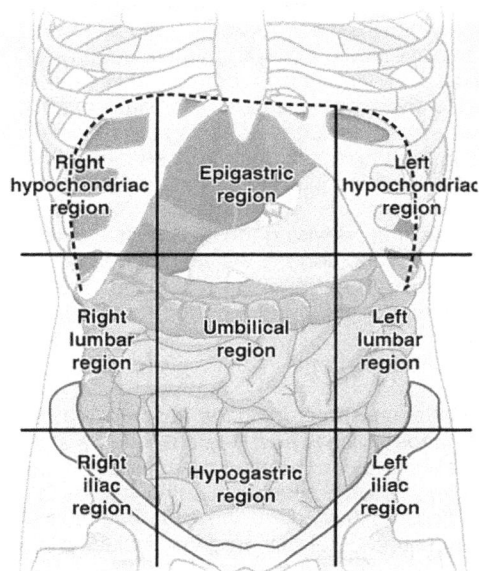

Figure 8: Nine regions of the abdominal area

① The right hypochondriac region is located on the right side of the epigastric region. This region includes the right lobe of the liver, gallbladder, upper right half of the kidneys, adrenal gland, and part of the duodenum. ② The epigastric region is located superior to the umbilical region and contains a segment of the descending (abdominal) aorta, inferior vena cava, pyloric end of the stomach, segment of the duodenum, most of the pancreas, and a portion of the liver. ③ The left hypochondriac region is located on the left side of the epigastric region. This region includes the part of the stomach, the spleen, the tail of the pancreas and part of the transverse and the descending colon. In addition the hypochondriac region also includes the upper half of the left kidney and the left adrenal gland. ④ The right lumbar region is located on the right side of the umbilical region and contains the ascending colon, lower half of the right kidney, and a segment of the small intestines that includes the duodenum and jejunum. ⑤ The umbilical region is located towards the center of the abdominal area right beneath the umbilicus. This region contains the omentum (a sheet of adipose tissue that is a part of the peritoneum) and the mesentery (folds of membranes that that supports the jejunum and ileum from the posterior abdominal wall). In addition, the umbilical regions also include a portion of the inferior vena cava, abdominal aorta, the lower part of the duodenum, and a segment of the small intestines that includes the jejunum and ileum. ⑥ The left lumbar region is located on the left side of the umbilical region and contains the descending colon, lower left half of the kidney, and parts of both the duodenum and jejunum. ⑦ The right iliac (inguinal) region is located on the right side of the hypogastric region and contains the cecum, appendix, lower end of the ileum, right ureter, and right spermatic cord (male) or right ovary (female). ⑧ The hypogastric (pubic) region is located below the umbilical region and contains a portion of the ilium, sigmoid colon, urinary bladder (when enlarged), ureters, the uterus (female), and ovaries (female). ⑨ The left iliac (inguinal) region is located on the left side of the hypogastric region and contains the sigmoid colon, left ureter, and left spermatic cord (males) or the left ovary (females).

BODY CAVITIES

Body cavities are spaces within the torso. Depending on the particular cavity, they may contain organs and/or fluids and may or may not possess external openings. In general there are two major areas that possess body cavities: the posterior and anterior body cavities.

The posterior (dorsal) cavities are located near the posterior portion of the body, where it is subdivided into two cavities: cranial and vertebral (spinal) cavities. The cranial cavity contains the brain and is formed from the flat bones of the skull. The vertebral cavity contains the spinal cord and is formed by the vertebral foramen within the vertebrae (Figure 9).

The cranial and vertebral cavities are continuous with one another and are lined by the meninges. The meninges are composed of three distinct membranes called the dura mater, arachnoid mater, and pia mater. The meninges are formed to protect the delicate nervous tissues that compose the brain and the spinal cord.

As the name implies, the anterior (ventral) cavities are located near the front of the body and are subdivided into various smaller cavities which contain either organs or bodily fluids (Figure 10). For example, the thoracic cavity is bordered inferiorly by the diaphragm and contains the lungs, a pleural cavity and the mediastinum. The pleural cavity is between the two serous membranes that surround the lungs. The two serous membranes are called the parietal pleura and the visceral pleura. These serous membranes are responsible for secreting pleural fluid into the pleural cavity. The pleural fluid has the same composition as blood serum and is intended to lubricate and protect the internal organs.

Pericardial cavity is located inferior/posterior to the mediastinum and is also formed from two serous membranes that surround the heart: parietal pericardium and the visceral pericardium. The pleural cavity contains a fluid known as pericardial fluid.

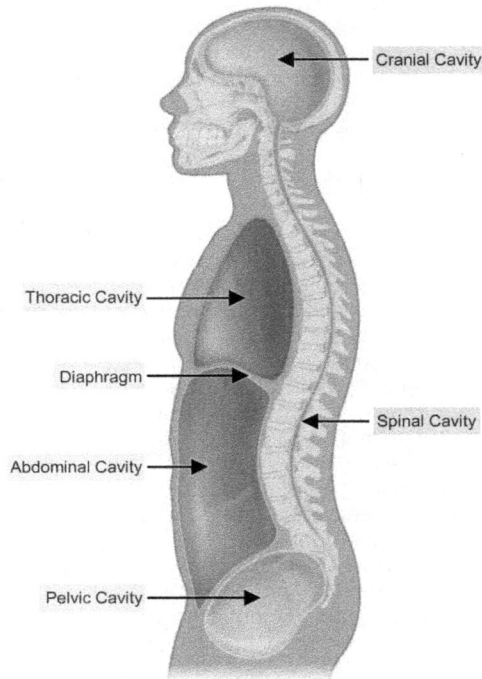

Figure 9: Lateral view of body cavities. Please note that the labels for the posterior cavities are in highlighted in red. Anterior cavities are highlighted in blue

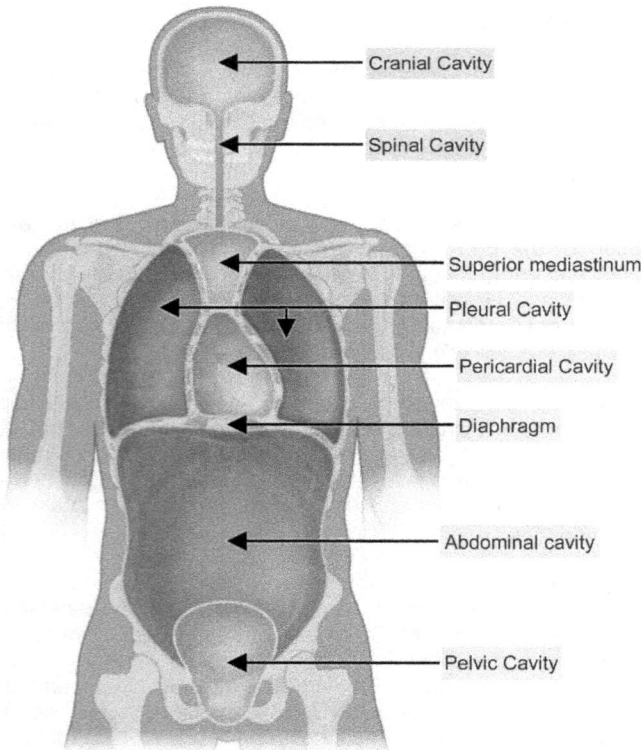

Figure 10: Anterior view of body cavities. Please note that the labels for the posterior cavities are in highlighted in red. Anterior cavities are highlighted in blue

The abdomino-pelvic cavity consists of two sub-cavities: the abdominal cavity and the pelvic cavity. These cavities are surrounded by two serous membranes, known as the parietal peritoneum and the visceral peritoneum. The space formed between the two membranes is called the peritoneal cavity. The abdominal cavity contains the stomach, liver, spleen, small intestines, and most of the large intestines. The pelvic cavity contains the bladder, sigmoid colon, rectum, and the internal reproductive organs. The peritoneal cavity is made up of the parietal and visceral peritoneum. Asites, a serious medical condition that is caused by peritoneal fluid buildup, may occur in the peritoneal cavity (Figure 11).

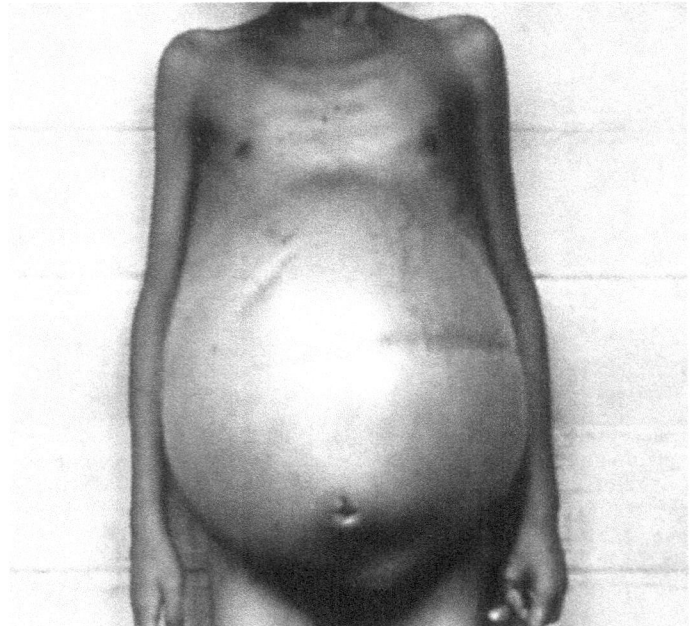

Figure 11: Photograph of a boy infected with *Schistosoma japonicum* showing signs of ascites, or a distended abdomen

NURSING CARE

When assessing any organ system it is important to understand its baseline functions or average homeostatic values. Therefore, any results obtained from analysis that deviate from the baseline may signify an illness or a disorder. It is with this concept in mind that the basic physical assessment is introduced which allows a medical professional to investigate the patient for signs of affliction. Basic physical assessment includes acquiring the medical history of a patient, recording the symptoms experienced by the individual, and a physical examination. These data will become part of the patient's medical records, and is used to determine the correct diagnosis, as well as, devising the best possible treatment plan. The following sections contain the standardized methodologies used in a physical examination after the recording of the individual's medical history and reported symptoms.

Auscultating Assessment

Basic physical assessment includes auscultating breath, heart, and bowel sounds which is best heard through a stethoscope. In determining abnormal auscultating breath sounds, one must first be familiar with various naturally occurring sounds in the thoracic areas, including regions above the collarbones and below the rib cage. Any sounds that are irregular, including decreased or absent breath sounds, and abnormal breath sounds, can aid in diagnosis. For example, absent or decreased breathing sounds could indicate that air or fluid has accumulated in or around the lungs (e.g. pneumonia, congested heart failure, pleural effusion etc.). Abnormal sounds could also point to increased thickness of the chest wall (i.e. soft tissue sarcomas, breast cancers), over-inflation of the lungs (i.e. emphysema) and reduced airflow to part of the lungs (i.e. chronic obstruction pulmonary disease). Some of the abnormal sounds include rales, which are slight clicking, bubbling, and rattling sounds that happen when air opens closed air spaces. Rhonchi are sounds that are similar to snoring and they occur when air is blocked in the large air passageways. Wheezes are sharp sounds produced by narrowing airways and could be heard during exhalations and stridor is a wheeze/whistle sound that is generated when there is a blockage of airflow in the trachea or the nasopharynx (Figures 12 and 13).

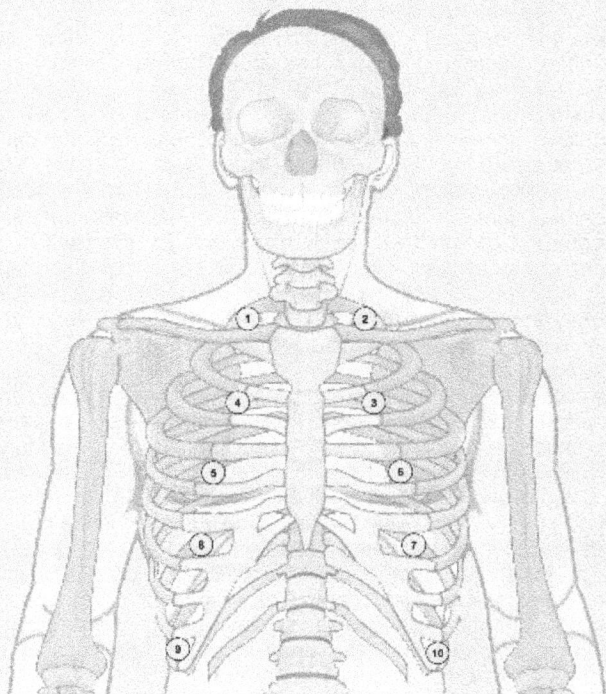

Figure 12: Lung auscultating locations anterior view

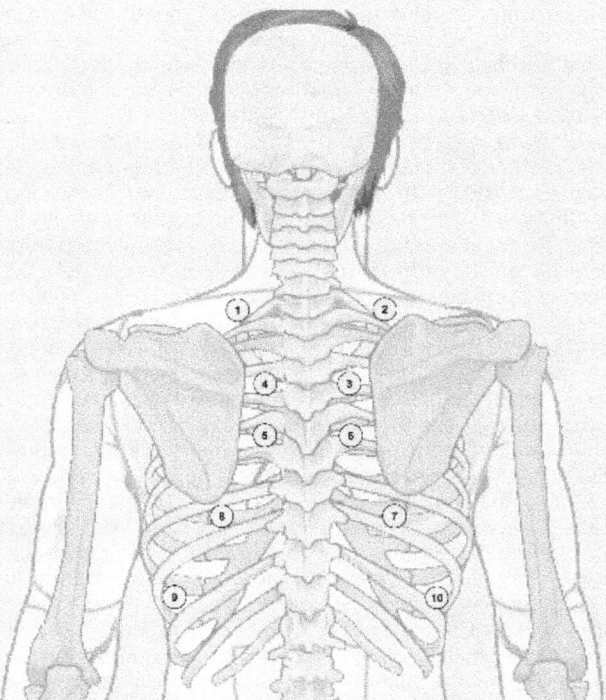

Figure 13: Lung auscultating locations posterior view

Respiratory patterns can also help with the diagnosis of a patient. For example, a healthy individual will demonstrate normal respiratory patterns (also known as eupnea) at approximately 12-

20 breaths per minute (b/m). However, a person suffering from tachypnea will exhibit breathing patterns of ~20 b/m. A person who is bradypenic (bradypnea) will express breathing patterns of less than <12 b/m. In contrast, a person who is hyperventilating will demonstrate a breathing pattern of greater than >20 b/m. A person who is demonstrating deep, gasping inspiration with a pause at full inspiration followed by a brief, insufficient release could be suffering from apneusis. An individual displaying progressively deeper, and at times faster respiration rates, followed immediately by a gradual decrease that results in apnea could be suffering from Cheyenne-Stokes respiration. Patients that are exhibiting rapid, deep, and labored breathing patterns could be suffering from Kussmaul's breathing (a common symptoms of diabetic ketoacidosis). Finally, if a person is displaying abnormal retention of air and having difficulties in exhalation, the individual may be suffering from air trapping, which is a symptom of emphysema.

Similarly, in determining auscultating heart sounds, a medical professional need to be familiar with the normal sounds of the heart. Anything that deviates from the baseline is considered to be abnormal. For example, a heart murmur is a whooshing, blowing, or rasping sound that could indicate irregular blood flow. There are several classifications of murmurs depending on the loudness that is exhibited. The loudness of the murmur is enumerated based on a Graded scale of I-VI. A Grade I murmur is extremely soft and is difficult to detect while a Grade VI murmur is loud and easily detected. In addition, a heart murmur can also be described based upon systole (contraction) or diastole (relaxation) phases of the heart. There are many possible causes of a heart murmur, but the most common cause is when a valve (tricuspid, bicuspid, or the semilunar valves) is not properly closed and backflow is occurring. Another possibility is when a narrowing (stenosis) at or around a heart valve has occurred (Figure 14).

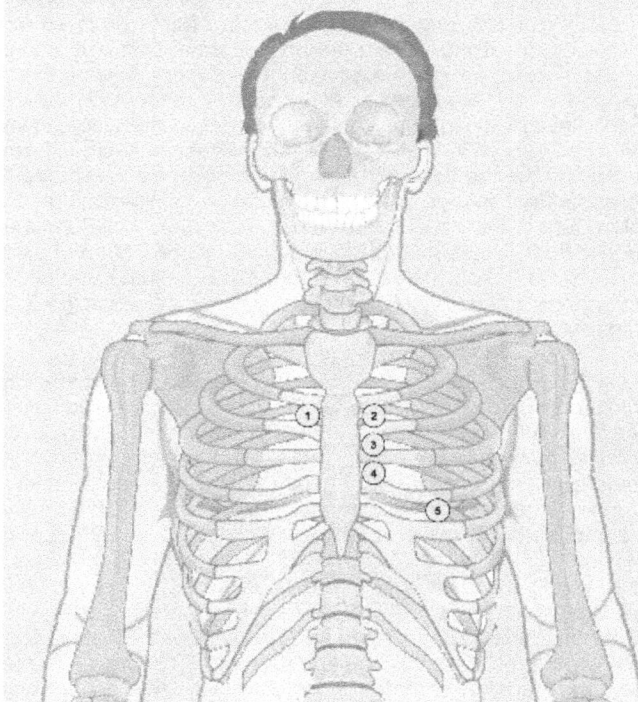

Figure 14: Illustration of cardiac auscultation locations. ① Aortic valve area, ② pulmonary valve area, ③ Erb's point, ④tricuspid valve area and ⑤mitral valve area

Auscultating bowel sounds occurs in the abdominal quadrants and they are made by peristalsis as food is being pushed through the digestive tract. It has been described that the sound heard in the digestive tract is similar to those of the echoing sounds heard through an empty pipe. Abnormal bowel sounds may indicate conditions such as ileus, or lack of intestinal activity.

Ileus could be caused by a rupture of the intestinal wall or may indicate necrosis of the intestinal tissues. Hypoactive bowel sounds can indicate a slowing of gastrointestinal activities which may indicate constipation, bowel obstructions (e.g. hernia, tumor etc.), paralytic ileus (decrease in nerve activities) or the patient may have ingested medications that decreased the movements in the digestive tract. Some of these medications that may induce a hypoactive bowel movement are opiates (e.g. codeine), anticholinergic, and phenothiazine. On the other hand, hyperactive bowel sounds indicate increased gastrointestinal activity. An increase in gastrointestinal activity may indicate that the patient is suffering from diarrhea, Crohn's disease, gastro-intestinal bleeding, or ulcerative colitis. Finally, a very high-pitched bowel sound may indicate early bowel obstruction.

Neuromuscular and Vascular Assessment

As a part of the basic physical assessment, medical staff should inspect the patient's arm and leg muscles for atrophy, tremors, fasciculation, or other abnormal movements. The strength of specific muscle groups should be tested by having client extend or flex individual joints against resistance (via the counter-forces) provided by examiner. The muscle groups to be examined include the brachials, carpals, femorals, crurals and tarsals.

The examiner should also palpate the joints carefully for symmetry, as well as, all muscle groups for size, symmetry, and tone. A range of motion for joint function, deformity, stability, contracture, and effusion should also be graded. The gait of the client should be examined for unsteadiness, irregular movements, or limping. Individual posture should be appraised for kyphosis (forward curvature of thoracic spine), lordosis (inward curvature of the lumbar spine), and scoliosis (lateral curving deviation of spine).

A capillary refill test should also be performed to determine the possibilities of dehydration and circulatory perfusion. Capillary refill test is a measurement of the amount of time required for a soft tissue under pressure to return to normal skin coloration. This test should be performed with the patient holding their hands above their heart before applying the pressure. Pressure is generally applied to the nail bed until it turns white (also known as blanched), which indicates that the blood has been forced from the tissue. Once the tissue has blanched, pressure is removed. If the patient possesses adequate circulation, normal coloration should return in less than two seconds. However, if blanch time is greater than two seconds, this may indicate reduction of skin turgor, which is often the result of circulatory volume depletion resulting from dehydration, peripheral vascular disease (PVD), hypothermia, or shock.

Assessing for pronator drift will aid in the evaluation of a patient's motor function. To perform this test, patients are required to be awake and able to follow directions. Pronator drift is a type of general neurological assessment that allows for the determination of neurologic deficiencies, comorbidity (also known as a patient with one or more additional disorders appearing simultaneously with a primary disorder), or acute neurologic deficits. Depending on the patient's general state of health, pronator drift is performed by first asking the patient to stand or sit, close their eyes and then extend both arms with palms facing upward. Please note that both extended arms should be level with one another. The patient should maintain this position for 20 to 30 seconds. If one of the arms starts to "pronate" or drift downward, the patient has just demonstrated pronator drift, which indicates an abnormal function of the corticospinal tract or the primary motor cortex. Pronator drift generally occurs in stroke patients or individuals with a cervical spine injury. If the patient starts to move an arm upward or laterally, this is an indication that the individual may suffer from a loss of proprioception (sense of position).

Vital Signs

A complete physical examination includes the evaluation and recording of bilateral arterial pulses at the femoral, popliteal, dorsalis pedis, and posterior tibial locations. While examining the pulse, the medical practitioner should also note the intensity, rate, rhythm, signs of blood vessel tenderness, tortuosity, or nodularity of the blood vessels exist. Palpation should be performed via the fingertips, while the intensity of the pulse should be graded on a scale of 0 to 4+. The grading scale 0 will indicate no palpable pulse, 1+ indicates a faint pulse, 2+ suggests a slightly diminished pulse when compared to the norm, 3+ indicates normal pulse and finally, 4+ indicating a bounding pulse.

Absence of a pulse may indicate blockage (occlusion) by a blood clot (thrombus), a detached, traveling intravascular mass (embolus), or a torn/damaged artery (dissection). Absence or diminished pulse in the legs along with intermittent muscular pain (claudication) may confirm a vascular condition. If the patient demonstrates a hypokinetic pulse (diminished pulse, 1+), it may suggest a low cardiac output due to shock, myocardial infarction, idiopathic dilated cardiomyopathy, valvular stenosis, pericardial tamponade, constrictive pericarditis, etc. In contrast, if the patient demonstrates a hyperkinetic pulse (a bounding pulse 4+) it indicates an abnormal increase of cardiac output due to anemia, anxiety, heart failure, aortic regurgitation, fever, chronic kidney disease, hyperthyroidism, etc.

Respiratory examination should begin by having the patient sit upright and exposed from the waist up. Make a general observation of the patient and determine if the individual is exhibiting signs of rapid breathing (tachypnoeic) or possesses any obvious signs of abnormalities. Please note that on average a person takes 18 to 22 breaths per minute. Examine the patient's hands to see if they appear normal in coloration or appear pink. Pink hands may indicate carbon dioxide retention. Request the patient to extend their arms and cock their wrists upward at a 90° angle for 30 seconds. A patient exhibiting irregular tremor (also known as CO_2 retention flap) may also indicate signs of CO_2 retention. Obtain the patient's radial pulse. If a bouncing pulse (leaping and forceful pulse that quickly disappears) is detected it may indicate CO_2 retention. Check the patient's tongue and note the coloration. If the color appears to be pale this may indicate anemia or central cyanosis (poor blood oxygenation in the lungs). Palpate the left supraclavicular node and see if it is enlarged. The left supraclavicular node drains the thoracic duct therefore an enlarged node may suggest metastatic cancer (cancer that have metastasized or spread). Palpate the chest of the patient. If the clavicles appear to be deviated, this may indicate a tumor or pneumothorax (abnormal collection of air or gas in the pleural space). Examine the individual's chest expansion by placing the hands firmly on the chest wall with your thumbs meeting in the midline. Request that the patient breathe in deeply and note the spread of the thumbs (normal spread is approximately 5 cm). Abnormal chest expansion may indicate lung or pleural disease. For example, patients with chronic obstructive pulmonary disease (COPD) will demonstrate an extremely low chest expansion. Chest percussion should be performed by first tapping the clavicle which will provide the resonance in the apex and then percuss normally for the entire lung field. There are four types of percussion sounds: resonant, hyper-resonant, and stony dull or dull. Hyper-resonant may indicate a collapsed lung while a dull sound suggests the presence of a solid mass under the surface. This solid mass may be an infection, effusion or a tumor. Place the medial edge of your hand on the patient's chest and check for tactile vocal fremitus. Tactile vocal fremitus indicates the vibrations that can be felt on the chest wall when the patient speaks. Generally, the patient is asked to repeatedly state the words "ninety-nine" during examination. In normal lungs, the vibrations felt on the bilateral sides of the chest should be similar. When the two sides of the chest are compared, increased vibrations may indicate pulmonary consolidation (which are lungs that are filled with liquid, swollen or hardening) while a decrease in vibrations may indicate pneumothorax. Use a stethoscope and auscultate both the anterior and posterior thorax region. Carefully listen for reduced breathing sounds, additional sounds such as crackles, wheezes, pleural rub or rhonchi.

The examination of blood pressure is an essential part of the physical examination. Blood pressure measurements can be taken from both arms and legs. Blood pressure measurements can be taken with the patient lying down, standing, or sitting. The results obtained from numerous easements should be averaged at the end of the exam in order to obtain an accurate result.

A normal result should be approximately 119 (systolic) over 79 (diastolic) mmHg. A patient with prehypertension will demonstrate a systolic measurement of 120 to 139 mmHg and a diastolic measurement of 80 to 89 mmHg. A patient with high blood pressure (hypertension) will have a systolic measurement of 140 mmHg or above and a diastolic measurement of 90 mmHg or above. Hypertension is subcategorized into two stages. For example, stage 1 will demonstrate a systolic value of 140 to 159 mmHg and a diastolic value of 90 to 99 mmHg. Stage 2 individual will demonstrate a systolic value of 160 mmHg or above and a diastolic value or 100 mmHg or above. Inversely, a patient with hypotension will demonstrate blood pressure below the normal values and will demonstrate poor circulation.

Pain is one of the most common reasons that people seek medical attention. It has been found that over 80 percent of all physicians' visit is because of acute pain and that nearly 80 percent of chronic pain sufferers are being undertreated. In November 1998, a nationwide initiative known as "pain as the 5th vital sign" was initiated by the Veterans Health Administration (VHA). This initiative requires the use of a Numeric Rating Scale (NRS) for all clinical encounters. The NRS is a standardized visual and verbal assessment tool based upon a 10 point scale with 0 indicating no pain and 10 representing the worst possible pain. This initiative requires that nurses access pain in their patients and maintain the results as a part of the medical records. This initiative expected that a score of 4 or higher will prompt a comprehensive pain evaluation and possible intervention by a physician. In addition to assessment of pain, nurses are also responsible for questioning the patient about other sensations such as paresthesia (tingling sensation in the fingers and toes) or absence of feelings.

Nursing Process

The nursing process is a systematic way of reasoning that guides client care. This process starts with the assessment of the client. The basic physical assessments as stated in the previous sections are just one of the many tools used in patient evaluation. Other aspects include clear and concise communication with the client where historical health data such as allergies and current medications are collected. Finally, considerations of any tests that may have been ordered by the physician should factor prominently into the evaluation of the client. Assessment is a continuous process that must be done before planning any nursing interventions.

Next, after the client has been assessed, a nursing diagnosis must be made. Please note that the nursing diagnosis is written in a specific format, and cannot contain a medical diagnosis. For instance, instead of asserting that the client has a bone fracture, a nursing diagnosis may state: "the patient possesses impaired physical mobility related to musculoskeletal impairment."

The next part of the nursing diagnosis must assert the evidence presented by the client such as "impaired physical mobility." It is at this point where the assessment data is used to support this statement. It is understood that part of the nursing diagnosis is based upon observation, so if we observe that this patient has a casted right ankle, and is using crutches to walk, the nursing diagnosis may state: "as evidenced by immobility of casted right ankle, and use of assistive devices to ambulate."

In summary, the complete nursing diagnosis would read:

"Impaired physical mobility related to musculoskeletal impairment as evidenced by immobility of casted right ankle, and use of assistive devices to ambulate."

Please note that sometimes a patient can simply be "at risk" for a complication such as bleeding for example due to anticoagulant therapy or some pathology. An "at risk for" diagnosis doesn't need an "as evidenced by" because the situation has yet transpired.

The nursing diagnosis will then be used to guide the planning for patient's treatment. In this case, what would a nurse do to aid the patient with impaired physical mobility? In other words, what nursing interventions should be performed in order to accomplish the necessary outcome? The outcomes must be realistic and have a time frame which the tasks can be accomplished. Using the previous case as an example, the nursing diagnosis would state: "instruct in safe use of walker or cane for ambulation." The nursing diagnosis would further add that the client should verbalize understanding of the situation, individual treatment regimen and safety measures. Finally, a time frame must be placed within the nursing diagnosis. For example, to end a nursing diagnosis statement, a nurse may inscribe: "by end of shift."

Now that planned nursing care has been devised, it is time to proceed with implementation. After the implementation has been initiated, an evaluation is required towards the end of the shift in order to determine if the planned care was successful for the client. Changes to the planned nursing care are part of the treatment and a nurse must be flexible in their plans and actions until the desired outcome has been achieved.

Complications of Immobility

Prevention is the key when focusing on the complications of immobility. It is understood that prevention is a large part of nursing care. Please note that multiple body systems can be affected by immobility and the following sections will describe some complications of immobility of the system, along with nursing interventions that can prevent them.

Integumentary

When a client is immobile there is an increased risk for skin breakdown. This breakdown can cause the formation of pressure ulcers due to a decrease in circulation. If immobility is accompanied by malnutrition and incontinence, the risk of skin breakdown increases. Clients confined to the bed often push themselves upright, which may result in the sheering of the skin. To prevent these complications the nurse should reposition the client every two hours, and protect pressure points such as boney prominences. A draw sheet should be used to pull the client up in the bed to reduce sheering and friction. Special mattresses can be used to redistribute body weight at intervals. In conjunction with good nutrition and wound clinic referrals, the risk of integument breakdown may be mitigated.

Elimination

Immobility may result in the reduction of gastrointestinal motility. As a result, the client may suffer from numerous gastrointestinal issues including constipation. For example, nursing interventions of constipation may include switching the client to a high fiber diet (dietary consultation may be necessary), encouraging the client to ambulate, increase fluid intake, vitamin supplementation and use stool softeners. A patient's appetite may become poor due to decrease in gastrointestinal activities; therefore, encouraging the client to increase their physical activities, when possible, may aid in the elimination this complication.

Urinary

The bladder has a tendency to retain urine when the client is immobilized in bed. Stagnant urine can result in the growth of bacteria, which in turn, may results in a urinary tract infection. Commonly, a client may avoid the intake of fluids to circumvent the use of a bed pan. Therefore, fluids should be encouraged and any signs and symptoms of urinary tract infection should be reported to the physician.

Respiratory

Acid-base imbalances can occur when clients are immobile due to incomplete chest expansion or atelectasis (collapse of a part or the entire lung). The client should be encouraged to take deep breaths, and cough every two hours. Use of an incentive spirometer is also helpful and should be encouraged. These interventions will help in the prevention of stasis of secretions and prevent complications that lead to respiratory acidosis, and infections such as pneumonia.

Cardiovascular

Poor peripheral circulation can result from immobility which leads to a risk for venous thromboembolism (VTE). VTE is a blood clot that is formed in a vein. A common type of VTE is a deep vein thrombosis (DVT), which is a blood clot formed within the deep veins of the leg. If this clot embolizes (breaks off) and travels in the blood it is called a venous thromboembolism (VTE). If VTE flows towards the lungs, it could result in pulmonary embolism (PE), a life-threatening blood clot in the lungs. Compression stockings, sequential depression devices, and a range of motion exercises can help prevent VTE. Nonetheless, the best solution to reduce VTE is to encourage mobility. The nurse should be on the watch for symptoms of VTE such as warmth, tenderness, redness, swelling, and reports of pain in the gastrocnemius (calf muscle). If any of the above stated symptoms are observed, the findings should promptly be reported to the physician.

QUESTIONS

1. Which of the following is not a sub-discipline of anatomy?
   a. Gross anatomy
   b. Cytology
   c. Histology
   d. Developmental anatomy
   e. Zoology

2. Which of the following is the sub-discipline of physiology that examines bodily functions in a diseased state?
   a. Animal physiology
   b. Pathophysiology
   c. Human physiology
   d. Cellular physiology
   e. Systemic physiology

3. Which of the following is known as the basic unit of matter?
   a. Cell
   b. Tissue
   c. Organs
   d. Atom
   e. Organism

4. Which of the following is (are) the particle (s) that construct the nucleus of the atom?
   a. Proton
   b. Electron
   c. Neutron
   d. Both a and c
   e. Both b and c

5. Which of the following is the positively charged particle (s) of the atom?
   a. Proton
   b. Electron
   c. Neutron
   d. Both a and c
   e. Both b and c

6. Which of the following is the negatively charged particle (s) of the atomic core?
   a. Proton
   b. Electron
   c. Neutron
   d. Both a and c
   e. Both b and c

7. Which of the following particle (s) of the atomic core possesses no charge?
   a. Proton
   b. Electron
   c. Neutron
   d. Both a and c
   e. Both b and c

8. It has been theorized that the electrons are held in their orbits by which of the following force generated by the nucleus?
   a. Electromagnetic force
   b. Gravity
   c. Hydroelectrical force
   d. Electrochemical force
   e. Centripetal force

9. What type of molecule is formed when two hydrogen atoms join together?
   a. Homonuclear molecule
   b. Heteronuclear molecule
   c. Isonuclear molecule
   d. Chemical compounds
   e. Perinucelar molecule

10. What type of molecule is formed when two hydrogen atoms and an oxygen atom joined together?
    a. Homonuclear molecule
    b. Heteronuclear molecule
    c. Isonuclear molecule
    d. Chemical compounds
    e. Perinucelar molecule

11. Which of the following is known as the smallest living organism and serves as the building block for complex multicellular creatures?
    a. Cell
    b. Tissue
    c. Organs
    d. Atom
    e. Organism

12. Which of the following is also referred to as the blue print from which all living organisms are derived?
    a. Ribose nucleic acid (RNA)
    b. Proteins
    c. Lipids
    d. Deoxyribose nucleic acid (DNA)
    e. Carbohydrates

13. Which of the following is the simplest and the most ancient living organism on earth?
    a. Eukaryotic cell
    b. Hepatocytes
    c. Monocytes
    d. Prokaryotic cell
    e. Macrophage

14. Which of the following type of cell could exist in unicellular or multicellular forms?
    a. Eukaryotic cell
    b. Hepatocytes
    c. Monocytes
    d. Prokaryotic cell
    e. Macrophage

15. Which of the following is (are) grouped under prokaryotes?
    a. Bacteria
    b. Achaea
    c. Eukaryotes
    d. Both a and b
    e. Both b and c

16. It is estimated that the human body is composed of approximately _____ cells?
    a. 73 billion prokaryotic
    b. 45 trillion prokaryotic
    c. 92 billion eukaryotic
    d. 73 trillion eukaryotic

17. Which of the following statement(s) best describes tissue?
    a. Tissue is defined as a collective of interacting cells with a similar embryonic origin
    b. Tissues are composed of prokaryotic cells
    c. Tissues are composed of eukaryotic cells
    d. Both a and b
    e. Both a and c

18. Human body is (are) composed of which of the following type(s) of tissue(s)?
    a. Epithelial tissue
    b. Connective tissue
    c. Muscular tissue
    d. Nervous tissue
    e. All of the above

19. Which of the following physiological organ system is involved in the removal of excess water and waste materials from the blood stream?
    a. Integument system

b. Musculoskeletal system
c. Respiratory system
d. Digestive system
e. Urinary system

20. Which of the following physiological organ system is formed to provide protection from abrasion, regulates temperature, and prevents uncontrollable water gain and water loss etc.?
a. Integument system
b. Musculoskeletal system
c. Respiratory system
d. Digestive system
e. Urinary system

21. Which of the following physiological organ system is formed to provide the necessary surface area for gas exchange?
a. Integument system
b. Musculoskeletal system
c. Respiratory system
d. Digestive system
e. Urinary system

22. Which of the following physiological organ system is formed to provide protection, support and movement?
a. Integument system
b. Musculoskeletal system
c. Respiratory system
d. Digestive system
e. Urinary system

23. Which of the following physiological organ system is formed to provide the necessary protection against bacteria or foreign entities that invade the organism?
a. Circulatory system
b. Immune system
c. Reproductive system
d. Endocrine system
e. Nervous system

24. Which of the following physiological organ system is involved in the distribution of nutrients and removes wastes from the tissues and cells?
a. Circulatory system
b. Immune system
c. Reproductive system
d. Endocrine system
e. Nervous system

25. Which of the following physiological organ system is the command and control center of the organism that is involved in the coordination of movements, organ functions, reception of stimuli and memory formation etc.?
a. Circulatory system
b. Immune system
c. Reproductive system
d. Endocrine system
e. Nervous system

26. Which of the following physiological organ system produces sperms or eggs which are necessary for propagation of the species?
a. Circulatory system
b. Immune system
c. Reproductive system
d. Endocrine system
e. Nervous system

27. Which of the following physiological organ system produces intercellular signal molecules (ligands) and releases them into tissue fluids or the bloodstream?
a. Circulatory system
b. Immune system
c. Reproductive system
d. Endocrine system
e. Nervous system
f. Both a and d
g. Both d and e

28. Which of the following term best describes the maintenance of a constant environment for the cells in the human body, and in turn, maintains a constant and stable environment for the body?
a. Negative feedback system
b. Positive feedback system
c. Up regulation
d. Down regulation
e. Homeostasis

29. In the human body the process of maintaining a stable internal environment is accomplished through which of the following physiological system?
a. Circulatory system
b. Immune system
c. Reproductive system
d. Endocrine system
e. Nervous system
f. Both a and d
g. Both d and e

30. Although out of order, the following statements provided describes the manner by which the human body maintains equilibrium. Use the space provided and place the listed statements in the correct sequence of order.

_____, _____, _____, _____

a. If a deviation from the norm occurred, the regulatory center of the brain will utilize the effectors to activate the endocrine system to secrete hormones or utilizing motor nerves to initiate counteraction to bring the body back to normal condition
b. Information collected form the external and internal environments are sent to the brain
c. Sensory receptor that monitors the external and internal environments
d. The information collected form the external and internal environments are analyzed and assessed by the brain

31. Place the correct answer in the spaces provided.

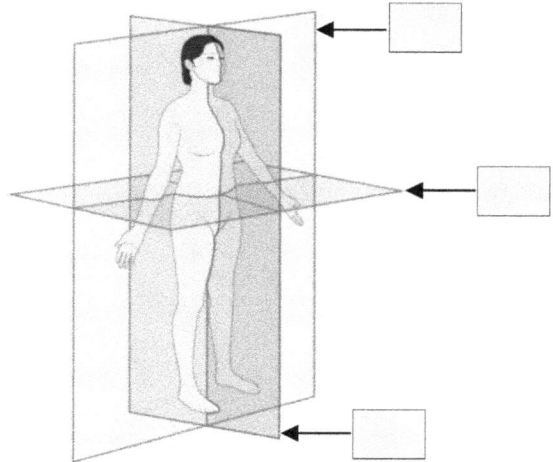

a. Femoral
b. Frontal
c. Transverse
d. Median
e. Oblique

32. Using the selection of external posterior anatomical terminologies provided, place the correct answer within the appropriate blanks.

_____ Armpit
_____ Forearm
_____ Head
_____ Heel of the foot
_____ Neck
_____ Point of the shoulder
_____ Upper arm
_____ Wrist

a. Acromial
b. Antebrachial
c. Axillary
d. Brachial
e. Calcaneal
f. Carpal
g. Cephalic
h. Cervical
i. Coxal
j. Cranial
k. Cubital
l. Dorsum
m. Femoral
n. Gluteal
o. Interscapular
p. Lumbar
q. Manus
r. Nuchal
s. Occipital

33. Using the selection of external posterior anatomical terminologies provided, place the correct answer within the appropriate blanks.

_____ Back of the hand
_____ Between the ribs and hips
_____ Buttocks
_____ Elbow
_____ Hand
_____ Head
_____ On the two sides of the spine
_____ Tail bone
_____ Thigh

a. Acromial
b. Antebrachial
c. Axillary
d. Brachial
e. Calcaneal
f. Carpal
g. Cephalic
h. Cervical
i. Coxal
j. Cranial
k. Cubital
l. Dorsum
m. Femoral
n. Gluteal
o. Interscapular
p. Lumbar
q. Manus
r. Nuchal
s. Occipital

34. Using the selection of external posterior anatomical terminologies provided,

place the correct answer within the appropriate blanks.

_____ Ankle

_____ Area between anus & genitals

_____ Back of the knee

_____ Back of the neck

_____ Between the hips

_____ Calf

_____ Chest

_____ Ear

_____ Lower back of the skull

_____ Posterior aspect of the elbow

_____ Shoulder blade

_____ Sole of the foot

_____ Spinal column

a. Acromial
b. Antebrachial
c. Axillary
d. Brachial
e. Calcaneal
f. Carpal
g. Cephalic
h. Cervical
i. Coxal
j. Cranial
k. Cubital
l. Femoral
m. Gluteal
n. Nuchal
o. Occipital
p. Olecranal
q. Otic
r. Perineal
s. Plantar
t. Popliteal
u. Sacral
v. Scapular
w. Sural
x. Tarsal
y. Thoracic
z. Vertebral

35. Using the selection of external anterior anatomical terminologies provided, place the correct answer within the appropriate blanks.

_____ Anterior body trunk inferior to ribs

_____ Anterior surfaces of the elbow

_____ Arm

_____ Armpit

_____ Between knee and ankle

_____ Cheek

_____ Collar bone

_____ Elbow

_____ Face

a. Abdominal
b. Acromial
c. Antebrachial
d. Antecubital
e. Axillary
f. Brachial
g. Buccal
h. Carpal
i. Cervical
j. Clavicular
k. Coxal
l. Cranial
m. Crural
n. Cubital
o. Digital
p. Dorsum
q. Facial
r. Femoral
s. Fibular
t. Frontal

36. Using the selection of external anterior anatomical terminologies provided, place the correct answer within the appropriate blanks.

_____ Finger or toe

_____ Forearm

_____ Forehead

_____ Head

_____ Hip

_____ Lateral side of the lower leg

_____ Neck region

_____ Point of the shoulder

_____ Thigh

_____ Top of the foot

_____ Wrist

a. Abdominal
b. Acromial
c. Antebrachial
d. Antecubital
e. Axillary
f. Brachial
g. Buccal
h. Carpal
i. Cervical
j. Clavicular
k. Coxal
l. Cranial
m. Crural
n. Cubital
o. Digital
p. Dorsum
q. Facial
r. Femoral
s. Fibular (peroneal)
t. Frontal

37. In anatomy, the human body is divided into eight sections. For example, when an individual is standing in the anatomical position, the _____ also known as the head is located at the top of the body and is visible either anteriorly or posteriorly.
a. Cephalic section (cranial section)
b. Cervical section
c. Thoracic section
d. Abdominal section
e. Pelvic section
f. Upper extremity sections
g. Lower extremity sections

38. In anatomy, the human body is divided into eight sections. For example, when

an individual is standing in the anatomical position, the _____ also known as the neck begins below the head, ends at the thorax, and is visible either anteriorly or posteriorly.
a. Cephalic section (cranial section)
b. Cervical section
c. Thoracic section
d. Abdominal section
e. Pelvic section
f. Upper extremity sections
g. Lower extremity sections

39. Using the selection of external anterior anatomical terminologies provided, place the correct answer within the appropriate blanks.

_____ Breast

_____ Chin

_____ Ear

_____ Eye socket

_____ Great toe

_____ Groin

_____ Hand

_____ Mouth

_____ Nose

_____ Palm of the hand

a. Hallux
b. Inguinal
c. Mammary
d. Manus
e. Mental
f. Nasal
g. Oral
h. Orbital (orbit)
i. Otic
j. Palmar
k. Patellar
l. Pectoral
m. Pedal
n. Pelvic
o. Pollex
p. Pubic
q. Sternal
r. Tarsal
s. Thoracic
t. Umbilical

40. Using the selection of external anterior anatomical terminologies provided, place the correct answer within the appropriate blanks.

_____ Ankle

_____ Anterior knee (kneecap)

_____ Breast bone

_____ Chest (entire)

_____ Chest (upper)

_____ Foot

_____ Genital

_____ Navel

_____ Pelvis

_____ Thumb

a. Hallux
b. Inguinal
c. Mammary
d. Manus
e. Mental
f. Nasal
g. Oral
h. Orbital (orbit)
i. Otic
j. Palmar
k. Patellar
l. Pectoral
m. Pedal
n. Pelvic
o. Pollex
p. Pubic
q. Sternal
r. Tarsal
s. Thoracic
t. Umbilical

41. Which of the following quadrant includes the duodenum, jejunum and ilium, left ureter, sigmoid colon, segment of the descending colon and the bladder if distended? In addition this quadrant also include lower lobe of left kidney, left ureter, left spermatic cord, left ovary and left fallopian tube and the enlarged uterus in pregnant women.
a. Right upper quadrant
b. Right lower quadrant
c. Left upper quadrant
d. Left lower quadrant

42. Which of the following quadrant includes the cecum, section of the ascending colon, part of the duodenum, jejunum and ilium? In addition, this quadrant will also include the appendix, the extended bladder, right ureter, right spermatic cord, right ovary, right fallopian tube, the lower lobe of the right kidney and the enlarged uterus in pregnant women.
a. Right upper quadrant
b. Right lower quadrant
c. Left upper quadrant
d. Left lower quadrant

43. Which of the following quadrant includes right lobe of the liver, gallbladder, head of the pancreas, a segment of the ascending and transverse colon and pylorus? In addition this quadrant also includes the hepatic flexure (the sharp bend between the ascending and the transverse colon), upper lobe of the right kidney, right adrenal gland and duodenum.
a. Right upper quadrant
b. Right lower quadrant
c. Left upper quadrant
d. Left lower quadrant

44. Which of the following quadrant includes the left lobe of liver, stomach, section of the transverse and descending colon? In addition, this quadrant also includes the splenic flexure, pancreas, upper lobe of the left kidney, left adrenal gland and spleen.
a. Right upper quadrant

b. Right lower quadrant
c. Left upper quadrant
d. Left lower quadrant

45. Which of the following region contains a portion of the ilium, sigmoid colon, urinary bladder (when enlarged), ureters, the uterus (female), and ovaries (female)?

a. Epigastric region
b. Hypogastric (pubic) region
c. Left hypochondriac region
d. Left iliac (inguinal) region
e. Left lumbar region
f. Right hypochondriac region
g. Right iliac (inguinal) region
h. Right lumbar region
i. Umbilical region

46. Which of the following region contains a segment of the descending (abdominal) aorta, inferior vena cava, pyloric end of the stomach, segment of the duodenum, most of the pancreas, and a portion of the liver?
a. Epigastric region
b. Hypogastric (pubic) region
c. Left hypochondriac region
d. Left iliac (inguinal) region
e. Left lumbar region
f. Right hypochondriac region
g. Right iliac (inguinal) region
h. Right lumbar region
i. Umbilical region

47. Which of the following region contains the ascending colon, lower half of the right kidney and a segment of the small intestines that include the duodenum and jejunum?
a. Epigastric region
b. Hypogastric (pubic) region
c. Left hypochondriac region
d. Left iliac (inguinal) region
e. Left lumbar region
f. Right hypochondriac region
g. Right iliac (inguinal) region
h. Right lumbar region
i. Umbilical region

48. Which of the following region contains the cecum, appendix, lower end of the ileum, right ureter, right spermatic cord (male) and right ovary (female)?
a. Epigastric region
b. Hypogastric (pubic) region
c. Left hypochondriac region
d. Left iliac (inguinal) region
e. Left lumbar region
f. Right hypochondriac region
g. Right iliac (inguinal) region
h. Right lumbar region
i. Umbilical region

49. Which of the following region contains the descending colon, lower left half of the kidney, and parts of both duodenum and jejunum?
a. Epigastric region
b. Hypogastric (pubic) region
c. Left hypochondriac region
d. Left iliac (inguinal) region
e. Left lumbar region
f. Right hypochondriac region
g. Right iliac (inguinal) region
h. Right lumbar region
i. Umbilical region

50. Which of the following region contains the omentum (a sheet of adipose tissue that is a part of the peritoneum) and the mesentery (a double layer of peritoneum that supports the jejunum and ileum from the posterior abdominal wall)? In addition, this region also includes the inferior vena cava, abdominal aorta, the lower part of the duodenum, and a segment of the small intestines that include the jejunum and ileum.
a. Epigastric region
b. Hypogastric (pubic) region
c. Left hypochondriac region
d. Left iliac (inguinal) region
e. Left lumbar region
f. Right hypochondriac region
g. Right iliac (inguinal) region
h. Right lumbar region
i. Umbilical region

51. Which of the following region contains the sigmoid colon, left ureter, left spermatic cord (males) and the left ovary (females)?
a. Epigastric region
b. Hypogastric (pubic) region
c. Left hypochondriac region
d. Left iliac (inguinal) region
e. Left lumbar region
f. Right hypochondriac region
g. Right iliac (inguinal) region

h. Right lumbar region
i. Umbilical region

52. Which of the following region includes the right lobe of the liver, gallbladder, upper right half of the kidneys, adrenal gland, and part of the duodenum?
a. Epigastric region
b. Hypogastric (pubic) region
c. Left hypochondriac region
d. Left iliac (inguinal) region
e. Left lumbar region
f. Right hypochondriac region
g. Right iliac (inguinal) region
h. Right lumbar region
i. Umbilical region

53. Which of the following region includes the stomach, spleen, the tail of the pancreas, part of the transverse and the descending colon? In addition this region also includes the upper half of the left kidneys and the adrenal gland.
a. Epigastric region
b. Hypogastric (pubic) region
c. Left hypochondriac region
d. Left iliac (inguinal) region
e. Left lumbar region
f. Right hypochondriac region
g. Right iliac (inguinal) region
h. Right lumbar region
i. Umbilical region

54. Using the selection of body orientation terminologies provided, place the correct answer within the appropriate blanks.

_____ Above

_____ Away from the midline (side)

_____ Back of or backside of the body

_____ Below

_____ Closer to the trunk of the body

_____ In front of / most forward

_____ Towards the backside of the body

_____ Towards the head

_____ Towards the head

_____ Towards the midline or middle

a. Superior
b. Inferior
c. Anterior
d. Posterior
e. Medial
f. Lateral
g. Cephalad
h. Cranial
i. Caudal
j. Proximal
k. Distal
l. Superficial
m. Intermediate
n. Deep
o. Dorsal
p. Ventral
q. Unilateral
r. Bilateral
s. Ipsilateral
t. Contralateral

55. Using the selection of body orientation terminologies provided, place the correct answer within the appropriate blanks.

_____ Away from the surface of the body

_____ Further away from the trunk of the body

_____ In between

_____ On both sides

_____ On one side

_____ On opposite sides

_____ On the same side

_____ Towards the surface of the body

a. Superior
b. Inferior
c. Anterior
d. Posterior
e. Medial
f. Lateral
g. Cephalad
h. Cranial
i. Caudal
j. Proximal
k. Distal
l. Superficial
m. Intermediate
n. Deep
o. Unilateral
p. Bilateral
q. Ipsilateral
r. Contralateral

56. The posterior (dorsal) cavities are located near the posterior portion of the body, where it is subdivided into two cavities. Which of the following cavities contains the brain?
a. Abdominal
b. Cranial
c. Diaphragm
d. Pelvic
e. Spinal
f. Thoracic

57. The posterior (dorsal) cavities are located near the posterior portion of the body, where it is subdivided into two cavities. Which of the following cavities contains the spinal cord?
a. Abdominal

    b. Cranial
    c. Diaphragm
    d. Pelvic
    e. Spinal
    f. Thoracic

58. Which of the following is (are) the membrane(s) that surrounds the brain and the spinal cord?
    a. Arachnoid mater
    b. Dura mater
    c. Epithelial tissue
    d. Pia mater
    e. Answers are a, b and d

59. The anterior (ventral) cavities are located near the front of the body and are subdivided into various smaller cavities which contain either organs or bodily fluids. Which of the following anterior cavity contains the lungs, a pleural cavity and the mediastinum?
    a. Abdominal
    b. Diaphragm
    c. Pelvic
    d. Pericardial
    e. Peritoneal
    f. Pleural
    g. Superior mediastinum
    h. Thoracic

60. The anterior (ventral) cavities are located near the front of the body and are subdivided into various smaller cavities which contain either organs or bodily fluids. Which of the following anterior cavity is formed between two serous membranes (parietal pleura and the visceral pleura) that surround the lungs?
    a. Abdominal
    b. Diaphragm
    c. Pelvic
    d. Pericardial
    e. Peritoneal
    f. Pleural
    g. Superior mediastinum
    h. Thoracic

61. The anterior (ventral) cavities are located near the front of the body and are subdivided into various smaller cavities which contain either organs or bodily fluids. Which of the following anterior cavity is formed two serous membranes that surround the heart: parietal pericardium and the visceral pericardium?
    a. Abdominal
    b. Diaphragm
    c. Pelvic
    d. Pericardial
    e. Peritoneal
    f. Pleural
    g. Superior mediastinum
    h. Thoracic

62. The anterior (ventral) cavities are located near the front of the body and are subdivided into various smaller cavities which contain either organs or bodily fluids. Which of the following cavity is located between the parietal peritoneum and the visceral peritoneum?
    a. Abdominal
    b. Diaphragm
    c. Pelvic
    d. Pericardial
    e. Peritoneal
    f. Pleural
    g. Superior mediastinum
    h. Thoracic

63. The anterior (ventral) cavities are located near the front of the body and are subdivided into various smaller cavities which contain either organs or bodily fluids. Which of the following cavity contains the stomach, liver, spleen, small intestines and most of the large intestines?
    a. Abdominal
    b. Diaphragm
    c. Pelvic
    d. Pericardial
    e. Peritoneal
    f. Pleural
    g. Superior mediastinum
    h. Thoracic

64. The anterior (ventral) cavities are located near the front of the body and are subdivided into various smaller cavities which contain either organs or bodily fluids. Which of the following cavity contains the bladder, sigmoid colon, rectum and the internal reproductive organs?
    a. Abdominal
    b. Diaphragm
    c. Pelvic
    d. Pericardial
    e. Peritoneal
    f. Pleural
    g. Superior mediastinum
    h. Thoracic

65. The anterior (ventral) cavities are located near the front of the body and are

subdivided into various smaller cavities which contain either organs or bodily fluids. Which of the following cavity is made up of the parietal and visceral peritoneum?
    a. Abdominal
    b. Diaphragm
    c. Pelvic
    d. Pericardial
    e. Peritoneal
    f. Pleural
    g. Superior mediastinum
    h. Thoracic

66. _____ is a serious medical condition that is caused by peritoneal fluid buildup
    a. Abdominal
    b. Asites
    c. Diaphragm
    d. Pelvic
    e. Pericardial
    f. Peritoneal
    g. Pleural
    h. Superior mediastinum

ANSWERS

1.	e	2.	b	3.	d
4.	d	5.	a	6.	b
7.	c	8.	a	9.	a
10.	d	11.	a	12.	d
13.	d	14.	a	15.	d
16.	d	17.	e	18.	e
19.	e	20.	a	21.	c
22.	b	23.	b	24.	a
25.	e	26.	c	27.	g
28.	a	29.	g	30.	c, b, d, a
31.	b, c, d	32.	c, b, g(j), e, h, a, d, f	33.	l, p, n, k, q, g(j), o, l, m
34.	x, r, t, n, u, w, y, q, o, p, v, s, z	35.	a, c, f, e, m, g, j, n, q	36.	a
37.	b	38.	o, c, t, l, k, s, i, b, r, p, h	39.	c, e, i, h, a, b, d, g, f, j
40.	r, k, q, s, l, m, p, t, n, o	41.	d	42.	b
43.	a	44.	c	45.	b
46.	a	47.	h	48.	g
49.	e	50.	i	51.	d
52.	f	53.	c	54.	a, f, d, b, j, c, d, h, g, e
55.	n, k, m, p, o, r, q, l	56.	b	57.	e
58.	e	59.	h	60.	f
61.	d	62.	e	63.	a
64.	c	65.	e	66.	b

REFERENCES

Hickey J. The Clinical Practice of Neurological and Neurosurgical Nursing. 6th ed. Philadelphia, PA: Wolters Kluwer Health/Lippincott Williams and Wilkins. 2009

Rank W. Simplifying neurological assessment. Nursing Made Incredibly Easy! 2010. 8: pp. 15-19

Moran JF. Clinical Methods: The History, Physical, and Laboratory Examinations. 3rd ed. Boston: Butterworths. 1990. Chapter 17

Moran JF. Clinical Methods: Examination of the Extremities. 3rd ed. Boston: Butterworths. 1990. Chapter 30

Jevon P. Chest examination - Part 1 - chest palpation. Nursing Times, 2006. 102: pp. 26

Mularski RA, White-Chu F, Overbay D, Miller L, Asch SM, Ganzini L. Measuring Pain as the 5th Vital Sign Does Not Improve Quality of Pain Management. J Gen Intern Med., 2006; 21: pp. 607–612

Seeley RR, Stephens TD, Tate P. Anatomy and Physiology 6th Edition. McGraw-Hill, New York, New York. 2003

Tate P. Seeley's Principles of Anatomy & Physiology 1st Edition. McGraw-Hill, New York, New York. 2009

Saladin KS. Anatomy & Physiology. The Unity of Form and Function 6th Edition. McGraw-Hill, New York, New York. 2010

PHOTO AND GRAPHIC BIBLIOGRAPHY

1. Figure 1: Graphic designed by P.Y.P. Jen. Copyright ©
2. Figure 2: Graphics designed and arranged by PYP Jen
3. Figure 3: Graphic designed by PYP. Jen. Copyright ©
4. Figure 4: Graphic designed and released by Open Stax College, Connexions, reprint permission granted under the Creative Commons Attribution 3.0 Unported license
5. Figure 5: Graphic designed and released by Open Stax College, Connexions, reprint permission granted under the Creative Commons Attribution 3.0 Unported license
6. Figure 6: Graphic designed and released by Open Stax College, Connexions, reprint permission granted under the Creative Commons Attribution 3.0 Unported license.

7.   Figure 7: Graphic designed and released by Open Stax College, Connexions, reprint permission granted under the Creative Commons Attribution 3.0 Unported license.
8.   Figure 8: Graphic designed and released by Open Stax College, Connexions, reprint permission granted under the Creative Commons Attribution 3.0 Unported license.
9.   Figure 9: Graphic designed and released by Open Stax College, Connexions, reprint permission granted under the Creative Commons Attribution 3.0 Unported license.
10. Figure 10: Graphic designed and released byOpen Stax College, Connexions, reprint permission granted under the Creative Commons Attribution 3.0 Unported license.
11. Figure 11: The photo was taken in 1986 and released by Centers for Disease Control and Prevention. Public domain photo
12. Figure 12: Graphic designed and released by Open Stax College, Connexions, reprint permission granted under the Creative Commons Attribution 3.0 Unported license.
13. Figure 13: Graphic designed and released by Open Stax College, Connexions, reprint permission granted under the Creative Commons Attribution 3.0 Unported license.
14. Figure 14: Graphic designed and released byOpen Stax College, Connexions, reprint permission granted under the Creative Commons Attribution 3.0 Unported license.

CHAPTER TWO

ELEMENTS AND SUBATOMIC PARTICLES

Matter is anything that takes up space, has mass and is composed of elements. There are approximately 92 naturally occurring elements and 26 laboratory synthesized elements. Put simply, as of 2014, there are 118 confirmed elements shown in the periodic table (Appendix I).

An element is defined as a substance that cannot be broken down into another form through ordinary chemical reactions. In other words, elements, naturally occurring or otherwise, are the building blocks that combine to form the physical world. The elements are listed by their atomic symbol, which is formally recognized by the International Union of Pure and Applied Chemistry (IUPAC). Atomic symbols stand for the abbreviated name of each particular element. For example, living organisms are primarily composed of six elements also recognized by their acronym *CHNOPS* formed through their atomic symbols. These six elements are carbon (C), hydrogen (H), nitrogen (N), oxygen (O), phosphorus (P), and sulfur (S) (Table 1).

Table 1: Examples of atomic symbols

Element	Atomic Symbols
Carbon	C
Hydrogen	H
Nitrogen	N
Oxygen	O
Phosphorus	P
Sulfur	S

The atom is composed of protons, electrons and neutrons. The proton is a subatomic particle with one positive electric charge (fundamental unit 1.602×10^{-19} Coulomb) and a mass of 938.3 MeV/c^2 (1.6726×10^{-27} kg). A Coulomb (C) is a unit of electrical charge and is defined as the amount of electricity transported per second via one ampere of current. A proton is approximately 1836 times the mass of an electron. The electrons possess a negative electric charge (-1.6×10^{-19} Coulombs) and a mass of about 0.51 MeV/c^2 (9.11×10^{-31} kg). The third subatomic particle is the neutron, which has no net electric charge and a mass of 939.573 MeV/c^2 (1.6749×10^{-27} kg) (Table 2).

Table 2: Charge of subatomic particles

Atomic Particles	Net Charge
Proton	+1
Neutron	0
Electron	-1

The atom is composed of a dense central core containing positively charged proton(s) and neutral neutron(s), which are slightly greater in mass than the proton. Circling the nucleus in orbit(s) is(are) negatively charged electron(s). It is understood that electrons are bounded to their specific orbit by electromagnetic forces generated by the nucleus.

The atomic mass of the element is equal to the total mass of neutrons and protons within the atom. Since the electrons are so minute (their weight is relatively insignificant), their mass is generally ignored (Table 3; Appendix I).

The atomic number is used to distinguish one element from another and it corresponds to the electric charge of the nucleus. Put simply, the atomic number is the number of protons within the element (Table 4; Appendix I).

In a neutral atom, the positive charge of the proton is equal to the negative charge of the electron. Therefore, the opposite charges cancel each other out and the atom remains neutral. For example, if the atomic number of carbon (C) is six, there would be six protons located within the nucleus of the atom while there would be six electrons in orbit around the nucleus. The positive charges of the protons is cancelled out by the negative charges of the electrons and the carbon atom remains neutral (Table 5; Appendix I).

Table 3: Examples of atomic mass

Element	Atomic Mass
Hydrogen	1.0079
Carbon	12.011
Potassium	39.0983
Magnesium	24.305
Calcium	40.078

Table 4: Examples of atomic number

Element	Atomic Number
Hydrogen	1
Carbon	6
Potassium	19
Magnesium	12
Calcium	20

Table 5: Number (#) of proton and electrons

Element	Atomic #	# Protons	# Electrons
Hydrogen	1	1	1
Carbon	6	6	6
Potassium	19	19	19
Magnesium	12	12	12
Calcium	20	20	20

ISOTOPES

It is the atomic numbers that makes the elements found within the periodic table unique. However, on occasion, the same element may possess the same atomic number (same number of protons) but different atomic mass due to variations in the number of neutrons. These alternative forms of the same element are called isotopes. For example, hydrogen has three isotopes with the same atomic number but has different atomic masses. These three hydrogen isotopes are protium (also known as hydrogen), deuterium, and tritium. Each of these isotopes possesses one proton, but differs in the number of their neutrons. For example, Protium (^{1}H) has no neutron, deuterium (^{2}H) has one neutron, and tritium (^{3}H) has two neutrons. Therefore, these isotopes of hydrogen have an atomic mass of one, two, and three respectively. Each of these isotopes has one electron to balance the charge of the one proton (Figure 1). Experiments have shown that some isotopes are radioactive while others are quite stable even under extreme laboratory condition. For example, both protium and deuterium are stable isotopes while only tritium is radioactive.

Presently, more than 1,000 radioactive isotopes have been discovered. Only ~50 of these radioactive isotopes are found naturally, while the remaining ~950 are synthesized in laboratory experiments. These radioactive isotopes are also referred to as radionuclides.

Radioactive isotopes have unstable nuclei due to the excess number of neutrons. The nuclei of these radioactive particles will undergo radioactive decay or dissipate excess energy in the form of alpha, beta, and gamma rays until it becomes a stable element. Radioactive decay is measured based upon its half-life. A half-life is defined as the amount of time it takes for half of these unstable isotopes to break down into a more stable form. As an example, tritium possesses a half-life of approximately

12.3 years, where half of the atoms will break down into helium-3 (a non-radioactive isotope of helium with two protons and one neutron).

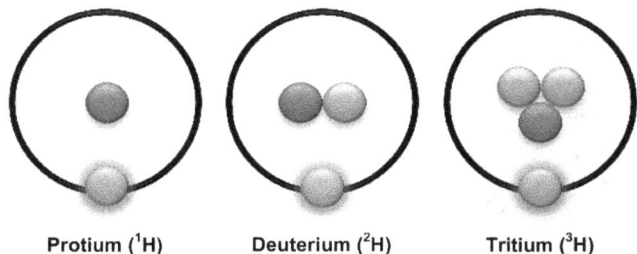

Figure 1: Isotopes of hydrogen. Note: proton (red), electron (yellow) and neutrons (blue)

Carbon 14 (^{14}C) is the most commonly used radioactive isotope in science. For example, ^{14}C is commonly employed in a process known as carbon dating, where this isotope is used to estimate the age of a fossil or archeological artifact. ^{14}C is unstable and undergoes radioactive decay. In time, one of its excess neutrons will split into a proton and an electron. The newly formed electron is quickly ejected from the nucleus and into orbit in the L shell, while the proton is retained, which will give this atom the atomic number of a nitrogen atom with a total count of seven protons and seven electrons (Figure 2).

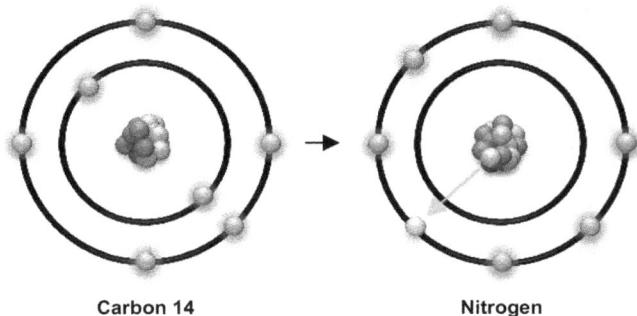

Figure 2: Illustration of ^{14}C decay. Note that the excess neutron splits into a proton and an electron. The electron is quickly ejected from the nucleus and into orbit within the L shell while the newly formed proton remains within the nucleus

It has been calculated that the half-life of C^{14} is approximately 5,730 years, where approximately half of these isotopes found within fossils or artifacts are broken down to form N^{14} (Figure 3).

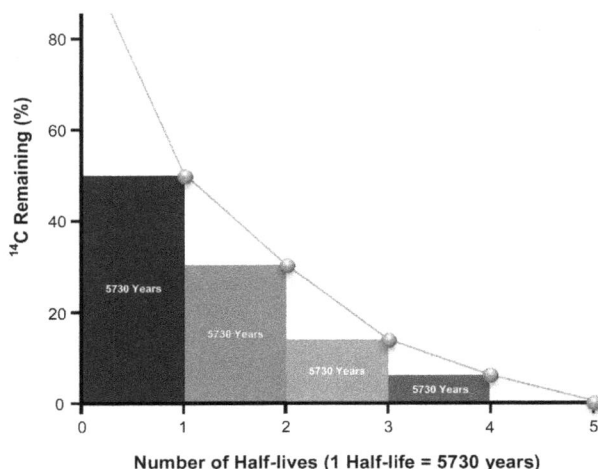

Figure 3: The half-life of C^{14}

It is known that living organisms accumulate small amounts of ^{14}C as they assimilate non-isotopic carbon (^{12}C) through photosynthesis or consumption of other organisms. When the organism is alive, ^{14}C is continuously decaying and being released as ^{14}N into the atmosphere. Nonetheless, while the organism is alive, ^{14}C isotopes, along with normal non-isotopic carbon (^{12}C) atoms, are constantly being incorporated through their metabolism. However, when the organism dies the replenishment of ^{14}C ceases. Since all organisms are estimated to contain similar percentages of ^{14}C at the time of death, this knowledge forms the baseline of the dating experiment. By using the predetermined baseline, scientists examine the ratio of ^{14}C radiation remaining within the fossil in comparison to the non-isotopic carbon (^{12}C). From this comparison the scientist can determine the approximate age of the fossil up to 40,000 years old.

Since ^{14}C has a relatively short half-life, scientists use other commonly found isotopes to determine the age of older fossils. Potassium (K) 40 is commonly used for fossils that date older than 20,000 to 50,000 years (the ratio of ^{40}K to argon 40 (^{40}Ar) is used to determine the approximate fossil age).

NUCLEAR MEDICINE

Nuclear medicine is a branch of medical science that utilizes radiation emitted from isotopes for diagnosis or in the treatment of disease. When nuclear medicine is used during medical diagnosis, radionuclides (radioactive isotopes) are introduced into the body via intravenous injection, inhalation (via the respiratory tract), or orally. These radioactive tracers have a short half-life and are designed to emit small amounts of gamma (γ) radiation from the internal structures of the body. The distinct advantage of nuclear imaging over the traditional x-ray techniques is that soft tissues and bone can be imaged and scrutinized by the attending physician.

During the early years of nuclear medicine, single photons were detected by a gamma camera, which viewed organs from various angles and stored the images in a database. This imaging system, called single photon emission computed tomography (SPECT), allows the computer to construct a two dimensional (2D) or three dimensional (3D) image of the body. Once the imaging process is completed, the collected information is then compiled and digitally enhanced before being viewed by a physician on a monitor for indications of abnormal conditions. Commonly used isotopes in SPECT are technetium 99m (Tc 99m; half-life six hours), iodine-131 (I-131; half-life eight days), and thallium-201 (T1-201; half-life three days) (Figure 4).

More recently, the development of positron emission tomography (PET), provides the physician with a more precise method for imaging. PET is a functional imaging technique that produces a 3D image of physiological process within the body. PET utilizes isotopes produced in a cyclotron. A cyclotron is an instrument used to make short half-life radioisotopes through hyper-acceleration of hydride atoms. One of the most commonly used cyclotron created isotope is fluorine-18, which has proven to be the most accurate method of detecting and evaluating cancers, cardiac functions, and brain imaging. These isotopes are introduced into the body intravenously and allowed to be absorbed by the organs. For example, organ malfunction may be indicated if the isotope is either incompletely taken up (cold spot), or taken up in surplus (hot spot). Once absorption of these radionuclides is complete, they will accumulate within the target tissue and radioactive decay begins. As they decays, the radionuclides emit positrons, which promptly combines with an available electron, resulting in the simultaneous emission of two gamma rays in opposing directions. These dual gamma rays are then detected by a PET camera, which allows precise imaging of internal functions without invasive procedures. Commonly used isotopes in PET are fluorine-18 (F-18; half-life 109 minutes), nitrogen-13 (N-13; half-life 10 minutes), carbon-11 (C-11; half-life 20 minutes), and oxygen-15 (O-15; half-life two minutes) (Figure 5).

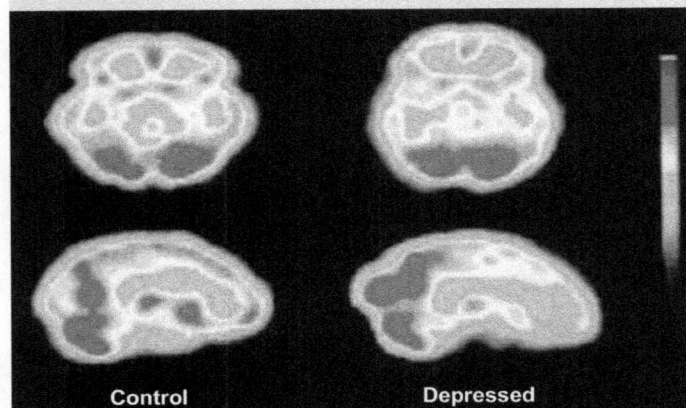

Figure 4: SPECT images from a depressed patient showing characteristic hypofrontality relative to a healthy control subject

Control   Depressed

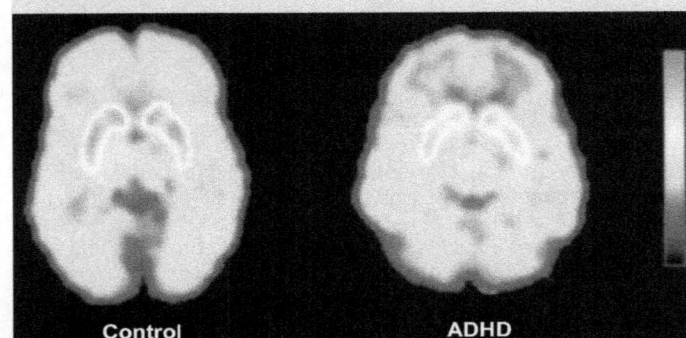

Control   ADHD

Figure 5: PET scan show that patients with ADHD had lower levels of dopamine transporters in the nucleus accumbens, a part of the brain's reward center, than control subjects

A newer procedure combines PET with computed X-ray tomography (CT) scans to provide two overlapping digitized images. This new diagnostic technique is estimated to provide 30% better imaging for diagnosis, when compared with the more traditional methods previously described (Figure 6).

Along with the advancement in techniques of using radionuclides in diagnostic imaging, therapeutic applications of radionuclides have equally made great strides. The use of β-emitting radiopharmaceuticals for pain relief in patients with widespread skeletal metastases is now widely employed. Beta (β) particles are subatomic particles equivalent to electrons that are ejected from the nuclei of some radioactive atoms. These β-emitters are designed to distribute highly energetic electrons several millimeters in the surrounding tissues. These radiopharmaceuticals are capable of providing simultaneously relief of pain to multiple areas of the skeleton. In comparison to the traditional external beam radiation therapy (EBRT), radiopharmaceuticals are capable of performing the tasks without significant soft-tissue toxicity and/or various technical complications. β-emitting radiopharmaceuticals are generally delivered intravenously and are designed to rapidly proceed to sites of active bone reaction and remodeling. Excess radionuclides are promptly eliminated either through the urinary or gastrointestinal system. Examples of these radio-pharmaceuticals are Phosphorus-32 (P-32; half-life14.3 days), Strontium-89 (Sr-89; half-life of 50.5 days), rhenium-186 (Re-185; half-life of 3.7 days), and samarium-153 (sm-153; half-life of 1.9 days).

Radionuclides are effective in the treatment of cancers. For example, the treatment, as well as imaging, for thyroid cancer following thyroidectomy is the ingestion of Iodine-131 (I-131).

I-131 has been found to be an effective treatment of disease recurrence and metastases.

Recently an alpha (α) emitter, called radium-223 (Ra-223; half-life of 11.4 days), has been developed for killing tumor cells and/or reducing the tumor burden. α-particles are a type of ionizing radiation in the form of large subatomic fragments consisting of two protons and two neutrons (e.g. helium nucleus) ejected by the nuclei of some unstable atoms.

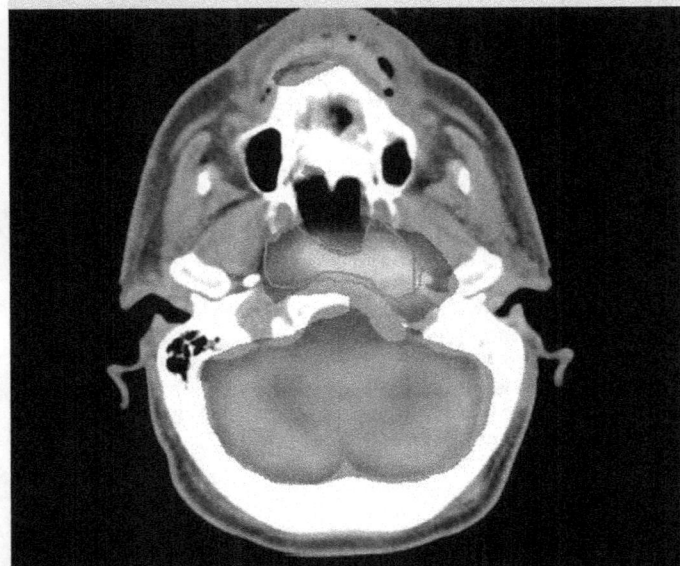

Figure 6: PET/CT image of a patient with nasopharyngeal undifferentiated carcinoma. The computed tomography gross tumor volume and the PET are highlighted with red and light blue contours

OCTET RULE

In 1913, Nobel Laureate Niels Bohr (1885-1962) proposed that electrons orbit around the nucleus (containing protons and neutrons) in concentric circles known as electron shells. His theory is based on the knowledge that even though all electrons have the same negative charge, not all electrons have the same amount of energy; therefore, depending on the specific energy levels of the electrons, they would inhabit different orbits circling the nucleus of an atom. Put simply, the higher the energy of an electron the further the orbit will be from the nucleus. Please examine Figure 7 for the electron distribution for the first three shells.

The diagram in Figure 8 shows the various electron shells envisioned by Bohr. The K shell which possesses the lowest electron energy level has only one shell (1s) and no subshell. The K shell can only accommodate 2 electrons. The L shell with higher energy level (in comparison to the K shell) possesses one shell (2s) and one subshell (2p). The 2s shell can accommodate 2 electrons while the subshell 2p will allow 6 additional electrons (2+4 electrons). The M shell will have 3s, 3p, and an additional shell, 3d. The 3s will accommodate 2 electrons, 3p will have 6 electrons (2+4 electrons), while 3d will have 10 electrons (6+4 electrons). Each additional electron shell will have one more subshell than the previous shell. In addition, each of the subshells will be able to accommodate 4 additional electrons. This gives the pattern of 2, 6, 10, 14, 18 and 22 electrons that will be able to fit into successive subshells.

The octet rule refers to the placement of electrons into electron shells. The ① lower levels of energy shells must be full before electrons are added to the next higher shell, and ② atoms with full outer energy shells are the most stable and do not chemically react under ordinary conditions. Inversely, atoms with incomplete outer shells readily react in an attempt to obtain a full outer shell format. Put simply, atoms will transfer or share electrons in an attempt to complete their outer electron shells and become more stable in the process.

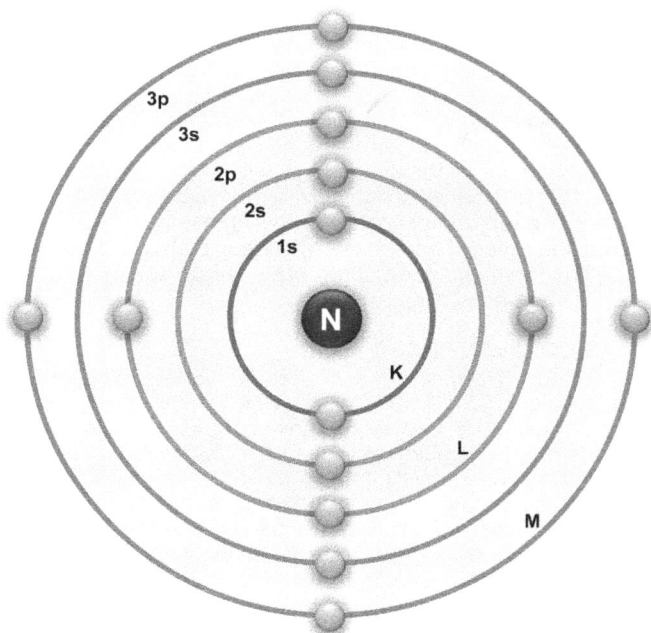

Figure 7: Illustration of Bohr's electron shells. Note that the K shell possesses only 1 shell (1s) with no subshell and can accommodate only 2 electrons. The L shell possesses 2s and 2p electron shell and subshell and can accommodate 2 and 6 electrons respectively. The same is shown for the M shell (3s and 3p) although the third subshell (3d) is not shown

Figure 8: Electron distribution of Bohr's electron shells

For example, hydrogen with only one electron in its outermost shell (K shell) is unstable and readily borrows an electron from neighboring atoms so that its single energy shell is complete with two electrons. In contrast, helium contains two electrons in its single shell, so it is stable and does not seek to share or transfer electrons with neighboring atoms. For this reason we refer to helium as an inert (noble) gas.

CHEMICAL BONDING

Chemical bonding occurs when two or more atoms interact with one another. There are several types of bonds that are essential to the creation of all things living and non-living on earth. The first of these bonds are ionic bonds, which involve the complete transfer of electron(s) from one atom to another. Ionic bonds occur only in inorganic molecules because this form of bonding does not provide the necessary platform for the creation of organic macromolecules (large organic molecules) (Figure 9).

The second type of bond are covalent bonds, which involve the

sharing of an electron(s) between two atoms. This type of bond exists in organic molecules where it allows for the formation of organic macromolecules.

A third type of chemical bond that is essential to all living organisms on earth is the hydrogen bond. Hydrogen bonds are weak bonds formed between polar covalent molecules that contain positively and negatively charged poles. For example, hydrogen bonds between water molecules are responsible for many of water's unique and life-enabling properties.

IONIC BOND

Ionic bonds are formed when an electron is transferred from one atom to another. For example, lithium (Li), having only one election in its outermost shell, would readily give away that single electron in order to form a more stable K shell. Li is also known as an electron donor. On the other hand, fluorine (F), with only seven electrons in its outer most (L) shell, would readily accept another electron to become stable. F is known as an electron acceptor (Figure 9).

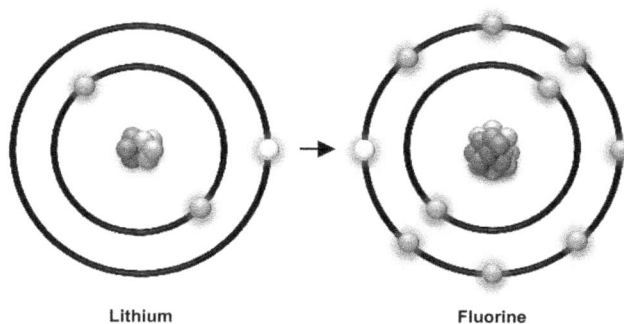

Lithium   Fluorine

Figure 9: Illustration of an ionic bond formed between lithium and fluorine to form the compound called lithium fluoride. Please note that lithium is the electron donor while fluoride is the electron acceptor. Lithium, by giving away an electron will possess the maximum number of electron in its K shell (1s) while fluorine by accepting an electron will complete its maximum number of electron in its L shell (2s and 2p) which is 8 electrons

The transfer of an electron results in the two atoms becoming ions or electrically charged atoms. For example, lithium, by giving up an electron will have an imbalance of 3 protons versus 2 electrons. This net imbalance will give lithium a positive (+) net charge. Fluorine, on the other hand, by accepting an electron will create an imbalance of 9 protons versus 10 electrons. This imbalance will give fluorine a net negative (-) charge.

It is insolution when these charges become significant. For example, when dissolved in water, lithium will become Li^+ or also known as a lithium cation (positive ion) while fluorine will become F^- or also known as a fluoride anion (negative ion).

NON-POLAR COVALENT BOND

Non-polar covalent bonds are formed when electron(s) is/are shared equally between two atoms. For example, hydrogen, with its single electron in its outermost shell (i.e. K shell), can become stable by adding a single electron to its outer shell. Therefore, naturally existing hydrogen is found in an H_2 format, where a pair of electrons is shared by two hydrogen molecules (H_2). The sharing of a single pair of electrons (two electrons) is called a single covalent bond and it is symbolized by a single line between two atomic symbols such as H-H. Other atoms, such as the oxygen atom, require two electrons to complete their outermost shells. Therefore, oxygen molecules tend to covalently bond together to share two pairs of electrons (a total of four electrons), which form a double covalent bond (Figure 10).

A double covalent bond is symbolized by a double line between the atomic symbols such as O=O. Finally, the nitrogen atom requires three electrons to complete its outermost shell; therefore, two nitrogen atoms tend to bind together and share three pairs of electrons (total of six electrons), resulting in a triple covalent bond. Triple covalent bond is symbolized by placing three lines between the atomic symbols such as N≡N.

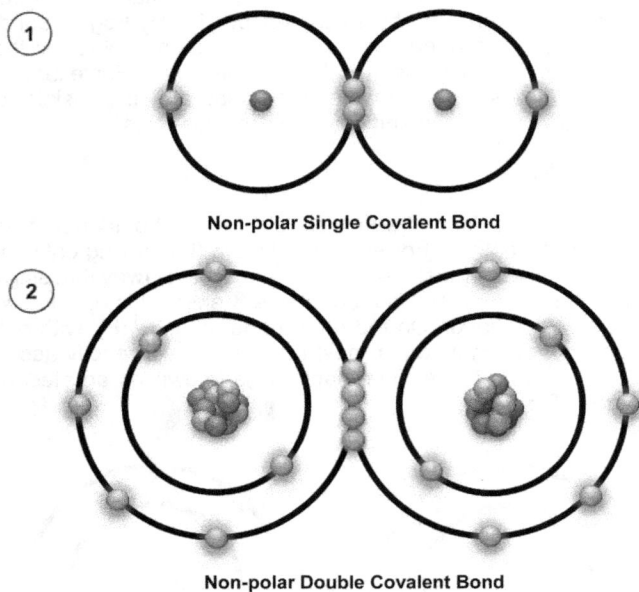

Non-polar Single Covalent Bond

Non-polar Double Covalent Bond

Figure 10: ① A single covalent bonded hydrogen sharing a pair of electrons and double covalent bonded oxygen ② sharing two pairs of electrons. Please note that the atoms involved in the electron sharing are equal, therefore not one atom possesses more affinity to gain unequal possession of the shared electrons. Put simply, the two examples given in the illustration are forming non-polar covalent bond

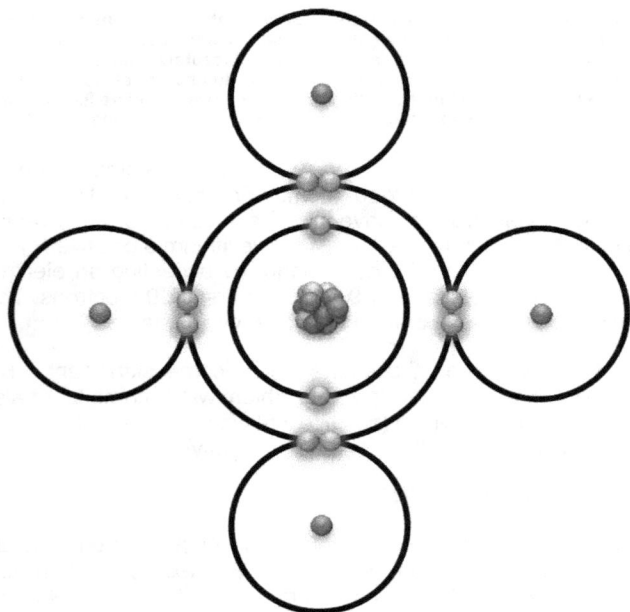

Figure 11: Graphic illustration of a methane (CH_4) molecule. Please note that the carbon atom possesses the ability to form four covalent bonds with other atoms or molecules. Additionally, carbon is a relatively weak atom and will share electrons equally with other atoms like hydrogen. Therefore, in the case of methane, this molecule is considered to be non-polar

Carbon is one of the most abundant elements and is the basic building block of all life on Earth. If we examine carbon's electron distribution, we will notice that within the L shell, there are 4 electrons in orbit (two election in the 2s subshell and two electrons in the 2p subshell). Based on the octet rule, carbon will need 4 additional electrons to become stable. For this reason, the carbon atom is capable of forming four covalent bonds with other

carbons, as well as, with a variety of other atoms or molecules. Because of this flexibility, there are approximately 10 million known carbon based compounds on Earth. This high abundance of the share abundance of carbon compounds, scientists have based an entire discipline, organic chemistry, towards the elucidation of this atom and the compounds that it forms.

Another unique quality of carbon is its rather weak affinity in sharing electrons with other atoms and molecules. For example, when carbon is sharing electrons with four other hydrogen forming CH_4 or otherwise known as methane gas, the sharing of the electrons are approximately equal. Consequently, scientists consider the covalent bond formed between carbon and hydrogen (as well as the bonds formed between carbon and other carbon atoms) as non-polar (Figure 11).

POLAR COVALENT BOND

Polar covalent bonds are formed when there is an unequal sharing of electrons between atoms. The unequal attraction of electrons by a particular atom within a molecule or compound is known as electronegativity. In polar covalent bonds, if one atom attracts electrons more than the other or if the electrons spend significantly more time orbiting one particular nucleus, then that atom is more electronegative than the other. Inversely, if the atom is shown to have less access to the electron comparatively to the other within the molecule, this atom is said to be electropositive.

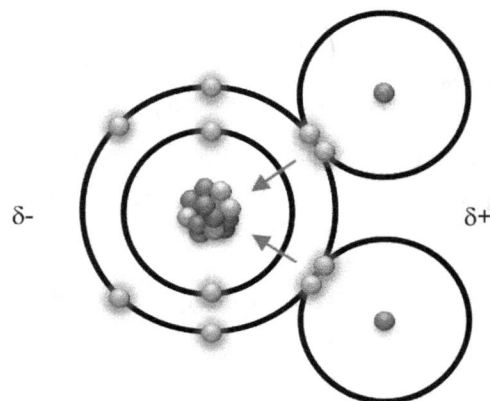

Figure 12: A polar covalent bonded water molecule (H_2O). Please note that oxygen has stronger attraction to electrons, indicated by the arrows demonstrating that the electrons are pulled closer towards its nucleus. Since electrons spend more time orbiting closer towards the nucleus of oxygen, it is therefore more electronegative δ-. Hydrogen, on the other hand, with fewer electrons orbiting around its nucleus is electropositive δ+

For example, in a water molecule (H_2O), oxygen, being a much bigger atom and possessing more protons within its nucleus, exerts more pull on the electrons. Due to oxygen's stronger attraction to electrons, it is therefore more electronegative than hydrogen. In contrast, hydrogen being a much smaller atom with only one proton in its nucleus, will exert only a limited amount of pull on the electrons. Due to the hydrogen's weaker attraction, it is therefore more electropositive to oxygen. Because of this unequal attraction, the water molecule is a polar molecule, which is defined as a molecule possessing two polar ends. On one end, the oxygen end, it is slightly negative, which is referred to as electronegative, and is shown with the symbol δ- (pronounced 'delta negative'). The opposite end of the water molecule, the hydrogen end, is slightly electropositive, δ+ (pronounced 'delta positive') (Figure 12).

HYDROGEN BOND

As a result of the polar covalent bond, an intermolecular force called hydrogen bond could be formed. Hydrogen bonding is created when the electro-negative (δ-) end of one molecule forms an attraction with the electropositive (δ+) end of another molecule. For example, water molecules have two polar ends, and each will form hydrogen bond with their polar opposite on other water molecules (Figure 13). This form of attraction between water molecules is referred to as cohesion. Being polar, water molecules can form hydrogen bond with other charged surfaces found in containers, pipes, or even blood vessels. This form of attraction between water molecule and charged surfaces is called adhesion.

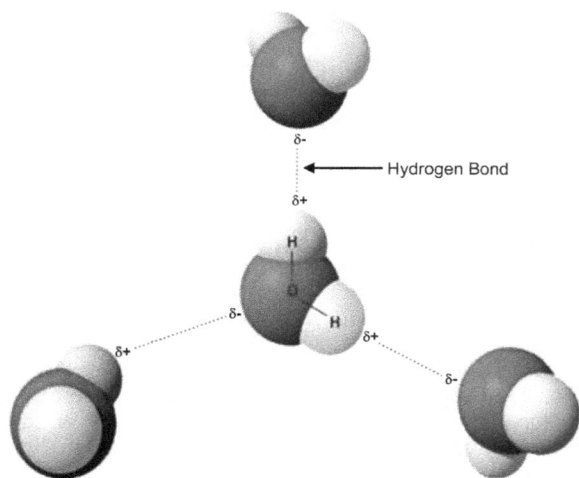

Figure 13: Hydrogen bonds formed between water molecules. Please note that the dotted lines shown in the illustration indicates hydrogen bonding and shows the attraction between polar opposite forces

The hydrogen bond is a weak bond, in comparison to both ionic and covalent bonding, but nonetheless it is essential component to the life sciences. It is understood that without hydrogen bond, life cannot exist. For instance, hydrogen bonding is partly responsible for the unique 3-dimensional structures that determine protein functions, which are essential to life.

ENERGY

Energy is defined as the capacity to do work. Work (W) on the other hand is defined as a force used to move matter. For example, a force of 50 Newtons (N) is applied to push an object 50 meters (m) will accomplish 2500 joules (J) of work.

$$W = Force \times Distance$$
$$W = 50N \times 50m$$
$$W = 2500J$$

Energy can be subdivided into two forms: potential energy and kinetic energy. Potential energy is stored energy, while kinetic energy is the energy of motion or a form of expended energy that is doing work. For example, if an archer draws a bow the potential energy is transferred from the arms of the archer through the draw string and into the bent bow limbs. However, when the string is released, the potential energy quickly transforms into kinetic energy which propels the arrow towards its target. Both potential and kinetic energy can also be used to describe mechanical and chemical energy, two forms of energy that are essential in the proper functioning of the human body.

Mechanical energy is the energy needed or generated by the movement of an object. Many of the functions of the human body require mechanical work. Functions such as movement of arms or legs, respiration, peristalsis, blood circulation, involve the movement of objects (kinetic mechanical energy) or the potential to move objects (potential mechanical energy).

Chemical energy is the potential of chemical substances to undergo a chemical reaction thereby releasing stored energy.

In addition, chemical energy is needed to transform reactants (substances involved in a chemical reaction) into other chemical substances necessary in maintaining homeostasis. For example, humans, like many other living organisms, require the consumption of food source in order to survive. The consumed food source contains potential energy that is released and used in myriad of controlled reactions after digestion and absorption.

ACID AND BASES

By definition, an acid is any substance that donates a proton in solution. For example, hydrochloric acid (HCl), readily loses a proton in the form of hydrogen ion (H^+) when placed in an aqueous solution (please note that hydrogen without its electron is simply a proton).

$$HCl \rightarrow H^+ + Cl^-$$

Inversely, a base is any substance that accepts a proton in solution. For example, sodium hydroxide (NaOH) is a base that readily dissociates into a sodium ion (Na^+) and hydroxide ion (OH^-) in solution. The hydroxide ion, in turn, will readily accept a hydrogen ion (H^+) or a proton to form water (Figure 11).

$$NaOH \rightarrow Na^+ + OH^-$$
$$OH^- + H^+ \leftrightarrow H_2O$$

Acid and bases are classified as strong or weak. A strong acid, such as HCl, is a substance that dissociates readily and completely in solution, releasing a proton (H^+). Similarly, a strong base, such as NaOH, is defined as a substance that dissociates readily and completely in solution and releases a hydroxide ion (OH^-). Simply, the more readily or completely an acid or base dissociates in solution, the stronger they are.

By definition, weak acids and bases only partially dissociate in solution. Only some of their protons (H^+) and hydroxide ions (OH^-) are released. An example of a weak acid is carbonic acid (H_2CO_3), which is the human blood buffer. Carbonic acid partially dissociates in solution (blood) and is always involved in reactions that are reversible.

$$CO_2 + H_2O \leftrightarrow H_2CO_3 \leftrightarrow HCO_3^- + H^+$$

Ammonia (NH_3) is an example of a weak base. Ammonia does not possess hydroxide ion but in an aqueous solution it reacts with the water molecules to form hydroxide ions. Ammonia partially dissociates in solution and its reaction with water is always reversible.

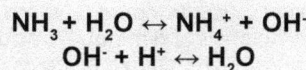

$$NH_3 + H_2O \leftrightarrow NH_4^+ + OH^-$$
$$OH^- + H^+ \leftrightarrow H_2O$$

pH SCALE

The pH scale is the manner by which acid and bases are measured. The pH scale ranges from 0 to 14 and is designed to show the concentration of protons (H^+) in solution (Figure 14). For example, pure water occasionally dissociates to form hydroxide ion (OH^-) and hydrogen ion (H^+, or a proton). Since pure water, on average, possesses equal numbers of protons and hydroxide ions, it is neutral and has a pH of 7.

$$H_2O \leftrightarrow OH^- + H^+$$

An acidic solution possesses more protons (H^+) in comparison to hydroxide ions (OH^-) in aqueous solution. Because of the higher concentration of hydrogen ions, an acidic solution will have a pH less than 7. In contrast, a basic solution will have more hydroxide ions in comparison to hydrogen ions. Because of the higher concentration of hydroxide ions, a basic solution will have a pH greater than 7 (Figure 14).

Concentration of H+	pH	Examples of Solutions
1/10,000,000	14	Liquid drain cleaner
1/1,000,000	13	Bleach
1/100,000	12	Soapy water
1/10,000	11	Ammonia
1/1,000	10	Milk of magnesia
1/100	9	Toothpaste
1/10	8	Baking soda
0	7	Pure water
10	6	Milk
100	5	Black coffee
1,000	4	Tomato juice
10,000	3	Orange juice
100,00	2	Lemon juice
1,000,000	1	Stomach acid
10,000,000	0	Battery acid

Figure 14: The pH scale

Calculating pH

Acids and bases are defined by the concentration of hydrogen ions (H+ mol/L) in solution. For example, if the concentration of H+ ions is 0.001mol/L, the pH of this solution is:

$pH = -\log (0.001)$
$pH = -\log (1 \times 10^{-3})$
$pH = - (-3)$
$pH = 3$

In contrast, if the concentration of H+ ions is 0.0000000001mol/L, the pH of this solution is:

$pH = -\log (0.0000000001)$
$pH = -\log (1 \times 10^{-10})$
$pH = - (-10)$
$pH = 10$

ACIDOSIS AND ALKALOSIS

On average the normal pH of the human blood is maintained between 7.35 and 7.45. However, when the pH value of body fluids is below 7.35, the condition that results is called acidosis. On the other hand, when the pH is above 7.45, the condition that results is known as alkalosis. Acidosis is a serious medical condition where the blood pH drops below 7.35. In humans, there are two forms of acidosis: respiratory acidosis and metabolic acidosis. Respiratory acidosis (also known as hypercapnic acidosis or carbon dioxide acidosis) is the result of the buildup of carbon dioxide (CO_2) within the system due to the respiratory tract's inability to remove excess CO_2. There are two forms of respiratory acidosis. Chronic respiratory acidosis takes place over an extended period of time and acute respiratory acidosis that onsets rapidly.

Respiratory acidosis may result from many different medical conditions. For example, a type of pathologic fracture of the thoracic vertebra called kyphosis may result in the decreased blood pH. Patients suffering from kyphosis demonstrate an over-curvature of the thoracic vertebra which can result in the bowing of the back in a slouching posture. This pathologic curving of the spine could place undue pressure upon the thoracic cavity and hinder respiration. Kyphosis is caused by a congenital condition or diseases such as tumor or osteoporosis etc. Similarly, respiratory acidosis may also be caused by a medical condition called scoliosis, which also results in the abnormal S-shaped curvature of the spine. Generally, this type of spinal deformity begins in the intervertebral discs causing distortions in the epiphyseal cartilage. Most often, scoliosis is idiopathic (cause unknown or arises spontaneously) although some may be caused by a congenital condition or neuromuscular disease. Respiratory acidosis can be caused by diseases of the respiratory tract, such as asthma and chronic obstructive lung disease (COPD). This condition is also often found in individuals that are chronically obese or are suffering from injuries to the chest and/or respiratory muscular weakness.

Metabolic acidosis results from the inability of the kidneys to excrete acid from the body. There are many ways in which metabolic acidosis can occur. For example, people suffering from diabetes (both type I and II) may develop diabetic ketoacidosis (DKA). DKA develops from a shortage of insulin and the body switches from using available glucose to utilizing fatty acids as a fuel source. The consequence of the switch in fuel source is a buildup of acidic ketones which the kidneys are unable to remove adequately. Hyperchloremic acidosis is a type of metabolic acidosis caused by the excessive loss of sodium bicarbonate ($NaHCO_3$) from the body. This condition usually occurs with individuals suffering from severe diarrhea. Lactic acidosis is a buildup of lactic acid within the system. This condition may be caused by excessive exercise, over drinking, liver failure, hypoglycemia, heart failure, severe anemia, or cancer. Proximal and distal renal tubular acidosis is caused by diseases associated with the kidneys.

Alkalosis is a medical condition where the body's pH is higher than average. There are two forms of alkalosis: respiratory alkalosis and metabolic alkalosis. Respiratory alkalosis is caused by the reduction of carbon dioxide in the body whereby a marked reduction of carbonic acid (blood buffer) is detected. Respiratory alkalosis may be caused by fever, diseases of the liver, hyperventilation, salicylate (e.g. aspirin) poisoning and/or lack of oxygen.

Metabolic alkalosis is generally caused by an increase in bicarbonates in the bloodstream. There are three different types of metabolic alkalosis. For example, hypochloremic alkalosis is caused by a lack of chloride which is generally associated with excessive vomiting. Hypokalemic alkalosis is generally the result of an excessive loss of potassium. This condition is generally associated with individuals taking diuretic medication. Finally, compensated alkalosis is generally seen in patients emending (correcting) from an acid-base imbalance where their bicarbonate and carbon dioxide levels remain abnormal.

ORGANIC MOLECULES

All organic molecules are covalently bonded and for the most part are extremely large molecules (also referred to as macromolecules) in comparison to inorganic molecules, where electrons are transferred from one atom to another and not shared. This ability to create large complex macromolecules is the result of the unique nature of carbon (C).

Carbon has four electrons in its outermost electron shell, meaning that it can gain four more electrons and become a more stable element based upon the Octet rule. This ability to share four electrons allows carbon to form four covalent bonds with other atoms or molecules, also referred to as functional groups. For example, the simplest organic compounds, composed of carbon and hydrogen, are known as hydrocarbons. The simplest of these hydrocarbons is called methane, a gas that has one carbon and four hydrogen atoms (CH_4) (Figure 8). By replacing one of the hydrogens with any of the functional groups, a different molecule is formed. Therefore, through the flexibility of the carbon atom any of the myriad of carbon based molecules that exist on Earth could be created. For example, by replacing a hydrogen atom with a chlorine atom will create a new molecule called chloromethane (CH_3Cl); replacing a hydrogen atom with a hydroxyl group (-OH) forms methanol (CH_3OH) (Figure 15).

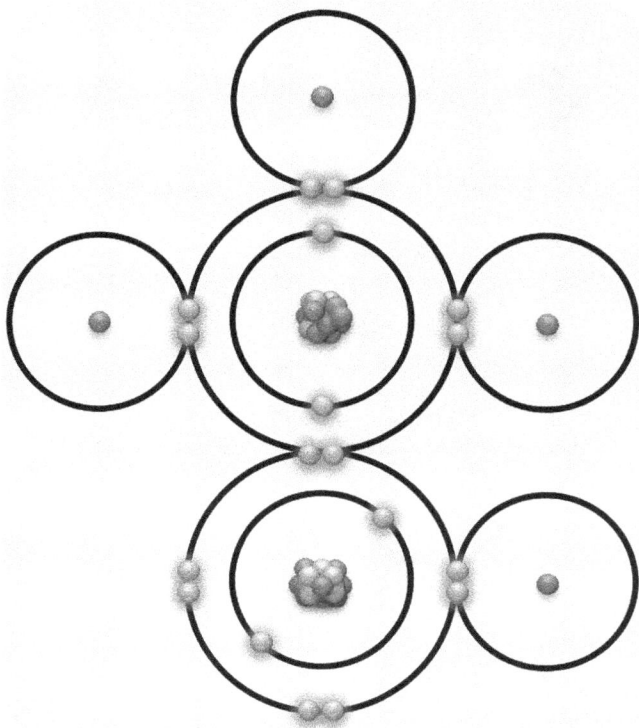

Figure 15: Graphic illustration of methanol (CH_3OH). Please note the flexibility of the carbon molecule (highlighted top) and its ability to form four covalent bonds with functional groups such as hydroxyl ion (highlighted bottom)

Carbon atoms may also form covalent bonds with other carbon atoms through the process of condensation synthesis, permitting the formation of an unbroken chain of carbon atoms. This "carbon backbone" is the fundamental characteristic of all organic macromolecules (Figure 16).

Condensation synthesis, also known as dehydration synthesis, is an organic chemical reaction where two molecules are joined together to create a large molecule with the loss of water (Figure 17).

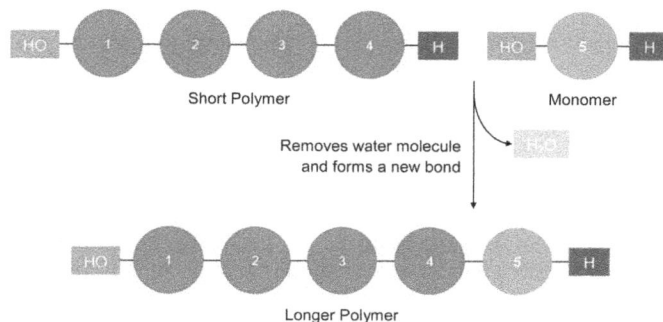

Figure 17: Condensation synthesis

The process of hydrolysis refers to the break down a larger molecule into smaller subunits through the addition of water. Please note that both hydrolysis and condensation synthesis require the support of organic catalysts, also known as enzymes, which increase the rate of the reaction (Figure 18).

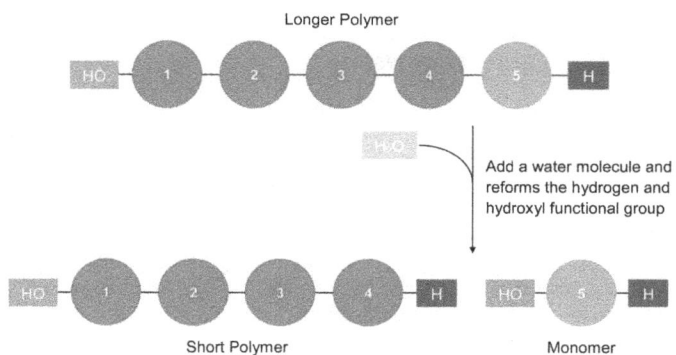

Figure 18: Hydrolysis

Organic macromolecules are classified into four major groups. These four groups are carbohydrates, lipids (including cholesterol and steroids), proteins and nucleic acids.

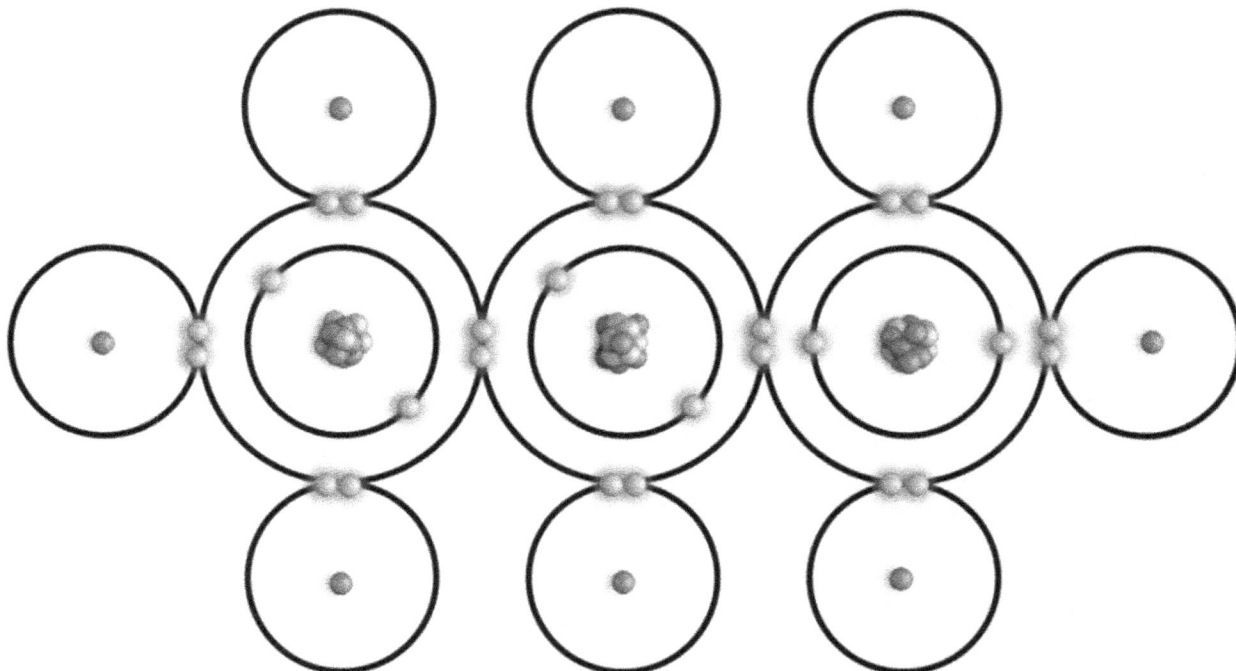

Figure 16: Graphic illustration of a propane molecule (C_3H_8). Please note the flexibility of the carbon molecule to form covalent bond with other carbon atoms to form macromolecules

Carbohydrates

The name carbohydrate is drawn together from the abbreviated word for carbon, carbo- and the abbreviated word for water, -hydrate. The reasoning behind this compilation of words is the molecular similarities of this macromolecule with water and its ability to dissolve in polar solvents, such as water. Put simply, carbohydrate is constructed out of a combination of carbon, oxygen, and hydrogen with the ratio of one carbon, two hydrogens and one oxygen (CH_2O, also known as hydrated carbon).

Carbohydrates, also referred to as sugars, are the major chemical energy source for most living things on earth. These complex molecules form a variety of structural components of living cells. For example, cellulose, the most abundant compound on earth, makes up most of the plant cell wall. This material, which makes up the majority of the composition of lumber, is what we use to build houses and furniture, as well as, ingesting as dietary fibers to prevent constipation. Other carbohydrates, such as starch and glycogen, are used by plants and animals respectively as storage sugars.

Carbohydrates are also present on the surfaces of plasma membranes and are responsible for some of its specific antigenic properties. For example, the antigenic properties of various polysaccharides located on the human red blood cells create the ABO blood types in primates.

Carbohydrates are subdivided based upon the number(s) of sugar components they contain. For example, a monomer (single unconnected molecule) of carbohydrate is called a monosaccharide and it is the basic building block of more complex sugars. Monosaccharides are a broad group of simple sugars. Some have a simplistic structure, possessing a carbon backbone of only three carbon atoms ($C_3H_6O_3$). These simple monosaccharides include aldotriose (glyceraldehyde) and ketotriose (dihydroxyacetone). On the other extreme, some monosaccharides, such as neuraminic acid, possess complex structures that contain a carbon backbone of nine carbon atoms ($C_9H_{17}N_1O_8$).

Examples of a typical monosaccharide commonly used in the biological world are pentose sugars (five carbon sugars) such as ribose and deoxyribose, the sugars found in RNA and DNA, respectively. Other examples of monosaccharides include those with a carbon backbone of six carbon atoms (hexose). Examples of these six carbon molecules are glucose, mannose, and fructose (Table 6).

Glucose is the principal source of energy for most eukaryotic organisms. This monosaccharide is transported to cells as an immediate source of energy. For instance, a patient in the hospital receiving nutrients intravenously is placed on a "glucose drip." Mannose (D-mannose) is hexose that is commonly prescribed for the prevention of urinary tract infections and the treatment of a metabolic disorder known as carbohydrate-deficient glycoprotein syndrome. Fructose, on the other hand, is also referred to as corn syrup and is found in most packaged foods and canned drinks. Presently, there are a lot of controversies surrounding the widespread use of this sweetener and its contribution to obesity.

Table 6: Terminology of sugars based upon the number of carbon in their carbon backbone

Carbon #	Term	Examples
3	Triose	Glyceraldehyde and dihydroxyacetone
4	Tetrose	Erythrose, threose, erythrulose
5	Pentose	Arabinose, ribose, lyxose, xylose, deoxyribose
6	Hexose	Glucose, mannose, fructose, galactose, talose
7	Heptose	Sedoheptulose, mannoheptulose
8	Octose	Synthetic monosaccharides
9	Nonose	Neuraminic acid, sialic acid
10	Decose	Synthetic monosaccharides

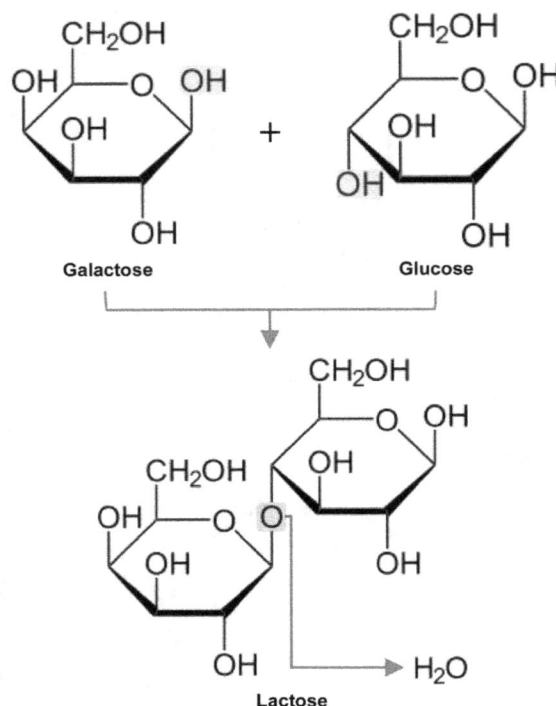

Figure 19: Graphic illustration of the formation of lactose. Please note that both glucose and galactose have the same molecular formula but their hydroxyl group is orientated differently at the fourth carbon

A disaccharide is a polymer (compounds with more than one monomer) that is constructed of the combination of two monomers through the process of condensation synthesis (Figure 17). For example, lactose, an animal sugar found in milk, is formed from the combination of glucose ($C_6H_{12}O_6$) and galactose ($C_6H_{12}O_6$) (Figure 19). Sucrose, on the other hand, is a disaccharide, commonly known as cane sugar or table sugar. Sucrose is a polymer formed from a glucose and fructose molecule. Maltose ,is another example of a disaccharide that is composed of two molecules of glucose. In the human digestive process, maltose is formed when the enzyme α-amylase, found in saliva, breaks down starch.

Polysaccharides are constructed through the combination of many monomers through the process of condensation synthesis. Starch, cellulose, and glycogen are the most common polysaccharides seen in the biosphere. Glycogen, for example, is a multibranched polysaccharide that consists of linear chains of glucose with various other chains of glucose, branching off approximately every 10 glucose residue. On average, a glycogen molecule consists of approximately 30 to 60 thousand glucose monomers. This polysaccharide serves as a form of energy storage in animals and fungi. In humans, glycogen is stored in the liver by hepatocytes (liver cells) and can be readily hydrolyzed to release glucose monomers into the blood stream.

In addition to the three most common polysaccharides mentioned previously, there are also unique sugars that are associated with genomes and genetic information. For example, ribose and deoxyribose are five carbon ring sugars and form the polymeric carbon backbone of deoxyribose nucleic acid (DNA) and ribose nucleic acid (RNA).

Lipids

Like carbohydrates, lipids are composed of carbon, hydrogen and oxygen. Nonetheless, the quantity of oxygen atoms in lipids is lower. In addition, elements like phosphorous and nitrogen are often included in the molecular structure of lipids. Unlike carbohydrates, which are hydrophilic (water loving) molecules, lipids are hydrophobic (water hating), and thereby insoluble in polar solvents like water. They can only be dissolved in non-polar solvents such as pentane, benzene, and toluene.

The family of lipids includes compounds such as fats, phospholipids, steroids, and eicosanoids (including prostaglandin, thromboxane, and leukotriene). Fats, for example, are energy storage molecules. In humans, fats in the form of fatty acids, are found in adipose tissue distributed throughout the body. The adipose tissue serves to provide an individual with a stored energy source, insulation from the environment, and padding to protect internal organs from trauma.

In general, lipids are composed of two types of molecules: fatty acids and glycerol. There are approximately 70 fatty acids that occur naturally. Fatty acids typically consists of carbon chains 14 to 22 atoms long and are soluble in water because they contain a polar carboxyl group (-COOH). Fatty acids can come in two forms: saturated or unsaturated. Saturated fatty acids contain no double bonds between the carbon atoms; on the other hand, unsaturated fatty acids have a mixture of double and single bonds (Figure 20).

The second type of molecule required to construct fats and oils is glycerol. Glycerol contains three polar hydroxyl groups (-OH) and is water soluble.

When a fat is formed, the carboxyl group of the fatty acids react with the hydroxyl group of glycerol through condensation synthesis (Figure 17). Because glycerol contains three hydroxyl groups, three fatty acids combine per glycerol molecule. It is for this reason that this type of fat is also known as a triglyceride. Once the carboxyl and hydroxyl functional groups are utilized in the condensation synthesis, the newly formed triglyceride is no longer polar, therefore insoluble in water. Triglycerides are the most common form of fat found in animal adipose tissue (Figure 21).

Similar to the formation of a triglyceride, the formation of a phospholipid involves the carboxyl group of two fatty acids reacting with the hydroxyl group of a glycerol through condensation synthesis. Instead of the attachment of the third fatty acid, a phosphate group will combine with the glycerol. The result is a slightly polar molecule, due to the phosphate, while the reminder of the molecule is hydrophobic. This unique property of phospholipid allows it to serve as the bulk of the cellular/plasma membrane in all living organisms (Figure 22).

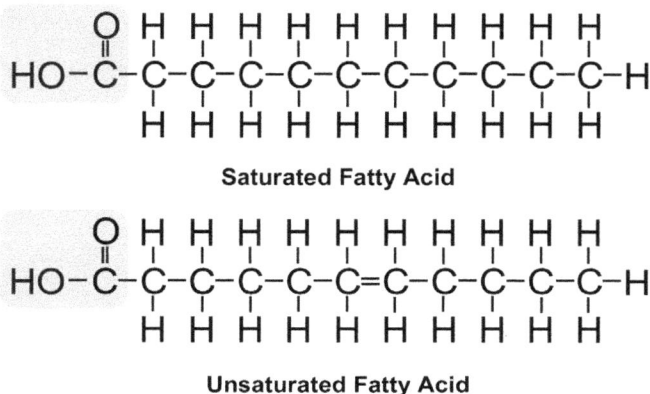

Saturated Fatty Acid

Unsaturated Fatty Acid

Figure 20: Illustration of saturated and unsaturated fatty acids. Please note that the carboxyl group is highlighted in red

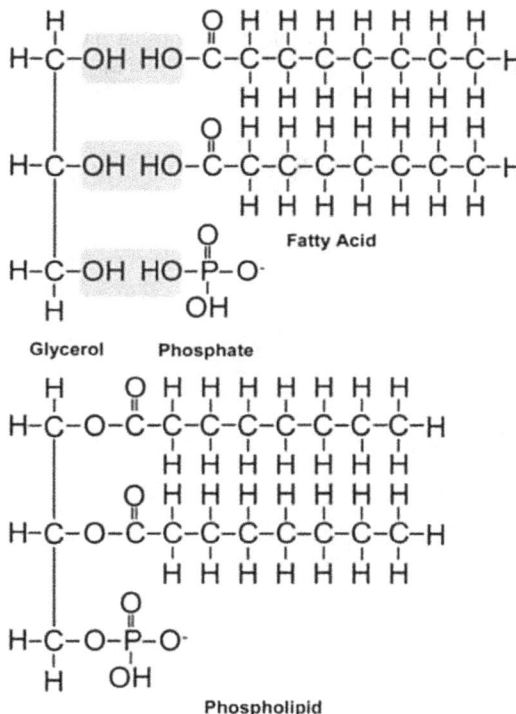

Glycerol   Fatty Acid

Triglyceride

Figure 21: Formation of a triglyceride through condensation synthesis

Glycerol   Phosphate

Phospholipid

Figure 21: Formation of a phospholipid through condensation synthesis

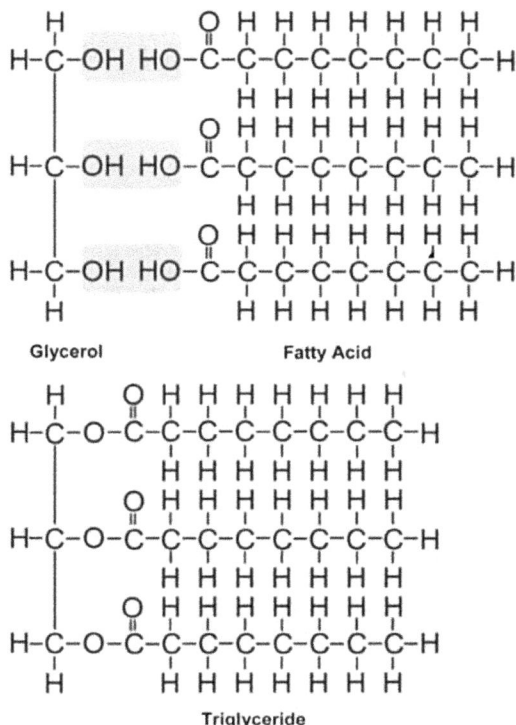

FATS AND HEALTH

Fats are contained in almost all foods we consume, and contrary to popular beliefs, some are beneficial to your health. It is important not to eliminate all fats from your diet, just be knowledgeable as to which ones are actually helpful. Presently, the medical community recommends two main types of potentially helpful dietary fats. Monounsaturated fat (MUFA) is a type of dietary fat that contains single double bonded carbons within its chemical structure and are found in a variety of foods and oils. Studies have shown that MUFAs improve blood cholesterol levels and decrease the chance of heart disease. In addition, medical research has demonstrated that MUFAs can actually aid in the control of blood sugar levels. Generally, MUFAs are liquid at room temperature but turn to a solid form when chilled. Olive and peanut oils are examples of ingredients that contain mono-unsaturated fats. Polyunsaturated fat (PUFA) is a type of dietary fat that contains more than one set of double bonded carbons within its chemical structure and is typically found in cooking oils (e.g. soybean, corn and sunflower oils) and fish (salmon, mackerel, herring, trout etc.). Scientific research has shown that consuming moderate proportion of PUFAs improves blood cholesterol levels, which may serve to decrease the risk of heart disease. Like MUFAs, PUFAs may also help in controlling blood sugar levels and decrease the risk of developing type 2 diabetes. Safflower, sunflower, corn, and omega-3 fish oils are examples of foods or ingredients that contains PUFA.

In contrast, there are two groups of fats that may be harmful to health. Saturated fat is a type of fat that contains no double bond within its chemical structure. Saturated fats are mainly found in animal sources such as meat and dairy products. Nonetheless, plant sources such as palm oil, palm kernel oil and coconut oil, also contain the potentially harmful fats. Research has shown that saturated fats raise low-density lipoprotein (LDL) cholesterol levels, which may increase the risk of cardiovascular disease, as well as chances of developing type 2 diabetes (Figure 23).

Transfats are also considered to be harmful. Trans fats, also known as trans-fatty acids, are found naturally in animal sources but are generally created during the processing of packaged foods. Hydrogen is added to liquid vegetable oils to make them more solid. Research has shown that saturated fats raise LDL cholesterol levels, which may increase the risk of cardiovascular disease (Figure 23).

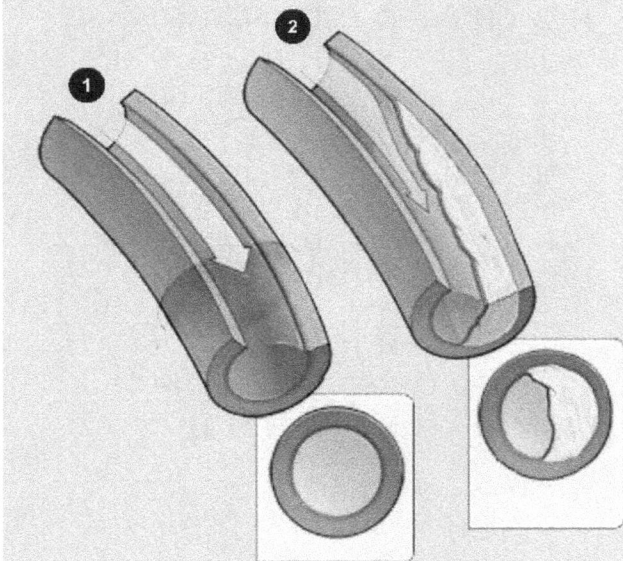

Figure 23: Coronary heart disease is a condition in which plaque builds up inside the coronary arteries

Steroids

Steroids are a type of lipid commonly found in nature and they are composed of four cycloalkane carbon ring structure. More specifically, steroids are formed from three cyclohexane (circular six-sided) rings and one cyclopentane (circular-five sided) ring. The rings are named A, B and C for the cyclohexane rings and D for the cyclopentane ring (Figure 24). Common steroid molecules include cholesterol, bile salts, testosterone, estrogen, and progesterone. Many steroids serve as hormones (signal molecules or ligands) within the human body. These hormones are responsible for the activation of various bio-physiological processes. For example, the cells found with the zona glomerulosa of the adrenal cortex secrete a steroid called aldosterone. Aldosterone is a type of mineralocorticoid that helps regulate sodium and potassium concentration of the blood

through the kidneys, thereby aiding in the maintenance of blood pressure.

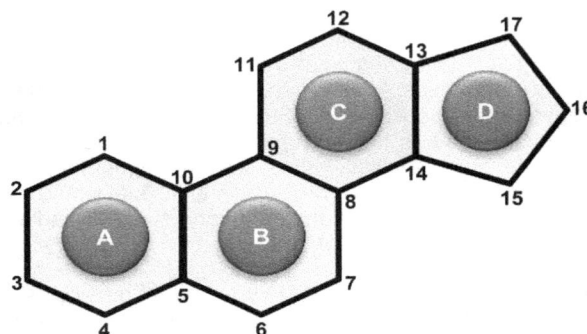

Figure 24: IUPAC standard lettering and numbering of the steroid ring on this gonane or steroid parental structure. This gonane could be modified through addition, subtraction or replacement of atoms to form any steroid in existence

Proteins

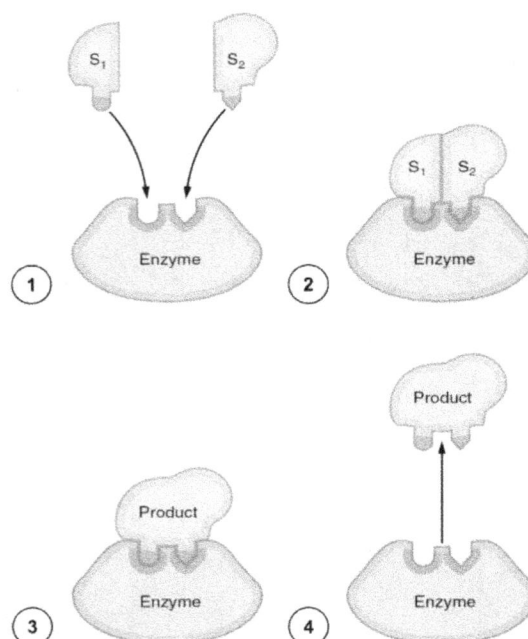

Figure 25: Illustration of an enzyme assisted reaction where the enzyme is used to speed up a reaction within living organisms. ① Substrates (reactants) S1 and S2 is configured into the active site of the enzyme. ② Active site is the location where the reactions between the substrates take place to create a ③ product. ④ Once the product is formed, it is released and the enzyme is free to perform another reaction. Please note that an enzyme does not become part of the products formed

Proteins are one of the major building blocks of living organisms and are among the most abundant organic molecules on earth. For example, proteins make up about 50% of the dry weight of most living organisms.

There are many different types of protein molecules. For example, keratin is one of the major components found in your hair and nails. Keratin is a hydrophobic fibrous structural protein that is formed to create a permeability barrier to prevent water seepage and loss. Additionally, enzymes are proteins which serve as organic catalysts that speed up chemical reactions in cells (Figure 25).

Many hormones are also made of proteins. A prime example is insulin, a signal molecule essential in controlling our blood sugar levels. Other examples of proteins found in living organisms are transport proteins, such as those that construct hemoglobin; contractile proteins, such as myosin found in muscles; immunoglobulins (e.g. antibodies); and various types of structural proteins found within cells and among the extracellular matrix of the body. No matter what type of protein or their particular function, all proteins follow the same blueprint where they are all polymers of amino acids, the building block of this molecule.

The basic component of protein are amino acids. Amino acids, as the name indicates, have an amino functional group (NH_2) and an acid group (carboxyl -COOH). There is also a third functional group, known as R-group, which binds to the central carbon (also known as the α-carbon) of the molecule. It is this R-group that allows for the differentiation of one amino acid from another (Figure 26).

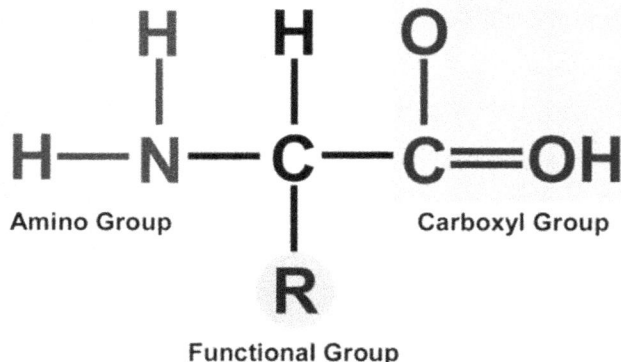

Figure 25: Standard structure of an amino acid

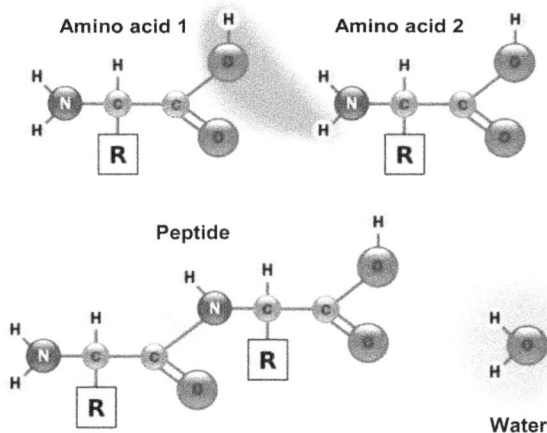

Figure 27: Formation of a peptide bond and the creation of a peptide through condensation synthesis

Based on the simplistic construct of amino acids, it is theoretically possible to create a large variety of amino acids simply by substituting one R-group with another. However, at present, there are only 21 naturally existing amino acids and these protein building blocks are used by all life forms and viruses on earth (Appendix II).

To form polymers, amino acids are joined together through a condensation reaction between the carboxyl group (-COOH) and the amino group ($-NH_2$), forming a type of covalent bond also known as a peptide bond (Figure 27). The sequence of amino acids that are joined together will determine the biological characteristics of the proteins formed. This is called the primary sequence, which in turn determines the overall structure of protein. Protein structure is critical for proper function; put simply, the shape determines the function. Since proteins are macromolecules containing hundreds, if not thousands, of amino acids, the number of different proteins that can be formed are astronomical.

Genetic Material

Nucleotides are monomers that form nucleic acids, a polymer. A nucleotide is composed of a nitrogenous (heterocyclic) base, a five-carbon (pentose) sugar ring (which could be either ribose of deoxyribose) and a phosphate group. Condensation synthesis forms between the phosphate group (PO_4H) located at the 5' end (5th carbon) of the pentose sugar with the hydroxyl group (-OH) at the 3' end (3rd carbon) of another pentose sugar. The formation of these covalent (ester) bonds creates the long linear chain of the sugar phosphate backbone that is either deoxyribonucleic acid (DNA) or ribonucleic acid (RNA).

There are 5 types of heterocyclic bases in existence and they are used as the source of genetic information by all living organisms, as well as, viruses on earth. The five heterocyclic bases are adenine (A), guanine (G), thymine (T) and cytosine (C) found in DNA, while RNA is composed of the same bases with the exception of uracil (U) in place of thymine. Both adenine and guanine are classified as purines because of their double ring structures, while thymine, cytosine and uracil are classified as pyrimidines because of their single ring structures (Figure 28).

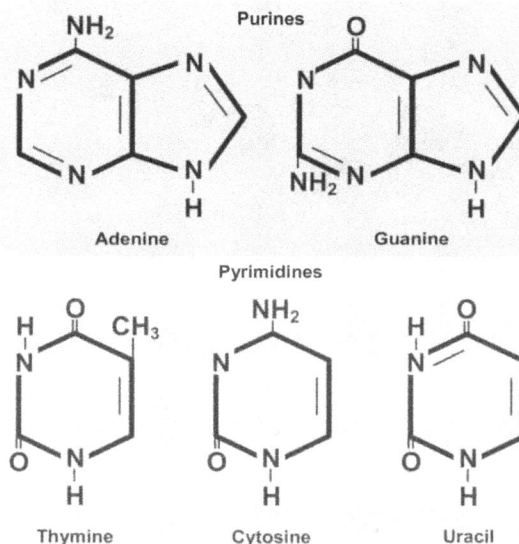

Figure 28: The five heterocyclic bases. Please note that uracil (U) is found exclusively in RNA while thymine (T) is found exclusively in DNA

Deoxyribonucleic Acid

The structure of DNA in living organisms, as well as some viruses, is composed of a double stranded parallel and anti-parallel format known as a double helix. The DNA double helix is composed of four distinct nitrogenous bases and a backbone composed of deoxyribose sugar (a pentose, five-sided sugar) bounded to phosphate groups.

Within the double helix of a DNA sequence, thymine (T) is always matched with adenine (A) on the opposite strand while cytosine (C) is always paired with guanine (G) on the opposite strand. The heterocyclic bases are interconnected through the formation of hydrogen bonds which provide a stable structure. For example, two hydrogen bonds are formed between adenine and thymine, while three hydrogen bonds are formed between cytosine and guanine (Figure 29).

The DNA sequences are the genetic instructions used in the maintenance and development by every living organisms, as well as some viruses (please note that some viruses use RNA as their genetic material). DNA sequences are similar to a blue print and dictates the organism's genotype (genetic makeup) and phenotype (appearance) by outlining the construction of the organism. As indicated previously, DNA molecules are double helix polymers, with backbones made of pentose sugars (deoxyribose) and phosphate groups joined by ester bonds. The strands of the double helix run in opposite directions: parallel (3' to 5') and anti-parallel (5' to 3'). The genetic codes for all living organisms are stored within the sequence of the heterocyclic bases (Figure 30).

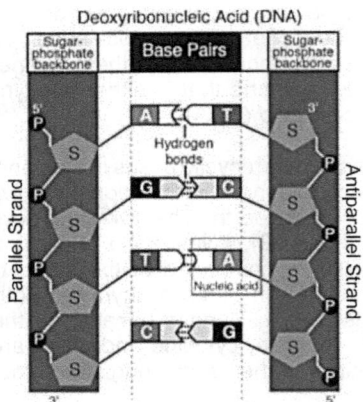

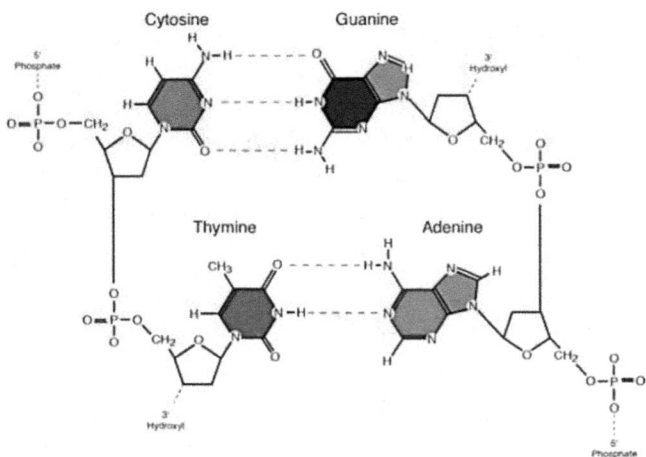

Within the nucleus of eukaryotic cells, DNA is organized through its interaction with histone proteins into structures called chromosomes. The replication of these chromosomes occurs prior to cellular division (mitosis and meiosis) (Figure 31). In contrast, prokaryotic cells (bacteria and archaea) store their DNA in the cytoplasm. The DNA structure of prokaryotes is a double stranded circular DNA, which we refer to a prokaryotic chromosome although no chromosome is actually formed.

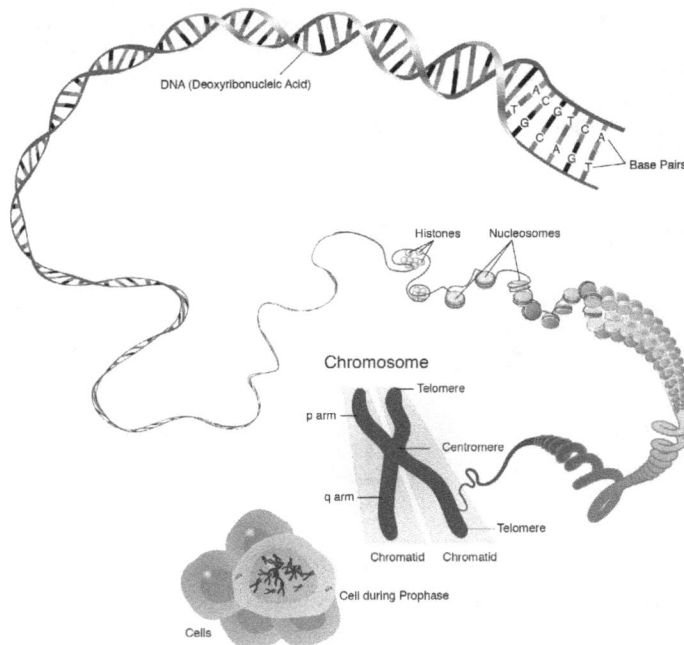

Figure 29: A DNA double helix. Please note that adenine (A) binds with thymine (T) while guanine (G) binds with cytosine (C)

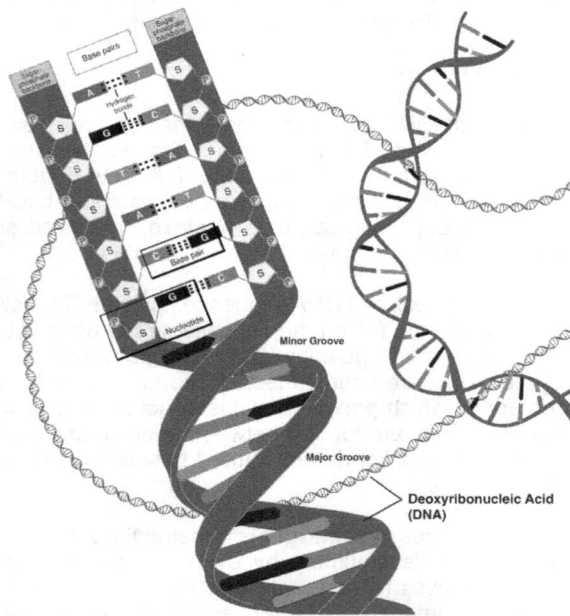

Figure 31: Prior to cellular division, replicated DNA interacts with the histone protein and is packed into a tightly wounded structure called a chromosome

Ribonucleic Acid

RNA, like DNA, exists in living organisms, as well as, some viruses. There are three types of RNA in eukaryotic cells: messenger RNA (mRNA), transfer RNA (tRNA) and ribosomal RNA (rRNA). mRNA is formed through the process of transcription. Transcription is the starting point for protein synthesis whereby a segment of the DNA molecule is copied into an mRNA sequence, which then leaves the nucleus and enters the cytoplasm. Within the cytoplasm, the mRNA proceeds to the ribosome to be translated into a polypeptide sequence. rRNA is formed in the nucleolus and constitutes 60% of the composition of the organelle known as the ribosome. Ribosomes are complex membrane-less molecular organelle that serves as the site for translation of mRNA. Finally, tRNA is a small RNA molecule that is responsible for transferring a specific amino acid to the site of protein synthesis, the ribosome.

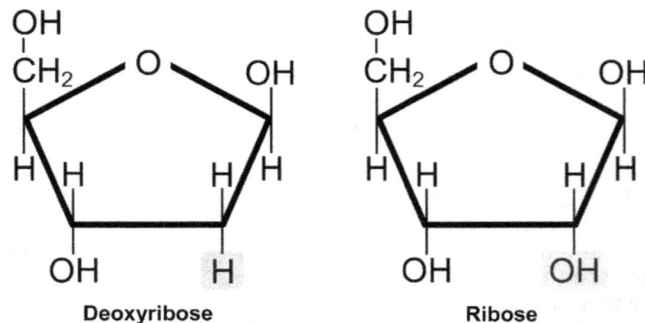

Figure 32: Deoxyribose and ribose sugars. Please note the subtle differences highlighted

With the exception of some double stranded RNA viruses, RNA exists in a single linear strand format. RNA is also composed of

Figure 30: A DNA double helix

nucleotides, but there are several differences when compared to its DNA counterpart. For example, the pentose sugar of RNA is called ribose, which is structurally different than deoxyribose. The 2' carbon on the pentose molecule of deoxyribose has two hydrogen atoms, while on ribose (at the same location) there is a hydroxyl group (-OH) and a hydrogen atom (Figure 33).

Another difference between RNA and DNA is that one of the heterocyclic bases thymine (which only exists in DNA) is replaced by uracil (U) as one of the pyrimidines (Figure 28). In an RNA sequence, adenine (A) is matched with uracil (U), while cytosine (C) is paired with guanine (G). Please note that the pairing between guanine and cytosine remains the same as in DNA.

Transcription

Within the nucleus, RNA is transcribed (copied) from a DNA template by a group of enzymes called RNA polymerases (Figure 32). Once transcription is completed, the mRNA proceeds through an editing process known as RNA splicing. Segments of non-coding RNA sequences, called introns are removed, while only the sequences necessary for protein synthesis are retained. These remaining coding sequences, called exons, are the final version of the mRNA. Once the editing is completed, the mRNA leaves the nucleus and proceeds to the ribosome, the site of protein synthesis, also known as translation. mRNA is coded in a three heterocyclic base format called a codon, where each codon represents one specific type of amino acid.

The second type of RNA is called transfer RNA (tRNA). tRNA is approximately 80 nucleotides in length and is capable of transferring a specific amino acid to a growing polypeptide chain in a ribosome. Each tRNA possesses a site for amino acid attachment located at its 3' end, while on the opposite side a three heterocyclic base sequence called an anticodon is located. The anticodon is intended to match the codon sequence of mRNA. Put simply, each of the codon found on the mRNA codes for one specific amino acid (Figure 34). For example, if the codons read CCU-UAA-CGU, there will be three tRNAs to match to the codon. The three tRNA's anticodon sequence will read ①

GGA ② AUU and ③ GCA (Appendix II).

tRNA

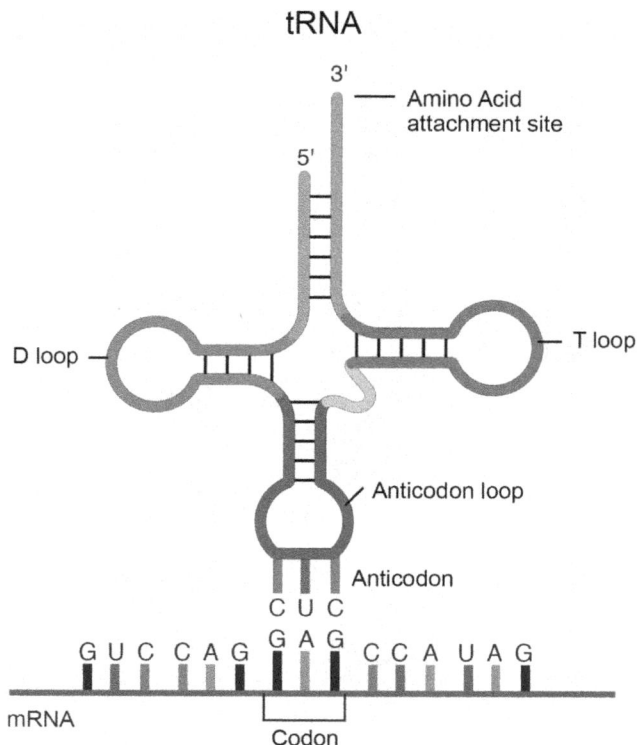

Figure 34: A transfer RNA (tRNA). Please note that the anticodon is matched with the codon from the messenger RNA (mRNA)

Lastly, the ribosomal RNA (rRNA) is the catalytic component of the ribosome. In eukaryotic cells, there are four known rRNA molecules and they are distinguished by molecular size: 18S, 5.8S, 28S, and 5S. The 18S, 5.8S and 28S rRNA are formed in the nucleolus, while the 5S rRNA is formed in the nucleoplasm; however, the final assembly of this molecule takes place in

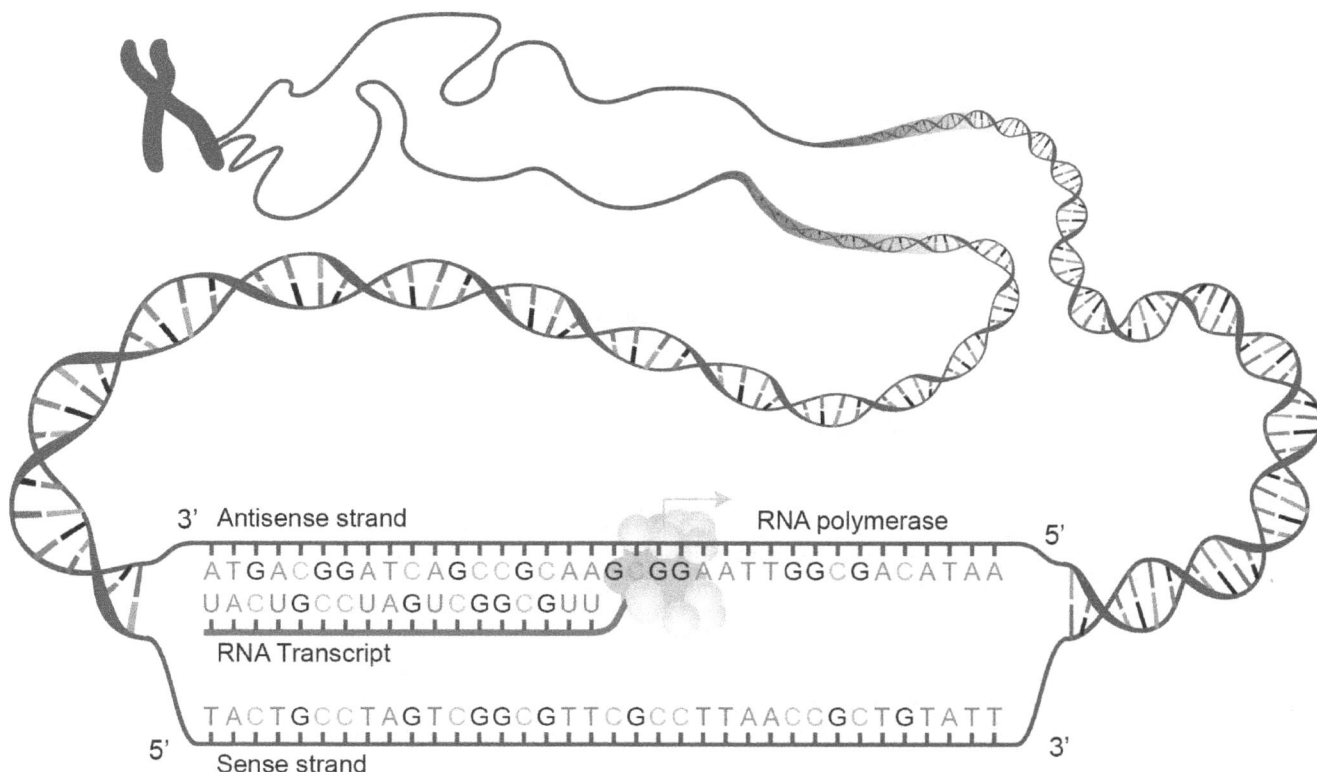

Figure 33: Graphic illustration of a DNA double helix and transcription

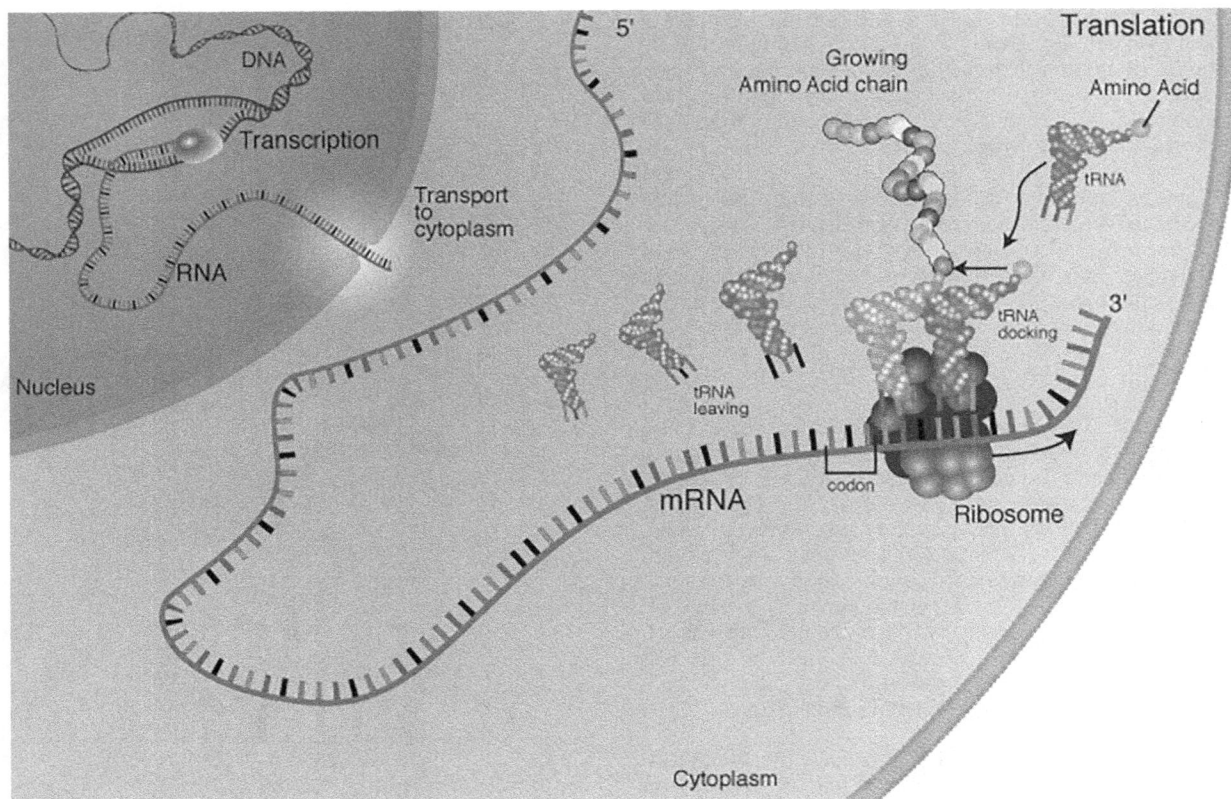

Figure 35: Graphic illustration of translation

the nucleolus. These rRNA are assembled along with various proteins in the cytoplasm to form two nucleoprotein structures: 60S (large subunit) and 40S (small subunit). Subsequently, the large and small subunits will combine to form a ribosome. The ribosome is designed as a site of translation (protein synthesis) for the mRNA molecule (Figure 35).

TRANSCRIPTION AND TRANSLATION

Proteins constitute over 50% of total dry cell mass. Therefore protein synthesis within both prokaryotic and eukaryotic cells is essential. The information used to create proteins is encoded within the DNA sequences found within the nucleus of a eukaryotic organism. In order for protein synthesis to take place, the information stored within the DNA must first be transcribed into messenger ribonucleic acid (mRNA) and subsequently translated into polypeptide at the ribosome. For example, if the 5' → 3' sequence of DNA (sense strand) is:

5' ATG-CCG-CAT-TTA-TTC-TAC-TAC-TAA 3'

The complimentary strand (antisense strand) would be (3' → 5'):

3' TAC-GGC-GTA-AAT-AAG-ATG-ATG-ATT 5'

The complimentary strand (antisense strand) will be used as a template to produce a mRNA sequence via transcription. Please note that the mRNA sequence is almost identical to the sense strand (5' → 3') of the DNA. The only difference is that thymine is replaced by uracil. This transcribed sequence is:

5' AUG-CCG-CAU-UUA-UUC-UAC-UAC-UAA 3'

Once the transcription is completed, the mRNA leaves the nucleus and proceeds into the cytoplasm to be translated by ribosomes (e.g. polyribosomes). Please note that each of the three heterocyclic base of the mRNA is called a codon. Each codon must match the anticodon found on the tRNA. The tRNA is responsible for bringing the correct amino acid to the ribosome;

the correct amino acid sequence is dictated by the mRNA. The tRNA and amino acid sequences are:

UAC	GGC	GUA	AAU	AAG	AUG	AUG	AUU
Met	Pro	His	Leu	Phe	Tyr	Try	Stop

SICKLE CELL ANEMIA

Sickle cell anemia is a general term for a group of related, chronic, genetic disorders caused by sickle shaped erythrocytes (red blood cells) (Figure 36).

In normal individuals, the erythrocytes are biconcave in shape and move easily throughout the blood vessels as they transport oxygen (O_2) and remove some carbon dioxide (CO_2). Normal adult erythrocytes are constructed out of hemoglobin (HBB^A) proteins. HBB^A, on the other hand, are constructed out of two alpha (α) and two beta (β) protein chains. Each of the chains is associated with a heme group (Figure 37).

The HBB^A makes up 95% of all hemoglobin in adults, while the remaining 5% consist of Hemoglobin A_2 ($α_2σ_2$) and Hemoglobin F ($α_2γ_2$), a type of fetal hemoglobin. A normal heme molecule consists of an iron atom at the center surrounded by a protein ring called a porphyrin. Porphyrins, are metalloproteins, which are proteins associated with an inorganic metal.

In individuals suffering from sickle cell anemia, the sickle cell hemoglobin (HBB^S) possesses two normal alpha (α) chains, while their two beta (β) chains are defective and prone to structural collapse. The genetic defect that caused the malfunctioning beta (β) chains are called point mutations or also known as base substitutions, where a single nucleotide of the β-globin gene is replace with another (Figure 38).

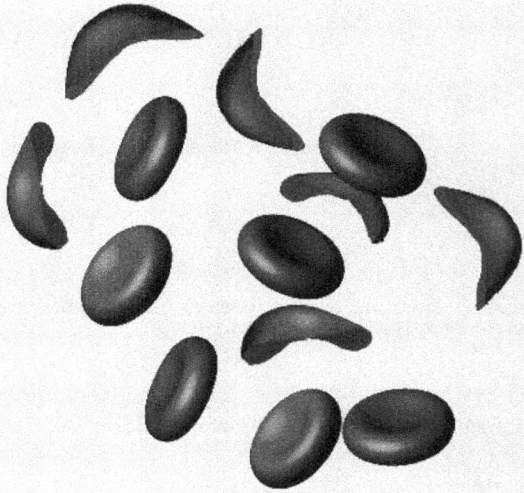

Figure 36: Sickle shaped cells which is a familiar occurrence with people suffering from sickle cell anemia

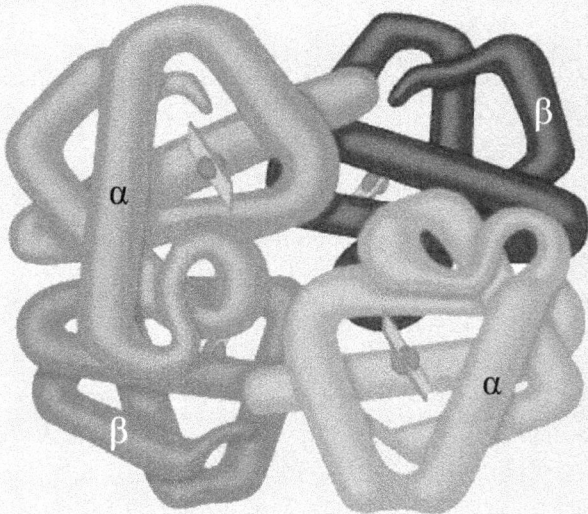

Figure 37: Hemoglobin (HBB^A). Please note that the heme molecule is illustrated by a simple red dot located within the alpha and beta protein chains

During the subsequent transcription and translation, the point mutation will result in the exchange of the amino acid glutamic acid, which exists in normal hemoglobin with valine, within the sickle cell beta (β) globin proteins on the hemoglobin. This mutated β chain is referred to as an S chain and this exchange of amino acids results in sickling.

Normal Hemoglobin (HBB^A)

5'	C	C	T	G	A	G	G	3'
3'	G	G	A	C	T	C	C	5'

Sickle Cell Hemoglobin (HBB^S)

5'	C	C	T	G	T	G	G	3'
3'	G	G	A	C	A	C	C	5'

Figure 38: Illustration demonstrate a single base substitution of HBB^S in comparison to HBB^A

In many forms of the disease, the structure of the erythrocytes (red blood cells) changes shape upon deoxygenation due to the polymerization of the abnormal hemoglobin. This process damages the cellular membrane of the erythrocytes, and can also cause a condition known as vaso-occlusion (Figure 39).

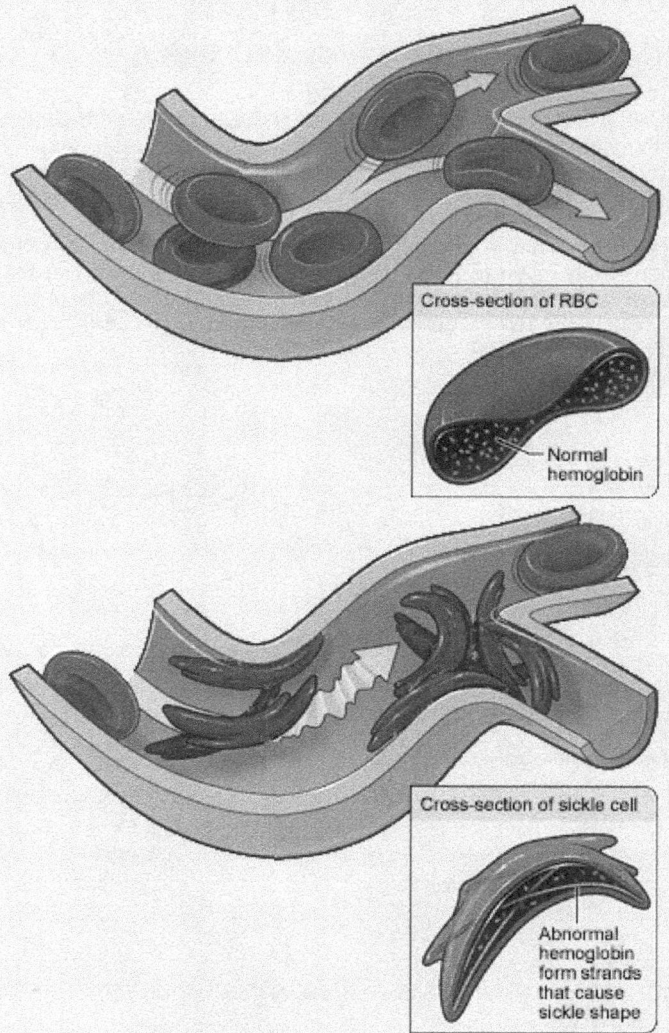

Figure 39: Illustration of sickle shaped cells which is a familiar occurrence with people suffering from sickle cell anemia

Vaso-occlusion is a serious medical condition where the erythrocytes become wedged in the branching of the blood vessels or become lodged in the smaller vessels. The blockage will deprive the tissues further downstream of oxygen and may cause serious damage to internal organs, such as spleen or liver and may also result in a stroke if the blockage occurs in the brain.

Adenosine Triphosphate

Adenosine triphosphate (ATP) is the most prominent energy currency of living organisms. Adenosine stands for adenine (a heterocyclic base) combined with ribose sugar, while triphosphate stands for three phosphates linked together via phosphate (covalent) bond (Figure 40).

The action of ATP is similar to that of a rechargeable battery where it can be used to power myriad of cellular functions and is recharged and reused through the process of phosphorylation and de-phosphorylation. Put simply, potential energy formed through cellular respiration is stored within the phosphate bond (also known as a high energy bond) located between the second and third phosphate (Figure 40).

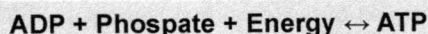

ADP + Phospate + Energy ↔ ATP

The energy stored within this energy molecule is subsequently released in the form of kinetic energy when the bond is hydrolyzed.

ATP ↔ ADP + Phospate + Energy

ATP is used to create intracellular messengers. For example, adenylate cyclase, a membrane bounded enzyme utilizes ATP to create cyclic adenosine monophosphate (cAMP) in order to activate other enzyme and prompt additional cellular processes.

There are other energy molecules similar to ATP that are commonly referred to as nucleoside triphosphates (NTPs) that exist within an organism. For example, NTPs include uridine triphosphate (UTP), guanosine triphosphate (GTP), and cytidine triphosphate (CTP).

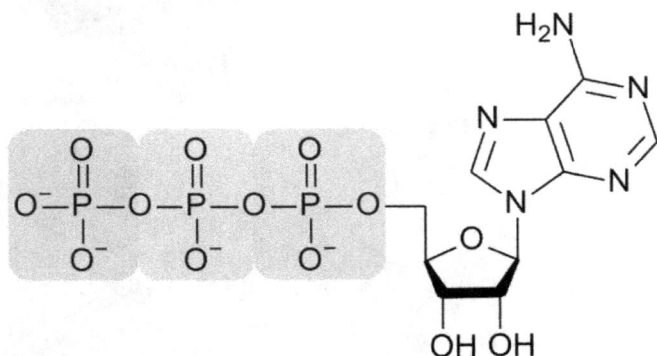

Figure 40: Adenosine triphosphate

QUESTIONS

1. How many naturally occurring elements are known to scientists?
   a. 26
   b. 50
   c. 87
   d. 92
   e. 118

2. How many elements have been synthesized in laboratories?
   a. 26
   b. 50
   c. 87
   d. 92
   e. 118

3. As of 2014, how many confirmed elements are known to scientific communities?
   a. 26
   b. 50
   c. 87
   d. 92
   e. 118

4. What is the atomic symbol for potassium?
   a. P
   b. K
   c. Na
   d. Ca
   e. Fe

5. What is the atomic symbol for iron?
   a. P
   b. K
   c. Na
   d. Ca
   e. Fe

6. What is the atomic symbol for phosphorous?
   a. P
   b. K
   c. Na
   d. Ca
   e. Fe

7. If the atomic number of carbon is 6, how many electrons does this element possess?
   a. 2
   b. 3
   c. 6
   d. 5
   e. 12

8. If the atomic number of argon is 18, how many electrons does this element possess?
   a. 6
   b. 8
   c. 10
   d. 12
   e. 18

9. The atom is composed of three subatomic particles: proton, neutron and electron. What is the net charge of electrons?
   a. Positive
   b. Negative
   c. Neutral

10. The atom is composed of three subatomic particles: proton, neutron and electron. What is the net charge of protons?
   a. Positive
   b. Negative
   c. Neutral

11. The atom is composed of three subatomic particles: proton, neutron and electron. What is the net charge of neutrons?
   a. Positive
   b. Negative
   c. Neutral

12. Which of the following terminology is defined as the amount of electricity transported per second via one ampere of current?
   a. Coulomb (C)
   b. Kilograms (Kg)
   c. Microliter (μl)
   d. Nanometer (nm)
   e. Newton (N)

13. If the atomic mass of magnesium is 24 (24.305 rounded number), how many neutrons are present within the nucleus of this element?
   a. 12
   b. 24
   c. 36
   d. 48
   e. 60

14. If the atomic mass of cobalt is 59 (58.933 rounded number), how many neutrons are present within the nucleus of this element?
   a. 24
   b. 32
   c. 46
   d. 58
   e. 74

15. If the atomic mass of aluminum is 27 (26.98 rounded number), how many neutrons are present within the nucleus of this element?
   a. 12
   b. 14
   c. 24
   d. 15
   e. 16

16. Which of the following hydrogen isotope possesses a single neutron within its nucleus?
   a. Cadium
   b. Deuterium
   c. Protium
   d. Titanium
   e. Tritium

17. Which of the following hydrogen isotopes possesses two neutrons within its nucleus?
   a. Cadium
   b. Deuterium
   c. Protium
   d. Titanium
   e. Tritium

18. Which of the following terms is defined as the amount of time it takes for half of these unstable isotopes to break down into a more stable element?
   a. Anions
   b. Cations
   c. Half-life
   d. Isotope
   e. Radioactive isotope

19. Tritium is an isotope that possesses a half-life of approximately 12.3 years where half of its numbers will break down into which of the following element?
   a. Carbon-14

b. Helium-3
c. Nitrogen-14
d. Potassium-40
e. Sodium-25

20. Which of the following is the most utilized isotope in science? This isotope is generally used to estimate the age of a fossil or archeological artifact.
   a. Carbon-14
   b. Helium-3
   c. Nitrogen-14
   d. Potassium-40
   e. Sodium-25

21. It has been calculated that the specific half-life of C^{14} is approximately 5730 years where approximately half of these isotopes found within fossils or artifacts are broken down to form which of the following elements?
   a. Carbon-14
   b. Helium-3
   c. Nitrogen-14
   d. Potassium-40
   e. Sodium-25

22. If the percentage of C^{14} in an artifact is approximately 30%, what is its estimated age?
   a. 5730 years
   b. 11460 years
   c. 17190 years
   d. 22920 years
   e. 28650 years

23. If the percentage of C^{14} in an artifact is approximately 15%, what is its estimated age?
   a. 5730 years
   b. 11460 years
   c. 17190 years
   d. 22920 years
   e. 28650 years

24. Potassium 40 (K40) is an isotope with a half-life of 1.2 billion years and is commonly used in radioactive dating of prehistoric fossils by paleontologists. Knowing that this isotope has an atomic weight of 40, how many neutrons does this isotope have?
   a. 18   e. 32
   b. 19   f. 35
   c. 20   g. 39
   d. 21   h. 50

25. Carbon 14 (C14) is an isotope with a half-life of 5700 years and is commonly used in radioactive dating of biological materials and/or antiquities discovered by archeologists. Knowing that this isotope has an atomic weight of 14, how many neutrons does it have?
   a. 5   e. 9
   b. 6   f. 10
   c. 7   g. 11
   d. 8   h. 12

26. Based on Bohr's theory, the electrons circles around the nucleus via various electron shells (or orbits) based upon their energy levels. Which of the following shells has the least amount of energy?
   a. K
   b. L
   c. M
   d. N
   e. O

27. Neon (Ne) is known as an inert gas. For which of the following reason or reasons is why neon does not react to another element and form a compound?
   a. It has 2 electrons in its outer most shell
   b. Neon does form other compounds. It donates up to 8 electrons for the process to take place
   c. It has eight electrons in its outer most shell
   d. Neon can't form another compound since it only have 10 protons within its nucleus
   e. Neon can't form another compound since it only have 10 neutrons within its nucleus

28. By using the octet rule, how many electrons does potassium (K) have in its outer most sub-shell?
   a. 1 electron in the 3d shell   e. 2 electron in the 3p shell
   b. 1 electron in the 3p shell   f. 3 electron in the 3d shell
   c. 1 electron in the 3s shell   g. 4 electron in the 3p shell
   d. 2 electron in the 3d shell   h. 5 electron in the 3s shell

29. What is the total number of electrons that exists for sulfur (S) in its outer most electron sub-shell?
   a. 1 electron in the 3d shell   e. 2 electron in the 3p shell
   b. 1 electron in the 3p shell   f. 3 electron in the 3d shell
   c. 1 electron in the 3s shell   g. 4 electron in the 3p shell
   d. 2 electron in the 3d shell   h. 5 electron in the 3s shell

30. What is the total number of electrons that exists for phosphorous (P) in its outer most electron sub-shell?

31.
   a. 1 electron in the 3d shell   e. 2 electron in the 3p shell
   b. 1 electron in the 3p shell   f. 3 electron in the 3p shell
   c. 1 electron in the 3s shell   g. 4 electron in the 3p shell
   d. 2 electron in the 3d shell   h. 5 electron in the 3s shell

31. What is the total number of electrons that exists for Boron (B) in its outer most electron sub-shell?
   a. 1 electron in the 2s shell   e. 2 electron in the 3p shell
   b. 1 electron in the 2p shell   f. 3 electron in the 3d shell
   c. 1 electron in the 3s shell   g. 4 electron in the 3p shell
   d. 2 electron in the 3d shell   h. 5 electron in the 3s shell

32. Nitrogen appears in nature as a stable compound N$_2$. Which of the following is the type of bond (s) formed in order to achieve this stable result?
   a. Ionic bond where three electrons are donated from one nitrogen to another
   b. Hydrogen bond, where the polarity of the atoms forms the attraction
   c. Covalent bond, where a single covalent bond is formed
   d. Covalent bond, where double covalent bonds are formed
   e. Covalent bond, where triple covalent bonds are formed
   f. Peptide bond, where the amino acids are sharing three pairs of electrons
   g. All of the above
   h. None of the above

33. Ionic bonds are formed when an electron is transferred from one atom to another. For example, sodium (Na) with the atomic number of _____ has _____ (number of electrons) in its outer most shell. This outer most shell is called _____. More specifically, sodium (Na) has_____ number of electrons in its outer most sub-shell which is called _____. On the other hand, chlorine (Cl) with the atomic number of _____ has _____ number of electrons in its outer most shell. This outer most shell is called _____. More specifically, chlorine (Cl) has_____ number of electrons in its outer most sub-shell which is called _____.

   Sodium (Na) having _____ electron (s) in its outermost shell would readily _____ (accept or donate) electron in order to become stable. On the other hand, chlorine (Cl), with _____ electron (s) in its outer most shell, would readily _____ (accept or donate) electron to become stable.
   a. Donate the electron   n. 19
   b. Accept the electron   o. K shell
   c. 1 electron   p. L shell
   d. 2 electron   q. M shell
   e. 3 electron   r. 2s
   f. 5 electron   s. 2p
   g. 6 electron   t. 3s
   h. 7 electron   u. 3p
   i. 11   v. 3d
   j. 12   w. 4s
   k. 14   x. 4p
   l. 16   y. 4d
   m. 17   z. 4r

34. Ionic bonds are formed when an electron is transferred from one atom to another. For example, lithium (Li) with the atomic number of _____ has _____ (number of electrons) in its outer most shell. This outer most shell is called _____. More specifically, lithium (Li) has _____ (number of electrons) in its outer most sub-shell which is called _____. On the other hand, fluorine (F) with the atomic number of _____ has _____ (number of electrons) in its outer most shell. This outer most shell is called _____. More specifically, fluorine (F) has _____ number of electrons in its outer most sub-shell which is called _____.

   Lithium (Li) having _____ electron (s) in its outermost shell would readily _____ (accept or donate) electron in order to become stable. On the other hand, fluorine (F), with _____ electron (s) in its outer most shell, would readily _____ (accept or donate) electron to become stable.
   a. Donate the electron   n. 19
   b. Accept the electron   o. K shell
   c. 1 electron   p. L shell
   d. 2 electron   q. M shell
   e. 3 electron   r. 2s
   f. 5 electron   s. 2p
   g. 6 electron   t. 3s
   h. 7 electron   u. 3p
   i. 1   v. 3d
   j. 2   w. 4s
   k. 3   x. 4p
   l. 9   y. 4d
   m. 17   z. 4r

35. Nitrogen appears in nature as a stable compound N$_2$. Which of the following is the type of bond (s) formed in order to achieve this stable result?
   a. Ionic bond where three electrons are donated from one nitrogen to another
   b. Hydrogen bond, where the polarity of the atoms forms the attraction
   c. Covalent bond, where a single covalent bond is formed
   d. Covalent bond, where double covalent bonds are formed
   e. Covalent bond, where triple covalent bonds are formed

36. H$_2$ is a relatively stable molecule formed through the _____ sharing of _____ (how many) electrons.

a. Polar covalent bond   e. 2
b. Nonpolar covalent bond   f. 4
c. Ionic bond   g. 6
d. Hydrogen bond   h. 8

37. Oxygen appears in nature as a stable compound O_2. Which of the following is the type of bond (s) formed in order to achieve this stable result?
    a. Ionic bond where three electrons are donated from one nitrogen to another
    b. Hydrogen bond, where the polarity of the atoms forms the attraction
    c. Covalent bond, where a single covalent bond is formed
    d. Covalent bond, where double covalent bonds are formed
    e. Covalent bond, where triple covalent bonds are formed

38. O_2 is a relatively stable molecule formed through the _____ (type of bone) where _____ (how many) electrons are shared.
    a. Polar covalent bond   e. 2
    b. Nonpolar covalent bond   f. 4
    c. Ionic bond   g. 6
    d. Hydrogen bond   h. 8

39. In a molecule that is formed by a polar covalent bond, electrons are pulled more closely to one atom than the other. This close attraction to one atom than the other creates two polar ends within this molecule. The polar 'end' where the electron(s) spend significantly more time orbiting is referred to as _____.
    a. Electropositive
    b. Pseudoelectoropositive
    c. Electroneutral
    d. Electronegative
    e. Electroreactive

40. In a molecule that is formed by a polar covalent bond, electrons are pulled more closely to one atom than the other. This close attraction to one atom than the other creates two polar ends within this molecule. The polar 'end' where the atom is shown to have less access to the electron (s) is called _____.
    a. Electropositive
    b. Pseudoelectoropositive
    c. Electroneutral
    d. Electronegative
    e. Electroreactive

41. Which of the following statement or statements is/are NOT true in regards to organic compounds (molecules)?
    a. They are composed of small number of atoms
    b. They are composed of large numbers of atoms
    c. They always have carbon and hydrogen
    d. They always have sodium and chloride
    e. They are formed through covalent bonding
    f. They are created with a carbon back bone
    g. Both a and d are incorrect
    h. Both b and f are incorrect

42. Water molecules are polar where they form _____ (type of bond) with other water molecules? This form of attraction is also referred to as _____.
    a. Hydrogen bond   e. Adhesive
    b. Electronegative   f. Ionic bond
    c. Electropositive   g. Non-polar covalent bond
    d. Cohesive   h. Polar covalent bond

43. Water molecules are polar where they form _____ (type of bond) with various charged surfaces? This form of attraction is also referred to as _____.
    a. Hydrogen bond   e. Adhesive
    b. Electronegative   f. Ionic bond
    c. Electropositive   g. Non-polar covalent bond
    d. Cohesive   h. Polar covalent bond

44. Energy is defined as the capacity to do work. Work (W) on the other hand is defined as force use to move matter. For example, if a force of 40 Newtons (N) is applied to push an object 150 meters (m), how much work is being performed?
    a. 4500 J
    b. 5000 J
    c. 5500 J
    d. 6000 J
    e. 6500 J

45. Energy is defined as the capacity to do work. Work (W) on the other hand is defined as force use to move matter. For example, if a force of 100 Newtons (N) is applied to push an object 100 meters (m), how much work is being performed?
    a. 100 J
    b. 1000 J
    c. 10000 J
    d. 100000 J
    e. 5000 J

46. Which of the following term is defined as potential energy?
    a. Stored energy
    b. Solar energy
    c. Chemical energy
    d. Energy of motion

47. Which of the following term is defined as kinetic energy?
    a. Stored energy
    b. Solar energy
    c. Chemical energy
    d. High energy bond
    e. Energy of motion

48. The energy needed to move arms or legs, respiration, peristalsis, blood circulation etc. is referred to as which of the following type of energy?
    a. Mechanical energy
    b. Solar energy
    c. Energy of motion
    d. Stored energy
    e. Chemical energy

49. The potential of chemical substances to undergo a chemical reaction thereby releasing stored energy is also referred to as which of the following type of energy?
    a. Mechanical energy
    b. Solar energy
    c. Energy of motion
    d. Stored energy
    e. Chemical energy

50. Which of the following is defined as a substance that donates a proton in solution?
    a. Acid
    b. Base
    c. Buffer
    d. Electron donor
    e. Electron acceptor

51. Which of the following is defined as a substance that accepts a proton or donates a hydroxyl group in solution?
    a. Acid
    b. Base
    c. Buffer
    d. Electron donor
    e. Electron acceptor

52. Which of the following is also known as the human blood buffer?
    a. HCl
    b. H_2SO_4
    c. H_2CO_3
    d. CO_2
    e. $NaOH$

53. Which of the following is considered as the homeostatic pH level of human blood?
    a. 2.8
    b. 3.9
    c. 6.2
    d. 7.4
    e. 8.6

54. Which of the following statement best describes a solution with the pH of 9?
    a. Acidic
    b. Basic
    c. Neutral

55. Which of the following statement best describes a solution with the pH of 3?
    a. Acidic
    b. Basic
    c. Neutral

56. Which of the following statement best describes a solution with the pH of 1?
    a. Acidic
    b. Basic
    c. Neutral

57. Which of the following statement best describes a solution with the pH of 7?
    a. Acidic
    b. Basic
    c. Neutral

58. .0000001 mol/L of hydrogen ion (H^+) in solution has a pH of _____.
    a. 4
    b. 5
    c. 6
    d. 7
    e. 8
    f. 9
    g. 10
    h. 11

59. .01 mol/L of hydrogen ion (H^+) in solution has a pH of _____.
    a. 2
    b. 4
    c. 6
    d. 8
    e. 9
    f. 10

60. When two monomers join together, the hydroxyl group is removed from one monomer and the hydrogen is removed from the other monomer forming water. What is this process called?
   a. Electrostatic bonding
   b. Ionic bonding
   c. Condensation synthesis
   d. Evaporation synthesis
   e. Hydrolysis

61. Water is added to a polymer and thereby reconstructing the hydroxyl and the hydrogen to each of the monomers. What is this process called?
   a. Electrostatic bonding
   b. Ionic bonding
   c. Condensation synthesis
   d. Evaporation synthesis
   e. Hydrolysis

62. The ability to create large complex macromolecules is the result of the unique nature of carbon (C). Carbon has _____ electrons in its outermost shell which means that it could form _____ (how many) covalent bond to become more stable.
   a. 1
   b. 2
   c. 3
   d. 4
   e. 5

63. Place the corresponding alphabet letter to the correct answer within the spaces provided

   _____ Amino acid
   _____ Carbohydrate
   _____ Fatty acid
   _____ Protein
   _____ Nucleic acid
   _____ Nucleotide
   _____ lipid
   _____ Glycerol

   a. Polymer
   b. Monomer

64. Sugars are classified based upon the number of carbons in their carbon backbone. Please place the corresponding alphabet letter to the correct answer within the spaces provided.

   _____ 3 carbon   a. Heptose
   _____ 4 carbon   b. Hexose
   _____ 5 carbon   c. Pentose
   _____ 6 carbon   d. Tetrose
   _____ 7 carbon   e. Triose

65. Which of the following is the monomer that is the principle source of energy for most eukaryotic organisms and it is this monosaccharide that is transported throughout the blood stream in the human body?
   a. Fructose
   b. Galactose
   c. Glucose
   d. Maltose
   e. Sucrose

66. Which of the following is the monomer that is also referred to as corn syrup and is found in most packaged foods and canned drinks?
   a. Fructose
   b. Galactose
   c. Glucose
   d. Maltose
   e. Sucrose

67. Disaccharide is a polymer (compounds with more than one monomer) that is constructed of the combination of two monomers through the process of _____. For example, lactose, an animal sugar found commonly in milk, is formed from the combination of _____ and _____ monomers.
   a. Condensation synthesis   e. Glucose
   b. Hydrolysis   f. Maltose
   c. Fructose   g. Sucrose
   d. Galactose   h. Sialic acid

68. Which of the following is (are) the term (s) describing sugar (s)?
   a. Carbohydrate   e. Glycogen
   b. Glycerol   f. Amino acids
   c. Fatty acids   g. Cyclopentane
   d. Steroids   h. Both a and e

69. DNA is composed of which of the following pentose sugar?
   a. Deoxyribose
   b. Galactose
   c. Glucose
   d. Maltose
   e. Ribose

70. RNA is composed of which of the following pentose sugar?
   a. Deoxyribose
   b. Galactose
   c. Glucose
   d. Maltose
   e. Ribose

71. Which of the following is the most commonly found form of stored sugars in animal which are composed of long chains of glucose?
   a. Fatty acids   e. Disaccharides
   b. Glycerol   f. Polysaccharides
   c. Glycogen   g. Deoxyribose
   d. Carbohydrates   h. Ribose

72. In humans, _____ are energy storage molecules stored in _____ (type of tissue) distributed throughout the body.
   a. Adipose tissue   e. Fat (fatty acids)
   b. Carbohydrate   f. Nucleic acids
   c. Connective tissue   g. Protein
   d. Epithelial tissue   h. Skeletomuscular tissue

73. Which of the following is (are) the monomer (s) of fats?
   a. Amino acid   e. Monosaccharide
   b. Disaccharide   f. Steroids
   c. Fatty acid   g. Both a and d
   d. Glycerol   h. Both c and d

74. Which of the following fatty acid contains no double bonds between the carbon atoms?
   a. Saturated fatty acid
   b. Unsaturated fatty acid

75. Which of the following fatty acid contains at least one double bond between the carbon atoms?
   a. Saturated fatty acid
   b. Unsaturated fatty acid

76. Steroid is a type of lipid commonly found in nature and is composed of four carbon ring. Three out of the four carbon rings is called _____ while the remaining carbon ring is called _____.
   a. Cyclodixane
   b. Cycloheptane
   c. Cyclohexane
   d. Cyclomonoxane
   e. Cyclopentane

77. Steroids commonly serve as signal molecules within the human body. What is the common name given to identify a signal molecule?
   a. Amino acid
   b. cholesterol
   c. Glucose
   d. Ligand
   e. Lipids

78. Proteins are composed of many basic building block components called _____. These basic building blocks are single unit or basic structure of a more complex molecule which is referred to as _____ (Hint: the common name for single unit).

   Like all other organic molecules, the basic components of the protein is differentiated (separated) from one another based upon their functional groups which is referred to specifically as _____ (Hint: what are these functional groups called in proteins) in protein based compounds. It is known that these basic components of proteins are formed through a type of covalent bond called _____ (specific name given to the covalent bond formed in proteins) which is created through an enzymatic process of removing a water molecule called _____.

   a. Amino acid   g. Polymer
   b. Monosaccharide   h. R-group
   c. Glycerol   i. Peptide bond
   d. Fatty acid   j. Disulfide bond
   e. Nucleic acid   k. Condensation synthesis
   f. Monomer   l. Hydrolysis

79. Which of the following is defined as an organic catalyst that is formed to speed up chemical reaction within cells?
   a. Enzyme
   b. Fatty acids
   c. Glycerol
   d. Nucleic acid
   e. Steroids

80. In RNA, a single ring nucleic acid structure is known as _____ which is represented by _____ and _____ heterocyclic base.

   In RNA a double ring nucleic acid structure is known as _____ which is represented by _____ and _____ heterocyclic base.
   a. Adenine   e. Purine
   b. Cytosine   f. Pyrimidine
   c. Guanine   g. Thymine
   d. Polypeptide   h. Uracil

81. In DNA, a single ring nucleic acid structure is known as _____ which is

represented by _____ and _____ heterocyclic base.

In DNA a double ring nucleic acid structure is known as _____ which is represented by _____ and _____ heterocyclic base.

a. Adenine
b. Cytosine
c. Guanine
d. Polypeptide
e. Purine
f. Pyrimidine
g. Thymine
h. Uracil

82. You are provided with the mRNA sequence. What is the tRNA sequence it is coding?

AUG UAC CCC UAC CUC GUU UAG

a. UAC AUG GGG AUG GAG CAA AUC
b. ATG TAC CCC TAC CTC GTT TAG
c. ATG AAC CGC TAC CTC GTT TAG
d. TAC ATG GGG ATG GAG CAA ATC
e. TAC ATC CCC TUA GAG CAA ATC
f. UAG ATG CCC AUG GAC CTT AUC
g. ATG CCC GGG TCG CGT GTT TAG
h. None of the above

83. You are provided with the mRNA sequence. What is the 3' DNA sequence it is coded from?

AUG UAC CCC UAC CUC GUU UAG

a. UAC AUG GGG AUG GAG CAA AUC
b. ATG TAC CCC TAC CTC GTT TAG
c. ATG AAC CGC TAC CTC GTT TAG
d. TAC ATG GGG ATG GAG CAA ATC
e. TAC ATC CCC TUA GAG CAA ATC
f. UAG ATG CCC AUG GAC CTT AUC
g. ATG CCC GGG TCG CGT GTT TAG
h. None of the above

84. You are provided with a 5' DNA sequence. What is the amino acid it is coding?

ATG TAC CCC TAC CTC GTT TAG

a. Met Val-Pro-Val-Leu-Val-Stop
b. Met Leu-Pro-Leu-Leu-Val-Stop
c. Met-Tyr-Leu-Pro-Val-Leu-Stop
d. Met-Glu-Leu-Glu-Val-Leu-Stop
e. Met-Gly-Glu-Pro-Pro-Val-Stop
f. Met-Arg-Glu-Val-Pro-Leu-Stop
g. Met-Tyr-Pro-Tyr-Leu-Val-Stop
h. None of the above

85. You are provided with an mRNA sequence. What is the complimentary tRNA sequence?

UAU AUA CUG GUU UAA AAG CAU ACA

a. ATA TAT GAC CAA ATT TTC GTA TGT
b. AUA AUU GAC CAA AUU UUC GUA UGU
c. AUA UAU GAC CAA AUU UUC GUA UGU
d. AUA UAU GAC CAA AUU UUC GUA UGU
e. AUA UAU GAG CAU AUU UUC GUA UGU
f. AUU UAU GAC CAA AUU UUC GUA UGU
g. UAU UAA CUG GUU UAA AAG CAU ACA
h. None of the above

86. You are provided with the 5' DNA sequence. What is the complimentary 3' DNA sequence?

ATG CGC TAT AAT AAG CCG ACA TGA

a. AUC GCG TTT AUT UUA CGC TTT UTC
b. AUG CGC UAU AAU AAG CCG ACA UGA
c. AUG CGG UAU AAU AAG CCG ACA UGA
d. TAC GCG ATA TTA TTC GGC TGT ACT
e. UAC GCG AAA UUA UUG CCG UGU ACU
f. UAC GCG AUA UUA UUC GGC UGU ACU
g. UUU UAU UAG GGC CGC UTA UUU AAA
h. None of the above

87. You are provided with a 3' DNA sequence. What is the 5' complementary DNA sequence?

ATT CCC GCG GAT ATT TCT GAC GGG

a. AUU CCC GCG GAU AUU UCU GAC GGG
b. TAA CCC CGC GTA AAA AGA CTG CCC
c. TAA GCG CGC CTA TTT AGA CTG GGG
d. TAA GGG CGC CTA TAA AGA CTG CCC
e. TAC GCG ATA TTA TTC GGC TGT ACT
f. UAC GCG AAA UUA UUG CCG UGU ACU
g. UAC GCG AUA UUA UUC GGC UGU ACU

88. You are provided with the 5' DNA sequence. What is the 3' DNA sequence it is coding?

ATG TTC ATT TCG CCG GCA TAG

a. AUG UUG AUU UCG CCG GCA UAG
b. AAT UUG AUU UCG CCG GCA UAC
c. TAC AAG TAA AGC GGC CGT ATC
d. TAC ATC TAA AGC GGC GGT TTC
e. UAC AAC UAA AGC GGC CGU AUC
f. UUC UUC UAA AGC GGC CGU AUC
g. TAC AAC TAU AGC GGC CGT ATC
h. TTC CCG UTT TAA AUU AAU GGG
i. None of the above

89. You are provided with the 5' DNA sequence. What is the tRNA sequence it is coding?

ATG TTC ATT TCG CCG GCA TAG

a. AUG UUG AUU UCG CCG GCA UAG
b. AAT UUG AUU UCG CCG GCA UAC
c. TAC AAC TAA AGC GGC CGT ATC
d. TAC ATC TAA AGC GGC GGT TTC
e. UAC AAG UAA AGC GGC CGU AUC
f. UUC UUC UAA AGC GGC CGU AUC
g. TAC AAC TAU AGC GGC CGT ATC
h. TTC CCG UTT TAA AUU AAU GGG
i. None of the above

90. You are provided with the 5' DNA sequence. What is the mRNA sequence it is coding?

ATG TTC ATT TCG CCG GCA TAG

a. AUG UUC AUU UCG CCG GCA UAG
b. AAT UUG AUU UCG CCG GCA UAC
c. TAC AAC TAA AGC GGC CGT ATC
d. TAC ATC TAA AGC GGC GGT TTC
e. UAC AAC UAA AGC GGC CGU AUC
f. UUC UUC UAA AGC GGC CGU AUC
g. TAC AAC TAU AGC GGC CGT ATC
h. TTC CCG UTT TAA AUU AAU GGG
i. None of the above

91. You are provided with the mRNA sequence. What is the 5' DNA sequence it is coded from?

AUG UAC CCC UAC CUC GUU UAG

a. UAC AUG GGG AUG GAG CAA AUC
b. ATG TAC CCC TAC CTC GTT TAG
c. ATG AAC CGC TAC CTC GTT TAG
d. TAC ATG GGG ATG GAG CAA ATC
e. TAC ATC CCC TUA GAG CAA ATC
f. UAG ATG CCC AUG GAC CTT AUC
g. ATG CCC GGG TCG CGT GTT TAG
h. None of the above

92. You are provided with the 5' DNA sequence. What is the amino acid sequence it is coding?

ATG TTC ATT TCG CCG GCA TAG

a. Met-Gly-Glu-Ser-Pro-Ala-Stop
b. Met-Tyr-Glu-Ser-Pro-Ala-Stop
c. Met-Arg-Ile-Ser-Pro-Ala-Stop
d. Met-Phe-Ile-Ser-Pro-Ala-Stop
e. Met-Ile-Leu-Ser-Pro-Ala-Stop
f. Met-Thr-Pro-Val-Pro-Ala-Stop
g. Met-Lys-pro-Val-Pro-Ala-Stop
h. All of the above
i. None of the above

93. _____ is the most common energy molecule of all living organisms on earth. This molecule consists of 3 polar functional groups called _____ connected through a type of covalent bond known as _____. This molecule is also connected to a pentose sugar called _____ and a base called _____. This base belongs to a group of double ringed compounds also known as _____.

The last of the three covalent bonds is also referred to as _____ which refers to the bond that needs to be broken in order to release the potential energy. Once this bond is enzymatically broken through a process known as _____ (Hint: adding water), the remaining molecule (possessing only 2 out of the 3 polar functional groups) is now referred to as _____ (Hint: 2 functional groups).

At times when the body is desperate for energy the 1 out of the two covalent bonds could also be broken to release some energy. Once this bond is broken, the remaining molecule (now only has one functional group) is referred to as _____ (Hint: single functional group).

1. Adenine
2. Adenosine di-phosphate (ADP)
3. Adenosine mono-phosphate (AMP)
4. Adenosine tri-phosphate (ATP)
5. Amino
6. Amino bond
7. Carboxyl
8. Carboxyl bond
9. Condensation synthesis
10. Cytosine
11. Cytosine di-phosphate (CDP)
12. Cytosine mono-phosphate (CMP)
13. Cytosine tri-phosphate (CTP)
14. Deoxyribose
15. Guanine
16. Guanine di-phosphate (GDP)
17. Guanine mono-phosphate (GMP)
18. Guanine tri-phosphate (GTP)
19. High energy bond
20. Hydrolysis
21. Ketones
22. Phosphate bond
23. Phosphates
24. Purine
25. Pyrimidine
26. Ribose
27. Sulfhydrl
28. Sulfhydrl bond
29. Thymine
30. Thymine di-phosphate (TDP)
31. Thymine mono-phosphate (TMP)
32. Thymine tri-phosphate (TTP)
33. Uracil
34. Uracil di-phosphate (UDP)
35. Uracil mono-phosphate (UMP)
36. Uracil tri-phosphate (UTP)

ANSWERS

1.	d	2.	a	3.	e
4.	b	5.	e	6.	a
7.	c	8.	e	9.	b
10.	a	11.	c	12.	a
13.	a	14.	b	15.	b
16.	b	17.	e	18.	c
19.	b	20.	a	21.	c
22.	b	23.	c	24.	d
25.	d	26.	a	27.	c
28.	a	29.	g	30.	f
31.	b	32.	e	33.	i, c, q, c, t, m, h, q, f, u, c, a, h, b
34.	k, c, p, c, r, l, h, p, f, s, c, a, h, b	35.	e	36.	e
37.	d	38.	b, f	39.	d
40.	a	41.	g	42.	a, d
43.	a, e	44.	e	45.	c
46.	a	47.	e	48.	a
49.	e	50.	a	51.	b
52.	c	53.	d	54.	b
55.	a	56.	a	57.	c
58.	d	59.	a	60.	c
61.	e	62.	d, d	63.	b, a, b, a, a, b, a, b
64.	e, d, c, b, a	65.	c	66.	a
67.	a, d, e	68.	h	69.	a
70.	e	71.	c	72.	e, a
73.	h	74.	a	75.	b
76.	c, e	77.	d	78.	a, f, h, I, k
79.	a	80.	f, b, h, e, a, c	81.	f, b, g, e, a, c
82.	a	83.	d	84.	g
85.	c	86.	d	87.	d
88.	c	89.	e	90.	a
91.	b	92.	d	93.	4, 23, 22, 26, 1, 24, 19, 20, 2, 3

REFERENCES

I.C. Baianu, Editor. Nuclear Medicine, Diagnostic Tomography and Imaging: X-ray, CAT, Scanning, PET, FFT Imaging and Microscopy. Web address: https:// archive.org/details/ NuclearMedicineDiagnosticTomographyAndImaging

International Atomic Energy Agency (IAEA). Cyclotron Produced Radionuclides: Principles and Practice. Technical Reports Series 465. Web Address: http://www-pub.iaea.org/MTCD/ publications/PDF/trs465_web.pdf

Cummings JH, Stephen AM. Carbohydrate terminology and classification. European Journal of Clinical Nutrition, 2007: 61, pp. S5–S18

Klug WS, Cummings MR, Spencer CA, Palladino MA. Concepts of Genetics, Ninth Edition. Pearson, Benjamin Cummings. San Francisco. 2009

Seeley RR, Stephens TD, Tate P. Anatomy and Physiology 6th Edition. McGraw-Hill, New York, New York. 2003

Tate P. Seeley's Principles of Anatomy & Physiology 1st Edition. McGraw-Hill, New York, New York. 2009

Saladin KS. Anatomy & Physiology. The Unity of Form and Function 6th Edition. McGraw-Hill, New York, New York. 2010

PHOTO AND GRAPHIC BIBLIOGRAPHY

1. Figure 1: Graphic designed by P.Y.P. Jen. Copyright ©
2. Figure 2: Graphic designed by P.Y.P. Jen. Copyright ©
3. Figure 3: The half-lives of C¹⁴. Graphic designed by P.Y.P. Jen. Copyright ©
4. Figure 4: SPECT image taken and released by OpenI, United States National Library of Medicine. Carhart-Harris RL et al., 2008. Psychopharmacology Unit, University of Bristol, Bristol, UK. Public domain photo
5. Figure 5: PET scan taken and released by Brookhaven National Laboratory. Office of Science of the U.S. Department of Energy (DOE). Public domain photo
6. Figure 6: PET/CT image taken and released by Deantonio L et al., 2008. Radiotherapy, University of Piemonte Orientale Amedeo Avogadro, Novara, Italy. Public domain photo
7. Figure 7: Graphic designed by P.Y.P. Jen. Copyright ©
8. Figure 8: Graphic designed by P.Y.P. Jen. Copyright ©
9. Figure 9: Graphic designed by P.Y.P. Jen. Copyright ©
10. Figure 10: Graphic designed by P.Y.P. Jen. Copyright ©
11. Figure 11: Graphic designed by P.Y.P. Jen. Copyright ©
12. Figure 12: Graphic designed by P.Y.P. Jen. Copyright ©
13. Figure 13: Graphic designed by Michal Maňas reprint permission granted under the Creative Commons Attribution 3.0 Unported license
14. Figure 14: Graphic designed by P.Y.P. Jen. Copyright ©
15. Figure 15: Graphic designed by P.Y.P. Jen. Copyright ©
16. Figure 16: Graphic designed by P.Y.P. Jen. Copyright ©
17. Figure 17: Graphic designed by P.Y.P. Jen. Copyright ©
18. Figure 18: Graphic designed by P.Y.P. Jen. Copyright ©
19. Figure 19: Graphic designed by P.Y.P. Jen. Copyright ©
20. Figure 20: Graphic designed by P.Y.P. Jen. Copyright ©
21. Figure 21: Graphic designed by P.Y.P. Jen. Copyright ©
22. Figure 22: Graphic designed by P.Y.P. Jen. Copyright ©
23. Figure 23: Graphic designed and released by US department of Health and Human Services. Public domain graphics
24. Figure 24: Graphic designed by P.Y.P. Jen. Copyright ©
25. Figure 25: Graphic designed and released by Open Stax College, Connexions, reprint permission granted under the Creative Commons Attribution 3.0 Unported license.
26. Figure 26: Graphic designed by P.Y.P. Jen. Copyright ©
27. Figure 27: Graphic designed by YassineMrabet 2007. Reprint permission granted under the Creative Commons Attribution 3.0 Unported license.
28. Figure 28: Graphic designed by P.Y.P. Jen. Copyright ©
29. Figure 29: Graphic designed and released by Darryl Leja 2010, National Human Genome Research Institute. Public domain graphics
30. Figure 30: Graphic designed and released by National Human Genome Research Institute. Public domain graphics
31. Figure 31: Graphic designed and released by Darryl Leja 2010, National Human Genome Research Institute. Public domain graphics
32. Figure 32: Graphic designed by P.Y.P. Jen. Copyright ©
33. Figure 33: Graphic designed and released byDarryl Leja 2010, National Human Genome Research Institute. Public domain graphics
34. Figure 34: Graphic designed and released by Darryl Leja 2010, National Human Genome Research Institute. Public domain graphics
35. Figure 35: Graphic designed and released by Darryl Leja 2010, National Human Genome Research Institute. Public domain graphics.
36. Figure 36: Graphic released by OpenI, United States National Library of Medicine. Graphic designed by Ferrell JE - J. Biol. 2009, Department of Chemical and Systems Biology, Stanford University School of Medicine. Public domain graphics
37. Figure 37: Graphic designed and released by Darryl Leja 2010, National Human Genome Research Institute. Public domain graphics
38. Figure 38: Graphic designed and released by P.Y.P. Jen Copyright ©
39. Figure 49: Graphic designed and released by Nationlal Institute of Health. Public domain graphics
40. Figure 40: Graphic designed and released by NEUROtiker. Public domain graphics

Appendix I: Periodic table. Graphic produced and released by National institute of Standards and Technology, United States Department of Commerce. Public Domain Graphics
Appendix II: Twenty one amino acids. Graphic designed by Dan Cojocari, Department of Medical Biophysics, University of Toronto 2011. Reprint permission granted under the Creative Commons Attribution 3.0 Unported license
Appendix III: The codon table. Graphic designed by P.Y.P. Jen. Copyright ©

PERIODIC TABLE
Atomic Properties of the Elements

NIST
National Institute of Standards and Technology
U.S. Department of Commerce

Physical Measurement Laboratory
www.nist.gov/pml

Standard Reference Data
www.nist.gov/srd

Frequently used fundamental physical constants

For the most accurate values of these and other constants, visit physics.nist.gov/constants

1 second = 9 192 631 770 periods of radiation corresponding to the transition between the two hyperfine levels of the ground state of ^{133}Cs

speed of light in vacuum	c	299 792 458 m s^{-1} (exact)
Planck constant	h	6.626 07 x 10^{-34} J s ($\hbar = h/2\pi$)
elementary charge	e	1.602 177 x 10^{-19} C
electron mass	m_e	9.109 38 x 10^{-31} kg
	$m_e c^2$	0.510 999 MeV
proton mass	m_p	1.672 622 x 10^{-27} kg
fine-structure constant	α	1/137.035 999
Rydberg constant	R_∞	10 973 731.569 m^{-1}
	$R_\infty c$	3.289 841 960 x 10^{15} Hz
	$R_\infty hc$	13.605 69 eV
Boltzmann constant	k	1.380 6 x 10^{-23} J K^{-1}

Legend:
- Solids
- Liquids
- Gases
- Artificially Prepared

Key (example entry)

Group 1 IA
Atomic Number
Symbol
Name
Standard Atomic Weight
Ground-state Configuration
Ground-state Level
Ionization Energy (eV)

58 — Atomic Number
$^1G_4^\circ$ — Ground-state Level
Ce — Symbol
Cerium — Name
140.116 — Standard Atomic Weight
[Xe]4f5d6s^2 — Ground-state Configuration
5.5386 — Ionization Energy (eV)

† Based upon ^{12}C. ( ) indicates the mass number of the longest-lived isotope.

‡ IUPAC conventional atomic weights; standard atomic weights for these elements are expressed in intervals; see iupac.org for an explanation and values.

For a description of the data, visit physics.nist.gov/data

NIST SP 966 (September 2014)

Period	Group 1 IA	IIA	IIIB	IVB	VB	VIB	VIIB	VIII			IB	IIB	IIIA	IVA	VA	VIA	VIIA	VIIIA 18
1	1 H Hydrogen 1.008‡ [1s] $^2S_{1/2}$ 13.5984																	2 He Helium 4.002602 1s^2 $^1S_0$ 24.5874
2	3 Li Lithium 6.94‡ 1s^{2}2s $^2S_{1/2}$ 5.3917	4 Be Beryllium 9.0121831 1s^{2}2s^2 $^1S_0$ 9.3227												5 B Boron 10.81‡ 1s^{2}2s^{2}2p $^2P_{1/2}^\circ$ 8.2980	6 C Carbon 12.011‡ 1s^{2}2s^{2}2p^2 $^3P_0$ 11.2603	7 N Nitrogen 14.007‡ 1s^{2}2s^{2}2p^3 $^4S_{3/2}^\circ$ 14.5341	8 O Oxygen 15.999‡ 1s^{2}2s^{2}2p^4 $^3P_2$ 13.6181	9 F Fluorine 18.998403163 1s^{2}2s^{2}2p^5 $^2P_{3/2}^\circ$ 17.4228	10 Ne Neon 20.1797 1s^{2}2s^{2}2p^6 $^1S_0$ 21.5645
3	11 Na Sodium 22.98976928 [Ne]3s $^2S_{1/2}$ 5.1391	12 Mg Magnesium 24.305‡ [Ne]3s^2 $^1S_0$ 7.6462												13 Al Aluminum 26.9815385 [Ne]3s^{2}3p $^2P_{1/2}^\circ$ 5.9858	14 Si Silicon 28.085‡ [Ne]3s^{2}3p^2 $^3P_0$ 8.1517	15 P Phosphorus 30.973761998 [Ne]3s^{2}3p^3 $^4S_{3/2}^\circ$ 10.4867	16 S Sulfur 32.06‡ [Ne]3s^{2}3p^4 $^3P_2$ 10.3600	17 Cl Chlorine 35.45‡ [Ne]3s^{2}3p^5 $^2P_{3/2}^\circ$ 12.9676	18 Ar Argon 39.948 [Ne]3s^{2}3p^6 $^1S_0$ 15.7596
4	19 K Potassium 39.0983 [Ar]4s $^2S_{1/2}$ 4.3407	20 Ca Calcium 40.078 [Ar]4s^2 $^1S_0$ 6.1132	21 Sc Scandium 44.955908 [Ar]3d4s^2 $^2D_{3/2}$ 6.5615	22 Ti Titanium 47.867 [Ar]3d^{2}4s^2 $^3F_2$ 6.8281	23 V Vanadium 50.9415 [Ar]3d^{3}4s^2 $^4F_{3/2}$ 6.7462	24 Cr Chromium 51.9961 [Ar]3d^{5}4s $^7S_3$ 6.7665	25 Mn Manganese 54.938044 [Ar]3d^{5}4s^2 $^6S_{5/2}$ 7.4340	26 Fe Iron 55.845 [Ar]3d^{6}4s^2 $^5D_4$ 7.9025	27 Co Cobalt 58.933194 [Ar]3d^{7}4s^2 $^4F_{9/2}$ 7.8810	28 Ni Nickel 58.6934 [Ar]3d^{8}4s^2 $^3F_4$ 7.6399	29 Cu Copper 63.546 [Ar]3d^{10}4s $^2S_{1/2}$ 7.7264	30 Zn Zinc 65.38 [Ar]3d^{10}4s^2 $^1S_0$ 9.3942	31 Ga Gallium 69.723 [Ar]3d^{10}4s^{2}4p $^2P_{1/2}^\circ$ 5.9993	32 Ge Germanium 72.630 [Ar]3d^{10}4s^{2}4p^2 $^3P_0$ 7.8994	33 As Arsenic 74.921595 [Ar]3d^{10}4s^{2}4p^3 $^4S_{3/2}^\circ$ 9.7886	34 Se Selenium 78.971 [Ar]3d^{10}4s^{2}4p^4 $^3P_2$ 9.7524	35 Br Bromine 79.904 [Ar]3d^{10}4s^{2}4p^5 $^2P_{3/2}^\circ$ 11.8138	36 Kr Krypton 83.798 [Ar]3d^{10}4s^{2}4p^6 $^1S_0$ 13.9996
5	37 Rb Rubidium 85.4678 [Kr]5s $^2S_{1/2}$ 4.1771	38 Sr Strontium 87.62 [Kr]5s^2 $^1S_0$ 5.6949	39 Y Yttrium 88.90584 [Kr]4d5s^2 $^2D_{3/2}$ 6.2173	40 Zr Zirconium 91.224 [Kr]4d^{2}5s^2 $^3F_2$ 6.6339	41 Nb Niobium 92.90637 [Kr]4d^{4}5s $^6D_{1/2}$ 6.7589	42 Mo Molybdenum 95.95 [Kr]4d^{5}5s $^7S_3$ 7.0924	43 Tc Technetium (98) [Kr]4d^{5}5s^2 $^6S_{5/2}$ 7.1194	44 Ru Ruthenium 101.07 [Kr]4d^{7}5s $^5F_5$ 7.3605	45 Rh Rhodium 102.90550 [Kr]4d^{8}5s $^4F_{9/2}$ 7.4589	46 Pd Palladium 106.42 [Kr]4d^{10} $^1S_0$ 8.3369	47 Ag Silver 107.8682 [Kr]4d^{10}5s $^2S_{1/2}$ 7.5762	48 Cd Cadmium 112.414 [Kr]4d^{10}5s^2 $^1S_0$ 8.9938	49 In Indium 114.818 [Kr]4d^{10}5s^{2}5p $^2P_{1/2}^\circ$ 5.7864	50 Sn Tin 118.710 [Kr]4d^{10}5s^{2}5p^2 $^3P_0$ 7.3439	51 Sb Antimony 121.760 [Kr]4d^{10}5s^{2}5p^3 $^4S_{3/2}^\circ$ 8.6084	52 Te Tellurium 127.60 [Kr]4d^{10}5s^{2}5p^4 $^3P_2$ 9.0097	53 I Iodine 126.90447 [Kr]4d^{10}5s^{2}5p^5 $^2P_{3/2}^\circ$ 10.4513	54 Xe Xenon 131.293 [Kr]4d^{10}5s^{2}5p^6 $^1S_0$ 12.1298
6	55 Cs Cesium 132.90545196 [Xe]6s $^2S_{1/2}$ 3.8939	56 Ba Barium 137.327 [Xe]6s^2 $^1S_0$ 5.2117	71 Lu Lutetium 174.9668 [Xe]4f^{14}5d6s^2 $^2D_{3/2}$ 5.4259	72 Hf Hafnium 178.49 [Xe]4f^{14}5d^{2}6s^2 $^3F_2$ 6.8251	73 Ta Tantalum 180.94788 [Xe]4f^{14}5d^{3}6s^2 $^4F_{3/2}$ 7.5496	74 W Tungsten 183.84 [Xe]4f^{14}5d^{4}6s^2 $^5D_0$ 7.8640	75 Re Rhenium 186.207 [Xe]4f^{14}5d^{5}6s^2 $^6S_{5/2}$ 7.8335	76 Os Osmium 190.23 [Xe]4f^{14}5d^{6}6s^2 $^5D_4$ 8.4382	77 Ir Iridium 192.217 [Xe]4f^{14}5d^{7}6s^2 $^4F_{9/2}$ 8.9670	78 Pt Platinum 195.084 [Xe]4f^{14}5d^{9}6s $^3D_3$ 8.9588	79 Au Gold 196.966569 [Xe]4f^{14}5d^{10}6s $^2S_{1/2}$ 9.2256	80 Hg Mercury 200.592 [Xe]4f^{14}5d^{10}6s^2 $^1S_0$ 10.4375	81 Tl Thallium 204.38‡ [Hg]6p $^2P_{1/2}^\circ$ 6.1083	82 Pb Lead 207.2 [Hg]6p^2 $^3P_0$ 7.4167	83 Bi Bismuth 208.98040 [Hg]6p^3 $^4S_{3/2}^\circ$ 7.2855	84 Po Polonium (209) [Hg]6p^4 $^3P_2$ 8.414	85 At Astatine (210) [Hg]6p^5 $^2P_{3/2}^\circ$ 9.31751	86 Rn Radon (222) [Hg]6p^6 $^1S_0$ 10.7485
7	87 Fr Francium (223) [Rn]7s $^2S_{1/2}$ 4.0727	88 Ra Radium (226) [Rn]7s^2 $^1S_0$ 5.2784	103 Lr Lawrencium (262) [Rn]5f^{14}7s^{2}7p $^2P_{1/2}^\circ$ 4.90	104 Rf Rutherfordium (267) [Rn]5f^{14}6d^{2}7s^2 $^3F_2$ 6.0	105 Db Dubnium (268) [Rn]5f^{14}6d^{3}7s^2 7.8	106 Sg Seaborgium (271) [Rn]5f^{14}6d^{4}7s^2 7.8	107 Bh Bohrium (272) [Rn]5f^{14}6d^{5}7s^2 7.7	108 Hs Hassium (270) [Rn]5f^{14}6d^{6}7s^2 7.6	109 Mt Meitnerium (276)	110 Ds Darmstadtium (281)	111 Rg Roentgenium (280)	112 Cn Copernicium (285)	113 Uut Ununtrium (284)	114 Fl Flerovium (289)	115 Uup Ununpentium (288)	116 Lv Livermorium (293)	117 Uus Ununseptium (294)	118 Uuo Ununoctium (294)

Lanthanides

57 La Lanthanum 138.90547 [Xe]5d6s^2 $^2D_{3/2}$ 5.5769	58 Ce Cerium 140.116 [Xe]4f5d6s^2 $^1G_4^\circ$ 5.5386	59 Pr Praseodymium 140.90766 [Xe]4f^{3}6s^2 $^4I_{9/2}^\circ$ 5.473	60 Nd Neodymium 144.242 [Xe]4f^{4}6s^2 $^5I_4$ 5.5250	61 Pm Promethium (145) [Xe]4f^{5}6s^2 $^6H_{5/2}^\circ$ 5.582	62 Sm Samarium 150.36 [Xe]4f^{6}6s^2 $^7F_0$ 5.6437	63 Eu Europium 151.964 [Xe]4f^{7}6s^2 $^8S_{7/2}^\circ$ 5.6704	64 Gd Gadolinium 157.25 [Xe]4f^{7}5d6s^2 $^9D_2^\circ$ 6.1498	65 Tb Terbium 158.92535 [Xe]4f^{9}6s^2 $^6H_{15/2}^\circ$ 5.8638	66 Dy Dysprosium 162.500 [Xe]4f^{10}6s^2 $^5I_8$ 5.9391	67 Ho Holmium 164.93033 [Xe]4f^{11}6s^2 $^4I_{15/2}^\circ$ 6.0215	68 Er Erbium 167.259 [Xe]4f^{12}6s^2 $^3H_6$ 6.1077	69 Tm Thulium 168.93422 [Xe]4f^{13}6s^2 $^2F_{7/2}^\circ$ 6.1843	70 Yb Ytterbium 173.054 [Xe]4f^{14}6s^2 $^1S_0$ 6.2542

Actinides

89 Ac Actinium (227) [Rn]6d7s^2 $^2D_{3/2}$ 5.3802	90 Th Thorium 232.0377 [Rn]6d^{2}7s^2 $^3F_2$ 6.3067	91 Pa Protactinium 231.03588 [Rn]5f^{2}6d7s^2 $^4K_{11/2}$ 5.89	92 U Uranium 238.02891 [Rn]5f^{3}6d7s^2 $^5L_6^\circ$ 6.1941	93 Np Neptunium (237) [Rn]5f^{4}6d7s^2 $^6L_{11/2}$ 6.2655	94 Pu Plutonium (244) [Rn]5f^{6}7s^2 $^7F_0$ 6.0258	95 Am Americium (243) [Rn]5f^{7}7s^2 $^8S_{7/2}^\circ$ 5.9738	96 Cm Curium (247) [Rn]5f^{7}6d7s^2 $^9D_2^\circ$ 5.9914	97 Bk Berkelium (247) [Rn]5f^{9}7s^2 $^6H_{15/2}^\circ$ 6.1978	98 Cf Californium (251) [Rn]5f^{10}7s^2 $^5I_8$ 6.2817	99 Es Einsteinium (252) [Rn]5f^{11}7s^2 $^4I_{15/2}^\circ$ 6.3676	100 Fm Fermium (257) [Rn]5f^{12}7s^2 $^3H_6$ 6.50	101 Md Mendelevium (258) [Rn]5f^{13}7s^2 $^2F_{7/2}^\circ$ 6.58	102 No Nobelium (259) [Rn]5f^{14}7s^2 $^1S_0$ 6.65

APPENDIX II

Twenty-One Amino Acids

⊕ Positive   ⊖ Negative
• Side chain charge at physiological **pH 7.4**

A. Amino Acids with Electricaly Charged Side Chains

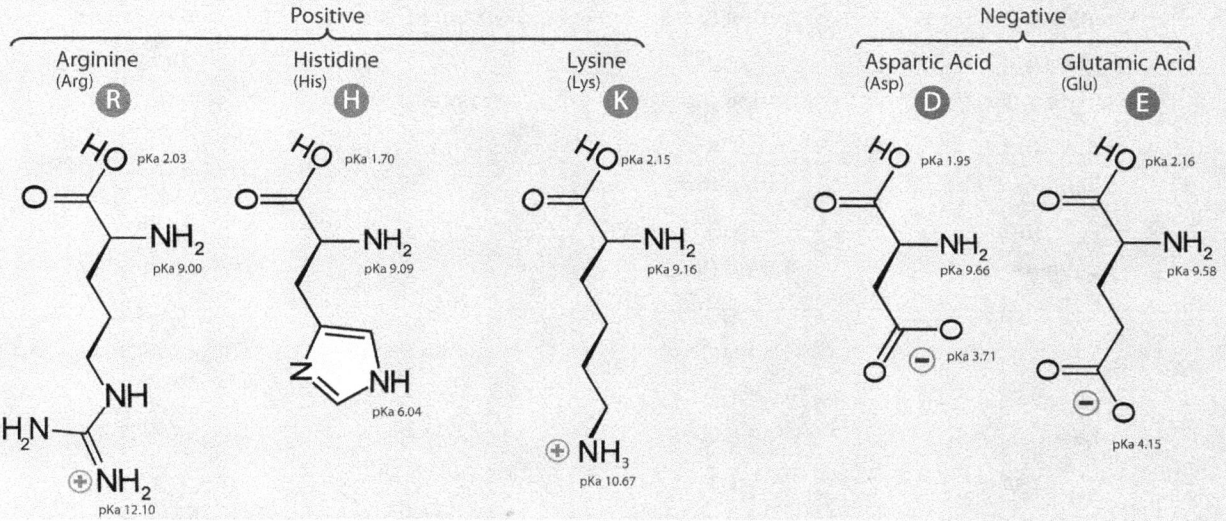

Positive

Arginine (Arg) **R**
pKa 2.03
pKa 9.00
NH
H_2N
⊕ NH_2
pKa 12.10

Histidine (His) **H**
pKa 1.70
pKa 9.09
N   NH
pKa 6.04

Lysine (Lys) **K**
pKa 2.15
pKa 9.16
⊕ NH_3
pKa 10.67

Negative

Aspartic Acid (Asp) **D**
pKa 1.95
pKa 9.66
⊖
pKa 3.71

Glutamic Acid (Glu) **E**
pKa 2.16
pKa 9.58
⊖
pKa 4.15

B. Amino Acids with Polar Uncharged Side Chains

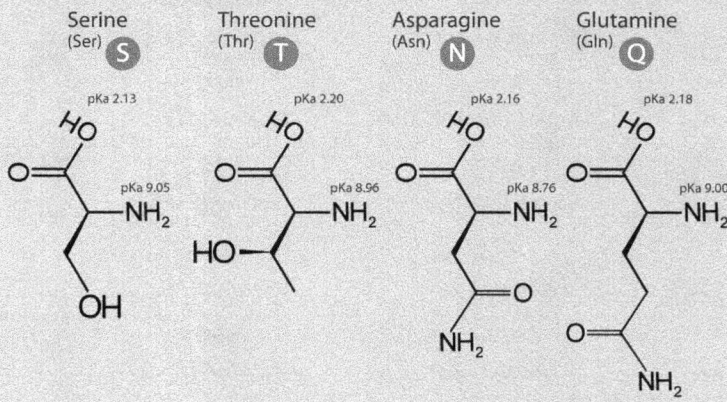

Serine (Ser) **S**
pKa 2.13
pKa 9.05
OH

Threonine (Thr) **T**
pKa 2.20
pKa 8.96
HO

Asparagine (Asn) **N**
pKa 2.16
pKa 8.76
NH_2

Glutamine (Gln) **Q**
pKa 2.18
pKa 9.00
NH_2

C. Special Cases

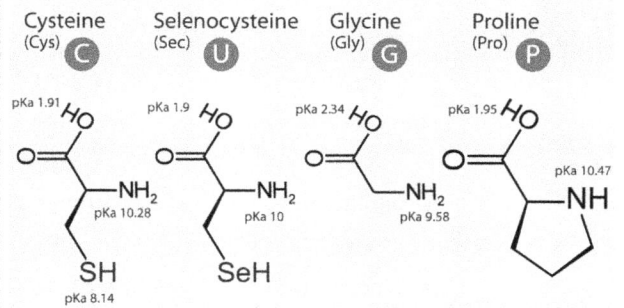

Cysteine (Cys) **C**
pKa 1.91
pKa 10.28
SH
pKa 8.14

Selenocysteine (Sec) **U**
pKa 1.9
pKa 10
SeH

Glycine (Gly) **G**
pKa 2.34
pKa 9.58

Proline (Pro) **P**
pKa 1.95
pKa 10.47
NH

D. Amino Acids with Hydrophobic Side Chain

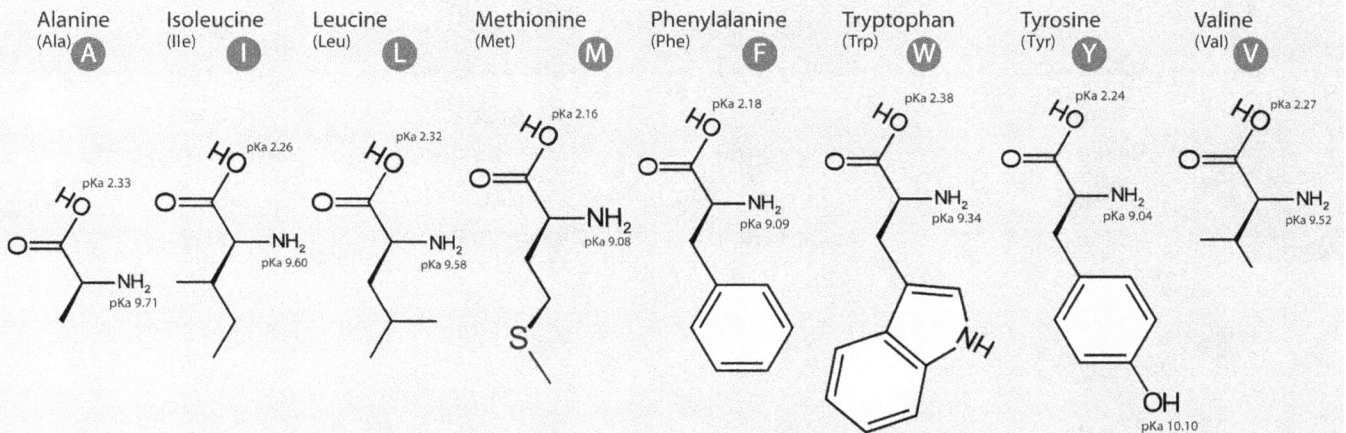

Alanine (Ala) **A**
pKa 2.33
pKa 9.71

Isoleucine (Ile) **I**
pKa 2.26
pKa 9.60

Leucine (Leu) **L**
pKa 2.32
pKa 9.58

Methionine (Met) **M**
pKa 2.16
pKa 9.08
S

Phenylalanine (Phe) **F**
pKa 2.18
pKa 9.09

Tryptophan (Trp) **W**
pKa 2.38
pKa 9.34
NH

Tyrosine (Tyr) **Y**
pKa 2.24
pKa 9.04
OH
pKa 10.10

Valine (Val) **V**
pKa 2.27
pKa 9.52

pKa Data: **CRC Handbook of Chemistry, v.** 2010

Dan Cojocari, Department of Medical Biophysics, **University of Toronto** 2009

	U	C	A	G	
U	UUU Phenylalanine (Phe)	UCU Serine (Ser)	UAU Tyrosine (Tyr)	UGU Cysteine (Cys)	U
	UUC Phenylalanine (Phe)	UCC Serine (Ser)	UAC Tyrosine (Tyr)	UGC Cysteine (Cys)	C
	UUA Leucine (Leu)	UCA Serine (Ser)	UAA Stop	UGA Stop	A
	UUG Leucine (Leu)	UCG Serine (Ser)	UAG Stop	UGG Tryptophan (Trp)	G
C	CUU Leucine (Leu)	CCU Proline (Pro)	CAU Histidine (His)	CGU Arginine (Arg)	U
	CUC Leucine (Leu)	CCC Proline (Pro)	CAC Histidine (His)	CGC Arginine (Arg)	C
	CUA Leucine (Leu)	CCA Proline (Pro)	CAA Glutamine (Glu)	CGA Arginine (Arg)	A
	CUG Leucine (Leu)	CCG Proline (Pro)	CAG Glutamine (Glu)	CGG Arginine (Arg)	G
A	AUU Isoleucine (Iso)	ACU Threonine (Thr)	AAU Asparagine (Asn)	AGU Serine (Ser)	U
	AUC Isoleucine (Iso)	ACC Threonine (Thr)	AAC Asparagine (Asn)	AGC Serine (Ser)	C
	AUA Isoleucine (Iso)	ACA Threonine (Thr)	AAA Lysine (Lys)	AGA Arginine (Arg)	A
	AUG Methionine (Met)	ACG Threonine (Thr)	AAG Lysine (Lys)	AGG Arginine (Arg)	G
G	GUU Valine (Val)	GCU Alanine (Ala)	GAU Aspartate (Asp)	GGU Glycine (Gly)	U
	GUC Valine (Val)	GCC Alanine (Ala)	GAC Aspartate (Asp)	GGC Glycine (Gly)	C
	GUA Valine (Val)	GCA Alanine (Ala)	GAA Glutamate (Glu)	GGA Glycine (Gly)	A
	GUG Valine (Val)	GCG Alanine (Ala)	GAG Glutamate (Glu)	GGG Glycine (Gly)	G

CHAPTER THREE

EUKARYOTIC CELLS

It is believed that the eukaryotic cells evolved from prokaryotic cells 1.6 to 2.0 billion years ago during the Mesoproterozoic Period. The distinctive feature that separates eukaryotes from prokaryotes is the unique development of membranous organelles. Like the divisions in a factory, the organelles within a eukaryotic cell perform specific roles in order to maintain normal cellular functions. Organelles are surrounded by a phospholipid bilayer similar in composition to the plasma membrane. Large organelles, such as the nucleus, and at times nucleolus, and vesicles, can be visible under low magnifications via a light microscope, while other organelles require special staining and high magnification, such as the scanning or transmission electron microscopes. It has been shown that some of the organelles are interconnected through intramembranous channels. For instance, the nuclear membrane is continuous with both the rough and smooth endoplasmic reticulum, while the remaining organelles exist independently within the cytoplasm (Figure 1).

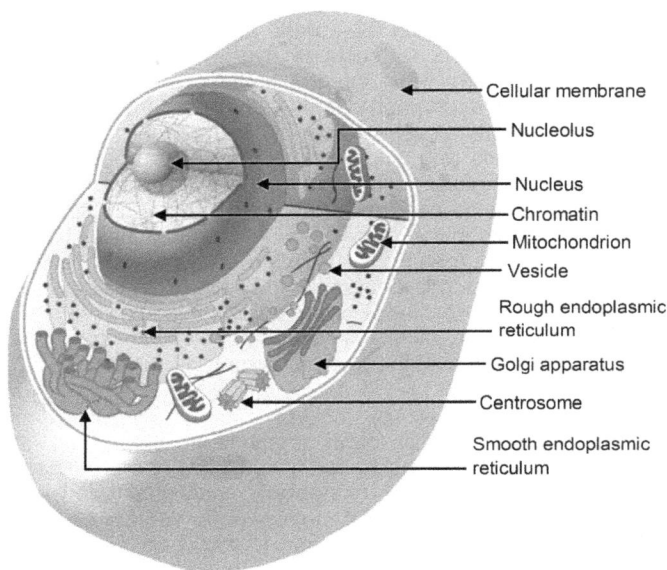

Figure 1: An eukaryotic cell and its organelles

Organelles such as the mitochondria and chloroplasts (found only in plant cells) are unique due to the fact that they do not share the same genetic materials as the cell that they inhabit. For example, the mitochondria possess their own DNA, which is similar to that found in prokaryotic cells. It is believed that the mitochondria, as well as the chloroplast, were at one time free-living organisms that were absorbed into the cytoplasm of prehistoric eukaryotic cells. Instead of consuming and digesting the smaller organisms, the eukaryotic cells and the prokaryote cells entered into a symbiotic relationship, (i.e. Endosymbiotic Theory), whereby both organisms benefited from their interactions. For example, the mitochondria provide the location for the production of ATP for the eukaryotic cells, while at the same time the eukaryotic cell provides the mitochondria with nutrients and a protective environment.

Simple observations will reveal that, on average, eukaryotic cells (10 to 100 μm in diameter) are much larger than prokaryotic cells (0.2 to 2.0 μm in diameter) and have the unique capability to form multicellular organisms. In contrast, prokaryotes can only maintain a single cellular status even though some are colonial. For this reason, animals and plants are made up of specialized eukaryotic cells conceived with specific functions. For instance, the human body is made of 200 different types of specialized cells working in conjunction to maintain a homeostatic environment.

Plasma Membrane

The plasma membrane is composed mostly of phospholipids arranged in a bilayer. This bilayer, also known as the cellular membrane, is separated into hydrophilic and hydrophobic regions due to the amphipathic nature of the phospholipids. The hydrophilic region of the cellular membrane is composed of the polar phosphate ends of the phospholipid. These polar ends of phospholipids, which comprise a minor portion of the entire molecule, are arranged to face the fluid filled medium (either towards the extracellular fluids or the cytoplasm of the cell). On the other hand, the non-polar ends of phospholipids consist of long chains of fatty acids that are closed within the inner layer of the bilayer away from any polar molecules. This dual property of the cellular membrane provides an effective barrier that keeps the internal and external environment separate (Figure 2).

Extracellular

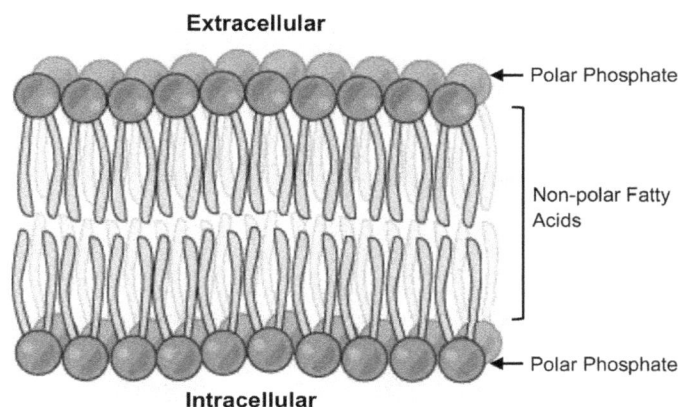

Figure 2: Illustration of the polar and non-polar layers of a phospholipid bilayer. Please note that the polar ends (shown in red) are facing the extracellular and intracellular fluid environments

The plasma membrane is a semipermeable membrane (also termed selectively permeable or differentially permeable), where it will allow certain molecules or ions to freely pass through, while preventing others from entering or exiting the cytoplasm. Without the assistance of active transport molecules (discussed later in this section), the manner by which substances pass through the cellular membrane is passive transport (Figure 3).

Passive transport is based on two properties that are common in biology, diffusion and osmosis. Diffusion is defined as the process of moving molecules from areas of high concentration to areas of low concentration based upon their concentration gradient (diffusion gradient). Lipid soluble substances such as oxygen (O_2), carbon dioxide (CO_2), and steroid-based molecules, can diffuse through the membrane freely based upon their diffusion gradient. Nonetheless, the rate of diffusion is based upon the magnitude of the concentration gradient. Put simply, the greater the concentration gradient, the greater the number of substances will flow from the higher concentration to lower concentration.

Osmosis, on the other hand, is the process by which water crosses from an area of high concentration to area of low concentration through a semipermeable membrane (e.g. the plasma membrane). Most often osmosis occurs through water channel proteins that exist in the membrane (discussed later in this section). The ability of water to move freely through the plasma membrane produces a problem for cells; this is especially true for multicellular creatures like humans since we are composed of approximately 80% water. For example, if the solute (dissolved substances) concentration within the cell is higher than its surrounding area, or inversely, the concentration of water is lower inside the cell than the surrounding environment, water will flow from its external environment into the cell. This hypotonic environment will result in the swelling of and possible bursting of the cell. Alternatively, if the solute concentration in the surrounding environment is higher than within the cell, or inversely, the concentration of water is higher inside the cell than the surrounding environment, water will flow from the cytoplasm out of the cell. In this hypertonic environment, crenation (shrinkage) of the cell occurs. In order to achieve homeostasis, the cell must be placed within an isotonic environment where the amount of water flowing into the cell is equal to the amount flowing out of the cell (Figure 4).

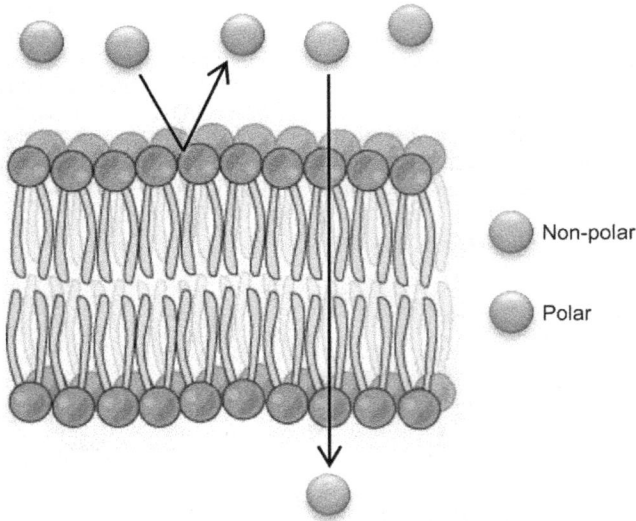

Non-polar

Polar

Figure 3: Illustration of the differentially permeable properties of a plasma membrane

Hypertonic	Hypotonic

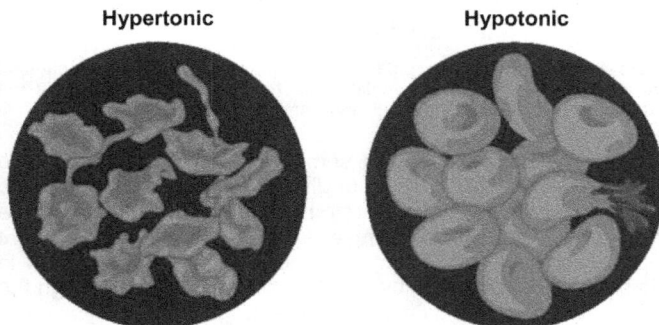

Figure 4: Red blood cells placed in a hypertonic (shriveled cells) and hypotonic (burst) environment

In summary, the permeability of the plasma membrane depends on many factors. For example, the specific osmotic pressure exerted on the cell, the concentration of solutes, solute size and shape, as well as, the ambient temperature may affect the membrane's permeability. In addition, the polarity of the molecule will also play a part in its permeability. For example, if the molecule is polar (hydrophilic), it will briefly enter the polar edges of the phospholipid bilayer before it is expelled by the nonpolar fatty acid layer. However, if the molecule is nonpolar (hydrophobic), it will face a slight repulsion by the polar phosphates. However, since the polar layer is minute, the nonpolar molecule will eventually pass through the phospholipid virtually unhindered (Figure 3).

Presently, the accepted concept for the proper organization of the plasma membrane is known as the fluid-mosaic model. This model theorizes that the membrane is relatively fluid, highly flexible, and has the capability to change shape and composition at times. It is theorized that the plasma membrane has the consistency of olive oil at body temperature (Figure 5).

Within the plasma membrane there are numerous other molecules such as proteins, glycolipids, and cholesterol randomly

embedded within to effectively shield the internal organelles from the external environment. For example, cholesterols, like phospholipids, are amphipathic molecules and possess both a hydrophilic and a hydrophobic portion. Cholesterol has a hydroxyl (-OH) group that aligns with the polar phosphates of the phospholipids while the remaining portion of the molecule inserts itself into the fatty acid portion of the membrane. The close interaction of cholesterol and phospholipids helps to stabilize the membrane, making it more fluid (Figure 5).

Structurally, there are two types of embedded proteins: peripheral and integral proteins. Peripheral proteins are found on the edges of the membrane and are positioned into the phospholipid bi-layer through covalent bonding, non-covalent interactions such as hydrogen bonding and/or anchored via cytoskeletons. The main function of peripheral proteins is to help stabilize the structural integrity of the plasma membrane; although if they are combined with a polysaccharide group, they are part of the cell's recognition molecules.

Integral proteins, also referred to as transmembrane proteins, are proteins that span the entire thickness of the plasma membrane. Like peripheral proteins, the integral proteins are anchored into the plasma membrane through covalent bonds, non-covalent interactions, such as hydrogen bonding, and/or via the cytoskeleton. Most integral proteins are glycoproteins which have their hydrophobic ends within the fatty acid layers of the plasma membrane (which are hydrophobic), while the hydrophilic regions (e.g. sugar groups) project out towards the intra- and extra-cellular fluid medium. There are numerous roles that integral proteins perform for the cell. Integral proteins may serve as cell recognition proteins, junctions (described below), transport proteins, membrane bounded enzymes and receptor proteins (described in later sections). Based on their molecular composition, each of these embedded molecules serves a specific purpose that maintains cellular homeostasis, as well as, the structural integrity of the plasma membrane (Figure 5).

Cell recognition molecules include various glycolipids and glycoproteins that are found embedded throughout the membrane. These polysaccharide-linked lipids and proteins are responsible for cell-to-cell recognition, individual-to-individual recognition, and recognition of molecules for cellular transport. The sugar groups that are attached to the lipids and proteins are called oligosaccharide or an antigen. Oligosaccharides are polysaccharide polymers that consists of three to nine monosaccharides in various combinations. It is the different combinations of these monosaccharides that provides the necessary identification markers for each of the cells. For example, cell-to-cell recognition is extremely important during embryonic development of an organism. The differences in the oligosaccharides allow the cells to differentiate and arrange themselves in proper locations within the developing body. Put simply, a skin cell will have a slightly different oligosaccharide marker when compared to a cardiac muscle cell. This difference prompts the skin cells and muscle cells to proceed to a different location as the body develops. Nevertheless, these subtle differences in antigens between the cells from the same body will not bring about an autoimmune response by the leukocytes (white blood cells), since they recognize all sugar types within the organism. However, if a particular cell is misplaced (i.e. a skin cell is misplaced among the cardiac muscle cells), the wayward cell would be quickly removed by the leukocytes (white blood cells).

In individual-to-individual recognition, the differences in oligosaccharides allow the body to recognize what is part of the body and what is alien. The best example of individual-to-individual response is visualized during blood transfusion when the wrong blood type is accidentally introduced into a patient's body. For example, an A-blood type individual could not accept blood from a B-blood type person because the differences in their antigen

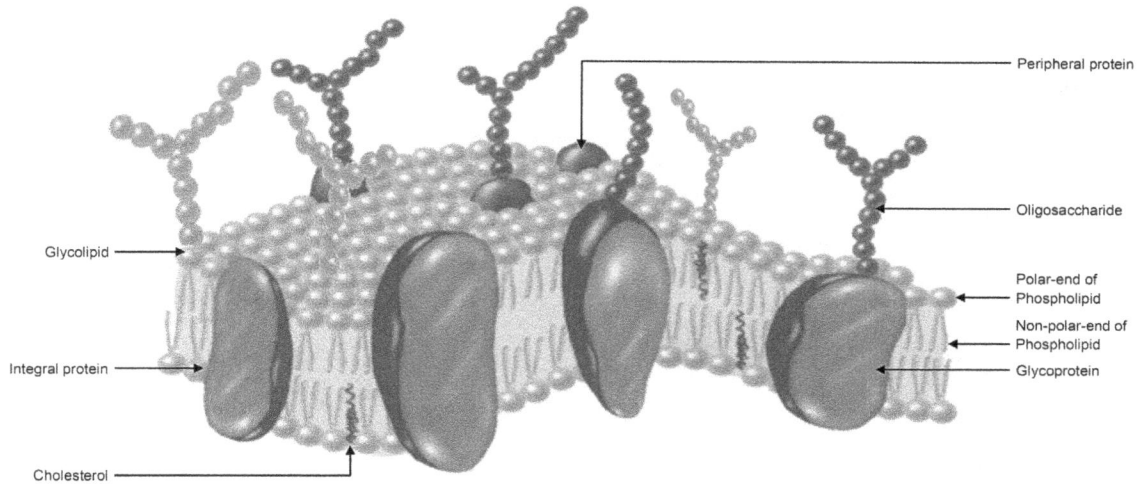

Figure 5: Illustration of a plasma membrane

will trigger an immediate immunoresponse by the leukocytes (Figure 6). For the same reason, a B-blood type individual will not be able to accept A-blood type blood. An O-blood type individual will possess no antigens on their red blood cells. O-blood type individuals will only be able to receive O-blood type blood and will reject all other blood types (e.g. A-, B- and AB-blood types). However, since O-type blood possesses no antigens, it could be transfused into individuals of all blood types. For this reason, O-blood type is referred to as the universal donor. On the other hand, a person who possesses an AB blood type will possess both A and B oligosaccharides and will be able to accept all transfusions. Donations, on the other hand, could only be given to another AB-blood type individual. Individuals with AB blood types are known as universal recipients.

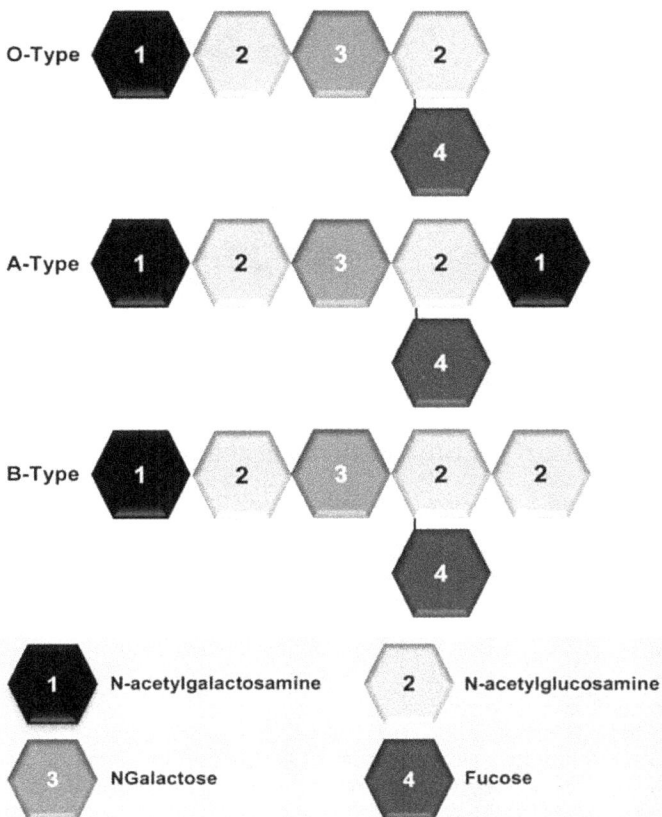

Figure 6: Illustration of the various oligosaccharides found on the membrane surface of red blood cells

An additional example of individual-to-individual recognition could be seen when the body is invaded by bacteria or viruses. The oligosaccharides exhibited by bacteria and viruses are foreign to the body's immune system, which will immediately trigger an immunoresponse.

Cellular Junctions

Cell-to-cell junctions are readily found between the membranes of eukaryotic cells. Most junctions are designed to link one cell with another, thereby allowing the formation of three-dimensional tissue structure, as well as, the maintenance of their proper function.

For example, tight junctions (also known as occluding junctions or zonula occludentes) generally form a belt-like loop around the apical lateral surfaces of epithelial tissues. Tight junctions form an impermeable barrier that prevents seepage of fluid and other materials, as well as preventing excessive movements between cells. For example, tight junctions found in the epithelial linings of the stomach prevent gastric juices (acids and enzymes) from seeping into and inadvertently digesting the underlining tissues. These junctions, located in the small intestines, prevent the nutrients from leaching in between the epithelial cells, limiting absorption only through the epithelial cells.

Tight junctions are deceivingly complex structures. It has been estimated that approximately 40 proteins are needed to form the protein complex, which in turn, is responsible for creating a single junction. The most notable of these proteins are the transmembrane proteins claudin, occludins, along with cytoplasmic scaffolding proteins ZO-1, ZO-2, and ZO-3 that creates links with the actin component of the cytoskeleton (Figure 7).

Adherens junctions (also known as anchoring junction, adhesion junctions, or zonula adherents) are found in cell types that are frequently exposed to mechanical stress (e.g. heart muscles, cardiac epithelium, skeletal muscles, and epidermis) and are responsible for maintaining physical contact between adjacent cells. One of the major component of adherens junction is the transmembrane linker proteins called cadherins (type I transmembrane proteins or also known as "classic" cadherins), which are responsible for forming bonds with other cadherins from adherens junctions with neighboring cells through homotypic (like molecule) interactions. These cadherins-cadherins interactions maintain a tightly regulated distance of 10 to 20 nm between the two plasma membranes. On the cytoplasmic side, cadherins couples with β-catenins, which in turn, are fastened into place by bundles of actin filaments and microtubules from the cytoskeleton.

Other proteins involved in the formation of adherens junctions include vinculin, a membrane-cytoskeletal protein that functions in linking both integrin adhesion molecules and cadherin to the actin cytoskeleton, and α-actinin which is necessary for the bundling of the actin filaments. Like the occluding junctions, anchoring junctions form an impermeable barrier that prevents seepage and minimizes cellular movements within tissues (Figure 8).

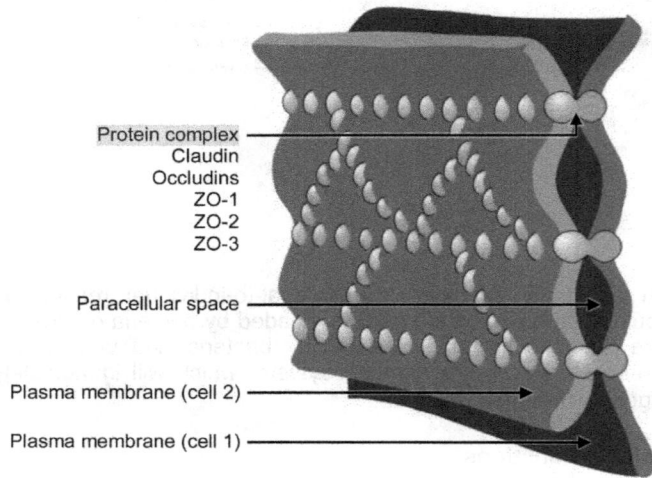

Figure 7: Illustration of a tight junction demonstrating the protein complex composed of a claudins, occludins, ZO-1, ZO-2, and ZO-3. Please note that the actin component of the cytoskeleton is not shown

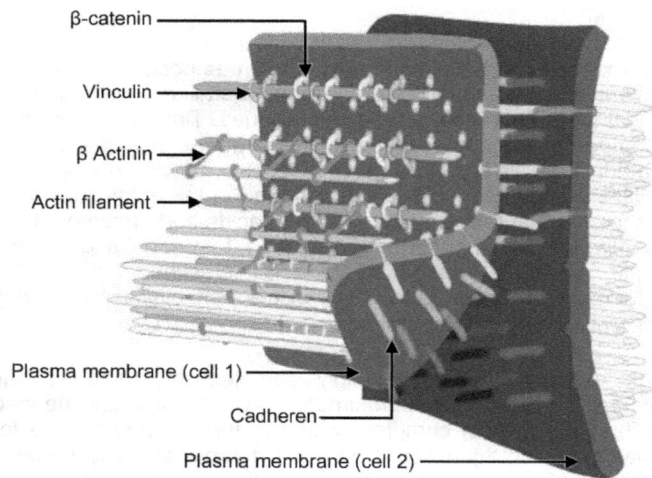

Figure 8: A adherens junction. Please note that the cadherens from plasma membrane (cell 1) is bounded to cadherens from plasma membrane (cell 2) via homotypic interactions. On the cytoplasmic side, β-catenin are responsible for connecting the junction to actin and microtubules

Desmosomes, also known as spot-weld junctions, provide strong adhesion between cells and are designed to withstand stresses, such as stretching, bending, twisting and compression, between cells of the epithelial tissue. Because of their ability to withstand high levels of stress, desmosomes are commonly found in stress prone tissues such as the epidermis (skin) and myocardium (heart muscles) (Figure 9).

On the extracellular side of the plasma membrane, various desmosomal cadherins, such as desmogleins, and desmocollins, engage in calcium-dependent adhesive interactions. On the cytoplasmic side, the desmosomal cadherins are linked to the

keratin intermediate filament partly facilitated by plakoglobin and plakophilins. It is understood that plakoglobins and plakopilins form the desmosomal (protein) plaques. In turn, the desmosomal plaques are connected to the keratin intermediate filament by desmoplakin, an intermediate filament binding protein. Desmoplakin also associates with both plakoglobin and plakophilins to complete the loop of interjunctional bonding (Figure 9).

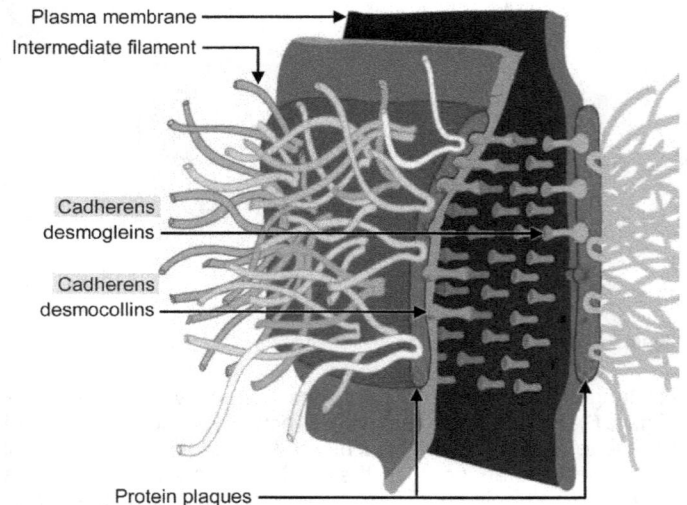

Figure 9: Illustration of a desmosome

Gap junctions, also known as communication junctions, are intercellular channel proteins that are found in nearly all cells (in association of solid tissues) in the human body. The gap junction is formed for cell-to-cell communications and allows the direct exchange of ions and molecules between cells (Figure 10).

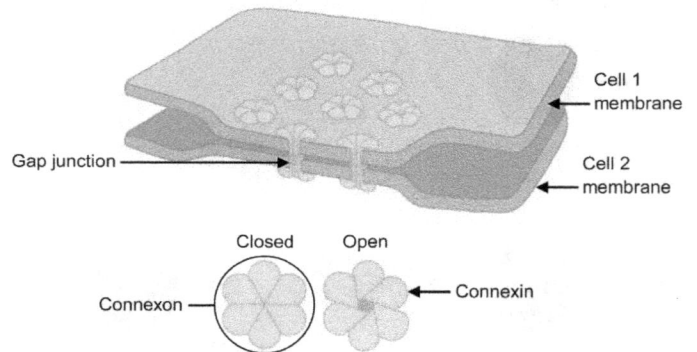

Figure 10: Illustration of a gap junction. Please note that a single gap junction is made from two protein channels, one from each adjacent cell

Gap junctions are formed from two protein channels located within each of the plasma membrane of adjacent cells. Each of the protein channels are aligned to form a continuous pore or aquaporin between the cells. These protein channels are made from the assembly of six transmembrane proteins called connexins to form a hexameric structure called connexon. The close interactions of the connexins allows no breaks to form between the transmembrane proteins while only a narrow, approximately 2 nm, corridor or "gap" is formed at the center of the hexameric structure.

In cardiac muscle cells and in nerves, gap junctions are essential in the spread of action potentials from one cell to the next. For example, in the cardiac muscle cells, the gap junctions located at the intercalated disks allow the electrical impulses to pass from one cell to the next thereby allowing the uniform contraction of the heart. In nerves, the gap junctions allow the electrical impulses to spread from one neuron to the next via synapses

(Figure 10).

Membrane Transport Proteins

Integral proteins are responsible for either passive or active transport. For example, channel proteins, which are composed of integral proteins, possess a corridor called pore (or aquaporin if the channel protein allows water transport) towards the midpoint of the protein structure. This corridor allows selected ions or polar molecules, as well as, water to pass through the plasma membrane based on their concentration gradient. For example, if sodium ions flow from areas of higher concentration to areas of lower concentration through the pore of a channel protein, this process occurs through the laws of diffusion. If water molecules flow from areas of higher concentration to areas of lower concentration through the aquaporin of a channel protein, this process is called osmosis.

Regulation of the passage of substances through the channel protein is based on size and charge. Put simply, if the substance is too large, it will not physically fit through the pore of channel protein. The regulation of charge molecule is slightly more complicated and involves the functional groups (R-groups) on the amino acids that construct the channel protein.

First, the hydrophilic R-groups of amino acids are arranged so that they are directed into the pore of the channel proteins. These R-groups would possess alternating charges depending upon the substances that they are allowing passage. For instance, R-groups of the channel proteins could be positively charged so that they attract anions. Alternatively, the R-groups could be negatively charged where they would attract cations. In addition, channel proteins may also be gated (Please examine Appendix II in Chapter Two).

There is a tremendous diversity among gated ion channels. For example, ion channels are controlled through different stimuli including voltage, temperature, pH, mechanical force, and ligands. Furthermore, the diversity of the gated ion channels also includes distinct properties, which cause the gated channels to open and close hastily, open quickly but close gradually, and open slowly and continue to stay open for an extended period of time. In addition, the biophysical properties of the gated channel may only allow ions to flow in one direction, or keep the channels that remain open all the time (leak channel).

Ion channels are large transmembrane polypeptides with many hydrophobic membrane-spanning domains anchored in lipid bilayer by cytoskeleton, where they perform a designated function. Most ion channels are very specific as to what is allowed through their pores. Potassium (K^+) channels, for instance, selectively allow K^+ ions to flow through but actively exclude Na^+ ions (or any other ions). Please note that both Na^+ and K^+ ions are positively charged (cations), therefore the selectivity is more complicated than simple charge or size exclusion. The specific mechanism of this selective exclusion baffled scientists for many years but recent discoveries have clarified the specific process. Recent research has shown that the exterior (the side facing the external environment) of the K^+ ion channel, near the opening of the pore, is negatively charged. This negative charge is generally accomplished by positioning negative charged amino acids (amino acids that possess negative charged R-groups) such as glutamic and aspartic acids near the opening. This negative charge serves to increase the effective concentration of cations near the outside of the pore (Figure 11).

It is understood that all ions are surrounded by water molecules due to hydrogen bonds. Therefore it is necessary to strip away the water molecules in order for the ions to fit through the pore. In the case of K^+ ions, the channel eliminates the water molecules by utilizing an amino acid R-group (functional group) called carbonyl oxygen ($C=O^-$). Carbonyl oxygen possesses a strong negative charge and is capable of stripping the water molecules surrounding the K^+ ions. However, this carbonyl oxygen is unable to strip away the water molecules surrounding Na^+ ions, since water molecules form much stronger bonds with Na^+ by comparison. Because of this inability to remove the water molecules surrounding Na^+ ions, the effective radius of this Na^+-water molecule complex is much wider than the K^+ ions.

As the ions continue down into the pore, they will be funneled single-file through a narrow center. At this narrowing, the larger diameter Na^+ ions (i.e. still surrounded by water) will be physically excluded from proceeding any further. In contrast, the much smaller diameter K^+ ions, having been stripped of its surrounding water molecules, will pass unimpeded. Once the K^+ ions pass through the narrow center, they will enter an area called the energy well. Like the opening of the pore, the energy well is also populated with amino acids with negatively charged R-groups. This negatively charged energy well attracts the positively charged K^+ ions and prevents them from diffusing backwards and return to the way they came.

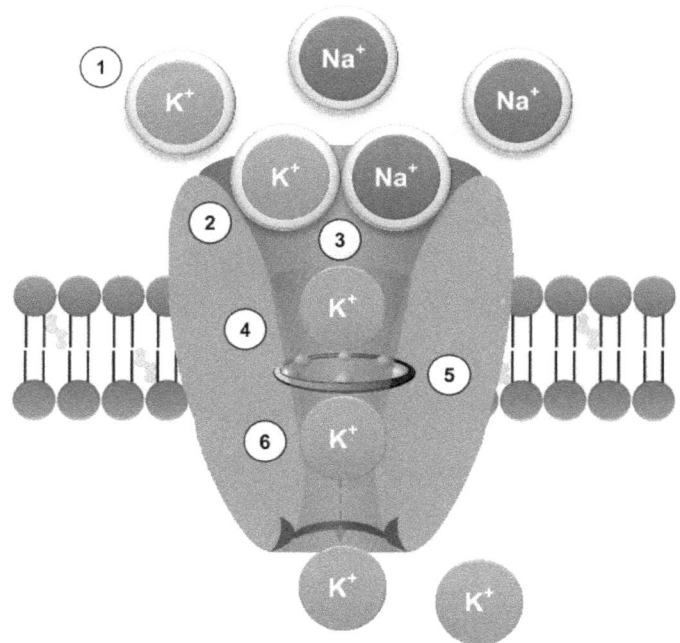

Figure 11: A Potassium ion channel. ① Both Na+ and K+ cations are surrounded by water molecules. ②The opening of the pore is negatively charged. ③ The negativity of the pore opening serves to increase the effective concentration of cations near the outside of the pore. ④ Carboxyl oxygen (C=O) strips the water molecules surrounding the K+ ions. ⑤The negatively charged energy well will attract K+ ion and prevent it from diffusing backwards. ⑥ The aggregated K+ ions will repulse one another and push each other towards the cytoplasm

As the first K^+ ion sits comfortably in the energy well, other K^+ ions will follow its progress down the pore. Since like-charges repulse one another, the latter K^+ ions will push the K^+ in front of them further into the pore and finally, based on their diffusion gradient, these ions will exit the channel protein pore and into the cytoplasm (Figure 11).

Receptor Proteins

There are two types of membrane bound receptor proteins: ionotropic and metabotropic receptors. The ionotropic receptors are transmembrane proteins that possesses a ligand binding site directly linked to the ion channels to form a single molecule entity. Ionotropic receptors are composed of four to five trans-membrane protein subunits, also known as multimers organized around a pore. These ionotropic receptors, are fast acting receptors commonly associated with neurotransmitters (signal molecules, ligands) such as γ-aminobutyric acid (GABA), glycine, serotonin, acetylcholine, etc. and are specific in the type of ions that are allowed to pass (Na⁺, K⁺, Ca⁺⁺ or Cl⁻) (Figure 12).

A G-protein consists of alpha (α), beta (β) and gamma (γ) subunits on the inner surface of the cell membrane. G-protein is activated when a signal molecule (ligand), such as a neurotransmitter or a hormone, attaches to its binding site. Please note that the receptor binding site is conformation-dependent (shape-dependent) so that only a ligand of a specific shape can bind with the receptor binding site. Once the ligand binds to the G-protein complex the receptor changes its shape and the three subunits (α, β and γ) separate. The β- and γ-subunits form a stable dimer (i.e. the joining of two molecules through intermolecular bonding) and are responsible for activating the α-subunit. As a result, guanosine diphosphate (GDP; spent energy molecule) bounded α-subunit is replaced with guanosine triphosphate (GTP; charged energy molecule). Subsequently, the α-subunit-GTP complex either causes the conformation to change for a channel protein or activation of an enzyme (Figure 13).

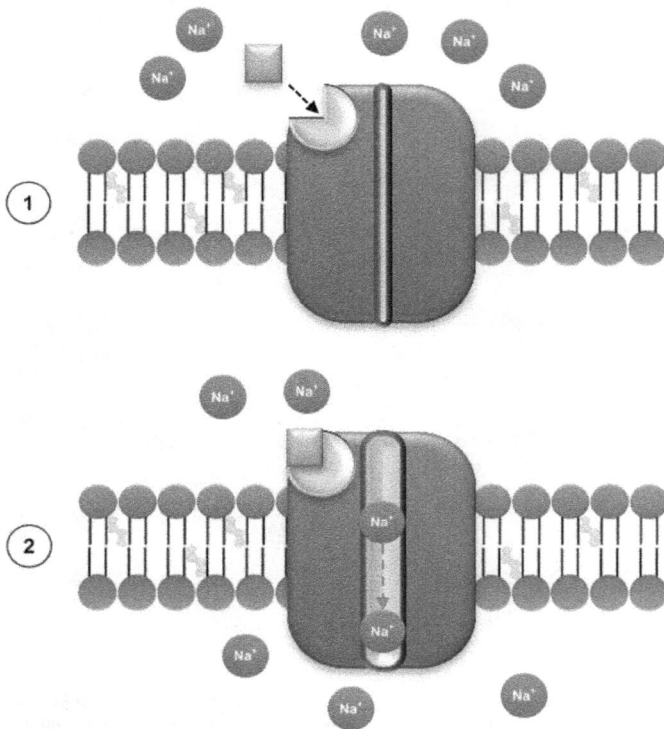

Figure 12: An ionotropic receptor. Please note that this receptor is a transmembrane protein that possesses a ligand binding site directly linked to the ion channels to form a single molecular entity. ① The pore is in the closed position prior to the binding of the ligand. ② Once the signal molecule binds to the ligand binding site the intermolecular interaction will result in the opening of the pore, thereby allowing Na⁺ ions to flow down its concentration gradient

For example, once the neurotransmitter acetylcholine (ACh) attaches to the binding site of a G-protein complex (e.g. muscarinic acetylcholine receptor, mAChR) this attachment will cause a conformational change whereby the subunits will separate. The β- and γ- dimer activate the α-subunit causing the α-subunit to exchange GDP with GTP. The α-subunit/GTP complex causes a conformation change in the sodium channel protein, thereby resulting in the opening of the pore allowing sodium (Na⁺) ions to flow down its concentration gradient (via diffusion) and enter the cytoplasm of a cell (Figure 13). In time, when an adequate amount of Na⁺ cations have diffused through the pore, the cell closes the gated channel. The process of returning the gated channel to its original conformation is the inverse reaction of the previously described process. First, the inter-cellular ligand acetylcholine has to be removed from the binding site on the G-protein by the enzyme acetylcholinesterase. Once it is removed, the α-subunit will hydrolyze the energy molecule GTP releasing a phosphate

molecule and energy thereby breaking the interaction with the gated channel. Once the connection between the α-subunit-GDP complex is severed, the gated channel will reconstitute its original conformation and prevent Na⁺ ions from entering the cytoplasm (Figure 14).

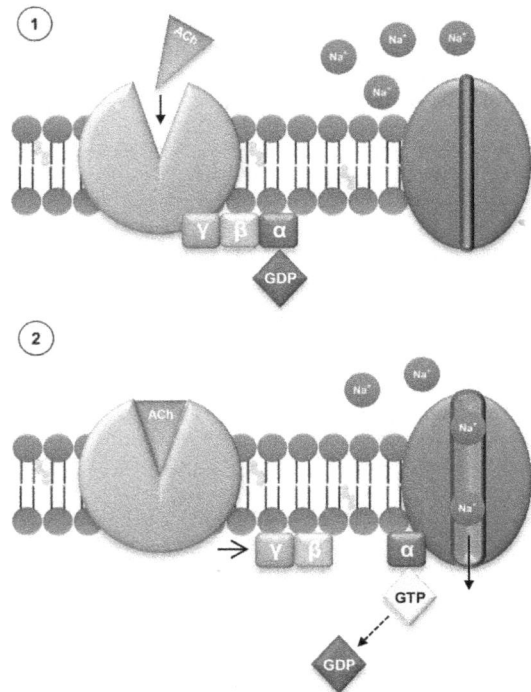

Figure 13: Illustration of a G-protein complex. ① Prior to the binding of ACh the gated channel is closed. ② Binding of ACh causes the α, β, and γ subunits to separate and GDP on the α subunit is exchanged with GTP. The α subunit-GTP complex causes a conformational change of the gated channel causing it to open, allowing Na⁺ ions to flow into the cell via their diffusion gradient

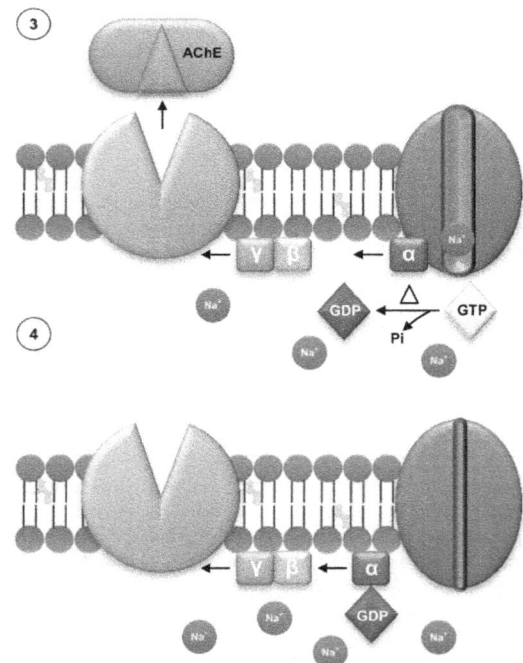

Figure 14: A G-protein complex. ③ ACh is removed from the G-protein binding site by the enzyme acetylcholinesterase (AChE). The removal of ACh causes the α-subunit to hydrolyze the energy molecule GTP thereby releasing a phosphate molecule (Pi) and energy. ④ Once the energy molecule GTP is hydrolyzed and converted into the spent energy molecule GDP, the interaction between the α subunit and the gated channel is severed thus returning the gated channel back to its original conformation (closed gate). The closed gate of the channel protein will in turn prevent any further diffusion of Na⁺ cations into the cytoplasm

An example of G-protein complex activation of an enzyme is the attachment of the hormone glucagon with its specific metabotropic receptors on the membrane of hepatocytes (i.e. liver cells). Glucagon is a hormone secreted by the alpha cell located in the Islets of Langerhans in the pancreas. opposing the actions of insulin, glucagon is responsible for increasing blood glucose concentration.

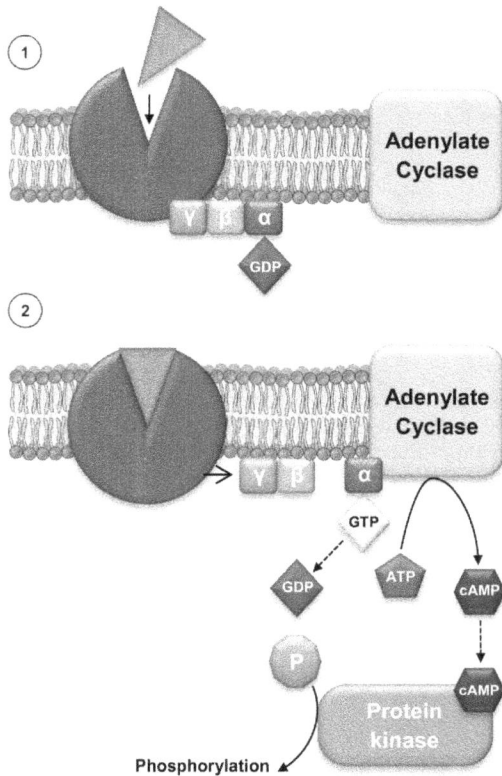

Figure 15: ① Prior to the binding of glucagon, the enzyme adenylate cyclase is not activated. ② The binding of glucagon causes α-, β- and γ-subunits to separate and GDP on α-subunit is exchanged with GTP. The α-subunit-GTP complex causes a conformational change of adenylate cyclase, activating the enzyme. Adenylate cyclase converts ATP into cAMP which then proceeds to activate protein kinase, an enzyme responsible for phosphorylating various intracellular enzymes

Once the hormone glucagon binds to the G-protein complex, the α-, β- and γ-subunits separate. The β- and γ-dimer activates the α-subunit causing the α-subunit to exchange GDP with GTP. The α-subunit/GTP complex causes a conformation change and subsequent activation of a membrane bounded enzyme called adenylate cyclase. Adenylate cyclase catalyzes the conversion of an ATP molecule into cyclic adenosine monophosphate (cAMP). cAMP is an intracellular (inside the cell) messenger and will in turn activate another enzyme, protein kinase. Protein kinase is responsible for transferring phosphate groups (i.e. process of phosphorylation) onto other enzymes, and therefore activating them. At the end of this cascade reaction, the hepatocytes are prompted to convert stored glycogen (stored sugar in the liver) into glucose and release it into the bloodstream (Figure 15).

Carrier proteins, also known as transporters, are involved in both the passive and active transport of molecules across the cellular membrane. Passive transport is called facilitated diffusion. For example, glucose, a large polar molecule, has a difficult time passing through the phospholipid bilayer of the plasma membrane. Nonetheless, glucose is essential for the survival of the cell and has to be transported into the cytoplasm. This problem is solved by using a glucose transporter. Once glucose binds to the transporter, the intermolecular interaction causes conformation change of this integral protein. The conformational change facilitates the movement of the glucose molecule towards the center of the integral protein. From that point, the glucose molecule flows down its concentration gradient into the cytoplasm of the cell (Figure 16).

In contrast to facilitated diffusion, active transport physically moves molecule(s) or ion(s) against its concentration gradient. Substances are moved from areas of low concentration to high concentration. It is understood that the function of active transport carrier protein depends upon the availability of ATP, the mainstay energy molecule of all living organisms.

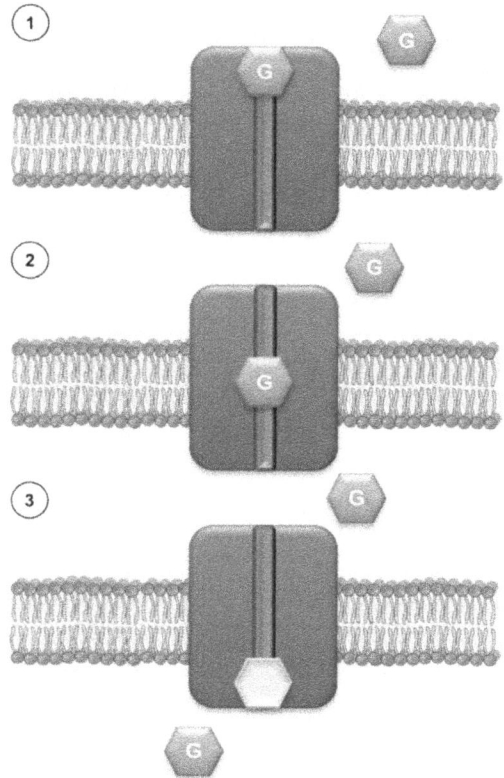

Figure 16: Illustration of facilitated transport. The binding of glucose (G) to the facilitated transport protein causes ① the change in the transport protein's conformation ② which allows the glucose molecule to be moved from the extracellular matrix and into ③ the cytoplasm of the cell

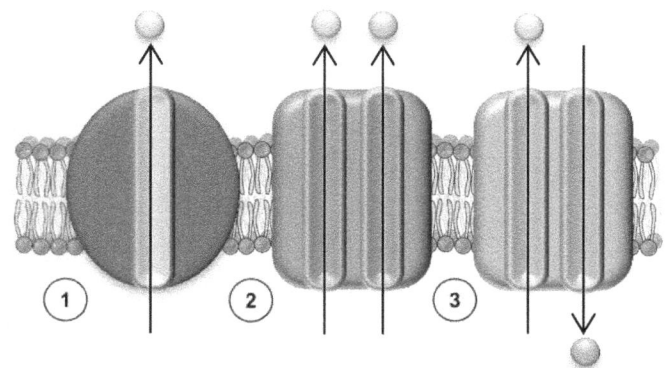

Figure 17: Illustration of active transport. ① Uniport, one substance is transported one direction. ② Symport, two substances are transported in the same direction. ③ Antiport, two substances are transported in opposite directions

There are three types of carrier protein based upon the amount and direction of ions or molecules they are transporting. For example, an uniport (uniporter) is capable of transporting one ion or molecule in one direction, in or out of the cell. A symport (symporter) transports two substances in the same direction, in or out of the cell. Finally, an antiport (antiporter) transports two substances in and out of the cell, in opposite directions. The most commonly referenced antiport is the sodium-potassium pump which are commonly found in the nervous and muscular tissues. Sodium-potassium pumps are responsible for restoring the membrane potential of both nervous and muscular tissues, so that additional action potentials can be generated.

Both symport and antiport are referred to as cotransporters, or carrier proteins, that move more than one molecule in one direction at the same time (Figure 17).

In general, there are two types of active transport. As described previously, primary or direct active transport is where energy in the form of ATP is used to move ions or molecules against their concentration gradient. Active transport transmembrane proteins are commonly found on the membrane of all cells within the human body and it is estimated that the energy requirements of primary active transport amounts to approximately 30% of the entire ATP supply of the cell. Primary active transport carrier protein complexes are an ideal example of ATPases. ATPases are enzymes that hydrolyze ATP to form ADP and release its energy.

Secondary active transport uses the kinetic energy (energy of motion) that is generated by a molecule, such as sodium, moving down its concentration gradient to drive its functions. One common example of secondary active transport is the reabsorption of glucose from the filtrate within the nephrons of the kidneys. In normal conditions, the concentration of Na^+ is high in the filtrate, while the concentration of glucose is comparatively high. Within the nephritic cells of the proximal convoluted tubule, also known as tubule cells, an Na^+ concentration gradient is generated through the actions of sodium-potassium pumps located on the basal domain (bottom) of the cell, which is adjacent to the blood vessels. The sodium-potassium pump sends three Na^+ ions out

of the cell and into the blood stream while moving two K^+ ions into the cell from the blood stream. On the apical domain (top) of the cell a sodium-glucose cotransporter is located. Because of the diffusion gradient generated for Na^+ ions by the sodium and potassium pump, Na^+ ions will readily diffuse down its concentration gradient into the cell. The kinetic energy generated by the movement of Na^+ ions into the cell allows the glucose molecule to be transported into the cell. Once inside the nephritic cells, glucose enters the interstitial fluid through facilitated diffusion via glucose transporter 2 (GLUT2) and reenters the bloodstream via diffusion (Figure 18).

Membrane Assisted Transport

Membrane assisted transport allows the movement of large substances or large amounts of substances in or out of the cell. All membrane assisted transport is receptor mediated and the receptors involved in this form of transport system are G-protein complexes. Put simply, the presence of a specific ligand or marker (antigen), attached to the membrane bound receptors (G-proteins), will trigger the activation of this form of transport (Figures 19 and 20).

There are two distinct types of membrane assisted transport: endocytosis and exocytosis. Endocytosis is transporting substances into the cell through the manipulating of the cellular membrane. Endocytosis is subdivided into two subtypes: phagocytosis and pinocytosis. Phagocytosis (defined as cell eating) occurs when generally large materials, such as foreign substances like bacteria, fungal spores, a damaged or dying cell, as well ass cellular debris are brought into the cell through the manipulation of the membrane and the formation of a phagosome. A phagosome is a vesicle containing the phagocytized material.

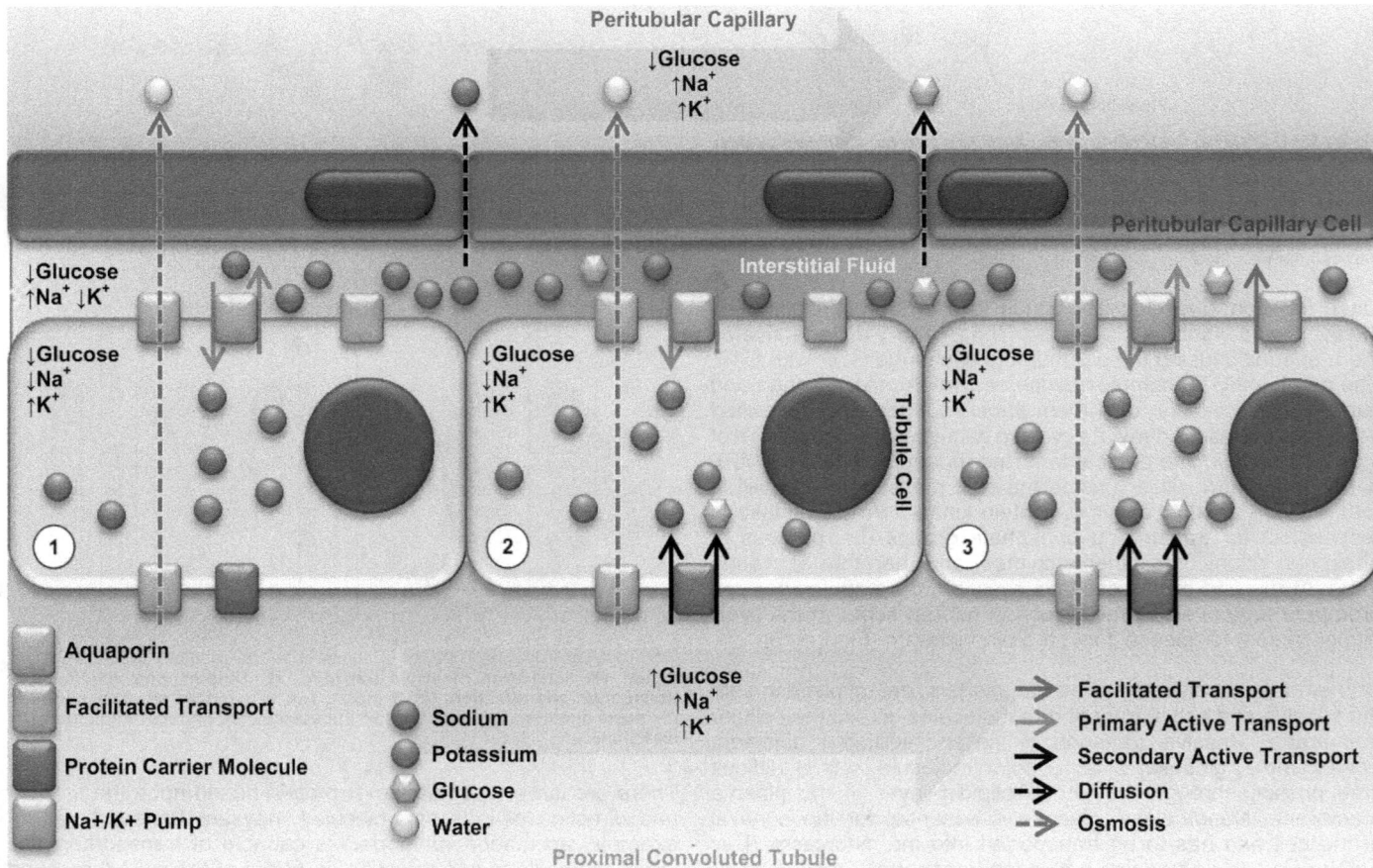

Figure 18: Primary active transport, secondary active transport and facilitated diffusion. ① Na^+/K^+ pump (primary active transport) will send out 3 Na^+ to the interstitial fluid and send in 2 K^+ into the cytoplasm of the tubule cell creating a Na^+ diffusion gradient from the filtrate in the proximal convoluted tubule. Please note that water flows readily back into the blood stream via osmosis. ② Using the kinetic energy generated by Na^+ flowing down its concentration gradient, glucose is brought onto the cytoplasm of the tubule cell using secondary active transport. ③ Glucose will be sent into the interstitial fluid via facilitated diffusion and eventually sent back into the peritubular capillary via facilitated diffusion (GLUT2 transporter)

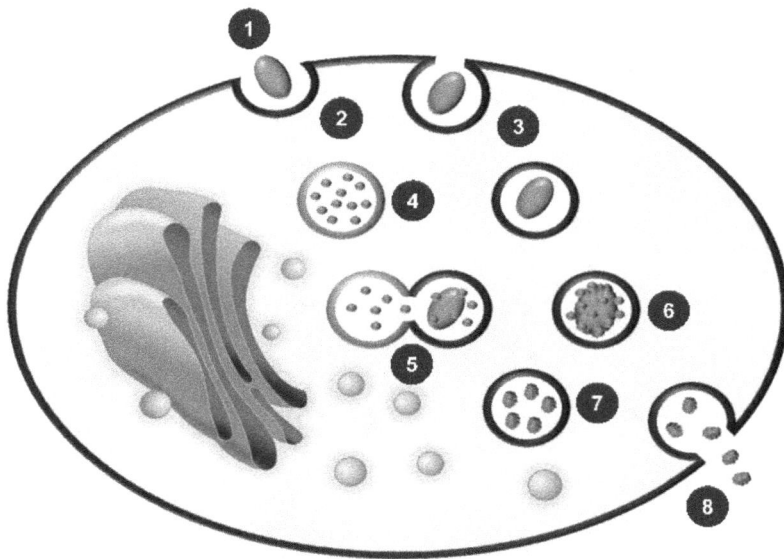

1	The cell phagocytized a foreign entity (e.g. bacterium)
2	An invagination of the cellular membrane accommodates the foreign object
3	A phagosome is formed
4	A lysosome containing digestive enzyme is formed from the Golgi apparatus
5	Fusion of the phagosome with the lysosome
6	Digestion of the foreign object
7	Waste products are formed
8	Waste products are expelled from the cell via exocytosis

Figure 19: Illustration of phagocytosis and the utilization of the cellular organelle, lysosome. Please note that all types of membrane assisted transport are receptor mediated

Please note that the phospholipid bilayer of the phagosome is formed from a small section of the phospholipid bilayer of the plasma membrane. In this manner the substances are absorbed without compromising the structural integrity of the plasma membrane (Figure 19).

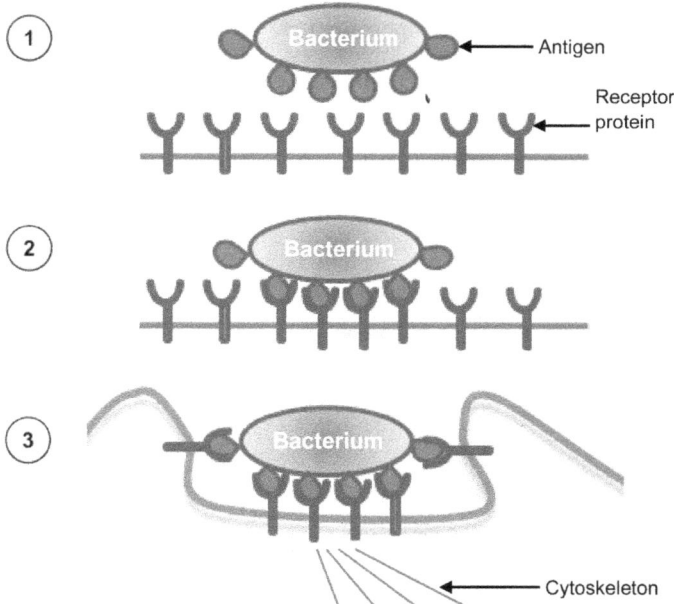

Figure 20: Receptor mediated endocytosis. ① Bacterium possesses a set of antigen that is alien to the immune system. ② Antigen activates the receptor protein. ③ Membrane invaginates via the manipulation of the cell membrane by cytoskeletons

For example, a type of leukocyte (white blood cell) called a macrophage, is involved in initiating the innate immune response. Macrophage is capable of distinguishing between healthy cells of the body and foreign materials, such as bacteria through the presence of mannose receptors (e.g. phagocytic receptors). These mannose receptors recognize the antigens located on the cellular membrane or cell walls of bacteria. Once the mannose receptor and bacterial antigen binds, these complexes induce the rearrangements in the actin cytoskeleton that lead to the internalization of the bacteria (phagocytosis) and quarantine them inside a phagosome. Once isolated within the cytoplasm of the macrophage, the bacteria will be further digested by the enzymes (lysozyme) contained within membrane bounded

organelles called lysosomes (Figure 20).

Pinocytosis (defined as cell drinking) is the uptake of smaller material into the cell through the initial activation of membrane surface receptors. Pinocytosis also involves the formation of a vesicle, which also includes some extracellular fluids. The manner by which a pinocytotic vesicle is created is identical to the formation of a phagosome.

Exocytosis is defined as the mass movement of materials out of the cell. Again, this form of transport involves a vesicle that contains materials that need to be released into the surrounding environment. Please note that when the phospholipid bilayer of the vesicle merges with the plasma membrane, the phospholipids of the vesicle will join their counterparts in the plasma membrane while the contents stored within the vesicle will subsequently be expelled. Again, the substances are released without compromising the structural integrity of the plasma membrane (Figure 21).

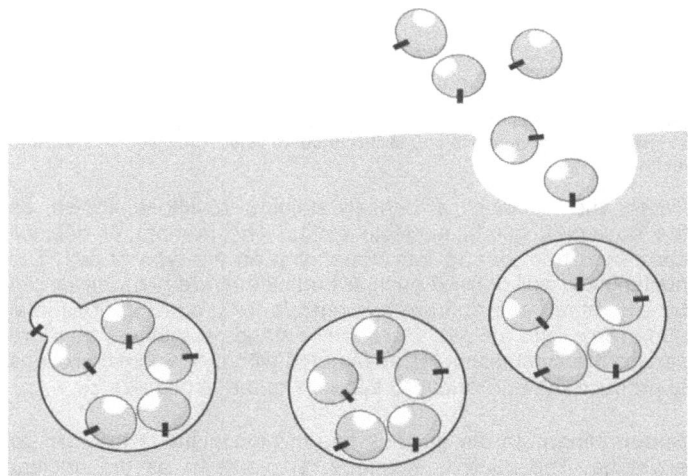

Figure 21: Illustration of exocytosis. Please note that the graphic demonstrates the process from the formation of the secretory vesicle to the final release of the substances into the interstitial space

ORGANELLES

Nucleus

Nucleus is often referred to as the command center of a eukaryotic cell and it contains the genetic information necessary for cellular division and protein synthesis. The genetic materials are suspended in a fluid matrix, known as nucleoplasm, and are organized as double helix deoxyribonucleic acid (DNA) molecules in association with histone proteins in the formation of chromosomes (Figures 22 and 23).

When the cell is not undergoing cellular division (e.g. mitosis) the genetic material is dispersed throughout the nucleus as delicate DNA filaments loosely coiled around histone proteins called chromatin. However, during cell division, the genetic material will consolidate first into chromomeres before finally being condensed into chromosomes (Figure 22).

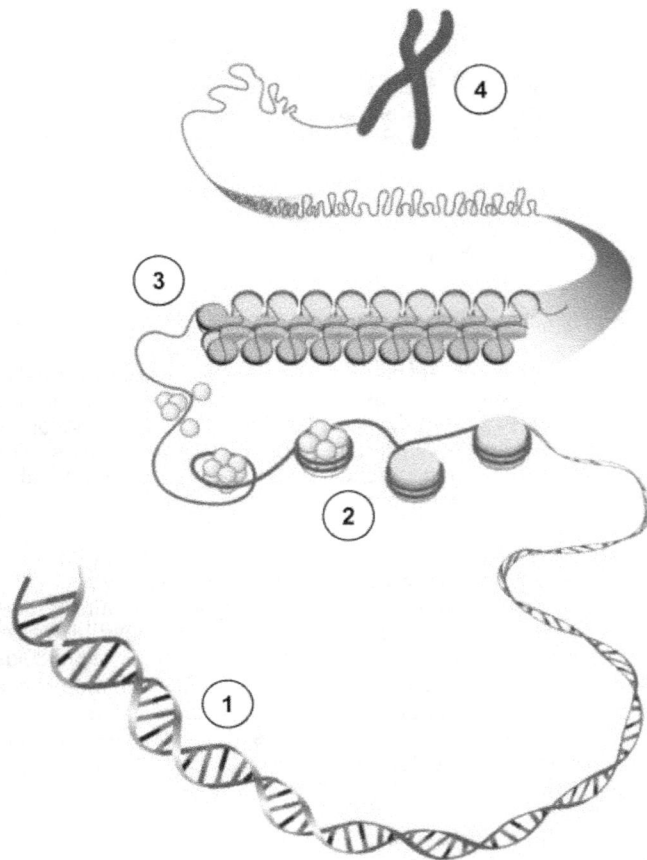

Figure 22: ① Illustration of DNA double helix before being ② wounded around histone proteins to form nucleo-somes. ③ The nucleosomes will interact with one another to form chromatin. During cellular division, ③ the chromatin will condense to create chromomere before finally ④ shaping into a chromosome

Within the nucleus, a densely-staining structure known as the nucleolus (plural nucleoli) exists. The number of nucleoli can range from one to four depending on the type of cell. The nucleoli are well defined nuclear bodies that are not surrounded by any membrane and are responsible for producing ribosomal ribonucleic acid (rRNA). Once rRNA stands are made, they are sent to the cytoplasm and assembled with cytoplasmic proteins to produce the 60S and 40S subunits of the ribosomes.

Surrounding both the nucleus and the nucleolus is the nuclear envelope. The nuclear envelope, also known as the nuclear membrane, is composed of a phospholipid bilayer similar to that of the plasma membrane. The nuclear envelope serves to separate the cell's genetic material from the surrounding cytoplasm, and acts as a barrier that prevents ions, solutes, and macromolecules from passing freely between the nucleus and the cytoplasm.

The nuclear envelope contains transmembrane proteins called

nuclear pores. These nuclear pores serve as gateways that allow substances to enter and exit the nucleus. For example, the rRNA strands that are produced in the nucleolus are exported (exported materials form the nucleus are called exportin) through the nuclear pores into the cytoplasm (Figure 23).

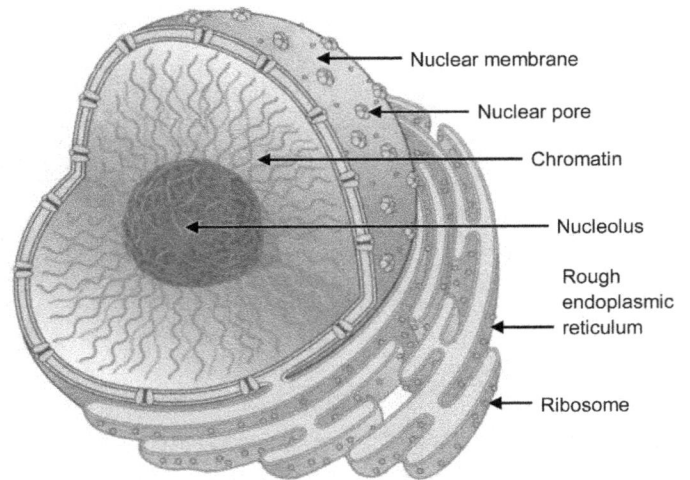

Figure 23: Illustration of a nucleus. Please note that the membrane structure of the rough endoplasmic reticulum (rER) is continuous with the nuclear membrane

Ribosome

Eukaryotic ribosomes are small, non-membrane bounded organelles made up of approximately 50 proteins and several long rRNAs intricately intertwined. Ribosomes are found floating freely in the cytoplasm, as well as, bound to the rough endoplasmic reticulum (Figure 24).

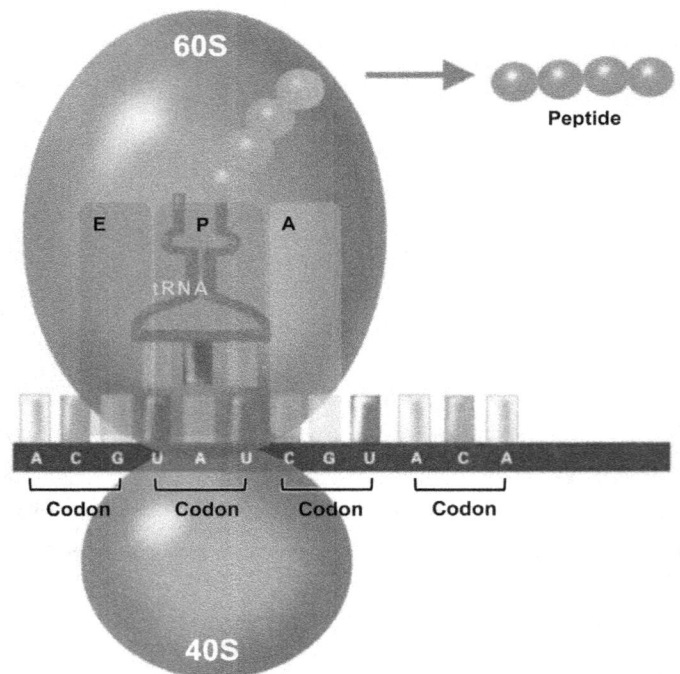

Figure 24: Illustration of a ribosome and its subunits

Ribosomes are the site of mRNA translation into polypeptide sequences (i.e. protein synthesis). A ribosome is made of two subunits: the large 60S subunit and the small 40S subunit. Once combined, the two subunits work as one to interact with mRNA and tRNA at three specific sites. The first site is also known as the A (aminoacyl-tRNA) site, is where the ribosome holds the tRNA, as well as, the amino acid that it is transporting. In addition, this is the location where the anticodon of the tRNA is matched with the codon of the mRNA. The second site is known as the P (peptidyl-tRNA) site, is where the tRNA unloads the amino acid, and allows the monomer to form peptide bonds with other amino acids (now polypeptide chain) that were brought prior to its arrival. The final site is known as the E (exit) site, where the now unloaded tRNA exits the ribosome (Figure 24).

The introduction of the mRNA from the nucleus into the cytoplasm triggers the binding of free floating ribosomes. It is common for hundreds, if not thousands or free-floating ribosomes to bind to the mRNA. These multiple bindings of ribosomes when viewed under electron microscopy are referred to as polyribosomes. Immediately after its attachment to the mRNA, the polyribosomes begin translation (protein synthesis), thereby producing hundreds, if not thousands, of the same polypeptide sequences coded by the mRNA. Once the polypeptide sequences are completed, it is inserted into the lumen of the rough endoplasmic reticulum (rER) via the docking protein, also known as the signal recognition particle (SRP) receptor for further processing.

Mitochondrion

The mitochondrion (plural mitochondria) is surrounded by an inner and an outer membrane both constructed out of phospholipid bilayers. The outer membrane is smooth in contour, while the inner membrane is highly folded and is also referred to as the christae. The christae serve to increase the surface area of the membrane on which cellular respiration can take place (Figure 25).

The mitochondrion is the site of aerobic cellular respiration where glucose is oxidized to produce adenosine triphosphate (ATP), carbon dioxide (CO_2) and cellular water (H_2O). For this reason, the mitochondrion is commonly referred to as the powerhouse of the cell. The enzymes located in the christae and the matrix (internal fluids of this organelle) are responsible for the breakdown of glucose and perform chemiosmosis through which most of ATPs are produced.

$$C_6H_{12}O_6 + 6O_2 \rightarrow 6CO_2 + 6H_2O + 36\text{-}38 \text{ ATP} + \text{heat}$$

The mitochondrion is one of the few organelles within the eukaryotic cell that possesses its own genetic material. Similar to the chloroplasts in plant cells, the mitochondrial DNA is circularly arranged and resembles that of the bacteria. It is believed that the mitochondrion exists in a symbiotic relationship with the eukaryotic cell (Figure 25).

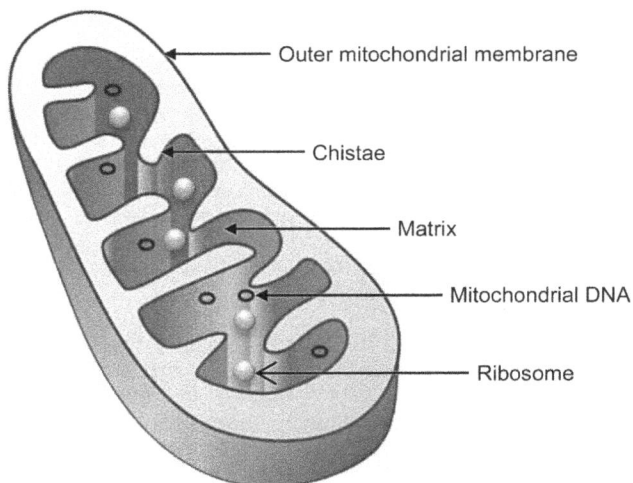

Figure 25: Illustration of a mitochondrion. Please note that this organelle possesses its own DNA and ribosomes that are similar to bacterial DNA and ribosome

Endoplasmic Reticulum

The endoplasmic reticulum (ER) consists of an intricate and complicated network of membranous channels and vesicles known as cisternae. The cisternae of the ER are continuous with the outer membrane of the nuclear envelope. In the eukaryotic cell, with the exception of erythrocytes and spermatozoa, there are two types of ER: rough endoplasmic reticulum (rER) and smooth endoplasmic reticulum (sER).

The rER is closely associated with ribosomes and obtained its name due to the peppered appearance because of bounded ribosomes at specific sites on its membrane, called translocons. The translocon is lined with a receptor called docking protein, also known as signal recognition particle (SRP) receptor, which is responsible for recognizing the signal peptides that are present on the N-terminus of most newly formed proteins as they are being made. This signal peptide triggers the insertion of the polypeptide sequence into the lumen; otherwise, the polypeptide sequence remains within the cytoplasm. Within the rER, the protein is folded and forms a multi-subunit complex under the scrutiny of a rER lumen chaperone molecule. The chaperone molecule helps the polypeptide interact with various enzymes. There are certain enzymes located within the lumen of the rER that are responsible for making modifications to the polypeptide by adding carbohydrate chains, thereby forming a glycoprotein. Other enzymes are also available in the lumen of this organelle to assist in forming higher levels of proteins by deleting or rearranging the amino acid sequences. Once the polypeptide is edited and deemed functional, the nearly completed product is then sent to the Golgi apparatus (Figure 26).

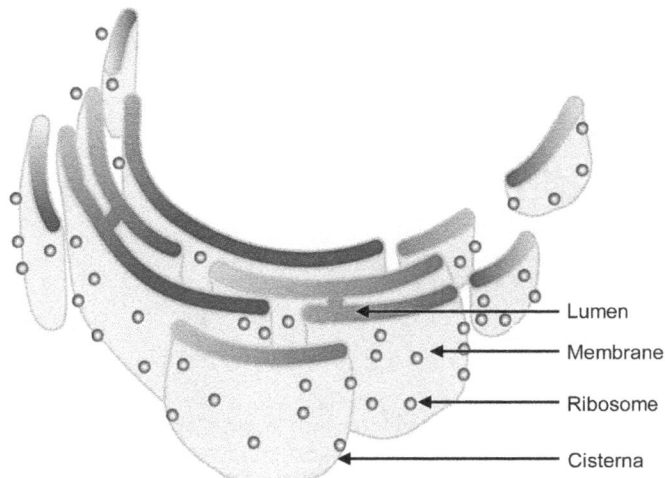

Figure 26: Rough endoplasmic reticulum. Please note the rough appearance is the result of ribosomes attaching to the rER membrane

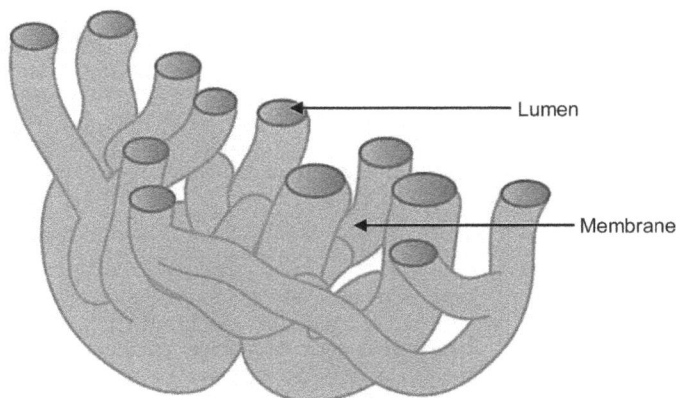

Figure 27: Smooth endoplasmic reticulum (sER)

The smooth endoplasmic reticulum (sER) is a separate organelle, although its membrane is continuous with that of the rER. The sER is thus named because there are no ribosomes attached to its membrane surface. In most cells, the amount of sER is relatively sparse. However, in cells that are responsible for synthesizing steroid based hormones, and lipids or responsible for catabolizing lipid-soluble (hydrophobic) molecules, the sER is quite abundant. For example, certain lipid-soluble drugs are delivered to the hepatocytes in the liver to be conjugated and made more water soluble so that they can be excreted by the kidneys. In muscle cells, a modified sER, known as sarcoplasmic reticulum, is responsible for storing intracellular calcium ions (Ca^{++}). Ca^{++} in muscle cells are essential component of muscle contraction (Figure 27).

Golgi Apparatus

Golgi apparatus is best described as the final processing center of products within the eukaryotic cell. The Golgi is comprised of membrane-bound stacks known as cisternae (approximately four to eight cisternae per Golgi apparatus per cell). Within the cisternae, various enzymes are available to modify substances. The Golgi apparatus is responsible for the addition of carbohydrates via glycosylation to proteins, thereby forming glycoproteins or proteoglycans. Additionally, it is also responsible for adding phosphates via phosphorylation to oligosaccharides of lysosomal glycoproteins, as well as, merge lipids generated in the sER with proteins to form lipoproteins (Figure 28).

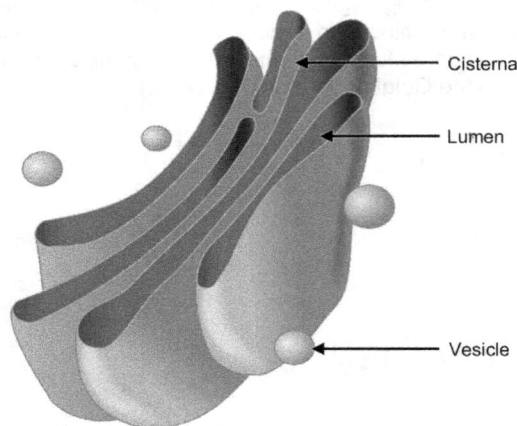

Figure 28: Golgi apparatus

For example, after a polypeptide is produced by the ribosome and "edited" (post translational modified) in the rER, the polypeptide then enters a cisterna before it pinches off and forms a vesicle. This vesicle then moves to the Golgi apparatus with the aid of cytoskeleton. Once the vesicle arrives at the Golgi apparatus, it fuses with the organelle's membrane before entering the lumen of its cisternae. Within the lumen of the Golgi apparatus, the polypeptide is ready for excretion by concentrating it with other identical polypeptides arriving from rER. The Golgi apparatus could also assemble the peptides with other functional products to form an organelle such as the lysosome and peroxisome. Once the final assembly of the products is completed, the Golgi apparatus pinches off a section of its cisternae and forms a functional intramembranous organelle or a secretory vesicle ready for exocytosis (Figure 28).

Intracellular Lipid Transport

The transport of lipids from the ER is accomplished through vesicular and non-vesicular transport mechanisms that are commonly found in mammalian cells. In vesicles, the hydrophobic substances produced in the sER are packaged within a phospholipid bilayer. Even though the fat-soluble substances stored within the bilayer are capable of diffusion through the

vesicle, it is the hydrophilic cytosol that maintains the integrity of the hydrophobic product/vesicle interaction and allows it to be delivered to the Golgi apparatus. Vesicular transport requires energy in the form of ATP and an intact cytoskeleton involved in the vesicular transport mechanism. Although this form of lipid transport mediates the bulk transport of many lipids, another method known as non-vesicular transport also plays an import role (Figure 29).

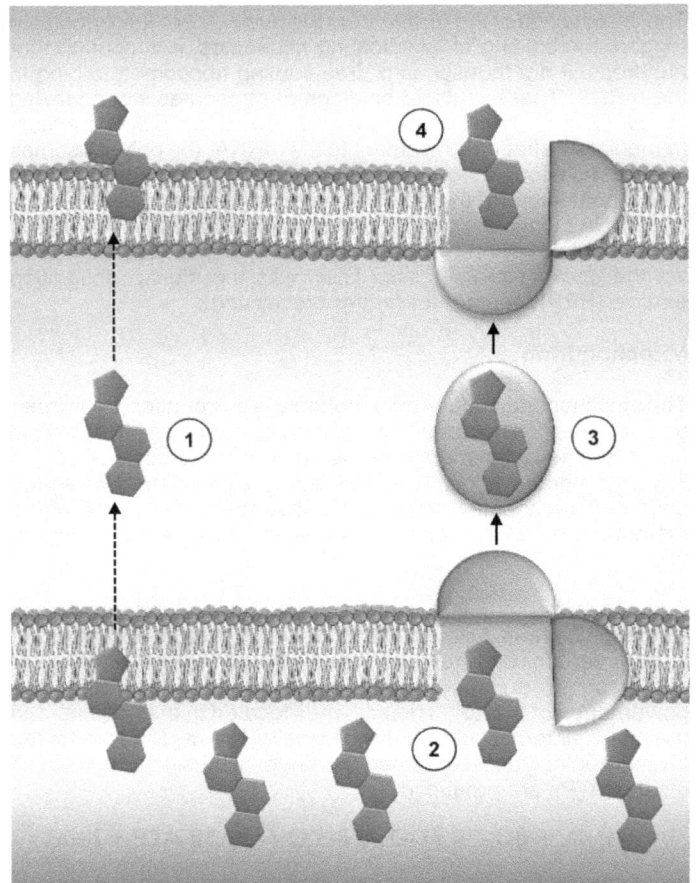

Figure 29: Intracellular lipid transport of steroid molecules. ① Spontaneous lipid transport is when a lipid monomer diffused through the sER and into the cytosol. ② Lipid-transfer proteins (LTPs) possess a "lid" which is opened subsequent to interacting to the donor membrane. LYP have a lipid-transfer domain that allows the isolation and storage of lipids during the cytosol transfer. ③ The lipid is being transported through hydrophilic aqueous cytosol. ④ The "lid" reopens after interacting with the target membrane allowing the delivery of the lipid into its lumen

Non-vesicular lipid transport between membranes occurs in one of two different manners. Spontaneous lipid transport is when a lipid monomer diffuses through the sER into the cytosol. Since a lipid is highly hydrophobic, its diffusion through an aqueous medium is extremely slow and insufficient to support the needs of the cell. Nonetheless, the speed of spontaneous lipid transport can be greatly increased if the transport occurs at various membrane contact sites. Membrane contact sites are defined as small cytosolic gaps of 10 to 20 nm between organelles, including the sER and the Golgi apparatus (Figure 29).

Lipid-transfer proteins (LTPs) are the second way lipids can be transported intracellularly. LTPs demonstrate specificity with various lipid types. LTPs contain a lipid-transfer domain which is a hydrophobic pocket within the protein that allows the isolation and storage of lipids during the cytosol transfer from a donor organelle to a target organelle (e.g. sER to the Golgi apparatus). Research has shown that the interaction of an LTP with membranes of organelles induces conformational changes, leading to displacement of a peptide "lid", which exposes the lipid-transfer domain. In addition, the "lid" is thought to penetrate the bilayer of the organelle and help scoop out the lipids from its lumen and transfer it to the hydrophobic cavity, thereby facilitating lipid absorption. Once the lipid is collected and stored, the "lid" will close and the LTP dissociates from the donor organelle. LTP then diffuses through the aqueous cytosol towards the target organelle. The transport is terminated by interaction with the target organelle's membrane which causes the "lid" to reopen and allows the delivery of the lipid (Figure 29).

Lysosomes

Lysosomes are membrane bound organelles (vesicles) that contain approximately 50 different degradative enzymes that are capable of breaking down polymers, such as: proteins (lysosomal proteases), carbohydrates (glycosidases), nucleic acids (nucleases), and lipids (lipases). Lysosomes are formed from the Golgi apparatus and serve as an intracellular digestive system. This organelle digests waste materials and food within the cell or brought into the cytoplasm via endocytosis. It is responsible for breaking down molecules into their simplest components so that they may be neutralized (i.e. bacteria, cellular debris, damaged and/or dying cells), and/or used in cellular maintenance. For example, leukocytes, such as macrophages, are programmed to seek out foreign bodies such as bacteria. Once a bacterium is located, the macrophage will internalize the organism via phagocytosis and quarantine it in a vesicle known as phagosome. Subsequently, the phagosome will fuse with lysosomes, which results in the degradation of the bacterium (Figure 19). Similarly, cellular debris, dead or damaged cells from the body, are also phagocytized and placed within a phagosome before being digested by the hydrolytic enzymes of the lysosome. This process of removing cellular debris is called autodigestion. In addition, lysosomes are also involved in a process known as autophagy where the organelles within a cell are replaced and renewed. In this process, a damaged organelle is enclosed within a vesicle called autophagosome and is then fused with lysosomes where it is promptly digested. It is understood that one of the main reasons for inflammation of injured tissue is caused by the rupture of lysosomes due to cellular damage. Put simply, once the enzymes stored within the lysosomes are inadvertently released into the surrounding environment, it could become toxic to other cells, and in turn, cause inflammation.

Lysosomal enzymes are all classified as acid hydrolases, which are enzymes that becomes active in acidic environments (pH ~5). Therefore, unlike most of the organelles within the cell, lysosomes maintain an acidic internal environment through the actions of proton pumps which actively transport hydrogen cations (H^+) into the lysosome from the cytosol.

Lysosomal Storage Disease

Lysosomal storage disease consists of approximately 30 different types of genetic conditions that are caused by mutations in the genes that encode the formation of one or more lysosomal enzymes. The inability to form proper functioning enzymes results in the buildup of undegraded materials within the cell which in turn could result in cellular hypertrophy and eventually cell death. For example, the most common of these disorders is Gaucher's disease, also known as glucocerebrosidase deficiency. Gaucher's disease is an autosomal recessive condition that commonly occurs in Ashkenazi Jews. This genetic disorder results in the malfunction of an enzyme called acid glucosylceramidase, which is responsible for breaking down β-glucosylceramide. The inability to breakdown this glycolipid causes the accumulation of β-glucosylceramide (a glycolipid) in various tissues (i.e. bone) and organs (i.e. spleen, liver, lungs and the brain). This buildup of glucosylceramide results in the hypertrophy of these tissues and organs which in turn affects their proper functions. Symptoms of this disease can vary widely, mostly depending on the location where glucosylceramide buildup is most pronounced. Nonetheless, this disease can manifest as enlargement of the spleen and liver, which results in severe abdominal pains. This disease can result in the accumulation of glucosylceramide in the bones which in turn could impede blood flow and result in the necrosis of bone cells (e.g. osteocytes) and increase the risk of fractures, death of pluripotent stem cells in the red marrow which may impede the formation of blood cells and platelets. Infrequently, this disease may affect the brain which may result in muscle rigidity, abnormal eye movements, dysphagia (swallow difficulties), and seizures. It has been reported that Gaucher's disease may increase the risk of Parkinson's disease and various blood cancers, such as myeloma, leukemia, and lymphoma. Being a genetic disorder, there are no cures, only treatment options for patients. These treatment options include enzyme replacement therapy, application of an oral medication called Miglustat which interferes with the production of glycolipids, and in turn prevents their buildup, and the administering of osteoporosis medication which aids in the reformation of weakened bones.

Peroxisomes

Peroxisomes are small enzyme containing organelles that are involved in the breakdown of fatty acids, amino acids, uric acids and hydrogen peroxide (H_2O_2). Hydrogen peroxide is a byproduct of fatty acid and amino acid catabolism and is extremely toxic to cells. Peroxisomes contain catalase, an enzyme that converts hydrogen peroxide to water and oxygen during peroxidation reactions. Peroxisomes are abundant in cells that metabolize lipids and steroids and detoxify alcohol. For example, peroxisomes are readily found in liver and kidney cells.

In addition to oxidative reactions, the peroxisomes are also the location for lipid biosynthesis. Substances such as cholesterol are synthesized both in the peroxisomes and in the sER. In many cell types found in the human body (cardiac and skeletal muscle cells, kidneys, lungs, and the brain etc.) peroxisomes are responsible for the formation of plasmalogen, a unique type of membrane phospholipid (e.g. glycerophospholipids). Plasmalogen is involved in maintaining the physical properties of the plasma membrane when transitioning from a gel to fluid state at various temperatures. In addition, these molecules are also involved in maintaining the proper functions of various membrane bounded integral proteins. In hepatocytes (liver cells), peroxisomes are involved in the formation of bile acids, a derivative of cholesterol found in bile that aid in fat absorption and modulate cholesterol levels.

Cytoskeleton

The cytoskeleton is a network of interconnected filaments and tubules that extend from the nucleus to the plasma membrane in eukaryotic cells. This intracellular structure has various functions such as maintaining cell shape and the movement of organelles. In addition, it provides the means to form cell to cell junctions. The cytoskeleton is constructed out of actin filaments, microtubules, and intermediate filaments (Figure 30).

Actin filaments form networks in the cytoplasm of the cell and act to support the plasma membrane and maintain the shape of the cell. Building and reconstructing the various actin filaments within the cytoplasm allows the amoeboid-like movements which allow white blood cells such as macrophage to venture throughout the human body. Actin filaments in conjunction with myosin provide the ability for the muscles to contract.

Microtubules are formed from alpha and beta tubulin protein subunits arranged in a hollow tubular structure. This structure supports the cytoplasm of the cell and is involved in cell division through the formation of kinetochore and spindle microtubules. In addition, microtubules forms the foundation of various organelles such as the centrioles, cilia, and flagella.

Intermediate filaments are a broad class of protein fibers that serve both structural and functional roles. Intermediate filaments function as tension-bearing cellular components to help maintain cell shape and rigidity. In addition, the intermediate filament serves to anchor organelles such as the nucleus (Figure 30).

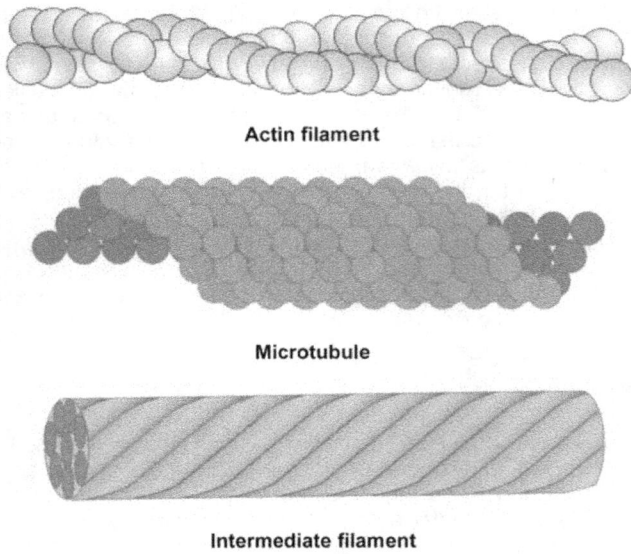

Actin filament

Microtubule

Intermediate filament

Figure 30: Illustration of the main components of the cytoskeleton

Centrosomes

Centrosomes are composed of two centrioles arranged at a 90° angle from one another. Centrioles are composed of 9 + 0 microtubule triplets. This microtubule arrangement is designed as nine microtubule filaments arranged in groups of threes. Centrioles are also the site to which other microtubules within the cell assemble and disassemble. For example, centrosomes are responsible for assembling the spindle and kinetochore microtubules which are essential during mitotic division. In addition, the centrioles within the centrosome give rise to basal bodies at the base of the flagella and cilia. The basal bodies direct the organization of microtubules and cilia and their microtubule counts are 9 + 0, identical in format as the centriole itself (Figure 31).

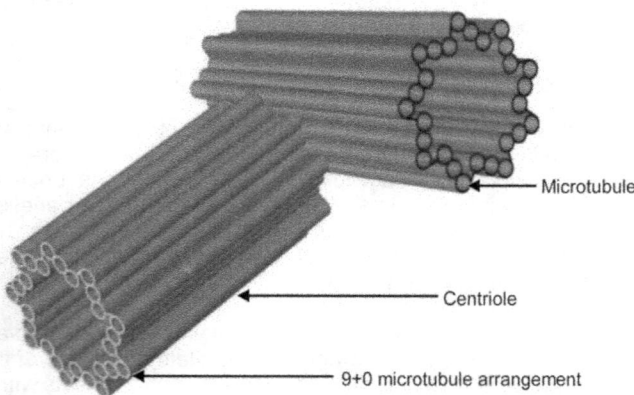

Microtubule

Centriole

9+0 microtubule arrangement

Figure 31: Illustration of a centrosome, the site of spindle and kinetochore microtubule assembly during mitosis. Please note that one centrosome is composed of two centrioles arranged at a 90° angle form one another

For decades, the existence of centrosome and the manner by which it divides within an eukaryotic cell had baffled scientists. Was centrosome duplication within dividing cells similar to that of mitochondria? As we mentioned previously, mitochondrial division is in many ways independent of the cell since it possesses its own DNA. Mitochondria, along with chloroplasts in plant cells, are believed to be coexisting with the eukaryotic cell in an endosymbiotic relationship.

It was only recently that this question was answered. After numerous experiments, no DNA has been found in the centrosome, therefore, it is concluded that centrosome duplication is directly linked to cellular division and not a semi-independent event. Although DNA was not found in the centrosome, RNA was discovered. This interesting experimental result leads to another question which will require further elucidation.

Cilia and Flagella

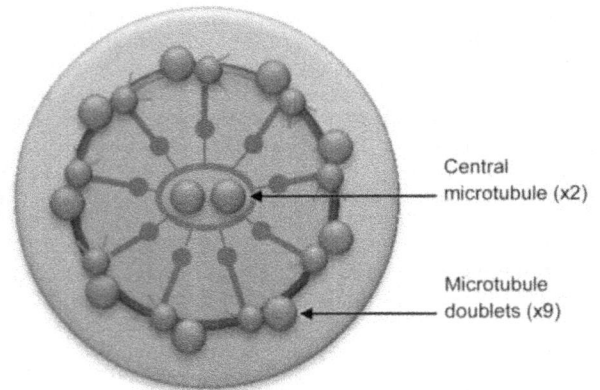

Central microtubule (x2)

Microtubule doublets (x9)

Figure 32: Illustration of 9+2 microtubule doublet arrangement of cilia and flagella

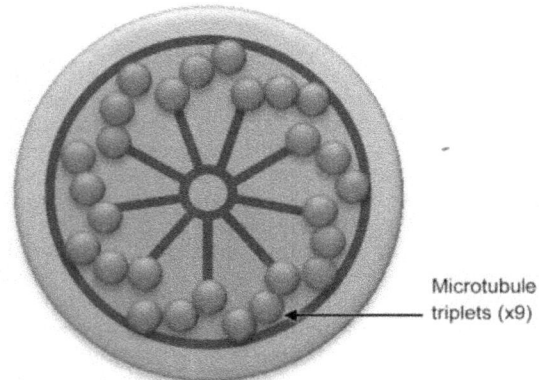

Microtubule triplets (x9)

Figure 33: Illustration of 9+0 microtubule triplet arrangement of the basal body

Cilia (Singular: cilium) and flagella (Singular: flagellum) are hair-like projections that facilitate movement. Cilia are found on the apical surface of pseudostratified columnar epithelial cells in the respiratory track. Their movements propel mucous away from the lungs. Flagella, on the other hand, provide movement for sperm.

Structurally, both flagella and cilia are arranged in a nine microtubule doublets around two single central microtubules (9 + 2 pattern) design (Figure 32). The centrioles within the centrosome give rise to basal bodies which form the foundation of the flagella and cilia. The basal bodies direct the organization of microtubules. Cilia and their microtubule counts are 9 + 0; identical in format as the centriole itself (Figure 33).

Microvilli

Microvilli (singular: microvillus) are non-motile structures that are generally located at the apical surface of cells. The microvillus is constructed of a core bundle of interlinking proteins surrounded by the cellular membrane. The core bundle of a microvillus is composed of ~19 actin filaments anchored at its basal end to horizontally oriented actin filaments through spectrin protein crosslinks. This crosslinking is also referred to as the terminal web. Within the length of the microvillus, the core bundle is also cross-linked by fimbrin and villin proteins. In comparison with fimbrin, villin is less organized and looser in crosslinks. Finally, the protein Myosin1A is responsible for laterally tethering the core bundle to the membrane of the microvillus (Figure 34).

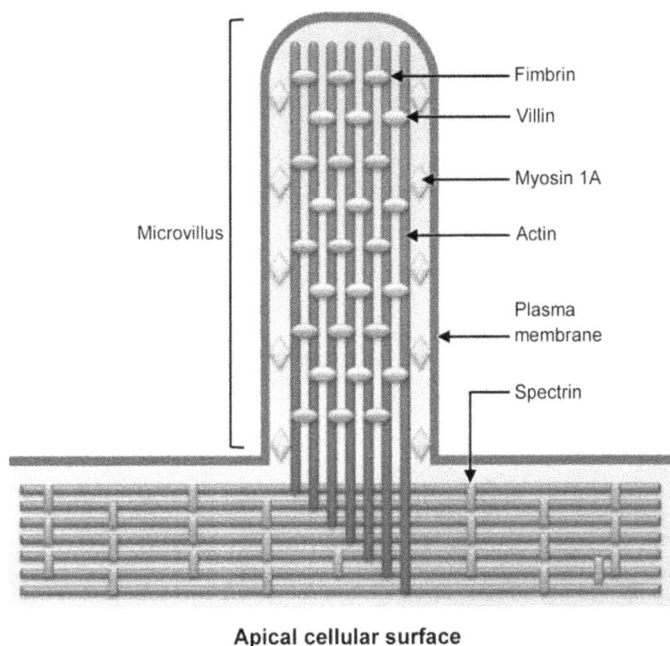

Apical cellular surface

Figure 34: Graphical illustration of a microvillus

These cellular membrane protrusions are responsible for increasing the surface area for absorption, secretion, adhesion, and mechanotransduction. For example, in the small intestines of the digestive track, microvilli are located upon epithelial folds known as villi. Microvilli are generally 0.08 µm in diameter and 1 µm long projections of the epithelial cellular surfaces. It is approximated that the microvilli increase the surface area for absorption and secretion of digestive track epithelial cells by approximately 600 fold in humans.

Leukocytes (white blood cells) possess a highly irregular cellular surface that contains many microvilli. These microvilli allow the leukocytes to initiate rolling adhesion on the vessel wall of fast flowing blood vessels. In addition, the microvilli allow the leukocytes to acquire foreign entities (e.g. bacteria) more quickly and efficiently.

In the case of stereocilia found within the inner ear, these microvilli are designed to increase surface area for mechanotransduction which is defined as transforming the pressure generated from sound waves into action potential).

QUESTIONS

1. Organelles are membrane bounded structures found within eukaryotic cells. Which of the following is the primary substance that the membranes of these organelles are constructed?
   a. Integral proteins
   b. Peripheral proteins
   c. Phospholipids
   d. Glycolipids
   e. Cholesterol

2. It is known that the nuclear membrane is continuous with which of the following organelle (s)?
   a. Peroxisome
   b. Ribosome
   c. Rough endoplasmic reticulum
   d. Smooth endoplasmic reticulum
   e. Both c and d

3. Which of the following organelle(s) found in the animal cell do not share the same genetic materials located within the nucleus?

   a. Chloroplasts
   b. Lysosome
   c. Mitochondria
   d. Peroxisome
   e. Ribosome
   f. Both a and c
   g. Both c and d

4. It is believed that the mitochondria (as well as the chloroplast and centrosome) are at one time free-living organisms that were absorbed into the cytoplasm of prehistoric eukaryotic cells. Instead of consuming and digesting the smaller organisms, the eukaryotic cells and the prokaryote cells entered into a form of agreement where both organisms will benefit from their interactions. What is the name given to this interaction?
   a. Big Bang theory
   b. Diffusion
   c. Endosymbiotic Theory
   d. Natural selection
   e. Osmosis

5. Which of the following type of cells can form multicellular organisms?
   a. Eukaryotes
   b. Plasmodium
   c. Prokaryotes
   d. Shigella
   e. Yersinia

6. The plasma membrane of both prokaryotic and eukaryotic cells is constructed out of which of the following compounds?
   a. Integral proteins
   b. Peripheral proteins
   c. Phospholipids
   d. Glycolipids
   e. Cholesterol

7. The plasma membrane is separated hydrophilic and hydrophobic regions due to its amphipathic nature. It is understood that the hydrophilic region of the cellular membrane is composed of _____.
   a. Charged portion of their sodium bounded region
   b. Non-polar ends of phospholipids are consists of long chains of fatty acids
   c. Polar phosphate ends of phospholipid
   d. Uncharged portion of their potassium formed regions
   e. Water loving portion of the glycerol

8. The plasma membrane is separated hydrophilic and hydrophobic regions due to its amphipathic nature. It is understood that the hydrophobic region of the cellular membrane is composed of _____.
   a. Charged portion of their sodium bounded region
   b. Non-polar ends of phospholipids are consists of long chains of fatty acids
   c. Polar phosphate ends of the phospholipid
   d. Uncharged portion of their potassium formed regions
   e. Water loving portion of the glycerol

9. Plasma membrane is semipermeable in nature of which is defined as _____.
   a. Allowing certain molecules or ions to freely pass through while preventing others from entering or exiting the cytoplasm
   b. Allowing everything to pass through
   c. Allowing nothing to pass through
   d. Allowing only water to pass through which preventing everything else from passage
   e. Creating a site of protein synthesis

10. Which of the following will pass through the plasma membrane freely?
   a. Amino acids
   b. Ca^{++}
   c. K^+
   d. Lipids
   e. Na^+

11. Which of the following is defined as the process of moving molecules from areas of high concentration to areas of low concentration based upon their concentration gradient?
   a. Active transport
   b. Diffusion
   c. Exocytosis
   d. Osmosis
   e. Phagocytosis

12. Diffusion belongs to which of the following form of cellular transport?
   a. Passive transport
   b. Active transport
   c. Membrane assisted transport
   d. Temperature dependent transport
   e. Substrate dependent transport

13. Which of the following is defined as the process by which water crosses from an area of high concentration to areas of low concentration through a semi-permeable membrane (plasma membrane)?
   a. Active transport

b. Diffusion
c. Exocytosis
d. Osmosis
e. Phagocytosis

14. Osmosis belongs to which of the following form of cellular transport?
a. Passive transport
b. Active transport
c. Membrane assisted transport
d. Temperature dependent transport
e. Substrate dependent transport

15. Which of the following term is defined as the concentration of solutes within the cell is higher than the surrounding environment? In this environment, water from the environment will continuously enter the cell causing it to swell and could lead to its destruction (e.g. bursting).
a. Active transport
b. Diffusion
c. Homeostasis
d. Hypertonic
e. Hypotonic
f. Isotonic
g. Osmosis
h. Both c and f

16. Which of the following term is defined as the concentration of solutes within the cell is lower than the surrounding environment? In this environment, water from the cell will continuously escape its cytoplasm causing it to shrivel and desiccate.
a. Active transport
b. Diffusion
c. Homeostasis
d. Hypertonic
e. Hypotonic
f. Isotonic
g. Osmosis
h. Both c and f

17. Which of the following term is defined as the concentration of solutes within the cell is equal to the concentration of solutes with the surrounding environment? In this environment, the cell is maintaining its tonicity.
a. Active transport
b. Diffusion
c. Homeostasis
d. Hypertonic
e. Hypotonic
f. Isotonic
g. Osmosis
h. Both c and f

18. The permeability of the plasma membrane is not affected by which of the following condition?
a. Ambient temperature
b. Cell size
c. Concentration of solutes
d. Shape of the solute
e. Size of the solute
f. Specific osmotic pressure exerted on the cell

19. Which of the following is the name given to the accepted concept for the proper function of the plasma membrane?
a. Fluid-mosaic model
b. Ambient temperature model
c. Solute size and shape model
d. Solute concentration model
e. Membrane charge model
f. Hydrophobic motion model

20. Which of the following is (are) amphipathic molecule(s)?
a. Cholesterol
b. Monosaccharides
c. Oligosaccharides
d. Peripheral proteins
e. Phospholipid
f. Both a and e
g. Both a and d

21. Which of the following compounds embedded into the plasma membrane is used to stabilize this semi-fluid barrier (making it less fluid)?
a. Cholesterol
b. Monosaccharides
c. Oligosaccharides
d. Peripheral proteins
e. Phospholipid
f. Both a and e
g. Both a and d

22. Integral proteins are transmembrane proteins that span the entire thickness of the plasma membrane. Like peripheral proteins, the integral proteins are anchored into the plasma membrane through which of the following interactions?
a. Adhesive interactions

b. Covalent and hydrogen bonds
c. Cytoskeletons
d. Ionic bonding
e. Both b and c
f. Both a and d

23. Most integral proteins are glycoproteins that have their hydrophobic ends located _____ while their hydrophilic ends are located _____ and _____.
a. Within the fatty acid layers of the plasma membrane
b. Projecting towards the intracellular fluid medium
c. Projecting towards the extracellular fluid medium
d. Among the nonpolar fatty acids of the plasma membrane
e. Adjacent to the peripheral proteins

24. Which of the following integral protein is involved in the transport of substances from areas of low concentration to areas of high concentration?
a. Channel protein
b. Carrier protein
c. Gated Channel protein
d. Enzyme protein
e. Receptor protein
f. Cell recognition protein
g. Adhesion protein
h. Desmosome

25. Which of the following is the name given to the polysaccharide chain that is involved in a cell-to-cell and an individual-to-individual recognition?
a. Antibody
b. Antigen
c. Disaccharide
d. Monosaccharide
e. Oligosaccharide
f. Both b and e

26. Which of the following is essential during embryonic or development stage of an organism?
a. Cell-to-cell recognition
b. Fatty acid catabolism
c. Individual-to-individual recognition
d. Ionic transport

27. In cell-to-cell recognition, the difference in the combinations of _____ (HInt: monomers joined to form an antigen), allows the cell to differentiate themselves in proper locations in the developing body. For example, a skin cell will have a slightly different oligosaccharide marker when compared to a cardiac muscle cell. If by chance a cell is misplaced within the human body _____ will be responsible for removing it immediately.
a. Antibody
b. Polysaccharide
c. Disaccharide
d. Monosaccharide
e. Oligosaccharide
f. Erythrocytes (red blood cells)
g. Leukocytes (white blood cells)
h. Both b and e

28. In individual-to-individual recognition, the difference in the combinations of _____ (HInt: monomers joined to form an antigen), allows the _____ (type of cells) to distinguish between what belongs to your body and what is foreign.
a. Antibody
b. Polysaccharide
c. Disaccharide
d. Monosaccharide
e. Oligosaccharide
f. Erythrocytes (red blood cells)
g. Leukocytes (white blood cells)
h. Both b and e

29. During a blood transfusion, an individual with A blood type can receive transfusions from which of the following blood type(s)?
a. A blood type
b. B blood type
c. O blood type
d. AB blood type
e. All of the above
f. Both a and c
g. Both b and c

30. During a blood transfusion, an individual with B blood type can receive transfusions from which of the following blood type(s)?
a. A blood type
b. B blood type
c. O blood type
d. AB blood type
e. All of the above
f. Both a and c
g. Both b and c

31. During a blood transfusion, an individual with AB blood type can receive transfusions from which of the following blood type(s)?
a. A blood type
b. B blood type
c. O blood type

d. AB blood type
e. All of the above
f. Both a and c
g. Both b and c

32. During a blood transfusion, an individual with O blood type can receive transfusions from which of the following blood type (s)?
a. A blood type
b. B blood type
c. O blood type
d. AB blood type
e. Both a and c
f. Both b and c

33. O blood type is also known as _____.
a. Universal cooperator
b. Universal donor
c. Universal prognosticator
d. Universal recipient
e. Universal sharer
f. Universal fluid

34. AB blood type is also known as _____.
a. Universal cooperator
b. Universal donor
c. Universal prognosticator
d. Universal recipient
e. Universal sharer
f. Universal fluid

35. Tight junctions (zonula occludentes) form an impermeable barrier that prevents seepage and tissue movements. These junctions generally exist on the _____ domain (side) of the cell.
a. Apical
b. Apical lateral
c. Basal
d. Basal lateral
e. Lateral
f. Oblique
g. Diagonal

36. Tight junctions (zonula occludentes) form an impermeable barrier that prevents seepage and tissue movements. When these junctions are found in the epithelial tissues of the stomach linings they prevent which of the following form occurring?
a. Preventing the seepage of acid that could digest the underlining tissue
b. Preventing the seepage of blood into the digestive tract
c. Preventing the loss of nutrients into the surrounding tissues
d. Preventing the hypercontraction of smooth muscles of the digestive tract
e. Preventing the influx of nutrients into the underlining tissues

37. Tight junctions are deceivingly complex structures and are estimated to be constructed out of ~40 different proteins. Which of the following is NOT one of these proteins?
a. Claudins
b. Cytoplasmic scaffolding proteins ZO-1
c. Cytoplasmic scaffolding proteins ZO-2
d. Cytoplasmic scaffolding proteins ZO-3
e. Occludins
f. β-catenins

38. Adherens junctions (zonula adherents) are found in various cell types that are frequently exposed to mechanical stress. Which of the following is not one of these tissues?
a. Cardiac muscles
b. Cardiac epithelial lining
c. Nervous tissue
d. Skeletal muscles
e. Epidermis (skin)

39. Adherens junctions (zonula adherents) are found in various cell types that are frequently exposed to mechanical stress. Which of the following is NOT one of these proteins?
a. "Classic" cadherins
b. Claudins
c. Vinculin
d. α-Actinin
e. β-catenins

40. In adherens junctions, the transmembrane linker proteins called cadherins (type I transmembrane proteins or also known as "classic" cadherins) are responsible for forming bonds with _____ through a type of interaction called _____. This interaction maintains a tightly regulated distance of 10-20 nm between the two plasma membranes.
a. "Classic" cadherens
b. Claudins
c. Heterotypic
d. Homotypic
e. Isotypic
f. Vinculin
g. α-Actinin
h. β-catenins

41. On the cytoplasmic side of adherens junctions, "classic" cadherins couples with β-catenins, which in turn, are fastened into place by bundles of _____ and _____ filaments (part of the cytoskeleton).
a. Actin
b. Intermediate filament
c. Kinesin
d. Microtubule
e. Myosin

42. In adherens junctions, which of the following is the membrane-cytoskeletal protein that functions in linking both integrin adhesion molecules and cadherin to the actin cytoskeleton?
a. "Classic" cadherens
b. Claudins
c. Heterotypic
d. Homotypic
e. Isotypic
f. Vinculin
g. α-Actinin
h. β-catenins

43. In adherens junctions, which of the following is the protein that is necessary for the bundling of the actin filaments?
a. "Classic" cadherens
b. Claudins
c. Heterotypic
d. Homotypic
e. Isotypic
f. Vinculin
g. α-Actinin
h. β-catenins

44. Which of the following is (are) desmosomal cadherens that is (are) enguaged in calcium-dependent adhesive interactions on the extracellular side of the junction?
a. Desmocollins
b. Desmogleins
c. Desmoplakin
d. Plakoglobin
e. Plakophilins
f. Vinculin
g. α-Actinin
h. β-catenins
i. Both a and b
j. Both d and e

45. On the cytoplasmic side, the desmosomal cadherins are linked to the keratin intermediate filament partly facilitated by which of the following protein (s)?
a. Desmocollins
b. Desmogleins
c. Desmoplakin
d. Plakoglobin
e. Plakophilins
f. Vinculin
g. α-Actinin
h. β-catenins
i. Both a and b
j. Both d and e

46. Which of the following protein (s) forms the desmosomal plaques?
a. Desmocollins
b. Desmogleins
c. Desmoplakin
d. Plakoglobin
e. Plakophilins
f. Vinculin
g. α-Actinin
h. β-catenins
i. Both a and b
j. Both d and e

47. The desmosomal plaques is connected to the keratin intermediate filament by which of the following intermediate filament binding protein (s)? This (these) protein (s) also associates with both plakoglobin and plakophilins to complete the loop of inter-junctional bonding.
a. Desmocollins
b. Desmogleins
c. Desmoplakin
d. Plakoglobin
e. Plakophilins
f. Vinculin
g. α-Actinin
h. β-catenins
i. Both a and b
j. Both d and e

48. Are gap junctions found in free floating cells such as the erythrocytes and leukocytes?
a. Yes
b. No

49. Communication junctions also known as_____ are extremely wide spread

and are frequently found in the cardiac, smooth muscle as well as neurons. These are like little intercellular channel proteins located mainly on the _____ (Hint: which side of the cell) domain of the cells which allow molecules to be passes from one adjacent cell to another. These junctions are built by six closely packed trans-membrane (integral) proteins. Each of these proteins are known as _____ while six of these proteins together forms a complex channel protein-like structure called _____.

a. Adhesion junctions
b. Apical
c. Basal
d. Cadherins
e. Catalase
f. Chondronectin
g. Collagen
h. Connexins
i. Connexons
j. Claudins
k. Cristae
l. Desmoplakins
m. Desmosomes
n. Elastin
o. Fibronectin
p. Gap junctions
q. Glycosaminoglycans
r. Histone
s. Integrin
t. Laminin
u. Lateral
v. Lysozyme
w. Occludins
x. Pakoglobins
y. Proteoglycan
z. Tight junctions

50. In the cardiac muscle cells, the gap junctions are found at a location known as _____.
a. Hypertonic disk
b. Hypotonic disk
c. Intercalated disk
d. Intercellular domain
e. Intracellular domain

51. Which of the following category of integral proteins are responsible for passive transport?
a. Carrier proteins
b. Cell recognition proteins
c. Channel proteins
d. Enzyme proteins
e. Receptor proteins
f. Both a and c
g. Both b and d

52. Channel proteins possess a corridor (passage way) that connects the cytosol with the extracellular fluids. Which of the following is the type of corridor that allows ions and other materials to pass through?
a. Aquaporin
b. G-protein complex
c. Pore
d. Sodium potassium pump
e. Stoma

53. Channel proteins possess a corridor (passage way) that connects the cytosol with the extracellular fluids. Which of the following is the type of corridor that allows water to pass through?
a. Aquaporin
b. G-protein complex
c. Pore
d. Sodium potassium pump
e. Stoma

54. Channel proteins possess a corridor (passage way) that connects the cytosol with the extracellular fluids. Which of the following is the type of corridor that allows water to pass through?
a. Aquaporin
b. G-protein complex
c. Pore
d. Sodium potassium pump
e. Stoma

55. If water molecules flow from areas of higher concentration to areas of lower concentration through the corridor of a channel protein, this process is called _____.
a. Diffusion
b. Hypertonic
c. Hypotonic
d. Isotonic
e. Osmosis

56. Which of the following restricts substances from passing through the channel protein?
a. Charge of the molecule
b. Density of a molecule
c. Shape of the molecule
d. Size of the molecule
e. Outer membrane of the molecules
f. Inner core ofthe molecules
g. Time
h. Both a and d

57. If the R-groups found at the entrance of the channel protein are positively charge, this channel protein will attract which of the following?
a. Anions
b. Cations
c. Hydrophilic molecules
d. Hydrophobic molecules

e. Isotopes
f. both a and c
g. both b and c
h. both d and e

58. If the R-groups found at the entrance of the channel protein are negatively charge, this channel protein will attract which of the following?
a. Anions
b. Cations
c. Hydrophilic molecules
d. Hydrophobic molecules
e. Isotopes
f. both a and c
g. both b and c
h. both d and e

59. Which of the following does NOT exert control over the opening of closing of gated channels?
a. Ligands
b. Mechanical force
c. pH
d. Temperature
e. Voltage
f. Weight

60. It has been reported that the exterior (the side facing the external environment) of the potassium ion channel, near the opening of the corridor, is _____ charged. This particular charge is generally accomplished by positioning _____ and _____ (hint: amino acids) near the opening of the channel protein. This charge serves to increase the effective concentration of _____ (select: cation or anion) at the entrance of the channel protein corridor and at the same time repulse _____ (select: cation or anion). Since both sodium and potassium are _____ (select: cations or anions), both these ions would be attracted to the opening of the channel protein. Being charged particles, both ions would have a cadre of water molecules surrounding them due to a weak attraction called _____. Therefore, it is necessary to strip away the water molecules from the ions before allowing them to enter the corridor.

The potassium ion channel eliminates the water molecules by utilizing an amino acid R-group called _____. This R-group possesses a strong negative charge and is capable of stripping the water molecules surrounding the potassium but unable to strip away water molecules surrounding sodium because of their stronger attraction to water. Because of this inability to remove the water molecules surrounding sodium ions, the effective radius of this sodium-water molecule complex is much too big thereby unable to fit through the narrow center of the corridor. Once the potassium ions pass through the narrow center it enters an area called _____. Like the opening of the ion channel, the narrow center is also populated by amino acids with _____ charged R-groups which will continue to attract potassium and prevent them from diffusing backwards to the extracellular matrix. As the first potassium ion sit comfortably in the energy well, other potassium ions will follow its progress down the pore. Since like-charges repulse one another, the latter potassium ions will push the potassium ions in front of them further into the pore and finally, based upon their diffusion gradient, these ions will exit the channel protein pore and into the cytoplasm.

a. Aldehyde
b. Alkene
c. Alkyne
d. Amide
e. Anion
f. Aspartic acids
g. Carbonyl oxygen
h. Carboxyl
i. Cation
j. Covalent bond
k. Energy well
l. Hydrogen bond
m. Hydrophobic interactions
n. Glutamic acid
o. Ionic bond
p. Ketone
q. Negative
r. Phosphodiester
s. Positive
t. Sulfide

61. Which of the following receptor protein is a type of transmembrane protein that possesses a ligand binding site directly linked to the ion channel?
a. Ametropic
b. Geotropic
c. Periotropic
d. Ionotropic
e. Metabotropic

62. Which of the following receptor protein is not a part of the channel protein? The opening and closing of this channel protein is dependent upon the activation of a G-protein complex.
a. Ametropic
b. Geotropic
c. Periotropic
d. Ionotropic
e. Metabotropic

63. Ionotropic receptors are composed of four to five trans-membrane protein subunits also known as _____ organized around a _____. Ionotropic receptors possess two functional domains. The first domain is referred to as the _____ and the second is called the _____ which creates the ion channel.
a. Aquaporin
b. Ligand binding site
c. Membrane spanning domain
d. Monomers
e. Multimers

f. Polymers
g. Pore

64. Ionotropic receptors are fast acting receptors commonly associated with neurotransmitters. Which of the following is NOT one of the ligands that associates with ionotropic receptors?
a. Acetylcholine
b. Glucagon
c. Glycine
d. Serotonin
e. γ-Aminobutyric acid (GABA)

65. In regards to a metabotropic receptor, a common name for the intercellular signal molecule (e.g. hormones and/or neurotransmitters) is called _____ where it is designed to bind to a specific receptor protein (Hint: a common switch) called _____.

The binding of the signal molecule causes a conformational change (change in the shape) of this protein thereby allowing three subunits also known as _____, _____ and _____ to separate. The _____ and _____ subunits form a stable dimer (joining of two molecules through intermolecular bonding) and are responsible for activating the _____ subunit. As a result, _____ (a spent energy molecule) bounded to the _____ is replaced with _____ (a charged energy molecule). Subsequently, the subunit-energy molecule complex either causes the conformation to change for a channel protein or activation of an enzyme.

Receptor protein opening a gated channel

This newly formed energy-subunit complex then comes into contact with a channel protein whereby causing it to alter in its conformation and opening a corridor also known as a _____ located at the center of the integral protein. This opening allows ions to flow through based upon their concentration gradient. Concentration gradient is based upon the theory of _____ which states that substances will flow from areas of _____ (Hint: high or low) concentration to areas of _____ (Hint: high or low) concentration. The above stated example of the gated channel protein is a part of the _____ (Hint: a type of transport that does not require energy) transport system.

Receptor protein activating a cascade reaction

This newly formed energy-subunit complex will come into contact with a membrane bounded enzyme. One of the most common membrane bounded enzyme is called _____ and its activation causes the decomposition of a (an) energy molecule called _____ (Hint: the most common energy molecule in the human body) into a secondary messenger molecule also known as a _____. A secondary messenger molecule is also commonly known as a (an) _____ which is defined as a signal molecule within the cell. This secondary messenger molecule in turn activates another enzyme called _____ which is responsible for transferring a phosphate group onto various unfinished products produced by the cell through a phosphate adding process known as _____. The addition of the phosphate effectively alters the shape of the unfinished products and thereby changes them into an active product.

Shutting down these reactions

After the necessary amount of ions has flowed through the membrane or adequate number of product has been formed through the cascade reaction, the channel protein and the membrane bounded protein has to be closed or turned off. First, the signal molecule will be removed from the receptor protein which causes _____ (a subunit) to break the _____ (Hint: a type of bond in an energy molecule) via a process known as _____ (Hint: adding water) and thereby releasing a (an) _____ (Hint: a functional group) and energy. The release of the functional group and energy causes the subunit and now spent energy molecule complex to break connection with the channel protein and the membrane bounded enzyme thereby closing the gated channel or shutting down the cascade reaction.

1.	Active	18.	GTP
2.	ADP	19.	High
3.	ATP	20.	High energy bond
4.	Alpha (α) subunit	21.	Hydrolysis
5.	Aquaporin	22.	Pericellular messenger
6.	Beta (β) subunit	23.	Intracellular messenger
7.	Condensation synthesis	24.	Ligand
8.	cAMP	25.	Lipase
9.	cCMP	26.	Low
10.	cGMP	27.	Osmosis
11.	cTMP	28.	Passive
12.	CDP	29.	Phosphate
13.	CTP	30.	Phosphorylation
14.	Diffusion	31.	Pore
15.	Gamma (γ) subunit	32.	Protease
16.	G-protein complex	33.	Protein kinase
17.	GDP	34.	Adenylate cyclase

66. Which of the following is a type of passive transport that moves substances through intermolecular interactions between the substrate and the transporter? For example, if a glucose binds to the transporter, the intermolecular interaction causes conformation change of this integral protein. This conformational change facilitates the movement of the glucose molecule towards the center of the integral protein. From that point, the glucose molecule flows down its concentration gradient into the cytoplasm of the cell.
a. Active transport
b. Diffusion
c. Facilitated diffusion
d. Membrane assisted transport
e. Osmosis

67. It is estimated that the energy requirement of primary active transport amounts to approximately _____ (%) percent of the entire ATP supply of the cell.
a. 10%
b. 20%
c. 30%
d. 40%
e. 50%

68. There are two types of active transport. Which of the following type of active transport utilizes energy in the form of ATP to move ions or molecules against their concentration gradient?
a. Facilitated diffusion
b. Membrane assisted transport
c. Osmosis
d. Primary active transport
e. Secondary active transport

69. There are two types of active transport. Which of the following type of active transport utilizes kinetic energy that is generated by a molecule (such as sodium) moving down its concentration gradient to propel its functions?
a. Facilitated diffusion
b. Membrane assisted transport
c. Osmosis
d. Primary active transport
e. Secondary active transport

70. Which of the following is a type of transport that physically moves molecule (s) or ion (s) against its concentration gradient where substances are moved from areas of low concentration to high concentration? It is understood that the function of active transport carrier protein depends upon the availability of ATP, the mainstay energy molecule of all living organisms.
a. Primary active transport
b. Diffusion
c. Secondary active transport
d. Facilitated diffusion
e. Membrane assisted transport

71. Which of the following best describes a carrier protein that is designed to transport 2 molecules (generally ions) in or out of the cell in the same direction?
a. Antiport
b. Cotransporter
c. Ligand gated channel
d. Mechanically gated channel
e. Symport
f. Uniport
g. Both a and c
h. Both b and e

72. Which of the following best describes a carrier protein that is designed to move one molecule in one direction in or out of the cell?
a. Antiport
b. Cotransporter
c. Ligand gated channel
d. Mechanically gated channel
e. Symport
f. Uniport
g. Both a and c
h. Both b and e

73. Which of the following best describes a carrier protein that is designed to move two or more molecules in the opposite direction in and out of the cell?
a. Antiport
b. Cotransporter
c. Ligand gated channel
d. Mechanically gated channel
e. Symport
f. Uniport
g. Both a and b
h. Both b and e

74. Which of the following occurs when large materials such as foreign substances like bacteria, fungal spores, a damaged or dying cell, as well as cellular debris is brought into the cell through the manipulation of the membrane and the formation of a phagosome?
a. Exocytosis
b. Phagocytosis
c. Pinocytosis
d. Primary active transport
e. Secondary active transport
f. Passive transport

75. Which of the following best describe (s) sodium and potassium pump?
    a. Active transport
    b. Antiport
    c. Ligand gated channel
    d. Mechanically gated channel
    e. Symport
    f. Uniport
    g. Voltage gated channel
    h. Both a and b
    i. None of the above

76. Which of the following mechanism triggers the activation of membrane assisted transport?
    a. Receptor proteins such as G-protein
    b. Temperature
    c. pH
    d. Electrical impulses
    e. Pressure
    f. Calcium ions
    g. Sodium ions
    h. Potassium ions

77. Which of the following occurs when the cell takes in many smaller materials into the cell through the manipulation of the plasma membrane?
    a. Exocytosis
    b. Phagocytosis
    c. Pinocytosis
    d. Primary active transport
    e. Secondary active transport

78. Which of the following occurs when a cell moves materials out of the cell? This form of transport involves a vesicle that contains materials needed to be released into the surrounding environment.
    a. Exocytosis
    b. Phagocytosis
    c. Pinocytosis
    d. Primary active transport
    e. Secondary active transport

79. The nucleus and its contents are separated from the cytoplasm of the eukaryotic cell by a _____ which is composed of phospholipids. When the cell is not dividing, the nucleus contains loosely coiled DNA strands called _____ which are suspended in a fluid filled matrix called _____. However when the cell is getting ready to divide, the DNA strands undergoes coiling and wrapping itself around a protein called _____ thereby converting itself into a rod-shaped structure called _____. Under the microscope, the _____ is the dark region of the nucleus (a small dark colored dot within the nucleus – this is not the Barr body). This dark spot marks the location where _____ is produced. After the production of this particular nucleic acid (nucleotide), the product leaves the nucleus through small channels within the membrane called _____.

    a. Actin
    b. Aquaproin
    c. Channel proteins
    d. Chromosome
    e. Chromatin
    f. Cristae
    g. Cytoplasm
    h. Cytosol
    i. DNA
    j. Docking proteins
    k. Exportin
    l. Globular
    m. Glycosaminoglycans (GAGs)
    n. Golgi apparatus
    o. Histone
    p. Importin
    q. Matrix
    r. Nuclear membrane
    s. Nuclear pore
    t. Nucleolus
    u. Nucleoplasm
    v. Ribosome
    w. rRNA
    x. rER
    y. sER
    z. Tubulin

80. Which of the following is the organelle responsible for translating mRNA?
    a. Mitochondria
    b. Nucleus
    c. Peroxisome
    d. rER
    e. Ribosome
    f. sER

81. The _____ is the main genetic sequence of all eukaryotic cells and it never leaves the nucleus. However, when the cell requires a protein product to be manufactured, a segment of the previously stated genetic sequence that codes for the protein is copied or transcribed into another genetic sequence called _____. This newly created genetic sequence is allowed to leave the nucleus through protein channels found in the nuclear membrane called _____. It is known that any substance that exits the nucleus are called _____. Once this transcribed genetic sequence enters the cytoplasm, it is immediately bounded (attached) to many (hundreds and/or thousands) organelles known as _____ (Hint: pleural form that indicates many of these organelles). These organelles are responsible for _____ the genetic sequence until hundreds or thousand copies of the same polypeptides are made. Once the polypeptide sequence is formed (via matching its anticodon to the codon etc.) the newly formed product is inserted into a protein structure (channel) called _____ located in the membrane of the organelle known as _____. This organelle is responsible for editing the polypeptide sequence within its lumen.

    a. Aquaporin
    b. Catabolism
    c. Centrioles
    d. Centrosome
    e. Channel protein
    f. Condensation synthesis
    g. Cytoplasm
    h. Cytosol
    i. DNA
    j. Docking protein (SRP)
    k. Endocytosis
    l. Exocytosis
    m. Exportin
    n. Golgi apparatus
    o. Hydrolysis
    p. Importin
    q. mRNA
    r. Nuclear plasm
    s. Nuclear pore
    t. Polyribosome
    u. Protein synthesis
    v. rER
    w. rRNA
    x. sER
    y. tRNA
    z. Vesicle

82. The ribosome binds or links to the rER at which of the following site?
    a. Attachment plaques
    b. Cadherins
    c. Claudins
    d. Desmoplakins
    e. Docking protein (SRP)
    f. Occludins
    g. Pakoglobins
    h. Transmembrane linker proteins

83. The mitochondria is an organelle that possesses double membranes. The outer membrane of this organelle is smooth while the inner membrane is highly folded. What is the name of this highly folded inner membrane?
    a. Chistae
    b. Matrix
    c. Nuclear envelope
    d. Plasma membrane
    e. Signal recognition particle

84. The mitochondrion is an organelle that possesses double membranes. The outer membrane of this organelle is smooth while the inner membrane is highly folded. What is the proposed reason behind the folding of the inner membrane?
    a. An artifact of inner enzymatic reactions
    b. Decrease surface area
    c. Increase structural endurance
    d. Increase surface area
    e. Modification of internal molecular sequence

85. Which of the following is the correct formula for cellular respiration?
    a. Maltose + CO_2 → Oxygen + carbon monoxide + ATP
    b. Glucose + ADP → water + CO_2 + ATP
    c. GDP + energy → GTP + water + CO_2
    d. Lactose + Oxygen → CO_2 + water + ATP
    e. Glucose + Oxygen + ADP → CO_2 + water + ATP
    f. G-protein + alpha subunit → channel + diffusion + ATP
    g. Glycogen + Oxygen → CO_2 + water + ATP
    h. None of the above

86. The endoplasmic reticulum (ER) consists of an intricate and complicated network of membranous channels called _____.
    a. Cisternae
    b. Matrix
    c. Chistae
    d. Microvilli
    e. Nuclear envelope

87. The rough endoplasmic reticulum (rER) is closely associated with the ribosome and obtained its name due to the peppered appearance of bounded ribosomes at specific sites on its membrane. These specific sites where ribosomes bind is called _____.
    a. Antiport
    b. Docking protein
    c. Translocon
    d. Uniport
    e. Synport

88. Which of the following is responsible for recognizing the signal peptides that are present on the N-terminus of most newly formed proteins as they are being made? This signal peptide triggers the insertion of the polypeptide sequence into the lumen, otherwise the polypeptide sequence remains within the cytoplasm.
    a. Antiport
    b. Claudins
    c. Signal recognition particle (SRP) receptor
    d. Translocon
    e. Uniport

89. Once the polypeptide from the ribosome is inserted into the lumen of the rER, it is escorted by which of the following molecule? This molecule escorts the polypeptide to interact with various enzymes.
    a. Chaperone molecule
    b. Signal recognition particle
    c. Chistae
    d. Matrix
    e. Translocon

90. Which of the following is (are) the type (s) of modification made to the polypeptide chains inside the rER lumen?
    a. Adding amino acid sequences to create a larger protein
    b. Adding carbohydrate chains to create glycoproteins
    c. Adding lipids to create a lipoprotein
    d. Adding steroids to create a sterol-protein
    e. Deleting or rearranging amino acid sequences

91. Which of the following organelle is responsible for producing steroid based hormones and lipids?
    a. Golgi apparatus
    b. Mitochondria
    c. Nucleus
    d. rER
    e. sER

92. Which of the following organelle is responsible for catabolizing lipid-soluble (hydrophobic) molecules?
    a. Golgi apparatus
    b. Mitochondria
    c. Nucleus
    d. rER
    e. sER

93. Which of the following is a modified form of sER that is responsible for storing calcium ions?
    a. Golgi apparatus
    b. Nucleus
    c. rER
    d. Sarcoplasmic reticulum
    e. sER

94. Golgi apparatus is best described as the final processing center of products within the eukaryotic cell. The Golgi is comprised of membrane-bound stacks known as _____.
    a. Cisternae
    b. Matrix
    c. Chistae
    d. Microvilli
    e. Nuclear envelope

95. Within the membrane bounded stacks of the Golgi apparatus, various enzymes are available to perform which of the following task(s)?
    a. Acetylation
    b. Carbonation
    c. Carboxylation
    d. Glycolsylation
    e. Phosphorylation
    f. Both d and e

96. In vascular transport of lipids, the hydrophobic substances produced in the sER are packaged within a phospholipid bilayer and are kept from simply diffusing through the membrane via which of the following process?
    a. Active transport maintains the integrity of the product/vesicle interaction
    b. Gated channels maintains the integrity of the product/vesicle interaction
    c. Hydrophilic cytosol maintains the integrity of the product/vesicle interaction
    d. Hydrophobic phospholipids maintains the integrity of the product/ vesicle interaction
    e. Protein interactions maintains the integrity of the product/vesicle interaction

97. Since lipid substances produced by the cell is hydrophobic, could this substance simply diffuse through the phospholipid bilayer membranes?
    a. Yes, hydrophobic substances will mix in hydrophobic substances
    b. No, hydrophobic and hydrophobic substances do not mix

98. The speed of spontaneous lipid transport can be greatly increased if the transport occurred at various membrane contact sites. Which of the following is the membrane contact site?
    a. Cytosolic gaps
    b. Gated channels
    c. Non-gated channels
    d. Symport
    e. Uniport

99. Which of the following is used for intracellular transport of lipids? This form of transport involves a transport protein complex that contains a hydrophic inner domain that stores the lipids.
    a. Lipid-transfer proteins
    b. Cytosolic gaps
    c. Chaperone molecule
    d. Signal recognition particle
    e. Translocon

100. _____ are membrane-bounded vesicles (an example of one of the many final products of the Golgi apparatus) that contain products that are very low in pH and contain ~50 different powerful hydrolytic digestive enzymes.

    For example, enzymes that are responsible for breaking down proteins are called _____. Enzymes that are responsible for breaking down carbohydrates are called _____. Enzymes that are responsible for breaking down nucleic acids are called _____. Enzymes that are responsible for breaking down

lipids are called_____.

In certain types of white blood cell, bacteria can be brought in via a membrane assisted transport process known as _____ (Hint: bringing in large objects) and isolated (quarantined) in a vesicle called _____. Once isolated, the bacterium is destroyed by the hydrolytic digestive enzymes stored in the above mentioned organelle. Similarly, cellular debris, dead or damaged cells from the body are also absorbed and isolated before being digested by the hydrolytic enzymes. This process of removing cellular debris, dead or damaged cells from the body is called _____. These membrane bounded enzymes are also used in a cellular repair process termed _____ where the organelles within a cell is replaced and renewed. . In this process, a damaged organelle is enclosed within a vesicle called _____ and is then promptly digested.

    a. Autodigestion   j. Lysosome
    b. Autophagosome   k. Multiphagosome
    c. Autophagy   l. Nucleases
    d. Autoprocessing   m. Peroxisome
    e. Exocytosis   n. Phagocytosis
    f. Glycosidases   o. Phagosome
    g. Golgi apparatus   p. Pinocytotic vesicle
    h. Lipases   q. Pinosytosis
    i. Lysosomal proteases   r. Ribosome

101. _____ are small enzyme containing organelles that are involved in the breakdown of fatty acids, amino acids uric acids and hydrogen peroxide (H_2O_2). Hydrogen peroxide is a byproduct of _____ and _____ catabolism and is extremely toxic to cells. This organelle contains _____, an enzyme that converts hydrogen peroxide to harmless _____ and _____ during a reaction known as _____.

In addition to oxidative reactions, these organelles are also the location for _____. For example, substances such as _____ are synthesized both in this organelle, as well as another organelle called _____.

    a. Amino acid   k. Monosaccharide
    b. Catalase   l. Nitrogen
    c. Catalytic reaction   m. Nucleic acid
    d. Cellular respiration   n. Oxygen
    e. Cholesterol   o. Peroxidation reactions
    f. Disaccharide   p. Peroxisomes
    g. Fatty acid   q. rER
    h. Hydroxide   r. Steroid
    i. Golgi apparatus   s. sER
    j. Lipid biosynthesis   t. Water

102. In many cell types found in the human body (cardiac and skeletal muscle cells, kidneys, lungs, and the brain etc.) peroxisomes are responsible for the formation a unique type of membrane phospholipid (e.g. glycerophospholipids) that is involved in maintaining the physical properties of the plasma membrane when transitioning from a gel to fluid state in various temperatures. In addition, these molecules are also involved in maintaining the proper functions of various membrane bounded integral proteins. What is this unique membrane phospholipid?
    a. Plasmalogen
    b. Cholesterol
    c. Alpha amylase
    d. Protein kinase
    e. Adenylate cyclase

103. Which of the following is the enzyme responsible for breaking hydrogen peroxide (H_2O_2) down into water?
    a. Adenylate cyclase
    b. Amylase
    c. ATPase
    d. Catalase
    e. Lipase
    f. Peptidase
    g. Phosphodiesterase
    h. Protein kinase

104. The cytoskeleton is a network of interconnected filaments and tubules that extend from the nucleus to the plasma membrane in eukaryotic cells. Which of the following is NOT a function of cytoskeleton?
    a. Maintaining cell shape
    b. Providing movement of the organelles
    c. Providing movement of the cells (forming flagella etc.)
    d. Forming cell-to-cell junctions
    e. Allows for cell-to-cell recognition

105. Which of the following is NOT a type of filament that is (are) included as cytoskeletons?
    a. Actin
    b. Glycosidase
    c. Intermediate filament
    d. Microtubules

106. Which of the following protein(s) are used in the construction of the microtubule?
    a. Alpha tubulin
    b. Beta tubulin
    c. Actin
    d. Keratin
    e. Intermediate filament

f. Both a and b
g. Both b and d

107. Which of the following is essential during cellular division?
a. Actin
b. Intermediate filament
c. Microtubules
d. Keratin

108. Which of the following best describes the microtubule composition of cilia?
a. 9+0 triplets
b. 9+1 triplets
c. 9+2 triplets
d. 9+0 doublets
e. 9+2 doublets

109. Which of the following have microtubule arrangements of 9+0 triplets and are known to be the center of microtubules regulation? This organelle is active during cellular division as well as creating basal bodies from which cilia and flagella may arise.
a. Centriole (centrosome)
b. Cilia
c. Endoplasmic reticulum
d. Nucelolus

110. Which of the following is involved in the increase of surface area for organs such as the small intestines?
a. Cilia
b. Flagella
c. Microvilli
d. Villus

111. The microvillus is constructed of a core bundle of interlinking proteins surrounded by the cellular membrane. It is understood that the core bundle of microvillus is composed of ~19 _____ anchored at its basal end to horizontally orientated _____ through _____. This crosslink is also referred to as the _____. Within the length of the microvillus, the core bundle is also cross-linked by two proteins called _____ and _____. Finally, the protein _____ is responsible for laterally tethers the core bundle to the membrane of the microvillus.
a. Actin filaments
b. Spectrin protein crosslinks
c. Terminal web
d. Fimbrin proteins
e. Villin
f. Myosin1A
g. Desmocollins
h. Desmogleins
i. Desmoplakin
j. Plakoglobin
k. Plakophilins
l. Vinculin
m. α-Actinin
n. β-catenins

ANSWERS

1.	c	2.	e	3.	c
4.	c	5.	a	6.	c
7.	c	8.	b	9.	a
10.	d	11.	b	12.	a
13.	d	14.	a	15.	e
16.	d	17.	f	18.	b
19.	a	20.	f	21.	g
22.	e	23.	a, b, c	24.	b
25.	f	26.	a	27.	d, g
28.	d, g	29.	e	30.	f
31.	e	32.	c	33.	b
34.	d	35.	b	36.	a
37.	f	38.	c	39.	b
40.	a, d	41.	a, d	42.	f
43.	g	44.	i	45.	j
46.	j	47.	c	48.	b
49.	p, u, h, i	50.	c	51.	f
52.	c	53.	a	54.	a
55.	e	56.	e	57.	a
58.	b	59.	f	60.	q, f, n, i, e, i, l, g, k, q
61.	d	62.	e	63.	e, g, b, c
64.	b	65.	24, 16, 4, 6, 15, 6, 15, 4, 17, 4, 18, 31, 14, 19, 26, 28, 34, 3, 8, 23, 33, 30, 4, 20, 21, 29	66.	C
67.	c	68.	d	69.	e
70.	a	71.	h	72.	f
73.	g	74.	b	75.	h
76.	a	77.	c	78.	a
79.	r, e, u, o, d, t, w, s	80.	e	81.	i, q, s, m, t, u, j, v
82.	e	83.	a	84.	d
85.	e	86.	a	87.	b
88.	c	89.	a	90.	b
91.	e	92.	e	93.	d
94.	a	95.	f	96.	c
97.	a	98.	a	99.	a
100.	j, i, f, l, h, n, o, a, c, b	101.	p, g, a, b, t, n, o, j, e, s	102.	a
103.	e	104.	e	105.	b
106.	f	107.	c	108.	e
109.	a	110.	c (d)	111.	a, a, b, c, d, e, f

REFERENCES

Alberts B, Johnson A, Lewis J, Raff M, Roberts K, Walter P. Molecular Biology of the Cell. 4th edition. New York: Garland Science. 2002

Meng W, Takeichi M. Adherens Junction: Molecular Architecture and Regulation. Cold Spring Harb Perspect Biol, 2009. 1: pp. 1-13

Nimigean CM, Allen TW. Origins of ion selectivity in potassium channels from the perspective of channel block. J. Gen Physiol, 2011. 137: pp. 405-413

Purves D, Augustine GJ, Fitzpatrick D, Katz LC, LaMantia A-S, McNamara JO, Williams SM. Neuroscience, 2nd Edition. Sunderland Massachusetts, Sinauer Associates 2001

Cooper GM. The Cell, a Molecular Approach, 2nd Edition. Sunderland Massachusetts, Sinauer Associates 2000

Kim K, Kim SO, Han J. Susceptibility of Lysosomes to Rupture Is a Determinant for Plasma Membrane Disruption in Tumor Necrosis Factor Alpha-Induced Cell Death Mol Cell Biol, 2003. 23: pp. 665–676

Braverman NE, Moser AB. Functions of plasmalogen lipids in health and disease. Biochim. Biophys. Acta, 2012. 1822: pp. 1442-1452

Goodenough DA, Paul DL. Gap Junctions. Cold Spring Harb Perspect Biol, 2009. 1: pp. a002576

Seeley RR, Stephens TD, Tate P. Anatomy and Physiology 6th Edition. McGraw-Hill, New York, New York. 2003

Tate P. Seeley's Principles of Anatomy & Physiology 1st Edition. McGraw-Hill, New York, New York. 2009

Saladin KS. Anatomy & Physiology. The Unity of Form and Function 6th Edition. McGraw-Hill, New York, New York. 2010

PHOTO AND GRAPHIC BIBLIOGRAPHY

1. Figure 1: Graphic designed and released by Darryl Leja 2010, National Human Genome Research Institute. Public domain graphics
2. Figure 2: Graphic designed and released by Open Stax College, Connexions, reprint permission granted under the Creative Commons Attribution 3.0 Unported license.
3. Figure 3: Graphic designed and released by Open Stax College, Connexions, reprint permission granted under the Creative Commons Attribution 3.0 Unported license.
4. Figure 4: Graphic designed and released by Ladyofhats 2008. Public domain graphics
5. Figure 5: Graphic designed and released by National Institute of Health. Public domain graphics
6. Figure 6: Graphic designed and released by P.Y.P. Jen. Copyright ©
7. Figure 7: Graphic designed and released by Ladyofhats 2008. Public domain graphics
8. Figure 8: Graphic designed and released by Ladyofhats 2008. Public domain graphics
9. Figure 9: Graphic designed and released by Ladyofhats 2008. Public domain graphics
10. Figure 10: Graphic designed and released by Open Stax College, Connexions, reprint permission granted under the Creative Commons Attribution 3.0 Unported license.
11. Figure 11: Graphic designed and released by Open Stax College, Connexions, reprint permission granted under the Creative Commons Attribution 3.0 Unported license. Paweł Tokarz, 2009. Public domain graphics
12. Figure 12: Graphic designed and released by P.Y.P. Jen. Copyright ©
13. Figure 13: Graphic designed and released by Open Stax College, Connexions, reprint permission granted under the Creative Commons Attribution 3.0 Unported license.
14. Figure 14: Graphic designed and released by Open Stax College, Connexions, reprint permission granted under the Creative Commons Attribution 3.0 Unported license.
15. Figure 15: Graphic designed and released by P.Y.P. Jen. Copyright ©
16. Figure 17: Graphic designed and released by P.Y.P. Jen. Copyright ©
17. Figure 18: Graphic designed and released by P.Y.P. Jen. Copyright ©
18. Figure 19: Graphic designed and released by Graham Colm, 2009. This graphic is licensed under the Creative Commons Attribution-Share Alike 3.0 Unported license.
19. Figure 20: Graphic released by National Institute of Health. The Human Genome Project. Eukaryotic Cell. The graphics was developed by Darryl Leja, 2010. Public domain graphic
20. Figure 21: Graphic designed and released by Miguelferig, 2011. Public domain graphic
21. Figure 22: GGraphic designed and released by Darryl Leja 2010, National Human Genome Research Institute. Public domain graphic
22. Figure 23: Graphic designed and released by Darryl Leja 2010, National Human Genome Research Institute. Public domain graphic
23. Figure 24: Graphic designed and released by Darryl Leja 2010, National Human Genome Research Institute. Public domain graphic
24. Figure 25: Graphic designed and released by Darryl Leja 2010, National Human Genome Research Institute. Public domain graphic
25. Figure 26: Graphic designed and released by Darryl Leja 2010, National Human Genome Research Institute. Public domain graphic
26. Figure 27: Graphic designed and released by Darryl Leja 2010, National Human Genome Research Institute. Public domain graphic
27. Figure 28: Graphic designed and released by Darryl Leja 2010, National Human Genome Research Institute. Public domain graphic
28. Figure 29: Graphic designed and released by Open Stax College, Connexions, reprint permission granted under the Creative Commons Attribution 3.0 Unported license.
29. Figure 30: Graphic designed and released by Zlir's 2011. Simon Caulton 2013. Boumphreyfr 2009. Creative Commons Attribution 3.0 Unported license.

30. Figure 31: Graphic designed and released by Darryl Leja 2010, National Human Genome Research Institute. Public domain graphic
31. Figure 32: Graphic designed and released by P.Y.P. Jen. Copyright ©
32. Figure 33: Graphic designed and released by P.Y.P. Jen. Copyright ©
33. Figure 34: Graphic designed and released by P.Y.P. Jen. Copyright ©

CHAPTER FOUR

EPITHELIAL TISSUE

Epithelial tissue forms the external covering of the organism, as well as, the internal lining of organs. Put simply, epithelial tissue covers the entire human body, both inside and out. Epithelial tissue is made up of densely arranged cells bound to one another through numerous cellular junctions previously described in Chapter Three.

In general, the cells of the epithelium possess an apical surface that is directed towards a lumen. Depending on the tissue type and its location in the human body, the apical surface may possess microvilli, which increases surface area, or cilia to provide movement. The epithelial cells also have a lateral surface that is connected by junctions with other cells of similar origin and a basal surface that is positioned above a loose connective tissue, the basement membrane (Figure 1).

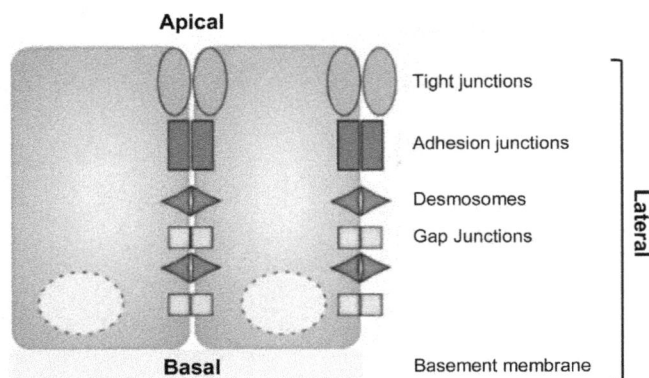

Figure 1: Illustrations of a cellular domain. The apical portion is facing the lumen, lateral is the site of connection between adjacent cells and basal portion is anchored to the basement membrane

The basement membrane, also known as the basal lamina, is a connective tissue layer formed from extracellular secretions (i.e. extracellular matrix) that isused to anchor the epithelial tissues, as well as, providing a medium for filtration and diffusion.

The major molecular components of basement membranes are collagen (collagen IV), laminin, entactin, fibronectin, proteoglycans (e.g. heparin sulfate and chondroitin sulfate), and hyaluronic acids. For example, collagen IV provides the physical support structure for the connective tissue and creates a scaffold for the other structural macromolecules to build on (Table I). Laminin is an adhesive glycoprotein that is a major structural component of the basement membrane. In addition, laminin interacts with integrin, an integral protein of plasma membrane, which helps to adhere the epithelial cells to the basement membrane. Fibronectin is another adhesive glycoprotein found in the basement membrane. Fibronectin also interacts with integrin and plays a role in affixing the epithelial cells to the basement membrane.

It is understood that many cellular responses are due to the interactions between integrin and various proteins within the basement membrane. For example, once integrin is attached to the proteins within basement membrane, this attachment will in turn affect the intracellular arrangement of the cytoskeleton. Put simply, these interactions will have definitive effect on cell adhesion, shape, migration, proliferation, and differentiation.

Entactin is a type of glycoprotein that serves as scaffolding that connects laminins and fibronectins to collagen in order to form a structural network. This structural network is essential in maintaining the structural integrity of the basement membrane

and also provides the necessary foundation for the attachment of the epithelial tissue (Figure 2).

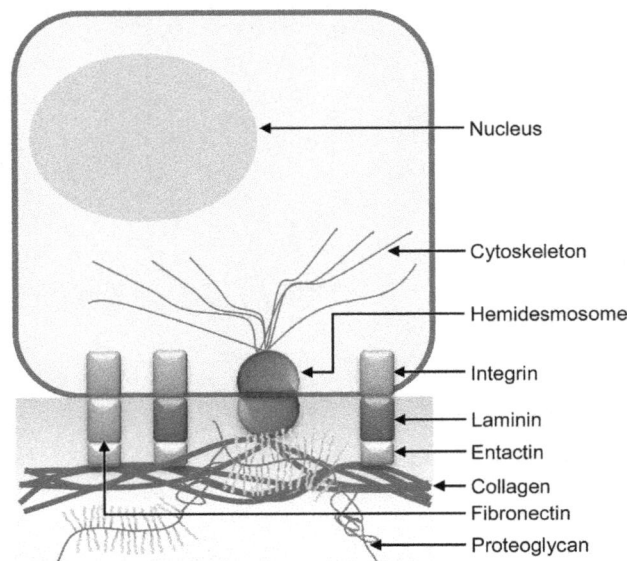

Figure 2: Illustration of the interaction between the basement membrane and epithelial cell. Please note that the adhesive glycoproteins laminin and fibronectin are bounded to the transmembrane protein integrin and through the interactions of enactin, both adhesive glycoproteins are connected to collagen IV fibers

Heparan sulfate, dermatan sulphate, and chondroitin sulfate are three types of glycosaminoglycans (GAGs) containing sulphate found within the basement membrane. These sulphate-containing GAGs are bounded covalently to a protein core (an integral protein) to form proteoglycans. Also located in the basement membrane are non-sulphate containing GAGs called hyaluronic acids. Unlike the proteoglycans described previously, hyaluronic acids are not associated with a protein core. Both sulphate-containing GAGs and non-sulphate containing GAGs are formed to provide structural support and a barrier for physical filtration for the basement membrane. Moreover, being negatively charged, these molecules repulse many anions, while simultaneously attract sodium ions (Na^+), which in turn attracts water molecules. Being water filled, the basement membrane provides resistance to compression, and serves as a diffusion/osmosis friendly medium for the exchange of waste materials, nutrients and gases (Figure 3).

The organization of the epithelial tissue can be as simple as a single layer of cells, or organized into several layers. For example, the endothelial lining of blood vessels consists of a single layer of cells attached to the basement membrane. In contrast, the thick skin of the human epidermis can consist of approximately 60 layers of skin cells affixed to the basement membrane. In general, epithelial tissues are divided into two groups: simple and stratified (Figure 4).

The simple epithelium is seen mostly in tissues formed for absorption, secretion, and gas exchange, where every single cell of the simple epithelium remains in contact with the basement membrane. On the other hand, stratified epithelium consists of multiple layers of cells situated on top of the basement membrane, while only the lowest layer of cells remains in contact with the basement membrane. Stratified epithelium is used in transportation and protection of underlining tissues and organs.

The categories of the epithelia are based upon the shape of the cells, as well as, the number of layers of cells situated above the basement membrane. We will examine the different categories below.

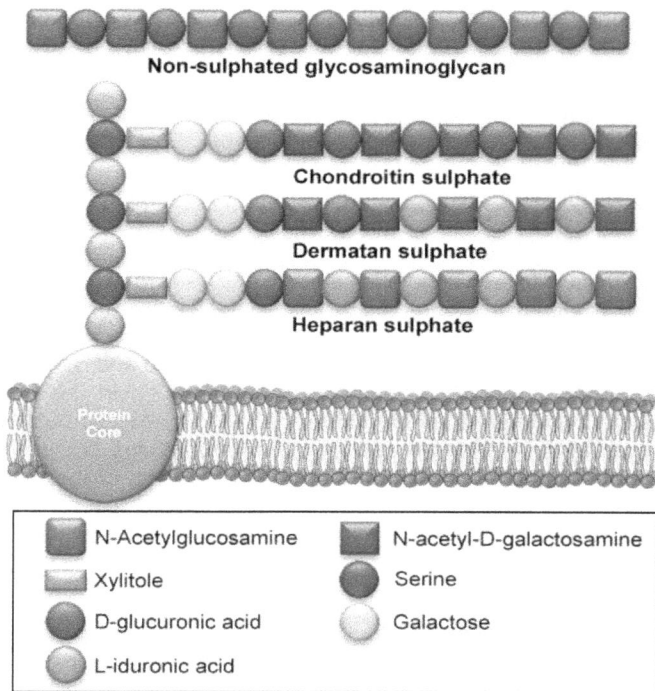

Figure 3: Illustration of the various structures of glycosaminoglycan and the formation of a proteoglycan. Please note that the hyaluronic acid (non-sulphated GAGs) is not linked to a protein core. Heparan sulphate, dermatan sulphate and chondroitin sulphate are connected to a protein core to form proteoglycan via a serine residue

Figure 4: Illustration of simple and stratified epithelium

Simple Squamous Epithelium

Simple squamous epithelium is composed of flat, thin cells that possess a low profile and large surface area. This type of epithelium is commonly found in areas of the body that perform gas exchange, require smooth surfaces for cellular, and fluid movement, involved in filtration, reabsorption of water, ions, and molecules, as well as, reducing friction between organs (Figure 5).

For example, simple squamous epithelium is found lining blood vessels where its smooth surface allows the erythrocytes to flow with minimal friction. The alveoli found at the terminus of the respiratory tract consist of simple squamous epithelium, which allows oxygen and carbon dioxide to diffuse freely in a process known as gas exchange. This epithelium located in the glomerulus of the kidneys and creates a barrier through which ions and small molecules, such as water, can pass through while keeping large molecules and cells away from the filtrate. In addition, this tissue located in the loop of Henle is involved in the reabsorption of ions and molecules during the formation of urine. Finally, simple squamous epithelium forms the serous pericardium and visceral pleura of the heart and lungs, respectively. These thin membranes secret a slippery fluid which lubricates the surfaces between the heart and lungs, which prevent damage that may result from friction (Figure 5).

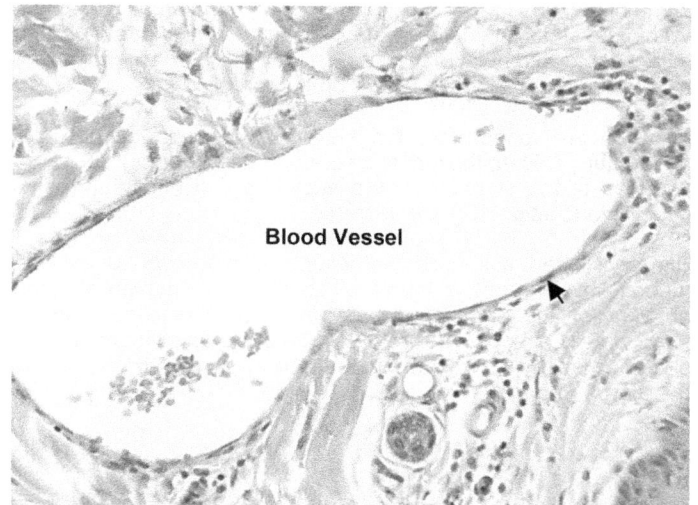

Figure 5: Simple squamous epithelium. Please note the graphic interpretation is shown above and the actual photomicrogaph is located below 400x

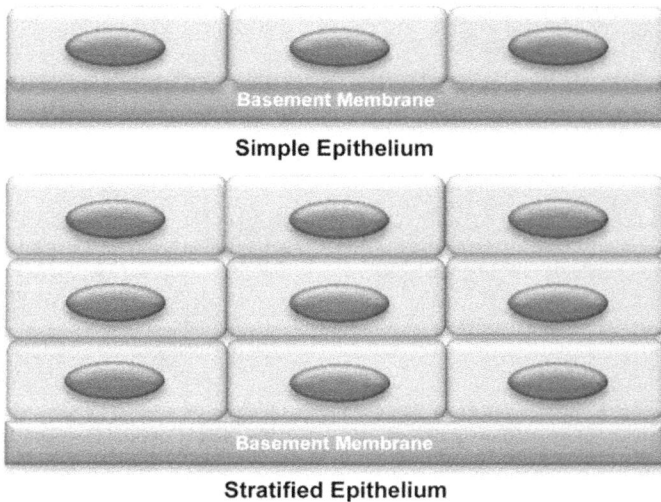

Simple Cuboidal Epithelium

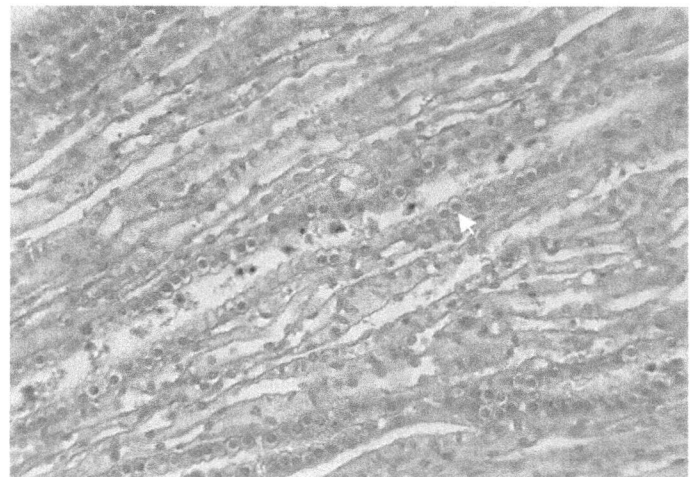

Figure 6: Simple cuboidal epithelium. Please note the graphic interpretation is shown above and the actual photomicrogaph is located below 400x

Simple cuboidal epithelium is composed of a single layer of cube-like cells that are capable of performing functions such as active transport, facilitated diffusion, secretion of substances, and/or protective roles within the human body. This tissue is commonly found in many small ducts and tubules (Figure 6).

For example, simple cuboidal epithelia are found in the renal (convoluted) tubules of the nephrons (located in the kidneys), thyroid, ovaries, as well as, lining the smaller ducts of sweat glands. In addition, this epithelial tissue makes up the germinal epithelia of the ovary and testes. The germinal epithelia of the ovaries are responsible for producing oocytes (eggs) and the sperm cells in the male testes (Figure 6).

Simple Columnar Epithelium

Simple columnar epithelium is most commonly found in organs when absorption occurs. Nonetheless, this type of epithelium is located in areas of the body that requires secretory functions and motility. For example, simple columnar epithelium is found in the highly absorptive surfaces throughout the small intestines. Within the small intestines, these cells are easily spotted through microscopic examinations of their apical cellular extensions or microvilli (also referred to as the brush border).

Microvilli are responsible for increasing the surface area for absorption. This epithelium is also found on the highly secretory surfaces of the stomach where enzymes and highly corrosive hydrochloric acid (HCl) are secreted.

Simple ciliated columnar epithelium (also known as ciliated mucous membrane) is found within the infundibulum and the ampulla of the fallopian tubes (oviduct). These rectangular shaped cells possess ~200 to 300 cilia on its apical surface. The cilia lined epithelium move the ovum (oocyte) or zygote towards the uterus (Figure 7).

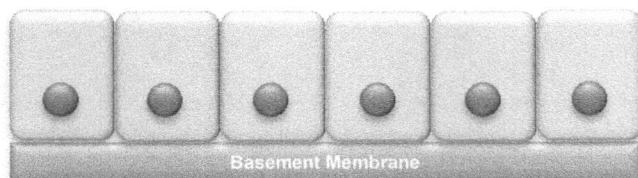

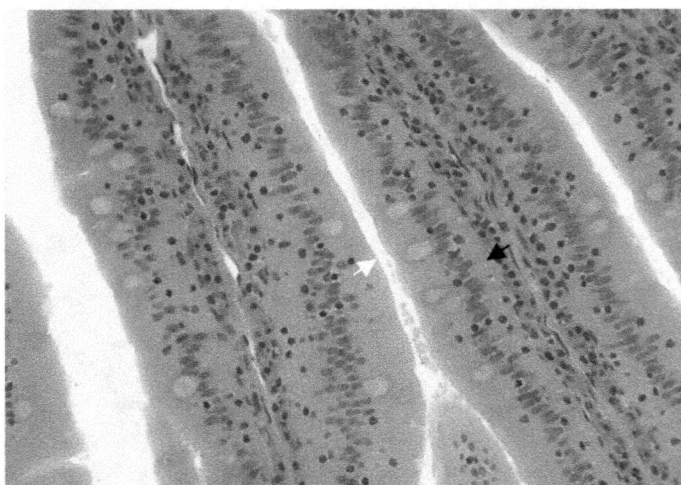

Figure 7: Simple ciliated columnar epithelium. Please note the graphic interpretation is shown above and the actual photomicrograph is located below 400x. The black arrow indicates the columnar cells and the white arrow indicates the cilia located on the apical surface of these cells

Pseudostratified Columnar Epithelium

Pseudostratified ciliated columnar epithelium is constructed out of irregularly distributed cells which may appear under the microscope as stratified tissue. However, appearances can be deceiving. This type of epithelium is actually a form of simple epithelium. Because of this visual misconception, the title

'pseudo' or false is applied to the name of this tissue.

Like all simple epithelium, every cell found in the pseudostratified columnar epithelium is situated upon the basement membrane. However, not all of the cells of the pseudostratified epithelium extend to the luminal (lumen) surface. The ones that do not extend to the lumen are the cells that are proceeding through constant mitosis to provide replacement cells for those surface luminal cells that have been lost or damaged.

Pseudostratified columnar epithelium lines the respiratory passage, vas deferens, and the epididymis. In the respiratory track, the pseudostratified ciliated columnar epithelia are responsible for moving mucous from the respiratory tract up towards the epiglottis. In the epididymis of the male reproductive track, the pseudostratified columnar epithelia with stereocilia (large non-motile microvilli) are formed to secrete and reabsorb epididymal fluids. Epididymal fluids are involved in the final maturation stage of sperm where the haploid sex cell obtains the ability to fertilize an ovum, as well as, gaining the ability of movement of its flagellum. In the vas deferens of the male reproductive track, the ciliated pseudostratified epithelia are used to propel sperm from the epididymis to the ejaculatory ducts (Figure 8).

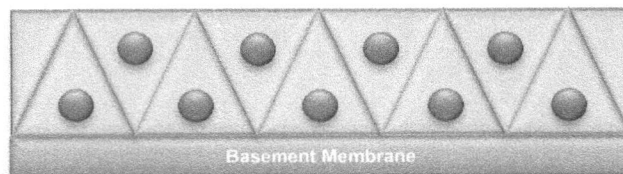

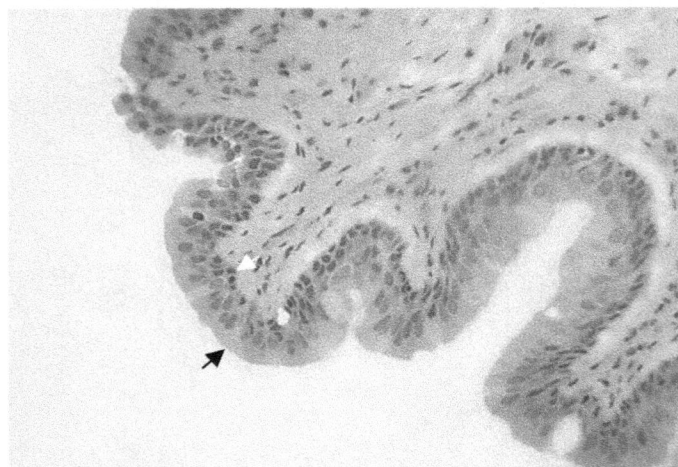

Figure 8: Pseudostratified ciliated columnar epithelium Please note the graphic interpretation is shown above and the actual photomicrogaph is located below 400x. The white arrow indicates the pseudostratified columnar cells and the black arrow indicates the cilia located on the apical surface of these cells

Stratified Squamous Epithelium

Stratified squamous epithelium consists of a variable number of cell layers that exhibit a transition from a cuboidal basal layer (the cell layer on top of the basement membrane) to flattened surface layers. The basal layer of cells is known to undergo continuous cellular division or mitosis. As these newly formed cells divide and are driven up towards the surface, many of them will become water proofed (keratinized) and dehydrated, which in turn reshapes the cells to become flat and thin. Eventually, these cells will replace the previous surface cells that have been damaged or lost due to abrasion.

Stratified squamous epithelium is designed to withstand abrasions, protect the tissues below from chemicals, and prevent excessive water gain or loss etc. Put simply, stratified squamous epithelium forms a protective outer covering of the human body.

Stratified squamous epithelia are further classified into either keratinized or non-keratinized depending upon the deposition of keratin, a non-soluble protein (a type of intermediate filaments) that are utilized to 'water-proof' the tissue. For example, the keratinized stratified squamous epithelia include the skin (epidermis), which consists of living cells in its deepest layer and dead cells on its upper-most surface (Figure 18). Non-keratinized epithelia, on the other hand, line the oral cavity, the vagina and anal canals. These epithelia are covered with fluids and consist of living cells in all layers (Figure 9).

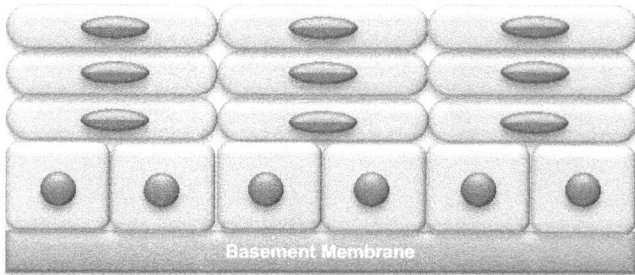

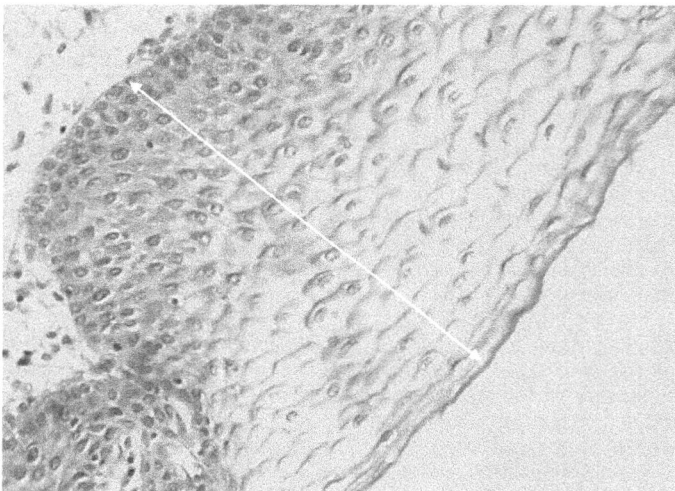

Figure 9: Stratified squamous epithelium. Please note the graphic interpretation is shown above and the actual photomicrogaph is located below 400x

Stratified Cuboidal Epithelium

Stratified cuboidal epithelium is a rare epithelial tissue that is generally constructed of two, or at the most, three layers of cuboidal cells. This type of epithelium is generally found in lining larger excretory ducts or exocrine glands, such as the salivary glands, mammary glands, sweat glands, ovarian follicles, and various ducts in the pancreas. Stratified cuboidal epithelium is designed for protection, secretion and is able to perform some absorption (Figure 10).

Stratified Columnar Epithelium

Stratified columnar epithelium is an atypical and rare type of epithelial tissue that is formed for protection, secretion, and may perform some absorption. This epithelium usually consists of several layers of cells, with the deeper layer cuboidal in shape while only the surface cells appears to be columnar.

Stratified columnar epithelium is found lining the salivary duct (also known as interlobular ducts) and is closely associated with goblet cells. This epithelium is also found in portions of the male urethra and the vas deferens. In females, this epithelium is located in the glandular ducts of the mammary gland and the

uterus. In both sexes, this epithelium is found in parts of the larynx and the ocular conjunctiva of the eye (Figure 11).

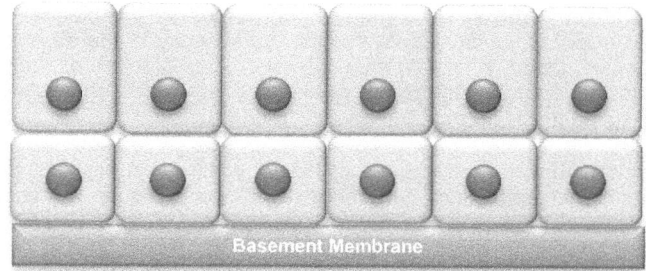

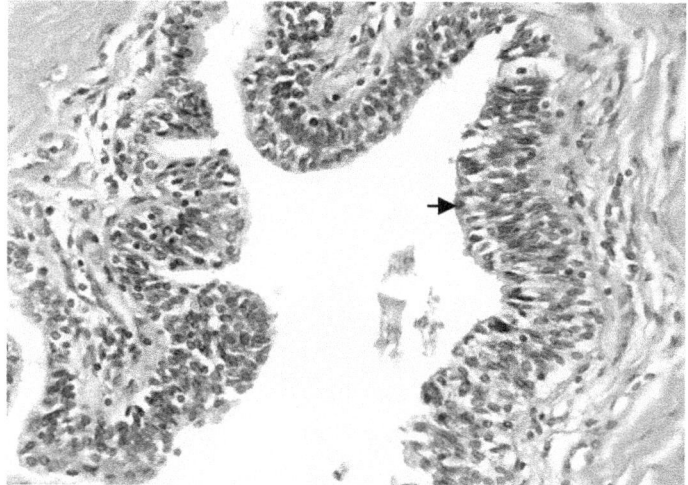

Figure 11: Stratified columnar epithelium. Please note the graphic interpretation is shown above and the actual photomicrogaph is located below 400X

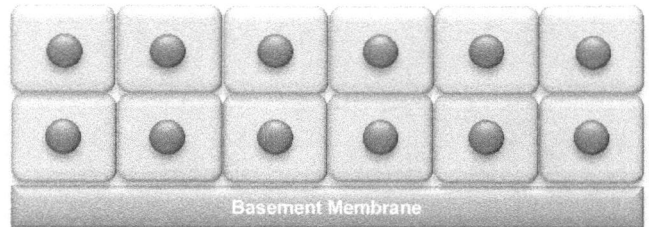

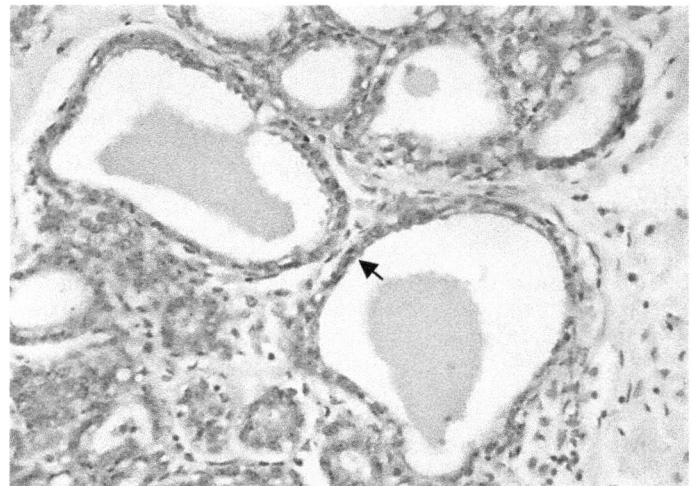

Figure 10: Stratified Cuboidal Epithelium. Please note the graphic interpretation is shown above and the actual photomicrogaph is located below 400X

Transitional Epithelium

Transitional epithelium is a form of stratified epithelium that is found in the urinary tract of mammals. Th cells found within this epithelial tissue is packed with elastic fibers, which allow it to withstand stretching. For example, the elastic fibers are formed to withstand the stretching/distending forces as the bladder or the ureters are being filled with urine. Additionally, numerous tight junctions could be found on the lateral domain of these epithelial cells, which makes the transitional epithelium watertight, as well as, being impervious to the toxicity of urine.

Depending on the volume of urine within the bladder the transitional epithelium can range in shape between stratified cuboidal or low columnar in its unstretched state to stratified squamous in its stretched state. Transitional epithelia can also be found in the ureters, ducts of the prostate, and the superior urethra (Figure 12).

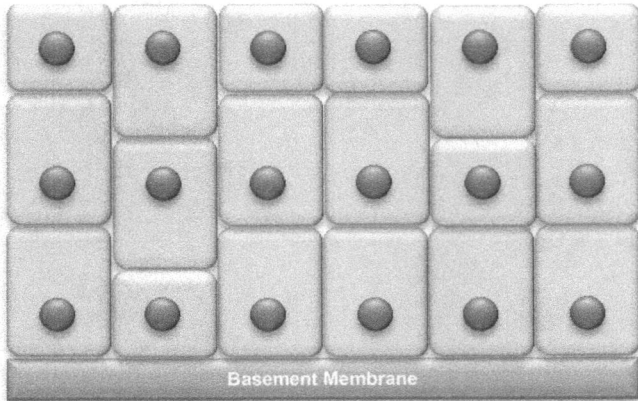

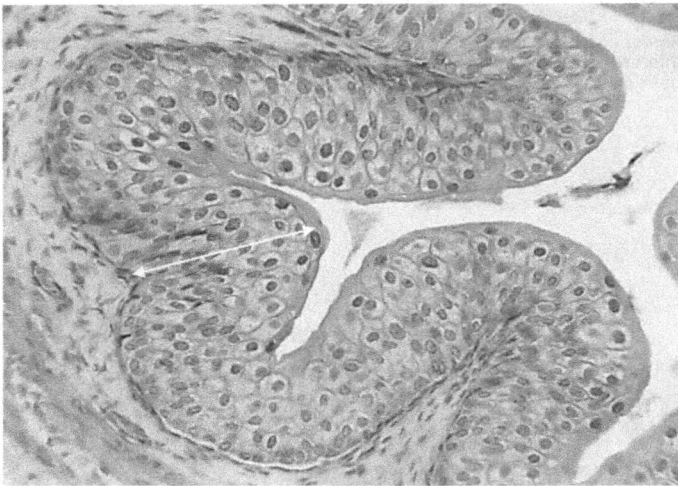

Figure 12: Transitional epithelium. Please note the graphic interpretation is shown above and the actual photomicrogaph is located below 400X

MUSCLES

Muscles, by definition, are specialized tissues that are designed to produce movements of the body, as well as, body parts. In mammals, there are three distinct types of muscle tissues: skeletal, cardiac and smooth.

Skeletal Muscle

Skeletal muscles are the most abundant muscles in the mammalian body and are a type of striated muscle. Striated muscles by definition are contractile tissues that possess alternating light and dark bands when view under a microscope.

In most areas of the human body, skeletal muscles are voluntary muscles and are subjected to conscious control. An example of involuntary skeletal muscle are the stapedius and tensor tympani muscles found in the middle ear.

Most often, the skeletal muscles are attached to two bones via tendons and its contraction will bring the bones closer together thereby providing bodily movement. On occasion, skeletal muscles are connected to one another via aponeurosis, a flat sheet-like connective tissue. This form of muscle connection is commonly seen on the human skull where the contraction of the muscle will play a part in speech formation and facial expressions (Figure 13).

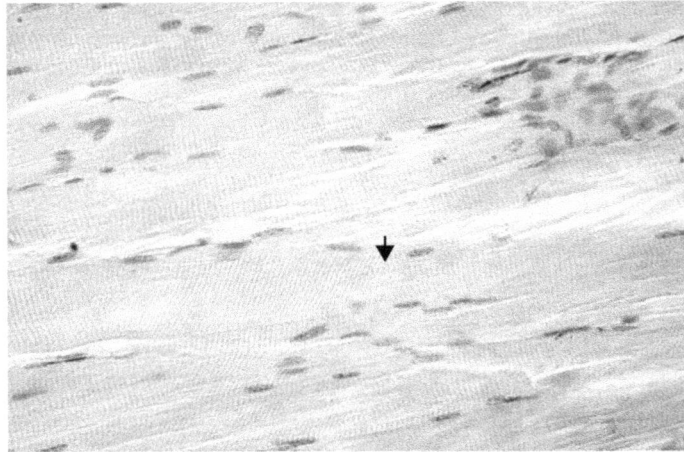

Figure 13: Photomicrograph of skeletal muscle 1000x. Please note that the striations are indicated by a black arrow

Cardiac Muscle

Cardiac muscles are striated muscles that also possess alternating light and dark bands when viewed under the microscope. Cardiac muscles are responsible for pumping blood throughout the body. However, unlike skeletal muscles, the cardiac muscle is an involuntary muscle and is not subjected to conscious control.

Cardiac muscle cells are interconnected end-to-end via the intercalated disks. Intercalated disks are the locations where numerous cellular junctions exist. For example, gap junctions which allow intercellular communications and facia adherens, a type of desmosome that prevents seepage of blood into the surrounding tissue, are also frequently found as an essential component of the intercalated disk (Figure 14).

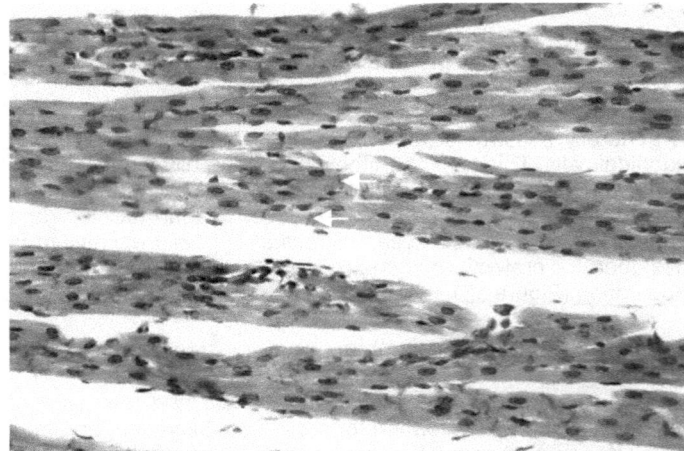

Figure 14: Photomicrograph of cardiac muscle 400x. Please note the intercalated disks are indicated by black arrows

Smooth Muscle

Smooth muscles are found in the internal organs, as well as, the blood vessels of the mammalian body. This muscle type does not contain striations and is involuntary, which is defined as a type of muscle that is not subjected to conscious control (Figure 15).

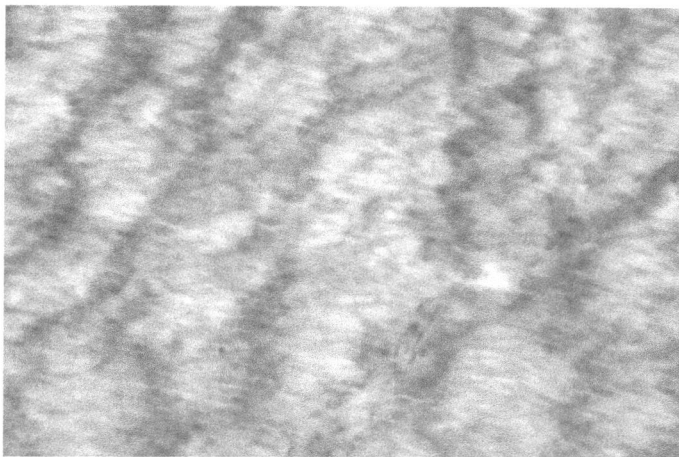
Figure 15: Photomicrograph of smooth muscle (no striations) 400x

CONNECTIVE TISSUE

Connective tissue is one of the four major tissue types and is the most abundant and widely distributed tissue in the human body. The connective tissue is responsible for providing cohesion and internal support for various tissues and organs. Although there are numerous different types of connective tissues, they all share some similar characteristics. For example, most connective tissues are composed of the extracellular secretions from cells, such as epithelial cells and fibroblasts (Figure 16).

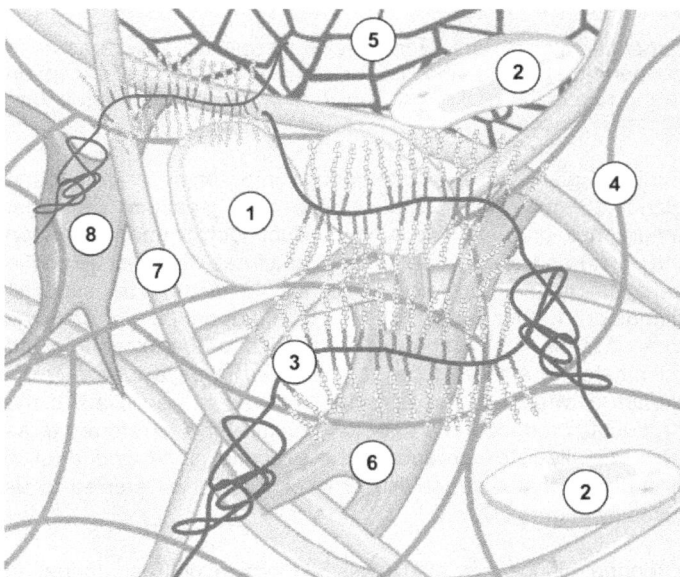
Figure 16: Illustration of connective tissue and its extracellular matrix. The graphics demonstrates: ① Adipocytes, ② macrophage, ③ glycosaminoglycans, ④ elastic fibers, ⑤ reticular fibers, ⑥ Fibroblasts, ⑦ collagen fibers, ⑧ mesenchymal cell

Extracellular Matrix

The extracellular matrix, also known as cell coat, is composed of protein fibers such as collagen, reticular fibers, and elastic fibers in varying percentage of concentrations dependent on the type of connective tissue. Collagen, for instance, is the most abundant glycoprotein in the animal kingdom. It is estimated that this glycoprotein makes up approximately 30% of the dry weight of the human body. Similar percentages can also be found with other animals. Collagen is an insoluble (hydrophobic) fibrous glycoprotein that is responsible for providing structural strength in the extracellular matrix. Collagen is known to be secreted by both fibroblasts (Figure 17) and epithelial cells.

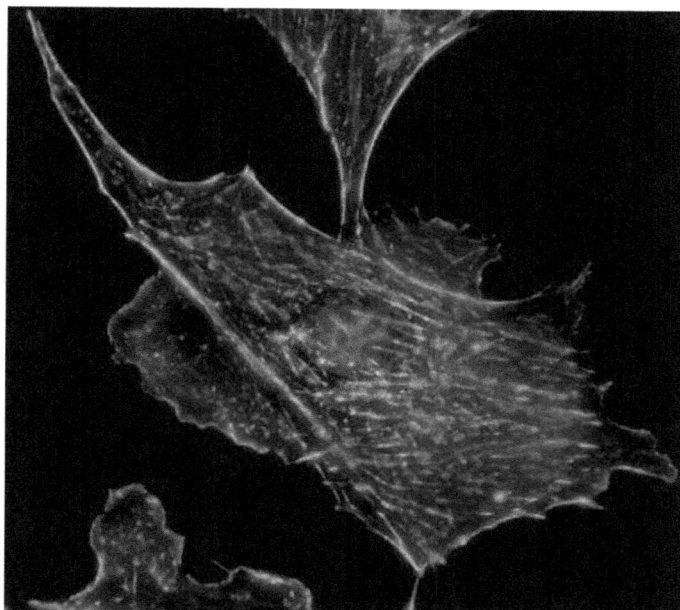
Figure 17: Photomicrograph of a fibroblast cell. Fibroblasts secrete collagen proteins that are used to maintain a structural framework for many tissues 1172x

In general, collagen molecules are assembled into collagen fibrils approximately 300 nanometers (nm) in length. It takes three collagen fibrils coiled together in a rope-like fashion to create a single collage fiber. Two of the collage fibrils will have identical molecular formula, referred to as α1 chains, while the third collage fibrils will have a slightly different molecular formula, which is labelled α2 chain. It has been calculated that each collage fibril chain is constructed out of approximately 1050 amino acids wound together in a triple helix format (Figure 18).

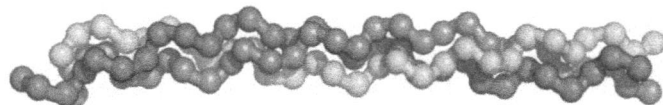
Collagen fibril
Figure 18: 3D graphic illustration of a collagen fibril. Please note that there are two α1 chains and one α2 chain

It was surprising to discover that all collagen fiber types have the same three-stranded helical structure. Presently, there are at least 21 types of collagen known. Nonetheless, most of the collage types found in the human body (~80-90%) consist of types I, II, and III (Table I).

Elastin is a highly hydrophobic protein composed of long sequences of glycine and proline amino acids, similar to that of collagen, but this protein is not glycosylated (i.e. not a glycoprotein). In addition to glycine and proline, elastin also contains alanine and lysine rich segments. Alanine provides this molecule with elasticity, while lysine forms cross-links with adjacent molecules. Elastin proteins, along with a structural glycoprotein called fibrillin, create elastin fibers. There are two forms of fibrillin (fibrillin 1 and fibrillin 2) and both are necessary to form the scaffolding (crosslink bonding) for the deposition of elastin protein, essential in the creation of elastic fibers. Elastin fibers provide resilience (or elasticity) for connective tissue. Both elastin and fibrillin are secreted by fibroblasts and epithelial cells.

The ground substances of the extracellular matrix also contain glycosaminoglycans (GAGs). GAGs are composed of long unbranched polysaccharides consisting of repeating disaccharide molecules. GAGs are assembled as proteoglycans and serve as a barrier, or filter, through which all substances entering and exiting the cells must pass. In addition, these proteoglycans possess a negative charge which attracts positively charged ions (cations) like sodium (Na^+). As a rule of thumb, where Na^+ goes, water will follow. For this reason, the extracellular matrix is composed mostly of water, providing resistance to compression, and a diffusion/osmosis friendly medium for the exchange of waste materials, nutrients and gases.

Table 1: List of collagen fiber types and the location where they are commonly found

Type	Location
I	Skin, tendon, bone, ligaments, dentin, interstitial tissues
II	Hyaline cartilage, vitreous humor
III	Reticular tissue, skin, muscle, blood vessels, bone
IV	Fetal tissues
V	Interstitial tissues, bone
VI	Cartilage, vitreous humor
VII	Basal laminas
VIII	Cornea
IX	Skin, bone, cartilage, tendons, and ligaments
X	Hyaline cartilage
XI	All cartilages
XII	Skin, tendon, bone, ligaments, dentin, interstitial tissues
XIII	Fetal bone, cartilage, skin, striated muscle, intestines
XIV	Bone and bone marrow
XV	Basement membrane
XVI	In association with Type I and II collagen
XVII	Hemidesmosomes, basement membrane
XVIII	Retinal structure and fetal neural tube closure
XIX	In association with Type I and II collagen
XX	Bone
XXI	Bone

Marfan Syndrome

Marfan syndrome is a human genetic mutation that is caused by the malformation of the connective tissues. It has been estimated that one in every five thousand individuals in the United States suffers from this condition. It has been hypothesized by many scientists that President Abraham Lincoln suffered from this disorder but the validity of this assumption has yet been verified.

Marfan syndrome can be either an autosomal dominant mutation or a spontaneous mutation that occurs on the fibrillin 1 (FBN1) gene, which encodes the connective tissue protein fibrillin 1. The FBN1 gene is located on chromosome 15, q arm and the 21st band. It is estimated that there is approximately a 25% chance that an individual may develop Marfan syndrome through spontaneous mutation. Fibrillin is a structural glycoprotein secreted by fibroblast and epithelial cells and is essential in the formation of elastic fibers.

Marfan syndrome displays a wide-range of effects on the human body, but most often it affects the connective tissue of the heart, blood vessels, eyes, bones, lungs, and meninges, which are the connective tissue covering of the spinal cord. The telltale sign of Marfan syndrome is its influence on the long bones which make the individual tall and lanky. Additional phenotypes of this condition include long arms, legs, fingers, toes, flat feet and rather flexible joints. Afflicted individuals may also suffer from scoliosis (i.e. S-shaped curvature of the spine) (Figure 19), pectus excavatum (i.e. a sunken rib cage) or pectus carinatum (i.e. a protruding rib cage). In addition, individuals suffering from this condition are subjected to lens dislocation of the eyes.

The most serious and life threatening complications of Marfan syndrome is its effect on the aorta (i.e. the main artery that originates from the left ventricle and extends down to the abdominal region), where it becomes structurally weak and prone to develop aortic aneurysm. Presently there is no cure for this genetic condition. Treating the symptoms of the disease is the only option.

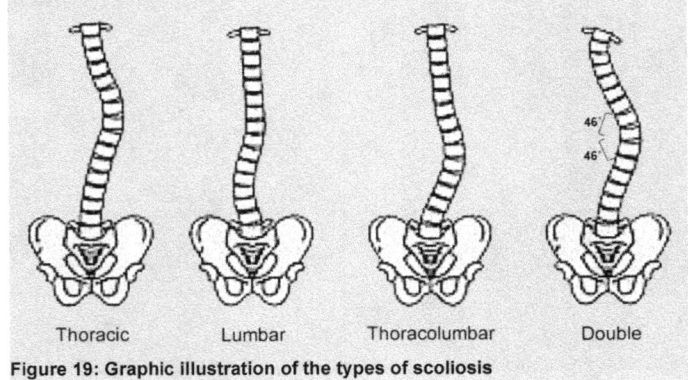

Figure 19: Graphic illustration of the types of scoliosis

Thoracic   Lumbar   Thoracolumbar   Double

GLANDS

Glands are secretory structures that are classified as either endocrine or exocrine depending on where they release their products. Endocrine glands secrete intercellular ligands (also known as hormones) directly into the blood stream, while exocrine glands secrete their products into a duct which eventually empties onto tissue surfaces or within cavities. The topic of endocrine glands will be covered in Chapter One, Principles of Anatomy and Physiology II. In this section attention will be focused on the exocrine glands.

Exocrine glands are generally multicellular structures with supporting epithelial and connective tissues. The only exception to the rule is the goblet cell, which are single celled glands that secretes mucous.

Exocrine glands are subcategorized into simple or compound glands. Simple glands are composed of glandular cells that empty their products into a single duct. Compound glands, on the other hand, are multi-branched structures where the exocrine cells empty their products into multiple ducts. The exocrine glands are further subcategorized based upon the type of ducts they possess. For example, if the product of the exocrine gland culminates in a small tube, it is referred to as a tubule. If the tubule is twisted at the end it is referred to as coiled. If the product is produced in a sac-like structure, this is referred to as acini. Finally if the product of the exocrine gland wound-up in a structure that is grape-like or in clusters, it is referred to as alveolar (Figure 20).

Exocrine glands are also classified based on the manner in which they secrete their products. For example, merocrine glands secrete their products either through active transport or via exocytosis. There are two types of cells belonging to the merocrine category: serous and mucous. For example, serous cells secrete solutions that contain water and a high concentration of enzymes. Examples of serous cells are the parotid glands (a type of salivary gland) in the oral cavity. Alternatively, the mucous cells secrete solutions that contain a mixture of water and high concentrations of a glycoprotein called mucin. Please note that mucous is made when water mixes with mucin. Submandibular and sublingual salivary glands, sweat glands, as well as, the exocrine portion of the pancreas are examples of merocrine glands (Figure 21).

Apocrine glands lose parts of their cells during secretion; therefore their released products possess both secretory products, as well as, cellular fragments. Mammary glands and the ceruminous glands (produces ear wax) are examples of apocrine glands

(Figure 22).

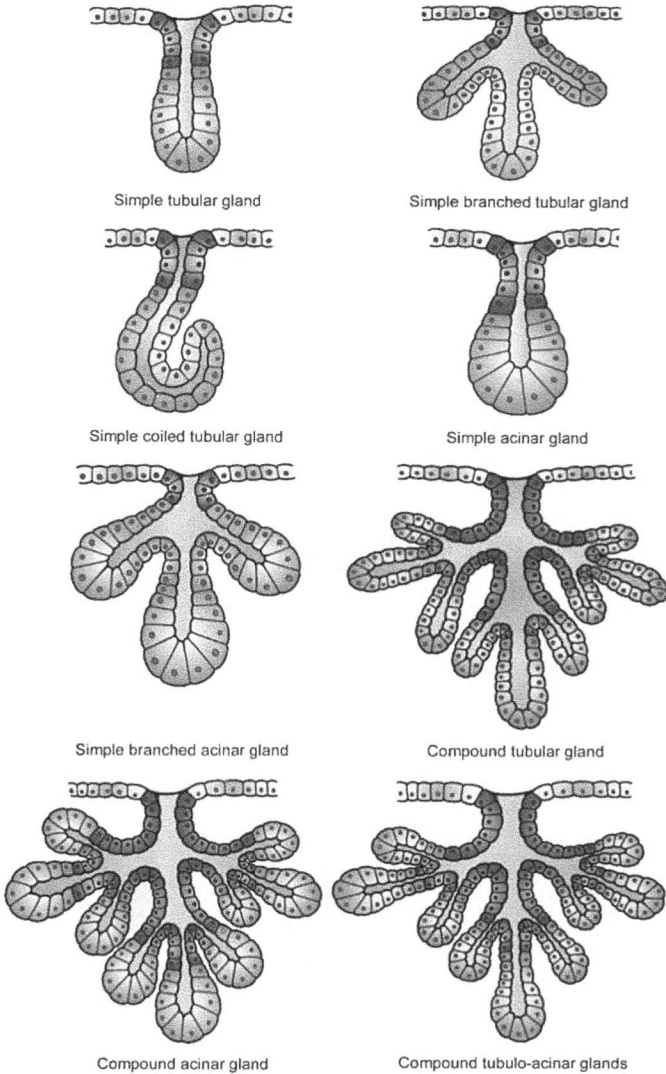

Simple tubular gland

Simple branched tubular gland

Simple coiled tubular gland

Simple acinar gland

Simple branched acinar gland

Compound tubular gland

Compound acinar gland

Compound tubulo-acinar glands

Figure 20: Illustrations of the different type of exocrine glands that exists in the human body. Please note that the goblet cell, a single cell exocrine gland, is not shown in the graphics

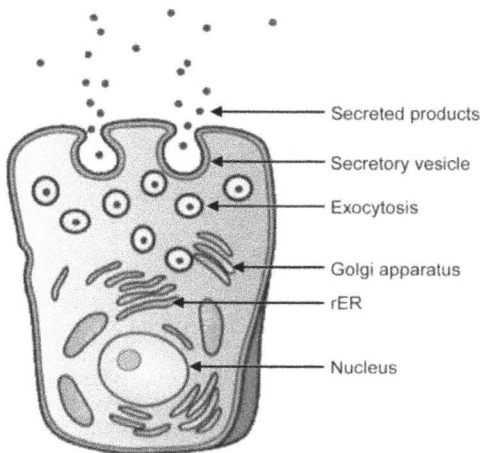

Secreted products

Secretory vesicle

Exocytosis

Golgi apparatus

rER

Nucleus

Figure 21: Illustration of a merocrine gland. Please note that the secretion is accomplished through exocytosis

Finally, holocrine glands possess cells that disintegrate completely during secretion. This form of secretion is a type of apoptosis. Apoptosis is defined as programmed cell death.

Example of this gland is the oily lubricants called sebum secreted by the sebaceous gland of the skin (Figure 23).

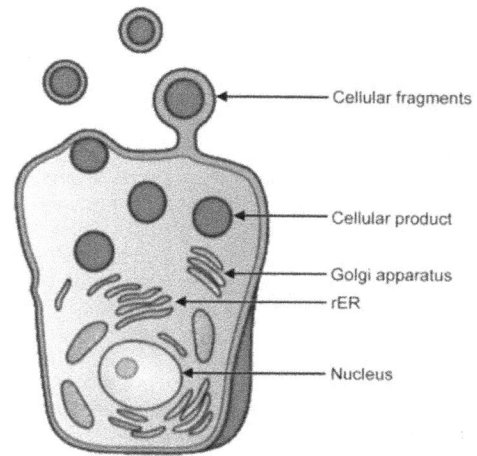

Cellular fragments

Cellular product

Golgi apparatus

rER

Nucleus

Figure 22: Illustration of an apocrine gland. Please note that fragments of the cell are secreted along with the cellular products

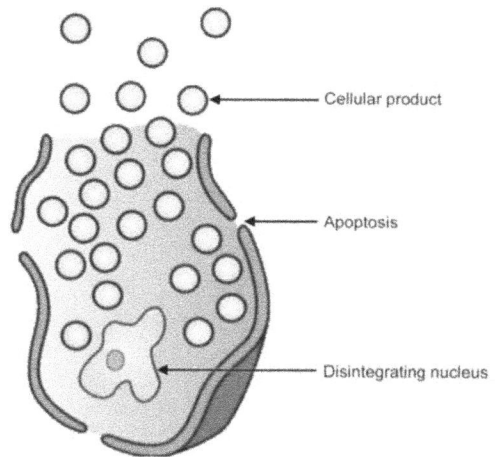

Cellular product

Apoptosis

Disintegrating nucleus

Figure 23: Illustration of a holocrine gland. Please note that the entire cell disintegrates (apoptosis) as the cellular products are being released

CONNECTIVE TISSUE CELLS

Connective tissue consists of two generalized grouping of cells. The first are referred to as fixed cells, which are present in specific areas of the human body in stable numbers. These fixed cells include fibroblasts, pericytes and mast cells. Wondering cells makes up the second connective tissue cell type. These cells are present in variable numbers depending upon the specific demands of the body. Wandering cells include leukocytes, such as macrophages, lymphocytes, neutrophils and eosinophils to name a few.

Fixed Cells

Fibroblasts (mature fibroblasts are called fibrocytes) are the most common fixed cells in connective tissue (Figure 17). Fibroblasts produce collagenous and elastic fibers, and secrete these into the matrix of connective tissues. Other fixed cells, such as pericytes, are found near the endothelial cells of the capillaries and venules. These cells are very similar to smooth muscle cells and share the same basal lamina with the endothelial cells of the capillaries and/or venules. Pericytes have the capability to differentiate into smooth muscle cells and/or endothelial cells after injury.

Mast cells are frequently found throughout the body and play an essential role in both inflammatory and allergenic responses. Mast cells contain numerous granules when stained. These granules contain heparin, histamine, serotonin, thromboxane, prosta-glandin, leukotriene, and cytokines.

Adipocytes, commonly known as fat cells, are filled with lipids and function as energy storage cells for the mammalian body. Adipocytes form the adipose tissue and are commonly found in the hypodermis and surround various organs, such as the kidneys. In addition to storing energy, the adipose tissue also provides physical protection for the internal organs, as well as, providing thermal insulation for the organism. There are two types of adipose tissue: white adipose tissue composed of unilocular fat cells (white fat cells), and brown adipose tissue composed of multilocular fat cells (brown fat cells). The unilocular fat cells (approximately 0.1 mm in diameter) contain a single lipid droplet surrounded by a layer of cytoplasm. Generally, the size of the fat droplet is so large that the nucleus and other organelles of the cells are flattened and located on the periphery of the cell mass and are diminished, since the primary function of the adipocytes is fat storage. On average, a human adult possesses approximately 30 billion fat cells (Figure 24).

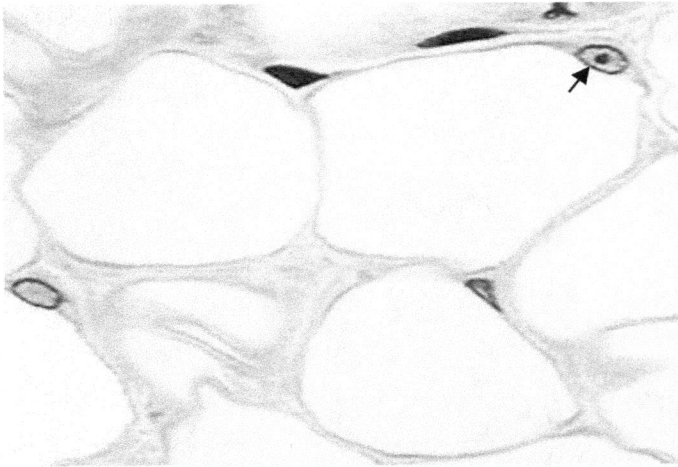

Figure 24: Unilocular adipocytes found within an adipose tissue. Please note that the nucleus (arrow) and various organelles are squeezed to the corner of the cell 1000x

Unlike the unilocular fat cells, the multilocular fat cells possess numerous lipid droplets scattered throughout its cytoplasm. The nucleus and organelles are located throughout the cytoplasm and are not squeezed by a single fat droplet to the periphery of the cell. Multilocular fat cells exist in abundance in infants and toddlers and are used to generate heat. Nonetheless, by adulthood, the multilocular fat cells are replaced by unilocular fat cells (Figure 25).

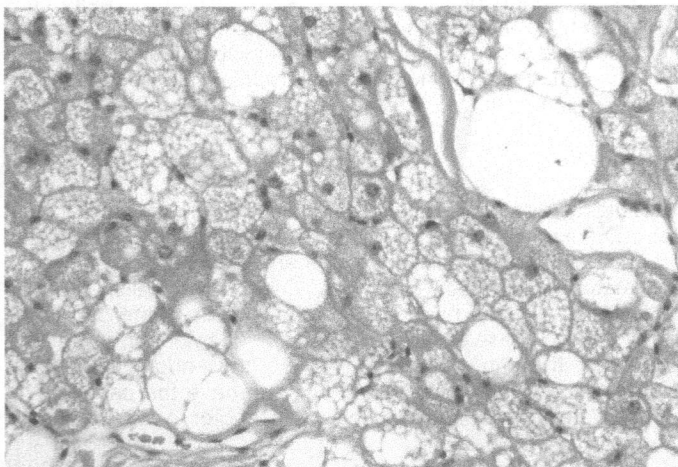

Figure 25: Photomicrograph of multilocular adipocytes. Please note that multilocular adipocytes are found in young children before being replaced by unilocular adipocytes when reaching adulthood 400x

Wandering Cells

Erythrocytes, also known as red blood cells, are anucleated (contain no nucleus) and biconcave in shape. Being anucleated, erythrocytes are unable to repair themselves when damaged and have to be replaced approximately every 120 days. The biconcavity of the erythrocytes provides more surface area for gas exchange.

Erythrocytes consist mainly of a complex globin protein and iron binding heme structure called hemoglobin (Hgb). The heme group allows oxygen molecules to form a temporary bond in the lungs which is then released down its concentration gradient in the tissues throughout the body. On its return, the globular protein portion of the erythrocyte can carry about 2% of carbon dioxide back to the lungs. The remaining carbon dioxide remains either in its original gaseous state, or is converted into carbonic acid (H_2CO_3), a human blood buffer. The formation of carbonic acid is accomplished through the enzymatic actions of carbonic anhydrase; a membrane-bound enzyme in the erythrocytes (Figure 26).

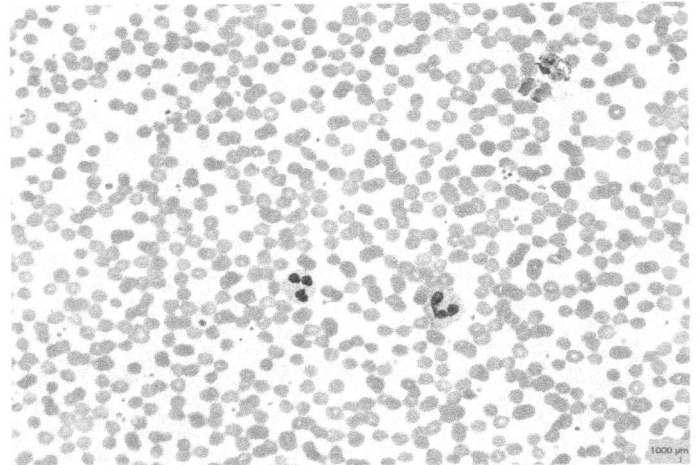

Figure 26: Photomicrograph of erythrocytes surrouding two neutrophils 1000x

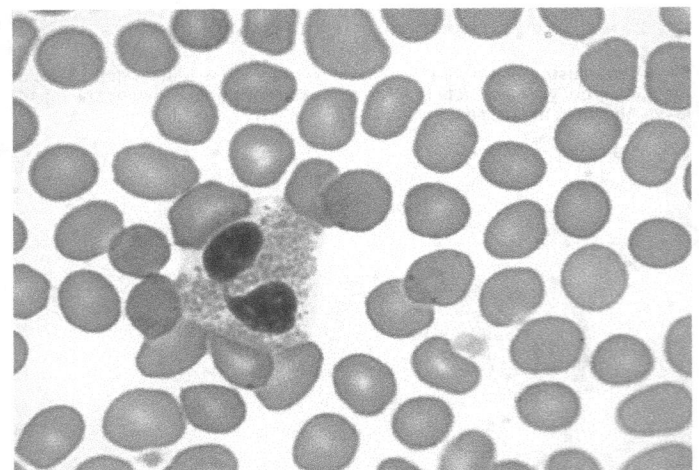

Figure 27: Photomicrograph of an eosinophil 1000x

Leukocytes, also known as white blood cells, are commonly found in connective tissue. They migrate from the blood vessels to the peripheral tissues of the body after maturation, especially to sites of injury or inflammation. There are two major groups of leukocytes: granulocytes and the agranulocytes. Granulocytes, when stained either by acid or Wright Staining solution, reveal small granules, which are actually vesicles, within their cytoplasm. Granulocytes include neutrophils, eosinophils and basophils (Figure 27). Agranulocytes, on the other hand, appear to contain no granules when stained. Examples of agranulocytes include monocytes and lymphocytes.

Monocytes will remain in circulation for 8 to 12 hours before leaving the blood vessels and migrating to sites of infection. Once it leaves circulation, this agranulocyte will differentiate into a macrophage or other macrophage related cells (e.g. dendritic cell, Kuffer cell etc.). For example, macrophages exist between transient and fixed cell mode. A macrophage is a mononuclear cell that is essential in the body's immune system and performs numerous functions within the human body. For example, macrophages remove damaged cells, necrotic cellular debris, as well as, pathogens. In addition, they also stimulate immunological cells and induce an immune response by secreting various cytokines (Figure 28).

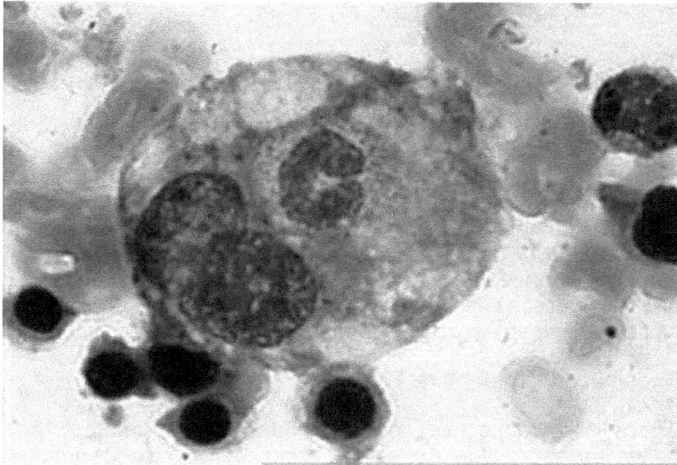
Figure 28: Photomicrograph of a macrophage taken at 1125x phagocytizing a monocyte and a lymphocyte

Another example of an agranulocyte are plasma cells. Plasma cells are mature B lymphocytes that are responsible for antibody production. Plasma cells are found in sites of chronic inflammation and sites with high risks of invasion by bacteria or foreign proteins (such as the lamina propria of the intestinal and respiratory tracts) (Figure 29).

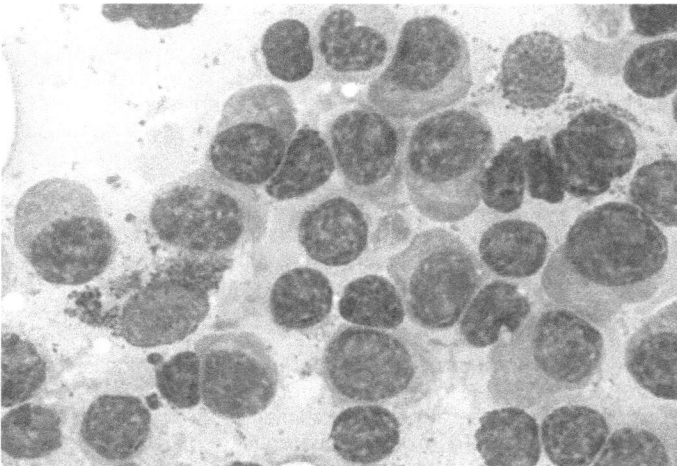
Figure 29: Photomicrograph of plasma cells 1000x

TYPES OF CONNECTIVE TISSUE

Connective tissue can be broadly separated into embryonic, general, and specialized connective tissues.

Embryonic Connective Tissues

Embryonic connective tissues can be divided into mucoid and mesenchymal connective tissues. Mucoid connective tissue is found during fetal development and is composed of ground substance with some protein fibers (collagen and reticular fibers), as well as primitive fibroblast cells. For example, mucoid connective tissue is characteristically found in the umbilical cord. This gelatinous connective tissue, also known as Wharton's jelly, supports embryonic blood vessels.

Mesenchymal connective tissues are constructed out of mesenchymal cells surrounded by an abundance of extracellular matrix, composed mainly of loosely constructed reticular fibers. Mesenchymal cells are generally oval or elongated shaped cells containing very little cytoplasm, with thin cytoplasmic processes protruding from their plasma membrane. Mesenchyme cells are embryonic stem cells found in the middle germ layer of the embryo, also known as the mesoderm. Mesoderm is responsible for developing into various types of connective tissue during fetal growth and continues to exist into adulthood, where they retain their capacity to differentiate into other connective tissue cells in response to injury.

General Connective Tissue

General connective tissue is subdivided into two subgroups: loose connective tissue and dense connective tissue. Dense connective tissue is further divided into dense regular and dense irregular connective tissues. General connective tissue is composed of fibroblasts or fibrocytes, when it reaches its adult or less active form. These fibroblasts are responsible for secreting extracellular matrix composed of collagen fibers, elastin fibers as well as reticular fibers.

Loose connective tissue includes loose areolar connective tissue which is found in the basement membrane of epithelial tissue, as well as, forming adventitia that surrounds and protects blood vessels and nerves. Loose connective tissue is composed of an extracellular matrix of loosely organized collagen and elastin fibers (Figure 30).

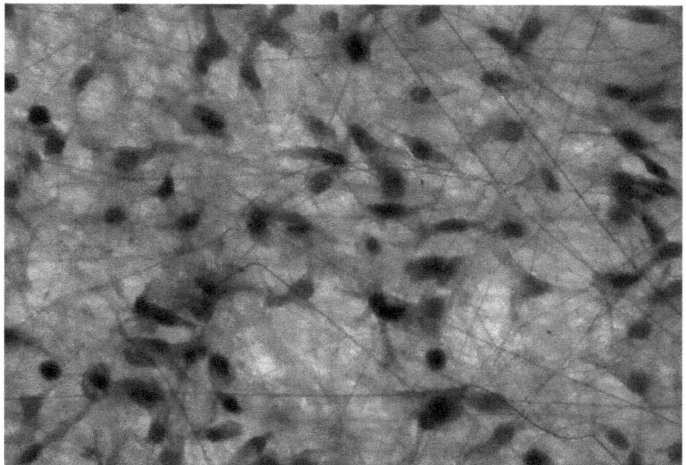
Figure 30: Photomicrograph of loose areolar connective tissue 400x

Adipose tissue is composed of fat cells also known as adipocytes. Adipose tissues provide the human body a location for storing fuel and offer insulation from the external elements. There are two types of adipose tissue: the white adipose tissue composed of unilocular fat cells (white fat cells) and the brown adipose tissue composed of multilocular fat cells (brown fat cells). The white fat cells compose adult adipose tissue while the brown fat cells compose the majority of fatty tissue found in infants. The brown fats of infants generate heat and provide the necessary energy for survival (Figures 24 and 25).

Reticular tissue is made out of a specialized type of fibroblasts called reticular cells. These reticular cells secrete an extracellular matrix composed of reticular fibers, also known as collage III fibers. Reticular tissues are commonly found around the liver and kidneys. They also form the internal structures that make up the bone marrow, spleen, thymus, and lymph nodes (Figure 31).

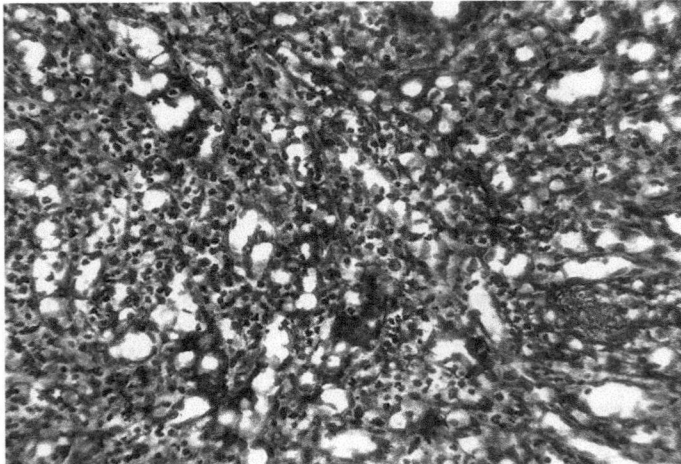

Figure 31: Photomicrograph of reticular tissue 400x

Dense connective tissues are divided into four subtypes depending upon the composition of their extracellular matrix. For example, dense regular connective tissue is composed of tightly packed and well organized mass of collagenous fibers and a fine network of elastic fibers secreted by fibroblast cells. This tissue is exceptionally strong and forms ligaments, aponeuroses, and tendons. Examples include connective tissue structures, such as tendons and aponeuroses, that allow muscles to be bounded to bones or to other muscles respectively (Figure 32).

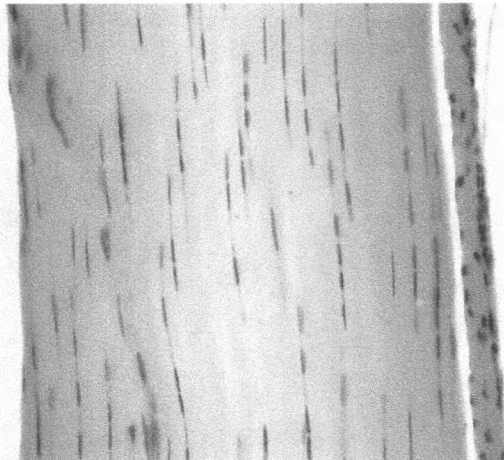

Figure 32: Photomicrograph of dense regular connective tissue 400x

Dense irregular connective tissue is mainly composed of collagen fibers with some elastic fibers. Again, these fibers are secreted by fibroblast cells. Dense irregular connective tissues are found in most fibrous connective tissue capsules of the body, the submucosa of the digestive track, as well as, the dermal layer (dermis) of the skin (Figure 33).

Dense regular elastic connective tissue is composed of parallel bundles of collagen fibers and large amounts of elastic fibers. This type of connective tissue is elastic and is capable of being stretched and returns to its original shape when the applied force is removed. Examples of structures that possess dense regular elastic connective tissue include the vocal folds, which allow an individual to modulate the flow of air being expelled from the lungs during phonation. In addition, the elastic ligaments found in the vertebrae are also composed of dense regular elastic connective tissue. These ligaments allow for limited amounts of back bending and stretching.

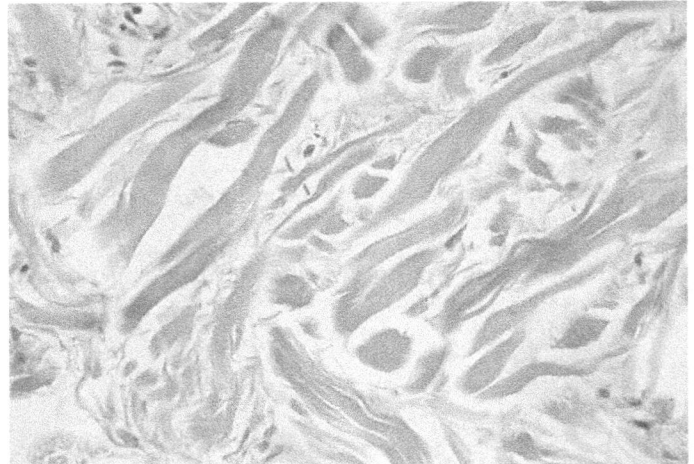

Figure 33: Photomicrograph of dense irregular connective tissue 400x

Dense irregular elastic connective tissue contains collagen fibers and large amounts of elastic fibers arranged in a random mesh-like orientation. This type of connective tissue is elastic and capable of being stretched in all directions but returns to its original shape when the applied force is removed. All elastic arteries possess this connective tissue. For example, the internal elastic lamina of the aorta is composed of this type of tissue.

Specialized Connective Tissue

Specialized connective tissue includes blood, cartilage and bones. Blood is a type of special fluid connective tissue composed of red blood cells (erythrocytes), white blood cells (leukocytes), and cellular fragments (platelets), suspended within a colloidal solution (a solution in which a material is evenly suspended in a liquid) called plasma. Plasma constitutes approximately 55% of total blood volume within the human body and is mainly composed of water (~91%). Dissolved within the fluid of plasma are hormones, sugars, ions (e.g. Na^+, Cl^-, Ca^{++} etc.), and dissolved gases such as carbon dioxide (CO_2). In addition, various proteins such as albumin, globulin, and fibrinogen are also found within the colloidal solution.

Blood is the primary method of transporting gases, nutrients and hormones to all tissues within the body, as well as removing wastes from the same tissues. Blood is an essential part of the circulatory system. Erythrocytes are blood cells specifically designed for oxygen transport.

Leukocytes, on the other hand, are responsible for fighting off infections which may be caused by bacteria or viruses. Leukocytes are classified by the way they stain and their appearance under the microscope. Please see the section entitled Wandering Cells for more information (Figures 26 to 29).

Cartilage is composed of cartilage producing cells called chondrocytes. These cells are located in spaces called lacunae surrounded by secreted extracellular matrix composed of collagen, elastin and ground substance rich in proteoglycan. While collagen fibers provide for resiliency and strength and elastic fibers allow it to be flexible, the most important component of the cartilage are the various proteoglycans found within the matrix. The main proteoglycan found in cartilage is called aggrecan which is a large molecule containing sizable numbers of chondroitin sulphates attached to a protein core. Like all forms of glycosaminoglycans (GAGs), chondroitin sulphate possesses a negative charge and attracts sodium cations, which in turn, attracts water. The ability of the proteoglycan to retain water allows cartilage to spring back to its original form after being compressed (Figure 34). There are three type of cartilages in the human body.

Hyaline cartilage is the most common and is composed of fine collagenous fibers. This cartilage is generally found at the ends of bones, providing protection against abrasion and pressure, and is also seen forming the soft parts of the nose, the supporting rings of the trachea and the bronchi found within the respiratory passages (Figure 34).

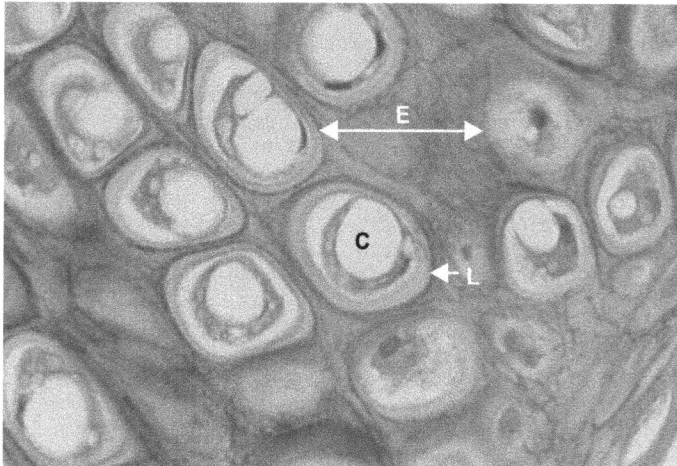

Figure 34: Photomicrograph of hyaline cartilage at 1000x magnification. Please note that (C) indicates the location of a chondrocyte, (L) indicates the encircling lacuna while (E) indicates the extracellular matrix of the cartilage

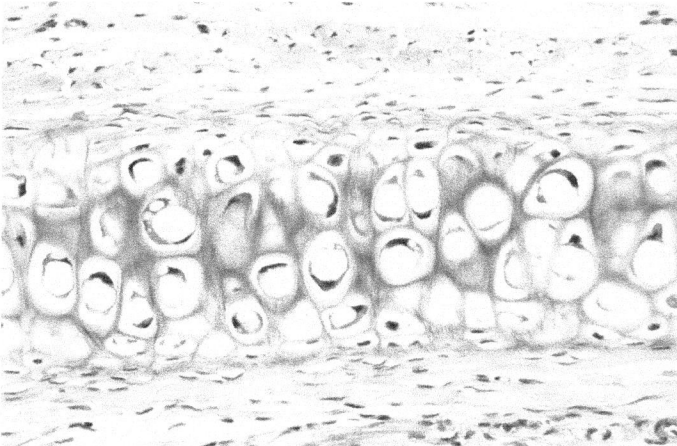

Figure 35: Photomicrograph of elastic cartilage 400x

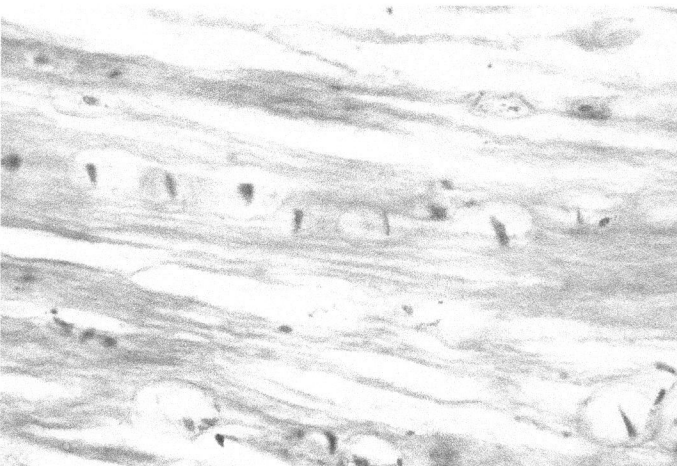

Figure 36: Photomicrograph of fibrocartilage 400x

Elastic cartilage is made up of a network of elastic fibers,

collagen, and proteoglycan. Elastic cartilage is responsible for stretching and recoiling to its original shape when the tension is removed. This type of cartilage maintains openings to structures and tubes. Examples of this connective tissue are found forming the framework of the external ear and parts of the auditory canal, as well as, providing structural support for the larynx (Figure 35).

Fibrocartilage is made up of tightly packed collagenous fibers in greater concentrations when compared to hyaline cartilage. The high collagen concentration makes fibrocartilage flexible, tough ,and gives it high tensile strength. Fibrocartilage is generally found between the vertebrae, also known as the intervertebral disk, and at sites connecting tendons and ligaments to bones (Figure 36).

Bones are composed of osteocytes (bone cells) and its secreted extracellular matrix of collagen and hydroxyapatite, a calcium and phosphate mineral deposit (Figure 37).

$$[Ca_{10}(PO_4)_6(OH)_2]$$

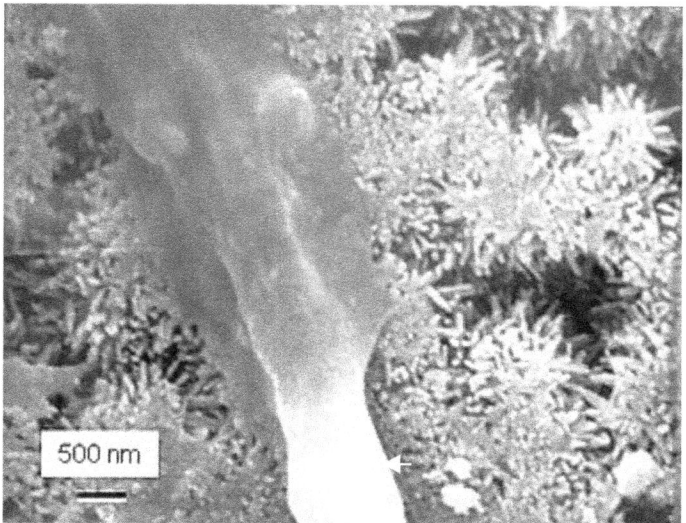

Figure 37: Scanning electron microscopy shows (arrow) the cytoplasmic extension of an osteoblast attaching to hydroxyapatite crystals

Bones are designed to form the endoskeleton of vertebrates and provide protection and support for the entire body. In addition, bones are the site of mineral storage (calcium and phosphate), a lever system for bodily movement and are partly responsible for blood cell formation (blood cell formation are formed by the stem cells found in the red marrow). There are 206 bones in the human body and 270 bones in an infant. The skeletal system is subdivided into appendicular and axial skeleton (Figure 38).

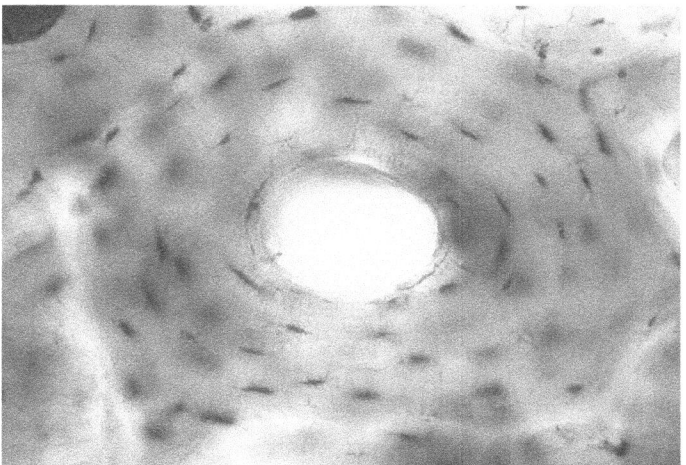

Figure 38: Photomicrograph of compact bone 400x

INTEGUMENT SYSTEM

Integument system is composed of the skin (cutaneous membrane) and its complimentary structures and glands. The integumentary system serves numerous functions. For example, it forms a protective covering against foreign materials, regulates body temperature, prevents excess water loss or gain, produces biochemical molecules (e.g. pheromones), and excretes waste materials. The dermis of the skin is also the attachment site for sensory receptors that detect pain, pressure, as well as temperature. Lastly, the skin is also the location for the formation of cholecalciferol or vitamin D_3, in a photoreactive process where the energy from ultraviolet radiation (UV) converts a precursor molecule called 7-dehydrocholesterol into vitamin D_3. Please note that cholecalciferol is biologically inert until it is further processed by the kidneys to the bioactive vitamin D, also known as calcitriol, or 1, 25-dihydroxy-cholecalciferol.

The skin, being the largest organ of the human body, is composed of many layers. The deepest layer, immediately superior to the muscles and skeletal structures, is called the hypodermis. The hypodermis, also known as the subcutaneous layer, is composed of connective tissues (mainly collagen and elastin fibers), blood vessels, and adipose tissues in a layer of subcutaneous fat. Visually, the hypodermal layer appears to be continuous with the dermis which rests immediately superior to its location (Figure 39).

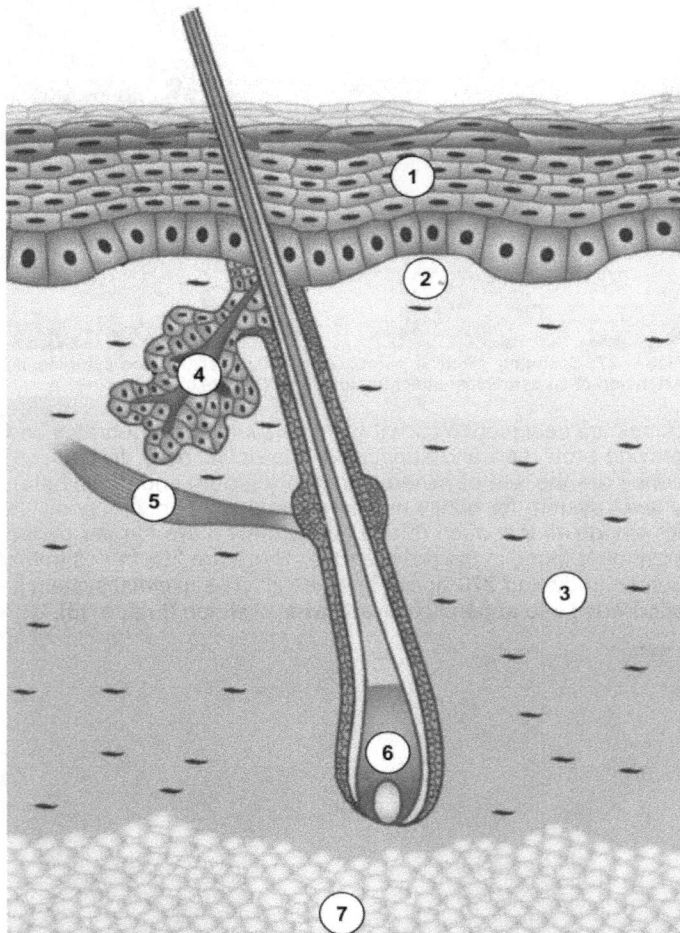

Figure 39: Illustration of the cutaneous membrane. ① Epidermis or stratified squamous epithelium, ② papillary layer of the dermis. ③ Reticular layer of the dermis, ④ sebaceous gland, ⑤ arrector pili muscle, ⑥ root of the hair, and ⑦ hypodermis

The dermis is a relatively thick layer of dense irregular collagenous connective tissue, its accompanying ground substances, smooth muscles (arrector pili muscles), nerves (autonomic and sensory

nerves), and blood vessels. The dermal blood vessels are designed to distribute nutrients and gases, as well as, take away cellular products and waste materials, to the epidermal layers of cells and are used for temperature regulation. For example, if the temperature gets too hot, the capillaries of the dermal layer will dilate, and allow more blood to flow, thereby allowing more heat to be dissipated through the skin. In addition, the eccrine sweat glands will produce its exocrine secretion, known as sweat, which is also designed to release heat from the body. In contrast, if the temperature is too low, the blood vessels will constrict and limit blood flow to the skin thereby limiting heat loss.

The dermis is composed of two layers. The inferior or the reticular layer, which rests immediately above the hypodermis, is where the roots of the hair rest, as well as the arrector pili muscles, sudoriferous (sweat) glands and sebaceous glands. The second or the superior layer of the dermis is called the papillary layer. The papillary layer is named after its dermal papillae, which are dermal contusions, that are responsible for tying down the basement membrane (basal lamina), and in turn, the epidermis (Figure 39).

The basement membrane, also known as the basal lamina, is composed of a layer of loose areolar connective tissue. The composition of this thin layer is mainly collagen, elastin, heparin sulfate, and adhesive glycoproteins, such as laminin and fibronectin. The collagen content of the basal lamina provides structural support, while the elastin allows some flexibility for movement. Heparin sulfate, on the other hand, is a proteoglycan, an essential component of the basement membrane that carries a strong negative charge. This electronegative charge repulses many anions, thereby acting as a filtration system. More importantly, heparin sulfate attract cations, such as sodium and in turn attracts polar molecules, such as water, which constitutes the fluid matrix of the basement membrane. Finally, adhesive glycoproteins, such as laminin and fibronectin, have a high affinity for a transmembrane protein located in the cellular membrane of the keratinocyte called integrin. In addition, entactin, a glycoprotein, connects laminins and fibronectin to collagen in order to form a structure network. These interactions help adhere the superior layer of epidermis to the basal lamina (Figures 2 and 40).

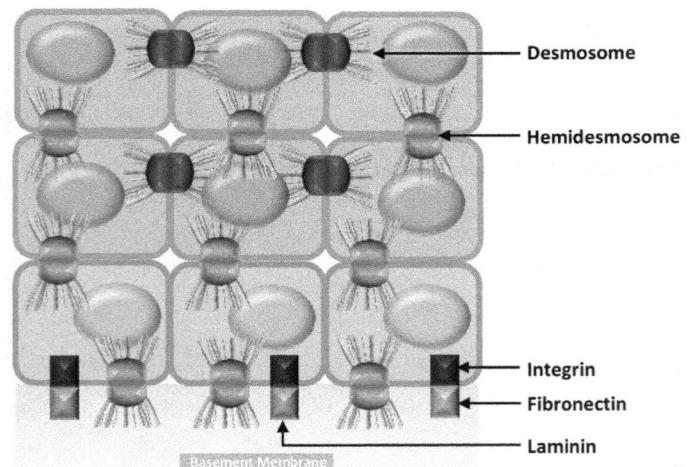

Figure 40: Illustration of the hemidesmosomes, desmosomes and adhesive glycoproteins within the stratum basale (the single layer above the basement membrane) and stratum spinosum

The epidermis is composed of stratified squamous epithelium and is subdivided into various layers. The most inferior layer (the layer superior to the basement membrane) is called the stratum basale. The stratum basale is a single layer of cuboidal and/or columnar cells that is anchored to the basement membrane via hemi-desmosomes and various transmembrane proteins and adhesive glycoproteins as described earlier. These cells are tied to one another though the formation of desmosomes on the lateral domain to prevent seepage and tissue movement (Figure 40).

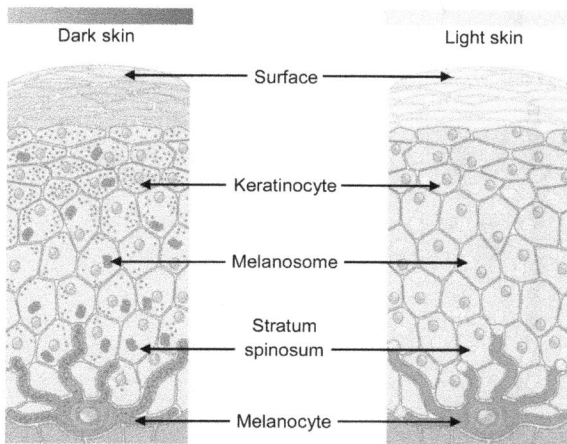

Figure 42: Coloration of the skin is dependent upon the amount of melanin produced by the skin pigment producing cells, melanocytes. Please note that the amount of melanin produced and packaged in the melanosomes (vesicles containing melanin) determines a person's skin color

The bulk of the cells located in the stratum basale are undifferentiated stem cells that proceed through mitosis approximately every 19 days to form new skin cells, also known as, keratinocytes. This continuous cellular division forms new cells which replace the dead skin cells that are desquamated, or flaked off, from the surface of the skin. It is estimated that the entire journey of the skin cells from the stratum basale to the surface of the skin takes approximately 40 to 56 days (Figure 41).

The second type of cell that exist in the stratum basal are melanocytes. Melanocytes are responsible for producing melanin pigments that create the coloration of the skin, as well as, functioning to shield the deeper tissues from ultraviolet (UV) radiation. Melanin pigments are produced in two primary forms: eumelanin is a black and brown pigment, while pheomelanin is a red pigment. Once the pigments are produced, they are packaged within a cellular vesicle called a melanosome and are subsequently absorbed by the newly divided keratinocytes via mitosis. All humans are born with near equal number of melanocytes; however, it is the activeness of these pigment cells that determine a person's skin color (Figures 41, 42 and 44). Finally, the third type of cells that are found in this layer are tactile (Merkel) cells which function as pressure receptors for sensory nerves.

The next layer is called stratum spinosum, which is situated immediately superior to the stratum basale. This layer is a continuation of the stratum basale but is composed of layers of newly divided keratinocytes (Figures 41 and 42). Within the stratum spinosum, these newly divided keratinocytes begin to produce lipid filled membrane coating vesicles called lamellar bodies, that are stored within the cytoplasm. Contained within the lamellar bodies are water proofing lipids called glucosylceramides (i.e. lipids with a glucose component), proteins, and enzymes. In addition, these keratinocytes will begin to form keratin fibers, which are also placed within the cytoplasm.

Also found but not limited to the stratum spinosum are dendritic (Langerhans) cells, which are modified macrophages. Dendritic cells are responsible for phagocytizing any foreign objects or organisms that may have penetrated the skin as a part of the primary immune response (Figure 43).

The next layer is called stratum granulosum, situated immediately superior to the stratum spinosum. This layer is a continuation of stratum spinosum. The cells within this layer perform two preprogrammed duties. First, these keratinocytes will secrete the glucosylceramides from the lamellar bodies via exocytosis. Glucosylceramides coat the plasma membrane of the keratinocytes and make the cell impermeable to water and by extension, prevent these cells from absorbing nutrients and water via diffusion and osmosis, respectively. Second, the keratinocytes with this layer will begin producing keratohyalin granules. Immediately after the formation of these granules, it begins to cross-link with keratin fibers (a type of intermediate filament of the cytoskeleton) within the cytoplasm forming keratin (a water-proof protein complex), as well as, promoting dehydration of the cell. The shear mass of keratohyalin granules and keratin physically crushes the organelles of the keratinocytes (Figures 41 and 44).

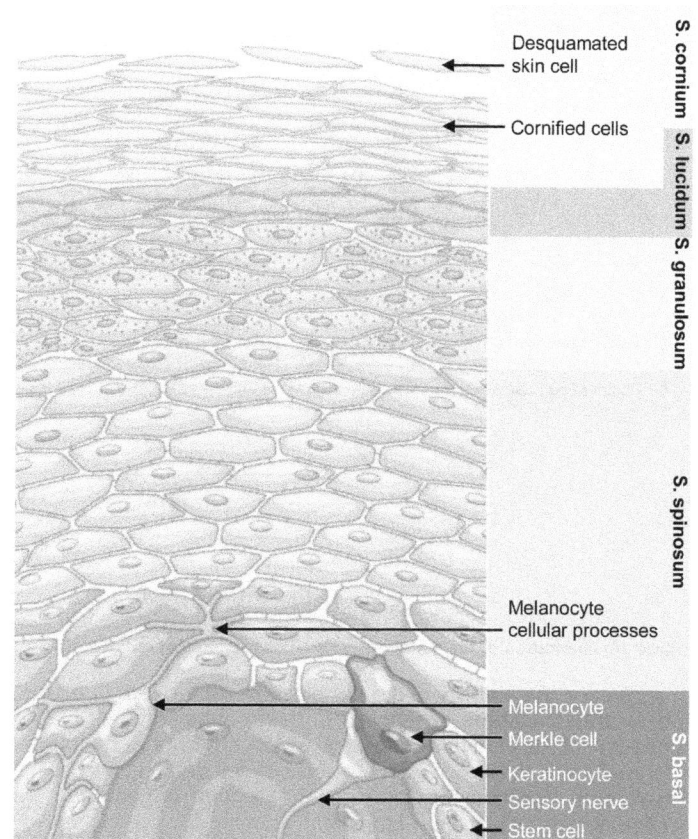

Figure 41: Illustration of the cells and layers of the skin

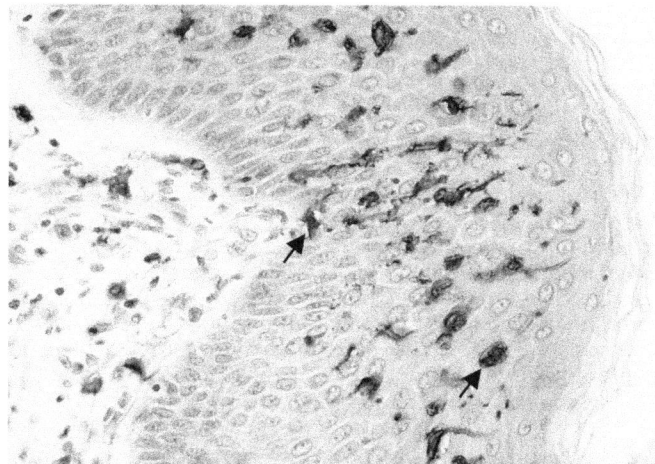

Figure 43: Section of skin showing large numbers of dendritic (Langerhans) cells in the epidermis using S100 immunoperoxidase stain. This photomicrograph was taken at 1000x magnification

THICK SKIN

Epidermis
- Stratum corneum
- Stratum granulosum
- Stratum basal
- Merkle cell

Dermis
- Papillary layer
- Superficial arterio-venus plexus
- Sweat duct
- Reticular layer
- Meissner corpuscle
- Apocrine sweat gland
- Deep arteriovenus plexus

Hypodermis
- Subcutaneous fat
- Nerve fibers
- Pacinian corpuscle

THIN SKIN

Epidermis
- Hair shaft
- Stratum corneum
- Stratum granulosum
- Stratum basal

Dermis
- Papillary layer
- Superficial arterio-venus plexus
- Arrector pili muscle
- Sebaceous gland
- Reticular layer
- Sweat duct
- Hair follicle
- Eccrine sweat gland
- Hair papilla

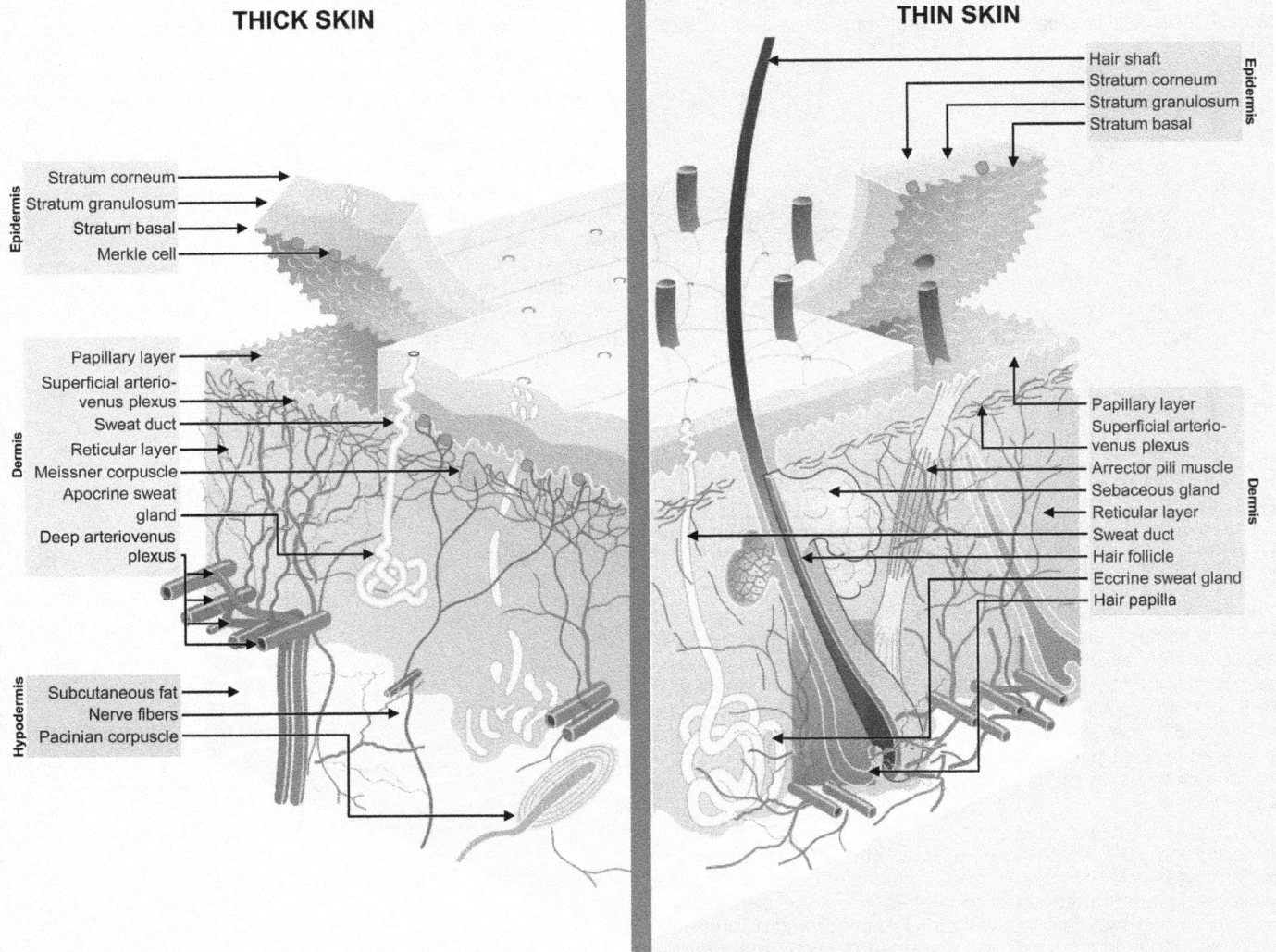

Figure 44: Illustration of thick and thin skin. Please note that only the thin skin possesses hair follicles

In summary, the inability of the keratinocytes to absorb nutrients, the destruction of the organelles of these cells, as well as, the formation of keratohyalin granules which prompts dehydration, eventually kills the cell. Since this form of cell death is part of the lifecycle of keratinocytes, we refer to this form of cell death as preplanned or apoptosis (Figure 41).

The next layer is called stratum lucidum (~30 layers of cells), situated immediately superior to the stratum granulosum. This layer is a continuation of stratum granulosum and only exists in the thick skin found on the finer tips, palmar, and plantar regions. At this point, the cells have effectively completed the apoptotic cycle and are, for all intents and purposes, dead. These dead cells are also called cornified cells (Figure 41).

The most superior layer is called the stratum corneum which consists of up to 30 layers of cornified cells completely filled with keratin. Keratin is a water-proof protein compound in the family of intermediate filaments of the cytoskeleton which makes the surface layer of the skin impermeable to water. The cells of the stratum corneum will continue to desquamate in a scale-like fashion called dander. These desquamated cells will have to be replaced constantly through the mitotic divisions of the cells within the stratum basale (Figure 41 and 44).

There are two distinct types of human skin located in specific locations on the body. The distinction between the two types is based upon the number of layers of cells it possesses, as well as the availability of hair follicles. For example, the thick skin is

found on the fingertips, palms of the hands and the soles of the feet. The thick skin is thus named because it possesses an extra layer called stratum lucidum. The thick skin does not possess hair follicles and is designed to withstand heavy pressure and abrasion. Thin skin, on the other hand, exists in the remaining areas of the body. This type of skin is thinner, more flexible and has hair follicles (Figure 44).

PRESSURE ULCERS

Pressure ulcers, also known as bedsores, are injuries to skin and underlying tissues which lead to cellular necrosis. There are six factors contributing to the formation of pressure ulcers. Pressure ulcers could be caused by external pressures resulting from boney prominences of the body such as the sacrum, heels, hips, and elbows coming into contact with a surface. This occurrence could be alleviated by using specialty beds or by placing specialty surfaces and paddings under the patient's boney prominences.

The intensity of the pressure is another factor where concentration of force applied may result in pressure ulcers. A tissue tolerance test should be performed to determine if an individual is prone to skin breakdown. Duration of pressure plays a part in pressure ulcer formation. Typically a hospital stay of three to five days could increase the likelihood of ulceration. Friction that occurs when two surfaces rub together contributes to skin break down. Shearing forces, such as, when the client tries to push themselves up in the bed, causing one layer of tissue to slide

over another will also contribute to skin breakdown.

Fragile skin is probably the major factor in the formation of pressure ulcers. Fragile skin may be caused by poor nutrition which could be indicated by low albumin levels in the blood stream. Minimal albumin levels indicates a lack of protein consumption by the individual, thereby indicating poor skin integrity and may prevent wound healing and make pressure ulcer formation occur more readily. Other nutritional factors include a deficiency in ascorbate. Ascorbate, also known as vitamin C, is involved in the process of collagen formation. A deficiency in vitamin C will result in fragile capillaries that could easily burst. Lack of capillary integrity will make blood delivery difficult and prevent the healing process. Iron is an essential mineral for the human body. Low iron levels will indicate a lack of hemoglobin development, reduced erythrocyte production and can contribute to the occurrence of iron deficiency anemia. Put simply, a lack of iron will prevent proper and efficient circulation of nutrients and oxygen whereby making healing difficult. Low zinc levels may also contribute to ulceration. Zinc is an essential mineral that is involved in the catalytic activities of over a hundred different enzymes in the human body. Zinc is involved in collagen formation and prevention of collagen breakdown in granulation tissues. Granulation tissue is a thin newly formed connective tissue filled with blood vessels and capillaries on the surface of a wound during the early stages of the healing process.

Fragile skin may also be caused by aging. With increasing age, the population of melanocytes decreases, making the skin more pale and translucent, thereby appearing thinner. Changes in the production of connective tissue, also known as elastosis, diminish the skin's elasticity and strength. Blood vessels in the dermis of the skin become fragile, which leads to bruising and dermal bleeding. The hypodermal fat layer thins thereby reducing its ability to provide insulation and padding. The combination of the changes in the aging skin increases the risk of pressure ulcer formation.

Accumulation of moisture will eventually cause ulceration. Skin that is in constant contact with urine or stool will eventually become fragile. The risk of moisture induced pressure ulceration is particularly high in immobile patients. Immobility may be caused by limited movement due to pain, musculoskeletal deficiencies, mental status, or other variables. The mental status of the patient may also contribute to the development of pressure ulcers. The mental status may include a confused patient who may not remember to reposition themselves, or a patient that may be unconscious, chemically sedated, or paralyzed.

Diseases, such as diabetes, may contribute to the development of fragile skin and increase the possibility of developing pressure ulcers. High blood glucose levels hinder circulation and are a risk factor in decreased skin integrity. Finally, obesity, as well as, patients undergoing corticosteroid therapy, may also contribute to the development of pressure ulcers.

Staging

Pressure ulcer staging describes the depth and severity of tissue injury. This assessment is important for the nurse to accurately identify. A stage 1 ulcer is an area of unblanchable redness. On darker skin tones the area may just look different from the surrounding area. Stage 2 pressure ulcers present as an ulcer that reaches the dermis, also referred to as partial thickness. The wound bed may be red/pink in appearance, or even presenting as a serum filled blister or an open/ruptured blister. Stage 3 is demonstrated by full thickness tissue loss which may extend as far as the subcutaneous fat layer. Some slough or yellow stringy tissue may also be present in this third stage, and there may also be undermining and tunneling. Stage 4 is demonstrated by full thickness tissue loss where the wound extends past the subcutaneous fat layer and exposes the underlying muscles,

tendons and/or bones. This stage can include some slough and or eschar (necrotized tissue), as well as, tunneling and undermining. Finally, an unstagable pressure ulcer is demonstrated by full thickness tissue loss in which the depth of the base of the ulcer is undetermined. This pressure ulcer is covered by slough and/or eschar in the wound bed (Figure 45).

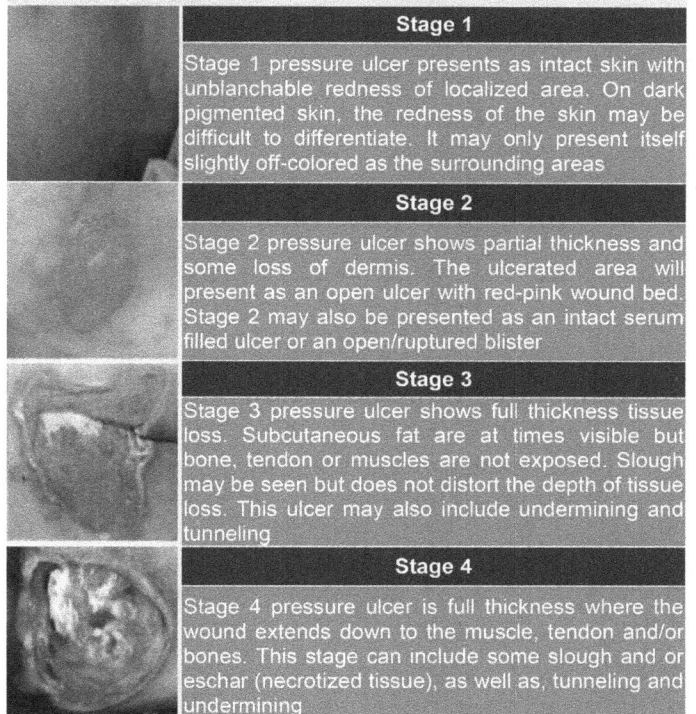

Stage 1
Stage 1 pressure ulcer presents as intact skin with unblanchable redness of localized area. On dark pigmented skin, the redness of the skin may be difficult to differentiate. It may only present itself slightly off-colored as the surrounding areas
Stage 2
Stage 2 pressure ulcer shows partial thickness and some loss of dermis. The ulcerated area will present as an open ulcer with red-pink wound bed. Stage 2 may also be presented as an intact serum filled ulcer or an open/ruptured blister
Stage 3
Stage 3 pressure ulcer shows full thickness tissue loss. Subcutaneous fat are at times visible but bone, tendon or muscles are not exposed. Slough may be seen but does not distort the depth of tissue loss. This ulcer may also include undermining and tunneling
Stage 4
Stage 4 pressure ulcer is full thickness where the wound extends down to the muscle, tendon and/or bones. This stage can include some slough and or eschar (necrotized tissue), as well as, tunneling and undermining

Figure 45: Pictorial examples of the stages of pressure ulcers

Pressure ulcers may demonstrate some form of drainage and there are four types of drainage known. There is serous drainage which is red in coloration similar to that of blood and serum. There is seroussanginous drainage which is pink in coloration. Sanguineous drainage is clear, while purulent drainage is thick, cloudy, foul smelling and may indicate infection. A client with purulent drainage may have other indicators of infection such as an elevated white blood cell count. If a wound is believed to be infected a wound culture should be done.

Assessment and Intervention of Wounds

The braden Scale is an assessment tool that the nursing staff uses to determine a client's risk for developing pressure ulcers. The scale identifies a series of risk factors such as level of immobility, potential for friction and shear forces, nutrition status, skin moisture and sensory perception. With the exception of "friction and shear" which is rated 1 to 3 points, the remaining categories are rated on a scale of 1 to 4 points, with the combined score of 23 points. If a patient possesses a combined score of 9 or less, this individual is at high risk of developing pressure ulcers. Inversely, if a patient has a combined score of 19-23, this individual is at no risk of developing pressure ulcers (Table 2).

In order to make certain that the comfort and safety of patients are not at risk, nurses are required to examine the condition of the patient's skin once upon their admission, as well as, at the beginning of every shift. Nurses must document any wound(s) detected based upon their location, size, stage, wound bed description (color, granulation, slough, eschar present), drainage (color, consistency, type, and amount), and wound edges. Wound edges and depth should be measured. In addition, nurses are required to determine if there are any undermining or tunneling present. A cotton tipped swab can be used to measure depth, tunneling, and undermining. If a wound appears to be beefy red and possesses an uneven red coating, this signifies that a granulation tissue (new growth of healthy tissue) is forming and this wound is in the process of healing. Nonetheless, slough (which is stringy and yellowish in appearance) or eschar may need to be debrided before dressing appropriately.

One of the methods for the prevention of wound development is nursing interventions. For example, nurses should turn an immobile patient every two hours in order to alleviate the constant external pressure upon boney prominences of the body. For clients who are semi-mobile, they should be educated about the importance of turning themselves every two hours, or shifting their weight every 15 minutes if sitting in a chair. Special beds, and pillows, and padding can be used for cushioning pressure points and boney prominences. Nurses should take great care in treating underlying illnesses, such as diabetes. As indicated previously, diabetic patients are highly susceptible to pressure ulcers. For clients with a nutritional deficit, nurses should regularly provide supplements such as parenteral or enteral feeds. In some cases a dietician may need to be consulted (Table 3).

Table 2: Example of a Braden scale that is used to evaluate patients' potential for developing pressure ulcers. Please note that 15-18 indicate mild risk of developing pressure ulcers, 12-14 indicates moderate risk of developing pressure ulcers and ≤ 11 indicates severe risk of developing pressure ulcers

Risk factors	1	2	3	4
Sensory perception	Completely limited	Very limited	Slightly limited	No impairment
Moisture	Constantly moist	Very moist	Occasionally moist	Rarely moist
Activity	Bedfast	Chair-fast	Walks occasionally	Walks frequently
Mobility	Completely immobile	Very limited	Limited	No limitation
Patient Nutrition	Very poor	Inadequate	Adequate	Excellent
Friction and Shear	problem	Potential problem	No apparent problem	NA

Table 3: Treatment of pressure ulcers at various stages

Stage	Treatment
Stage 1	Remove pressure and reduce friction and shearing, promote good nutrition, fluid and electrolyte balance, prevent moisture
Stage 2	Measures used in stage one, cleanse with sterile saline, semipermeable occlusive dressings, hydrocolloid wafers, or wet saline dressings
Stage 3	Measures used in stage one, and cleansing and debriding (provide analgesia before attempting), wound vac, surgical intervention when complications such as fistula arise or ulcer is not responding to treatment
Stage 4	Utilizes the same interventions as shown on stage 3

Wound Complications

Complications of wounds include infection, hemorrhage, dehiscence (when a wound, such as, a surgical incision opens back up), evisceration (when a wound such as a surgical incision opens back up and organs protrude), and fistulas (an opening occurs between two sites e.g. rectum and vagina).

Symptoms of wound infection include redness, swelling, warmth, and drainage at the site. Drainage may be purulent, which means it could be cloudy and foul smelling. Lab results may indicate an increased in white blood cell count. The physician may order a culture of the wound site to determine if the site is infected. Important points to remember about collecting wound cultures are that the wound should be cleaned first, and the swab and inside of the culture tube should only come into contact with the specimen collected, therefore special care must be taken not to contaminate the swab or inside of the tube by touching other objects or surfaces. Please note that even allowing the swab to touch the outside of the culture tube renders specimen contaminated. Infection is often managed with antibiotics.

Symptoms of hemorrhage may include an excessive amount of bloody drainage present in the drain or on the wound dressing. In addition, patients may also demonstrate symptoms of tachycardia, hypotension, and lab results may also indicate low hemoglobin and hematocrit. If a client's dressing indicates that the wound is bleeding, the nurse should not remove the dressing, but should rather reinforce the current dressing to prevent removing any clots that may have begun to form. The physician should be notified immediately if the nurse suspects that the client is hemorrhaging.

Wound dehiscence and evisceration after surgery are serious wound complications. If it occurs the nurse should immediately focus on taking the pressure off the site. This is accomplished by putting the client in a low Fowler's position which is slightly elevating the patient's head. Cover the dehisced or eviscerated wound with sterile moist dressings. The surgeon must be notified promptly and client must not be left alone.

Sterile Technique and Standard Precautions

The skin is the human body's first defense against pathogens, but it can also be a manner by which pathogens are transmitted. This is why hand hygiene is essential. As a nurse, hand hygiene will be one of the most important interventions in preventing spread of disease from client to self or from client to client. Plain soap and water, antiseptic handrubs, and utilization of surgical hand asepsis are ways to achieve hand hygiene. Surgical hand antisepsis is a method by which debris and transient microorganisms are removed. Surgical hand antisepsis is designed to minimize microorganisms and prohibit regrowth on the hands, nails and forearms. Nursing staff should dress in the correct sterilized attire before initiating the procedures for surgical hand antisepsis. Hair must be covered by a surgical hat, while scarves and other supplementary clothing should be removed. All jewelry must be removed to prevent any potential habitation of microorganisms beneath the jewelry and to avoid accumulation of scrub solution in their crevices. Finger nails must be clean and short and nail polish and artificial nails must not be worn (jewelry, artificial nails may harbor microorganisms). Hands must be clean and free from any breaks in the integrity of their skin. Minor abrasions should be covered with a waterproofed dressing. Individual with serious or infected wound should not scrub.

In some cases nurses need to wear personal protective equipment (PPE) such as gown, goggles, gloves, and mask to reduce the chances of becoming cross contaminated by a pathogen.

There are different levels of "transmission based precautions" medical staff use depending on the route by which a pathogen is spread. However, standard precautions are used for all clients regardless of the diagnosis. Standard precautions include the use of gloves anytime body fluids or blood are handled and possibly all PPEs (gown, goggles, gloves, and mask) if splashes of bodily fluids could occur.

Airborne precautions are used for diseases spread through the air such as TB, Varicella, and Rubeola. Airborne precautions require the use of a N95 mask and negative pressure room. Negative pressure rooms being used for purposes of containing airborne pathogens must remain with the doors sealed except when medical staff enter and leave the enclosure. This client gets a private room and a sign should be placed outside the door to alert all entering to wear the proper PPE. When the client is being transported within the facility they must don a surgical mask to prevent the spread of the infection.

Droplet precautions are for diseases spread through large particle droplets. Droplet transmission of pathogens involves the contact of infected particles of fluids (~5 μm) with the conjunctivae or the mucous membranes (nose or mouth) of unsuspecting individual. These droplets are generated from an infected individual usually during sneezing, coughing, or speech. Droplet transmission could also occur during suctioning and bronchoscopy. The diseases that may spread though droplet transmission include but not limited to, Rubella, Mumps, Diptheria, and Adenoviruses. For droplet precautions the proper attire includes a surgical mask, when standing within three feet of client. The room does not need to be a negative pressure room, but should be private. A sign and PPE set-up should be outside the door for potential visitors and medical staff.

Contact precautions are used for pathogens spread through direct or indirect contact (through a fomite such as something a client may have touched in the room). Diseases that could be transmitted through direct and/or indirect contact include, but not limited to, MRSA (Methicillin-resistant *Staphylococcus aureus*), VRE (vancomycin-resistant enterococci), VISA (Vancomycin Intermediate *Staphylococcus aureus*), and VRSA (Vancomycin-resistant *Staphylococcus aureus*). The proper PPE for contact precautions are gown and gloves.

Neutropenic precautions are unique because it is used to protect the client against pathogens others may bring to them. Clients with compromised immune systems (neutrophil count is below 500 cells/µL) will require this form of protection. These precautions include prohibiting the patient from consuming fresh fruit, while visitors of the client would not be allowed to bring fresh flowers. Sick visitors should not be permitted to enter, and healthy visitors must always don masks.

Asepsis is defined as being free from disease causing microorganisms. Medical and surgical asepsis are two methodologies that nurses follow when maintaining a clean and infection free environment. Medical asepsis is known as "clean technique" and it is employed in an effort to reduce the number of pathogens. Medical asepsis techniques require that medical personnel should routinely wash their hands before and after coming into physical contact with the client. Depending on the condition of the patient, sterile or "clean gloves" may be used to protect the caregiver, as well as, the client. The work surfaces and equipment used must be clean at all times especially when performing procedures such as placement of a nasogastric tube or rectal tube. Special handling techniques must be employed with linens and all other items that may come into contact with the patient in order to avoid the spread of germs.

Surgical asepsis is also known as "sterile technique". Whereas medical asepsis reduces the number of pathogens, sterile technique is meant to keep objects and areas completely free from pathogens. Antiseptic solutions, such as chlorhexidine gluconate and iodine, should be used to prepare areas of the skin for sterile procedures. Please note, when preparing the skin with antiseptic solutions, the solution should be applied starting from the center and move outwards to the periphery in a circular pattern. Like medical asepsis, surgical asepsis requires that medical personnel routinely wash their hands before and after coming into physical contact with the client.

Sterile gloves must be used to perform all procedures. There are two recommended techniques used by nurses for donning sterile gloves in order to prevent contamination. The close donning technique requires that the acquisition of the gloves does not include direct contact with the hands. For example, nurses are required to grip the edge of the sterile glove package through the gowns. The package holding the sterile gloves should be acquired no more than an inch from its corners; again, with the gown covering the fingers, gripping it before carefully opening the package to display the gloves. Use the right hand to remove the left glove and with the left hand palm up, fingers straight before laying the glove on your left wrist. Carefully grip the cuff with your left thumb before placing the right thumb inside the top cuff edge. Once the right thumb is in place, make a fist with your right hand and stretch the glove over your left fingertips. Make certain that all your left fingers are straight before pulling the glove down the glove over the entire hand. Repeat the above procedure to don the right glove.

The caregiver must always consider the package contaminated and should never touch the package once the sterile gloves are retrieved. Once the gloves are firmly in place, make certain that all surface powder be removed with either sterile water from a pour rinse, or a sterile wipe.

The open donning technique requires that the sterile glove package be gripped through the gown. Again, the package holding the sterile gloves should be acquired no more than an inch from its corners. With the gown covering the fingers, gripping it before carefully open the package to display the gloves. Unlike the close donning technique, the open donning technique does not restrict contact of the gloves with an individual's bare hands. The nurse should pick up the cuff of the right glove with the bare left hand and slide your right hand into the glove until you have a snug fit over the thumb joint and knuckles. Unfold the glove cuffs so that it is covering the gown sleeves. Please note that the bare left hand should only touch the folded cuff thereby keeping the rest of the glove sterile. To don the left glove, first slide your right fingertips into the folded cuff of the left glove. Carefully pull the glove over your fingertips, and fit your left hand into it until you have a snug fit over the thumb joint and knuckles. Again, unfold the glove cuff to cover your gown sleeves. It is important that the gloved fingertips do not touch the bare forearms or wrists. After donning sterile gloves, the surface powder must be removed from powdered gloves. Use either sterile water from a pour rinse or a sterile wipe. Once the gloves are snuggly in place make certain that the hands must never drop below waist level to remain sterile, and nothing non-sterile may be touched.

Sterile techniques require some forethought and planning. For example, the establishment of sterile field in preparation for surgery is an essential part of sterile techniques. All flat surfaces within the sterilized area must be dry and dust free before placing the sterile drapes. Sterile drapes should be used to cover surfaces or operative fields. Sterile drapes provide a barrier against pathogens, liquids, and particulate matters. Please note that drapes are only sterile at table level. Areas of the drapes that fall below the working surface and are out of the direct vision of the surgical team are not considered sterile. Similarly, any items that fall below the table level are considered unsterile. This rule applies to the portion of suction and irrigation tubing that is off the sterile field. In addition, if the drape is not wide enough to cover an entire surface, a one inch margin around the edge of the drape is considered unsterile.

Packages holding sterile equipment or materials are not considered sterile. Care must be taken opening packages for the sterile procedure. The unsterile package must never come into contact with the sterile field. In addition, all flaps of non-woven wrap must be prevented from coming into contact with the sterile field.

All surgical supplies must be opened as close to the surgical start time as possible. This methodology is to limit the amount of time that the sterilized surgical supplies are exposed to the environment. The potential of contamination increases with time when particles stirred up by movement of personnel could settle upon the flat surfaces within the sterile environment. Sterile surgical supplies shall be handled as little as possible, while all items added to sterile field shall be assessed for its packaging integrity and that the chemical indicators indicate a "pass" result before introducing them into the sterile field.

Burns

First degree burns (superficial thickness) involve only the epidermis. First degree burns appear pink/red in coloration, as well as being dry. This type of burn may demonstrate swelling and small blisters may also be present. Although painful, first degree burns will heal within two to three days and shouldn't leave scars. These minor burns should be treated by running cool water over the burned areas.

There are two types of second degree burns. The first is a superficial partial-thickness burn and involves the epidermis and upper 1/3 of the dermis. This type of second degree burn appears bright red or mottled. Swelling may be present, along with fluids weeping from the wound. Blisters will also be present that are more than 5mm in size (also known as bulla: singular; bullae: plural). The injured area will blanch (i.e. go pale) with pressure and are sensitive to air currents. Although this burn will be painful, the blisters will heal within 10 to 21 days while scarring should be minimal. The second type of second degree burn is a deep partial-thickness burn and involves the epidermis to the deeper portions of the dermis. Besides being bright red or mottled it may also have a waxy appearance and have wet or dry blisters. Nonetheless, the hair follicles will remain intact. These burns take about three weeks to heal, and are treated surgically with excision and grafting. It is important that burns don't become infected. A second degree burn that becomes infected can easily convert to a full thickness skin loss.

Third degree burns are also known as full-thickness burns. This type of burn involves the total destruction of the epidermis, dermis, as well as, all associated glands, hair follicles and nerves. These burns are white, brown or black, leathery in appearance and are associated with edema (swelling due to the buildup of fluids). Third degree burns will display various zones. For example, zone of coagulation is located towards the center and is the most damaged area. The tissue loss in this zone is irreversible. The zone of stasis surrounds the zone of coagulation. The tissue of this zone has suffered from decreased circulation but the damage is reversible. Finally, surrounding the zone of coagulation is the zone of hyperemia. This zone is the least damaged area and is the most salvageable (Figure 46).

Due to the damage to the sensory nerves, third degree burns are generally painless. Treatment includes wound closure by skin grafting for all burns larger than 4 cm. All wounds that are less than 4 cm are allowed to close by contraction.

There are four different types of skin grafts. An autograft is a permanent graph. Autografts utilize the client's own skin. A homograft is skin taken from a cadaver or donor, and is usually rejected by the patient on week two post-surgery. Heterografts are skin grafts acquired from an animal such as a pig. Heterografts may last 4-5 days before being rejected by the patient. Synthetic skin grafts are biosynthetic dressings. Synthetic skin grafts are usually rejected by the patient within 8 days post-surgery.

Emergency management of burns includes first stopping the burning process through chemical flush, or smothering of fire. Once the source of the burn is eliminated, airway (i.e. respiration, breathing) and circulation should be addressed. Neurological function should also be assessed for deficit. Remove clothing and/or jewelry that may become constrictive once swelling starts. If clothing is stuck to the skin it should not be pulled off, but removed through cutting. In cases of an electrical burn observe for signs of necrosis of organs between the entrance and exit wounds. Since the heart is one of the most affected organs during an electrocution, an ECG is necessary to monitor the electrical activities of the patient's heart.

An important assessment tool in the management of burns is the Rule of Nines. The Rule of Nines is a standardized technique that is used to promptly evaluate the amount of body surface area (BSA) that has been burned on a patient. This rule is only applied to partial thickness second degree burns and full thickness third degree burns.

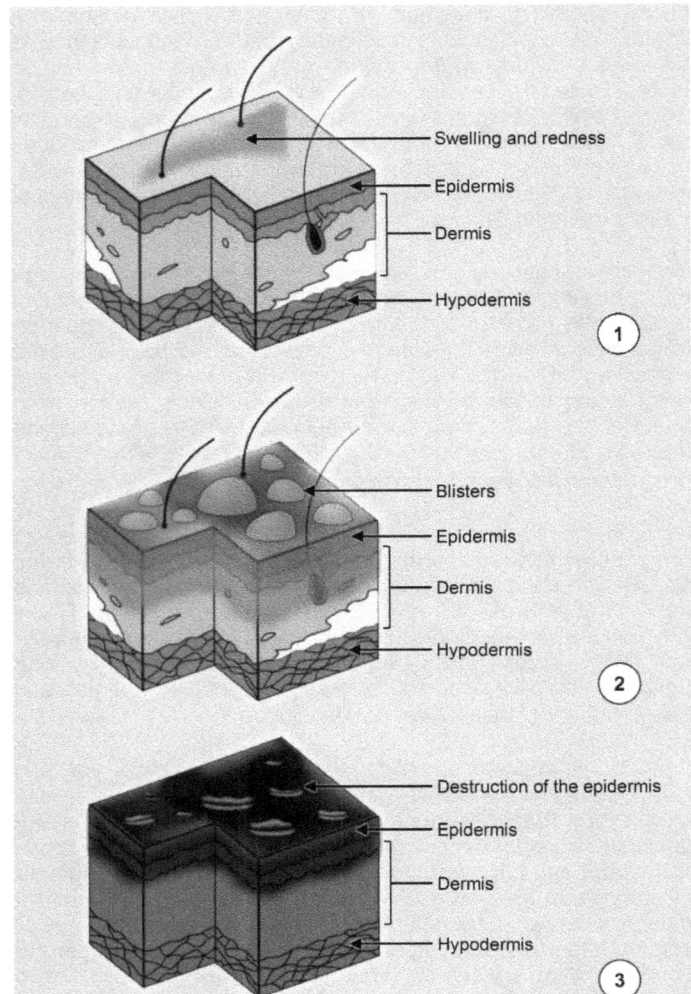

Figure 46: Illustration of burns. (a) First degree burn with swelling and redness. (b) Second degree burn with redness and blisters. (c) Third degree burn with the total destruction of the epidermis

For children add 0.5% to each leg and subtract 1% for the head for each year above one year old. For example, a 6-year old child would be +2.5% for each leg and -5% for the head. This formula should be used until the adult Rule of Nines values are reached. An alternative method to calculating the BSA is utilizing the size of the patient's palm, which is approximately 1% BSA. For example, if a burn area is approximately three palm surfaces, the burn would be roughly 3% BSA. This method can be used to estimate the BSA for both adults and children (Table 4). The Rule of Nines is essential to determine the fluid needs of a burn victim in order to prevent shock.

Table 4: Body surface area (BSA) of adult and child

Area	Adult	Child
Anterior head	4.5	9
Posterior head	4.5	9
Anterior torso	18	18
Posterior torso	18	18
Anterior leg	9	6.75
Posterior leg	9	6.75
Anterior arm	4.5	4.5
Posterior arm	4.5	4.5
Genitalia/perineum	1	1

The body responds to burns by vasoconstriction followed by vasodilation. Loss of the capillary seal occurs which causes third spacing. Third-spacing occurs when fluid transfers from the intravascular space into the interstitial areas. Third spacing leads to massive fluid and electrolyte shifts, therefore prompt fluid resuscitation is essential to survival and staying ahead of shock. Adequate urine output is not a good indicator of fluid status in burn victims. Normal urine output may be present for the first 24 hours even in presence of tachycardia. Decreased blood pressure is a late sign and signals the onset of burn shock. Three to five liters of fluid may be lost in the first 24 hours and peaks in the first 6 to 8 hours. This is why the most fluid is given within the first 8 hours of resuscitation.

Fluid and electrolyte balance intervention involve the utilization of two large bore IV's (20 gauge or less, even better if one is a central line) introducing fluids for resuscitation. The amount of fluids is calculated through Parkland formula which is 4 ml lactated ringers (LR) multiplied by total body surface area (TBSA) of percentage burned multiplied by the weight in kg (lbs/2.2). The first half of the calculated amount is delivered in the first 8 hours from the burn incident, and the remaining fluid is delivered in the next 16 hours. For example, if a person has burns in 45% of his body and weighs 75 kg then:

4 ml	x	TBSA %	x	Weight	=	Answer
4 ml	x	45%	x	75kg	=	13500 ml kg

Maintenance rate first 8 hrs = 6750 ml kg
Maintenance rate remaining 16 hrs = 6750 ml kg

Please note that children require an additional maintenance dose of fluid to compensate for insensible water loss and this solution should be D5 ¼ NS (a solution containing .20% sodium chloride and 5% dextrose in sterile water) for infants and D5 ½ NS (a solution containing .45% sodium chloride and 5% dextrose in sterile water) for children.

The maintenance formula is 100 ml/kg for the first 10 kg, 50 ml/kg for the second 10 kg, and 20 ml/kg for the remainder of weight in kg and then the total should be divided by 24 hours to arrive at the hourly maintenance rate. For example, if a child weighs 25 kg then:

First 10 kg	=	100 ml/kg x 10 kg	=	1000 ml
Second 10 kg	=	50 ml/kg x 10 kg	=	500 ml
Remaining kg	=	20 ml/kg x 5 kg	=	100 ml

Total maintenance rate = 1600 ml
Total maintenance rate per hour = 66.7 ml/hour

It is important to titrate the fluids depending on the child's response. In addition, it is essential to monitor serum electrolytes of the client.

Another important assessment is the client's airway and breathing. Inhalation injury is common due to direct heat or edema. Indicators of injury to the airway include hoarseness, stridor (abnormal, high-pitched sounds), problems swallowing, difficulty breathing, wheezing, rhonchi (coarse rattling respiratory sound) and coughing up carbon particles. Look for burns to face or neck, singed nasal hair, and bloody sputum as other indicators. Labs may indicate abnormal arterial blood gas (ABG) with decreased partial pressure of oxygen (PO_2), and increasing carboxyhemoglobin levels. Carboxyhemoglobin is a stable complex formed between carbon monoxide hemoglobin within the red blood cells. Keep in mind that clients exposed to carbon monoxide should have their O_2 level assessed through ABG examinations and not a non-invasive pulse oximeter. Non-invasive pulse oximeter may indicate an adequate O_2 level falsely since the carbon monoxide's presence in the blood mimics the presence of O_2. Early intubation is essential before swelling of the airway causes obstruction. If there is any suspicion that a person has a possible respiratory injury they should be observed for 24 hours for complications which may lead to acute respiratory failure.

Nurses should also assess for burns around the chest and neck that might restrict breathing once edema and tightness of skin manifests. Escharotomy may be needed to relieve the restrictiveness of tightening skin allowing the chest to expand.

Interventions pertaining to gas exchange and airway clearance are employed to aid in the respiration of patients. For example, turning the patient from side to side and/or side to back while allowing deep inhalation and strong cough for exhalation may help clear the airway for respiration. Incentive spirometer and suctioning may be used to aid respiration. Other equipment such as mechanical ventilation of humidified oxygen could also be employed.

Escharotomy may also be used during cases of full thickness burns. This medical procedure is designed to combat compartment syndrome, which is the constriction of deep rehydrated tissue by the leathery superficial eschar that has lost its elasticity. Escharotomy involves an incision made through the eschar in order to expose the fatty tissue below and promote chest expansion.

The lab results from severe burn patients may initially display low sodium and high potassium levels. Even so, potassium levels should be closely monitored as it may go to the opposite extreme (becoming too low), due to large fluid boluses in the resuscitation period. Lab results may also demonstrate that the patient's hemoglobin/hematocrit is too low. Nonetheless, at times, the patient's hematocrit may appear artificially high due to dehydration. Coagulation labs may also report irregularities. Coagulation factor levels tested include platelet count, fibrinogen, factors V, VIII, and IX. Patients may demonstrate high baseline levels of platelets, fibrinogen, and factor VIII while the removal rates of platelet and factor IX are significantly lower when compared with the average. Other systemic effects of burns include the gastrointestinal (GI) system, urinary system (renal function), and the immune system. GI system complications include bacterial translocation which is the translocation of viable bacteria from the gastrointestinal (GI) tract to extra-intestinal sites, such as the mesenteric lymph nodes, liver, spleen, kidney, and bloodstream. A burn patient may also suffer from paralytic ileus which is defined as the severe loss of intestinal peristalsis. It is understood that not all segments of the GI tract are equally susceptible to this condition. For example, small intestines appear to be relatively resistant and its activity generally returns within the first few hours. However, the motility of the colon may not return for 24 to 72 hours and requires intra-abdominal manipulations. Finally, curling ulcer of the duodenum may also form in patients with extensive superficial burns. This type of ulcer appears in individuals that have suffered extreme stress.

Acute renal failure (ARF) is a well-known complication of severe burns and is due to the low blood flow rate to the kidneys. ARF is defined as the rapid loss (< 2 days) of kidney's ability to remove waste and help balance fluids and electrolytes. ARF is an important factor that results in an increase in mortality rate in burn patients.

The immune system can become breached due to the compromise of the body's first line of defense which is the skin. Burn victims are more susceptible to infections and sepsis. Even though immunoresponses are activated in response to burns, in cases of severe burns, these responses are for a time suppressed. For example, neutrophil chemotaxis is hindered by decreased perfusion caused by hypovolemia (decrease in the volume of blood plasma) and the formation of small blood clots throughout the system also known as microthrombi. In addition, neutrophil activities are drastically reduced due to the reduction of oxygen delivery to the injured tissues. Furthermore, a drop in IgG antibodies could be detected as both B- and T-lymphocytes are depressed due to the injury.

Aseptic technique should be used when performing dressing changes. Protective isolation using neutropenic precautions should be implemented to protect the client. Topical antibacterial agents with regular changes in dressing should be implemented. When performing a dressing change, it is important to monitor the room temperature, as well as, to perform the task quickly to prevent chilling and possible hypothermia. Skin cultures taken from the wounds should be performed on a regular basis until instructed otherwise. Fluid blisters less than 2 cm are usually left intact. Larger fluid blisters (>2cm) are broken, drained and debrided (removal of dead or damaged tissue). Hydrotherapy is also done once or twice a day for 30 minutes each session. Minimize the amount of pain by administering IV morphine or other opioids. Additionally, it is important to watch for swellings and compartment syndrome where the eschar is causing the tightening of skin.

Prevention of scarring is also important and custom made elastic pressure garments can be worn for six months to a year after grafting. These garments are worn at all times except for when bathing. Tubular support bands can also be used to prevent scarring. They should be applied within five to seven days after grafting. Prevention of contractures after grafting is another concern. Splinting is used to control contractures where it immobilizes body parts.

Nutrition will be important during the healing process. A high protein and high calorie diet is essential during this hypermetabolic time. Carbohydrates and fats will spare the metabolism of proteins needed to generate wound healing. Vitamin, mineral supplements and caloric counts should be performed. Careful monitoring of the patient's weight and blood glucose level is essential. After extubation (removal of tubing) a dysphagia screen should be performed to prevent aspiration of food and liquids.

Long term rehabilitation involves teaching the client to engage in aerobic exercises for muscle and cardio-vascular strength. Teach the client to clean the skin with mild soap and apply an emollient several times a day.

Hair

Hair is a common characteristic in humans and other mammalian species. At 5th-6th months of gestation (fetal development), the fetus is covered in a layer of light, delicate hair called lanugo. At term, long course and pigmented terminal hairs replace lanugo on the scalp, eye brows, and eye lashes. In addition, vellus hairs, which are short non-pigmented and fine, replace the lanugo on the rest of the body. At puberty, the terminal hairs begin to grow and replace vellus hairs at the pubic and axillary regions.

In general, the hair is divided into the shaft which protrudes above the surface of the skin and the root, which exists in the dermis (reticular layer) of the skin. The hair shaft is composed of columns of dead keratinized epithelial cells arranged in three layers. The three layers are the medulla, cortex, and cuticle. The medulla composes the central axis of the hair and is constructed out of two to three layers of cells containing soft keratin.

The cortex forms the bulk of the hair and consists of cells containing hard keratin. The cuticle is constructed out of layers of hard keratin and it surrounds the cortex and forms the hair surface. The difference between soft and hard keratin is based on their sulfur content and the amount of cysteine disulfide crosslinks that are formed between the molecules. For example, soft keratin will have fewer cysteine disulfide crosslinks thereby making them soft, while hard keratin have numerous cysteine

disulfide crosslinks, making them hard and inflexible. The edges of the cuticle cells overlap one another creating an edge similar to that of shingles on a roof. These edges help to hold the hair follicle in place.

The hair follicle is a tubular invagination of the epidermis and the dermis. The hair follicle consists of the dermal root sheath (connective tissue root sheath) and the epithelial root sheath. The dermal root sheath surrounds the epithelial root sheath and is composed of the tissue of the dermis. The epithelial root sheath, on the other hand, is an extension of the epidermis and possesses all the layers previously described. Towards the bulb of the hair follicle, the epithelial root sheath narrows to form a single layer called the stratum basale, which consist of undifferentiated stem cells that provide for hair follicle growth, as well as, the stem cells necessary for dermal tissue repairs. In addition, this layer also contains melanocytes that produce the skin pigment melanin. It is melanin that provides for hair color.

The base of the root is expanded to form the hair bulb. The hair bulb contains a mass of undifferentiated cells at a location known as the matrix. These undifferentiated cells also provide for hair growth. At the base of the hair bulb, right beneath the stratum basale, the dermis of the skin protrudes into the bulb forming the hair papilla. The hair papilla is comprised of blood vessels that nourish the cells of the hair (Figure 47).

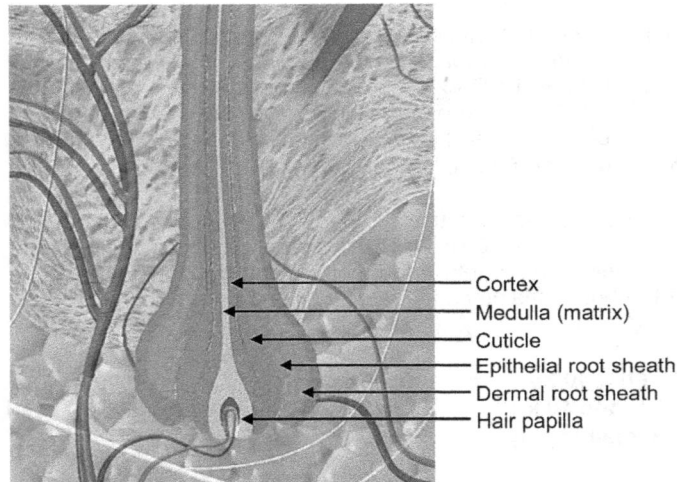

Cortex
Medulla (matrix)
Cuticle
Epithelial root sheath
Dermal root sheath
Hair papilla

Figure 47: Illustration of a hair follicle

Hair growth proceeds through two phases: the growth phase and the resting phase. It is known that the pattern of hair growth is dependent on the type of hair. For example, eyelashes grow for about 30 days and rests for 105 days while the scalp hair grows for three years and rests for one to two years.

Associated with the hair follicles is a type of smooth muscle called arrector pili muscle. Since, it is a smooth muscle, it is involuntary and not under conscious control. Nonetheless, the rise in adrenaline levels when a person is emotionally charged (e.g. angry or scared) will cause the arrector pili to contract, causing the hair to stand on its end. In addition, the contraction of the arrector pili muscles gives off the appearance of goose bumps.

The associated gland with the hair follicle is called the sebaceous gland. The sebaceous glands are a holocrine type of exocrine gland that consists of specialized epithelial cells that produce an oily secretion along with cellular debris called sebum. The sebaceous glands keep the skin and hair moist and waterproof.

Nails

The keratin cells produce nails in a stratified squamous format similar to that of the epidermis. The nail is composed of a layer of dead stratum corneum cells filled with hard keratin. The visible part of the nail is called the nail body and the part that lies under the skin is called the nail root. The nail root is held into place by a fold of skin called the nail fold, as well as, the underlying nail bed. A small portion of the nail fold extends on the nail body, and this area is called eponychium or the cuticle. The nail root is constructed distally of the nail matrix which consists of the mitotically active stem cells located in the stratum basale. A small part of the nail matrix called the lunula is visible generally beneath the cuticle. The lunula is a pale, crescent shaped area that could equate to the various subsequent stratum of the skin (e.g. stratum spinosum and granulosum). Unlike the hair, nails do not proceed into a resting phase but grow constantly until death (Figure 48).

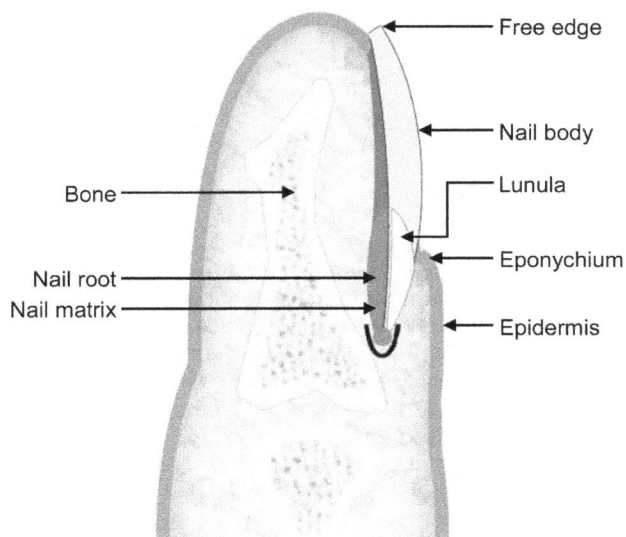

Figure 48: Graphic illustration of the nail and finger

Glands of the Skin

The major glands found within the dermal layer of the skin are the sebaceous and the sweat glands. Sebaceous glands can be either simple or compound alveolar glands that are connected to the superficial parts of the follicle via a duct (Figure 20). As mentioned previously, a sebaceous gland is a type of holocrine exocrine gland where the entire secretory cell disintegrates as their products are being released. The exocrine product of the sebaceous gland is called sebum, which is an oily white substance, high in lipid content. Sebum is formed to lubricate the hair and skin, preventing desiccation (Figure 44).

There are four types of sweat glands, also known as sudoriferous glands, found within the human body. For example, eccrine sweat glands appear throughout the body but are most prevalent in the forehead, neck, back, palms of the hands, and soles of the feet. Eccrine sweat glands are exocrine glands that consist of a ball-shaped coil in the deeper reticular layer of the dermis or superficial subcutaneous layer. The coils are lined with sweat-secreting epithelial cells and the sweat produced is carried away via a sweat duct to the pore, which is located on the surface of the skin. Eccrine sweat glands are prompted to produce their products when the body temperature starts to rise above homeostatic levels. Put simply, the sweat helps the body release heat and helps return the core temperature back to normal levels. In addition, the eccrine sweat glands could also release some of the similar waste products that the kidneys release. It has been estimated that there are 3 to 4 million eccrine sweat glands found in the adult human skin (Figure 44).

Apocrine sweat glands are less common when compared with

eccrine sweat glands and only appear in the groin, anal, areola and axilla regions. Apocrine sweat glands may also be found in bearded areas of the adult male. However, based on recent research, there is a racial component to the appearance of this gland. For example, apocrine glands are absent from or rarely found in the axillary regions of Asians.

In comparison to the eccrine sweat glands, the lumen of the apocrine sweat glands are much wider and they produce a thicker form of sweat. The thickness of the apocrine sweat is due to the high fatty acid content of the secretions. Apocrine sweat glands are the human version of scented glands. They are prompted to produce and secrete their exocrine product when the person becomes sexually aroused, upset, frightened, or in pain. It is assumed that they possess pheromone secretions (Figure 44).

Other minor populations of modified sweat glands may also be found in the skin. For example, the ceruminous glands are modified eccrine sweat glands found in the auditory canal of the ears. These glands produce an exocrine product called cerumen or ear wax. The mammary glands are modified apocrine sweat glands, located within the breasts responsible for milk production.

TEMPERATURE REGULATION

The maintenance of a constant body temperature is essential in the survival of an individual since most of the enzymes within the human body catalyze chemical reactions in narrow temperature parameters. Put simply, if the temperature rises or drops below preset levels, the enzymes that produce the products the body needs will begin to malfunction.

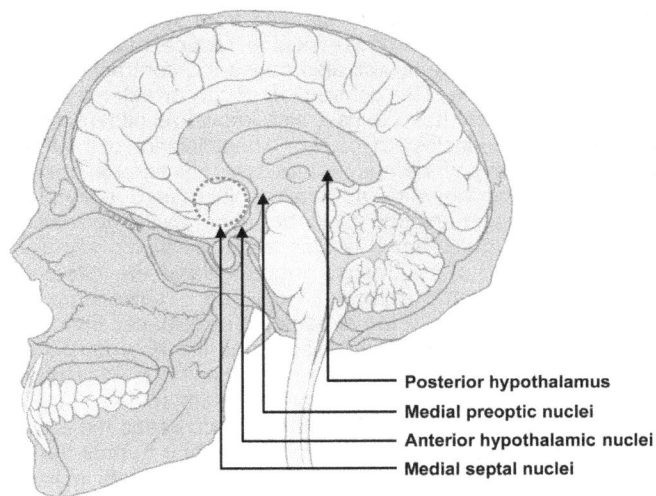

Figure 49: Temperature regulation centers which includes medial preoptic nuclei, anterior hypothalamic nuclei, neurons in the adjoining medial septal nuclei (POAH) and the posterior hypothalamus

The normal body temperature set-point is primarily determined by the activity of neurons in the medial preoptic nuclei, anterior hypothalamic nuclei, and the neurons in the adjoining medial septal nuclei. Cooperatively, these three regions form the preoptic-anterior hypothalamus (POAH). The second region for thermoregulation is located in the posterior hypothalamus. For example, the information regarding temperature is transmitted from the temperature receptors, also known as peripheral thermo-receptors, found in the skin and mucous membranes to the temperature-regulating centers in the POAH. In addition, internal temperature sensors, also known as central thermo-receptors, convey information regarding the core body temperature (e.g. blood temperature), which includes temperature information from the hypothalamus itself, to the POAH. These signals are then combined in the posterior hypothalamus, analyzed and integrated. In turn, necessary controls are triggered to increase

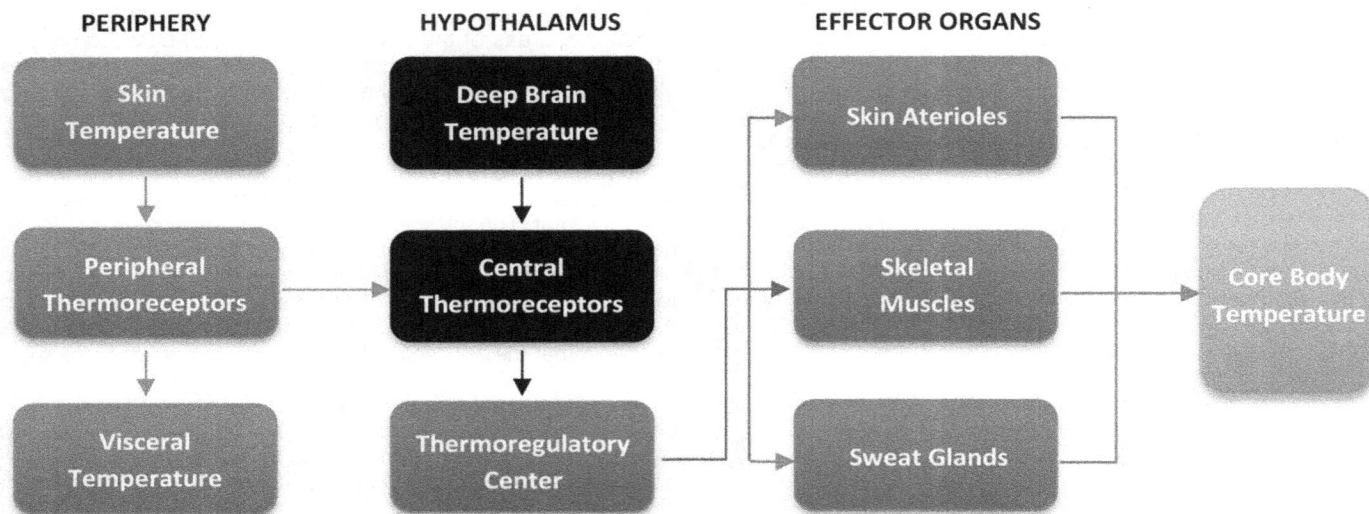

Figure 50: Diagrammatic view of thermoregulation within the human body

or decrease heat producing processes, as well as initiating heat conservation or release mechanisms of the body (Figure 49).

For example, if the temperature of the body is detected to be lower than the average 37.0 °C (98.6 °F), the thermoregulation centers in the POAH, in conjunction with the posterior hypothalamus, will trigger responses that promote heat gain. Immediately the sympathetic centers are activated where the norepinephrine, also known as noradrenaline, released from sympathetic fibers constricts blood vessels of the skin to reduce heat loss. The release of norepinephrine will also cause the contraction of the arrector pili muscles. The contraction of the arrector pili will in turn cause the body hairs to rise, also known as piloerection, in an effort to trap heat close to the skin. The activation of the POAH and posterior hypothalamus will cause the adrenal medulla to release epinephrine or adrenaline. Increased epinephrine secretion will trigger an increase in thermogenesis. The shivering center in the hypothalamus is also activated which stimulates the motor centers of the brainstem and initiates the involuntary contraction of skeletal muscles causing shivering and generating heat. Finally, the hypothalamus releases thyrotropin releasing hormone (TRH) which activates the anterior pituitary gland to release thyroid stimulating hormone (TSH). TSH triggers the thyroid to release thyroxin (T_4) and triiodothyronine (T_3) which in turn increases metabolism and thereby increases the amount of heat that the body generates. It is interesting to note that the negative feedback system in response to cold also activates various compensatory behavioral responses such as huddling, increased voluntary physical activity, sheltering next to a heat source and/or wearing warm clothing. Nonetheless, if proper body temperature adjustments are not made, and if the body temperature drops to or below 34.4° C (94° F), hypothermia will inevitably occur (Figures 49 and 50).

In contrast, if the temperature of the body increases above the pre-set values, the thermoregulation centers in the POAH, in conjunction with the posterior hypothalamus, will trigger responses that promote heat loss. For example, the thermoregulation centers will once again activate the autonomic nervous system and prompt the bulbous secretory coils of the eccrine sweat glands of the skin to produce and secrete sweat. It is understood that the excretion of sweat allows for heat loss through evaporate cooling (Figure 50).

Although the nervous innervation of the sweat gland is not completely understood, evidence for cholinergic (parasympathetic) rather than the adrenergic (sympathetic) neurotransmitters being responsible for triggering the process.

For example, Shibasaki *et al.* (2005) shows that the process of sweating could be initiated through the binding of acetylcholine to muscarinic receptors on the eccrine sweat gland. This evidence was found when atropine (a muscarinic-receptor antagonist) was administered both locally and systemically which greatly impeded and/or completely abolished sweating during a thermal challenge experiment. In addition to triggering sweating, the cholinergic nerves also cause the arrector pili muscles to relax, thereby causing the hairs on the skin to lay flat against the skin. The relaxation of the arrector pili muscles, and by extension the body hairs, will prevent heat from being trapped by the layer of still air between the hairs and increasing heat loss by convection (Figure 50).

Finally, the cholinergic nerves also trigger arteriolar vasodilation (the smooth muscle walls of the arterioles relax allowing increased blood flow through the artery). Vasodilation allows the blood flow to increase to the superficial capillaries in the skin, thereby increasing heat loss by convection and conduction (Figure 50).

Please note that most animals can't sweat efficiently. Horses and humans are two of the few animals capable of sweating to eliminate excess heat from the body core. For example, dogs and cats have sweat glands only on the pads of their feet and they dissipate excess heat through panting. It is known that dogs and cats possess large surface areas within their lungs where expulsion of air promotes heat loss.

QUESTIONS

1. Which of the following form the external coverings and the internal lining of organs or the organism?
   a. Connective tissue
   b. Epithelial tissue
   c. Nervous tissue
   d. Muscular tissue

2. Which part of the epithelial cell is directed at the lumen?
   a. Apical
   b. Basal
   c. Lateral
   d. Oblique

3. Which domain of the epithelial tissue is anchored to the basememt membrane?
   a. Apical
   b. Basal
   c. Lateral
   d. Oblique
   e. Diagonal

4. Which of the following membrane protrusions located on the epithelial cell is formed to increase surface area for secretion of absorption?
   a. Cilia
   b. Flagella
   c. Microvilli
   d. Villi

5. Which of the following membrane protrusions located on the epithelial cell is formed to provide movement?
   a. Cilia
   b. Flagella
   c. Microvilli
   d. Villi

6. Which of the following molecular components of the basement membrane is formed to provide physical support and creates a scaffold for the other structural macromolecules for built upon?
   a. Collagen IV
   b. Entactin
   c. Fibronectin
   d. Hyaluronic acid

7. Which of the following is a type or are types of adhesive glycoproteins that interact with integrin, an integral membrane protein?
   a. Collagen IV
   b. Entactin
   c. Fibronectin
   d. Hyaluronic acid
   e. Laminin
   f. Proteoglycans
   g. Both c and e

8. Which of the following basement membrane glycoprotein(s) is (are) formed to aid in affixing the epithelial cells to the basement membrane?
   a. Collagen IV
   b. Entactin
   c. Fibronectin
   d. Hyaluronic acid
   e. Laminin
   f. Proteoglycans
   g. Both c and e

9. Which of the following protein once bounded to its counterpart in the basement membrane is responsible for the intracellular arrangement of cytoskeletons and in turn will have definitive effect on cell adhesion, shape, migration, proliferation, and differentiation?
   a. Collagen IV
   b. Entactin
   c. Fibronectin
   d. Integrin

10. Which of the following serves as a scaffolding protein and connects the adhesive glycoproteins to collagen in order to form a structure network? This structural network is essential in maintaining the structural integrity of the basement membrane but also provide the necessary foundation for the attachment of the epithelial tissue.
    a. Collagen IV
    b. Entactin
    c. Fibronectin
    d. Hyaluronic acid

11. Heparan sulfate, dermatan sulphate and Chondroitin sulfate are three types of _____ that contains sulphate that are found within the basement membrane.
    a. Chondroitin
    b. Glycosaminoglycan
    c. Keratin
    d. Keratohyalin granules

12. When glycosminoglycans are bounded to a protein core they are referred to as a _____.
    a. Collagen IV
    b. Entactin
    c. Fibronectin
    d. Hyaluronic acid
    e. Proteoglycan

13. Which of the following is a non-sulfate containing GAG?
    a. Collagen IV
    b. Entactin
    c. Fibronectin
    d. Hyaluronic acid

14. Both sulfate and non-sulfate containing GAGs are _____ charged where they repulse _____ (type of ions) and attract _____ (type of ions) like sodium. The attraction towards sodium in turn will attract _____ which will enable the basement membrane to resist compression and serves as a diffuse/osmosis

friendly medium for the exchange of waste materials, nutrients and gases.
   a. Anions
   b. Cations
   c. Negative
   d. Neutral
   e. Positive
   f. Water

15. _____ consist of a single layer of irregular shaped cells that are commonly found in the _____ (Hint: name the particular system that this epithelial tissue is often found). These cells normally possess mobile hair like projections called _____. Scattered along this epithelial tissue are numerous mucous secreting cells called _____. It is understood that these mucous secreting cells belong to the _____ type of exocrine gland that is responsible for producing a protein compound called _____. This compound when mixed with water will result in the formation of mucous.

_____ consist of many layers of flattened cells (~60 layers depending upon the location). This tissue is commonly found in _____ (Hint: name one organ that this epithelial tissue is often found) and forms the protective layer (water proof layer) that protects the internal systems from foreign materials, viruses, bacteria and various chemicals.

_____ consist of several layers of cells that are specifically designed to stretch but also forms a barrier to protect from seeping wastes into the surrounding tissues and organs. This epithelial tissue is commonly found in the _____ (Hint: name one organ that this epithelial tissue is commonly found).

_____ consists of a single layer of square shaped cells that are involved in absorption and secretion of glandular substances. This epithelial tissue is commonly found in the _____ (Hint: name one gland that this epithelial tissue is commonly found), a gland that is formed to secrete hormones.

_____ consists of a single layer of rectangular shaped cells and lines most glands/organs found in the _____ (Hint: name the particular system that this epithelial tissue is often found). Because of their thickness, these cells are responsible for secreting enzymes/acids, absorption and protection. It is known that many of these rectangular shaped cells possess specialized projections called _____ which are designed to _____. It is understood that these projections are located on the _____ (Hint: the particular domain) of the cell.

_____ are comprised of a thin (single layer) flattened cells that are involved in _____ which is defined as the process of transporting substances from areas of high concentration to areas of low concentration. These epithelia are commonly found in the _____ (Hint: name a particular organ system).

_____ consist of several layers of elongated rectangular-shaped cells which are found in the _____ (Hint: name one organ that this epithelial tissue is commonly found).

_____ consist of two or three layers of square shaped cells that provide protection to various ducts in the human body. For example, this epithelial tissue is designed to line the ducts of the _____ (Hint: name one gland that this epithelial tissue is commonly found).

1.	Active transport	31.	Hydrogen bond
2.	Anionic	32.	Increase surface area
3.	Apical domain	33.	Integrin
4.	Apocrine	34.	Ionic bond
5.	Basal domain	35.	Laminin
6.	Basement membrane	36.	Lateral domain
7.	Calcium	37.	Loose areolar CT
8.	Cationic	38.	Magnesium
9.	Chloride	39.	Maltose
10.	Cilia	40.	Mammary gland
11.	Circulatory system	41.	Merocrine
12.	Collagen	42.	Microvilli
13.	Covalent bond	43.	Mucin
14.	Dense irregular CT	44.	Potassium
15.	Dense irregular elastic CT	45.	Provide movement
16.	Dense regular CT	46.	Pseudostratified columnar
17.	Dense regular elastic CT	47.	Respiratory system
18.	Digestive system	48.	Simple columnar
19.	Diffusion/osmosis	49.	Simple cuboidal
20.	Elastic cartilage	50.	Simple squamous
21.	Elastic fiber	51.	Small
22.	Epidermis	52.	Sodium
23.	Fibrocartilage	53.	Stratified columnar
24.	Fibronectin	54.	Stratified cuboidal
25.	Goblet cells	55.	Stratified squamous
26.	Heparan sulfate	56.	Thyroid gland
27.	Holocrine	57.	Transitional
28.	Hyaline cartilage	58.	Urinary bladder
29.	Hydrogen	59.	Vas deferens

16. There are three types of muscle fibers. The _____ is the muscle of physical movement (e.g. biceps, triceps etc.) This muscle gives a physical appearance of striations thereby also known as a type of striated muscle.

The _____ is also a type of striated muscle and it provides the contractility of our heart. The cells of this heart muscle are connected end-to-end via the _____.

The _____ contains no striations and they comprise the bulk of the muscles of our internal organs.
   a. Skeletal muscle
   b. Cardiac muscle
   c. Smooth muscle
   d. Intercalated disk
   e. Both a and b
   f. Both b and c

17. Which of the following muscles are the most abundant in the human body?
   a. Skeletal muscle
   b. Cardiac muscle
   c. Smooth muscle
   d. Intercalated disk
   e. Both a and b
   f. Both b and c

18. Which of the following muscles are striated muscles?
   a. Skeletal muscle
   b. Cardiac muscle
   c. Smooth muscle
   d. Intercalated disk
   e. Both a and b
   f. Both b and c

19. Which of the following are involuntary muscles?
   a. Skeletal muscle
   b. Cardiac muscle
   c. Smooth muscle
   d. Intercalated disk
   e. Both a and b
   f. Both b and c

20. Which of the following is (are) voluntary muscle(s)?
   a. Skeletal muscle
   b. Cardiac muscle
   c. Smooth muscle
   d. Intercalated disk
   e. Both a and b
   f. Both a and c

21. Commonly, the skeletal muscles will connect to bones of other muscles via which of the following to provide movement for the body?
   a. Aponeurosis
   b. Bone
   c. Cartilage
   d. Connective tissue
   e. Epithelial tissue
   f. Nervous tissue
   g. Tendon
   h. Both a and g

22. In certain locations of the body (e.g. the cranium), the skeletal muscles will be attached to other skeletal muscles via a flat sheet-like connective tissue called _____.
   a. Aponeurosis
   b. Bone
   c. Cartilage
   d. Connective tissue
   e. Epithelial tissue
   f. Nervous tissue

23. Intercalated disks contain numerous junctions. Which of the following is (are) the type(s) of junction(s) most often found in this location?
   a. Gap junction
   b. Tight junction
   c. Adhesion junction
   d. Facia adherens
   e. Both a and d
   f. Both a and b

24. Which of the following has been stated to be the most abundant glycoprotein in the animal kingdom?
   a. Collagen
   b. Elastin
   c. Eleidin
   d. Entactin
   e. Fibronectin

   f. Integrin
   g. Laminin

25. Which of the following is an insoluble (hydrophobic) fibrous glycoprotein that is responsible for providing structural strength for the extracellular matrix?
   a. Collagen
   b. Elastin
   c. Eleidin
   d. Entactin
   e. Fibronectin
   f. Integrin
   g. Laminin

26. In the extracellular matrix, collagen is produced and secreted by which of the following cell(s)?
   a. Endothelial cells
   b. Epithelial cells
   c. Erythrocytes
   d. Fibroblasts
   e. Leukocytes
   f. Both a and d
   g. Both b and d

27. Generally, collagen molecules are assembled into _____ which are ~300 nm in length.
   a. Collagen fiber
   b. Collagen fibrils
   c. Elastin
   d. Integrin
   e. Laminin
   f. Linea alba

28. It takes _____ (how many) collagen fibrils to assemble into a collagen fiber? Two of the collagen fibril will have the same molecular formula and they are referred to as _____ while the third fibril (which has a slightly different formula) is called _____.
   a. 1
   b. 2
   c. 3
   d. 4
   e. α1
   f. α2
   g. α3
   h. α4

29. Presently, how many different types of collagen are known to scientists?
   a. 5
   b. 10
   c. 16
   d. 21
   e. 30

30. The majority of collagen found within the human body is (are) which of the following type(s). Please circle all correct answers; you may circle more than one answer.
   a. Collagen I
   b. Collagen II
   c. Collagen III
   d. Collagen IV
   e. Collagen V
   f. Collagen X
   g. Collagen XI
   h. Collagen XXI

31. Similar to the composition of collagen, elastin is composed of long sequences of which of the following amino acid(s)?
   a. Asparagin
   b. Glutamic acid
   c. Glycine
   d. Leucine
   e. Methionine
   f. Proline
   g. Both a and c
   h. Both c and f

32. Which of the following amino acid provides the elastin with elasticity?
   a. Alanine
   b. Asparagin
   c. Glutamic acid
   d. Glycine
   e. Leucine
   f. Lysine
   g. Methionine
   h. Proline

33. Which of the following amino acid found in elastin creates crosslink with

various other molecules?
a. Alanine
b. Asparagin
c. Glutamic acid
d. Glycine
e. Leucine
f. Lysine
g. Methionine
h. Proline

34. Which of the following form an elastic fiber?
a. Collagen protein
b. Elastin protein
c. Fibrillin
d. Fibrin
e. Fibrinogen
f. Both a and c
g. Both b and c

35. Which of the following protein(s) is (are) responsible for forming crosslinks with elastin proteins?
a. Fibrillin 1
b. Fibrillin 2
c. Collagen
d. Fibrin
e. Fibrinogen
f. Both a and b
g. Both b and c

36. Elastin is produced and secreted by which of the following cell type(s)?
a. Endothelial cells
b. Epithelial cells
c. Erythrocytes
d. Fibroblasts
e. Leukocytes
f. Both a and d
g. Both b and d

37. Which of the following is a type of gland that secretes its product directly into blood stream?
a. Endocrine glands
b. Exocrine glands

38. Which of the following is a type of gland that secretes its product directly into a duct?
a. Endocrine glands
b. Exocrine glands

39. Which of the following exocrine gland secretes its product into a single duct?
a. Simple gland
b. Compound gland
c. Multiple gland
d. Complex gland

40. Which of the following exocrine gland secretes its product into a multiple ducts?
a. Simple gland
b. Compound gland
c. Multiple gland
d. Complex gland

41. Which of the following exocrine gland's secreted products culminates in a small single tube?
a. Compound acinar
b. Compound tubular
c. Compound tubulo-acinar
d. Simple acinar
e. Simple branched acinar
f. Simple branched tubular
g. Simple coiled tubular
h. Simple tubular

42. Which of the following exocrine gland's secreted products culminates in a small single tube that is twisted at the end?
a. Compound acinar
b. Compound tubular
c. Compound tubulo-acinar
d. Simple acinar
e. Simple branched acinar
f. Simple branched tubular
g. Simple coiled tubular
h. Simple tubular

43. Which of the following exocrine gland's secreted products is produced in a sac-like structure which culminates in a small single tube?
a. Compound acinar

b. Compound tubular
c. Compound tubulo-acinar
d. Simple acinar
e. Simple branched acinar
f. Simple branched tubular
g. Simple coiled tubular
h. Simple tubular

44. Which of the following exocrine gland's secreted products is produced in multiple sac-like structures? Hint: no more than 3 sac-like structures.
a. Compound acinar
b. Compound tubular
c. Compound tubulo-acinar
d. Simple acinar
e. Simple branched acinar
f. Simple branched tubular
g. Simple coiled tubular
h. Simple tubular

45. Which of the following exocrine gland's secreted products culminates in a multiple small tubes?
a. Compound acinar
b. Compound tubular
c. Compound tubulo-acinar
d. Simple acinar
e. Simple branched acinar
f. Simple branched tubular
g. Simple coiled tubular
h. Simple tubular

46. Which of the following exocrine gland secretes its products via exocytosis?
a. Merocrine glands
b. Apocrine gland
c. Holocrine gland
d. Mucous cells
e. Serous cells

47. Which of the following type(s) of exocrine cell(s) produce(s) and secrete(s) a protein/water rich compound?
a. Tubular glands
b. Apocrine gland
c. Holocrine gland
d. Mucous cells
e. Serous cells
f. Both a and b
g. Both d and e

48. Which of the following exocrine cell produces and secretes an enzyme/water rich compound?
a. Tubular glands
b. Apocrine gland
c. Holocrine gland
d. Mucous cells
e. Serous cells
f. Both a and b
g. Both d and e

49. Which of the following protein is needed to form mucous?
a. Mucin
b. Collagen
c. Elastin
d. Fibrinogen
e. Alpha amylase

50. Which of the following exocrine gland loses parts of its cell during secretion therefore its released products possess both secretory products, as well as cellular fragments?
a. Merocrine glands
b. Apocrine gland
c. Holocrine gland
d. Mucous cells
e. Serous cells

51. Which of the following exocrine gland completely disintegrates through a form of apoptosis during secretion?
a. Merocrine glands
b. Apocrine gland
c. Holocrine gland
d. Mucous cells
e. Serous cells

52. Which of the following is (are) example(s) of apocrine exocrine glands?
a. Ceruminous glands
b. Mammary glands
c. Parotid gland
d. Sebaceous gland

e.　Sweat gland
f.　Both a and b
g.　Both c and e

53. Which of the following is (are) example(s) of merocrine exocrine glands?
    a.　Ceruminous glands
    b.　Mammary glands
    c.　Parotid gland
    d.　Sebaceous gland
    e.　Sweat gland
    f.　Both a and b
    g.　Both c and e

54. Which of the following is (are) example(s) of holocrine exocrine glands?
    a.　Ceruminous glands
    b.　Mammary glands
    c.　Parotid gland
    d.　Sebaceous gland
    e.　Sweat gland
    f.　Both a and b
    g.　Both c and e

55. Which of the following is (are) fixed (connective tissue) cells?
    a.　Fibroblasts
    b.　Pericytes
    c.　Mast cells
    d.　Adipocytes
    e.　All of the above

56. Which of the following is a type of cell that possesses the capability to differentiate into smooth muscle cells and/or endothelial cells after injury?
    a.　Fibroblasts
    b.　Pericytes
    c.　Mast cells
    d.　Adipocytes
    e.　All of the above

57. Which of the following is a type of cell that contains heparin, histamine, serotonin, thromboxane, prostaglandin, leukotriene and cytokines within its cytoplasm and plays an important role in the inflammation process?
    a.　Fibroblasts
    b.　Pericytes
    c.　Mast cells
    d.　Adipocytes
    e.　All of the above

58. Which of the following is a type of cell that is filled with lipids and function as energy storage cells for the mammalian body?
    a.　Fibroblasts
    b.　Pericytes
    c.　Mast cells
    d.　Adipocytes
    e.　All of the above

59. Which of the following is known as adult fat cells?
    a.　Brown fat cell
    b.　Multilocular fat cells
    c.　Unilocular fat cells
    d.　White fat cell
    e.　Both a and b
    f.　Both c and d

60. Which of the following is a type of fat cell that exists in young children?
    a.　Brown fat cell
    b.　Multilocular fat cells
    c.　Unilocular fat cells
    d.　White fat cell
    e.　Both a and b
    f.　Both c and d

61. Which of the following cell is anucleated?
    a.　Chondrocyte
    b.　Erythrocyte
    c.　Fibrocyte
    d.　Leukocyte
    e.　Osteocyte

62. What is the life expectancy of an erythrocyte?
    a.　~50 days
    b.　~60 days
    c.　~90 days
    d.　~120 days
    e.　~160 days

63. Which of the following cell cannot repair itself when damaged and has to be

removed and replaced?
    a.　Chondrocyte
    b.　Erythrocyte
    c.　Fibrocyte
    d.　Leukocyte
    e.　Osteocyte

64. Erythrocytes consist mainly of a complex globin protein and iron structure called _____.
    a.　Elastin
    b.　Fibrin
    c.　Fibrinogen
    d.　Heme
    e.　Hemoglobin

65. Carbonic acid is produced by which of the following membrane bounded protein?
    a.　Adenylate phosphate
    b.　Alpha amylase
    c.　Carbonic anhydrase
    d.　Protein kinase

66. Which of the following is also known as human blood buffer?
    a.　Carbonic acid
    b.　Sodium hydroxide
    c.　Hydrochloric acid
    d.　Sulfuric acid
    e.　Magnesium phosphate

67. Which of the following is (are) known as granulocyte(s)?
    a.　Neutrophil
    b.　Eosinophil
    c.　Basophil
    d.　All of the above

68. Which of the following is (are) known as agranulocyte?
    a.　Eosinophil
    b.　Lymphocyte
    c.　Monocyte
    d.　Neutrophil
    e.　Both b and c

69. Which of the following embryonic connective tissue, also known as Wharton's jelly, is formed to support umbilical blood vessels?
    a.　Mucoid
    b.　Mesenchymal
    c.　Loose connective tissue
    d.　Dense connective tissue

70. Which of the following is a type of embryonic stem cell found in the middle germ layer of the embryo, also known as the mesoderm?
    a.　Mucoid
    b.　Mesenchymal
    c.　Loose connective tissue
    d.　Dense connective tissue

71. Which of the following is the type of connective tissue that composes the basement membrane?
    a.　Dense irregular
    b.　Dense irregular elastic
    c.　Dense regular
    d.　Dense regular elastic
    e.　Loose areolar
    f.　Reticular tissue

72. Which of the following is the type of connective tissue that composes the adventitia?
    a.　Dense irregular
    b.　Dense irregular elastic
    c.　Dense regular
    d.　Dense regular elastic
    e.　Loose areolar
    f.　Reticular tissue

73. Dense connective tissues consist of closely packed collagenous fibers and a fine network of elastic fibers. There are four types of dense connective tissue. For example:

    The _____ is composed of mostly thick matrix of elastic fibers with very little collagen. This type of tissue if found in the walls of arteries

    The _____ is composed of collagen fibers (majority) and some elastic fibers. It is found in most fibrous connective tissue capsules of the body as well as the submucosa of the digestive tract

    The _____is composed of mostly thick matrix of elastic fibers with very little collagen. This type of tissue if found in the vocal cords

The _____ is composed mainly of collagen fibers with very little elastic fibers. This type of tissue is found in tendons, aponeuroses (e.g. linea alba) and ligaments
a. Loose areolar connective tissue
b. Dense irregular connective tissue
c. Dense regular connective tissue
d. Dense irregular elastic connective tissue
e. Dense regular elastic connective tissue

74. Which of the following connective tissue is composed mainly of collagen III fibers and are commonly found around the liver, kidneys and form the internal structures that make up the bone marrow, spleen, thymus and lymph nodes?
a. Dense irregular
b. Dense irregular elastic
c. Dense regular
d. Dense regular elastic
e. Loose areolar
f. Reticular tissue

75. _____ is a type of connective tissue that is mainly composed of collagen and elastic fibers (in varying concentrations depending on the specific type) embedded in a gel-like ground substance. There are 3 different types of this kind of connective tissue.

The _____ (**Answer 1**) is generally found at the ends of bones and provides protection against abrasion and pressure between the bones. In addition it is also found in the soft parts of the nose and the supporting rings of the respiratory passage (e.g. trachea, bronchi etc.).

The _____ (**Answer 2**) consists of network of elastic fibers and provides the framework for the external ear and the larynx.

The _____ consists of many collagenous fibers and very small amounts of elastic fibers. This type of connective tissue is extremely strong and serves as the cushion between the vertebrae. The _____ (**Answer 3**) is the specific name of this connective tissue between the vertebrae.
a. Cartilage
b. Ligaments and tendon
c. Bone
d. Blood
e. Elastic cartilage
f. Fibrocartilage
g. Hyaline cartilage
h. Intervertebral disk

76. Which of the following is also known as bone cells?
a. Chondrocyte
b. Erythrocyte
c. Fibroblast
d. Leukocyte
e. Osteocytes

77. Bone cells secrete both collagen and _____ to create the extracellular matrix or the bulk materials that create bones.
a. Hydroxyapatite
b. Hyaline cartilage
c. Fibrocartilage
d. Elastic cartilage
e. Reticular tissue

78. Lowest Layer of skin

The human skin is composed of multiple layers of interlining tissues. The lowest layer of the skin (the layer superior to muscles and bones etc.) is called _____ (**Answer 1**) and is composed of blood vessels, loose connective tissues and an energy storage and thermo-regulating fat layer called _____. This fat tissue is composed of mature fat cells called _____.

Middle layer of skin

On the superior side of the previously mentioned layer (**Answer 1**) is another distinct layer of skin tissue called _____ (**Answer 2**). This middle layer of skin is subdivided into two distinct layers. For example, the inferior sub-layer is called _____ (Hint: provide the specific name for the layer right above **Answer 1**). This layer consists mostly of irregular intertwining _____ fibers which are designed to provide support. These protein fibers do not stretch but are flexible and have high tensile strength. The specific name given to this type of connective tissue is _____ (Hint: provide the specific name for this type of connective tissue).

The superior sub-layer (second sub-layer) is called _____. This layer is named after the _____ which are structure contusions that are responsible for tying down the superior (adjacent) layer of connective tissue, and of course, the upper skin.

This middle layer of skin is responsible for providing blood supplies (e.g. nutrients and gases) to all superior skin cells, as well as used to regulate body temperature. In addition, there are numerous exocrine glands located within this layer. For example, one type of oil-secreting gland that exists within this layer is called _____. These specialized epithelial cells are responsible for producing an oily secretion along with cellular debris called _____ (Hint: name the product secreted) which is created in an effort to keep the surface of the skin (the layer that is exposed to the environment) as well as hair shafts moist and water proof. These glands are generally found immediately adjacent to the hair shafts and are categorized as a part of _____ type of exocrine glands. Numerous nerves are also found in this particular layer of skin. Most of the nerves that are located in this layer are sensory in origin and allow an individual to sense pain, temperature, pressure etc. These sensory nerves are also commonly referred to as _____ (Hint: going towards the brain).

The human hair is composed of the same type of epithelial cells that constructs the upper layer of skin. The base of the hair follicle is called _____ which is also located within (**Answer 2**). A particular type of smooth muscle called _____ is attached to the hair follicles and is responsible for raising and lowering the shaft. This smooth muscle is innervated by a particular involuntary nervous system called _____. This particular nervous system is subdivided into two major divisions. The first division is called _____ and produces a type of neurotransmitter called acetylcholine that causes the relaxation of this muscle. The second division is called _____ and produces a type of neurotransmitter called noradrenaline which causes the contraction of the above mentioned muscle.

Also existing in the middle layer of the skin are sweat glands which are also referred to as _____ (Hint: provide the scientific name for this glandular epithelium). These glands consist of a ball-shaped coil structure and are lined with sweat-secreting epithelial cells. These cells produce their product (e.g. sweat) and release them directly into the sweat duct and are released onto the surface of the skin via an opening called the _____.

There are various different types of sweat glands. For example, the _____ is a type of sweat gland that only produces its product when the person becomes upset, frightened or in pain. This is what is typically termed "cold sweat" and these glands are commonly found in the armpits, groin and along the hair follicles of the scalp. The _____ is a modified sweat gland that is locate within the breasts and secretes milk and _____ is a modified sweat gland that is located in the ear and produces waxes. The most common type of sweat gland is called _____ and it appears throughout the body but most prevalent in the forehead, neck, and back. This gland produces sweat during exercise and when the external temperature becomes too hot.

1.	Adipocytes	37.	Keratin
2.	Adipose tissue	38.	Mammary glands
3.	Afferents	39.	Melanocytes
4.	Alpha-amlylase	40.	Melanosome
5.	Apocrine	41.	Merocrine
6.	Apocrine sweat glands	42.	Mitosis
7.	Apoptosis	43.	Mucin
8.	Arrector pili	44.	Osmosis
9.	Autonomic nervous system	45.	Papillary layer
10.	Basement membrane	46.	Parasympathetic
11.	Ceruminous glands	47.	Phagocytosis
12.	Collagen	48.	Pheomelanin
13.	Cornified cells	49.	Pinocytosis
14.	Dendritic (Langerhans) cells	50.	Pore
		51.	Reticular layer
15.	Dense irregular	52.	Root
16.	Dense irregular elastic	53.	Sebaceous gland
17.	Dense regular	54.	Sebum
18.	Dense regular elastic	55.	Simple columnar
19.	Dermal papillae	56.	Simple cuboidal
20.	Dermis	57.	Simple squamous
21.	Desquamate	58.	Stem cells
22.	Diffusion	59.	Stratified columnar
23.	Eccrine sweat glands	60.	Stratified cuboidal
24.	Efferents	61.	Stratified squamous
25.	Elastin	62.	Stratum basal
26.	Epidermis	63.	Stratum corneum
27.	Eumelanin	64.	Stratum granulosum
28.	Exocytosis	65.	Stratum lucidum
29.	Glucosylceramides	66.	Stratum spinosum
30.	Holocrine	67.	Sudoriferous glands
31.	Hypodermis	68.	Sympathetic
32.	Keratinocytes	69.	Tactile (Merkel) cells
33.	Keratohyalin granules	70.	Thick skin
34.	Lamellar bodies	71.	Thin skin
35.	Loose areolar	72.	Ultraviolet (UV)
36.	Keratin fibers		

79. Upper layer of skin

On the superior side of the previously mentioned layer of skin (**Answer 2**) a thinly constructed connective tissue layer known as _____ (**Answer 3**). This layer is consists of a type of loosely arranged fluid filled connective tissue called _____ and it is the site of attachment for the upper most layers of skin as well as allowing diffusion of nutrients and gases from **Answer 2**.

The upper most layer of the human skin, also known as the _____, is composed of _____ type of epithelial tissue. This upper layer of skin is composed of many different sub-layers in chronological progression towards the surface of the body.

The first sub-layer immediately above **Answer 3** (the previous mentioned loosely arranged connective tissue layer) is called _____. This first layer consists of a single layer of cells. It is understood that there are three types of cells located within this layer. The most predominant type of cell is called _____ (Hint: pre-skin cells) which is constantly undergoing a type of cellular division called _____ to form skin cells called _____. This active division is responsible for forming new cells that replace the dead and old cells that flaked off or _____ at the surface of the skin which later produces the dander found around your home.

The second type of cells that exist in the first sublayer are _____ which are responsible for producing skin pigments known as _____ which creates the coloration of the skin, as well as functions to shield the deeper tissues from a form of cancer causing radiation called _____. It is understood that this skin pigment comes in two primary forms in humans. The _____ is a black and brown pigment while _____ is a red pigment. Once these pigments are produced, they are packaged within a cellular vesicle called _____ and are subsequently absorbed by the newly divided skin cells called _____ (cell divided via mitosis).

The third type of cells that are found in this first sub-layer are _____ which function as pressure receptor for sensory nerves.

The next (superior) layer is called _____. Within this layer, the newly divided skin cells begin producing lipid filled membrane coating vesicles called _____ (also known as membrane coating vesicles) and stored within the cytoplasm. Contained within these membrane coating vesicles are water proofing lipids called _____, proteins and enzymes. In addition, these skin cells will begin forming a water proofing fibers called _____ that are also placed within the cytoplasm.

Also found but not limited to this second sub-layer is a type of white blood cells called _____ which are modified macrophages. These cells are responsible for _____ or "to eat" any foreign objects or organisms that may have penetrated the skin as a part of the primary immune response.

The next (superior) layer is called _____ which is a continuation of the previous layer of skin cells. The cells within this layer perform two preprogrammed duties. First, these skin cells will secrete the water proofing material called _____ from the membrane coating vesicle called _____ via a membrane assisted transport process called _____. This water-proofing material will coat the plasma membrane of the skin cells and make the cell impermeable to water and by extension, prevent these cells from absorbing nutrients and water via a passive transport process called _____ (absorbing nutrient and ions) and _____ (absorbing water) respectively. Second, the skin cells with this layer will begin producing _____ (**Answer 4**) which will immediately form cross-links with the previously mentioned water proofing protein fibers called _____ within the cytoplasm forming _____, a water-proof protein complex which is a part of the intermediate filament of the cytoskeleton, as well as promoting dehydration of the cell. In addition, the sheer numbers of the **Answer 4** will crush the organelles within the skin cells.

In summary, the inability of the skin cells to absorb nutrients, the destruction of the organelles, as well as the dehydration of these cells cause it to die. Since this form of cell death is part of the lifecycle of skin cells, we refer to this form of cell death as preplanned or _____.

The next (superior) layer is called _____. This layer only exists in the _____ which are found on the finger tips, palmar and plantar regions. At this point, all the skin cells are dead or have become _____ (another term for dead skin cells).

The most superior layer is called the _____ which consists of up to ~30 layers of dead skin cells The cells of this layer will continue to flake off in a scale-like fashion called dander.

1. Adipocytes
2. Adipose tissue
3. Afferents
4. Alpha-amlylase
5. Apocrine
6. Apocrine sweat glands
7. Apoptosis
8. Arrector pili
9. Autonomic nervous system
10. Basement membrane
11. Ceruminous glands
12. Collagen
13. Cornified cells
14. Dendritic (Langerhans) cells
15. Dense irregular
16. Dense irregular elastic
17. Dense regular
18. Dense regular elastic
19. Dermal papillae
20. Dermis
21. Desquamate
22. Diffusion
23. Eccrine sweat glands
24. Efferents
25. Elastin
26. Epidermis
27. Eumelanin
28. Exocytosis
29. Glucosylceramides
30. Holocrine
31. Hypodermis
32. Keratinocytes
33. Keratohyalin granules
34. Lamellar bodies
35. Loose areolar
36. Keratin fibers
37. Keratin
38. Melanin pigments
39. Melanocytes
40. Melanosome
41. Merocrine
42. Mitosis
43. Mucin
44. Osmosis
45. Papillary layer
46. Parasympathetic
47. Phagocytosis
48. Pheomelanin
49. Pinocytosis
50. Pore
51. Reticular layer
52. Root
53. Sebaceous gland
54. Sebum
55. Simple columnar
56. Simple cuboidal
57. Simple squamous
58. Stem cells
59. Stratified columnar
60. Stratified cuboidal
61. Stratified squamous
62. Stratum basal
63. Stratum corneum
64. Stratum granulosum
65. Stratum lucidum
66. Stratum spinosum
67. Sudoriferous glands
68. Sympathetic
69. Tactile (Merkel) cells
70. Thick skin
71. Thin skin
72. Ultraviolet (UV)

80. The medial preoptic nuclei, anterior hypothalamic nuclei, and the neurons in the adjoining medial septal nuclei belong to which of the following region?
a. Superior hypothalamus
b. Lateral hypothalamus
c. Posterior hypothalamus
d. Postoptic hypothalamus
e. Preoptic-anterior hypothalamus
f. Both a and e
g. Both c and e

81. The normal body temperature set-point is primarily determined by the activity of neurons in which of the following region(s)?
a. Superior hypothalamus
b. Lateral hypothalamus
c. Posterior hypothalamus
d. Postoptic hypothalamus
e. Preoptic-anterior hypothalamus
f. Both a and e
g. Both c and e

82. Which of the following transmits sensory information to the POAH?
a. Cardio thermo-receptor
b. Central thermo-receptors
c. Dermal thermo-receptors
d. Peripheral thermo-receptors
e. Vascular thermo-receptor
f. Both a and b
g. Both b and d

83. Which of the following is responsible for transmitting temperature information from the skin and mucous membranes to the temperature-regulating centers in the POAH?
a. Cardio thermo-receptor
b. Central thermo-receptors
c. Dermal thermo-receptors
d. Peripheral thermo-receptors
e. Vascular thermo-receptor

84. Which of the following is responsible for transmitting temperature information from the core body temperature (e.g. blood temperature), which includes temperature information from the hypothalamus itself, to the POAH?
a. Cardio thermo-receptor
b. Central thermo-receptors
c. Dermal thermo-receptors
d. Peripheral thermo-receptors
e. Vascular thermo-receptor

85. Which of the following location is responsible for analyzing, integrating and control the body's heat production process?
a. Anterior hypothalamus
b. Lateral hypothalamus
c. Posterior hypothalamus
d. Postoptic hypothalamus
e. Preoptic-anterior hypothalamus
f. Both a and e
g. Both c and e

86. If the temperature of the body is detected to be lower than the average 37.0 °C (98.6 °F), the thermoregulation centers in the _____, in conjunction with the _____, will trigger responses that promote heat gain. Immediately, the _____ (a nervous system) are activated which in turn activates one of its two

subdivisions called _____ (sympathetic or parasympathetic). The activation of this subdivision will cause the _____ (sympathetic or parasympathetic) nerves to secrete its neurotransmitter called _____. This neurotransmitter causes the constriction of the blood vessels in the skin to reduce heat loss and causes the contraction of a type of smooth muscle found in the dermis called _____, causing the body hairs to rise. The actions which cause the rise of the body hairs is also known as _____ which traps heat closer to the skin. In addition, the activation of this subdivision also causes _____ to be secreted from adrenal medulla which increases thermogenesis. A shivering center in the hypothalamus is also activated which stimulates the motor centers of the brainstem and causes the involuntary contraction of _____, causing shivering and generating heat. Finally, the hypothalamus releases a hormone called _____ which activates the anterior pituitary gland to release _____. This hormone triggers the thyroid to release two hormones called _____ and _____which in turn increases metabolism and thereby increases the amount of heat that the body generates.

a.	Acetylcholine	o.	Posterior hypothalamus
b.	Anterior hypothalamus	p.	Postoptic hypothalamus
c.	Apocrine sweat gland	q.	Preoptic-anterior hypothalamus
d.	Arrector pili	r.	Sebaceous gland
e.	Autonomic nervous system	s.	Skeletal muscle
f.	Cardiac muscle	t.	Smooth muscle
g.	Convection	u.	Somatic nervous system
h.	Eccrine sweat gland	v.	Sympathetic
i.	Adrenaline	w.	Thyroid stimulating hormone (TSH)
j.	Lateral hypothalamus		
k.	Levator ani	x.	Thyrotropin releasing hormone (TRH)
l.	Noradrenaline		
m.	Parasympathetic	y.	Thyroxin (T_4)
n.	Piloerection	z.	Triiodothyronine (T_3)

87. In contrast, if the temperature of the body increases above the pre-set values, the thermoregulation centers in the _____, in conjunction with the _____, will trigger responses that promote heat loss. Immediately, the _____ (a nervous system) are activated which in turn activates one of its two subdivisions called _____ (sympathetic or parasympathetic). The activation of this subdivision will cause the _____ (sympathetic or parasympathetic) nerves to secrete its neurotransmitter called _____. The release of this neurotransmitter will cause the bulbous secretory coils of the _____ (a type of sweat gland) of the skin to produce and secrete sweat. It is understood that the excretion of sweat allows for heat loss through evaporate cooling. The previously secreted neurotransmitter also causes the relaxation of a type of smooth muscle found in the dermis called _____ which causes the body hairs to relax and lay flat against the skin. The relaxation of the body hairs prevents heat from being trapped by the layer of still air between the hairs and increasing heat loss by _____.

a.	Acetylcholine	o.	Posterior hypothalamus
b.	Anterior hypothalamus	p.	Postoptic hypothalamus
c.	Apocrine sweat gland	q.	Preoptic-anterior hypothalamus
d.	Arrector pili	r.	Sebaceous gland
e.	Autonomic nervous system	s.	Skeletal muscle
f.	Cardiac muscle	t.	Smooth muscle
g.	Convection	u.	Somatic nervous system
h.	Eccrine sweat gland	v.	Sympathetic
i.	Epinephrine (adrenaline)	w.	Thyroid stimulating hormone (TSH)
j.	Lateral hypothalamus		
k.	Levator ani	x.	Thyrotropin releasing hormone (TRH)
l.	Noradrenaline		
m.	Parasympathetic	y.	Thyroxin (T_4)
n.	Piloerection	z.	Triiodothyronine (T_3)

88. The fetus is covered in which of the following type of hair?
a. Course hairs
b. Fine hairs
c. Lanugo
d. Terminal hairs
e. Vellus hairs

89. At term the early fetal hair at the scalp, eye brows and eye lashes is replaced by which of the following type of hair?
a. Course hairs
b. Fine hairs
c. Lanugo
d. Terminal hairs
e. Vellus hairs

90. At term the early fetal hair on the rest of the body (excluding scalp, eye brows and eye lashes) is replaced by which of the following type of hair?
a. Course hairs
b. Fine hairs
c. Lanugo
d. Terminal hairs
e. Vellus hairs

91. In general, the hair is divided into the _____ which protrudes above the surface of the skin and the _____ which exists in the _____ layer of the skin.

a. Epidermis
b. Hypodermis
c. Papillary layer of the dermis
d. Reticular layer of the dermis
e. Root
f. Shaft

ANSWERS

1.	b	2.	a	3.	b
4.	c	5.	a	6.	a
7.	g	8.	g	9.	d
10.	b	11.	b	12.	e
13.	d	14.	e, a, b, f	15.	46, 47, 10, 25, 41, 43, 55, 22, 57, 58, 49, 56, 48, 18, 42, 32, 3, 50, 19, 11(47), 53, 59, 54, 40
16.	a, b, d, c	17.	a	18.	e
19.	f	20.	a	21.	h
22.	a	23.	e	24.	a
25.	a	26.	g	27.	b
28.	c, e, f	29.	d	30.	a, b, c
31.	h	32.	a	33.	f
34.	g	35.	f	36.	g
37.	a	38.	b	39.	a
40.	b	41.	h	42.	g
43.	d	44.	e	45.	b
46.	a	47.	g	48.	e
49.	a	50.	b	51.	c
52.	f	53.	g	54.	d
55.	e	56.	b	57.	c
58.	d	59.	f	60.	e
61.	b	62.	d	63.	b
64.	e	65.	c	66.	a
67.	d	68.	e	69.	a
70.	b	71.	e	72.	e
73.	e, c, f, d	74.	f	75.	a, g, e, f, h
76.	e	77.	a	78.	31, 2, 1, 20, 51, 12, 15, 45, 19, 53, 54, 30, 3, 52, 8, 9, 46, 68, 67, 50, 6, 38, 11, 23
79.	10, 35, 26, 61, 62, 58, 42, 32, 21, 39, 38, 72, 27, 48, 40, 32, 69, 66, 34, 29, 36, 14, 47, 64, 29, 34, 28, 22, 44, 33, 32, 37, 7, 65, 70, 13, 63	80.	e	81.	e
82.	g	83.	d	84.	b
85.	c	86.	q, o, e, v, v, l, d, n, i, s, x, w, v, z	87.	q, o, e, m, m, a, h, d, g
88.	c	89.	d	90.	e
91.	f, e, d				

REFERENCES

Shibasaki M, Wilson TE, Crandall CG. Neural control and mechanisms of eccrine sweating during heat stress and exercise. J Appl Physiol, 2005. 100: pp. 1692-1701

Cullen JJ, Murray DJ, Kealey GP. Changes in coagulation factors in patients with burns during acute blood loss. J Burn Care Rehabil, 1989. 10: pp. 517-522

Berg RD. Bacterial translocation from the gastrointestinal tract. Adv Exp Med Biol, 1999. 473: pp 11-30

Kirksey TD and Moncrief JS Jr.: Gastrointestinal complications in burns. Am J Surg, 1968. 116: 627-633

Woods JH, Erichkson LW, Condon RE, Schulte WJ, Sillin LFL. Postoperative ileus: A colonic problem? Surgery, 1978. 84: 527-532

Holm C, Hörbrand F, von Donnersmarck GH, Mühlbauer W. Acute renal failure in severely burned patients. Burns, 1999. 25: pp. 171-178

Robins EV.Immunosuppression of the burned patient. Crit Care Nurs Clin North Am, 1989. 1: pp. 767-74

Seeley RR, Stephens TD, Tate P. Anatomy and Physiology 6th Edition. McGraw-Hill, New York, New York. 2003

Tate P. Seeley's Principles of Anatomy & Physiology 1st Edition. McGraw-Hill, New York, New York. 2009

Saladin KS. Anatomy & Physiology. The Unity of Form and Function 6th Edition. McGraw-Hill, New York, New York. 2010

PHOTO AND GRAPHIC BIBLIOGRAPHY

1. Figure 1: Graphic designed by P.Y.P. Jen. Copyright ©. Wroblewski LE, 2011. U.S. National Library of Medicine. Public domain graphics
2. Figure 2: Graphic designed by P.Y.P. Jen. Copyright ©
3. Figure 3: Graphic designed by P.Y.P. Jen. Copyright ©
4. Figure 4: Graphic designed by P.Y.P. Jen. Copyright ©
5. Figure 5: Graphic designed and photomicrograph taken by P.Y.P. Jen. Copyright ©
6. Figure 6:. Graphic designed and photomicrograph taken by P.Y.P. Jen. Copyright ©
7. Figure 7: Graphic designed and photomicrograph taken by P.Y.P. Jen. Copyright ©
8. Figure 8: Graphic designed by P.Y.P. Jen. Copyright ©. Photomicrograph released by G. Matteo. (2012) U.S. National Library of Medicine. Public domain photo
9. Figure 9: Graphic designed and photomicrograph taken by P.Y.P. Jen. Copyright ©
10. Figure 10: Graphic designed and photomicrograph taken by P.Y.P. Jen. Copyright ©
11. Figure 11: Graphic designed and photomicrograph taken by P.Y.P. Jen. Copyright ©
12. Figure 12: Graphic designed and photomicrograph taken by P.Y.P. Jen. Copyright ©
13. Figure 13: Photomicrograph taken by Ganímedes 2013 at 1000x magnification. This photo is licensed under the Creative Commons Attribution-Share Alike 3.0 Unported license
14. Figure 14: Photomicrograph taken by P.Y.P. Jen at 400x magnification. Copyright ©
15. Figure 15: Photomicrograph taken by P.Y.P. Jen at 400x magnification. Copyright ©
16. Figure 16: Graphic released by Open Stax College, Connexions, reprint permission granted under the Creative Commons Attribution 3.0 Unported license.
17. Figure 17: Photomicrograph taken at 1172x and released by National Institute of Health. The Human Genome Project. Eukaryotic Cell. The graphics was developed by Darryl Leja NHGRI. Public domain photo
18. Figure 18: Graphic released by Vossman, 2005. Reprint permission is granted to copy, distribute and/or modify this document under the terms of the GNU Free Documentation License, Version 1.2
19. Figure 19: Graphic released by US National Institute of Health. Public domain graphic
20. Figure 20: Graphic released by Holly Fischer, 2013. This file is licensed under the Creative Commons Attribution 3.0 Unported license
21. Figure 21: Graphic released by Holly Fischer, 2013. This file is licensed under the Creative Commons Attribution 3.0 Unported license
22. Figure 22: Graphic released by Holly Fischer, 2013. This file is licensed under the Creative Commons Attribution 3.0 Unported license
23. Figure 23: Graphic released by Holly Fischer, 2013. This file is licensed under the Creative Commons Attribution 3.0 Unported license
24. Figure 24: Photomicrograph taken by P.Y.P. Jen at 1000x. Copyright ©
25. Figure 25: Photomicrograph taken at 400x and uploaded by Nephron and permission is granted to copy, distribute and/or modify this image under the terms of the GNU Free Documentation License Version 1.2
26. Figure 26: Photomicrograph taken at 400x by Erhabor Osaro, 2014. This photo is licensed under the Creative Commons Attribution-Share Alike 4.0 Unported license
27. Figure 27: Photomicrograph taken at 1000x by HeNe, 2014. This photo is licensed under the Creative Commons Attribution-Share Alike 3.0 Unported license
28. Figure 28: Photomicrograph taken at 1125x and released by Koenjo~commonswiki on June 2009. Public domain photo
29. Figure 29: Photomicrograph taken at 1000x and released by Wartak SA, 2008. Baystate Medical Center/Tufts University School of Medicine. U.S. National Library of Medicine. Public domain photo
30. Figure 30: Photomicrograph taken at 400x and released by P.Y.P. Jen. Copyright ©
31. Figure 31: Photomicrograph taken at 400x and released by P.Y.P. Jen. Copyright ©
32. Figure 32: Photomicrograph taken at 400x and released by P.Y.P. Jen. Copyright ©
33. Figure 33: Photomicrograph taken at 400x and released by P.Y.P. Jen. Copyright ©
34. Figure 34: Photomicrograph taken at 1000x and released by P.Y.P. Jen. Copyright ©
35. Figure 35: Photomicrograph taken at 400x and released by P.Y.P. Jen. Copyright ©
36. Figure 36: Photomicrograph taken at 400x and released by P.Y.P. Jen. Copyright ©
37. Figure 37: Scanning electron microscopy (50000x) taken by Newman ME, 2007. US Department of Commerce. Public domain photo
38. Figure 38: Photomicrograph taken at 400x and released by P.Y.P. Jen. Copyright ©
39. Figure 39: Graphic released by Wong DJ and Chang HY, 2013. This file is licensed under the Creative Commons Attribution 3.0 Unported license
40. Figure 40: Graphics by P.Y.P. Jen. Copyright ©
41. Figure 41: Graphic released by Open Stax College, Connexions, reprint permission granted under the Creative Commons Attribution 3.0 Unported license.
42. Figure 42: Graphic released by Open Stax College, Connexions, reprint permission granted under the Creative Commons Attribution 3.0 Unported license.
43. Figure 43: Photomicrograph taken at 1000x magnification by Haymanj. This file is licensed under the Creative Commons Attribution 3.0 Unported license
44. Figure 44: Graphic illustration released by Madhero88 and M.Komorniczak, 2012. This file is licensed under the Creative Commons Attribution 3.0 Unported license
45. Figure 45: Photographs taken by Babagolzadeh. This file is licensed under the Creative Commons Attribution 3.0 Unported license
46. Figure 46: Graphic released by BruceBlaus. 2014. This file is licensed under the Creative Commons Attribution 3.0 Unported license
47. Figure 47: Graphic released by KDS444, 2014. This file is licensed under the Creative Commons Attribution 3.0 Unported license
48. Figure 48: Graphic released by PJ Lynch, 2006. Permission for reprint is granted under Creative Commons Attribution 2.5 License 2006
49. Figure 49: Graphic released by P.Y.P. Jen. Copyright ©

CHAPTER FIVE

SKELETAL SYSTEM

The skeletal system is composed of bones, cartilage, tendons ,and ligaments. These components perform a variety of functions within the human body. For example, the skeletal system provides the framework for the body; it reinforces the softer tissues and serves as points of attachment for most skeletal muscles. The interaction between bone, cartilage, tendons and ligaments provides the necessary leverage and support for bodily movement. In addition, the bones provide physical protection for many of the body's internal organs, as well as, serving as sites to store minerals such as calcium (Ca) and phosphorus (P). When needed, the bones can be prompted to release these minerals into the blood stream. The red marrow, found within the medullary space of bones, is composed of hematopoietic cells (cells that give rise to all blood cells) and is the site of blood cell and platelet production. The yellow marrow, also located in the medullary spaces, consists mainly of adipose tissue and is one of the many sites of energy storage within the human body.

The human skeletal system is comprised of the axial and the appendicular skeletons. The axial skeleton is so named because it forms the central axis of the body. The axial skeleton consists of the skull, vertebral column, ribs and the sternum (Figure 1).

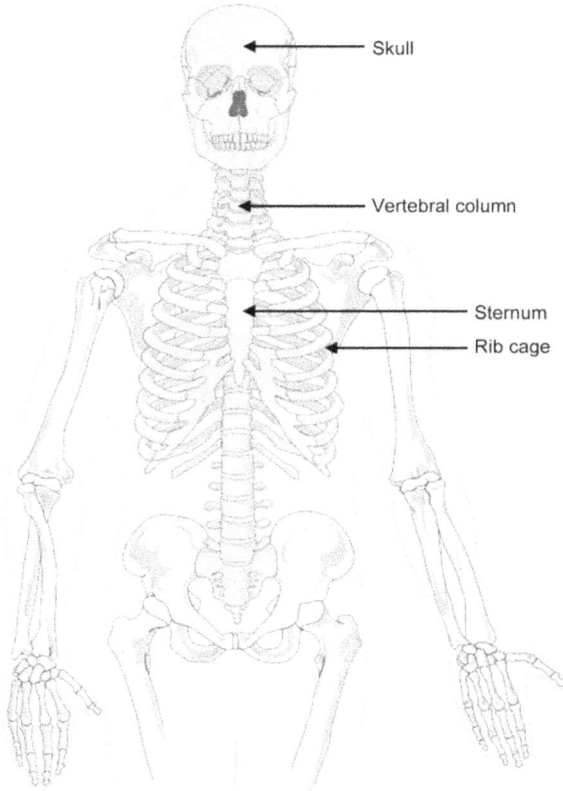

Figure 1: Illustration of the axial skeleton

The skull consists of 22 different bones, including the three smallest bones of the body; incus, stapes, and malleus, also known as the auditory ossicles. The remaining bones of the skull can be divided into cranial bones, which are designed to protect the brain, and facial bones, which serve to provide structural support for sensory organs (e.g. eyes, nose), as well as, forming the hinged jaw, necessary for the first stage of digestion.

The human appendicular skeleton is made up of 126 bones (out

of 206 bones total found in the adult human body) and consists of the skeletal structures that form the girdles and the limbs. The superior limbs are attached to the pectoral girdle, while the inferior limbs are attached to the pelvic girdle. The appendicular skeleton serves two major functions. The first is to provide physical protection for the underlining organs of the circulatory, digestive, urinary, and reproductive systems, as well as, portions of the endocrine system. The second is to provide the necessary attachment sites for tendons and ligaments at inter-skeletal articulations (e.g. joints) thereby making physical movement possible (Figure 2).

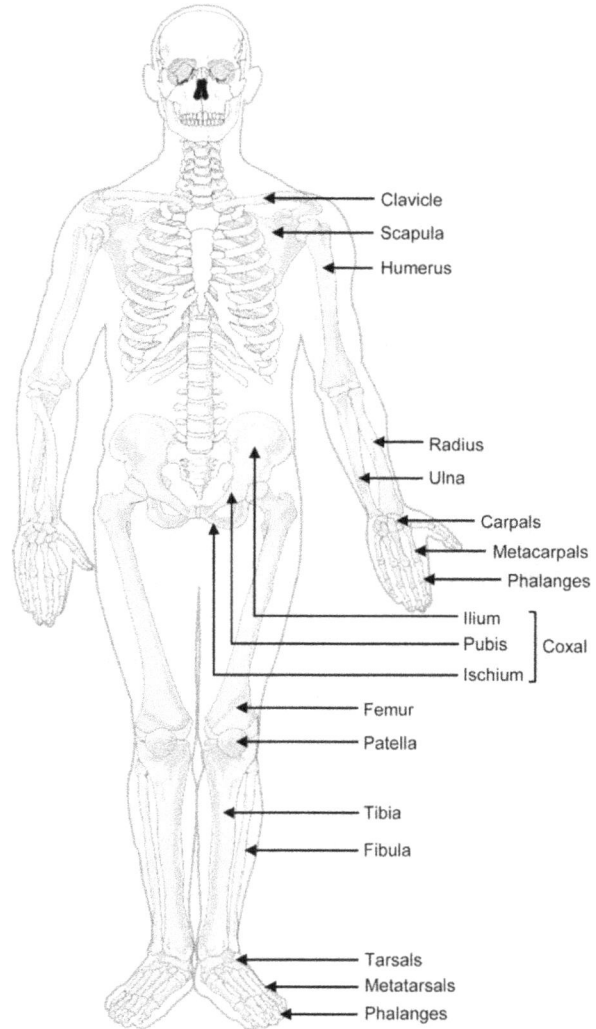

Figure 2: Illustration of the appendicular skeleton

HISTOLOGY OF BONES

Bone Cells

Mesenchymal stem cells give rise to osteogenic cells (also known as osteogenic, osteoprogenitor cells, osteochondrial progenitor cells) which can become osteoblasts or chondroblasts. Both osteoblasts and chrondroblasts will later mature into osteocytes and chondrocytes, respectively. In bones, mesenchymal stem cells are found in the embryonic tissue during development, and in adults they are found within the medullary space of bones. Developed mesenchymal stem cells, such as the osteogenic cells, are found within the dense irregular connective tissues of the periosteum and the thin endosteum enclosing the marrow space. Mesenchymal stem cells also can form chondroblasts (chondrocytes once matured), adipoblasts (adipocytes once matured), and fibroblasts (fibrocytes once matured) (Figure 3).

Osteocytes are mature bone cells residing within spaces inside the bone matrix called lacunae. Osteocytes have numerous cellular processes known as filopodia

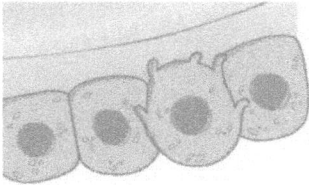

Osteoblasts are immature bone cells that are capable of secreting organic and inorganic bone matrix. In adults, osteoblasts are found in the endosteum and the periosteum of bones and are responsible for bone remodeling or repairs

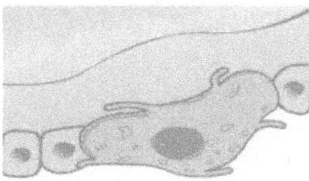

Osteogenic cells are developed from mesenchymal cells. It is the precursor cells that will mature to form osteoblasts and later osteocytes. In adults, osteogenic cells are found in the endosteum and the periosteum of bones

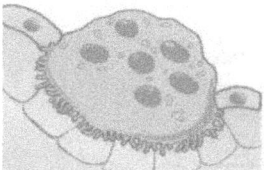

An osteoclast is a modified macrophage that is capable for dissolving organic and inorganic bone matrix. Osteoclasts are involved in bone remodeling and/or repairs

Figure 3: Illustration of the cells of the bone

The term ossification, also known as osteogenesis, is defined as the manner by which bone is produced by osteoblasts. As mentioned previously, bone marrow mesenchymal stem cells can develop into osteogenic cells and later mature into osteoblasts. With regards to the periosteum, the already formed osteogenic cells will mature into osteoblasts when repairs or remodeling of bone is required. Put simply, osteoblasts are specialized mesenchymal cells that are responsible for the synthesis and mineralization of bone during development and later in bone remodeling. Under microscopic examination, osteoblasts possess many ribosomes, an extensive array of endoplasmic reticulum (mainly rough ER), and Golgi apparatus within their cytoplasm. The osteoblasts possess numerous cellular processes that interconnect other osteoblasts in a network. These interconnected osteoblasts communicate with one another via gap junctions.

Osteoblasts are responsible for producing proteins such as collagen, proteoglycans, glycoproteins, and releasing them via exocytosis. These secreted unmineralized protein-based products are referred to as osteoid tissue. In addition, osteoblasts produce high concentrations of calcium and phosphate before releasing them in a matrix vesicle. The matrix vesicle is formed through the protrusion of the osteoblast plasma membrane through budding. Once released, the calcium and phosphate are enzymatically converted into hydroxyapatite $[Ca_{10}(PO_4)_6(OH)_2]$, which is the main inorganic component of bone matrix. Once enough bone matrix is secreted and the osteoblast is molded within a lacunae (a chamber of created by their own secretions) and when their cellular processes (also known as filopodia) are surrounded by a tunnel-like structure called canaliculi, it is no longer referred to as an osteoblast but rather a mature bone cell

called osteocyte.

Osteoclasts are modified macrophages that are capable of reabsorbing or breaking down of bone. Reabsorption of bone relies upon the secretion of the enzyme cathepsin K and hydrogen ions (H^+). The H^+ acidifies the area beneath the osteoclast, where the breakdown of the bone is taking place, to dissolve the hydroxyapatite (minerals), and then cathepsin K breaks down the protein component of the bone matrix. Once the bone matrix is broken down, many of the minerals, such as calcium and phosphate, are released into the blood stream, while the protein component that has been hydrolyzed is reabsorbed into the osteoclast via endocytosis.

Bone Matrix

It has been discovered that the majority of organic ground substances found in bones are collagenous. The most predominant being Type I collagen with trace amounts of type III, V, IX, XII, XIV, XIX, XX, and XXI collagen. Minor populations of proteins found within bone matrix include the blood plasma protein albumin and α2-HS-glycoprotein, which have tendencies to bind to hydroxyapatite due to the interactions between their acidic properties, carboxyl group (-COOH) and albumin's amino groups ($-NH_2$). It is believed that blood serum-derived proteins, such as albumin, aid in the process of matrix mineralization, while α2-HS-glycoproteins and other trace amounts of non-collagenous proteins, such as growth factors (e.g. insulin-like growth factors I and II), are believed to help regulate osteoblast proliferation.

Other non-collagenous proteins or protein based structures, such as proteoglycans and γ-carboxylated (gla) proteins, also exist in bone. For example, proteoglycans, such as small leucine-rich proteoglycans (SLRPs), are involved in various aspects of bone development. SLRPs are involved in bone cell proliferation and remodeling, as well as, organic matrix and mineral deposition. Evidence shows a correlation between SLRP depletion and degenerative diseases, such as osteoporosis.

Osteocalcin, a vitamin K dependent γ-carboxylated (gla) protein, is the most abundant non-collagenous protein in bone. Recent evidence suggests that osteocalcin is a dual functioning protein that is essential in directing the proper deposition and organization of hydroxyapatite in bones and as a hormone (intercellular ligand) that prompts the β-cells found within the Islet of Langerhans of the pancreas to increase insulin secretion. For example, it has been shown that bone depletion, as a result of disease or heritable condition, causes an increase in circulating osteocalcin levels, which in turn brings about an increase in insulin secretion, and the subsequent increase of fat deposition in the adipose tissues.

Other non-collagenous protein includes osteonectin and bone sialoprotein, which also occupies a substantial proportion of non-collagenous proteins in bones. For example, osteonectin, an adhesive glycoprotein is readily found within the bone matrix. Osteonectin possesses a high affinity for both hydroxyapatite and collagen fibers, which allows them to initiate mineralization and mineral crystal formation within bones. Bone sialoprotein is a highly glycosylated and sulphated phosphoprotein, possessing the capability to bind hydroxyapatite and transmembrane integrins, and is believed to be the protein core (protein nucleator) that allows for hydroxyapatite crystal formation.

A typical human bone is approximately 60% mineral, 30% organic matrix, 7% water and 3% lipids. The mineral component of bone is mostly comprised of hydroxyapatite, along with trace amounts of magnesium, carbonate, and free phosphate (not associated with hydroxyapatite). Bone mineral is responsible for mechanical rigidity and load-bearing strength while the organic matrix provides elasticity and flexibility.

Hydroxyapatite minerals found in bones are very small, (~200 Å) in diameter and more soluble in comparison to the hydroxyapatite crystals that exist in natural geologic formation. This increase in solubility allows the bone hydroxyapatite to be easily constructed and degraded by osteoclasts depending upon systemic needs. As mentioned previously, bone sialoprotein is required to form its nuclear core for hydroxyapatite crystallization. Osteonectin, the adhesive glycoprotein, is involved in initiating mineralization and mineral crystal formation within bones. Additionally, osteocalcin is responsible for directing the proper deposition and organization of hydroxyapatite.

CLASSIFICATION OF BONES

Woven and Lamellar Bone

Depending on the organization patterns of the collagen fibers (type I, IX, and XII) within the osteoid (bone matrix), the human bone is divided into woven and lamellar bone. The woven bone contains fewer collagen fibers that are arranged randomly in haphazard directions. In addition, hyaline cartilage or calcified hyaline cartilage is often found within the matrix of the woven bone, thereby making them physically and mechanically weaker in contrast to the lamellar bone (Figure 4).

In higher vertebrates, the existence of the woven or primary bone is believed to fulfill a temporary role during development before being replaced by the more durable and organized lamellar, secondary bone. The process of replacing woven bone with lamellar bone is called bone remodeling. For example, the developing human tibia is mostly woven bone at birth but will be gradually replaced by lamellar bone within the following twelve months. Another example is seen during fracture repair. Approximately three weeks post fracture, a callus is formed at the site of injury. A callus is a mass of tissue that forms at the fractured site which connects the fractured ends of the bones together and helps to immobilize the injured site. The callus contains large quantities of woven bone. As the healing process proceeds, the woven bone is gradually reabsorbed by osteoclasts and replaced by much stronger lamellar bone.

Lamellar, secondary bone is the type of bone found in adult humans. This type of bone possesses a regular parallel alignment of collagen fibers into sheets or lamellae that are approximately 5 micrometers (μm) thick. Lamellar bone is highly organized in concentric sheets and is mechanically strong. Nonetheless, when compared with woven bone, the lamellar bone has a much lower proportion of osteocytes versus bone matrix.

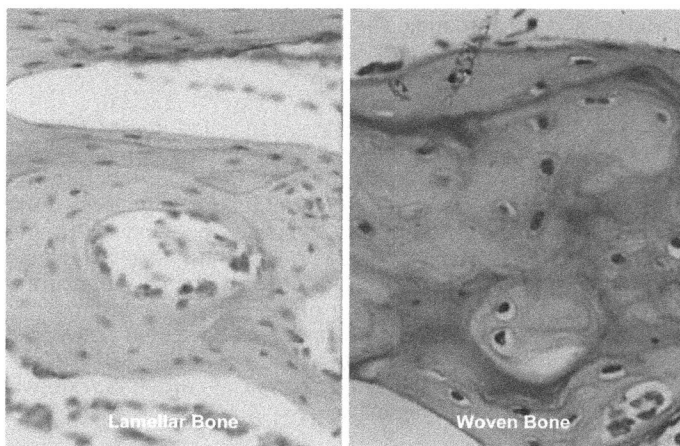

Figure 4: H&E stain sections of lamellar bone (left) which is concentrically organized and woven bone (right) which are mixed with cartilage and calcified cartilage tissues

Cancellous and Compact Bone

Both woven and lamellar bone are also categorized based on the volume of bone matrix in association with the amount of space present within the bone. For example, cancellous bone, also known as trabecular bone or spongy bone, has relatively less bone matrix and more space within the structure of the bone when compared with compact bone, also known as the cortical bone. Put simply, the cancellous bone is less dense, physically weaker, softer, and more flexible. In contrast, the compact bone possesses more bone matrix and less space within the bone. In other words, the compact bone is denser, physically stronger, harder, and a lot less flexible.

It is estimated that cancellous bone makes up approximately 20% of the entire skeletal structure. It is typically found at the ends of long bones or lining the medullary space. Cancellous bones are composed of interconnecting bone rods or plates called trabeculae. The trabeculae are thin strands of bone tissue measuring between 50 to 500 μm in diameter. The surface of the trabeculae is covered by a thin layer of tissue call endosteum, which is composed mostly of osteoblasts, few osteoclasts and osteogenic cells. Surrounding the trabeculae, within the intertrabecular spaces, there are numerous blood vessels, capillaries, nerves and red marrow which are involved in hematopoiesis. Additionally, lymphatic vessels could also be located within the intertrabecular spaces.

A cross section of the trabeculae will show numerous lacunae embedded within the concentric circular patterns of the lamellae. Each lacuna will house one mature bone cell, an osteocyte (Figure 5).

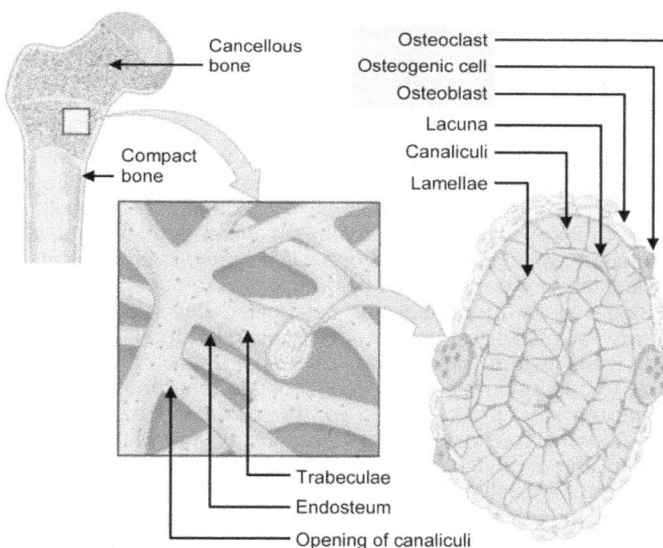

Figure 5: Illustration of a cancellous bone. Please note that the bone is constructed out of intertwining plates of trabeculae and that the layer of osteoblasts, osteoclasts and osteogenic cells (highlight in blue) forms the endosteum

These osteocytes are interconnected to one another via filopodia, which are located within the tiny canals or canaliculi. Since no blood supply is directly found within the trabeculae, the osteocytes must obtain their nutrients and perform gas exchange through the filopodia. For example, the osteocytes that are located on the periphery of the trabeculae will connect with the cells of the endosteum (osteoclasts, osteoblasts, and osteogenic cells), as well as, open to the capillaries surrounding the trabeculae (Figure 5).

Nutrients and gases will flow through the canaliculi openings of the trabeculae through the gap junctions located at the end

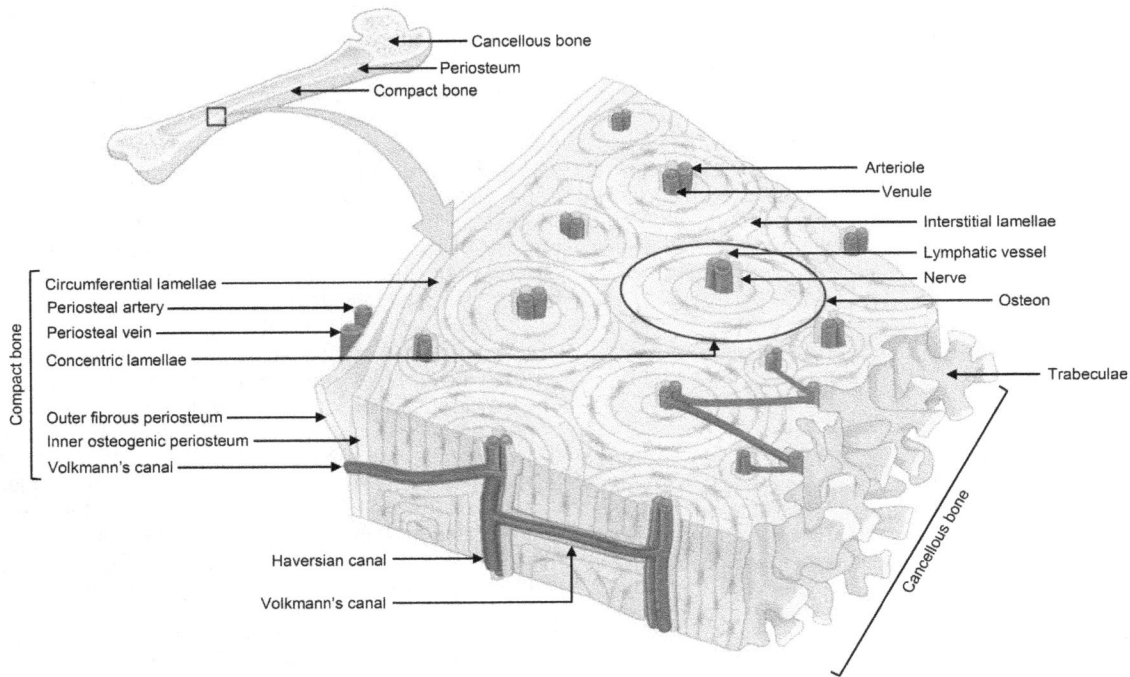

Figure 6: Cross sectional view of compact bone and cancellous bone lining the medullary space. Please note that the Haversian canal allows the blood vessels, nerves and lymphatic vessels to course vertically while the Volkmann's canal also contains blood vessels, nerves, and lymphatic's connects one osteon with another

of filopodia. From there, the nutrients and gases will proceed into the osteocytes before being shared with other networked osteocytes, each interconnected by filopodia and their terminal junctions. Inversely, waste disposal and elimination of carbon dioxide will proceed in the opposite direction.

The trabeculae are structured to deal with the stress and weight that are placed upon it. In addition, the trabeculae will resist bending and stretching. If the stress placed upon the cancellous bone is changed due to injury or an increase in weight, the trabeculae within cancellous bone will be remodeled (realigned) to meet the new demands. For example, this remodeling process is triggered by the trabecular osteocytes through a process known as mechanosensation. Stretching and bending of the bone and, in turn, osteocytes, causes the flow of canalicular fluids to change in response. It is believed that the rapid fluxes of calcium ions, prostaglandin E2, cyclooxygenase 2, kinases, Runx2, and nitrous oxide across the gap junctions at the terminal ends of the filopodia stimulate the transmission of information between the networked osteocytes, and the cells of the endosteum. Once triggered, the osteoclast will begin removing existing bone structures that are providing inadequate supports, while osteoblasts will proceed in and begin manufacturing new trabeculae formations to satisfy demand.

The spaces within the cancellous bone are filled with bone marrow and blood vessels. For example, the cancellous bone frequently contains red marrow where hematopoiesis, or blood cell formation, takes place.

Compact bone by comparison is a much denser, stronger and rigid structure and is estimated that this type of bone constitutes approximately 80% of the entire skeletal structure. The matrix of the compact bone is filled with organic ground substances and inorganic compounds like hydroxyapatite with relatively few spaces in between. In many instances the compact bone forms a shell around cancellous bone to provide added support especially in the long bones of the arms and the legs.

Unlike the cancellous bone, blood vessels directly enter the compact bone in horizontal tunnels called Haversian (central or osteon) canals. Surrounding the Haversian canal are lamellae forming concentric circles (4 to 20 concentric lamellae) constructed of bone matrix with numerous lacunae dispersed within. Based on morphology, there are three types of lamellae that exists in the compact bone. The outer surfaces of the compact bone are formed by circumferential lamellae arranged parallel to the surface of the bone. The bulk of the compact bone is formed by concentric lamellae and in between the circumferential lamellae are interstitial lamellae, which are remnant of circumferential lamellae that still remained after bone remodeling (Figure 6).

Found inside the lacunae are mature bone cells called osteocytes. Like the osteocytes in the cancellous bone, these found in the compact bone also possess numerous cellular processes, known as filopodia, running within canaliculi. The combination of the Haversian canal, lamellae, lacunae, osteocytes and canaliculi are referred to as a Haversian system or an Osteon. Each of the osteon is interconnected through the Volkmann's (perforating) canals which are perpendicular to the long axis of compact bone. The Volkmann's canals are simply tunnel branches carrying artery and veins from the periostium.

Due to the accessibility of the Haversian canal, the acquisition of nutrients and gases for the osteocytes of the compact bone is slightly different than those found in cancellous bone. The osteocytes located at the inner layer of the lamellae will extend its filopodia to the edge of the canaliculi opening within the Haversian canal. Subsequently, nutrients and gases will diffuse through the gap junctions located at the terminal ends of the filopodia and enter the cell body of the osteocyte. In turn, this osteocyte will share the nutrients and gases with the networked osteocytes located within the outlying layers of the lamellae. Inversely, waste disposal and elimination of carbon dioxide will proceed in the opposite direction.

Undifferentiated osteogenic cells, osteoblasts and osteoclasts will always remain in the cellular layer of periosteum and endosteum adjacent to mature osteoblasts. In addition, a thin layer of osteoblasts, osteoclasts and osteogenic cells exists in the lining of Haversian canals and are thus in position to rapidly become osteoblasts in case of needed remodeling or repair.

Shapes of Bones

The 206 bones of the human body are further classified based upon their shapes. For example, if the bones are longer than they are wide, then they are classified as long bones. If they are as long as they are wide, they are grouped as short bones. Flat bones are flat and slightly curved. Finally those that do not fit into any of the above categories are categorized as irregular bones.

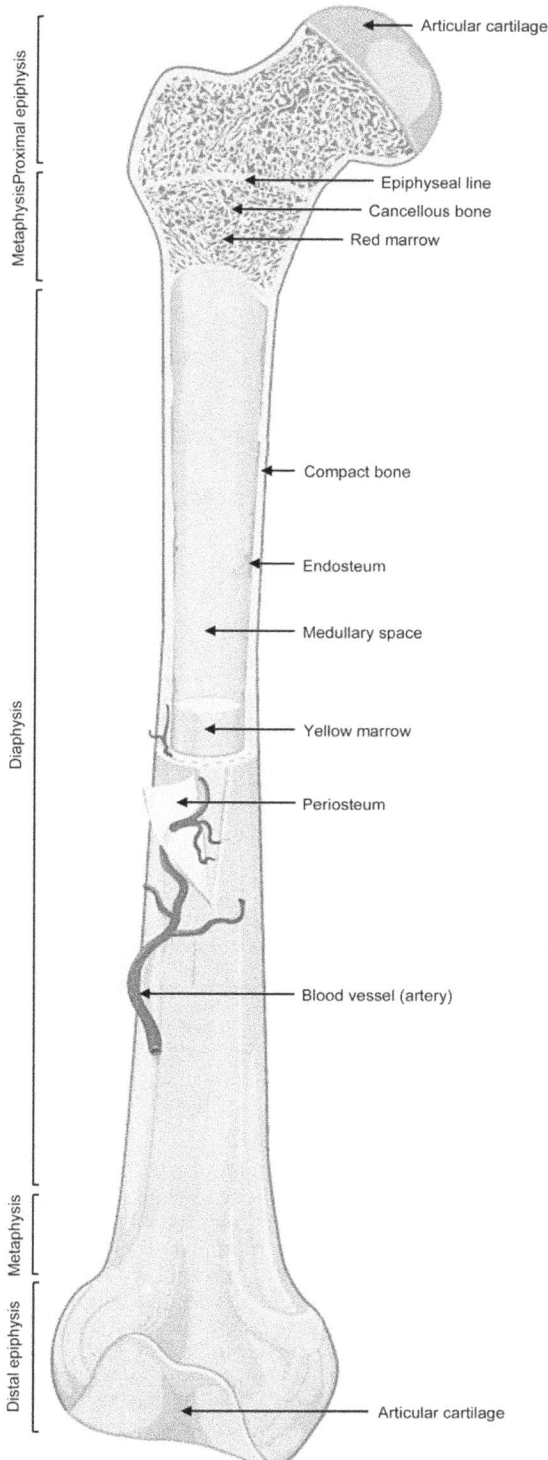

Figure 7: Illustration of a long bone represented by the largest long bone in the human body, the femur. The cancellous bone exists in the proximal and distal epiphysis, while the compact bone constitutes the majority of the structure. Please note that the lining the medullary space is not shown

The long bone is constructed with two enlarged ends called the epiphysis and a long shaft called the diaphysis. The epiphysis is composed mainly of cancellous bone with its margins lined with compact bone for additional support. The terminal ends of the epiphysis are generally rounded and are designed to articulate with adjacent bones, forming joints. The surface of the epiphysis is covered with a thin hyaline cartilage layer called the articular cartilage which is designed to prevent friction between bones during movement. If a long bone is still growing, a layer of hyaline cartilage called epiphyseal plate, also known as growth plate, would be visible towards diaphysis of the long bone. This epiphyseal plate is the location where the process of metaphysis or bone growth takes place. Once adulthood is reached and the individual stops growing, the cartilage of the epiphyseal plate will be replaced by compact bone forming the epiphyseal line. As mentioned previously, the red marrow is found within the intertrabecular spaces of the cancellous bone. The red marrow is the location of hematopoiesis (Figure 7).

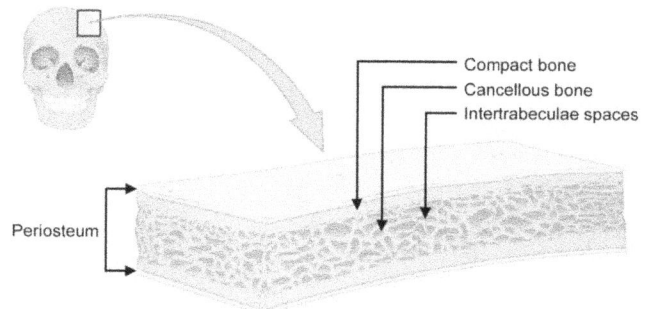

Figure 8: Illustration of a flat bone. Please note that the intertrabeculae spaces contains the red marrow

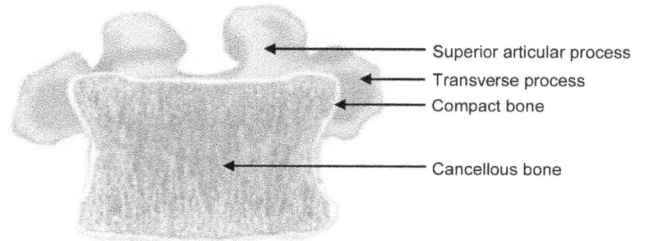

Figure 9: Illustration of an irregular bone, represented by a coronal plane of a lumbar vertebra. Please note that the cancellous bone forms the bulk of the bone, while the compact bone forms its external borders

Epiphyses are separated by a long shaft called diaphysis. The diaphysis is composed mainly of compact bone with cancellous bone existing only towards the medullary space. It is surrounded by a dense irregular connective tissue called the periosteum, containing numerous nerves, blood vessels, lymphatic vessels, osteoblasts, osteoclasts, and osteogenic cells. Within the medullary space, lining the cancellous bone of the diaphysis, is a thin cellular membrane called the endosteum. The endosteum is composed of layers of osteoblasts, osteoclasts and osteogenic cells. Within the medullary space, the marrow exists. During fetal development the medullary space of the diaphysis is filled with red marrow, but a conversion to yellow marrow, a tissue made out of loose areolar connective tissue and adipose tissue, occurs shortly after birth. Depending on the long bone, the medullary space may be completely converted to yellow marrow after reaching adulthood. Nonetheless, certain long bones, such as the femur, will still retain red marrow towards the distal ends (towards the epiphysis), while the remaining medullary space is filled with yellow marrow (Figure 7).

As the name states, the flat bone is plate–like bone with built in curvatures. Examples of the flat bones are mainly found in the skull. The constructing of a flat bone resembles that of a sandwich where the superficial and the deep layers are lined with compact bone, and sandwiched in between is a layer of cancellous bone also known as the dipolë. The compact bones are lined with periosteum while the cancellous bone is filled with red marrow used in hematopoiesis (Figure 8).

Short and irregular bones are similar in construction. They both possess a cancellous bone core with compact bone forming the external borders to provide additional strength. Neither short nor irregular bones are elongated and do not possess a diaphysis. Nonetheless, these bones do have epiphyseal plate which allows for bone growth (Figure 9).

BONE DEVELOPMENT

There are two patterns of fetal bone development in humans: intramembranous and endochondral ossification. The major difference between the two forms of ossification are the types of tissue from which the bone is derived. Nonetheless, there are some similarities. For example, both processes are initiated by mesenchyme tissue and mesenchymal cells. In addition, both patterns involve the initial formation of woven bone.

Intramembranous Ossification

The development of the flat bones (bones of the skull, sternum, clavicle, pelvis, ribs, and the diaphysis of the clavicle) or intramembranous bones begins in approximately the fifth week prenatal by forming a mesenchymal membrane at the site of the future bone. The mesenchyme tissue is composed of mesenchymal cells dispersed within delicate collage fibers richly supplied with capillaries. Intramembranous ossification begins when the mesenchymal cells start to cluster and form ossification centers. These mesenchymal cells enlarge and differentiate into osteogenic cells before developing into osteoblasts. Immediately, the osteoblasts begin secreting osteoid (e.g. bone matrix) on the collagen fibers forming the numerous trabeculae of woven bone. The secreted osteoid tissue traps the osteoblasts and their filapodia in lacunae and canaliculi, respectively, thereby creating osteocytes. The diameter of the trabeculae grow as more osteoblasts begin depositing osteoid on the delicate collagen fibers. As the trabeculae begin to link together, the future structure of cancellous bone begins to form. At this point, a thin layer of osteoblasts, osteoclasts, and osteogenic cells establishes on the surface of the trabeculae, creating the endosteum. As stated previously, there are many intertrabeculae spaces within a cancellous bone. It is within these spaces that blood vessels begin to condense and form red marrow. During this time, the mesenchymal cells that are superficial and deep towards the cancellous bone differentiate into osteoblasts and begin to form the compact bone. After numerous bone remodeling and reformation events, the woven bone is slowly converted into lamellar bone, which contributes to the final form of the flat bone (Figure 10).

Ossification of flat bones begins in approximately the eighth week prenatal development when the osteoblasts begin to deposit hydroxyapatite into their extracellular matrix. As ossification continues intramembraneously, the mesenchyme tissues that exist superficial and deep within the flat bone remain non-calcified and form the periosteum (fibrous connective tissue) (Figure 10).

Centers of ossification are defined as the locations of the mesenchymal membrane where intramembranous ossification begins. As the fetus grows, the center of ossification expands with the oldest areas of ossification, located at the center of the development, while the newly ossified tissues are located at the edges. At term, the fetus still possesses numerous locations on the skull that have yet to be transformed from osteoid tissue into bone (no hydroxylapatite has been added). These so-called soft spots are called fontanels. These fontanels will finally become bone when the child matures to approximately two years of age.

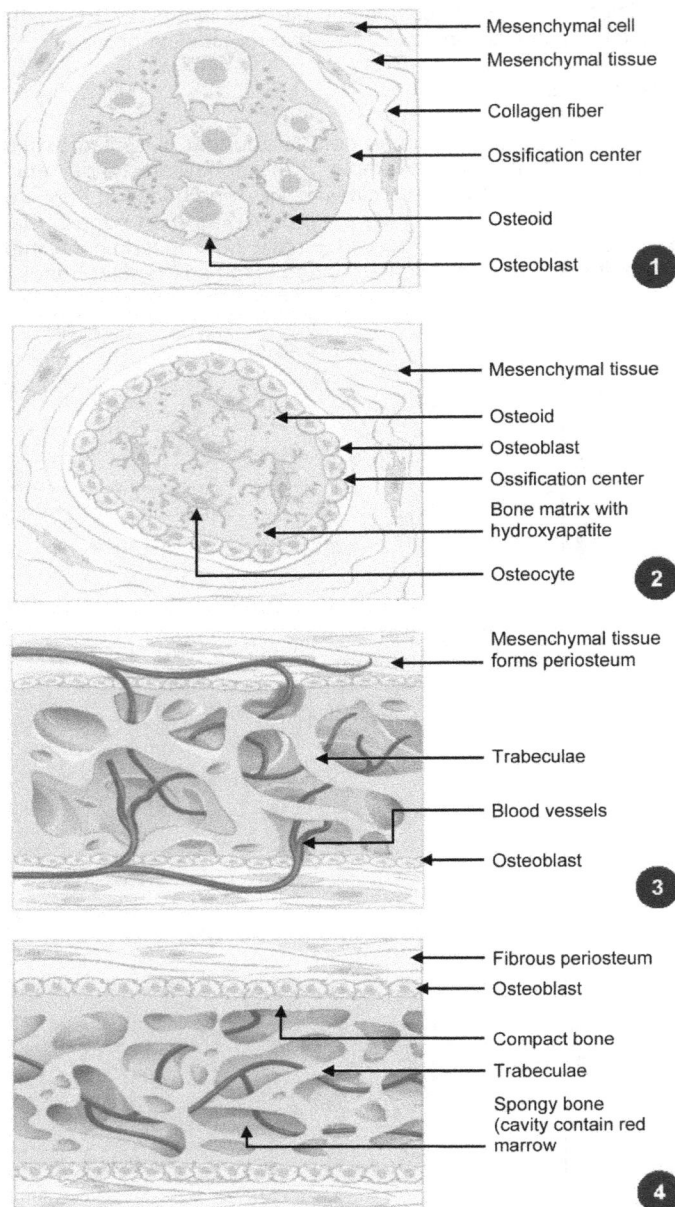

Labels: Mesenchymal cell; Mesenchymal tissue; Collagen fiber; Ossification center; Osteoid; Osteoblast — ①

Labels: Mesenchymal tissue; Osteoid; Osteoblast; Ossification center; Bone matrix with hydroxyapatite; Osteocyte — ②

Labels: Mesenchymal tissue forms periosteum; Trabeculae; Blood vessels; Osteoblast — ③

Labels: Fibrous periosteum; Osteoblast; Compact bone; Trabeculae; Spongy bone (cavity contain red marrow) — ④

Figure 10: Illustration of intramembranous ossification. ① Mesenchymal cells are grouped into clusters to form ossification centers which subsequently mature into osteoblasts. ② Osteoblasts will secrete osteoid tissue into their surroundings which effectively traps these osteoblasts. Osteoblasts that are surrounded by osteoid tissue become mature bone cells known as osteocytes. ③ The trabeculae matrix and periosteum form and ④ Compact bone develops superficially to the trabecular bone and crowded blood vessels condense to form red marrow

Endochondral Ossification

Endochondral ossification is defined as the formation of bone within cartilage. This type of bone development encompasses most of the bones in the human skeletal system. The development of endochondral bone begins in approximately the fourth week prenatal development when the mesenchymal cells enlarge and form osteogenic cells. Soon after, the osteogenic cells develop into immature cartilage cells called chondroblasts. This occurs because of the availability of growth factors, Pax1 and Scleraxis. Immediately after their formation, the chondroblasts begin producing a hyaline cartilage model in the shape of the future bone. As the chondroblasts become surrounded by cartilage or embedded inside lacunas, it is now referred to as mature cartilage cells or chondrocytes. With the exceptions to where the future joint will be, the remaining cartilage model is surrounded

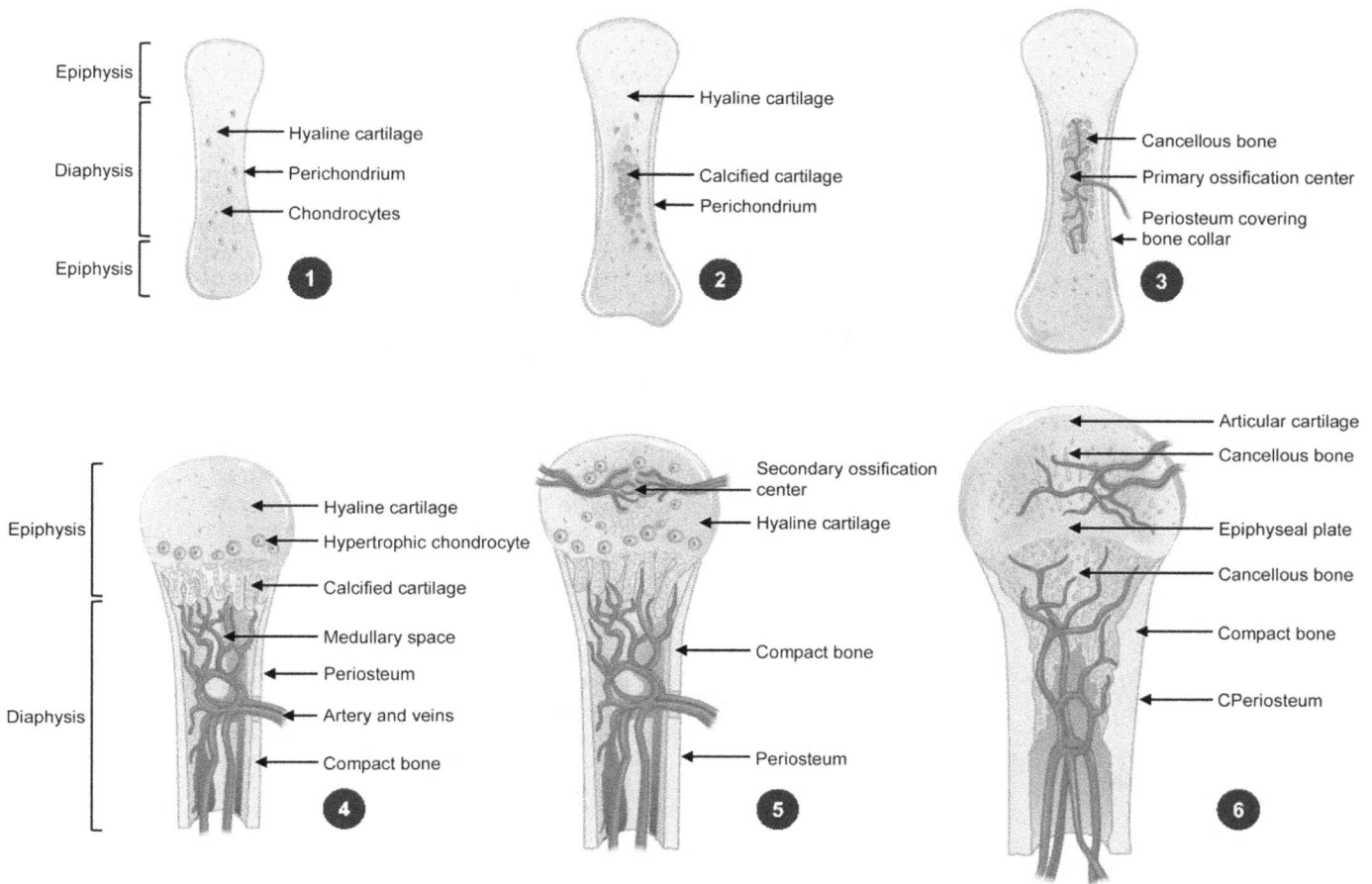

Figure 11: Illustration of endochondrial ossification. ① Mesenchymal cell will differentiate into chondroblasts and form a cartilage model of the future. Once the chondroblasts are trapped within lacunae, they are now referred to as chondrocytes. ② The chondrocytes will divide and become hypertrophic before creating calcified cartilage. Once the calcified cartilage is made, the cells proceed into apoptosis leaving behind emptied lacunae. ③ Blood vessels, osteoblasts and osteoclasts will penetrate the calcified cartilage and a primary ossification center develops. ④ Chondrocytes swell (hypertrophy) and divide at the epiphyseal ends of the future bone. ⑤ Calcified cartilage begins to form in the epiphyseal ends of the future bone. Chondrocytes proceed into apoptosis leaving behind emptied lacunae. Blood vessels, osteoblasts and osteoclasts will penetrate the calcified cartilage and a secondary ossification center develops. ⑥ Cartilage will remain at the epiphyseal plate and articular cartilage is formed

by an undeveloped mesenchyme tissue is now referred to as the perichondrium.

The mesenchymal cells within the perichondrium enlarge and mature into osteogenic cells and soon develop into osteoblasts. Immediately after their development, the osteoblasts begin producing a bone collar surrounding the diaphysis. As soon as the bone collar is completed around the shaft of the future bone, the mesenchymal tissue that surrounds the diaphysis is no longer referred to as a perichondrium. Since this tissue is now surrounding the bone collar, it is now referred to as periosteum.

The bone collar is an important support structure for the developing cartilage model. Like scaffolding surrounding a weakened structure, the bone collar will prevent bending or twisting of the future bone, allowing it to grow properly. In addition, this bone collar will be remodeled to become compact bone. At this point, some of the osteogenic cells in the periosteum will develop into osteoclasts.

While the formation of the bone collar is taking place, chondrocytes towards the center of the cartilage model become hypertrophic chondrocytes; they begin to divide rapidly and enlarge. In addition, these cells begin to increase their deposition of hyaline cartilage. This combination of rapid division, hypertrophy and deposition of additional cartilage, results in the widening and lengthening (growth) of the cartilage model.

Soon after, these newly divided and hypertrophied chondrocytes

begin to produce and release bone matrix vesicles which initiate the formation of hydroxyapatite into their surrounding matrix, effectively trapping themselves within water insoluble calcified cartilage lacunae. Without any connection to the blood vessels and with direct diffusion and osmosis made impossible by the water insoluble hydroxyapatite, these chondrocytes proceed on the path of apoptosis. As these cells deteriorate and die, they leave behind emptied enlarged hydroxyapatite hardened lacunae.

After the disintegration of the chondrocytes, blood vessels, osteoblasts, and osteoclasts begin invading the calcified cartilage. It is interesting to note that the excavating abilities of the osteoclast are not needed to allow blood vessels to enter the calcified cartilage. Nonetheless, the manners by which blood vessels invade cartilage and calcified cartilage require further elucidation.

Once settled on the surface of the calcified cartilage, the osteoblasts will begin forming numerous trabeculae of woven bone. The secreted bone matrix will trap the osteoblasts and their filapodia in lacunae and canaliculi respectively, creating osteocytes. The diameter of the trabeculae begins to grow as more osteoblasts begin secreting more bone matrix. As the trabeculae begin to link together, the future structure of woven cancellous bone begins to form. The formation of cancellous bone is referred to as the primary ossification center. It is during this time that the blood vessels begin to condense and form red marrow within the intertrabecular spaces of the cancellous bone.

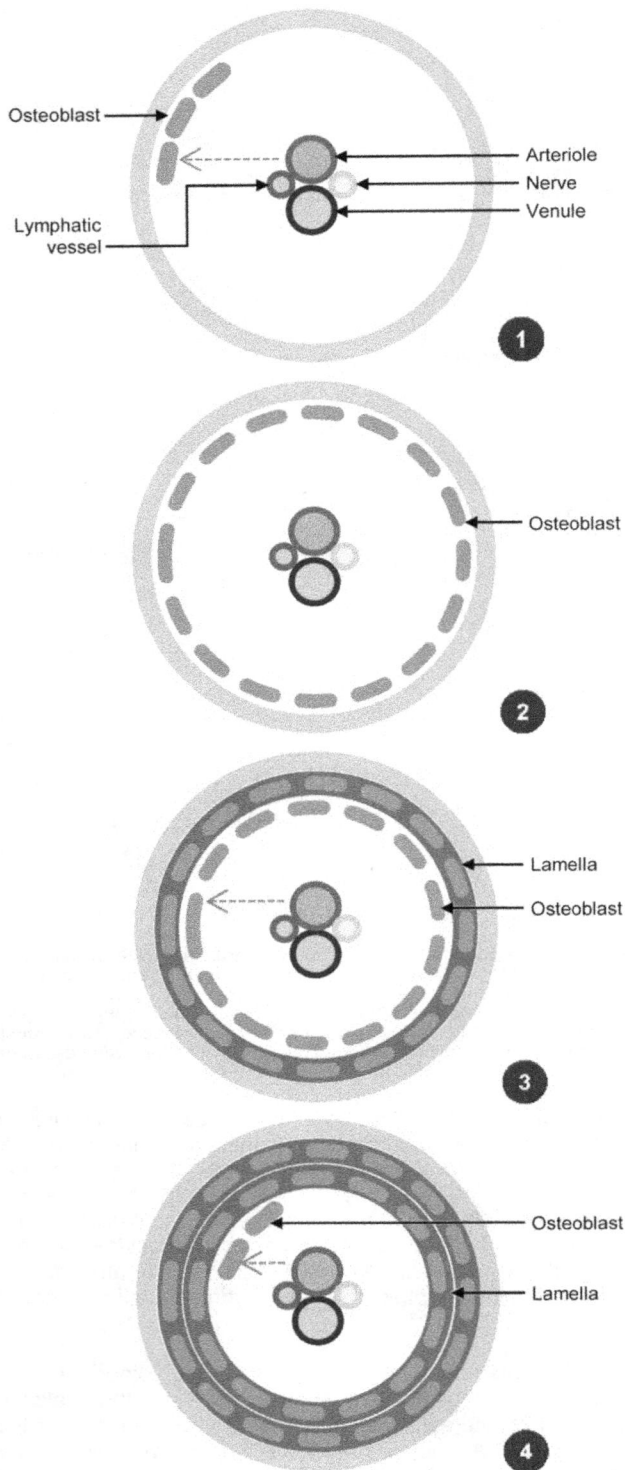

Figure 12: An illustration of compact bone creation. ① The osteoclasts will tunnel through both the bone collar and some cancellous bone while the blood vessels, nerves, lymphatic vessels and osteoblasts will follow. The osteoblasts will start to line the calcified surfaces of the tunnel. ② The osteoblasts will form the endosteum and begin secreting bone matrix. ③ A new layer or lamella will form while newly arrived osteoblasts will line the surface of the new lamella. ④ Again, the osteoblast will secrete bone matrix and another lamella will be created

to remove the cancellous bone towards the center of the growing bone to form a medullary space (Figure 11).

About one month before birth, the secondary ossification centers begins to form in the calcified cartilage areas of both the proximal and distal epiphysis of the tibia, humerus, and femur. If all three secondary ossification centers appear on a radiograph, the fetus is consider to be full term. It is important to note that the formation of other secondary ossification centers will continue as the child grows. For example, the secondary ossification center located on the medial epiphysis of the clavicle will remain with the individual until approximately 20 years of age (Figure 11).

The manner by which secondary ossification centers are formed is the same as the formation of the primary ossification centers; where blood vessels, osteoblasts, and osteoclasts begin invading the calcified cartilage. Once settled on the surface of the calcified cartilage, the osteoblasts will form trabeculae of woven bone. The bone matrix that is secreted will trap the osteoblasts within its lacunae turning it into a mature bone cell or osteocyte. As the trabeculae begin to link together, the future structure of woven cancellous bone begins to take shape. The major difference between primary and secondary ossification centers is that the secondary ossification center does not form a medullary space within the epiphysis (Figure 11).

Also during this time, towards the surface of the bone, in an area adjacent to the periosteum, the osteoclasts will begin a process of bone remodeling that transforms the bone collar, and some peripheral cancellous bone, into compact bone. For example, osteoclasts will begin tunneling through the bone collar, as well as, dissolving/remodeling some of the superficial trabeculae of the cancellous bone. The horizontal tunnels formed will become the Volkmann's canal while the vertical tunnels formed will become the Haversian canal. As the osteoclasts are tunneling through, blood vessels, lymphatic vessels, and nerves begin growing into the newly formed tunnels. In addition, osteoblasts and osteogenic cells from the periosteum also begin to follow the osteoclasts into the newly formed tunnels. A thin layer of osteogenic cells and osteoblasts will line the new tunnel forming an endosteum. The osteoblasts of the newly formed endosteum will begin to deposit bone matrix thereby trapping the osteoblast and its filopodia in a calcified matrix, turning the immature bone cells into osteocytes. This deposition of the bone matrix will create a concentric lamella. This process will repeat itself once more as the osteogenic cells of the endosteum will differentiate into osteoblasts and begin secreting another layer of bone matrix. This will create another concentric lamella. This process will continue until an osteon is created (Figure 12).

The exchange of cartilage with woven bone will continue until nearly all of the cartilage has been replaced. The remaining hyaline cartilage region, known as the epiphyseal disk, will remain until the adult stops growing (~20 year of age) and become the epiphyseal line. As mentioned previously, the process of replacing woven bone with lamellar bone is called bone remodeling. For example, the woven bone of the tibia will be replaced by lamellar bone between birth and one year of age.

Bone Growth in Length

Bone growth in both length and width will continue until approximately 20 years of age. The lengthening of bone occurs at the epiphyseal plate superior to the metaphysis which is a transitional region between the epiphysis and the diaphysis (Figure 7). The epiphyseal plate is divided into four zones (Figure 13).

The first is called the zone of reserve cartilage, also known as the zone of resting cartilage. It is the furthest zone from the primary marrow space. This zone is composed of chondrocytes embedded within hyaline cartilage. The cells within this zone

The cartilage model will continue to grow as more chondrocytes divide and proceed to hypertrophy. The bone collar surrounding the diaphysis will continue to lengthen to accommodate the newly formed areas of the cartilage model. Regions of calcified cartilage will also continue to expand, while calcified cartilage begins to form in the epiphysis. It is at this time that the osteoclasts begin

demonstrate no sign of development.

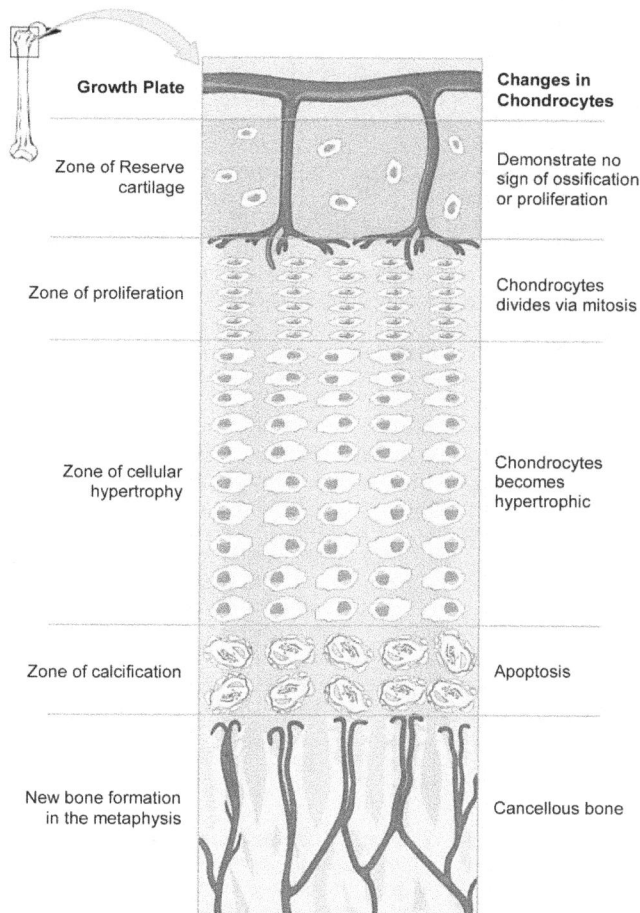

Figure 13: Illustration of an epiphyseal plate and its zones

The second is called the zone of proliferation. This zone is closer to the primary marrow space and here the chondrocytes are dividing, creating more cells, and more lacunae. Please note that it is the dividing chondrocytes that add length to the bone.

The third is the zone of cellular hypertrophy. This zone is even closer to the primary marrow space and the newly divided chondrocytes become hypertrophic (increase in the volume of the cell), which adds both length and width to the growing bone.

Finally, the zone of calcification is where the chondrocytes begin to deposit hydroxyapatite into their surrounding matrix, thereby effectively trapping themselves within water insoluble calcified cartilage lacunae. Without any connection to the blood vessels, and with direct diffusion and osmosis made impossible by the water insoluble hydroxyapatite, these chondrocytes proceed on the path of apoptosis. As these cells deteriorate and die, they leave behind emptied enlarged hydroxyapatite hardened lacunae.

After the disintegration of the chondrocytes, blood vessels, osteoblasts, and osteoclasts begin invading the calcified cartilage. Once settled on the surface of the calcified cartilage, the osteoblasts will begin forming the numerous trabeculae of woven bone. The secreted bone matrix will trap the osteoblasts and their filapodia in lacunae and canaliculi respectively, thereby creating osteocytes. The diameter of the trabeculae begin to grow as more osteoblasts begin secreting more bone matrix. As the trabeculae begin to link together, the future structure of woven cancellous bone starts to form (Figure 13).

As the individual ages, the woven bone will be replaced by lamellar bone. In addition, much of the cancellous bone will be remodeled to form compact bone, adding strength and hardness. Please examine the previous section entitled Endochondrial Ossification for more information.

Bone Growth in Width: Appositional Growth

Appositional growth occurs beneath the periosteum and this is the manner by which long bones grow in width and the flat bones grow in thickness. Appositional growth occurs on the surfaces of older bones where the osteoblasts (from the maturation of osteogenic cells via the periosteum) will lay down bone matrix to form a new layer of bone with series of ridges and grooves. Please note that since this new bone formation is occurring beneath the periosteum, the newly formed tissue is readily lined by the periosteum. The osteoblasts that formed these ridges are now referred to as osteocytes, since they and their filopodia are now trapped within lacunae.

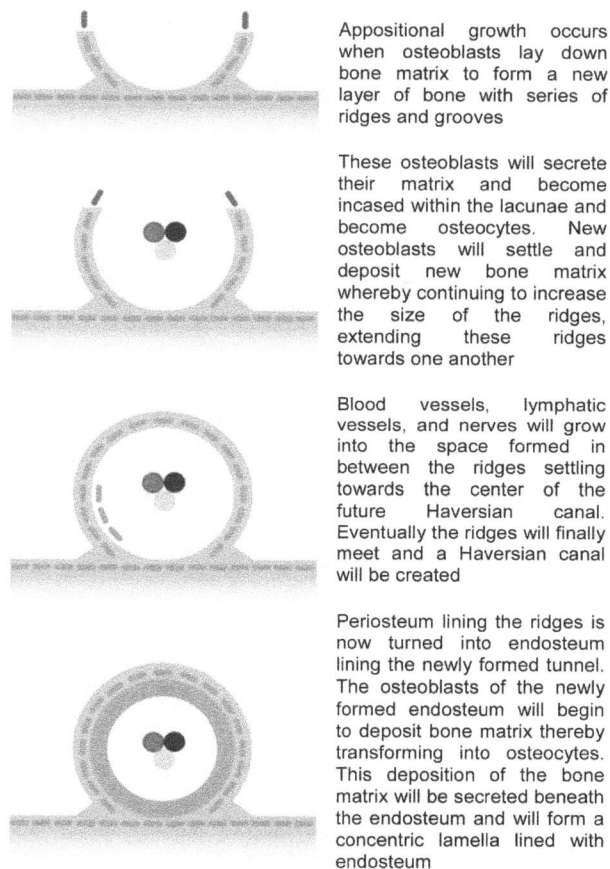

Appositional growth occurs when osteoblasts lay down bone matrix to form a new layer of bone with series of ridges and grooves

These osteoblasts will secrete their matrix and become incased within the lacunae and become osteocytes. New osteoblasts will settle and deposit new bone matrix whereby continuing to increase the size of the ridges, extending these ridges towards one another

Blood vessels, lymphatic vessels, and nerves will grow into the space formed in between the ridges settling towards the center of the future Haversian canal. Eventually the ridges will finally meet and a Haversian canal will be created

Periosteum lining the ridges is now turned into endosteum lining the newly formed tunnel. The osteoblasts of the newly formed endosteum will begin to deposit bone matrix thereby transforming into osteocytes. This deposition of the bone matrix will be secreted beneath the endosteum and will form a concentric lamella lined with endosteum

Figure 14: Illustration of appositional growth

As the individual ages, the osteoblasts will continue to increase the size of the ridges, extending these ridges towards one another. At this time blood vessels, lymphatic vessels, and nerves will grow into the space formed in between the ridges settling towards the center of the future Haversian canal. Eventually the ridges will finally meet and a Haversian canal will be created. What was previously the periosteum lining the ridges is now turned into endosteum lining the newly formed tunnel. The osteoblasts of the newly formed endosteum will begin to deposit bone matrix, thereby trapping the osteoblast and its filopodia in calcified matrix turning the immature bone cell into osteocytes. This deposition of the bone matrix will be secreted beneath the endosteum and will form a concentric lamella lined with endosteum. The osteogenic cells of the endosteum will once again differentiate into osteoblasts and begin secreting another layer of bone matrix. This will create another concentric lamella. This process will continue until an osteon is created (Figure 14).

Osteoporosis

Osteoporosis is condition that is caused by the thinning of the bone matrix resulting in the weakening of bones. This condition is generally caused by low calcium levels in the blood stream resulting in an excessive reabsorption of bone matrix in an attempt to restore homeostatic levels of calcium in the body.

The loss of bone mass results in a bone mineral density (BMD; amount of mineral matter per square centimeter of bone) and causes the bones to become porous and weak which make them prone to deformity and fracture. It is known that women suffer from this condition two and a half times (2.5x) more often than men. This highly imbalanced occurrence may be attributed to many factors. One reason that women tend to suffer from osteoporosis may be attributed to pregnancy since the fetus will satisfy its need for calcium from the mother's body. Another attributed factor is the decrease in the production of estrogen as a woman increases in age. Estrogen is a hormone that provides secondary female characteristics, as well as, helping to maintain calcium balance. Estrogen is known to inhibit the effects of parathyroid hormone (PTH; which causes osteoclasts to break down bone and release calcium into the blood stream) and thereby maintaining bone health. This situation becomes dire when women enter menopause, and estrogen production falls below desirable levels. Most often, the degeneration of bone occurs in the cancellous bones of the forearm and in the vertebrae. The degeneration of the vertebrae could cause a decrease in height or worse, kyphosis, a condition caused by the collapse of the back bone, thereby resulting in a forward curvature of the spine or a hump (Figure 15).

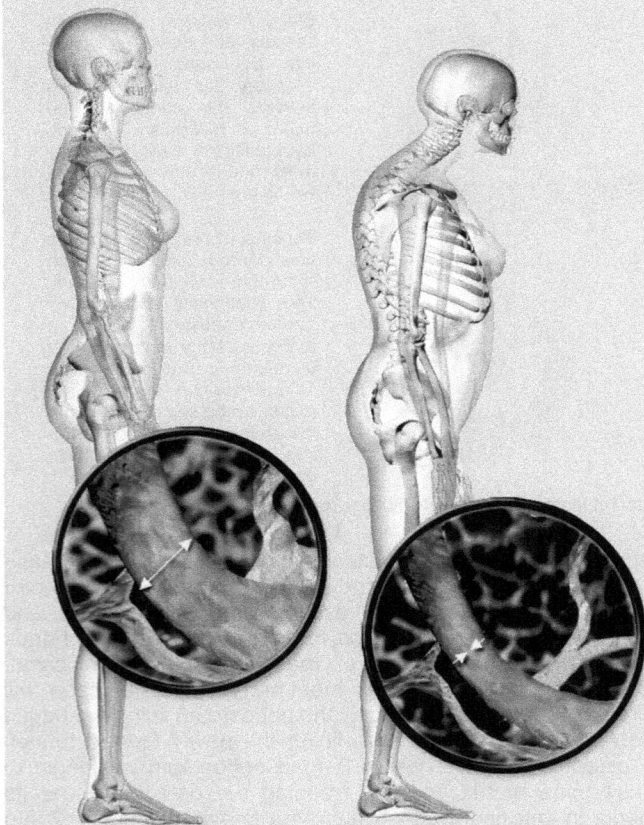

Figure 15: Illustration of kyphosis

Recently, it has been determined that there is a genetic component in osteoporosis. It is estimated that approximately 60% person's peak bone mass is genetically determined and the remaining 40% can be attributed to environmental factors. For example, if a person is predetermined to have low bone mass and does not ingest enough calcium in their diet, this combination of circumstances would make this individual more susceptible to osteoporosis.

There are many treatments for osteoporosis. For example, increase in dietary calcium and vitamin D can promote new bone formation. It is recommended that an individual needs to consume approximately 1000 to 1500 mg of calcium and 800 IU (international unit) or approximately 20 µg of vitamin D daily to generate a positive effect. Increase in exercise routine could also help prevent bone loss and increase in bone formation. In postmenopausal women, hormone replacement therapy (HRT) will replenish the reduction in natural estrogen production with synthetic estrogen. Nonetheless, the side effects of HRT may include increase risk in heart attack, stroke, blood clot, breast and uterine cancer. Recently, new group of drugs called selective estrogen receptor modulators (SERMs) have been developed. These drugs target estrogen receptors and produce the desirable bone regeneration without many of the dangerous side effects of HRT. There are three type of SERMs: tamoxifen citrate (Nolvadex and Soltamox), Evista and Fareston. Statins may be used, which will increase osteoblast activities, while inhibiting cholesterol synthesis. Calcitonin (drug name, miacalcin) is antagonistic hormones of PTH which inhibit osteoclasts activity and prevents bone degeneration. Finally, bisphosphonates can also be taken to prevent further degeneration of bones. Bisphosphonates inhibit the activities of osteoclasts thereby preventing further bone deterioration.

BONE FRACTURES

Broken bones, also referred to as bone fractures, are rather common occurrence. On average, a person will suffer two fractures during their lives. Generally, a fracture is caused when physical force exerted to the bone is higher than the bone is capable of coping. The risk of fracture in many ways is dependent on the individual's age. Put simply, the younger a person is the more likely that individual will place themselves in a position where damage to the bone can occur. Nonetheless, childhood fractures are generally less complicated and heal quickly. Although less frequent, adult fractures are more complicated and take longer to heal. Adult bones are more brittle and an individual is more likely to suffer fracture during a fall.

There are many types of fractures but in general fractures are classified into four categories. For example, a simple or closed fracture is when the bone breaks but there is no open wound caused by the broken bone on the skin. A compound fracture or open fracture is when there is an open wound on the skin with or without a protruding piece of fractured bone exposed. A complete fracture is when the bone is broken into two or more pieces. Incomplete fracture is when a bone cracks but is not broken all the way through (Figure 16).

Based on the categories of the break, and at times the age of an individual, bone fractures are also be classified into numerous subtypes. For example, an oblique fracture is a complete fracture that may be either compound of simple. The term oblique means that the fracture line is oblique to the long axis of the bone and the two fragments are in the same plane without spiraling. Oblique fractures are generally caused by over bending of a long bone (Figure 17).

Transverse fractures are a complete fracture that can be either compound or simple. The term transverse implies that the fracture line is at a right angle to the long axis of the bone. Most often this type of fracture is caused by bending forces (Figure 17).

Spiral fractures, also called torsion fractures, are a complete fractures that are either compound or simple. The term spiral means that the fracture line is spiraling along the long axis of the bone. This type of fracture is generally caused by torsional twisting or rotational forces (Figure 17).

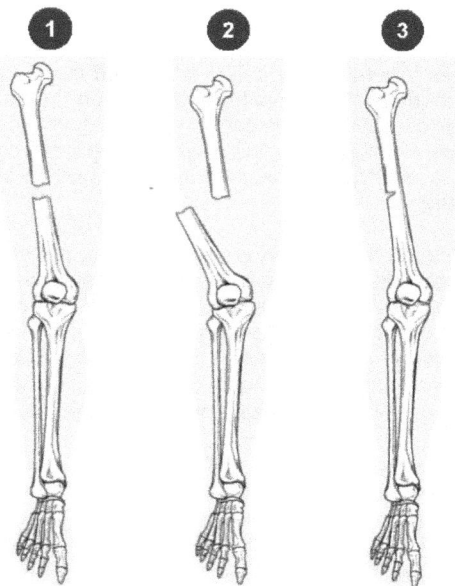

Figure 16: Illustration of the general categories of bone fractures. ① A complete fracture that is also a simple fracture. ② A complete fracture that is also a compound fracture. ③ An incomplete fracture that is also a simple fracture

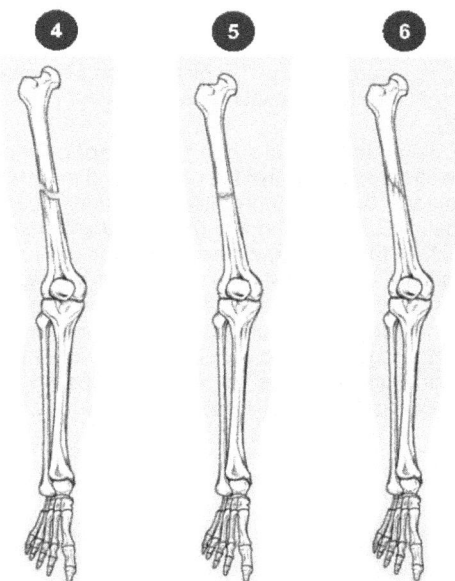

Figure 17: Illustration of bone fractures. ④ An oblique fracture. ⑤ A transverse fracture. ⑥ A spiral fracture

Comminuted fracture is a complete fracture that is either compound or simple. Comminuted fracture denotes that at least three fracture fragments have resulted from the break and the fracture lines interconnect. The fracture lines that form the comminuted fracture may be oblique, transverse, or spiral. Comminuted fractures are generally caused by high-impact trauma, such as an automobile accident (Figure 18).

Impacted fracture, also known as buckle fracture, is a simple fracture. An impacted fracture is where a bony fragment, generally compact bone, is driven into cancellous bone. This type of fracture is common in sky diving accidents (Figure 18).

Multiple fracture implies a complete fracture that is either compound or simple. The term multiple implies that at least three fracture fragments occurred in a single bone, however, their fracture lines do not interconnect and the individual fracture lines could be of any shape. Nonetheless, multiple fracture is generally describing two completely independent fractures that occurred on the same bone (Figure 18).

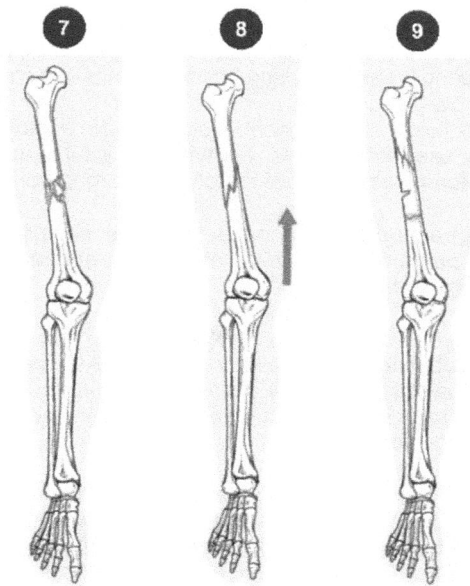

Figure 18: Illustration of bone fractures. ⑦ A comminuted fracture. ⑧ An impact fracture. ⑨ A multiple fracture

Compression fracture is a simple fracture that occurs when the cancellous bone collapses on itself. Typically, this type of fracture occurs in the vertebra following trauma and tend to be stable. This type of fracture heals in place, but shortening of the individual will occur (Figure 19).

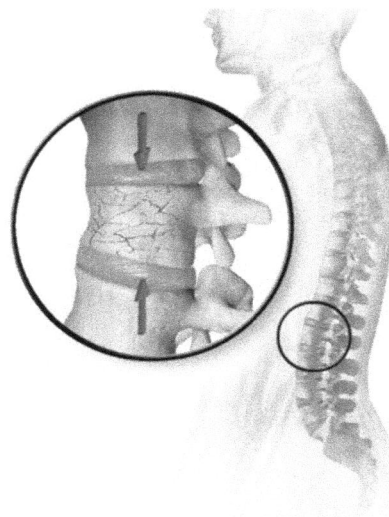

Figure 15: Kyphosis

A pathologic fracture is a simple fracture resulting from a disease that weakens the bones such as osteoporosis, tumor or cyst growth in the bones, osteomalacia (a condition caused by flawed mineralization of the bones), Paget's disease (a chronic disorder that is caused by the abnormal bone breakdown and development), or osteogenesis imperfecta (a congenital bone disorder typified by structurally weak bones that are prone to fracture). Additionally, pathologic fractures can also be caused by myeloma (plasma cell cancer), as well as, metastatic lung, and breast cancer. Pathological fractures are commonly seen in the spine of the elderly, in a condition called kyphosis, which is a deformity that causes the upper spine to bend forward, creating a hump (Figure 15).

A greenstick fracture is an incomplete simple fracture that occurs in infants and children. Since the bones of infants and children are mainly composed of woven bone, which is made of high percentage of cartilage, this type of incomplete fracture results in bones bending and cracking but does not break completely.

Stress fractures are an incomplete, simple fracture that are commonly seen in athletes. This type of fracture results from a repetitive force being applied to bones for a long period of time.

A hairline fracture is an incomplete simple fracture where the bone fragments remain in alignment, while appearing on x-ray film as a fine line.

The fissure fracture is an incomplete fracture that extends from the surface, but not through a long or flat bone. Generally this type of fracture is still covered by an intact periosteum; therefore the bone still retains its shape. The fracture lines may be transverse, oblique, spiral, longitudinal, or radiating from a central point.

Depression fractures may be a simple or complete fracture that occurs in the flat bones of the skull. This type of fracture is represented by multiple fracture lines, all intersecting and there is often an indention in the area of the fracture. Depression fractures generally occur in the calvarium (the portion of a skull including the braincase but excluding the lower jaw and facial portion), the maxilla, or the frontal bone.

BONE REPAIR

On average, a person will suffer two fractures in a lifetime; therefore bone repair is an essential part of life. It has been determined that there are four main phases of repair following fracture.

The hematoma stage, generally occurs during the first two weeks post-injury, is initiated after hemorrhage is formed due to vascular injury in and around a damaged bone. Subsequently, a hematoma is formed. A hematoma is a localized mass of blood released from damaged blood vessels (i.e. hemorrhage), but are confined within an organ or space. Hematoma results in a blood clot forming to prevent excessive bleeding. However, the formation of the blood clot will result in anoxia (lack of oxygen)

and poor or lack of nutrients being supplied to the damaged bone. This will result in the death of the bone tissue within the immediate area (Figure 20).

The callus stage occurs when a callus is formed at the damaged site. A callus is defined as a collection of tissue that forms at the fractured site, which is designed to connect the broken ends of the bones together. The callus is defined by its location. For example, an internal callus is formed between the broken bones and the marrow cavity and is constructed of hyaline cartilage and woven bone. An external callus is where a collar of cartilage and woven bone external to the fracture is formed to stabilize the break (Figure 20).

Callus formation follows various steps. First, blood vessels will begin to grow into the hematoma, which results in the infusion of macrophages and osteoclasts to the area. The macrophages are responsible for removing cellular debris, while the osteoclasts are responsible for extricating bone debris. Soon after the introduction of macrophage and osteoclasts, an infusion of T- and B-lymphocytes triggers inflammation of the tissues.

This inflammation is intended to fight off any potential infections. In addition, an infusion of fibroblasts will result in the production of collagen and other extracelluar materials. The production of collagen will form a granulation tissue, also known as aprocallus. After the formation of the procallus, osteoprogenitor cells from the periosteum will differentiate into chondroblasts and enter the procallus. Chondroblasts will form hyaline cartilage and eventually become trapped within the lacunae and become chondrocytes (Figure 20).

Callus ossification occurs when chondrocytes begin depositing hydroxyapatite into its extracellular matrix, which results in these cells proceeding into apoptosis. After the death of these cells, only empty lacunae remain. The ossification of the callus occurs approximately four to six weeks post-injury

Bone remodeling is the final step of bone repair. During this final stage the woven bones formed in the internal and external callus will be replaced by the much stronger laminar compact bone. First the osteoclasts will began removing/reforming the boney matrix which is followed by osteoprogenitor cells differentiating into osteoblasts. Osteoblasts will then enter the repair zone and begin forming lacunae by depositing the bone matrix of compact bone. The osteon from opposite sides of the break will extend across the damaged area to anchor or "peg" the break together. Adequate strength in bones usually develops six months after injury, however the complete healing process may not occur until

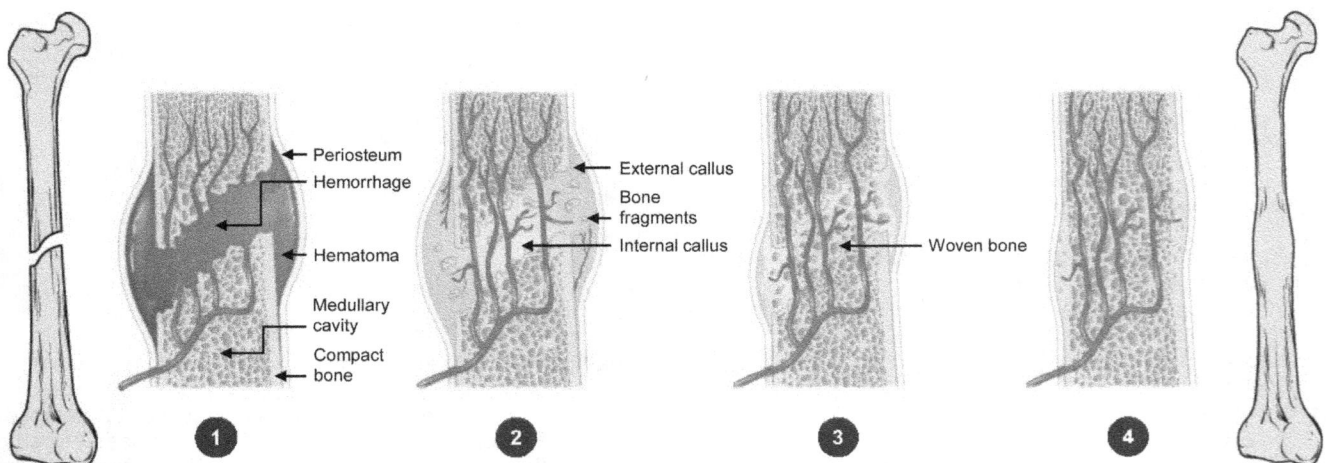

Figure 20: Graphic illustration of bone repair. ① The formation of a hematoma from a hemorrhage. ② The formation of an internal and external callus. ③ The formation of woven cancellous bone and ④ remodeled and healed bone

SKELETAL SYSTEM

A healthy skeletal system is maintained through various means. For example, bones are the major storage areas for calcium (Ca^{++}) ions in the human body. Ca^{++} ions are used to create hydroxyapatite by osteoblasts during bone formation, while Ca^{++} ions are released from bones when osteoclasts remodel bones. When Ca^{++} ions levels are low in circulation, parathyroid hormones are released by the parathyroid glands (located on the posterior side of the thyroid gland), which increases osteoclast activities causing more bone matrix to be broken down and increasing Ca^{++} ions being released into the blood stream. In contrast, when the levels of Ca^{++} ions are too high in circulation, the parafollicular cells, also located in the thyroid gland, release the hormone calcitonin to inhibit osteoclasts activities and increase osteoblast activities. Increased osteoblast activity will result in an escalation of Ca^{++} ions absorption from the blood stream to be used in bone formation. The absorption of Ca^{++} ions will in turn lower the levels of Ca^{++} ions in circulation (Figure 21).

The main sources of calcium for the human body are ingestion and absorption of this cation through the digestive system. It is understood that milk, dairy products, meat, and certain vegetables contain high levels of calcium. These food sources are mechanically broken down through mastication (chewing) and chemically processed in the stomach. The hydrochloric acid secreted by the parietal cells in the stomach converts calcium into Ca^{++} ions, which is then absorbed by the small intestines.

Ca^{++} ions are absorbed in the small intestine through distinct mechanisms found primarily in the duodenum and upper jejunum segments of the small intestines. The absorption processes involves three steps. First, the entry of the Ca^{++} ions across the brush border (i.e. microvilli) is accomplished through a calcium selective ion channel, CaT1. Second, the intracellular diffusion of the Ca^{++} ions within the epithelial cells of the small intestines is mediated by two cytosolic calcium-binding proteins calbindinD (9k) and CaBP. Finally, Ca^{++} ions are excreted from the basal domain of epithelial cells, which in turn enters circulation. The excretion of Ca^{++} ions is mediated by an enzyme CaATPase. It is known that the biosynthesis of CaT1 and CaBP are vitamin D dependent processes. Therefore, low vitamin D levels will hinder the production of CaT1 and CaBP, which in turn hinders the absorption of Ca^{++} ions.

Additional factors involved in bone maintenance include bone stress, and in females, estrogen. For example, bones are structured to deal with the stress and weight placed upon them. Changes in stress (e.g. increase in body weight) will trigger the remodeling of the bone to meet the new demands. It is understood that the remodeling process is triggered by the stretching and bending of the osteocytes through mechanosensation, a process where mechanical stimuli are transduced into neuronal impulses. An individual that is immobile will have weaker bones, since no stress has been placed on them to induce remodeling. Lack of bone remodeling will create weaker bones that are unable to meet the demands of stress.

The hormone estrogen also exerts its effects on bones within the body. Osteoblasts possess two intracellular estrogen receptors, ERa and ERb. When estrogen binds to the receptors, osteoblast synthesis of the paracrine secretion interleukin 6 is blocked. It is known that interleukin 6 is a potent stimulator of osteoclast activities and promotes the breakdown of bones. Therefore, the inhibition of interleukin 6 will prevent further bone breakdown. However, if the estrogen levels are low as in the cases of women proceeding into menopause, the osteoclasts activities will remain high because of the continuing secretion of interleukin 6, which may contribute to the development of osteoporosis. In addition, estrogen also regulates osteoclast apoptosis. If estrogen levels are low, osteoclasts will remain viable for longer periods of time and continue to actively breakdown bone.

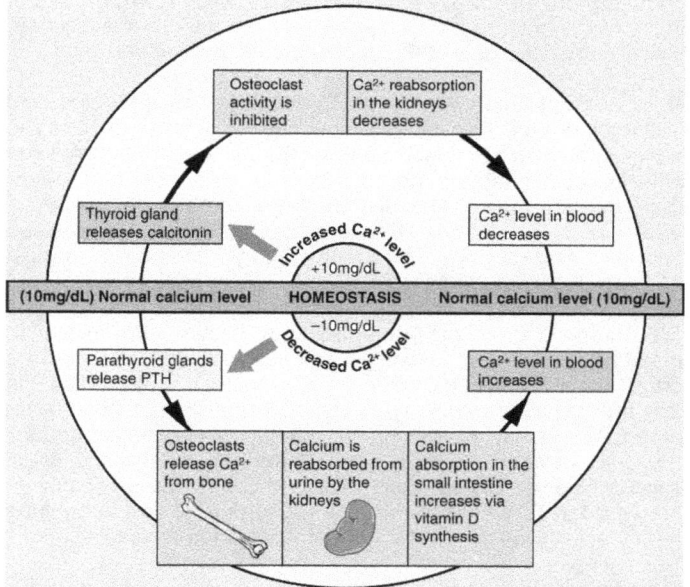

Figure 21: Illustration of calcium homeostasis

Bone Injury

When assessing a bone fracture, a nurse should look for the 5 P's: pain, paralysis, paresthesia, pallor (unhealthy pale appearance), and pulselessness. When observing the area of injury the nurse may also see deformity, shortening of the extremity, hear crepitus (grating, crackling or popping sounds), localized edema, and ecchymosis (escape of blood into the tissues from ruptured blood vessels).

When the assessment reveals the existence of a fracture in an emergency situation, the following interventions should be implemented. The extremity should be immobilized or splinted. The proximal or distal joint should be supported (e.g. an arm sling). If it is a compound fracture, the injury should be covered with a sterile dressing or clean dressing if sterile dressing isn't available. The client presenting with an open fracture should receive a tetanus shot and intravenous broad spectrum antibiotics for the first 48 hours.

Casting

Fiberglass casts are lighter, more durable, and take less time to prepare than plaster casts. Plaster casts, on the other hand, are more delicate, and take more time to dry, around 24 to 72 hours. Nonetheless, plaster casts will create a more accurate mold than fiberglass cast. During the setting process these casts will emit heat. It is necessary for the attending nurse to inform the client about this sensation in advance. While the cast is setting, it is advisable to handle it with palms of the hands and avoid direct contact with fingers to avoid deformation of the setting plaster.

Compartment syndrome is a complication that should be assessed when a client is using a cast. Compartment syndrome is a serious condition that is caused by increased pressure on muscle compartment(s), which may lead to neuromuscular damage, as well as, impeding adequate circulation. The swelling that may result in compartment syndrome is commonly caused by traumatic accidents such as car crash, crushing injury, or through surgery. Within the deep fascia between the muscles there are limited amounts of space where muscle fibers, nerves and blood vessels exist. These limited spaces are called compartments.

It is known that deep fascia does not expand and any swelling within the compartment will result in increased pressure which will increase stress on the muscle fibers, blood vessels, and nerves. If the pressure becomes too high, blood flow within the compartment will be limited or impeded, leading to permanent injury to the muscle and nerves. If the pressure is not alleviated, it may cause necrosis of the tissues within the compartment.

Compartment syndrome is characterized by deep, throbbing, and unrelenting pain. The pain will continue to increase despite the administration of opioids. Objective data will reveal compromised neurovascular status such as coolness, paleness, and longer capillary refill time to the distal extremity, these all indicate a decrease in circulation. This complication must be addressed promptly, as it can cause permanent injury. The physician must be immediately notified.

Discourage the client from inserting objects into the cast. Inserting objects into a confined space will only increase the pressure that is placed on the compartments. It is also advisable that the fractured extremity be elevated the first 24 hours after casting. Elevation will aid in blood drainage and relief of potential pressure buildup within the compartments. Finally, the client should also be encouraged to flex the affected extremity to prevent disuse syndrome. Disuse syndrome is cause by long term inactivity which results in the deterioration of tissues.

External and Internal Fixator

The purpose of fracture fixation is to stabilize the fractured bone and to facilitate healing of the injured bone. There are various ways that fractures are treated. For example, conservative fracture treatment relies on providing external support for fractures. This has been the manner by which fractures have been treated for most of the 19th and 20th centuries. Conservative fracture treatment consists of using braces (to limit range of motion of a joint), slings (to support and immobilize the injured area), splints (to stabilize an injured body part), or casts (to hold broken bones in place while it heals) in an effort to stabilize and restore bone alignment.

With the advent of sterile surgical procedures and the subsequent reduction of infection, external and internal fixators are now commonly used to set and stabilize fractured bones. External fixators are designed on the principle of splinting. There are three standard forms of external fixators: standard uniplanar fixator, ring fixator, and a hybrid fixator. Standard uniplanar fixator consists of series of screws that are inserted through the near cortex, medullary canal, and into the far cortex of the bone (to avoid penetration of the muscle) and are then connected to an external rod for stability (Figure 22).

Ring fixator also known as a wire-ring fixator, is commonly used in bone fractures that require deformity correction due to its ability to apply compression, control distraction, and orientation in three dimensions. Ring fixator is designed with thin wires passing through the fractured bone while they are held under tension by an external frame. There are two standard types of wire fixators. The closed ring fixators immobilize bone fractures and provide the most stability. The open ring fixators are generally used near joints to provide some controlled movements (Figure 22).

The hybrid fixator utilizes the techniques for both standard uniplanar and ring fixation. This form of fixator is commonly seen in the treatment of proximal and distal tibial fractures that tend to be close to the joint.

Internal fixator is generally used to treat comminuted and multiple fractures and it requires a surgical procedure. The surgeon will transect a wound and reposition the fragments of bone into their normal alignment. They are held together with special cobalt, chrome, stainless steel, or titanium implants (screws, plates, rods/nails, and wires) to reduce the chances of developing an allergic reaction from the patient's immune system. For example, screws are simple devises that could be used as a stand-alone product to hold bone fragments together or in conjunction with other implements such as plates. Screws come in many designs or sizes depending on the type of fracture. Screws can be left in place even after the wound has completely healed. Plates are simply internal splints that are designed to hold the broken pieces of bone together. The plates are fastened to the fragmented bones with screws. Plates can also be left in place even after the wound has completely healed. Rods/nails are commonly used in long bone (e.g. femur and tibia) fractures to hold the fractured pieces of bone together. These rods/nails are generally inserted through the medullary cavity of the bone. Once the rod/nail is in place, screws are inserted at each end to hold it in place, as well as, keeping the fracture from shortening or rotating. Rods/nails can be left in place even after the wound has completely healed. Wires are often used to immobilize pieces of fractured bone that are too small to be fixed with screws. Depending on the circumstances, wires may be removed after the bone has healed or left in place.

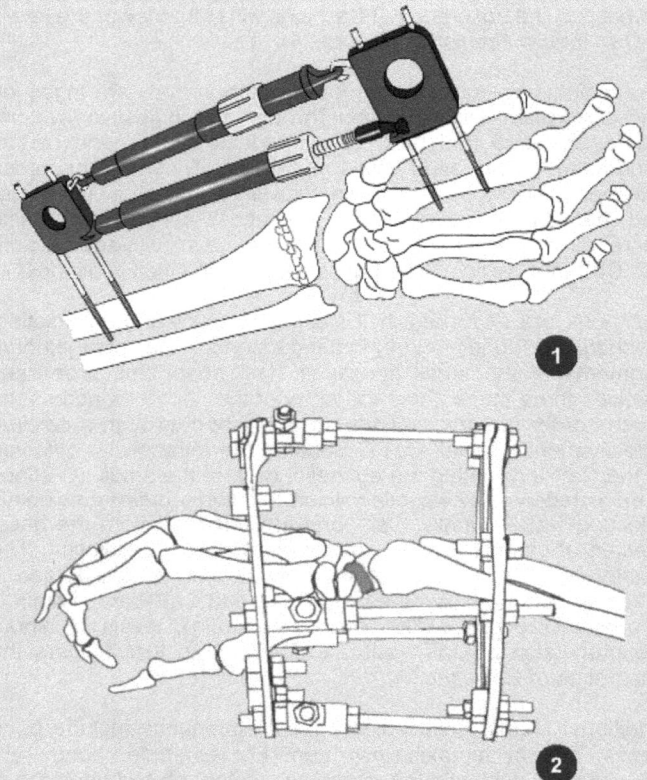

Figure 22: Illustration of ① uniplanar fixator (Dynawrist external fixator) and ② ring fixator (LIizarov device)

Nursing care for a client with conservative fracture treatment, external fixator or internal fixator must always involve keeping the damaged extremity elevated. Nurses should examine the patient's neurovascular status every two hours and assess the distal portion of the extremity (below the injury) for any changes in the color, temperature, sensation and capillary refill. If an external or internal fixator is used, the nurse should also evaluate the pin site for signs of infection such as swelling, redness, tenderness and drainage. The pin sites should be cleaned with chlorhexidine solution once or twice daily. Any signs of loosening pins should be immediately reported to the physician.

Traction

Traction is a form of short-term treatment of fractures and dislocations that are unable to be treated by casting. Traction is used to decrease muscle spasm, reduce pain and maintain the extremity in proper alignment. There are two types of traction commonly used in treatment of patients: skin traction (referred to as Buck's Traction), and skeletal traction.

Skin traction is generally used for femoral, acetabular, and hip fractures, as well as, lower back (lumbar) pains. Skin traction allows the fractured bone to be stabilized until doctors can agree on the final treatment plan.This traction method functions through the application of tape and foam strips to the patient's skin directly below the fracture in the direction and magnitude to obtain its desired effect. The weight which can be applied to skin traction is limited to the tolerance of the skin. Generally, the weight applied is restricted to less than 10% of the individual's body weight (up to a maximum of 10 pounds or 4.5 kg). Additional weights will not provide faster results. Experience has shown that a weight greater than 4.5 kg will result in the irritation of superficial skin layer.

In time, as soft tissue and muscle of the damaged area begin to relax, the amount of weight applied may change to maintain the desired pulling (traction) force. After the application of traction is completed, the amount of counter-traction forces will require consideration. Counter-traction is a force that is directed in an opposite direction which is designed to further stabilize the wound.

There are three main types of skin traction: Buck's extension traction, Russell's traction, and Dunlop's traction. Buck's traction is a type of traction where traction force is exerted in a single plane. Buck's traction is designed to partially or temporarily immobilize a fracture while awaiting surgical fixation. For example, subsequent to a lower leg injury, Buck's traction is applied through the utilization of foam rubber padded straps which are placed with the foam surface against the skin on each side of the affected leg. Loop the tape beyond the sole of the foot and place spreader to the distal end of the tape to prevent undue pressure exerted along the side of the foot. Place cast paddings along the malleolus (distal end) and proximal side of fibula which will prevent pressure sores from forming and the necrosis of the skin. With the leg slightly elevated, elastic bandage is applied in a spiral fashion over the traction tape. The application of the elastic bandage should begin from the ankle until it completely covers the wound. The elastic bandage is in constant contact with client's skin therefore the integrity of the skin is an important aspect of nursing care. The traction bandage should therefore be kept free from wrinkles and the skin should be inspected every shift. Pressure behind the knee to the peroneal nerve should be avoided to prevent nerve damage. The nurse should help the client maintain proper body alignment by assisting with shifting of weight every two hours, but the client should not be turned from side to side. Special mattresses come in handy in this situation to protect boney prominences. A sheepskin pad should also be placed under the injured leg to ease friction. If a foam boot is used with Buck's traction, make certain that the heel of the affected leg is firmly placed into the heel of the boot before securing the leg with Velcro straps. Weights are added to the rope fixed to the spreader or footplate of the foam boot before passing the rope over a pulley fastened to the end of the bed.

Russell's traction is generally used for treating fractures of the tibial plateau. A system of suspension and traction pull is used in this form of traction. Just like the Buck's traction, adhesive strips are applied while the knee is suspended in a sling. On one end, the rope is attached to the spreader or footplate of the foam boot while on the other end the rope passes over a pulley which is attached to an overhead bar and is then directed to a system of three pulleys at the foot of the bed. The three pulley system will provide both an upward and a forward pull to the injured leg.

Dunlop's traction is generally used for fractures of the upper extremities. For example, if an individual suffers a fracture humerous, Dunlop's traction will be used to immobilize and stabilize the arm. With the arm supinated, adhesive strips are applied to the abducted humerous. Counter-weight is hung from the upper part of the arm to aid in the alignment of the proximal and distal fragments of the fracture. Longitudinal traction will then be applied to the supinated forearm with the elbow angled at approximately 45°.

Skeletal traction is used as a temporary measure when a spanning external fixator or a long leg splint is unavailable. Skeletal traction utilizes a large K-wire or a strong Steinmann pin placed into the bone distal to the fracture and then connected to an appropriate stirrup with approximately 7 to 11 kg (15 to 25 pounds) weight. In certain circumstances, it may be required to apply counter-weight and the foot of the bed may need to be elevated.

Hip Fracture

When assessing a hip fracture, the most common complaint by the patient is severe pain located in the hip or groin area. The leg of the patient may also appear shortened or abducted and externally rotated. Hip fracture is treated with surgery and temporary traction. A common type of hip fracture is femoral neck fracture, also known as intracapsular fracture. This type of fracture is generally associated with stress fracture by individuals participating in strenuous athletic activities or high impact car accidents. Nonetheless, there are cases of elderly individuals with osteoporosis that developed this form of fracture. Frequent complication associated with femoral neck fracture is avascular necrosis (AVN). AVN is a pathologic process, which results from disruption in the blood supply to the bone. At present, AVN of the hip is not well understood but it does cause femoral head ischemia, which results in the death of osteocytes and the marrow and eventually brings about the collapse of the necrotic segment. Less commonly, an extracapsular fracture occurs at the trochanteric, intertrochanteric, and subtrochanteric regions. Unlike the intracapsular fractures, extra-capsular fractures do not interfere with femoral head blood supply and patients will not be at risk for developing AVN.

To manage hip fracture, the nurse should provide pain relief, and utilize proper techniques when moving the client to prevent further injury. The client should only be turned on the unaffected side with use of an abduction pillow between the knees. Physical activities should be encouraged when directed by the physician. Venous thromboembolism (VTE) is a clot that forms within the veins and its prevention is important and is generally controlled through the use of prescribed anticoagulants and compression devices. Monitoring a client's intake and output can give nurses clues about internal hemorrhage. For example, inadequate output may be an indicator of such a problem. Intravenous antibiotics may be ordered to prevent infection.

Hip Replacement

Hip dislocation is a possible complication in patients that have received a hip replacement. Therefore, it is essential that the nurse efforts in the prevention of this potential event. Nurses should avoid removing any stress on the joint for the first 8 to 12 weeks. Prevention of this complication is achieved by teaching the client to keep the affected hip abducted at all times and not to flex the hip more than 90°. The head of the bed should be at no more than a 60° angle. A fracture pan should be used for toileting, and raised toilet seats should also be utilized. The client should also be instructed not to cross their legs and keep their feet flat on the floor when extricating from their bed. Put simply, the patient should never dangle their legs along the side of the bed before standing. The client should not bend at the waist while standing. Signs and symptoms of a dislocated hip are swelling, pain, shortening of the leg, abnormal rotation, loss of movement, and popping sensation.

Drainage should also be monitored for amount, type, color, and odor. Any abnormalities can indicate an infection. More than 500 mL in the first 24 hour period post-surgery and more than

30 ml of drainage in an eight hour period by day two would be classified as excessive drainage. Prevention of heel pressure ulcers and thromboembolism are also essential parts of nursing care provided for hip replacement patients. Other complications associated with hip replacement that require close observation include heterotopic ossification (bone tissue forms outside of the bone), avascular necrosis, and loosening of the prosthesis.

Pelvic Fracture

Pelvic fracture includes any damage caused to the hip bone (the three fused bones of the ilium, ischium, and pubis), sacrum, and coccyx. The most common causes of this type of fracture are falls or vehicle accidents. Most pelvic fractures are non-life threatening and may be treated without surgery. However, serious pelvic fracture can cause damage to internal organs and produce significant internal bleeding. It is known that a serious pelvic fracture could result in the loss of four to five liters of blood, which may cause shock in patients.

Pelvic fractures can be assessed by observing bruising, tenderness, swelling, numbness, and tingling over the pubic area. These signs of tissue damage could also be seen on the proximal part of the thighs. The client may complain of pain or discomfort during weight bearing. A computed tomography (CT) scan may be performed to determine the extent of the fracture. A voiding cystogram may be done to determine if any damage has occurred to the urinary system, and this should be performed before insertion of a Foley catheter. Please note that the voiding cystogram uses contrast dye injected into the bladder. Aside from potential bladder injury, assessment should also be performed for any injury done to the rectum, intestines, and other abdominal organs.

An unstable pelvic fracture requires surgery, and open reduction internal fixation (ORIF). Stable fractures are managed through bed rest, and ankle and leg exercises and anti-embolic hose are used to prevent the complications of immobility. Defecation may pose a problem for patients with pelvic fracture and this situation could be alleviated through the increased fluid and fiber intake, as well as, the use of stool softeners to prevent straining. Pain management is also important. Swelling near and at the site of the fracture place pressure on nerves and exacerbates pain. Bone that is in the process of healing aches as the osteoblasts rebuilds the pelvic bone. Anti-inflammatory medications prescribed by the physician should reduce the swelling and pain. Pharmacological pain relief includes opioid medications that treat the pain itself, as well as, muscle relaxers that provide indirect pain relief by loosening the taut muscles. A fracture pan should be used in both unstable and stable pelvic fracture. To assess neurovascular status in these types of injury, the nurse should inspect the lower extremities for signs of neurovascular compromise.

Total Knee Replacement

After total knee replacement surgery, the client's neurovascular status will need to be assessed using the techniques mentioned earlier. The dressing around the knee should be assessed for bleeding. If there is blood on the dressing a circle should be drawn around it, this will help the nurse keep track of how much more bleeding has occurred since the last assessment. The drainage of the surgical site should be monitored for the type and amount of drainage. Drainage more than 400 mL in a 24 hour period is considered abnormal and the physician should be notified. Attending nurse should assess for signs and symptoms of infection.

Continuous passive motion (CPM) may be ordered for the client. However, the settings are controlled by the physical therapist and the nurse should not adjust it. The nurse may remove and apply the device. The knee immobilizer should be inspected for a proper fit.

Amputation

Amputations are commonly performed on lower extremities due to peripheral vascular disease (PVD). Complications of this surgery are infection, hemorrhage, skin breakdown, and joint contracture. To avoid joint contractures the residual limb should never be placed on a pillow, it should remain in a neutral alignment. It may be elevated briefly after surgery, but never for prolonged periods. The client should also be discouraged from sitting for extended durations and when sitting, the patient should be instructed to keep their legs close together. Range of motion should be performed frequently and the residual limb should be handled gently. Swelling can be controlled through compression dressings. When performing wound care aseptic technique should be used.

Phantom limb pain is a common problem with limb amputation. Phantom pain generally exhibits as sharp, aching, burning, and/or cramping pains. In time the phantom pain will subside, however, there are some medications available to ease the discomfort. For example, tricyclic antidepressants and anticonvulsants are common medications prescribed by physicians to manage this form of pain.

Bone Disorders

Osteoporosis is a reduction in bone density and structure. It is understood that Caucasian and Asian females are at the greatest risk. These risks are attributed to non-modifiable factors such as familial history for the disease. The pathology of osteoporosis can cause spontaneous fractures, kyphosis, loss in height, mid-thoracic pain, respiration difficulties, and a rounded abdomen. This disease can be managed through weight bearing exercises 3 days a week for about 30 minutes a session such as walking. Dietary modification may also prove beneficial. For example, patients should be encouraged to ingest a diet rich in calcium (1000 to 1200 mg daily) and vitamin D (800 to 1000 IU for those age 50 years and older).

Osteomalacia is a condition that causes the softening of bone due to inadequate mineralization (lack of hydroxyapatite). Signs and symptoms for this condition include pain, tenderness, bowing of bones, and pathological fracture. Laboratory examinations may reveal hypocalcemia. The client may waddle or limp when walking, and may develop kyphosis. Nurses should handle these clients gently when turning or repositioning, and use pillows to support them. Like osteoporosis, patients suffering from osteomalacia should also be encouraged to ingest a diet high in calcium and vitamin D.

Osteomyelitis is an acute or chronic infection of the bone usually caused by bacteria in the species of *Staphylococcus*, *Enterobacter*, *Streptococcus*, and *Haemophilus* etc. Typically this condition is spread through contamination from bone surgery. Signs and symptoms of the disease include chills, high fever, rapid pulse, general malaise, pain unrelieved by analgesics, swelling, and tenderness at the infected site. Typical treatment involves aggressive administration of intravenous antibiotics coupled with around the clock dosing for three to six weeks. Additional antibiotics may be taken orally for up to three months. Surgery may be needed for infections that do not respond to antibiotic.

Sterile technique should be used for dressing changes for patients suffering from osteomyelitis. The affected limb should be handled gently. Immobilization and proper alignment of the extremity should be performed.

R.I.C.E.

In the case of minor injury such as a knee or ankle sprain (injury to a ligament and tendons), contusions (a soft tissue injury, and a hematoma may develop) and strains (pulled muscle or tendon), the nurse could relieve pain, swelling and promote healing with methods specified by the acronym R.I.C.E. R.I.C.E. stands for Rest, Ice, Compression, and Elevation. For example, the client should be advised to rest and protect the injured or sore area. Using ice or a cold pack should reduce pain and swelling. Apply the ice or cold pack for a period of 10 to 20 minutes, three or more times a day for two to three days. If the swelling has been reduced, apply heat to the tender areas. Compress (wrap) the injured areas with elastic bandages and elevate the injured areas to help decrease the pain and swelling.

Morse Fall Scale

It is understood that numerous situations can increase the risk of falling. The Morse fall scale is an assessment tool often used to assess clients with mobility issues. During this assessment process, the client is assessed for a series of criteria that are each assigned a numerical value. For example, 0 to 24 is issued to an individual not at risk, 25 to 45 is given to an individual who is at low to moderate risk of a fall, and a score of 46+ is given to a patient that is at a high risk of falling. Nursing intervention has to be performed for a patient with a high score or at a high risk of falling. An example of these interventions would be non-skid socks, keeping the call light within arms-length, and keeping the bed in the low position with two to three side rails up.

Pediatric Fractures

The most common fractures in children are plastic/bowing (bone bent but not broken), buckle/torus (compressed porous bone; appears as raised or bulging projection at fracture line), greenstick (bone angulated beyond limits of bending, an incomplete fracture), and complete fracture (divides bone fragments). Spiral fractures are not common but if a child is presented with this diagnosis, this should be treated as an indicator of abuse. In the cases of spiral fracture, the nurse should assess the situation very carefully to determine the origin of the injury and in some states Social Services should be contacted immediately.

Neonates typically heal within two to three weeks; early childhood four weeks; later childhood six to eight weeks; and adolescence 8 to 12 weeks. The priority nursing intervention here is perfusion. The goal is to regain alignment and length, retain alignment, restore function, and prevent further injury. Medical management is through closed reduction and casting.

QUESTIONS

1. Which of the following is not a part of the skeletal system?
   a. Bones
   b. Cartilages
   c. Ligaments
   d. Muscles
   e. Tendons

2. Which of the following is not a function of the skeletal system?
   a. Provide physical protection for the internal tissues and organs
   b. Providing a location for immunoreaction
   c. Providing a site for blood cells and platelets production
   d. Providing a site for energy storage
   e. Proving leverage for bodily movements
   f. Serving as storage sites for calcium & phosphorus

3. Which of the following is composed of the skull, vertebral column, ribs and the sternum?
   a. Appendicular skeleton
   b. Axial skeleton

4. Which of the following is composed of the bones that construct the girdles and the limbs of the body?
   a. Appendicular skeleton
   b. Axial skeleton

5. How many bones construct the adult human skull?
   a. 10
   b. 12
   c. 18
   d. 20
   e. 22

6. The three smallest bones in the human body are the incus, stapes, and malleus which are also known as _____.
   a. Flat bones
   b. Round bones
   c. Irregular bones
   d. Long bones
   e. Auditory ossicles
   f. Both c and e

7. Excluding the skull, how many bones construct the appendicular skeleton?
   a. 126
   b. 150
   c. 167
   d. 200
   e. 206

8. Which of the following is a type of embryonic tissue that gives rise to osteogenic cells?
   a. Hematopoietic stem cells
   b. Hepatocytes
   c. Mesenchymal cells
   d. Osteochondrial progenitor cell
   e. Osteoprogenitor cells

9. The osteogenic cells will give rise to which of the following type (s) of cell (s)?
   a. Chondroblasts
   b. Hepatocytes
   c. Osteoblasts
   d. Osteochondrial progenitor cell
   e. Osteoprogenitor cells
   f. Both a and c
   g. Both d and e

10. An immature cartilage cell is called _____.
    a. Chondroblast
    b. Chondrocyte
    c. Hepatocytes
    d. Osteoblast
    e. Osteocyte

11. An immature bone cell is called _____.
    a. Chondroblast
    b. Chondrocyte
    c. Hepatocytes
    d. Osteoblast
    e. Osteocyte

12. Mesenchymal cells could form which of the following type (s) of cell (s)?
    a. Fibroblasts
    b. Adipoblasts
    c. Chondroblasts
    d. Osteoblasts
    e. All of the above

13. Which of the following term (s) is (are) defined as the manner by which bone is produced?
    a. Hematopoiesis
    b. Ossification
    c. Osteogenesis
    d. Osteoporosis
    e. Both b and c

14. Under the microscope, osteoblast possesses numerous cellular processes which interconnect with other osteoblast cellular processes. This interconnection of cellular processes allows for cellular communication. Which of the following type of junction allows for cellular communications to take place?
    a. Adhesion junction
    b. Desmosomes
    c. Gap junctions
    d. Hemidesmosomes
    e. Tight junction

15. Osteoblasts are responsible for producing proteins such as collagen, proteoglycans glycoproteins and releasing them via exocytosis. These unmineralized protein-based products that are secreted by osteoblast are referred to as the _____.
    a. Elastic tissue
    b. Lymphoid tissue
    c. Neuromuscular tissue
    d. Osteoid tissue

e. Reticular tissue

16. Osteoblasts produce high concentrations of calcium and phosphate and release them in a matrix vesicle. This matrix vesicle is formed through the protrusion of the osteoblast plasma membrane via a process known as _____.
    a. Budding
    b. Endocytosis
    c. Exocytosis
    d. Phagocytosis
    e. Pinocytosis

17. Osteoblasts produce high concentrations of calcium and phosphate and release them in a matrix vesicle. Once released, the calcium and phosphate is enzymatically converted into which of the following compounds?
    a. Calcium carbonate
    b. Calcium phosphate
    c. Hydroxyapatite
    d. Mesenchymal tissue
    e. Osteoid tissue

18. Under the microscope, osteoblast possesses numerous cellular processes which interconnect with other osteoblast cellular processes. What is the name given to these cellular processes?
    a. Filopodia
    b. Monopodial
    c. Parapodia
    d. Pseudopodia
    e. Stylopodia

19. Once adequate amounts of bone matrix are secreted, the immature bone cells are molded within its own creation. What is the name given to the chamber which the bone cells resides?
    a. Canaliculi
    b. Haversian's canal
    c. Lacunae
    d. Lamellae
    e. Volkmann's canal

20. Once adequate amounts of bone matrix are secreted, the immature bone cells are molded within its own creation. What is the name given to the tubular structure that the cellular processes are embedded?
    a. Canaliculi
    b. Haversian's canal
    c. Lacunae
    d. Lamellae
    e. Volkmann's canal

21. Once adequate amounts of bone matrix are secreted, the immature bone cells are molded within its own creation. Once embedded, these immature bone cells are now considered to be mature. What is the name given to indicate a matured bone cell?
    a. Chondrocytes
    b. Osteoblasts
    c. Osteocytes
    d. Osteogenic cells
    e. Osteoprogenitor cells

22. What is the name given to mature cartilage cells?
    a. Chondrocytes
    b. Osteoblasts
    c. Osteocytes
    d. Osteogenic cells
    e. Osteoprogenitor cells

23. Which of the following is a modified macrophage that is capable of resorbing or breakdown of bone?
    a. Chondroblasts
    b. Osteoblasts
    c. Osteoclasts
    d. Osteocytes
    e. Osteogenitor cells

24. Which of the following substances is (are) secreted during bone reabsorption?
    a. Alpha amylase
    b. Cathepsin K
    c. Hydrogen ions (H⁺)
    d. Osteonectin
    e. Sialoprotein
    f. Both a and c
    g. Both b and c

25. Which of the following is responsible for breaking down hydroxyapatite?
    a. Alpha amylase
    b. Cathepsin K
    c. Hydrogen ions (H⁺)
    d. Osteonectin

e. Sialoprotein
f. Both a and c
g. Both b and c

26. Which of the following is responsible for breaking down the protein component of the bone matrix?
    a. Alpha amylase
    b. Cathepsin K
    c. Hydrogen ions (H⁺)
    d. Osteonectin
    e. Sialoprotein
    f. Both a and c
    g. Both b and c

27. Which of the following is the most abundant collagenous compound found in bones?
    a. Type I collagen
    b. Type II collagen
    c. Type III collagen
    d. Type IV collagen
    e. Type V collagen
    f. Type VI collagen
    g. Type VII collagen

28. Which of the following is a blood serum-derived protein that aids in the process of matrix mineralization?
    a. Albumin
    b. Bone sialoprotein
    c. Fibrin
    d. Fibronectin
    e. Globulin
    f. Osteocalcin
    g. Osteonectin
    h. Small leucine-rich proteoglycans
    i. α2-HS-glycoprotein

29. Which of the following is believed to be a type of growth factor that helps to regulate osteoblast proliferation?
    a. Albumin
    b. Bone sialoprotein
    c. Fibrin
    d. Fibronectin
    e. Globulin
    f. Osteocalcin
    g. Osteonectin
    h. Small leucine-rich proteoglycans
    i. α2-HS-glycoprotein

30. Which of the following bone matrix substance is involved in cell proliferation, organic matrix and mineral deposition, and remodeling?
    a. Albumin
    b. Bone sialoprotein
    c. Fibrin
    d. Fibronectin
    e. Globulin
    f. Osteocalcin
    g. Osteonectin
    h. Small leucine-rich proteoglycans
    i. α2-HS-glycoprotein

31. In individuals with osteoporosis, which of the following bone matrix protein also demonstrates a sign of depletion?
    a. Albumin
    b. Bone sialoprotein
    c. Fibrin
    d. Fibronectin
    e. Globulin
    f. Osteocalcin
    g. Osteonectin
    h. Small leucine-rich proteoglycans
    i. α2-HS-glycoprotein

32. Which of the following is a dual functioning protein that is essential in directing the proper deposition and organization of hydroxyapatite in bones and as a hormone (intercellular ligand) that prompts the β-cells of the found within the Islet of Langerhans of the pancreas to increase insulin secretion?
    a. Albumin
    b. Bone sialoprotein
    c. Fibrin
    d. Fibronectin
    e. Globulin
    f. Osteocalcin
    g. Osteonectin
    h. Small leucine-rich proteoglycans
    i. α2-HS-glycoprotein

33. An increase in which of the following bone matrix protein in circulation is known to bring about an increase in insulin secretion and the subsequent increase of fat deposit in the adipose tissues?
    a. Albumin
    b. Bone sialoprotein
    c. Fibrin
    d. Fibronectin
    e. Globulin
    f. Osteocalcin
    g. Osteonectin
    h. Small leucine-rich proteoglycans
    i. α2-HS-glycoprotein

34. Which of the following is a type of adhesive glycoprotein that is responsible for initiating mineralization and mineral crystal formation within bones?
    a. Albumin
    b. Bone sialoprotein
    c. Fibrin
    d. Fibronectin
    e. Globulin
    f. Osteocalcin
    g. Osteonectin
    h. Small leucine-rich proteoglycans
    i. α2-HS-glycoprotein

35. Which of the following is believed to be the protein core (protein nucleator) that allows for hydroxyapatite crystal formation?
    a. Albumin
    b. Bone sialoprotein
    c. Fibrin
    d. Fibronectin
    e. Globulin
    f. Osteocalcin
    g. Osteonectin
    h. Small leucine-rich proteoglycans
    i. α2-HS-glycoprotein

36. Bone mineral is responsible for mechanical rigidity and load-bearing strength while the organic matrix provides elasticity and flexibility. Which of the following makes up the bulk of the bone minerals?
    a. Carbonate
    b. Hydroxyapatite
    c. Iron
    d. Magnesium
    e. Osteocalcin
    f. Osteonectin
    g. Phosphate
    h. Sialoprotein

37. Which of the following is believed to form the protein core for bone matrix formation?
    a. Albumin
    b. Bone Sialoprotein
    c. Carbonate
    d. Hydroxyapatite
    e. Iron
    f. Magnesium
    g. Osteocalcin
    h. Osteonectin
    i. Phosphate

38. Which of the following is involved in the initiating mineralization and mineral crystal formation within bones?
    a. Carbonate
    b. Hydroxyapatite
    c. Iron
    d. Magnesium
    e. Osteocalcin
    f. Osteonectin
    g. Phosphate
    h. Sialoprotein

39. Which of the following is responsible for directing the proper deposition and organization of hydroxyapatite?
    a. Carbonate
    b. Hydroxyapatite
    c. Iron
    d. Magnesium
    e. Osteocalcin
    f. Osteonectin
    g. Phosphate
    h. Sialoprotein

40. Which of the following type of bone possesses fewer collagen fibers that are arranged randomly in haphazard directions? In addition, hyaline cartilage or calcified hyaline cartilage is often found within the matrix of this bone thereby making them physically and mechanically weak by comparison to the lamellar bone.
    a. Flat bone
    b. Cancellous bone
    c. Compact bone
    d. Irregular bone
    e. Lamellar bone
    f. Long bone
    g. Woven bone

41. In higher vertebrates, the existence of the _____ is believed to fulfill a temporary role during development before being replaced by the more durable and organized _____. The process of replacing woven bone to lamellar bone is called _____.
    a. Bone modification
    b. Bone remodeling
    c. Cancellous bone
    d. Compact bone
    e. Flat bone
    f. Irregular bone
    g. Lamellar bone
    h. Long bone
    i. Woven bone

42. Which of the following type of lamellar bone possesses relatively less bone matrix and more space within the structure, which makes them less dense, physically weaker, softer and more flexible?
    a. Cancellous bone
    b. Compact bone
    c. Long bone
    d. Irregular bone

43. Which of the following type of lamellar bone is denser, physically stronger, harder and a lot less flexible?
    a. Cancellous bone
    b. Compact bone
    c. Long bone
    d. Irregular bone

44. Cancellous bones are composed of interconnecting bone rods or plates called _____.
    a. Canaliculi
    b. Endosteum
    c. Filopodia
    d. Lacunae
    e. Lamellae
    f. Osteoblasts
    g. Osteoclasts
    h. Osteocytes
    i. Trabeculae
    j. Both f and h
    k. Both f and g

45. The surface of the bone rods or plates is lined with which of the following tissue?
    a. Canaliculi
    b. Endosteum
    c. Filopodia
    d. Lacunae
    e. Lamellae
    f. Osteoblasts
    g. Osteoclasts
    h. Osteocytes
    i. Trabeculae
    j. Both f and h
    k. Both f and g

46. Which of the following type of cell (s) makes up the bulk of the tissue that lines the bone rods or plates?
    a. Canaliculi
    b. Endosteum
    c. Filopodia
    d. Lacunae
    e. Lamellae
    f. Osteoblasts
    g. Osteoclasts
    h. Osteocytes
    i. Trabeculae
    j. Both f and h
    k. Both f and g

47. What exists within the intertrabecular spaces?
    a. Nerves
    b. Osteoblasts
    c. Osteoclasts

d. Osteocytes
e. Red marrow
f. Yellow marrow
g. Both a and e
h. Both a and f

48. Which of the following is responsible for hematopoiesis?
a. Nerves
b. Osteoblasts
c. Osteoclasts
d. Osteocytes
e. Red marrow
f. Yellow marrow
g. Both a and e
h. Both a and f

49. A cross section of the trabeculae will show concentric circular patterns of the _____ while each _____ will house one mature bone cell called _____.
a. Canaliculi
b. Endosteum
c. Filopodia
d. Lacunae
e. Lamellae
f. Osteoblasts
g. Osteoclasts
h. Osteocytes
i. Trabeculae
j. Both f and h
k. Both f and g

50. The mature bone cells also known as _____ are interconnected to one another via cellular processes called _____ which are located within the tiny canals called _____.
a. Canaliculi
b. Endosteum
c. Filopodia
d. Lacunae
e. Lamellae
f. Osteoblasts
g. Osteoclasts
h. Osteocytes
i. Trabeculae
j. Both f and h
k. Both f and g

51. Since there are no blood supplies directly found within the trabeculae, the mature bone cells called _____ must obtain their nutrients and perform gas exchange through their interconnected cellular processes called _____ connected to the capillaries surrounding the trabeculae. For example, nutrients and gases will flow through the canaliculi openings of the trabeculae via the _____ (a type of junction) located at the end of the cellular processes. From there, the nutrients and gases will proceed into the initial bone cells before being shared with other networked bone cells.
a. Adherens junction
b. Canaliculi
c. Endosteum
d. Filopodia
e. Gap junction
f. Lacunae
g. Lamellae
h. Osteoblasts
i. Osteoclasts
j. Osteocytes
k. Tight junction

52. It is known that the trabeculae are structured to deal with the stress and weight that is placed upon it. If the stress placed upon the cancellous bone is changed due to injury or an increase in weight, the trabeculae will be remodeled (realigned) to meet the new demand. This remodeling process is triggered by the trabecular osteocytes through a process known as _____.
a. Chemosensation
b. Ligand sensation
c. Mechanosensation
d. Photosensation
e. Propriosensation
f. Voltage sensation

53. The process of bone remodeling is initiated when the bone (along with its osteocytes) is stretched and bent which in turn causes the flow of fluid found within the cellular processes called _____. This fluid contains various substances such as _____, _____, _____, _____, _____, _____ across a type of junction called _____ which are found at the ends of the interconnecting cellular processes. Once triggered, the _____ (a modified macrophage) will begin removing existing bone structures that are providing inadequate support while _____ (immature bone cells) will proceed in and

begin manufacturing new trabeculae formations to satisfy demand.
a. Adherens junctions
b. Calcium ions
c. Canalicular fluids
d. Cyclooxygenase 2
e. Gap junction
f. Kinases
g. Nitrous oxide
h. Osteoblasts
i. Osteoclasts
j. Osteocytes
k. Prostaglandin E2
l. Runx2
m. Tight junction

54. Unlike the cancellous bone, the compact bone possesses direct blood supplies. For example, the _____ are horizontal canals while _____ are vertical canals. Both these canals contain blood vessels, nerves and lymphatic vessels.
a. Canaliculi
b. Filapodia
c. Haverisan
d. Lacunae
e. Lamellae
f. Volkmann

55. Surrounding the osteon canal are concentric circles of deposited bone matrix called?
a. Canaliculi
b. Filapodia
c. Haverisan
d. Lacunae
e. Lamellae
f. Volkmann

56. In humans, there are three types of concentric circles created by bone matrix surrounding the osteon canal. The outer surfaces of the compact bone are formed by _____ (**Answer A**) arranged parallel to the surface of the bone. The bulk of the compact bone is formed by _____ and in between **Answer A** are _____ (remnant of **Answer A**) that still remained after bone remodeling.
a. Haversian canal
b. Volkmann canal
c. Canaliculi
d. Lacunae
e. Concentric lamellae
f. Circumferential lamellae
g. Interstitial lamellae

57. Which of the following makes up the Haversian system?
a. Canaliculi
b. Haversian canal
c. Lacunae
d. Lamellae
e. Osteocytes
f. All of the above
g. None of the above

58. In compact bone, which of the following cell (s) exists in the cellular layer of the periosteum?
a. Fibroblasts
b. Fibrocytes
c. Osteoblast
d. Osteoclasts
e. Osteocyte
f. Both a and b
g. Both c and d

59. What is the name given it undifferentiated future bone cells (type of adult stem cells) that also exist in the periosteum (not mentioned in the previous question)?
a. Fibroblasts
b. Fibrocytes
c. Osteogenic cells
d. Osteoblast
e. Osteoclasts
f. Osteocyte

60. Which of the following type of cell (s) exists within the lining of the Haversian canals tand are therefore immediately become available when repairs or remodeling need to take place?
a. Osteocytes
b. Osteogenic cells
c. Osteoclasts
d. Osteoblasts
e. Both a and b
f. Both c and d

g. Answers b, c and d

61. What is the name given to the type of bones that are longer than they are wide?
    a. Cancellous bone
    b. Compact bone
    c. Flat bone
    d. Irregular bone
    e. Long bone
    f. Short bone

62. What is the name given to the type of bones that are as long as they are wide?
    a. Cancellous bone
    b. Compact bone
    c. Flat bone
    d. Irregular bone
    e. Long bone
    f. Short bone

63. What is the name given to the type of bones that are flat and slightly curved?
    a. Cancellous bone
    b. Compact bone
    c. Flat bone
    d. Irregular bone
    e. Long bone
    f. Short bone

64. What is the name given to the type of bones that are variable/unequal in shape?
    a. Cancellous bone
    b. Compact bone
    c. Flat bone
    d. Irregular bone
    e. Long bone
    f. Short bone

65. The long bone is constructed with two enlarged ends called the _____ and a long shaft called the _____.
    a. Articular cartilage
    b. Cancellous bone
    c. Compact bone
    d. Diaphysis
    e. Elastic cartilage
    f. Epiphysis
    g. Fibrocartilage
    h. Hyaline cartilage

66. The enlarged ends of the long bone are composed mainly of _____ type of bone with its margins lined with _____ for additional support.
    a. Articular cartilage
    b. Cancellous bone
    c. Compact bone
    d. Diaphysis
    e. Elastic cartilage
    f. Epiphysis
    g. Fibrocartilage
    h. Hyaline cartilage

67. It is known that the surfaces of the enlarged ends of the long bones are covered with a thin connective tissue called _____ which is a type of cartilage called _____. This cartilage is designed to prevent friction between bones during movement.
    a. Articular cartilage
    b. Cancellous bone
    c. Compact bone
    d. Diaphysis
    e. Elastic cartilage
    f. Epiphysis
    g. Fibrocartilage
    h. Hyaline cartilage

68. The shaft of the long bone is composed mainly of _____ type of bone while only towards the medullary cavity is it lined with _____ type of bone.
    a. Articular cartilage
    b. Cancellous bone
    c. Compact bone
    d. Diaphysis
    e. Elastic cartilage
    f. Epiphysis
    g. Fibrocartilage
    h. Hyaline cartilage

69. Within the shaft of the long bone, lining the medullary cavity, there is a cellular membrane called _____.
    a. Endochondrium
    b. Endosteum
    c. Perichondrium
    d. Periosteum
    e. Epichondrium

70. The shaft of the long bone is surrounded by a dense irregular connective tissue called _____ which contains numerous nerves, blood vessels, lymphatic vessels, osteoblasts, osteoclasts and osteogenic cells.
    a. Endochondrium
    b. Endosteum
    c. Perichondrium
    d. Periosteum
    e. Diaphysis

71. During fetal development the medullary space of the diaphysis is filled with _____ but a conversion to _____, a tissue made out of loose areolar connective tissue and adipose tissue, occurs shortly after birth.
    a. Nerves
    b. Osteoblasts
    c. Osteoclasts
    d. Osteocytes
    e. Red marrow
    f. Yellow marrow
    g. Both a and e
    h. Both a and f

72. Flat bones are mainly found in which of the following anatomical locations?
    a. Clavicle
    b. Femur
    c. Radius
    d. Skull
    e. Tibia
    f. Both a and d
    g. Both b and e

73. The construct of a flat bone resembles that of a sandwich where the superficial and the deep layers are lined with _____ (type of bone) and sandwiched in between is a layer of _____ also known as the_____.
    a. Cancellous bone
    b. Compact bone
    c. Dipolë
    d. Irregular bone
    e. Long bone
    f. Short bone
    g. Both a and b
    h. Both d and f

74. Which of the following undersized bone(s) do not possess a diaphysis?
    a. Cancellous bone
    b. Compact bone
    c. Dipolë
    d. Irregular bone
    e. Long bone
    f. Short bone
    g. Both a and b
    h. Both d and f

75. The development of the flat bone or intramembranous bone begins approximately the fifth week prenatal by forming a _____ (embryonic tissues) at the site of the future bone. This embryonic tissue is composed of _____ (**Answer A**) dispersed within delicate collage fibers richly supplied with capillaries. Intramembranous ossification begins when **Answer A** begins to cluster to form _____ (an area where ossification begins). These **Answer A's** enlarge and differentiate into _____ (early form of embryonic bone cell) before developing into _____ (immature bone cell; **Answer B**). Immediately, **Answer B** begin secreting _____ (bone matrix) on the collagen fibers forming numerous _____ (bone rods or plates) of woven bone. The secreted bone matrix traps the **Answer B** and their cellular processes called _____. The cell body of **Answer B** is trapped within _____ while the cellular processes are trapped with _____ thereby creating mature bone cells called _____. The diameter of the bone rods begins to grow as more **Answer B** lays down more bone matrix upon delicate collagen fibers. As the bone rods begin to link together, the future structure of _____ begins to form. At this point, a thin layer of _____ (immature bone cell), _____ (modified macrophage) and _____ (early form of bone cell) forms upon the surface of the trabeculae to create a thin delicate tissue called _____. These **Answer A's** that are superficial and deep towards the cancellous bone differentiate into immature bone cell and begin to form _____ (a type of bone). After numerous bone remodeling and reformation, the woven bone is slowly converted into lamellar bone which contributes to the final form of the flat bone.
    a. Canaliculi
    b. Cancellous bone
    c. Compact bone
    d. Endosteum
    e. Filapodia
    f. Fontanels
    g. Hydroxyapatite
    h. Lacunae
    i. Mesenchymal cells
    j. Mesenchymal membrane
    k. Ossification centers
    l. Osteoblasts
    m. Osteoclasts
    n. Osteocytes
    o. Osteonectin
    p. Osteogenic cells
    q. Osteoid
    r. Periosteum
    s. Trabeculae
    t. Center of ossification

76. Ossification of flat bones begins approximately eighth week prenatal when the immature bone cells began to deposit _____ (calcium phosphate crystals) into their extracellular matrix. As the ossification continues intramembraneously, the embryonic tissues that exist superficial and deep within the flat bone remain non-calcified and become form the _____ (fibrous connective tissue).

_____ is defined as the locations of the embryonic tissues where intramembranous ossification begins. As the fetus grows, this area where ossification begins expands with the oldest area of ossification located at the center of the development while the newly ossified tissues are located at its edges. At term, the fetus still possesses numerous locations on their skull that have yet transformed into bone (no calcium phosphate crystals have been added). These so-called soft spots are called _____ and it will be approximately two years of age before these soft spots finally become bone.

a. Canaliculi
b. Cancellous bone
c. Compact bone
d. Endosteum
e. Filapodia
f. Fontanels
g. Hydroxyapatite
h. Lacunae
i. Mesenchymal cells
j. Mesenchymal membrane
k. Ossification centers
l. Osteoblasts
m. Osteoclasts
n. Osteocytes
o. Osteonectin
p. Osteogenic cells
q. Osteoid
r. Periosteum
s. Trabeculae
t. Center of ossification

77. Which of the following is defined as bone formation within cartilage?
a. Endochondrial ossification
b. Extrachondrial ossification
c. Intermembranous ossification
d. Intramembranous ossification

78. The development of endochondral bone begins approximately the fourth week prenatal where the _____ (embryonic cell) enlarges to form _____ (early form of embryonic bone cell). Soon after, these embryonic bone cells will develop into immature cartilage cells called _____ through the availability of growth factors called _____ and _____. Immediately after their formation, these immature cartilage cells will begin producing a cartilage model in the shape of the future bone. It is known that the cartilage model is made out of _____ (type of cartilage).

a. Bone collar
b. Chondroblasts
c. Chondrocytes
d. Dipolë
e. Diaphysis
f. Epiphysis
g. Elastic cartilage
h. Fibrocartilage
i. Hyaline cartilage
j. Lacunae
k. Mesenchymal cells
l. Mesenchymal tissue
m. Osteoblasts
n. Osteoclasts
o. Osteocytes
p. Osteogenic cells
q. Pax1
r. Perichondrium
s. Periosteum
t. Prostaglandin E2
u. Runx2
v. Scleraxis

79. As these immature cartilage cells become surrounded by the cartilage or embedded/surrounded by _____, it is now referred to as mature cartilage cells or _____. With the exception to where the future joint will be, the remaining cartilage model is surrounding by an undeveloped embryonic tissue called _____ or also known as _____ (surrounding cartilage).

These embryonic cells within the undeveloped embryonic tissue surrounding the cartilage model begin to enlarge and mature into _____ (early form of embryonic bone cell) and soon mature into _____ (immature bone cell). Immediately after its development these immature bone cells begin producing a layer of ossified tissue surrounded the shaft of the bone. It is known that this layer of ossified tissue is called _____ while the shaft of the future bone is called _____. As soon as this layer of ossified tissue is completed around the shaft of the future bone, the embryonic tissue that surrounds the shaft is now referred to as a _____ (surrounding the bone).

a. Bone collar
b. Chondroblasts
c. Chondrocytes
d. Dipolë
e. Diaphysis
f. Epiphysis
g. Elastic cartilage
h. Fibrocartilage
i. Hyaline cartilage
j. Lacunae
k. Mesenchymal cells
l. Mesenchymal tissue
m. Osteoblasts
n. Osteoclasts
o. Osteocytes
p. Osteogenic cells
q. Pax1
r. Perichondrium
s. Periosteum
t. Prostaglandin E2
u. Runx2
v. Scleraxis

80. Which of the following is an important support structure for the developing cartilage model of future bone?
a. Bone collar
b. Canaliculi
c. Lacunae
d. Lamellae
e. Perichondrium

f. Periosteum

81. While the formation of the bone collar is taking place, _____ (mature cartilage cells) towards the center of the cartilage model become _____ as they began to divide rapidly and enlarge. In addition, these cells began to increase in their deposition of cartilage. The combination of rapid division, hypertrophy and deposition of additional cartilage results in the widening and lengthening (growth) of the cartilage model.

Soon after, these newly divided and enlarged mature cartilage cells chondrocytes begin to produce and release _____ which initiates the formation of _____ (calcium phosphate crystals) into their surrounding matrix thereby effectively trapping themselves within water insoluble calcified cartilage enclosure also known as _____. Without any connection with the blood vessels and with direct diffusion and osmosis made impossible by the water insoluble calcium phosphate crystals, these mature cartilage cells begin a process of programmed cell death or _____. As these cells deteriorate and die, they leave behind emptied water insoluble calcified cartilage enclosure.

a. Apoptosis
b. Bone collar
c. Bone matrix vesicles
d. Chondroblasts
e. Chondrocytes
f. Dipolë
g. Diaphysis
h. Epiphysis
i. Elastic cartilage
j. Fibrocartilage
k. Hyaline cartilage
l. Hydroxyapatite
m. Hypertrophic
n. Hypotrophic
o. Lacunae
p. Mesenchymal tissue
q. Osteoblasts
r. Osteoclasts
s. Osteocytes
t. Osteogenic cells
u. Pax1
v. Perichondrium
w. Periosteum
x. Prostaglandin E2
y. Runx2
z. Scleraxis

82. After the disintegration of the mature cartilage cells, blood vessels, _____ (immature bone cells) and _____ (modified macrophage) begin invading the calcified cartilage. Once settled on the surface of the calcified cartilage, these immature bone cells will begin forming numerous _____ (bone rods or plates) of woven bone. The secreted bone matrix will trap the immature bone cells and their cellular processes called _____ in _____ (enclosing the cell) and _____ (enclosing the cellular process) respectively thereby creating mature bone cells called _____. The diameter of the bone rods or plates begins to grow as more immature bone cells begin secreting more bone matrix. As these bone rods or plates begin to link together, the future structure of woven cancellous bone begins to form. The formation of cancellous bone is referred to as the _____. It is during this time when the blood vessels begin to condense and form red marrow within the spaces within the cancellous bone.

1. Apoptosis
2. Bone collar
3. Bone matrix vesicles
4. Canaliculi
5. Chondroblasts
6. Chondrocytes
7. Dipolë
8. Diaphysis
9. Epiphysis
10. Elastic cartilage
11. Fibrocartilage
12. Filapodia
13. Hyaline cartilage
14. Hydroxyapatite
15. Hypertrophic
16. Hypotrophic
17. Lacunae
18. Mesenchymal tissue
19. Osteoblasts
20. Osteoclasts
21. Osteocytes
22. Osteogenic cells
23. Pax1
24. Perichondrium
25. Periosteum
26. Primary ossification center
27. Prostaglandin E2
28. Runx2
29. Scleraxis
30. Trabeculae

83. The manner by which secondary ossification centers are formed is the same as the formation of the primary ossification center where blood vessels, osteoblasts and osteoclasts began invading the calcified cartilage. The major difference between primary and secondary ossification centers is that the secondary ossification center does not form a _____ within the epiphysis.
a. Medullary center
b. Dipolë
c. Canaliculi
d. Lacunae
e. Filapodia

84. Bone growth in length occurs at the _____?
a. Epiphyseal plate
b. Metaphysis
c. Primary ossification center
d. Secondary ossification center
e. Zone of reserve cartilage
f. Zone of proliferation
g. Zone of cellular hypertrophy
h. Zone of calcification

85. Which of the following is defined as a transitional region between the cartilaginous head (epiphysis) of the developing bone and the developing diaphysis?

a. Epiphyseal plate
b. Metaphysis
c. Primary ossification center
d. Secondary ossification center
e. Zone of reserve cartilage
f. Zone of proliferation
g. Zone of cellular hypertrophy
h. Zone of calcification

86. Which of the following zone is where the mature cartilage cells begin to deposit bony materials (calcium phosphate crystals) around themselves and thereby cutting off their supplies of oxygen and nutrients which cause them to die?
a. Epiphyseal plate
b. Metaphysis
c. Primary ossification center
d. Secondary ossification center
e. Zone of reserve cartilage
f. Zone of proliferation
g. Zone of cellular hypertrophy
h. Zone of calcification

87. Which of the following zone is when the cells within show no sign of ossification and remain resting cartilage?
a. Epiphyseal plate
b. Metaphysis
c. Primary ossification center
d. Secondary ossification center
e. Zone of reserve cartilage
f. Zone of proliferation
g. Zone of cellular hypertrophy
h. Zone of calcification

88. Which of the following zone is when the mature cartilage cells begin to enlarge? These enlargements of the cells cause the lengthening of the bones.
a. Epiphyseal plate
b. Metaphysis
c. Primary ossification center
d. Secondary ossification center
e. Zone of reserve cartilage
f. Zone of proliferation
g. Zone of cellular hypertrophy
h. Zone of calcification

89. After which of the following zone is when the modified monocyte (phagocytic cells) begins to invade the newly formed ossified cartilage and dissolves them? Blood vessels and nerves from the primary marrow follow the newly dug tunnels. In addition, the immature bone cells also invade and establish themselves around blood vessels and start to deposit bone material in a concentric circular pattern.
a. Epiphyseal plate
b. Metaphysis
c. Primary ossification center
d. Secondary ossification center
e. Zone of reserve cartilage
f. Zone of proliferation
g. Zone of cellular hypertrophy
h. Zone of calcification

90. Which of the following is defined as the manner by which bones grow in width?
a. Appositional growth
b. Dipolë
c. Epiphyseal plate
d. Metaphysis
e. Primary ossification center
f. Scleraxis
g. Secondary ossification center
h. Zone of calcification

91. The formation of ridges and grooves on the surfaces of older bones and immediately beneath the periosteum is the manner by which bones grows in width. As additional osteoblasts settle and extend the ridges towards one another, blood vessels, lymphatic vessels, and nerves will grow into the space formed in between the ridges settling towards the center of this future _____ (a passageway).
a. Canaliculi
b. Dipolë
c. Filapodia
d. Haversian canal
e. Lacunae
f. Lamellae
g. Medullary center
h. Volkmann's canal

92. Which of the following is a type of simple fracture that is resulting from a

disease such as (osteoporosis, tumor etc.) which causes weaknesses and the eventual collapse of bones? This condition generally causes a condition such as kyphosis which the fracture causes the spine to bend forward.
a. Comminuted fracture
b. Greenstick fracture
c. Impacted fracture
d. Oblique fracture
e. Pathologic fracture
f. Spiral fracture
g. Stress fracture
h. Transverse fracture

93. Which of the following is a type of incomplete simple fracture that occurs in infants and children? This type of fracture results in bones bending and cracking but does not break completely since the bones of infants and children are mainly composed of cartilage.
a. Comminuted fracture
b. Greenstick fracture
c. Impacted fracture
d. Oblique fracture
e. Pathologic fracture
f. Spiral fracture
g. Stress fracture
h. Transverse fracture

94. Which of the following is a type of simple fracture that is also known as buckle fracture? This type of fracture is result from one end of the bone is driven into another.
a. Comminuted fracture
b. Greenstick fracture
c. Impacted fracture
d. Oblique fracture
e. Pathologic fracture
f. Spiral fracture
g. Stress fracture
h. Transverse fracture

95. Which of the following is a type of complete fracture that could be either compound of simple and is at times referred to as torsion fracture? This type of fracture results from bones being twisted apart.
a. Comminuted fracture
b. Greenstick fracture
c. Impacted fracture
d. Oblique fracture
e. Pathologic fracture
f. Spiral fracture
g. Stress fracture
h. Transverse fracture

96. Which of the following is a type of complete fracture that could be either compound of simple? This type of fracture results from the bones breaking at approximately 45° sloped angle.
a. Comminuted fracture
b. Greenstick fracture
c. Impacted fracture
d. Oblique fracture
e. Pathologic fracture
f. Spiral fracture
g. Stress fracture
h. Transverse fracture

97. Which of the following is a type of fracture that could be either compound or simple? This type of fracture denotes that at least three fracture fragments have resulted from the break and the fracture lines interconnect. The fracture lines that form this type of fracture could be oblique, transverse, or spiral.
a. Comminuted fracture
b. Greenstick fracture
c. Impacted fracture
d. Multiple fracture
e. Oblique fracture
f. Pathologic fracture
g. Spiral fracture
h. Stress fracture
i. Transverse fracture

98. Which of the following is a type of fracture that is a complete fracture that could be either compound or simple? This type of fracture involves at least three fracture fragments occurred in a single bone, however, their fracture lines do not interconnect and the individual fracture lines could be of any shape.
a. Comminuted fracture
b. Greenstick fracture
c. Impacted fracture
d. Multiple fracture
e. Oblique fracture
f. Pathologic fracture

g. Spiral fracture
h. Stress fracture
i. Transverse fracture

ANSWERS

1.	d	2.	b	3.	b
4.	a	5.	e	6.	f
7.	a	8.	c	9.	f
10.	a	11.	d	12.	e
13.	e	14.	c	15.	d
16.	a	17.	c	18.	a
19.	c	20.	a	21.	c
22.	a	23.	c	24.	g
25.	c	26.	b	27.	a
28.	a	29.	i	30.	h
31.	h	32.	f	33.	f
34.	g	35.	b	36.	b
37.	a	38.	f	39.	e
40.	g	41.	i, g, b	42.	a
43.	b	44.	i	45.	b
46.	k	47.	e (g)	48.	e
49.	e, d, h	50.	h, c, a	51.	j, d, e
52.	c	53.	c, b, k, d, f, l, g, e, i, h	54.	c, f
55.	e	56.	f, e, g	57.	f
58.	g	59.	c	60.	g
61.	e	62.	f	63.	c
64.	d	65.	f, d	66.	b, c
67.	a, h	68.	c, b	69.	b
70.	d	71.	e, f	72.	f
73.	b, a, c	74.	h	75.	j, i, k, p, l, q, s, e, h, a, n, b, l, m, p, d, c
76.	g, r, t, f	77.	a	78.	k, p, b, q, v, i
79.	j, c, l, r, p, m, a, e, s	80.	a	81.	e, m, c, l, o, a
82.	19, 20, 30, 12, 17, 4, 21, 26	83.	a	84.	a
85.	b	86.	h	87.	e
88.	g	89.	h	90.	a
91.	d	92.	e	93.	b
94.	c	95.	f	96.	d
97.	a	98.	d		

REFERENCES

Clarke B. Normal Bone Anatomy and Physiology. CJASN November 2008, vol. 3 no. Supplement 3, S131-S139

Plantalech L, Guillaumont M, Vergnaud P, Leclercq M, Delmas PD. Impairment of gamma carboxylation of circulating osteocalcin (bone gla protein) in elderly women. J Bone Miner Res. 1991, 6: pp. 1211-1216

Hoang QQ, Sicheri F, Howard AJ, Yang DSC. Bone recognition mechanism of porcine osteocalcin from crystal structure. Nature 2003, 425, 977–980

Ganss B, Kim RH, Sodek J. Bone sialoprotein. Crit Rev Oral Biol Med. 1999, 10, 79-98.

Nikitovic D, Aggelidakis J, Young MF, Iozzo RV, Karamanos NK, Tzanakakis GN. The biology of small leucine-rich proteoglycans in bone pathophysiology 2012, J. Bio chem. 287, 33926-33933 NM Hancox. Biology of Bone. Cambridge University Press, 1972. pp. 106

Rubin CT, Lanyon LE. Osteoregulatory nature of mechanical stimuli: Function as a determinant for adaptive bone remodeling. 1987, J. Orthop. Res. 5, 300– 310

Jorgensen NR, Teilmann SC, Henriksen Z, Civitelli R, Sorensen OH, Steinberg TH. Activation of L-type calcium channels is required for gap junction-mediated intercellular calcium signaling in osteoblastic cells. 2003, J. Biol. Chem. 278, 4082– 4086

Gilbert SF. Developmental Biology. Osteogenesis: The Development of Bones.6th edition. Sunderland (MA): Sinauer Associates. 2000

Zhuang L, Peng JB, Tou L, Takanaga H, Adam RM, Hediger MA, Freeman MR. Calcium-selective ion channel, CaT1, is apically localized in gastrointestinal tract epithelia and is aberrantly expressed in human malignancies. Lab Invest. 2002, 82: pp. 1755-1764

Seeley RR, Stephens TD, Tate P. Anatomy and Physiology 6th Edition. McGraw-Hill, New York, New York. 2003

Tate P. Seeley's Principles of Anatomy & Physiology 1st Edition. McGraw-Hill, New York, New York. 2009

Saladin KS. Anatomy & Physiology. The Unity of Form and Function 6th Edition. McGraw-Hill, New York, New York. 2010

PHOTO AND GRAPHIC BIBLIOGRAPHY

1. Figure 1: Graphic designed and released by MR Villarreal, 2009. Public domain graphics
2. Figure 2: Graphic designed and released by MR Villarreal, 2009. Public domain graphcis
3. Figure 3: Graphic designed and released by Open Stax College, Connexions, reprint permission granted under the Creative Commons Attribution 3.0 Unported license.
4. Figure 4: photomicrograph taken and released by Matos MA, Araújo FP, Paixão FB. Histomorphometric evaluation of bone healing in rabbit fibular osteotomy model without fixation. J Orthop Surg Res, 2008. OpenI National Institutes of Health. Public domain photo
5. Figure 5: Graphic designed and released by Open Stax College, Connexions, reprint permission granted under the Creative Commons Attribution 3.0 Unported license
6. Figure 6: Graphic designed and released by Open Stax College, Connexions, reprint permission granted under the Creative Commons Attribution 3.0 Unported license
7. Figure 7: Graphic designed and released by Open Stax College, Connexions, reprint permission granted under the Creative Commons Attribution 3.0 Unported license
8. Figure 8: Graphic designed and released by Open Stax College, Connexions, reprint permission granted under the Creative Commons Attribution 3.0 Unported license
9. Figure 9: Graphic designed and released by US Department of Health and Human Services. Public domain graphics
10. Figure 10: Graphic designed and released by Open Stax College, Connexions, reprint permission granted under the Creative Commons Attribution 3.0 Unported license
11. Figure 11: Graphic designed by P.Y.P. Jen. Copyright ©
12. Figure 12: Graphic designed and released by Open Stax College, Connexions, reprint permission granted under the Creative Commons Attribution 3.0 Unported license
13. Figure 13: Graphic designed by P.Y.P. Jen. Copyright ©
14. Figure 14: Graphic designed and released by BruceBlaus. This file is licensed under the Creative Commons Attribution 3.0 Unported license
15. Figure 15: Graphic designed and released by Open Stax College, Connexions, reprint permission granted under the Creative Commons Attribution 3.0 Unported license
16. Figure 16: Graphic designed and released by Open Stax College, Connexions, reprint permission granted under the Creative Commons Attribution 3.0 Unported license
17. Figure 17: Graphic designed and released by Open Stax College, Connexions, reprint permission granted under the Creative Commons Attribution 3.0 Unported license
18. Figure 18: Graphic designed and released by BruceBlaus. This file is licensed under the Creative Commons Attribution 3.0 Unported license
19. Figure 19: Graphic designed and released by Open Stax College, Connexions, reprint permission granted under the Creative Commons Attribution 3.0 Unported license
20. Figure 20: Graphic designed and released by Open Stax College, Connexions, reprint permission granted under the Creative Commons Attribution 3.0 Unported license
21. Figure 21: Graphic designed and released by Krukhaug Y, Ugland S, Lie SA, Hove LM. External fixation of fractures of the distal radius Acta Orthop, 2009. Bumbaširević M, Tomić S, Lešić A, Bumbaširević V, Rakočević Z, Atkinson HD - J Orthop The treatment of scaphoid nonunion using the Ilizarov fixator without bone graft, a study of 18 cases Surg Res, 2011. OpenI beta, United States National Library. Public domain graphics

CHAPTER SIX

MUSCULAR SYSTEM

The muscular system is composed of skeletal, cardiac, and smooth muscles. Through muscular contractions, this organ system is responsible for most of the movement of the body both externally and internally.

The skeletal muscles are the most abundant muscles in the body, and their contractions result in bodily movements, such as walking, running, and the ability to grasp and manipulate objects. Additionally, skeletal muscles are also responsible for producing body heat (i.e. shivering), respiration, and the manipulation of sounds.

Cardiac muscles are responsible for the movement of blood through the body and are essential in providing the vehicle for gas and nutrient exchange. Cardiac muscles are self-synchronizing in coordinating cardiac contraction, also known as autorhythmic. Additionally, cardiac muscles are myogenic, or capable of initiating their own contractions, and do not require any outside input in order to function. Nonetheless, the cardiac muscles are influenced by the autonomic nervous system, as well as, the endocrine system.

The smooth muscles are responsible for propelling and mixing foods in the digestive tract, controlling exocrine and endocrine secretions, and regulating the amount of blood flowing through the blood vessels. It is understood that the smooth muscles are directly controlled by the autonomic nervous system, as well as, being heavily influenced by the endocrine system.

Muscle Fibers

A muscle cell, also known as a muscle fiber, is capable of contraction, or the ability to forcefully become shorter. Inversely, a muscle cell is also capable of relaxation, which is defined as the passive process of muscle elongation. In order for muscle cells to contract and relax, muscles fibers are constructed to be extendable (i.e. stretched beyond its resting length), as well as, elastic (i.e. ability to recoil after being stretched). It is understood that all muscle fibers are excitable which indicates that they are capable of responding to external stimuli from the nervous and endocrine systems.

The plasma membrane of muscles fiber is called the sarcolemma and it plays an essential role in skeletal and cardiac muscle function. In addition to maintaining cellular integrity, the sarcolemma is involved in responding to neurotransmissions, the propagation of action potentials, and excitation-contraction coupling (i.e. action potential causes the contraction of a muscle fiber). For example, in relatively regular intervals throughout the skeletal and cardiac muscle cell, the sarcolemma invaginates to form transverse tubules (t-tubules), which wrap around and extend deep into the muscle fiber. These t-tubules are responsible for spreading electrical impulses throughout and deep within the muscle fibers.

The sarcolemma surrounds the sarcoplasm, which is the cytoplasm of the muscle fiber. Within the sarcoplasm, nucleus, or nuclei as in the case of skeletal muscles, as well as, organelles such as mitochondria, centrosomes, rough endoplasmic reticulum, ribosomes, and the Golgi apparatus exist. With the exception of the sarcoplasmic reticulum, which is a modified form of smooth endoplasmic reticulum, the remaining organelles perform their normal functions (see Chapter Three for more information regarding organelle function).

The sarcoplasmic reticulum is a specialized form of smooth endoplasmic reticulum, which stores and releases large amounts of calcium ions (Ca^{++}). The enlarged ends of sarcoplasmic reticulum are called terminal cisternae, which collect and concentrate calcium ions from the sarcoplasm before passing them down for storage within the modified organelle. Through electron microscopy, it has been observed that the terminal cisternae of the sarcoplasmic reticulum are in close association with t-tubules, forming a contact region called the diad in cardiac muscle cells and triad in skeletal muscle cells (Figure 1).

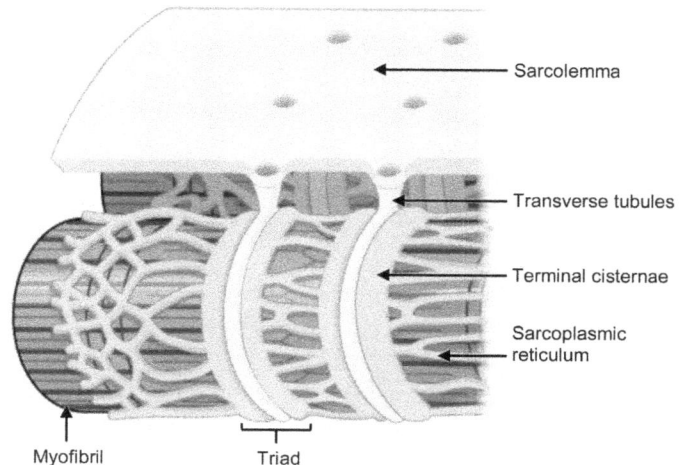

Figure 1: Illustration of the interactions between various structures within a skeletal muscle cell. Please note that the transverse tubules are simply invaginations of the sarcolemma

Also found within the sarcoplasm are numerous myofibrils which are thread-like structures measuring one to three µm in diameter that extend the entire length of the muscle fiber. The bulk of the myofibril is constructed out of actin and myosin myofilaments, as well as, various regulatory proteins (e.g. troponin and tropomyosin) and structural proteins (e.g. nebulin and titin).

Myofilaments, Regulatory and Structure Proteins

The structure of the myofibril is mainly composed of actin and myosin myofilaments. For example, actin myofilaments, also known as thin myofilaments, are created from fibrous actin (F-actin), which is a helix of uniformly oriented monomers that has a diameter of seven to eight nm. F-actin is constructed out of two globular actins (G-actin; 42 kD), twisted together to create a polymer. Once polymerized, the F-actin is capped and stabilized by a protein known as beta (β) actinin. Each G-actin subunit has an active site called a myosin binding site, that allows myosin binding to occur during muscle contraction (Figure 2).

Myosin myofilaments, also known as thick myofilaments, are motor proteins. Depending on the type of muscle, different isoforms (or types of molecular arrangements of myosin) exists. For example, in smooth muscle the myosin myofilaments are referred to as the p-light chain, which possesses a different molecular formula and structural orientation when compared to myosin myofilaments found in skeletal and cardiac muscles. In skeletal muscle, there are approximately 400 myosin molecules joined together to form a thick myofilament. These molecules are maintained in bundles through its interactions with a clamp protein (C protein), an M-line protein, as well as, the hydrophobic interactions of the myosin molecules themselves with the surrounding hydrophilic environment.

Each of the myosin molecules resembles the shape of a golf club. The head of the myosin molecule is called heavy meromyosin, while the shaft of the molecule is called light meromyosin. The two sections of the myosin molecule are separated at the location known as the trypsin hinge, which is formed by a proteolytic enzyme called trypsin. The trypsin hinge is the location where the heavy meromyosin angles sharply (~45°) downward from the main axis of the thick myofilament. The heavy meromyosin contains two binding sites. The first is called the actin binding site and is a deep cleft that is formed to connect with the active site (i.e. myosin binding site) of the G-actin. The second binding site is composed of an enzyme called adenylpyrophosphatase or ATPase. ATPase is used to hydrolyze the energy molecule adenosine triphosphate (ATP) to adenosine diphosphate (ADP), releasing a phosphate functional group (P_i), and the stored energy. Subsequently, the myosin motor protein converts the released energy from the ATP molecule into the energy of motion (Figure 3).

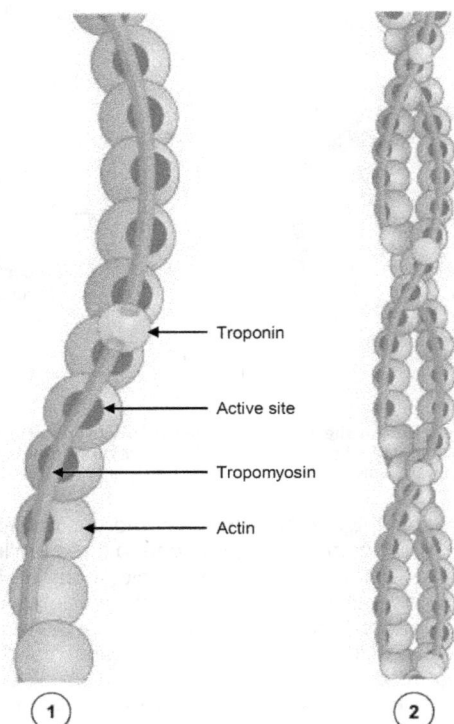

Figure 2: Illustration of actin molecules. ① Illustrates a G-actin molecule with its active site and associated regulatory proteins. ② Illustrates an actin myofilament or a F-actin, which is formed from two intertwining G-actin molecules

Tropomyosin is a regulatory protein, which is found coursing alongside of the actin molecules. Tropomyosin is a long, rod-like helically-intertwined molecule. In a relaxed muscle, each tropomyosin molecule spans a length of seven G-actin residues. This regulatory protein is strategically located so that it blocks part of the binding site between actin and myosin. Put simply, this protein serves as an "on/off" switch, where in the "on" position it allows the full contact of myosin and actin, thereby allowing the formation of a cross bridge. However, when it is at the "off" position, it will still allow some actin and myosin interaction, but will not allow the formation of a cross bridge (Figure 2).

Troponin is another regulatory protein located between the thin and thick myofilaments and is attached to tropomyosin. Troponin is composed of three protein subunits: troponin I, troponin T, and troponin C, and each possess their own unique functions. Troponin T (TnT) is responsible for binding troponin to tropomyosin. Troponin I (TnI) inhibits the interaction of myosin with actin. Troponin C (TnC) is a calcium ion binding protein. For example, once the motor nerve signals the skeletal muscle to contract, Ca++ ions are immediately released from the sarcoplasmic reticulum into the cytoplasm. Ca++ ions will bind with TnC, which in turn causes the entire troponin molecule to undergo a conformational change. This conformational change causes the tropomyosin to switch to the "on" position, thereby exposing the actin active site. Once the active site is exposed,

the actin binding site of the heavy meromyosin will immediately bind to the active site of the actin myofilament, forming a cross bridge. In order for relaxation to occur the Ca++ ion concentration must decrease, thereby allowing troponin to reform its original conformation and return tropomyosin back into the "off" position (Figure 2).

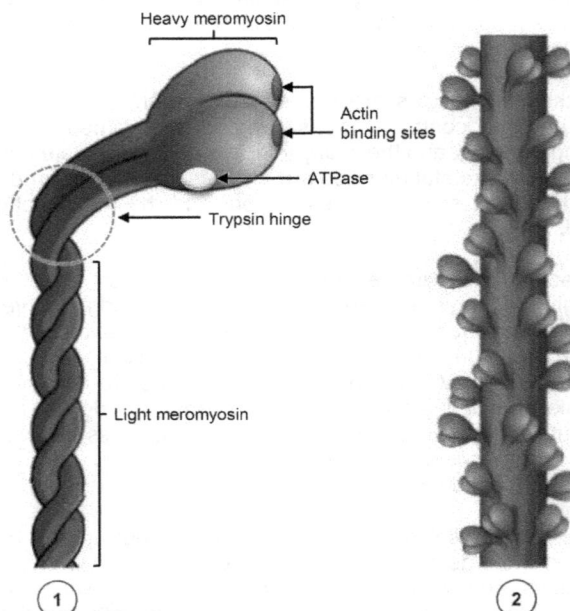

Figure 3: Graphic illustration of myosin molecules. ① Illustrates a single myosin molecule with its actin binding sites and ATPase. ② Illustrates a myosin myofilament (thick myofilament) with ~400 myosin molecules bundles together

Titin is one of the largest proteins known consisting of approximately 27,000 amino acids. Titin is an elastic protein that extends along the myosin myofilament and maintains the spatial structure of the myofibril, as well as, stabilizing the position of the contractile filaments. Like a spring, titin is capable of being compressed as the muscle contracts (i.e. shortening of the sarcomere) and is responsible for returning the contracted/shortened myofibrils back to their original relaxed shape once the contraction is terminated (Figure 4).

Nebulin is an enormous 600 to 900 kDa filamentous non-elastic structural protein that courses along the side of the actin (thin) myofilaments. Like scaffolding, nebulin is constructed to align and regulate the spatial orientation of the thin myofilament, as well as, the alignment of the Z disks. In addition, nebulin is also involved in the regulation of muscle contraction by enhancing calcium ion sensitivity and cycling kinetics of cross bridge formation. Evidence of the functions of nebulin is demonstrated in humans through a group of genetic disease called Nemaline Myopathy (NM). NM is a structural congenital myopathy that is passed down from previous generations via an autosomal dominant and/or recessive manner. In patients suffering from NM with nebulin mutation (NM-NEB), an approximate 60% reduction in contractile force has been reported. Additionally, the skeletal muscles of these individuals demonstrate a leftward-shift of the force-sarcomere length relation to NM-NEB muscle fibers. It is postulated that the cause of the reduction in the force of contraction in NM-NEB skeletal muscles is most likely due to the shorter and non-uniform thin filament lengths, when compared to control muscles.

Troponin Test

As the name states, the Troponin Test is designed to measure the levels of troponin in the bloodstream. More specifically, it is designed to measure the serum levels of both troponin T (TnT) and troponin

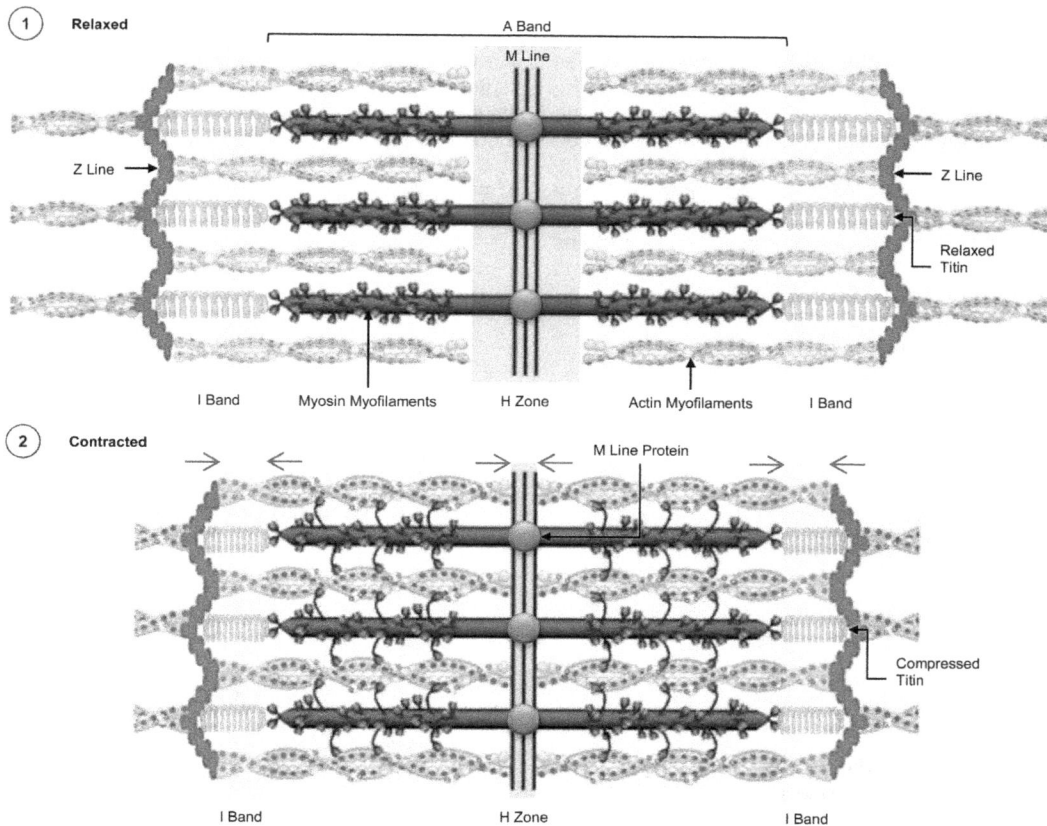

Figure 4: Illustration of the functional unit of the muscle fiber, the sarcomere. The sarcomere stretches from Z line to Z line and is composed of actin and myosin myofilaments, as well as, various other structural and regulatory proteins. Please note that the titin molecules and their interaction with contracting sarcomere. ① Titin molecules extend the entire length of the myosin myofilament and are ② compressed as the sarcomere shortened. Titin molecules are partly responsible for returning the sarcomere back to its original shape

I (TnI) subunits. It is understood that TnT and TnI are released when the cardiac muscle is damaged as in the case of myocardial infarction (i.e. heart attack). The amount of TnT and TnI that are released into the bloodstream is proportional to the seriousness of the cardiac muscle damage (Figure 5).

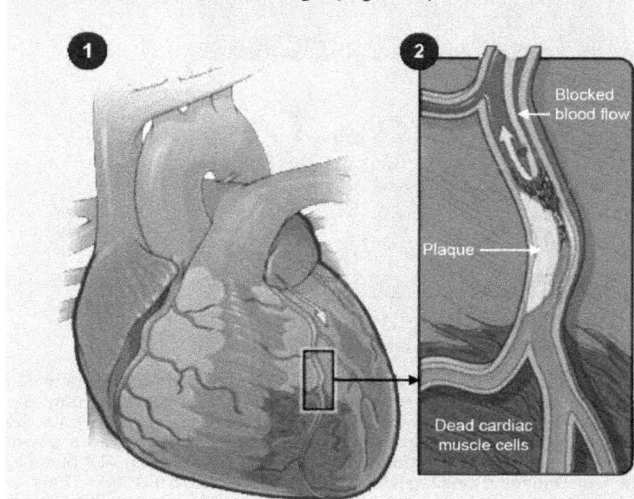

Figure 5: Illustration of a myocardial infarction. Please note that the blockage occurred at the ① anterior interventricular artery. ② Demonstrates the obstruction created by plaque formation which results in the backflow of blood

The Troponin Test is generally ordered by the attending physician for patients with chest pains. This test will be performed at least two to three times within 16 hours of the initial examination. The Troponin Test is used to allow the attending physician to determine if the individual is suffering from a myocardial infarction, as well as, assessing the amount of damage already

caused. This test is usually performed with other cardiac marker tests for CPK isoenzymes and/or myoglobin.

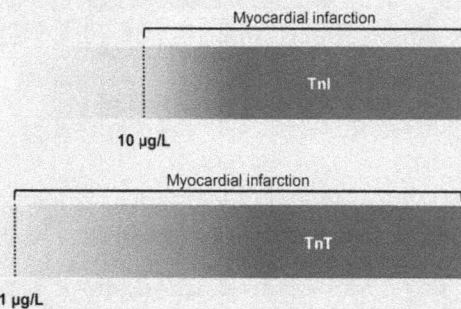

Figure 6: The levels of TnI and TnT. If the test results are >10 µg/L for TnI and >0.1 µg/L for TnT, the individual has suffered some cardiac muscle damage

Test results are normal when the TnI is less than 10 µg/L and TnT is between 0 to 0.1 µg/L. However, if the test results are greater than (>) 10 µg/L for TnI and greater than (>) 0.1 µg/L for TnT, then the individual is suffering from a myocardial infarction. An increase in the troponin level above the limit set for normal, no matter how slight, will indicate that damage has been done to the heart. It is estimated that most patients suffering from a myocardial infarction will demonstrate increased TnI and TnT levels within 6 hours of the primary event (Figure 6).

Sarcomere

A sarcomere extends from Z-line (Z-disk) to Z-line and is the basic contractile unit of the myofibril. Z-lines are constructed out of a filamentous protein network and are formed as an attachment site for actin myofilaments along with other regulatory and structural proteins. Other components of the sarcomere include I-bands also known as light bands. I-bands, are attached to the Z-line and consist of mostly actin myofilaments. A-bands, also known as dark bands, consist of both myosin and actin myofilaments. The H-zone is a location where I-bands and A-bands do not overlap and consists of mostly myosin myofilaments. The M-line is located at the center of the myosin myofilaments and it consists of M-line proteins that are used to hold the ~400 myosin molecules in a tight bundle (Figures 4 and 7). Please note, in an effort to provide a simple explanation of the sarcomere, various structure proteins (titin and nebulin), as well as, regulatory proteins (troponin and tropomyosin) are purposely omitted within the figures.

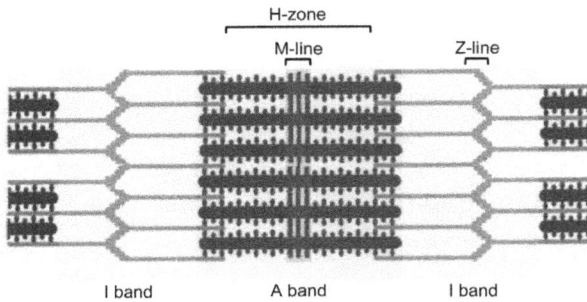

Figure 7: Illustration of a sarcomere

Sliding Filament Theory

The force created by contracting muscle is called muscle tension and this force is created through the interaction of actin and myosin myofilaments, which are powered by ATP. During contraction of a muscle, the length of the muscle fiber shortens, however, the actin and myosin myofilaments remain the same length.

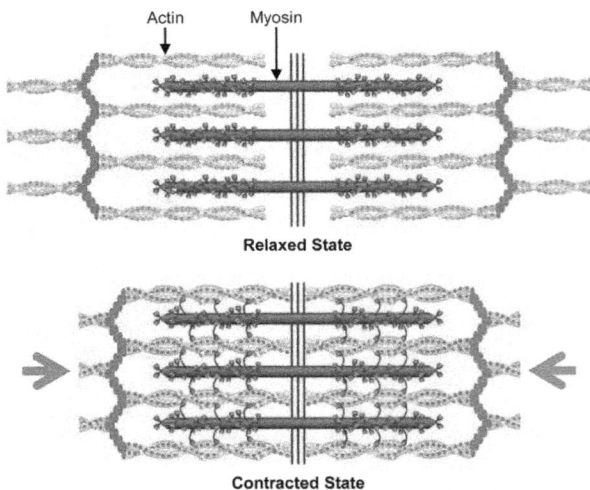

Figure 8: Illustration of a contracting sarcomere based upon the sliding filament theory. Please note it is the interaction of the actin binding site of the heavy meromyosin with the active site of the actin myofilament that forms the cross bridge. Subsequently, it is the release of energy from ATP that resulted in the "bending" of the heavy meromyosin causing a power stroke

This intriguing interaction was explained by two papers published by Dr. A.F. Huxley and colleagues in 1954. Using high resolution microscopy techniques, they described the state of actin and myosin myofilaments at different stages of contraction in muscle fibers. They observed that the A-band, which contains both actin and myosin myofilaments, remained at constant length during contraction while the I-band, which is composed of actin myofilaments, changed its length as the muscle fibers contracted. This observation allowed Huxley and colleagues to propose the sliding filament theory. In summary, this ground breaking theory simply proposed that the actin and myosin

myofilaments slide past one another, thereby pulling the Z-lines closer together, and in turn, creating muscle tension. Since actin is attached to the Z-lines, any movement of the actin myofilament sliding over myosin myofilament would result in a shortening of the sarcomere. Sequentially, the shortening of the sarcomere will result in a reduction of length in the myofibril. Since the myofibril stretches from end to end within the muscle fiber, any shortening of the myofibril will result in the contraction of the muscle and thereby creating muscle tension (Figures 4 and 8).

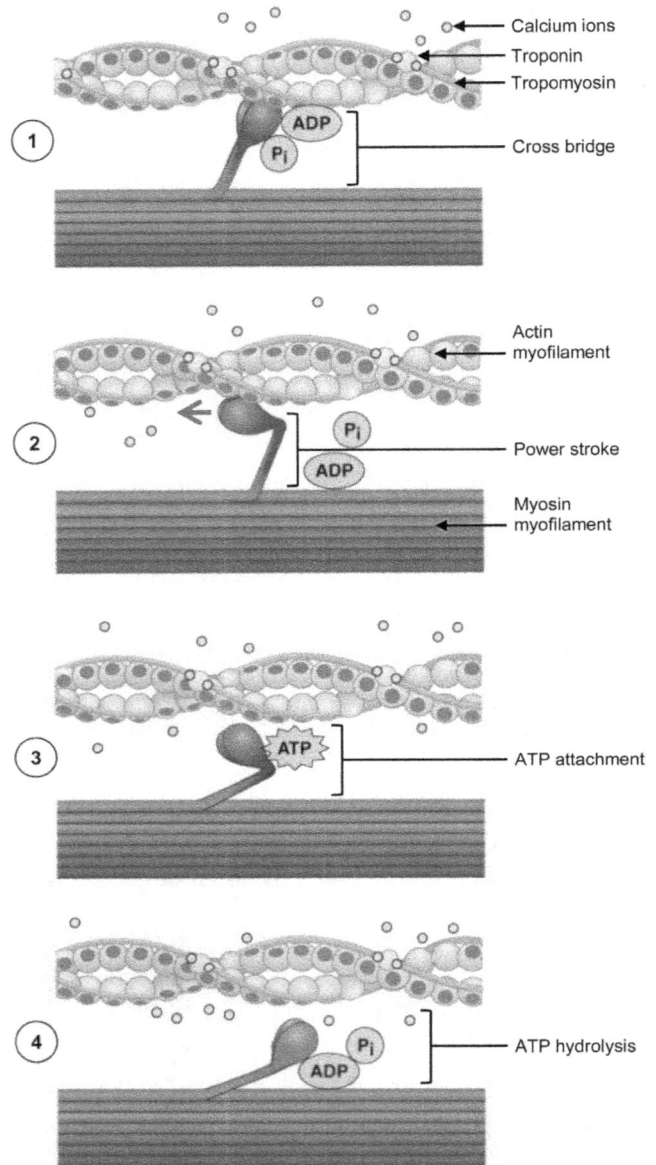

Figure 9: ① Calcium ions bind to troponin which in turn moves tropomyosin away from the active site. The exposed active site allows heavy meromyosin to bind thereby forming a cross bridge. ② The energy stored within the heavy meromyosin from the previous hydrolysis of ATP causes a power stroke. Subsequently, the remaining ADP and P_i is released. ③ ATP binds to the heavy meromyosin causing it to detach from the active site. ④ ATPase of the heavy meromyosin hydrolyzes the ATP causing it to return to its original position while the energy from the hydrolysis of ATP is stored

Ever since this theory's inception and introduction in 1954, the sliding filament theory has withstood decades of scrutiny from scientists and is now considered to be the foundational theory of muscle contraction. In recognition of his accomplishments, Dr. A.F. Huxley was awarded the Nobel Prize in Medicine in 1963.

Since the theory's proposal, more information has been added. One of these important additions is the manner by which myosin myofilaments act upon actin myofilaments to produce muscle tension. Scientists have demonstrated that the heavy meromyosin possesses multiple hinged elements called the subfragment 1 region (S1 region). These hinge elements are triggered to bend when ATP is hydrolyzed and a phosphate functional group and energy are released. This bending explains the manner by which heavy meromyosin 'walks' along actin myofilaments (Figure 9).

For example, the binding of calcium ion (Ca^{++}) to TnC will result in a conformational change of the regulatory protein, troponin. This conformational change of troponin will move tropomyosin away from the active site, thereby allowing the heavy meromyosin to reach forward and attach, via actin binding site, to the active site on the actin myofilaments and create a cross bridge. Once the cross bridge is formed, the energy stored from the hydrolysis of ATP causes the heavy meromyosin to bend, thus resulting in the shortening of the sarcomere. This bending action of the heavy meromyosin is also referred to as a power stroke. Immediately after the power stroke, a new ATP binds to the heavy meromyosin causing it to detach from the active site. The ATPase of the heavy meromyosin hydrolyzes the ATP causing it to return to its original position, while the energy from the hydrolysis of ATP is stored. If the calcium ions are still attached to the troponin and the active site is still exposed, the heavy meromyosin will reach forward once more and bind to another active site as the process repeats. This bind, bend and release process is also known as myosin-actin cycling (Figure 9).

SKELETAL MUSCLE

Skeletal muscles are organized from skeletal muscle fibers surrounded by deep fascia, also known as, muscular fascia. Deep fascia is constructed from a layer of dense irregular connective tissue, which helps to physically protect the skeletal muscle fibers, as well as, serving to compartmentalize individual muscle fibers, and to form groups of muscle fibers. In addition, the deep fascia will taper at the ends and blend with other connective tissues to form tendons or create a flat sheet-like connective tissue structure called aponeurosis.

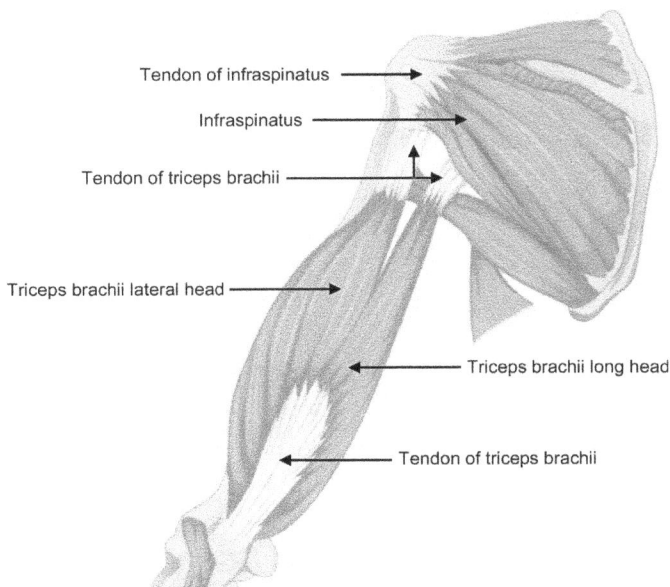

Figure 10: Triceps brachii, infraspinatus, and their associated tendons

Tendons are constructed from a tough band of fibrous connective tissue that generally connects muscle to bone via collagenous fibers. These cartilaginous fibers are called Sharpey fibers, and extend into the matrix of the bone to provide the necessary anchor for muscle attachment. Tendons serve to transmit force generated by the muscles, while simultaneously helping to position the limbs and/or acting to make movement more efficient (Figure 10). Nonetheless, tendons may also help connect muscles to organs. As in the case of the muscles that

control eye movement, the six ocular muscles, also known as extrinsic muscles of the eye, are connected directly to the orbs via tendons.

These ocular muscles along with their associated tendons are responsible for the movement of the eyes. For example, the superior oblique muscle depresses the eyes when moving from side to side. The inferior oblique muscle is opposite to the function of the superior oblique muscle where it elevates the eyes when moving from side to side. The superior rectus muscle moves the eye upward and secondarily rotates the top of the eye toward the nose, while the inferior rectus muscle moves the eye downward and secondarily rotates the top of the eye away from the nose. Finally, the lateral rectus muscle moves the eyes away from the nose and the medial rectus muscle moves the eyes towards the nose (Figure 11).

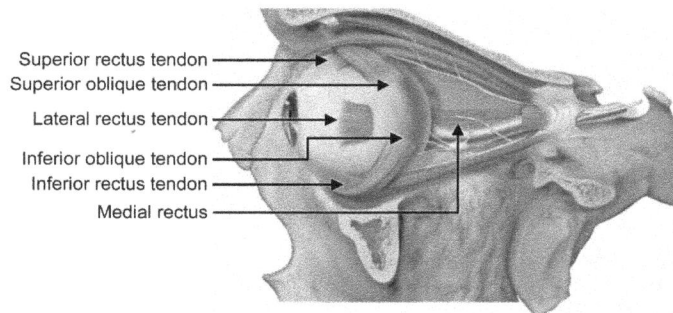

Figure 11: Illustration of the six muscles that control eye movement. Please note that each of the extreinsic muscles are directly connected to the eye via tendons

An aponeurosis, on the other hand, is a flat sheet-like structure that connects muscle to muscles or muscles to bones. Just like the tendon, an aponeurosis is composed of dense fibrous connective tissue. For example, the galea aponeurotica is an aponeuosis that is located at the top of the skull and connects frontalis muscle anteriorly to the occipitalis posteriorly (Figure 12).

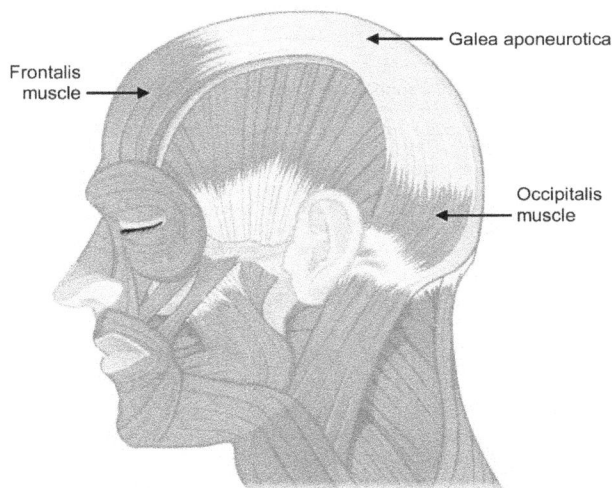

Figure 12: Illustration of galea aponeurotica, an aponeurosis, that connects frontalis with occipitalis muscles

All skeletal muscles of the body are surrounded by a layer of deep fascia. The deep fascia is richly innervated with nociceptors (i.e. sensory receptors that are capable of detecting pain), proprioceptors (i.e. sensory receptors that detects the position of the body in relation to balance), chemoreceptors (i.e. sensory receptors that detects the changes in pH), thermoreceptors (i.e. sensory receptors that detects changes in temperature), and changes in external and internal pressure, as well as, mechanoreceptors (i.e. sensory receptors that detects vibrations or changes in pressure). Additionally, the deep fascia provides a pathway for motor nerves and blood vessels to reach the muscle fibers.

The outer most layer of the deep fascia is called epimysium. Internally, the epimysium branches inward to form the perimysium, which divides the muscle into bundles of muscle fibers called muscle fasciculi. The perimysium will further branch into endomysium which forms a connective tissue layer surrounding each individual muscle fiber (Figure 13).

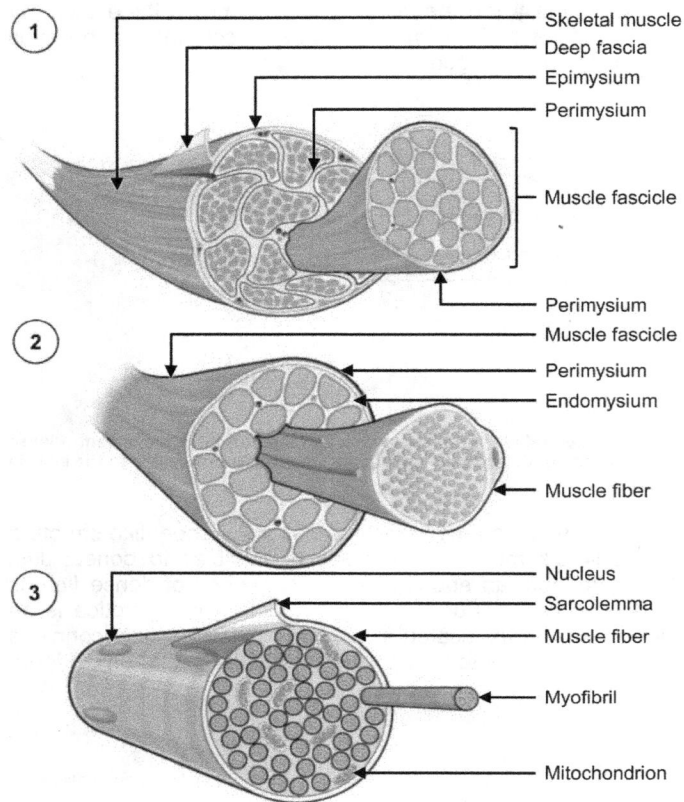

Figure 13: Illustration of connective tissue surrounding the skeletal muscle. ① **Illustrates the muscle surrounded by deep fascia and tapers at its end to form a tendon. Please note that the outer most layer of deep fascia is called epimysium which extends inward to form perimysium.** ② **The perimysium divides the muscle into smaller compartments called the muscle fascicle. The perimysium further extends inward to form endomysium which surrounds each muscle fiber.** ③ **Within the sarcoplasm of the muscle fiber are numerous myofibril and mitochondria. Please note that skeletal muscle fibers are multinucleated**

Neuromuscular Junction

In order to have fine motor control over skeletal muscles, each skeletal muscle fiber is directly coordinated by the central nervous system (CNS). For example, the primary motor cortex of the cerebral cortex determines the particular movement that the body needs to make and sends a signal via the corticospinal pathway to the cerebral peduncle, located in the midbrain. At the cerebral peduncle, these axons form the medullary pyramids that contain the corticospinal tracts on the ventral surface of the brainstem. At the level of the caudal medulla, the corticospinal tract splits into two tracts. Approximately 90% of axons cross over to the contralateral side at the pyramidal decussation, forming the lateral corticospinal tract. The lateral corticospinal tract continues down the spinal cord until it reaches the target segments in the spinal cord. The corticospinal tracts synapse with the cell bodies of somatic motor neurons, which are located in the anterior horn of the spinal cord, and in turn, project their axons via the ventral root directly to the skeletal muscles. The remaining 10% of axons

that do not cross at the caudal medulla constitute the anterior corticospinal tract. These nerves will continue down the spinal cord until they reach the target segment of the spinal cord, then they cross over to the contralateral side and innervate motor neurons or interneurons in the anterior horn (Figure 14).

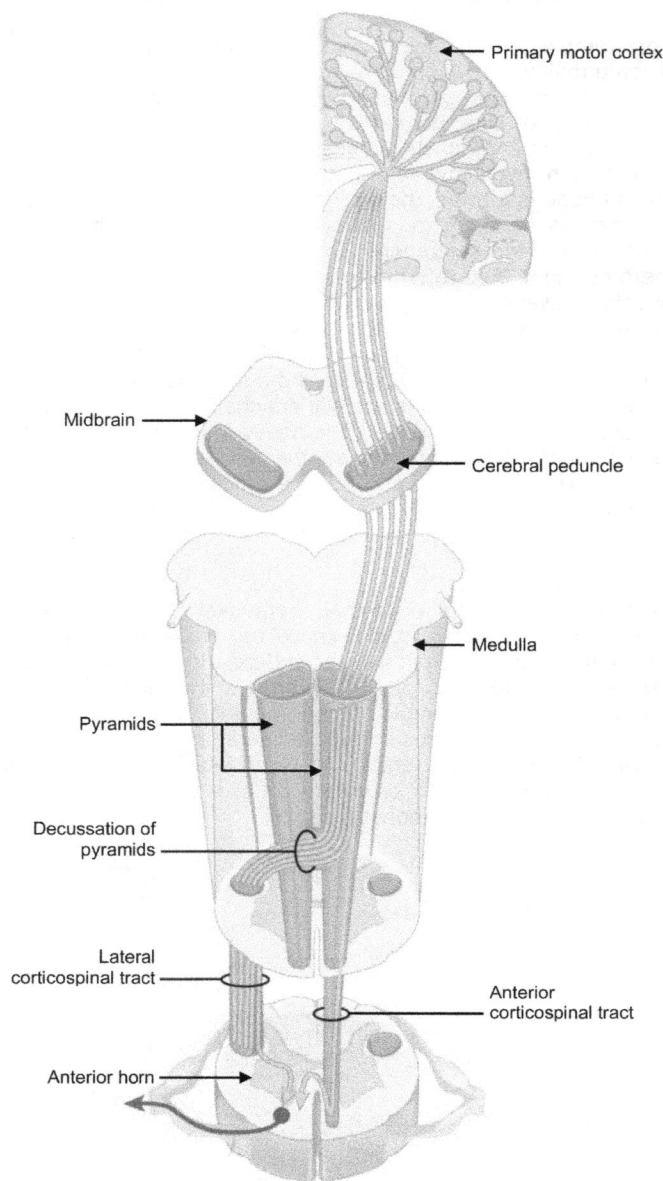

Figure 14: Illustration of primary motor cortex's direct control over skeletal muscles. Please note that the decision for skeletal muscle movement is made in the primary motor cortex which sends its signal via the corticospinal pathway to the cerebral peduncle where the corticospinal tracts of medullary pyramids are formed. The corticospinal tracts split into lateral and anterior corticospinal tract before reaching the anterior horn of the spinal cord segment where it synapses with a motor nerve. The signal is then transmitted by the motor nerve axon to the skeletal muscle fiber

Each skeletal muscle fiber is innervated by a single motor axon from a somatic motor neuron. This same axon may also form collaterals (i.e. axonal branches) and innervate other muscle fibers to exert a unified control over their contractions. All muscle fibers innervated by the same axon are called a motor unit.

The axonal terminals of motor nerves form a synapse called neuromuscular junction with a muscle fiber. At the neuromuscular junction, the sarcolemma of the skeletal muscle cell invaginates to accommodate the axonal terminal. This invagination is called the motor end plate. Interestingly, the neurolemma (nerve membrane) and the sarcolemma do not come into direct physical contact. As a result, a tiny fluid-filled space is formed between the nervous and muscular membranes. This tiny fluid-filled space is called the synaptic cleft and is the location where neurotransmitters, such as acetylcholine, are released via exocytosis (Figure 15).

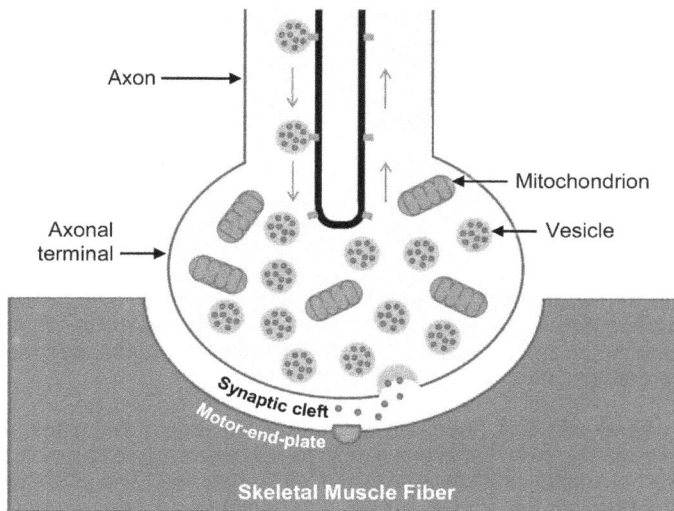

Figure 15: Illustration of a neuromuscular junction. Please note that the axonal terminal possesses numerous mitochondria and secretory vesicles containing neurotransmitters or neuropeptides. The mitochondria is needed to produce the necessary ATP needed to power the release of neurotransmitter or neuropeptides via exocytosis into the synaptic cleft

Membrane Potential

At rest and unstimulated, the skeletal muscles are polarized. Polarization is defined as a difference in electrical charge due to the uneven distribution of sodium (Na^+), potassium (K^+) and calcium (Ca^{++}) cations. This uneven distribution is created by the actions of the sodium-potassium pump (Na^+/K^+ pump) and calcium pumps (Ca^{++} pump) located in the sarcolemma. Na^+/K^+ pump is an active transport integral membrane protein, which sends out three Na^+ cations from the sarcoplasm and sends two K^+ cations from the extracellular matrix into the sarcoplasm. Ca^{++} pumps, on the other hand, are also an active transport molecule that sends Ca^{++} cations into the extracellular matrix.

An unstimulated skeletal muscle fiber keeps the concentration of Na^+ and Ca^{++} cations high in the extracellular matrix of the muscle cell. Since Na^+ and Ca^{++} are cations, the net charge in the extracellular matrix is positive. It is important to note that the high concentration of Na^+ and Ca^{++} in the extracellular matrix creates a diffusion gradient for these cations to flow into the sarcoplasm. In contrast, a resting skeletal muscle cell will maintain a high K^+ concentration in the sarcoplasm. However, due to the large number of negatively charged functional groups found on various contractile, regulatory, and structural proteins within the sarcoplasm, the net charge within the sarcoplasm will remain negative. Again, it is important to note that the high concentration of K^+ cations in the sarcoplasm creates the necessary diffusion gradient for these cations to flow outward into the extracellular matrix. On average, the charge difference of an unstimulated skeletal muscle cell is approximately -85 mV, which is also referred to as the resting membrane potential (Figure 16).

Once the resting membrane potential is created (e.g. the skeletal muscle cell is polarized), the skeletal muscle fiber is poised and ready to depolarize or to generate an action potential. An action potential is defined as the transient alteration of the transmembrane voltage where the membrane potential in the sarcoplasm becomes more positive in comparison to the membrane potential outside of the sarcolemma. This change in membrane potential is caused by alterations in the permeability of the sarcolemma through the opening of ion channels whereby specific ions are allowed to flow through.

In general, there are three groups of ion channels found in the

sarcolemma. The first group includes the ligand gated Na^+ ion channels and voltage gated Na^+ ion channels, which are involved in the depolarization of the skeletal muscle fiber. The second group includes the voltage gated Ca^{++} ion channels, which are involved in triggering muscle contraction. Finally the third group, which consists of the voltage gated K^+ ion channels, is involved in the repolarization of the muscle fiber. Repolarization is the process of restoring the membrane potential.

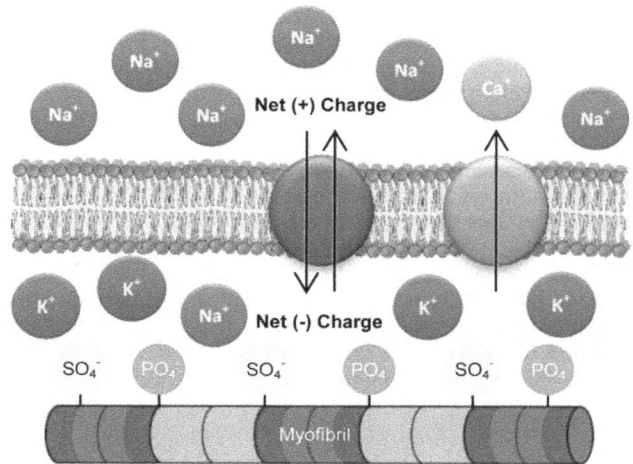

Figure 16: The uneven distribution of Na^+, K^+ and Ca^{++} cations creates a resting potential of -85 mV. Due to the high concentration of functional groups like PO_4^- and SO_4^- in the sarcoplasm, the net charge remains negative even though there is a high concentration of K^+. Please note that the Na^+/K^+ pump will send 3 Na^+ cations into the extracellular matrix while bringing in 2 K^+ cations into the sarcoplasm, while Ca^{++} pumps will send 1 Ca^{++} cations at a time out into the extracellular matrix

Ligand gated Na^+ channels are controlled by the binding of ligands, such as the neurotransmitter acetylcholine (ACh). There are two types of membrane bound receptor proteins: metabotropic and ionotropic receptors. The metabotropic receptors are not formed as a part of channel proteins. Instead, the opening and closing of the channel protein is dependent upon a guanine nucleotide-binding protein (G-protein) complex. The G-protein is responsible for the ligand mediated control of various functions of the cell. For example, ligand gated channels are controlled by a G-protein complex where the activation of the receptor protein alters the cellular membrane's permeability or activates certain enzymes. Put simply, a G-protein complex is an "on/off switch" that can activate or discontinue certain cellular actions.

In smooth and cardiac muscles, but not in skeletal muscles, the metabotropic receptor is the muscarinic acetylcholine receptor (mAChR). mAChR consists of alpha (α), beta (β), and gamma (γ) subunits on the inner surface of the cell membrane. Once ACh binds with mAChR, the receptor protein will undergo a conformation change and the three subunits (α, β and γ) will separate. The β and γ subunits form a stable dimer (joining of two molecules through intermolecular bonding) and is responsible for activating the α subunit. As the α subunit is separating a spent energy molecule, called guanosine diphosphate (GDP), is replaced with guanosine triphosphate (GTP). Consequently, the α subunit/GTP complex causes a conformation change in a K^+ ion channel protein, thereby resulting in the opening of the pore and allowing K^+ ions to diffuse down its concentration gradient, exiting the sarcoplasm of the muscle fiber.

The process of returning the ligand gated K^+ channel to its original conformation is an inverse reaction of the previously described process. First, the ACh has to be removed from the binding site on the mAChR by the enzyme acetylcholinesterase. Once the ligand is removed, the α subunit will hydrolyze the energy molecule GTP, releasing a phosphate molecule and energy, thereby breaking the interaction with the gated channel. Once the connection between the α subunit-GDP complex is severed, the ligand gated channel will reconstitute its original conformation and prevent K$^+$ ions from exiting the cytoplasm. (Please see Chapter three for more information in regards to the receptor and channel proteins).

The ionotropic receptors are transmembrane proteins that possess a ligand binding site directly linked to the ion channels forming a single molecular entity. Ionotropic receptors are composed of four to five transmembrane protein subunits, also known as multimers, organized around a pore. These ionotropic receptors, are fast acting receptors commonly associated with neurotransmitters such as γ-aminobutyric acid (GABA), glycine, serotonin, acetylcholine, etc., and are specific in the type of ions that are allowed to pass through its pore (Na$^+$, K$^+$, Ca^{++}, or Cl$^-$).

Ionotropic receptors possess two functional domains. The first domain is referred to as the ligand binding site and the second is called the membrane-spanning domain which creates the ion channel. For example, once the neurotransmitter binds to the ligand binding site, the intermolecular action will result in a change in the conformation of the membrane-spanning domain or the protein multimers. The resulting change in conformation will cause the pore to open, allowing ions to flow down the concentration gradient via diffusion.

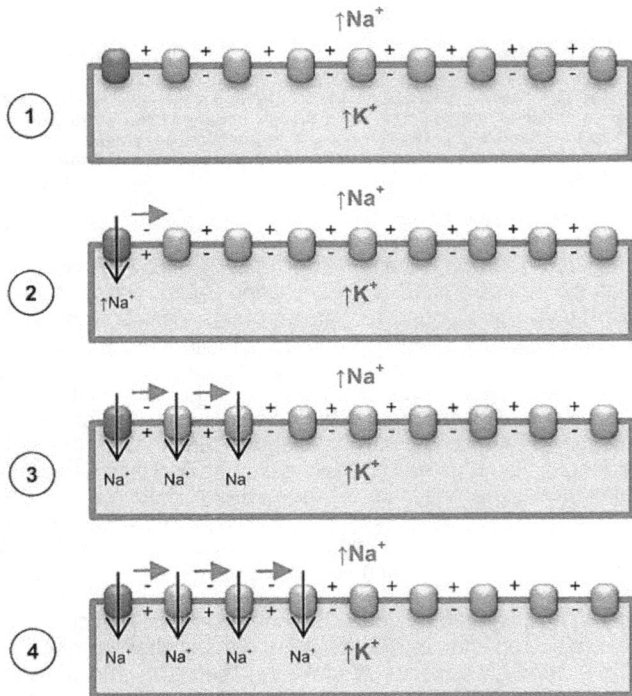

Figure 17: An illustration of the formation of an action potential. Please note that ① the pore of the ligand gated Na+ channel (first channel) will open after the binding of ACh which allows the initial influx of Na$^+$ and the generation of an electric impulse (arrow). ② Subsequently, the electric impulse will spread down the membrane by causing the first voltage gated Na$^+$ channel to open which in turn will create another electric impulse and ③ opening another voltage gated Na$^+$ channel. ④ Like falling dominos, another electric impulse will be generated whereby causing other voltage gated Na$^+$ channel to open

An example of ionotropic receptor associated with a Na$^+$ channel commonly found in skeletal muscles is the nicotinic acetylcholine receptor (nAChR). As mentioned previously, the binding of ACh to the ligand binding site results in a conformational change of the membrane-spanning domain, which in turn, allows Na$^+$ cations to flow down their concentration gradient and into the sarcoplasm. The influx of Na$^+$ cations will immediately alter the transmembrane voltage adjacent to the ligand gated Na$^+$ ion channel where the membrane potential in the sarcoplasm becomes more positive in comparison to the membrane potential outside of the sarcolemma. This switch of membrane charge will create an electrical impulse. If this impulse is sufficient to reach

the threshold (set at -58 mV), then the voltage gated Na$^+$ ion channel will open (Figure 17).

The threshold is the minimum amount of stimulus required to create an action potential and it is an all-or-none response. If the electrical impulse is inadequate (subthreshold stimulus) then it is too weak to instigate an action potential and the membrane quickly reestablishes the ionic imbalance. However, if the stimulus reaches the threshold then an action potential will be created.

The opening of the voltage gated Na$^+$ ion channel pore allows more Na$^+$ cations to flow into the cell thereby generating an additional electrical impulse. Like a multitude of dominoes toppling over one by one, each generated electrical impulse will cause another voltage gated Na$^+$ ion channel pore to open and a subsequent influx of more Na$^+$ cations. This process will continue until the concentration of Na$^+$ in the extracellular matrix and the sarcoplasm has reached equilibrium (Figure 17).

In time, when an adequate amount of Na$^+$ ions have diffused through the pore, the skeletal muscle cell closes the ligand gated Na$^+$ channels and voltage gated Na$^+$ channels. The process of returning the ligand gated Na$^+$ channels to their original conformation is an inverse reaction of the previously described process. First, the ACh has to be removed from the binding site on the nAChR by acetylcholinesterase. Once the ligand is removed, the membrane spanning domain will return to its original shape, which results in the closing of the pore, thereby discontinuing any further influx of the Na$^+$ cations.

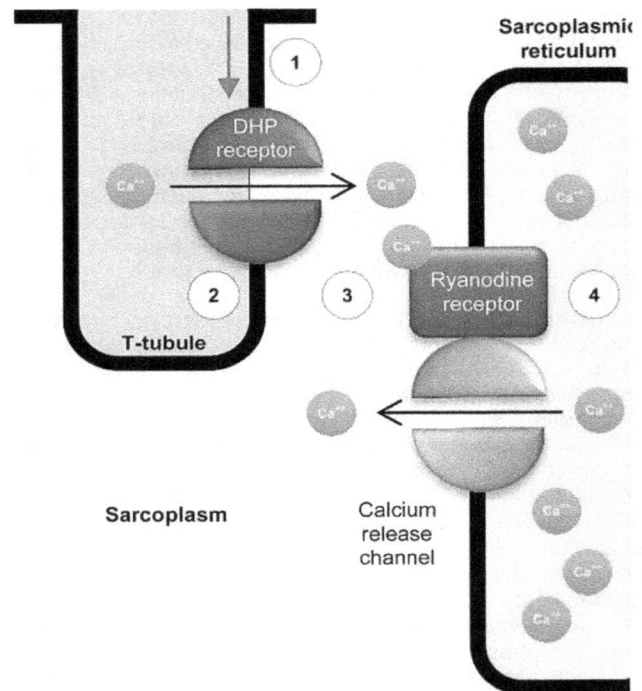

Figure 18: Illustration of the interactions between DHP receptor, ryanodine receptor and calcium release channel. ① The generated action potential activates the DHP receptor which causes this voltage gated Ca^{++} channel to open. ② Ca^{++} cations from the extracellular matrix to flow into the cell. ③ The Ca^{++} cations bonds to ryanodine receptor causing it to activate. ④ The activated ryanodine receptor trigger the opening of the calcium release channel thereby allowing the Ca^{++} stored within the sarcoplasmic reticulum to be released into the sarcoplasm

Repolarization is defined as the return of the membrane potential to its resting values. After the Na$^+$ ions have reached equilibrium and both ligand and voltage gated Na$^+$ channels have closed, the voltage gated K$^+$ channels are triggered to open. Once these channels open, the K$^+$ cations, which are kept at high concentrations inside the sarcoplasm, rush down their concentration gradient and into the extracellular matrix. This efflux of K⁺ cations results in the repolarization of the membrane potential.

In skeletal muscles, voltage gated Ca⁺⁺ channels are composed of two parts and are involved in the release of Ca⁺⁺ cations into the sarcoplasm. The first part is the dihydropyridine receptors (DHP receptors) in the transverse tubules (t-tubules), located in the triad. The second part involves ryanodine receptors and the calcium release channels located on the sarcoplasmic reticulum.

It is understood that the DHP receptor is a voltage sensing Ca⁺⁺ ion channel, which is opened in the presence of an electrical impulse. For example, the action potential, generated by the influx of Na⁺, will flow down the sarcolemma and the t-tubules as it opens up more and more voltage gated Na⁺ channels. At the triad, or the diad of cardiac muscle, the electrical impulse will activate the DHP receptor, thereby causing these calcium channels to open and allow Ca⁺⁺ cations from the extracellular matrix to flow into the t-tubules. The influx of Ca⁺⁺ cations will then bind directly onto the underlying ryanodine receptors, located on the sarcoplasmic reticulum. The attachment of Ca⁺⁺ cations activate the ryanodine receptors, which in turn, causes the adjacent calcium release channels to open, and result in the release of stored Ca⁺⁺ cations into the sarcoplasm. The process of releasing Ca⁺⁺ cations into the sarcoplasm is called a muscle impulse (Figure 18).

Skeletal Muscle Excitation-Contraction Coupling

Excitation-contraction coupling is the way by which an action potential causes the contraction of a muscle fiber. A detailed explanation of this process would be a compilation of nearly all that has been discussed within this chapter. In order to facilitate a better understanding of the methodology of skeletal muscle contraction, a bullet point description of the sequence of events is provided in three sections: polarization, depolarization and repolarization.

Polarization

- At rest and unstimulated, the skeletal muscles are polarized. Polarization is defined as a difference in electrical charge due to the uneven distribution of sodium (Na⁺), calcium (Ca⁺⁺) and potassium (K⁺) cations
- Na⁺ and Ca⁺⁺ cations are kept in high concentrations in the extracellular matrix of the skeletal muscle cells, which create a net positive charge external to the sarcolemma
- K⁺ cations are kept at a high concentration in the sarcoplasm
- However, due to the negativity of the charged functional groups, such as sulfates (SO_4^-) and phosphates (PO_4^-), found as a part of sarcoplasmic proteins, the interior of the muscle cell has a net negative charge regardless of the high concentration of the positively charged K⁺ cations (Figure 16)
- The difference between the charges in the sarcoplasm and the extracellular matrix is -85 mV (Figure 19)
- The primary motor cortex of the cerebral cortex determines a particular movement that the body needs to make and sends its signal (action potential) via the corticospinal pathway to the cerebral peduncle, located in the midbrain. The cerebral peduncle forms the medullary pyramids that contain the corticospinal tracts, which continue the transmission of the action potential. At the level of the caudal medulla, the corticospinal tract splits into two tracts: the lateral corticospinal tract (90% of the axons) and the anterior corticospinal tract (10% of the axons) (Figure 14)
- Both corticospinal tracts continue down the spinal cord until they synapse with the cell bodies of somatic motor neurons, located in the anterior horn of the spinal cord. The motor neurons project their axons, via the ventral root, directly to the skeletal muscles. Please note that each skeletal muscle fiber is innervated directly by motor nerves and/or collaterals from motor nerves, forming motor units

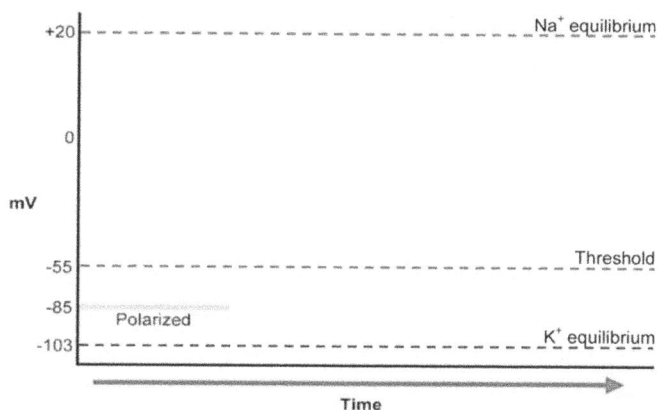

Figure 19: Illustration showing the membrane potential of an unstimulated skeletal muscle (yellow line). Please note that at -85 mV the muscle fiber is polarized

- At the neuromuscular junction, the axonal terminal of the motor nerve releases vesicles, containing ACh, via exocytosis into the synaptic cleft and is allowed to diffuse through the fluid filled medium to the nAChR (ionotropic receptor) located on the motor-end-plate (Figure 15)
- ACh binds with the nAChR causing a conformational change in the multimers, thereby resulting in the opening of the pore
- The opening of the pore allows Na⁺ cations to flow down their concentration gradient and into the sarcoplasm
- The influx of Na⁺ cations will create a "localized" electrical impulse
- The threshold of the skeletal muscle all-or-none response is -58 mV
- If the electrical impulse generated by the initial influx of Na⁺ cations is sufficient to reach or surpass the threshold (≥ -58 mV), then the adjacent voltage gated Na⁺ ion channel will open, causing other voltage gated Na⁺ ion channels to open
- An action potential will be generated as the voltage gated Na⁺ ion channels open. At this point the skeletal muscle cell is depolarized (Figure 20)

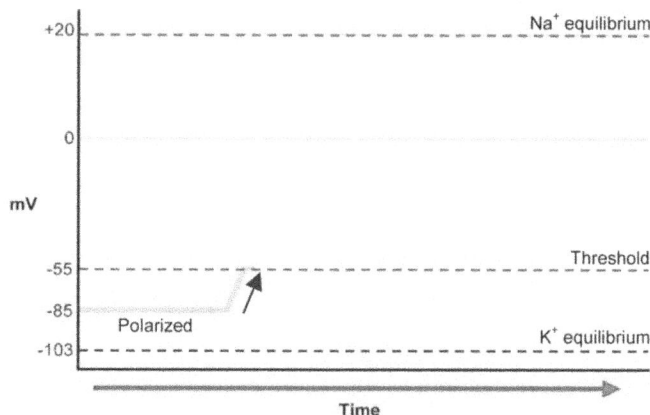

Figure 20: Illustration showing the membrane potential of a stimulated skeletal muscle (yellow line). Please note that the stimulation has risen pass the threshold and the all-or-none criteria for depolarization has been satisfied

Depolarization

- As mentioned previously, the opening of the voltage gated Na⁺ ion channel pores allows more Na⁺ cations to flow into the cell, thereby generating additional electrical impulse. Additional impulses will cause other voltage gated Na⁺ channels to open, thereby resulting in the spread of the action potential (Figure 17)
- The action potential will spread down the sarcolemma as well as the t-tubules
- At the triad, the generated action potential activates the DHP receptor which causes this voltage gated Ca⁺⁺ channel

to open. Ca⁺⁺ cations, kept at high concentrations in the extracellular matrix, begin to flow into the sarcoplasm of the cell (Figure 18)

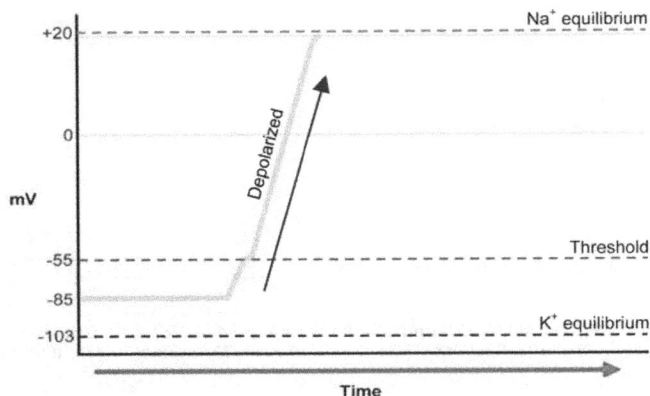

Figure 21: Illustration showing the membrane potential of a depolarized skeletal muscle (yellow line). Please note that at +20 mV the Na⁺ concentrations has reached equilibrium and that the skeletal muscle fiber has completely depolarized

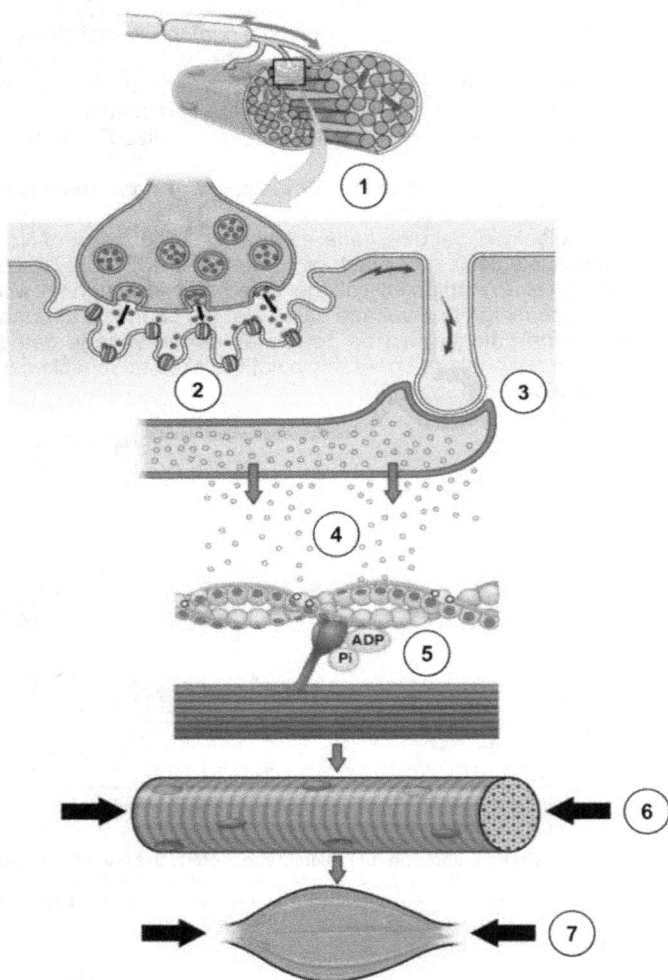

Figure 22: Illustration of excitation-contraction coupling. ① Action potential arrives at the neuromuscular junction which causes AChE to be released via exocytosis. ② AChE diffuses cross the synaptic cleft and binds with nAChR and initiates an action potential. ③ Action potential travels to the t-tubules and activates the DHP receptor which in turn causes the sarcoplasmic reticulum to release Ca⁺⁺. ④ Ca⁺⁺ is released into the sarcoplasm and ⑤ binds with troponin, which in turn moves tropomyosin away from the active site. Cross bridges are formed between the myosin and actin myofilament and actin-myosin cycling begins. ⑥ Actin-myosin cycling results in the shortening of the sarcomere. The shortening of the sarcomere causes the shortening of the myofibril. ⑦ The shortening of the myofibril results in muscle contraction

- The Ca⁺⁺ cations bind to ryanodine receptors causing them to activate and trigger the opening of the calcium release channels, thereby allowing the Ca⁺⁺ cations stored within the sarcoplasmic reticulum to be released into the sarcoplasm. The release of calcium into the sarcoplasm is called a muscle impulse (Figure 18)
- At this stage the skeletal muscle is completely depolarized and the membrane potential is approximately +20 mV (Figures 21 and 22)
- Ca⁺⁺ cations released from the sarcoplasmic reticulum will bind with troponin C (TnC), which in turn will cause the entire troponin molecule to undergo a structural conformational change. Please note that troponin has two other components: troponin T (TnT) and troponin I (TnI). TnT is responsible for binding troponin to tropomyosin, while troponin I (TnI) inhibits the interaction of myosin with actin
- This conformational change causes troponin to move the tropomyosin so that it now exposes the actin active sites, or switch to the "on" position
- Once the active site is exposed, the actin binding site of the heavy meromyosin will immediately bind, forming a bond called the cross bridge
- Since there are no net movements at this time, this tight binding between the active site on the actin myofilament and the actin binding site on the heavy meromyosin places the skeletal muscle in a rigor state
- Once the cross bridge is formed, the energy stored from the previous hydrolysis of ATP causes the heavy meromyosin to bend at the S1 region, resulting in the shortening of the sarcomere
- This bending action of the heavy meromyosin is also referred to as a power stroke
- Immediately after the power stroke, a new ATP binds to the heavy meromyosin causing it to change its conformation and detach from the active site
- The second binding site located on the heavy meromyosin is called ATPase. The ATPase hydrolyzes the newly attached ATP causing it to return to its original conformation, while the energy from the hydrolysis of ATP is once again stored
- If the Ca⁺⁺ cation is still attached to the troponin and the active site is still exposed, the heavy meromyosin will reach forward and bind to another active site, repeating the process
- This bind, bend and release process is also known as myosin-actin cycling (Figure 22)

Repolarization

- Skeletal muscles cannot be in the excited state for long, because it would result in muscle fatigue. Therefore, relaxation is a must. The relaxation of the skeletal muscle involves the process of repolarization
- The muscle cell is completely depolarized at +20 mV
- Acetylcholinesterase (AChE), present at the synaptic cleft, begins to break down the ACh, eliminating the stimulus form the nAChR complex
- Once the neurotransmitter ACh is removed, the multimers return to their original conformation, closing the pore
- At +20 mV, Na⁺ cations have reached their equilibrium; therefore no net movement of ions will flow across the membrane. Because there are no net movement of Na⁺, there will be no electrical impulse generated
- Lack of electrical impulse will cause the voltage gated Na⁺ channels to close
- At +20 mV, the voltage gated K⁺ channels are triggered to open. K⁺ cations flow down their concentration gradient and leave the sarcoplasm to enter the extracellular matrix of the skeletal muscle fiber. The process of repolarization of the skeletal muscle cell begins
- As more K⁺ cations leave the cell, the membrane potential becomes more negative. Please note that the negativity within the sarcoplasm is caused by the abundance of negatively charged functional groups, such as SO_4^- and PO_4^-
- Eventually, the voltage drops to -103 mV, which is also

known as hyperpolarization

- At hyperpolarization, the voltage gated K^+ channels closes, while the Na^+/K^+ pumps and the Ca^{++} pumps are switched on (Figure 23)

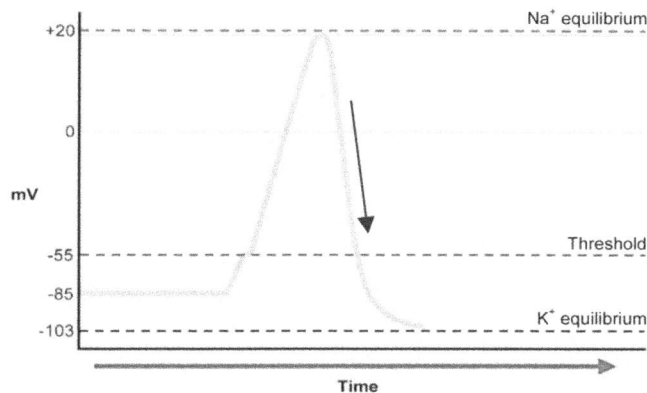

Figure 23: Illustration showing the membrane potential of a repolarizing skeletal muscle (yellow line). Please note that at -103 mV K^+ cations have reached equilibrium and that the muscle fiber is hyperpolarized. At -103 mv, the voltage gated K^+ channels are closed and the Na^+/K^+ pump is switched on

- Na^+/K^+ pump is an antiport active transport membrane bounded protein that sends 3 Na^+ cations out into the extracellular matrix and 2 K^+ cations into the sarcoplasm
- Ca^{++} pumps, on the other hand, send a single Ca^{++} cation at a time from the sarcoplasm and back into the extracellular matrix
- Please note that active transport requires energy molecules, such as ATP, to move substances against their concentration gradient
- The Na^+/K^+ pump resets the membrane potential back to -85 mV (Figure 24)

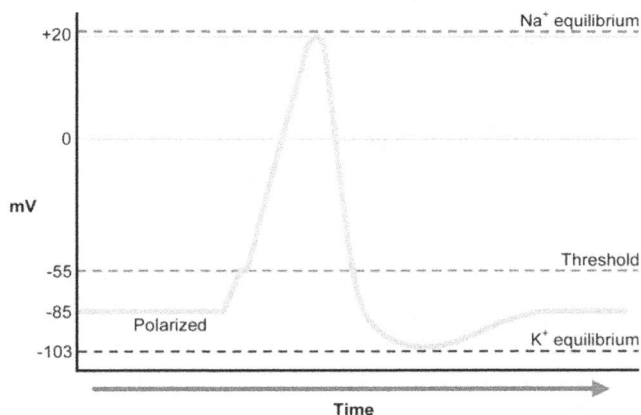

Figure 24: Illustration showing the membrane potential of a skeletal muscle fiber that has repolarized and returned to being polarized at -85 mV. Again, Na^+ cations are kept at a high concentration in the extracellular matrix, while the K^+ cations are kept at a high concentration within the sarcoplasm of the skeletal muscle cell

Muscle Relaxation

- Please note that the process of muscle relaxation occurs in conjunction with repolarization
- The elimination of the neurotransmitter ACh, and the initiation of repolarization causes the terminal cisternae to collect, concentrate and reabsorb Ca^{++} cations to be stored in the sarcoplamsmic reticulum
- The reabsorption of Ca^{++} cations is an energy dependent step (i.e. ATP is needed). If energy is not available, Ca^{++} cations will not be reabsorbed and tetany may result. Please examine the section on tetany for more information
- Ca^{++} cations are also removed from the DHP receptor located at the triad, which closes the pore within the voltage gated Ca^{++} channels and deactivates ryanodine receptors

- Ca^{++} cations are released/reabsorbed from the TnC, a troponin subunit, returning the regulatory molecule, troponin, back to its original conformation
- Once troponin is returned back to its original conformation, it will move tropomyosin back to its original position. Please note that tropomyosin is responsible for blocking the active site of actin
- With the active site blocked, the connection with the actin binding site, located on the heavy meromyosin, and the active site on the actin myofilament is severed. The cross bridge formed between the myosin and actin myofilament is broken
- At this time, the heavily compressed spring-like elastic protein, titin, will release its potential energy and push the myosin and actin myofilaments back to their original shape
- In turn, with the myosin and actin myofilaments return to their original shape, the sarcomere returns back to its relaxed state
- With the sarcomere back to its relaxed state, the myofibril is consecutively returned to its relaxed state. The final result is muscle relaxation (Figure 25)

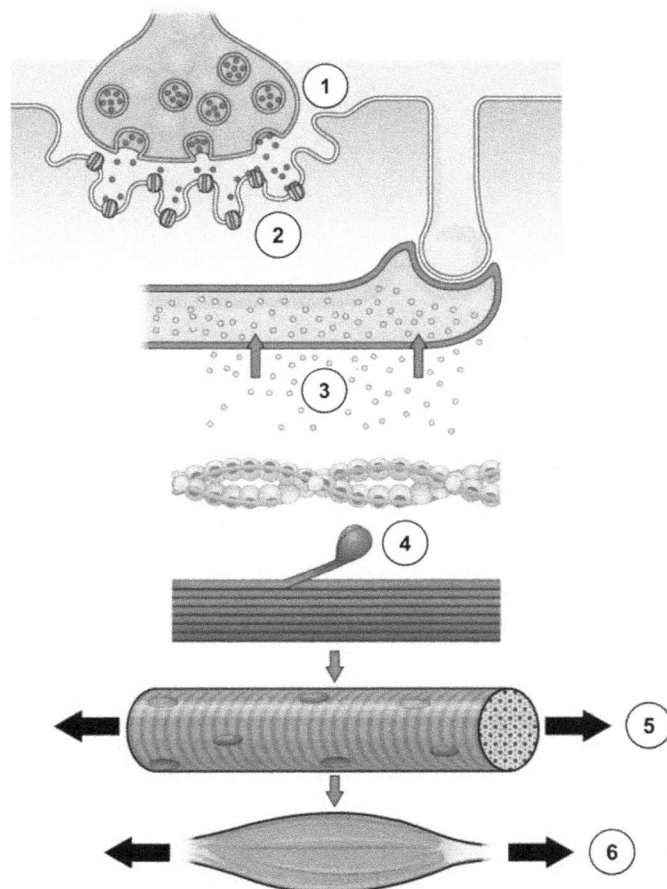

Figure 25: An illustration of skeletal muscle relaxation. ① AChE removes ACh, which results in the conformational change of the multimers, thereby closing the pore. ② Lack of an action potential causes the voltage gated Na^+ channels to close. ③ DHP receptor turns "off", causing the reabsorption of Ca^{++} by the terminal cisternae and subsequent storage in the sarcoplasmic reticulum. ④ Removal of Ca^{++} from TnC, causes the troponin to return to its original shape. Regaining its shape, troponin moves tropomyosin back to its original conformation. In its original position, tropomyosin covers the active site of actin myofilament and severs the cross bridge. ⑤ The sarcomere and myofibril return to its relaxed state. ⑥ muscle relaxation

Length-tension relationship

Length-tension relationship of muscle is defined as the correlation between the length of the muscle fiber and the force that it generates. It is understood that the length of the muscle fiber has a direct relationship to the position of the actin and myosin myofilaments within the sarcomeres.

As stated in the previous section, when a skeletal muscle fiber contracts, the actin binding site of heavy meromyosin attaches to the active site of the actin myofilament to form cross-bridges. Immediately after the formation of cross bridges, the energy stored from the previous hydrolysis of ATP causes the heavy meromyosin to bend at the S1 region, resulting in the shortening of the sarcomere. This bending action of the heavy meromyosin is also referred to as a power stroke and once completed, a new ATP binds to the ATPase (second binding site) located on the heavy meromyosin causing it to change its conformation and detaching it from the active site. Subsequently, ATPase hydrolyzes the newly attached ATP causing it to return to its original conformation, while the energy from the hydrolysis of ATP is once again stored. If the Ca^{++} cations are still attached to the troponin, it will continue to remove the tropomyosin away from the active site of actin, thereby exposing the active site. Since the active site is exposed, the heavy meromyosin will reach forward and bind it as the process repeats. This bind, bend and release process is also known as myosin-actin cycling, which causes the thick filament to slide over the thin filament. The sarcomere shortens as the thick and thin filaments slide over one another, creating muscle tension.

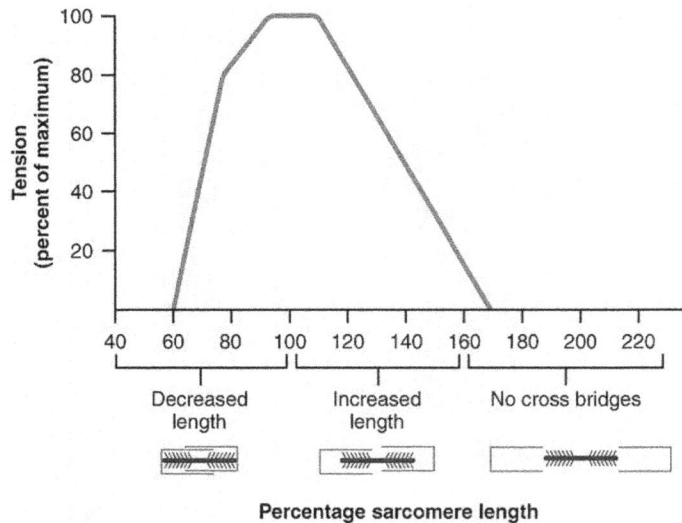

Figure 26: An illustration of the length-tension relationship of a sarcomere

Based on calculations, the ideal length of a sarcomere that will produce a maximal tension occurs between 80 to 120 percent of its resting length. Experiments have demonstrated that when an isolated muscle fiber is stretched to the point of minimal overlap of actin and myosin myofilaments (greater than 120 percent) and then stimulated by an electrode to contract, the muscle tension (force) generated is minimal. In other words, the thick and thin filaments do not overlap sufficiently, which results in less tension being produced. When an isolated muscle fiber is stimulated to contract at maximum overlap of the myofilaments (less than 80 percent), the muscle tension (force) produced is also minimal. Put simply, when there is too much overlap and the H-zones and I bands have narrowed, there would be nowhere for the thick and thin filaments to go, thereby diminishing the muscle tension generated (Figure 26).

Muscle Twitch

A muscle twitch is defined as the contraction of a muscle in response to a single motor neuron stimulus that causes an action potential in one or more muscles or motor units. Depending on the muscle type, the duration of a single twitch may last for a few milliseconds to a few hundreds of milliseconds. Using a myogram, the tension produced is shown to have three distinct phases: a lag phase, a contraction phase and a relaxation phase (Figure 27).

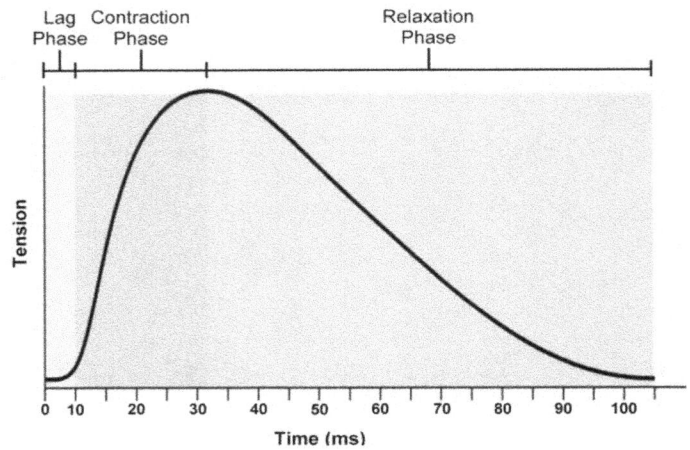

Figure 27: Illustration of a muscle twitch

The time between the application of a stimulus and the beginning of contraction is called the lag phase. This first phase includes the initiation of an action potential and its propagation along the sarcolemma and the transverse tubules. The stimulation of the DHP receptor at the triad and the subsequent stimulation of the ryanodine receptor cause the release of Ca^{++} cations from sarcoplasmic reticulum and the generation of a muscle impulse. The released Ca^{++} cations bind to TnC, causing a conformational change in troponin and the removal of tropomyosin from the active site of actin. The exposure of the active site allows the actin binding site of the heavy meromyosin to bind and form a cross bridge.

The contraction phase occurs when the energy stored from the previous hydrolysis of ATP causes the heavy meromyosin to bend at the S1 region, resulting in the formation of a power stroke and a subsequent shortening of the sarcomere. Immediately after the initial power stroke, a new ATP binds to the heavy meromyosin causing it to change its conformation and detach from the active site. The ATPase hydrolyzes the newly attached ATP causing it to return to its original conformation, while the energy from the hydrolysis of ATP is once again stored and the heavy meromyosin will reach forward and bind to another active site as the myosin-actin cycling process repeats.

The relaxation phase occurs when ACh is eliminated from the receptor protein (nAChR) and the initiation of repolarization process. During the repolarization of the muscle fiber, the terminal cisternae collects, concentrates, reabsorbs Ca^{++} cations and stores them in the sarcoplamsmic reticulum. The removal of Ca^{++} cations from the TnC causes the troponin to return to its original conformation, thereby returning tropomyosin back to its original position, blocking the active site. With the active site blocked, the cross bridge forms between the myosin, and actin myofilament is broken. At this time, the heavily compressed spring-like elastic protein titin will release its potential energy and push myosin and actin myofilaments back to its original relaxed state.

Strength of Muscle Contraction

There is a wide range of muscular responses, ranging from very weak to very strong, depending on the particular circumstances. For instance, the force required to lift a single can of soft drink is much less than the force required to lift a case of soft drinks. It is understood that there are two manners by which the force generated by the muscle can be increased. First is multiple motor unit summation, which involves the increased number of muscle fibers contracting. Second is called multiple wave summation, which involves increasing the force of contraction of the muscle fibers.

Multiple motor unit summation involves the progressive activation of a muscle through the incremental recruitment of motor units. Motor unit consists of a single motor neuron and the muscle fibers that it innervates through its collaterals. These muscle fibers are dispersed in part or throughout the muscle. Therefore, if a single motor unit is stimulated to contract, it will result in a weak, but evenly distributed muscle contraction. A muscle is composed of numerous motor units and the force of muscle contraction is dependent on the number of motor units stimulated. Hence, in a graded response, the more motor units that are stimulated, the summation of their contraction will in turn increase the force generated.

Experiments have demonstrated the graded response of muscles by applying external electrical stimuli in increasing intensity. For example, a sub-threshold stimulus is an electrical stimuli that is not strong enough to reach the threshold (-58 mV) and is thereby unable to generate an action potential. As the external electrical stimuli increases, a threshold stimulus will be reached, allowing an action potential to be generated in a single motor unit axon, in turn, causing the muscle fibers within the motor unit to contract. Further increasing the external electrical stimuli will create sub-maximal stimuli that will produce additional action potentials in numerous axons innervating different motor units. The action potentials generated by these additional axons will induce more muscle fibers from different motor units to contract. Additional increase in the external electrical stimuli will create a maximal stimulus, which will activate all the axons of all motor units of the muscle. Consequently, all of the muscles fibers of all motor units will contract, generating a maximum amount of contraction. Further increase in the external electrical stimuli will create supra-maximal stimuli, which will result in no additional effects since all the motor units are already activated (Figure 28).

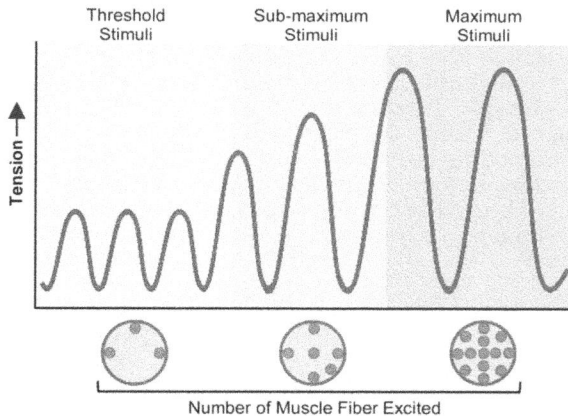

Figure 28: Illustration of threshold, sub-maximum and maximum stimuli

The second manner by which force is generated by a muscle can be increased is through multiple wave summation. Multiple wave summation is the correlation between the frequency of stimuli (i.e. action potentials) from motor nerves and the subsequent increase of tension in skeletal muscle (Figure 29).

For example, if the frequency of stimuli from a motor nerve is low, the innervated muscle fibers will have enough time to contract (i.e. exhibit a muscle twitch) and then relax completely. However, if the frequency of the stimuli from the motor nerve is increased, the previously contracted muscle fibers will not have the time to relax completely before a second contraction is initiated. Since, skeletal muscles do not have to completely relax between contractions; this second contraction will be added, or summarized, with the previous contraction, increasing the muscle tension. This manner of muscle stimulation and contraction is called incomplete tetanus. At the molecular level, incomplete tetanus occurs because the additional stimulus prompts more Ca^{++} cations to be released and in turn, become available to activate additional sarcomeres, even when the muscle is only partially relaxed. Nonetheless, if the frequency of stimuli

becomes so frequent, the contractions will become continuous, as the relaxation phase disappears completely. This continuous contraction is called complete tetanus, where the concentration of Ca^{++} cations in the sarcoplasm is so high that virtually all of the sarcomeres have formed cross bridges and shortened. Put simply, in complete tetanus, the muscle has reached maximum contraction, where no further contraction will be possible (Figure 27).

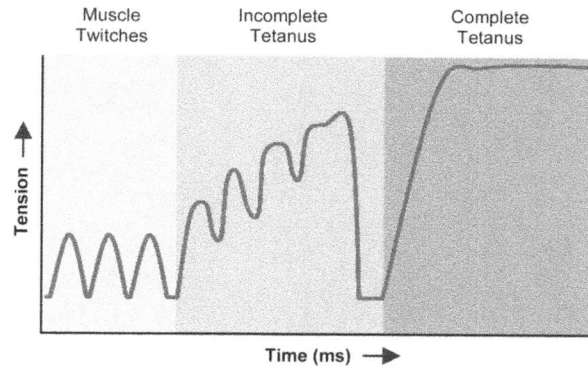

Figure 29: Illustration of muscle twitches, incomplete tetanus and complete tetanus

Treppe

Treppe, also known as the staircase effect, occurs when a skeletal muscle has been at rest for a prolonged period and then activated to contract. If a rested muscle is stimulated at maximum level but at a low frequency (allowing time for the muscle to relax), the initial contractions generate about 50 percent the force when the same muscle is stimulated the second time. In turn, if the skeletal muscle is stimulated the third time, the contraction would be stronger than the second and so forth, until finally the maximum contraction is reached (Figure 30).

One theory for this staircase effect is due to the availability of Ca^{++} cations in the sarcoplasm. For example, when the muscle is stimulated the first time, the stored Ca^{++} cations are released into the sarcoplasm. Before the Ca^{++} cations are completely absorbed back into the sarcoplasmic reticulum a second stimulus causes more Ca^{++} cations to be released into the sarcoplasm. An increase in Ca^{++} levels makes contraction more efficient. This situation repeats itself during the third and fourth stimuli until all of the stored Ca^{++} cations are released into the sarcoplasm which causes the muscle to reach maximum efficiency. Treppe is also demonstrated during the warm-up prior to aerobic or anaerobic exercise. An athlete would notice that the initial contraction of their muscle is weaker, and it is not until the muscle warmed-up before maximum efficiency is achieved (Figure 30).

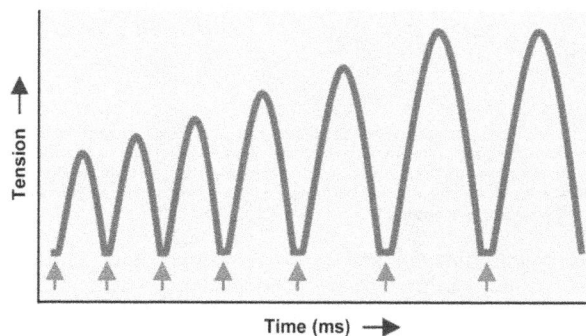

Figure 30: Illustration of treppe. Please note that the red arrows indicates a motor nerve stimulation

Types of Muscle Contractions

Muscle tension is defined as the force exerted on an object by a contracting muscle, while the force exerted on the muscle by an object (weight) that an individual is trying to lift is called load. Put simply, muscle tension and the load are opposing forces. Therefore, in order for the muscle fiber to move a load, the tension must be greater than load.

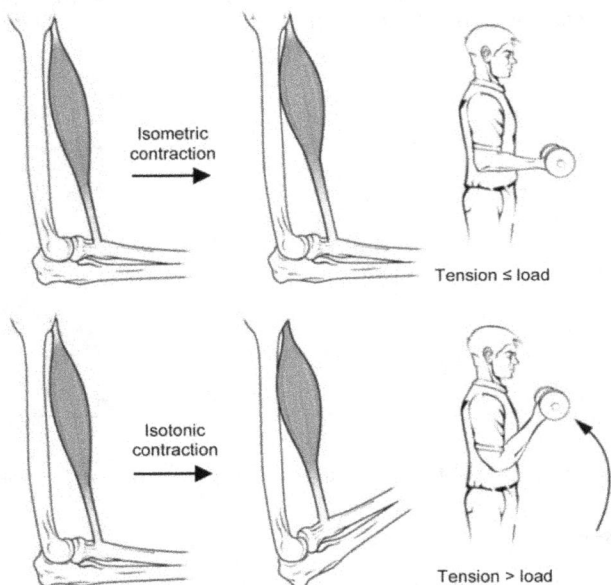

Figure 31: Isometric and isotonic contractions. Isometric contraction occurs when the tension is less or equal to the load. Typically, the tension will increase while the muscle length remains the same. Isotonic contraction occurs when the tension exerted is greater than the load. Typically, the tension will remain the same while the muscle length shortens

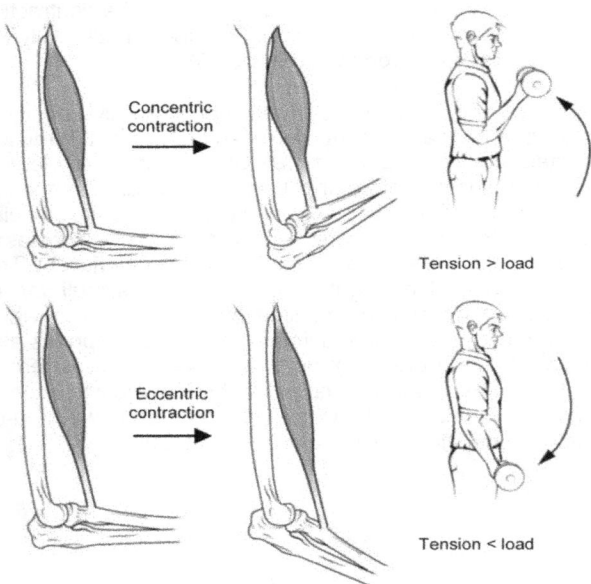

Figure 32: Concentric and eccentric contractions. Concentric contraction occurs when the tension is greater than the load. The tension will increase as the muscle length shortens. Eccentric contraction occurs when the tension exerted is less than the load. The tension will remain the same while the length of the muscle increases

The interactions between actin and myosin myofilaments generate muscle tension during myosin-actin cycling. Although muscle tension implies contraction or shortening of the sarcomere, the force created could also cause the muscle to lengthen or remain the same length. For example, during isometric contractions, the length of the muscle remains the same but the amount of tension increases. Isometric contractions are typically seen when muscle supports a load in steady position or struggles to move a supported load that is greater than the tension. During such a contraction, the bound cross bridges do not move. Isometric

contractions are commonly found in the muscles of the back (i.e. latissimus dorsi) which are responsible for an individual's posture (Figure 31).

Isotonic contractions occur when the tension exerted is greater than the load. During isotonic contraction, the tension increases briefly during the initial phase just to overcome the load but remains the same for the remainder of the contraction because additional force is no longer needed. Nonetheless, in this type of contraction, the length of the muscle does shorten. Muscles of the fingers and the upper limb typically demonstrate isotonic contractions (Figure 31).

Concentric contractions occur when an increase in tension also results in the shortening of the muscle. This type of contraction occurs when the tension generated is greater than the load, and is best demonstrated in flexing the biceps brachii when attempting to lift weights. Finally, eccentric contractions occur when the muscle lengthens while the tension generated remains the same. This type of contraction occurs when a load on a muscle is greater than the tension generated. The best example of this form of contraction is when a weight is being slowly lowered resulting in the lengthening of the biceps brachii, while no additional force is generated (Figure 32).

AEROBIC RESPIRATION

Aerobic respiration uses oxygen to break down glucose to produce ATP, water, and the waste product carbon dioxide (CO_2). Aerobic respiration involves four distinct reactions, acting together to produce a potential net gain of 36 to 38 ATPs. The four steps are glycolysis, transition reaction, the Krebs cycle (also known as citric acid cycle), and the electron transport chain.

In addition to the various enzymes that are involved in aerobic respiration there are three important molecules that are essential for reactions to take place. For example, coenzyme A (CoA) is involved in the oxidation of pyruvate to form Acetyl coenzyme A (AcetylCoA) in order for it to proceed into the Krebs cycle. Nicotinamide adenine dinucleotide (NAD^+) is an oxidized form of an electron carrier coenzyme that can be reduced by accepting two electrons to form NADH + H^+. Flavin adenine dinucleotide (FAD^+), is a second type of an oxidized electron carrier coenzyme that can be reduced by accepting four electrons to form $FADH_2$ + H_2^+.

The first stage in aerobic respiration involves the reaction called glycolysis and this reaction takes place in the cytosol of all cells. Unlike the remaining reactions in aerobic respiration, glycolysis does not require oxygen and is at times referred to as anaerobic glycolysis. The initial step of glycolysis involves the energizing of a molecule of glucose by investing two molecules of ATP to form fructose 1,6-bisphosphate. Once energized, the fructose 1,6-bisphosphate breaks down into two molecules of glyceraldehyde 3-phosphate before being enzymatically converted into two molecules of pyruvate, as well as 4 molecules of ATP (net gain of two ATP) and two molecules of NADH (reduced form of NAD^+) (Figure 33).

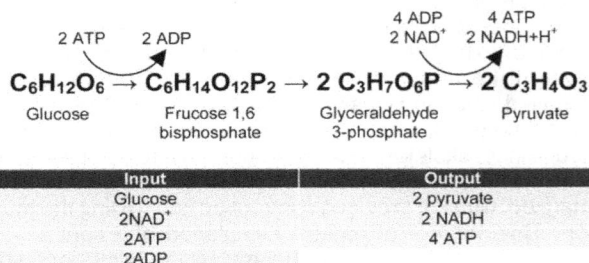

$$2 ATP \quad 2 ADP \qquad\qquad 4 ADP \quad 4 ATP$$
$$\qquad\qquad\qquad\qquad\qquad 2 NAD^+ \quad 2 NADH + H^+$$
$$C_6H_{12}O_6 \rightarrow C_6H_{14}O_{12}P_2 \rightarrow 2\ C_3H_7O_6P \rightarrow 2\ C_3H_4O_3$$
Glucose   Frucose 1,6   Glyceraldehyde   Pyruvate
bisphosphate   3-phosphate

Input	Output
Glucose	2 pyruvate
2NAD$^+$	2 NADH
2ATP	4 ATP
2ADP	

Figure 33: Glycolysis

The second stage of aerobic respiration is called the transition reaction. The transition reaction is an oxygen dependent step that uses the two pyruvates generated from glycolysis to form acetyl-CoA, releasing NADH, and CO_2. This reaction is also referred to as a link-reaction because it provides a connection between glycolysis and the Krebs cycle (Figure 34).

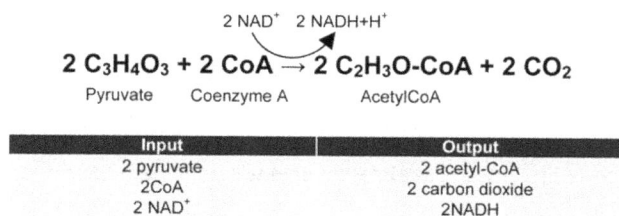

$$2\ NAD^+ \quad 2\ NADH+H^+$$

$$2\ C_3H_4O_3 + 2\ CoA \rightarrow 2\ C_2H_3O\text{-}CoA + 2\ CO_2$$

Pyruvate   Coenzyme A   AcetylCoA

Input	Output
2 pyruvate	2 acetyl-CoA
2CoA	2 carbon dioxide
2 NAD$^+$	2NADH

Figure 34: Transition reaction

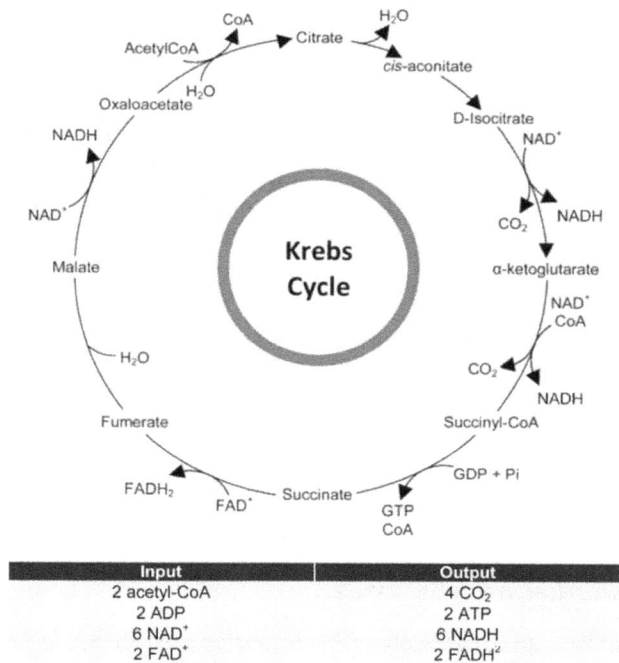

Figure 35: Illustration of the Krebs cycle

Input	Output
2 acetyl-CoA	4 CO$_2$
2 ADP	2 ATP
6 NAD$^+$	6 NADH
2 FAD$^+$	2 FADH^2

The Krebs cycle, also known as the citric acid cycle, is the third stage of aerobic respiration. The Krebs cycle is a series of oxygen dependent chemical reactions that oxidizes the acetyl groups released by CoA to form two ATP, six NADH, and two FADH$_2$. For example, acetyl-CoA enters the cycle after the enzyme citrate synthase consolidates it with oxaloacetate through the process of aldol condensation to create citrate. Please note that CoA is not part of the Krebs cycle, it is only a transfer molecule that aids in the initialization of the reaction and can be used again in future reactions. Once citrate is formed, it is immediately processed by the enzyme aconitase to form cis-aconitate and then forming D-isocitrate. Through oxidation and decarboxylation, D-isocitrate is processed by the enzyme isocitrate dehydrogenase to release an electron (reducing NAD$^+$ to NADH), a molecule of CO_2, and a product called α-ketoglutarate. Immediately after its formation, α-ketoglutarate is processed by the enzyme α-ketoglutarate dehydrogenase along with CoA to create succinyl-CoA, as well as, releasing another electron (reducing NAD$^+$ to NADH) and CO_2. Succinyl-COA is phosphorolated by succinyl-CoA synthetase to form succinate and a molecule of guanosine triphosphate (GTP). Once succinate is formed, it is oxidized by succinic dehydrogenase to form fumarate and releasing two elections (reducing FAD$^+$ to FADH$_2$). Fumarate is processed through hydration to form malate by the enzyme fumarase. Subsequently, malate is oxidized by the enzyme malate dehydrogenase to form oxaloacetate and releases an electron (reducing NAD$^+$ to NADH). Finally, the process repeats itself with another acetyl group being processed with oxaloacetate (Figure 35).

All the electrons obtained from glycolysis, the transitional reaction and the Krebs cycle are transported to the electron transport chain by NADH and FADH$_2$, electron carriers. The electron transport chain, also known as chemiosmotic phosphorylation, is where the bulk of the ATPs are produced in aerobic respiration.

The electron transport chain is initiated when electrons are passed from NADH to NADH-ubiquinoneubiquinone oxidoreductase, an integral protein electron carrier, and FADH$_2$ pass their electrons to succinate dehydrogenase, a peripheral protein electron carrier. Once the two initial membrane bounded electron carriers are reduced, they pass their electrons to the next carrier protein called ubiquinone (CoQ). Immediately, ubiquinone transfers the electrons down to cytochrome c oxidioreductase, which then pass the electrons down to cytochrome c oxidase. Cytochrome c oxidase then transfers the electrons to reduce molecules of O$_2$ into water within the mitochondrial matrix (Figure 36).

The purpose of transferring electrons from one membrane bound carrier to another is to use the high energy given off by the electrons to power a series of proton pumps which send protons (H$^+$) out of the matrix space and into the intermembrane space of the mitochondria. By sending the H$^+$ ions into the intermembrane space, a diffusion gradient is created (e.g. concentration of H$^+$ is high in the intermembrane space while H$^+$ is kept at low concentration in the matrix of mitochondria). Energy is released when H$^+$ ions diffuse down their concentration gradient back into the mitochondrial matrix through the membrane bounded enzyme ATP synthase. The energy released by the H$^+$ ions flowing down their concentration gradient powers ATP synthase to phosphorylate ADP into ATP. An ATP Synthase can phosphorylate up to 100 molecules of ADP per second (Figure 36). .

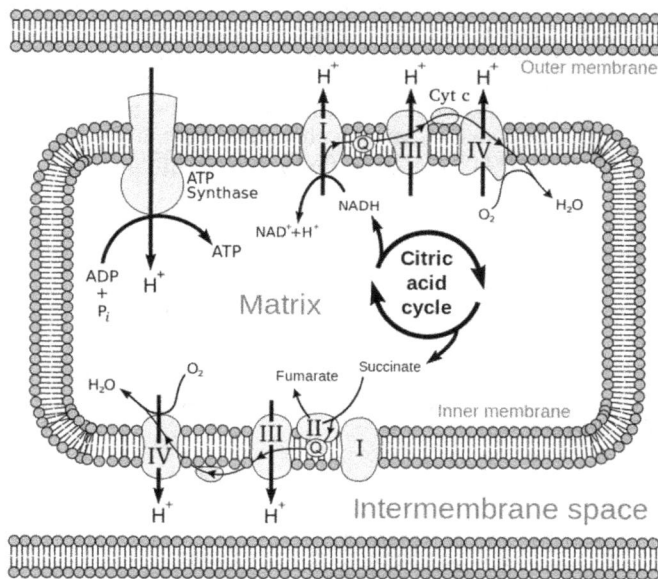

Figure 36: The electron transport chain found within the chistae of the mitochondria. This chain is initiated when electrons are passed from the NADH to NADH-ubiquinoneubiquinone oxidoreductase (I), and FADH$_2$ pass their electrons to succinate dehydrogenase (II). Once the two initial membrane bounded electron carriers are reduced, they pass their electrons to the next carrier protein called ubiquinone (CoQ). CoQ transfers their electrons down to cytochrome c oxidioreductase (III), which then pass the electrons down to cytochrome c (Cyt c). Cytochrome c then transfers the electrons to Cytochrome c oxidase (IV) where they reduce molecules of O$_2$ into water within the mitochondria matrix. Please note that the H$^+$ gradient generated is used to power ATP synthase to phosphorylate ADP into ATP

ENERGY MOLECULES

ATP is the energy molecule of choice for all three types of musculature. Nonetheless after a prolonged period of muscle stimulation, ATP molecules will be depleted and efforts have to be made to replenish the energy molecule supplies.

There are numerous kinds of energy molecules that exists within the muscle cell. For example, nucleoside triphosphates (NTPs) are energy molecules that provide energy and phosphate group for phosphorylations. An NTP includes a ribose pentose sugar and a heterocyclic base, which constitutes a nucleoside, as well as, three phosphates. NTPs found in the human body are adenosine triphosphate (ATP), guanosine triphosphate (GTP), cytidine triphosphate (CTP), 5-methyluridine triphosphate (m^5UTP), and uridine triphosphate (UTP). A second group of energy carrying molecules consists of energy molecules with deoxyribose as their pentose sugar. These deoxyribose nucleoside triphosphates (dNTP) with the exception of deoxyuridine triphosphate (dUTP) are used as building blocks for deoxyribose nucleic acid (DNA). dNTPs found in the human body are deoxyadenosine triphosphate (dATP), deoxy-guanosine triphosphate (dGTP), deoxycytidine triphosphate (dCTP), deoxythymidine triphosphate (dTTP) and deoxyuridine triphosphate (dUTP).

During emergencies the body may use NTPs and dNTPs to reform ATPs in order to replenish their supplies. This replenishment process is carried out by the enzyme nucleoside diphosphokinase.

Another way in which the supply of ATP is regenerated is the use of creatine phosphate (CP), also known as phosphocreatine. CPs are obtained from the ingestion of animal proteins and/or produced by the liver. Once ingested or produced, CP is then transferred via the circulatory system to the muscle fibers for storage. Just like NTPs, CPs are used only in emergencies. By using the enzyme creatine kinase, the energy stored in CPs can be transferred to reform ATP molecules.

Due to extensive usage of skeletal muscles, large amounts of ATPs are consumed, which leads to equal amounts of ADPs being generated. The final source of emergency energy transfer in the human body is found in the spent energy molecules of ADPs. This process is accomplished by the enzyme adenylate kinase, where a phosphate located on the ADP could be transferred to reform an ATP. The removal of a phosphate from ADP will create an adenosine monophosphate (AMP) molecule (Figure 37).

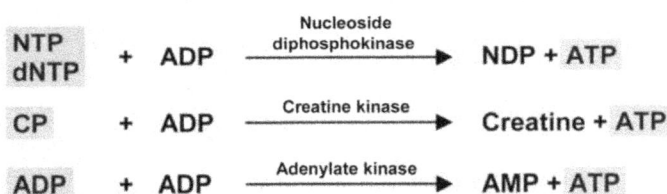

Figure 37: Illustration of various manners by which ATP is replenished within the human body

TETANY

Tetany is a condition of involuntary hypercontracted muscle that trails a prolonged period of repetitive and summed muscle stimulation. Tetany may be caused by the depletion of ATP, hypocalcaemia (low calcium levels) or bacterial infections. For example, one form of tetanic stimulation initially raises sarcoplasmic calcium and depletes the available ATPs, as well as, other energy molecules that aid in maintaining normal ATP levels. The outcome of this type of tetany is a hypercontracted muscle with Ca^{++} cations continuously being bounded to TnCs which will cause troponins to maintain its altered conformational state. With no ATP available to sequester the Ca^{++} cations back into the terminal cisternae of the sarcoplasmic reticulum, this altered conformational state of troponins will maintain tropomyosins in the "on" position, whereby continuing to expose the active sites of actin myofilament. The continuous exposure of

the active site will result in the inability of the myofibril to break actin-myosin cross bridges and causes the skeletal muscle to remain in the rigor state.

In an effort to compensate for the extreme situation, the mitochondria will be recruited to remove the excess Ca^{++} from the sarcoplasm by pumping Ca^{++} from the sarcoplasm and into its matrix. This action by the mitochondria will only temporarily remove the excess and in effect remove the Ca^{++} cations bounded to the TnC. This temporary removal of Ca^{++} cations allows the troponin to return to its original conformation and, in turn, allows the tropomyosin to move into the "off" position. Once the cross bridge is broken, the skeletal muscle will assume a flaccid state.

The situation repeats itself when additional ATPs and other energy molecules are created but fall far short of adequate amounts. Put simply, as soon as the newly produced ATPs and energy molecules are utilized the muscle falls back into the rigor state when Ca^{++} cations cannot be reabsorbed by the sarcoplasmic reticulum and subsequently back into flaccid state when the mitochondria removes Ca^{++} cations from the sarcoplasm.

Hypocalcaemia is a condition caused by low serum Ca^{++} levels in the blood. Hypocalcaemia is a biochemical abnormality that results in numerous life threatening complications, one of which is tetany. In primary care, the most common cause of hypocalcaemia is vitamin D deficiency. Vitamin D facilitates calcium absorption from the digestive tract and lack of vitamin D will result in inadequate Ca^{++} absorption, and in turn, low serum Ca^{++} levels. Other causes of hypocalcaemia are a lack of secreted parathyroid hormone (PTH). PTH is produced and released by the chief cells found in the parathyroid gland and is responsible for increasing serum Ca^{++} concentrations. Therefore, a lack of PTH will result in low serum Ca^{++} levels.

Excessive secretions of aldosterone by the adrenal cortex may also result in low serum Ca^{++} levels. Aldosterone is a mineralcorticoid that influences the secretion of Ca^{++} cations by the kidneys. Therefore, high levels of aldosterone will result in the increase in the secretion of Ca^{++} by the kidneys which will result in low serum Ca^{++} levels.

Alkalosis, an excessively alkaline (basic) state of bodily fluids resulting from persistent vomiting and rapid breathing, may also cause hypocalcaemia. The increase in the pH is due to the loss or the decrease in the production of carbonic acids. It is believed that the increase in alkalinity causes plasma proteins, such as albumin, to bind to Ca^{++}, reducing the amount of free Ca^{++} in the blood stream.

Low calcium levels in the bloodstream increase the permeability of the neurolemma of motor nerves, causing a progressive depolarization, which increases the possibility of action potentials. If the plasma Ca^{++} levels decrease to less than 50% of the normal value of 9.4 mg/dl, action potentials may be spontaneously generated, causing contraction of peripheral skeletal muscles. Continuous stimulation of the muscle limits the ability for the sarcoplasmic reticulum to requester Ca^{++}, allowing this cation to accumulate in the sarcoplasm. If intracellular Ca^{++} is not removed, the cross bridge will not be severed therefore the muscle remains in the rigor state or tetany.

Tetany can also be caused by *Clostridium tetani* (*C. tetani*), a spore forming, Gram positive anaerobic bacterium that causes an acute and often fatal disease called tetanus. Symptoms of tetanus are characterized by generalized rigidity (e.g. lockjaw) and convulsive spasms. The spores of *C. tetani* generally enter the body through a puncture or cut before germinating and multiplying in the relatively anaerobic environment of the dermis. Tetanus toxins will be produced by *C. tetani* and spread via in the circulatory and lymphatic systems. Tetanus toxins act upon both peripheral and central nervous system. For example, the tetanus toxin blocks the release of γ-aminobutyric acid (GABA). GABA is a neurotransmitter that inhibits the actions of motor neurons. Lack of GABA being released will result in the hyperactivity of motor neurons, which results in violent spastic paralysis.

ANAEROBIC RESPIRATION

Anaerobic respiration, also known as anaerobic glycolysis (or glycolysis as seen in Figure 33), involves the first stage of cellular respiration. Anaerobic respiration does not require oxygen and involves the breakdown of glucose to generate two ATP molecules and two molecules of lactic acid as metabolic waste. Lactic acid is a weak acid and has a tendency to lose its hydrogen cation to form lactate (sodium lactate). Anaerobic respiration is an inefficient reaction to gain ATP. However, it is a necessary alternative step when oxygen is scarce or unavailable. For example, during a heavy regimented exercise, when the demand for energy is high and the availability of oxygen is low (after the oxygen stored in the myoglobin is consumed), glucose is broken down and oxidized to pyruvate, which is then processed into lactic acid through anaerobic respiration. Additional energy can be supplied through the creatine phosphates stored in the muscle fibers but that source of energy is limited. Lactic acid can only be excreted from the skeletal muscle cells via facilitated diffusion with the aid of transport proteins found in the sarcolemma. When faced with an overproduction of lactic acid, this relatively slow removal process cannot keep up with demand and is overwhelmed. Rather quickly, the concentration of lactic acid will build up within the sarcoplasm. The combination of the demand of ATP, lack of oxygen, and the buildup of lactic acid and lactate creates a situation known as oxygen debt (Figure 38).

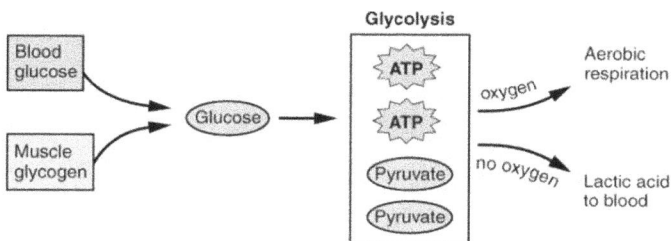

Figure 38: Illustration of various manners by which ATP is replenished within the human body

Once the heavy exercise is over and the consumption of ATP and oxygen catches up to post exercise levels, this would mark the time when the oxygen debt must be repaid. First, the ATP generated through aerobic respiration has to replenish the creatine phosphate used during the exercise. Second, the buildup of lactic acid must be allowed to leave the skeletal muscle cell and enter circulation. The lactic acid reaches the liver and is processed by gluconeogenesis, where the lactic acid is converted back into glucose. Once glucose is formed, it is either processed to form glycogen and stored within the hepatocytes, or released back into circulation and used in aerobic respiration by other cells within the body. Third, the myoglobin that was used during exercise must be recharged with oxygen so that it can be reused.

TYPES OF SKELETAL MUSCLE FIBERS

There are three types of skeletal muscle fibers based upon their resistance to fatigue. The slow-twitch oxidative fibers (Type I fibers) are most resistant to fatigue and are generally found in muscle groups used in maintaining form and posture, walking or lifting of heavy objects. The term "oxidative" indicates that these fibers require oxygen so that they can proceed through aerobic cellular respiration and obtain the highest amounts of ATP possible per molecule of glucose. In comparison to the other types of skeletal muscle fibers, slow-twitch oxidative fibers possess more mitochondria, are richly supplied by blood vessels and capillaries, but are slow to generate an action potential or

to contract. These fibers have the smallest diameter among the three types, allowing faster diffusion of oxygen, since there is not a lot of distance to cover. These fibers are described as red fibers because they contain large amounts of myoglobin.

Like hemoglobin, myoglobin is a heme based metalloprotein that is designed to store extra oxygen, which allows muscles to function at elevated levels of activity for a prolonged period of time. Unlike hemoglobin, which consists of four protein chains (α1, α2, β1, and β2 protein subunits) and four heme groups, myoglobin is a 153 amino acid protein with a single heme group at the core of the molecule. Nonetheless, myoglobin possesses a stronger affinity for oxygen than hemoglobin, which allows the transfer of oxygen from the circulating hemoglobin to myoglobin for storage (Figure 39).

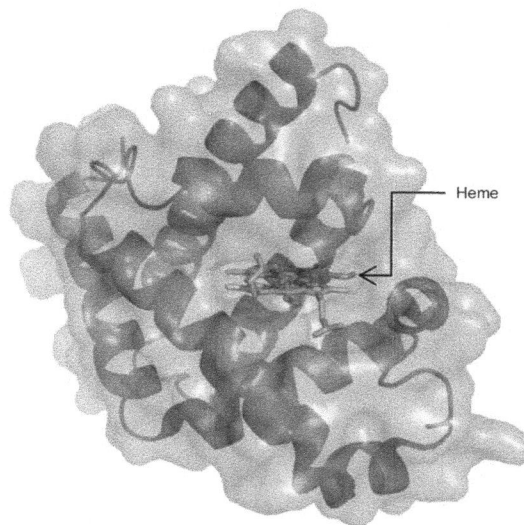

Figure 39: Molecular structure of myoglobin. Please note that a single heme group exists at the core of the molecule

Fast twitch oxidative fibers (Type IIA) have slightly larger cells when compared with slow-twitched oxidative fibers. Like the slow-twitched fibers, these fibers require oxygen for aerobic cellular respiration and to obtain the highest amounts of ATP possible per molecule of glucose. These fibers have good amounts of mitochondria, some myoglobin, well supplied with blood vessels and capillaries, and are faster in generating action potentials and contractions in comparison with the slow-twitch oxidative fibers.

Fast-twitch glycolytic fibers (Type IIX) possess the largest diameter of all three types of muscle fibers and have the ability to generate action potentials and contract two to three times faster than the slow-twitch oxidative fibers. Nonetheless, these fibers are the fastest to fatigue because they produce the least amount of ATP. Fast-twitch glycolytic fibers are explosive fibers that are commonly used in sprinting or leaping activities. These fibers rely solely on ATPs that are formed from glycolysis, which produces two ATPs in total. Since they depend on glycolysis, these fibers are accustomed to anaerobic respiration and the resulting lactic acid build up (i.e. hence muscle fatigue and oxygen debt). In comparison, fast-twitch glycolytic fibers have the least amount of mitochondria in the sarcoplasm, are supplied by fewer blood vessels and capillaries, but have high concentration of glycogen, which is used as their emergency fuel source. In addition, these fibers have very low myoglobin content, therefore they appear white under microscopic examination.

MYOGLOBIN TEST

Both skeletal muscle fibers and cardiac muscle fibers possess myoglobin as a reservoir for oxygen. This is also the reason that myoglobin

is released into the blood stream when these muscles are damaged. The serum myoglobin test measures the amount of myoglobin in the blood. It is known that myoglobin appears in the blood more rapidly than troponin and can aid in the diagnoses of a heart attack in the earliest stages. In addition to the diagnosis of a heart attack, this test could also aid in the diagnosis of rhabdomyolysis. Rhabdomyolysis may result from trauma, use of recreational drugs such as cocaine, heroin, or PCP. Genetic muscular diseases, such as muscular dystrophy, seizures, and severe dehydration, could also cause rhabdomyolysis.

For example, the content of myoglobin in the blood stream will begin to increase two to three hours after a heart attack and peaks approximately 12 hours subsequent to the cardiac muscle damage. The myoglobin in the blood will be filtered and excreted by the kidneys. Nonetheless, high amount of myoglobin may damage the kidneys and result in kidney failure.

SMOOTH MUSCLE

Smooth muscles are involuntary muscles that are under the influence of the endocrine and autonomic nervous systems. Smooth muscles are readily found in internal organs such as the genital, gastrointestinal, and urinary tracts, as well as, the muscular layers (e.g. tunica media) of the blood vessels.

There are two types of smooth muscles: visceral and multiunit. The visceral smooth muscles, as the name states, are found in most internal organs of the body. Visceral smooth muscles are generally organized in a sheet-like fashion and are interconnected via numerous gap junctions. It is understood that these gap junctions allow action potential to transmit from cell to cell via electrical synapses (see Chapter Three for more information), which in turn, allows these smooth muscle fibers to function as a single unit or unitary smooth muscle. Unlike the visceral smooth muscles, the multiunit smooth muscles function independently. Although these smooth muscle fibers also exist in sheets (i.e. blood vessels), small bundles (i.e. arrector pili) or individual cells found in the capsule of the spleen, it has been reported that the multiunit smooth muscle fibers possess very few gap junctions; therefore, making the electrical synapses, which are necessary for them to function as a single unit, an impossibility.

Unlike the striated muscles (cardiac and skeletal muscles) which measured approximately 10 to 100 μm, the smooth muscle fibers are much smaller, generally measured about 2 to 10 μm. Most noticeably, the smooth muscles have no visible striations. Nonetheless, smooth muscles are capable of isometric contractions with similar forces generated to those seen in striated muscles. However, the rates of contraction are much slower by comparison.

There are no myofibrils in the smooth muscle fiber. The actin and p-light chain (isoform of myosin that does not possess a trypsin hinge) myofilaments in a smooth muscle fiber are loosely organized and are scattered throughout the sarcoplasm. It is understood that the actin myofilaments of smooth muscle cells are connected to dense bodies and dense areas of the sarcolemma. The dense bodies and areas are primarily composed of a-actinin (a Z-line/disk protein) and non-contractile intermediate filaments (e.g. desmin and vimentin), which form a cytoskeleton anchoring structure. The dense bodies and areas are considered to be the smooth muscle equivalent to the Z-line/disk found in striated muscles. Put simply, when the actin and p-light chain slide past one another (i.e. sliding filament theory), this interaction will cause the length of the myofilament to shorten and in turn, cause the dense bodies and areas to be pulled closer together resulting in contraction.

Smooth muscle fibers do not possess transverse tubules (t-tubules) while the concentration of sarcoplasmic reticulum is drastically less by comparison to striated muscles. In other words, the smooth muscle cells have fewer Ca^{++} cations stored. To make up for the low amounts of Ca^{++} cations, the sarcolemma of the smooth muscle cells have numerous gated Ca^{++} channels and calcium pumps. It is known that the Ca^{++} pumps actively transport Ca^{++} cations from the sarcoplasm to maintain a high concentration gradient of Ca^{++} cations in the extracellular matrix. The gated Ca^{++} channels, on the other hand, will open and close based on the attachment of ligands, voltage or mechanical induction. For example, when the gated Ca^{++} channels are triggered to open, the Ca^{++} cations will flow passively from areas of high concentration (in the extracellular matrix) through the open pore and into the sarcoplasm.

Unlike the striated muscles, smooth muscles do not possess the regulatory protein troponin. Instead, smooth muscle fibers possess a regulatory protein called caldesmon. In addition, smooth muscle fibers possess an isoform of myosin called p-light chain. For example, when the smooth muscle is trigger to contract, either through the binding of hormones (endocrine system) or transmissions by the autonomic nerves, Ca^{++} cations will be released either by the sarcoplasmic reticulum or via diffusion through the various Ca^{++} gated channels found in its sarcolemma. Once the level of Ca^{++} cations in the sarcoplasm is increased, these cations will bind with a sarcoplasmic protein, calmodulin to create calcium/calmodulin (CaCM) complex. CaCM is an activation protein complex which is responsible for initiating the contraction of smooth muscle cells. First, CaCM will bind with caldesmon (a troponin replacement) which in turn will cause it to move tropomyosin, exposing the active site of actin. Secondly, CaCM will bind with myosin light chain kinase (MLCK) and activate the enzyme. Once activated, MLCK will hydrolyze ATP into ADP and begin transferring the phosphate group (phosphorylation) to the p-light chain. As indicated previously, the p-light chain does not possess a trypsin hinge; therefore, it is not angled away from the midline of the light meromyosin. Through the process of phosphorylation by MLCK, the addition of a phosphate group will cause the heavy meromyosin to angle, allowing its actin binding site to form a cross bridge with the active site of actin. Once the cross bridge is formed, the energy stored from the hydrolysis of ATP by the ATPase found on the heavy meromyosin causes it to bend thus resulting in smooth muscle contraction. This bending action of the heavy meromyosin is also referred to as a power stroke. Immediately after the power stroke, a new ATP binds to the heavy meromyosin causing it to detach from the active site. ATPase of the heavy meromyosin hydrolyzes the ATP causing it to return to its original position while the energy from the hydrolysis of ATP is stored. If the Ca^{++} cations are still attached to the activation protein complex calmodulin, the process will continue to repeat itself.

The contractile process ends when the hormone is removed from its binding site or the cell is signaled to do so by autonomic nerves. Once signaled, an enzyme called myosin phosphatase will remove the phosphate from the p-light chain, returning it to its original conformation, breaking the cross bridge. In addition, Ca^{++} cations will immediately be actively transported out of the cell via Ca^{++} pumps, reducing the sarcoplasmic Ca^{++} levels. By reducing the Ca^{++} cation levels in the sarcoplasm, the Ca^{++} will be removed from the CaCM, deactivating this complex. The deactivation of CaCM will cause the caldesmon to return to its original conformation which in turn will move the tropomyosin back to its original shape, blocking the active site of actin.

QUESTIONS

1. Which of the following is (are) striated muscle(s)?
    a.   Smooth muscle
    b.   Skeletal muscle
    c.   Cardiac muscle
    d.   Facia adherens
    e.   Arrector pili
    f.   Both a and c
    g.   Both b and c

2. Which of the following type of muscle(s) is (are) responsible for bodily

movements such as walking, running, and respiration?
a. Smooth muscle
b. Skeletal muscle
c. Cardiac muscle
d. Facia adherens
e. Arrector pili
f. Both a and c
g. Both b and c

3. Which of the following muscle(s) is (are) responsible for manipulating of sounds (i.e. speech etc.), producing body heat (i.e. shivering), as well as enabling the ability to grasp and manipulate objects?
a. Smooth muscle
b. Skeletal muscle
c. Cardiac muscle
d. Facia adherens
e. Arrector pili
f. Both a and c
g. Both b and c

4. Which of the following muscle(s) is (are) responsible for the movement (e.g. pumping) of blood throughout the system and are essential in providing the vehicle for gas and nutrient exchange?
a. Smooth muscle
b. Skeletal muscle
c. Cardiac muscle
d. Facia adherens
e. Arrector pili
f. Both a and c
g. Both b and c

5. Which of the following is defined as self-synchronizing in an effort to coordinate a cardiac contraction?
a. Autonomic
b. Autorhythmic
c. Myogenic
d. Parasympathetic
e. Peristalsis
f. Slow wave
g. Sympathetic

6. Which of the following is defined as a type of muscle that can initiate its own contraction without the input of nerves?
a. Autonomic
b. Autorhythmic
c. Myogenic
d. Parasympathetic
e. Peristalsis
f. Slow wave
g. Sympathetic
h. Ectopic

7. Which of the following muscle(s) is (are) responsible for propelling and mixing foods in the digestive tract, controlling exocrine and endocrine secretions and regulating the amount of blood flows through the blood vessels?
a. Smooth muscle
b. Skeletal muscle
c. Cardiac muscle
d. Facia adherens
e. Arrector pili
f. Both a and c
g. Both b and c

8. Smooth muscles are directly controlled by which of the following system or systems?
a. Circulatory system
b. Digestive system
c. Endocrine system
d. Integument system
e. Autonomic nervous system
f. Both c and e
g. Both b and d

9. Which of the following is (are) protein (s) commonly found in all type (s) of muscles within the human body?
a. Actin
b. Albumin
c. Globulin
d. Glycosaminoglycan
e. Myosin
f. Pectin
g. Proteoglycan
h. Both a and e

10. What is the name given to the plasma membrane of a muscle fiber?
a. Sarcolemma
b. Sarcoplasm
c. Sarcoplasmic reticulum
d. Neuroplasm
e. Neurolemma
f. Chromtophilic substance

11. What is the name given to the cytoplasm of a muscle fiber?

a. Sarcolemma
b. Sarcoplasm
c. Sarcoplasmic reticulum
d. Neuroplasm
e. Neurolemma
f. Chromatophilic substance

12. The plasma membrane of the striated muscle cells invaginates at regular intevals to form _____.
a. Diad
b. Sarcoplasmic reticulum
c. Terminal cisternae
d. Transverse tubules
e. Triad
f. Tropomyosin
g. Troponin

13. Which of the following is the name given to the modified form of a sER?
a. Diad
b. Sarcoplasmic reticulum
c. Terminal cisternae
d. Transverse tubules
e. Triad
f. Tropomyosin
g. Troponin

14. The modified form of the sER found in muscle cells are formed to perform which of the following function?
a. Collect and concentrate calcium ions
b. Generate ATP
c. Perform phagocytosis
d. Produce steroids
e. Protein synthesis
f. Store calcium ions
g. Store sodium ions

15. The enlarged ends of the modified form of the sER found in muscle cells are called _____.
a. Diad
b. Sarcoplasmic reticulum
c. Terminal cisternae
d. Transverse tubules
e. Triad
f. Tropomyosin
g. Troponin

16. These enlarged ends of the modified form of the sER found in muscle cells are constructed to perform which of the following function?
a. Collect and concentrate calcium ions
b. Generate ATP
c. Perform phagocytosis
d. Produce steroids
e. Protein synthesis
f. Store calcium ions
g. Store sodium ions

17. The enlarged ends of the modified sER forms close associations with the regular invaginations of its cellular membrane. What is the name given to this association in cardiac muscles?
a. Diad
b. Sarcoplasmic reticulum
c. Terminal cisternae
d. Transverse tubules
e. Triad
f. Tropomyosin
g. Troponin

18. The enlarged ends of the modified sER forms close associations with the regular invaginations of its cellular membrane. What is the name given to this association in skeletal muscles?
a. Diad
b. Sarcoplasmic reticulum
c. Terminal cisternae
d. Transverse tubules
e. Triad
f. Tropomyosin
g. Troponin

19. Which of the following protein is NOT found in a myofibril?
a. Actin
b. Albumin
c. Myosin
d. Nebulin
e. Titin
f. Tropomyosin
g. Troponin

20. Actin myofilaments, is also known as _____.
a. A bands
b. F-actin
c. G-actin
d. M-line
e. Thick filaments

f. Thin filaments
g. Z-line/disk

21. Myosin myofilaments, is also known as _____.
a. A bands
b. F-actin
c. G-actin
d. M-line
e. Thick myofilaments
f. Thin myofilaments
g. Z-line/disk

22. Actin myofilaments are created from which of the following protein polymer?
a. A bands
b. F-actin
c. G-actin
d. M-line
e. Thick myofilaments
f. Thin myofilaments
g. Z-line/disk

23. An actin monomer is also called _____.
a. A bands
b. F-actin
c. G-actin
d. M-line
e. Thick myofilaments
f. Thin myofilaments
g. Z-line/disk

24. Each G-actin subunit has an active site called _____ that allows myosin binding during muscle contraction.
a. Myosin binding site
b. Actin binding site
c. ATPase
d. GTPase
e. Thick filaments
f. Thin filaments
g. Z-line/disk

25. Which of the following is a motor protein?
a. Actin
b. Albumin
c. Myosin
d. Nebulin
e. P-light chain
f. Titin
g. Tropomyosin
h. Troponin

26. Myosin myofilaments of smooth muscles are also called _____.
a. Actin
b. Albumin
c. Myosin
d. Nebulin
e. P-light chain
f. Titin
g. Tropomyosin
h. Troponin

27. How many myosin molecules does it take to form a myosin myofilament?
a. 100
b. 200
c. 300
d. 400
e. 500
f. 600
g. 700

28. Various myosin molecules are joined together through the interactions of which of the following molecule(s)?
a. Actin
b. C protein
c. M-line protein
d. Myosin
e. Nebulin
f. Titin
g. Both a and d
h. Both b and c

29. Each of the myosin molecules resembles the shape of a golf club. The head of the myosin molecule is called _____ while the shaft of the molecule is called _____.
a. Light meromyosin
b. Heavy meromyosin
c. Trypsin hinge
d. Troponin
e. ATPase
f. F-actin
g. G-actin

30. The two sections of the myosin molecule are separated at the location known

as the _____ which is formed by a proteolytic enzyme called trypsin. This location allows the head of the myosin molecule to angle sharply (~45°) downward from the main axis.
a. Light meromyosin
b. Heavy meromyosin
c. Trypsin hinge
d. Troponin
e. ATPase
f. F-actin
g. G-actin

31. The head of the myosin molecule contains two binding sites. The first is called _____ which is a deep cleft that is formed to connect with the _____ of the G-actin. The second binding site is an enzyme called _____. This enzyme is used to _____ (breakdown) the energy molecule _____ into _____ and releasing a phosphate functional group (P_i) and the stored energy.
a. Actin binding site
b. Active site (myosin binding site)
c. Adenylpyrophosphatase
d. ADP
e. AMP
f. ATP
g. cAMP
h. Hydrolyze

32. Which of the following is a regulatory protein that is found coursing alongside of actin molecules? This protein is a long, rod-like helically-intertwined molecule. This regulatory protein is strategically located so that it blocks part of the binding site between actin and myosin.
a. Diad
b. Sarcoplasmic reticulum
c. Terminal cisternae
d. Transverse tubules
e. Triad
f. Tropomyosin
g. Troponin

33. Which of the following is a regulatory protein located between the thin and thick myofilaments and is attached to tropomyosin. This protein serves to move tropomyosin away from the actin/myosin binding sites.
a. Diad
b. Sarcoplasmic reticulum
c. Terminal cisternae
d. Transverse tubules
e. Triad
f. Tropomyosin
g. Troponin

34. Troponin is composed of three protein subunits. Which of the following is responsible for binding troponin to tropomyosin?
a. Actin binding site
b. Diad
c. Myosin binding site
d. TnC
e. TnI
f. TnT
g. Triad

35. Troponin is composed of three protein subunits. Which of the following is responsible for inhibiting the interaction of myosin with actin?
a. Actin binding site
b. Diad
c. Myosin binding site
d. TnC
e. TnI
f. TnT
g. Triad

36. Troponin is composed of three protein subunits. Which of the following is a calcium ion binding protein?
a. Actin binding site
b. Diad
c. Myosin binding site
d. TnC
e. TnI
f. TnT
g. Triad

37. Which of the following is one of the largest protein known consisting of approximately 27 thousand amino acids? It is an elastic protein that extends along the myosin myofilament and maintains the spatial structure of the myofibril as well as stabilizes the position of the contractile filaments. Like a spring, this elastic protein is capable of being compressed as the muscle contracts and is responsible for snapping the contracting myofibrils back to its original shape once the contraction is terminated.
a. Actin binding site
b. Diad
c. Myosin binding site
d. Nebulin
e. TnC
f. TnI
g. TnT

h. Triad
i. Tropomyosin
j. Troponin
k. Titin

38. Which of the following is a non-elastic structural protein that courses along the side of the actin myofilaments? Like scaffolding, this protein is constructed to align and regulate the spatial orientation of the thin myofilament.
a. Actin binding site
b. Diad
c. Myosin binding site
d. Nebulin
e. TnC
f. TnI
g. TnT
h. Triad
i. Tropomyosin
j. Troponin
k. Titin

39. Which of the following is a protein structure of the myofibril where I bands connects?
a. A bands
b. Deep fascia
c. H-zone
d. M-line
e. Sarcomere
f. T-zone
g. X bands
h. Z-line/disk

40. Which of the following is the name given to a single contractile unit of myofibril that extends from one z-line to another?
a. I bands
b. A bands
c. Sarcomere
d. myoglobin
e. H-zone
f. M-line
g. Sarcolemma
h. Sarcoplasm

41. Which of the following is a part of the myofibril that consists of a thickening of the myosin filaments and contains a protein that is used to tie / cap the myosin molecules together?
a. A bands
b. H-zone
c. I bands
d. M-line
e. Sarcomere
f. T-zone
g. Z-line/disk
h. None of the above

42. Which of the following is (are) segment (s) of the myofibril that consists of myosin filaments only?
a. M-line
b. Z-line/disk
c. H-zone
d. Z-bands
e. I bands
f. A bands
g. Both a and d
h. Both a and c

43. Which of the following is a part of the myofibril and is also known as the light band which consist of only actin filaments
a. A bands
b. H-zone
c. I bands
d. M-line
e. Sarcolemma
f. Sarcomere
g. Z-line/disk
h. None of the above

44. Which of the following is a part of the myofibril that is also known as dark bands?
a. A bands
b. H-zone
c. I bands
d. M-line
e. Sarcolemma
f. Sarcomere
g. Z-line/disk
h. None of the above

45. Which of the following is constructed out of a layer of dense irregular connective tissue which serves to physically protect the skeletal muscle fibers, as well as to compartmentalize individual muscle fibers and form groups of muscle fibers?
a. A bands
b. Deep fascia
c. H-zone

d. M-line
e. Sarcomere
f. T-zone
g. X bands
h. Z-line/disk

46. _____ is a part of the myofibril that consists of a thickening of the myosin filaments. This thickening of the myosin filament is consists of another protein structure that is used to tie/cap approximately 400 myosin molecules together to form a myosin filament

_____ is also known as the calcium ion storage site in muscle cells

_____ is the general term given to the connection between somatic motor neuron and muscle fibers? This area of connection is also referred to as synapses

_____ is the name given to a segment of the myofibril that stretches from one z-line to another z-line

_____ and _____ is the name given to portion of the contractile unit that possesses only myosin

_____ is the structural protein that is attached to the z-disk. This structural protein is designed to run along the thin filament and help keep it aligned

_____ is the term used to describe the in-folding (invagination) of the muscle cellular membrane and is designed to spread action potential evenly throughout the muscle cell

_____ is the term used to describe the in-folding of the muscle cellular membrane that is designed to accommodate the terminal end of the motor neuron axon

a. A bands
b. B (beta) – actinin
c. C-protein
d. Flavin adenine dinucleotide
e. H-zone
f. I bands
g. M-line
h. Motor end plate
i. Myoglobin
j. Myoglobin
k. Nebulin
l. Neuromuscular junction
m. Nicotinamide adenine dinucleotide
n. Prime mover
o. Sarcolemma
p. Sarcomere
q. Sarcoplasm
r. Sarcoplasmic reticulum
s. Synaptic cleft
t. Synergist
u. Terminal cisternae
v. Titin
w. Transverse tubules
x. Tropomyosin
y. Troponin-C
z. Trypsin hinge

47. This layer of dense irregular connective tissue which serves to physically protect the skeletal muscle fibers will taper at the ends and blend with other connective tissues to form _____ or create a flat sheet-like connective tissue structure called _____.
a. Adipose tissue
b. Aponeurosis
c. Elastic cartilage
d. Fibrocartilage
e. Hyaline cartilage
f. Ligaments
g. Sharpey fibers
h. Tendon

48. Tendons are formed to connect muscle to bone via cartilaginous fibers. These cartilaginous fibers are also known as _____ which extends into the matrix of the bone to provide the necessary anchor for muscle attachment.
a. Adipose tissue
b. Aponeurosis
c. Elastic cartilage
d. Fibrocartilage
e. Hyaline cartilage
f. Ligaments
g. Sharpey fibers
h. Tendon

49. Which of the following muscle of the eye depresses the eyes when moving from side to side?
a. Inferior oblique
b. Inferior rectus
c. Lateral rectus
d. Lateral oblique
e. Medial rectus
f. Superior oblique
g. Superior rectus

50. Which of the following muscle of the eye elevates the eyes when moving from side to side?
a. Inferior oblique
b. Inferior rectus
c. Lateral rectus
d. Lateral oblique
e. Medial rectus
f. Superior oblique
g. Superior rectus

51. Which of the following muscle of the eye moves the eye upward and secondarily rotates the top of the eye toward the nose?
    a. Inferior oblique
    b. Inferior rectus
    c. Lateral rectus
    d. Lateral oblique
    e. Medial rectus
    f. Superior oblique
    g. Superior rectus

52. Which of the following connective tissue structure is richly innervated with sensory receptors?
    a. A bands
    b. Deep fascia
    c. H-zone
    d. M-line
    e. Sarcomere
    f. T-zone
    g. X bands
    h. Z-line/disk

53. Which of the following muscle of the eye moves the eye downward and secondarily rotates the top of the eye away from the nose?
    a. Inferior oblique
    b. Inferior rectus
    c. Lateral rectus
    d. Lateral oblique
    e. Medial rectus
    f. Superior oblique
    g. Superior rectus

54. Which of the following muscle of the eye moves the eyes away from the nose?
    a. Inferior oblique
    b. Inferior rectus
    c. Lateral rectus
    d. Medial rectus
    e. Superior oblique
    f. Superior rectus

55. Which of the following is the name given to sensory receptors that provide information regarding pain?
    a. Chemoreceptors
    b. G-protein complex
    c. Mechanoreceptors
    d. Nociceptors
    e. Proprioceptors
    f. Receptor proteins
    g. Thermo-receptors

56. Which of the following is the name given to sensory receptors that provide information regarding pressure?
    a. Chemoreceptors
    b. G-protein complex
    c. Mechanoreceptors
    d. Nociceptors
    e. Proprioceptors
    f. Receptor proteins
    g. Thermo-receptors

57. Which of the following is the name given to sensory receptors that provide information in regards to the position of the body in relations to balance?
    a. Chemoreceptors
    b. G-protein complex
    c. Mechanoreceptors
    d. Nociceptors
    e. Proprioceptors
    f. Thermo-receptors

58. Which of the following is the name given to sensory receptors that provide information regarding to internal and external temperature?
    a. Chemoreceptors
    b. G-protein complex
    c. Mechanoreceptors
    d. Nociceptors
    e. Proprioceptors
    f. Receptor proteins
    g. Thermo-receptors

59. Which of the following is the name given to sensory receptors that provide information regarding to pH?
    a. Chemoreceptors
    b. G-protein complex
    c. Mechanoreceptors
    d. Nociceptors
    e. Proprioceptors
    f. Receptor proteins
    g. Thermo-receptors

60. Which of the following is a flat/broad fibrous sheet of connective tissue that forms connection with adjacent skeletal musculature?
    a. Aponeuroses
    b. Dense irregular connective tissue
    c. Dense regular connective tissue
    d. Endomysium

e. Epimysium
f. Fascia
g. Perimysium
h. Both c and g

61. The outer most layer of the deep fascia is called _____.
    a. Aponeurosis
    b. Endomysium
    c. Epimysium
    d. Muscle fasciculi
    e. Perimysium
    f. Sarcolemma
    g. Sarcoplasm
    h. Tendon

62. Internally, the outer most layers of the deep fascia branches inward to form _____ which divides the muscle into bundles of muscle fibers called _____. This inner branching of the deep fascia will further separate into _____ which forms a connective tissue layer surrounding each individual muscle fiber.
    a. Endomysium
    b. Epimysium
    c. Muscle fasciculi
    d. Perimysium
    e. Sarcolemma
    f. Sarcoplasm
    g. Tendon

63. In order to possess fine motor control over skeletal muscles, each skeletal muscle fiber is directly controlled by the _____ (which nervous system). For example, the _____ of the cerebral cortex determines the particular movement that the body needs to make and sends its signal via the _____ to the _____ (**Answer A**) located in the midbrain. At **Answer A**, these axons form the _____ that contain the corticospinal tracts on the ventral surface of the brainstem. At the level of the _____, the corticospinal tract splits into two tracts. Approximately 90% of the axons cross over to the contralateral side at the _____, forming the _____ (**Answer B**). **Answer B** continues down the spinal cord until it reaches the target segments in the spinal cord before connecting with the cell bodies of somatic motor neurons which is located in the _____ of the spinal cord, which in turn projects their axons via the ventral root directly to the skeletal muscles. The remaining 10% of the axons that do not cross at the caudal medulla constitute the _____. These nerves will continue down the spinal cord until they reach the target segment of the spinal cord before crossing over to the contralateral side and innervate motor neurons or interneurons in the _____.
    a. Anterior corticospinal tract
    b. Anterior horn
    c. Autonomic nervous system
    d. Caudal medulla
    e. Central nervous system
    f. Cerebral peduncle
    g. Corticospinal pathway
    h. Lateral corticospinal tract
    i. Lateral horn
    j. Medullary pyramids
    k. Peripheral nervous system
    l. Posterior horn
    m. Primary motor cortex
    n. Pyramidal decussation

64. In order to provide an unified control over movement, each skeletal muscle fiber could be innervated by _____ or _____.
    a. Collaterals form a somatic motor neuron
    b. Dendritic spines from a somatic motor neuron
    c. Parasympathetic nerve fibers form the autonomic nervous system
    d. Single motor axon from a somatic motor neuron
    e. Single sensory axon from a somatic motor neuron
    f. Sympathetic nerve fibers form the autonomic nervous system
    g. The endocrine system

65. Which of the following is defined as the axonal branching?
    a. Axonal terminals
    b. Cerebral peduncle
    c. Collaterals
    d. Dendritic spines
    e. Interneurons
    f. Intranuerons
    g. Pyramidal decussation

66. All muscles innervated by the same axon are called _____.
    a. Aponeuroses
    b. Endomysium
    c. Epimysium
    d. Fascia
    e. Motor neuron
    f. Motor unit
    g. Perimysium

67. Which of the following is a part of the nerve fiber and is designed to help increase surface area of nerve impulse transmission?
    a. Axolemma
    b. Axon
    c. Axonal hillock
    d. Axoplasm
    e. Collaterals
    f. Dendrites
    g. Dendritic spines
    h. Initial segment

68. Which of the following is the general term given to the synapses (connection) between somatic motor neuron and muscle fibers?

a. Beta- actinin
b. H-zone
c. M-line
d. Motor end plate
e. Neuromuscular junction
f. Synaptic cleft
g. Troponin-C
h. Trypsin hinge

69. At the location where the motor nerves connect with the sarcolemma of the skeletal muscle fiber, the sarcolemma of the skeletal muscle cell invaginates to accommodate the axonal terminal. This invagination is called the _____.
a. B (beta) - actinin
b. H-zone
c. M-line
d. Motor end plate
e. Neuromuscular junction
f. Synaptic cleft
g. Troponin-C
h. Trypsin hinge

70. At the location where the motor nerves connect with the sarcolemma of the skeletal muscle fiber the neurolemma (nerve membrane) and the sarcolemma do not come into direct physical contact. As a result a tiny fluid-filled space is formed in between the nervous and muscular membranes. This tiny fluid-filled space is called _____.
a. B (beta) - actinin
b. H-zone
c. M-line
d. Motor end plate
e. Neuromuscular junction
f. Synaptic cleft
g. Troponin-C
h. Trypsin hinge

71. Which of the following is defined as the manner by which an action potential causes contraction of a muscle fiber?
a. Depolarized
b. Excitation-contraction coupling
c. Motor end plate
d. Neuromuscular junction
e. Polarized
f. Synaptic cleft

72. At rest and unstimulated, the skeletal muscles are _____ which is defined as a difference in electrical charge due to the uneven distribution of three types of cations, the _____, _____ and _____. It is known that _____ and _____ are kept in high concentration in the extracellular matrix of skeletal muscle cell which creates a net _____ charge external to the sarcolemma. _____ cations are kept at a high concentration in the sarcoplasm. However, due to the negativity of the charged functional groups such as _____ and _____ as functional groups within the sarcoplasmic proteins, the interior of the muscle cell has a net _____ charge regardless of the high concentration of the positively charged cations. It has been calculated that the difference between the charges in the sarcoplasm and the extracellular matrix is _____ mV.

1.	-58 mV	23.	mAChR
2.	-103 mV	24.	Mechanically gated K⁺ ion channel
3.	-85 mV	25.	Medullary pyramids
4.	-30 mV	26.	Motor-end-plate
5.	Acetylcholine	27.	nAChR
6.	ADP	28.	Negative
7.	AMP	29.	Neuromuscular junction
8.	Anterior corticospinal tract	30.	Neuropeptide Y
9.	Anterior horn	31.	Nitric oxide
10.	ATP	32.	Noradrenaline
11.	Cerebral peduncle	33.	Phosphates (PO₄⁻)
12.	Collaterals	34.	Polarized
13.	Corticospinal tracts	35.	Positive
14.	Dendritic spines	36.	Posterior horn
15.	GDP	37.	Primary motor cortex
16.	GMP	38.	Sodium (Na⁺)
17.	GTP	39.	Potassium (K⁺)
18.	Lateral corticospinal tract	40.	Calcium (Ca⁺⁺)
19.	Lateral horn	41.	Sulfates (SO₄⁻)
20.	Ligand gated Ca⁺⁺ channel protein	42.	Synaptic cleft
21.	Ligand gated K⁺ channel protein	43.	Voltage gated Na⁺ ion channel
22.	Ligand gated Na⁺ channel protein	44.	α
		45.	β
		46.	γ

73. The _____ of the cerebral cortex determined a particular movement that the body needs to make and sends its signal (action potential) via the corticospinal pathway to the _____ located in the midbrain which forms the _____ that contain the _____ (**Answer A**) which continue the transmission of the action potential. At the level of the caudal medulla, **Answer A** splits into two tracts: _____ (90% of the axon) and _____ (10% of the axon). Both of these tracts continue down the spinal cord until it synapse with the cell bodies of somatic motor neurons is located in the _____ of the spinal cord. The motor neurons project their axons via the ventral root directly to the skeletal muscles. It is understood that each of the skeletal muscle fibers are innervated directly by motor nerves or branches also known as _____ from motor nerves to form

motor units.

1.	-58 mV	23.	mAChR
2.	-103 mV	24.	Mechanically gated K⁺ ion channel
3.	-85 mV	25.	Medullary pyramids
4.	-30 mV	26.	Motor-end-plate
5.	Acetylcholine	27.	nAChR
6.	ADP	28.	Negative
7.	AMP	29.	Neuromuscular junction
8.	Anterior corticospinal tract	30.	Neuropeptide Y
9.	Anterior horn	31.	Nitric oxide
10.	ATP	32.	Noradrenaline
11.	Cerebral peduncle	33.	Phosphates (PO₄⁻)
12.	Collaterals	34.	Polarized
13.	Corticospinal tracts	35.	Positive
14.	Dendritic spines	36.	Posterior horn
15.	GDP	37.	Primary motor cortex
16.	GMP	38.	Sodium (Na⁺)
17.	GTP	39.	Potassium (K⁺)
18.	Lateral corticospinal tract	40.	Sulfates (SO₄⁻)
19.	Lateral horn	41.	Synaptic cleft
20.	Ligand gated Ca⁺⁺ channel protein	42.	Voltage gated Na⁺ ion channel
21.	Ligand gated K⁺ channel protein	43.	α
22.	Ligand gated Na⁺ channel protein	44.	β
		45.	γ

74. At the _____, the axonal terminal of the motor nerve releases vesicles containing neurotransmitters called _____ via exocytosis into the _____ and allowed to diffuse through the fluid filled medium to the _____ (hint: a type of iontropic receptor) located on the invagination of the sarcolemma that accommodates the axonal terminal called _____. Once the neurotransmitter binds with the iontropic receptor, it will cause a conformational change in the protein subunits called _____, and in turn, causing the pore of the channel to open. Immediately, _____ cations begin to diffuse down its concentration gradient and enter the sarcoplasm of the muscle fiber. The influx of these cations will create a "localized" electrical impulse. It is known that the threshold of the skeletal muscle all-or-none response is _____ (mV). If the electrical impulse generated by the initial influx of cations is sufficient to reach or surpass the threshold, then the _____ will open (allowing more ions to diffuse down its concentration gradient) and an action potential will be generated. At this point the skeletal muscle cell is depolarized.

1.	-58 mV	23.	mAChR
2.	-103 mV	24.	Mechanically gated K⁺ ion channel
3.	-85 mV	25.	Medullary pyramids
4.	-30 mV	26.	Motor-end-plate
5.	Acetylcholine	27.	Multimers
6.	ADP	28.	nAChR
7.	AMP	29.	Negative
8.	Anterior corticospinal tract	30.	Neuromuscular junction
9.	Anterior horn	31.	Neuropeptide Y
10.	ATP	32.	Nitric oxide
11.	Cerebral peduncle	33.	Noradrenaline
12.	Collaterals	34.	Phosphates (PO₄⁻)
13.	Corticospinal tracts	35.	Polarized
14.	Dendritic spines	36.	Positive
15.	GDP	37.	Posterior horn
16.	GMP	38.	Primary motor cortex
17.	GTP	39.	Sodium (Na⁺)
18.	Lateral corticospinal tract	40.	Potassium (K⁺)
19.	Lateral horn	41.	Sulfates (SO₄⁻)
20.	Ligand gated Ca⁺⁺ channel protein	42.	Synaptic cleft
21.	Ligand gated K⁺ channel protein	43.	Voltage gated Na⁺ ion channel
22.	Ligand gated Na⁺ channel protein	44.	α
		45.	β
		46.	γ

75. The action potential will spread down the sarcolemma, as well as, the _____. At the _____ of the skeletal muscle, the generated action potential activates the _____ (**Answer A**) located on the modified sER called _____. **Answer A** is a channel protein called _____ and its resulting activation allows _____ cations which is kept at high concentrations in the extracellular matrix begins to flow into the sarcoplasm of the cell. These cations will bind to _____ (a receptor protein existing on the modified sER) causing it to activate and trigger the opening of the _____ thereby allowing cations stored within the modified sER to be released into the sarcoplasm. The release of these cations into the sarcoplasm is called a _____. At this stage the skeletal muscle is completely depolarized and the membrane potential is approximately set at _____.

1.	+20 mV	22.	Ligand gated K⁺ channel
2.	-103 mV	23.	Ligand gated Na⁺ channel
3.	-55 mV	24.	Light meromyosin
4.	-58 mV	25.	Mechanically gated Na⁺ channel
5.	-85 mV	26.	Muscle impulse
6.	Actin binding site	27.	Myosin binding site
7.	ADP	28.	Myosin-actin cycling
8.	AMP	29.	Na⁺
9.	ATP	30.	Power stroke
10.	ATPase	31.	Rigor state
11.	Cal⁺⁺	32.	Ryanodine receptor
12.	Calcium release channe		

143

33. DHP receptor
34. Diad
35. GDP
36. GMP
37. GTP
38. GTPase
39. Heavy meromyosin
40. K⁺
41. Ligand gated Ca⁺⁺ channel

42. S1 region
43. Sarcoplasmic reticulum
44. Transverse tubules
45. Triad
46. Tropomyosin
47. Troponin C (TnC)
48. Troponin I (TnI)
49. Troponin T (TnT)
50. Voltage gated Ca⁺⁺ channel
51. Voltage gated Na⁺ channel

76. These cations released into the sarcoplasm from the modified sER will bind with _____ (specific binding site located on a regulatory protein) which in turn causes the entire regulatory protein to undergo structural conformational change. Please note that this regulatory protein has two other components. The _____ is responsible for binding this protein to another regulatory protein that is formed to prevent the formation of cross bridges while _____ is formed to inhibit the interaction of myosin with actin. Once the entire regulatory protein has undergone structural conformational change, it moves another regulatory protein called _____ (regulatory protein that is formed to prevent the formation of cross bridges) out of the way of actin and myosin and thereby exposes the actin active site or switch into the "on" position. Once the active site is exposed, the _____ of the head of the myosin molecule also known as _____ will immediately bind thereby forming a bond called the cross bridge. Since there are no net movements at this time, this tight binding between the active site on actin and myosin places the skeletal muscle in a _____.

1. +20 mV
2. -103 mV
3. -55 mV
4. -58 mV
5. -85 mV
6. Actin binding site
7. ADP
8. AMP
9. ATP
10. ATPase
11. Ca⁺⁺
12. Calcium release channel
13. DHP receptor
14. Diad
15. GDP
16. GMP
17. GTP
18. GTPase
19. Heavy meromyosin
20. K⁺
21. Ligand gated Ca⁺⁺ channel

22. Ligand gated K⁺ channel
23. Ligand gated Na⁺ channel
24. Light meromyosin
25. Mechanically gated Na⁺ channel
26. Muscle impulse
27. Myosin binding site
28. Myosin-actin cycling
29. Na⁺
30. Power stroke
31. Rigor state
32. Ryanodine receptor
33. S1 region
34. Sarcoplasmic reticulum
35. Transverse tubules
36. Triad
37. Tropomyosin
38. Troponin C (TnC)
39. Troponin I (TnI)
40. Troponin T (TnT)
41. Voltage gated Ca⁺⁺ channel
42. Voltage gated Na⁺ channel

77. Once the cross bridge is formed, the energy stored from the previous hydrolysis of _____ causes the heavy meromyosin to bend at the _____ thus resulting in the shortening of the sarcomere. This bending action of the head of the myosin molecule is also referred to as a _____. Immediately after the movement, a new energy molecule will bind to the _____ or the 'head' of the myosin, thereby causing it to change its conformation and thereby detach from the active site. The second binding site located on the head of the myosin molecule is called _____. This binding site will hydrolyze the newly attached energy molecule called _____ causing it to return to its original conformation while the energy released through hydrolysis is once again stored. This bind, bend and release process is also known as _____.

1. +20 mV
2. -103 mV
3. -55 mV
4. -58 mV
5. -85 mV
6. Actin binding site
7. ADP
8. AMP
9. ATP
10. ATPase
11. Ca⁺⁺
12. Calcium release channel
13. DHP receptor
14. Diad
15. GDP
16. GMP
17. GTP
18. GTPase
19. Heavy meromyosin
20. K⁺
21. Ligand gated Ca⁺⁺ channel

22. Ligand gated K⁺ channel
23. Ligand gated Na⁺ channel
24. Light meromyosin
25. Mechanically gated Na⁺ channel
26. Muscle impulse
27. Myosin binding site
28. Myosin-actin cycling
29. Na⁺
30. Power stroke
31. Rigor state
32. Ryanodine receptor
33. S1 region
34. Sarcoplasmic reticulum
35. Transverse tubules
36. Triad
37. Tropomyosin
38. Troponin C (TnC)
39. Troponin I (TnI)
40. Troponin T (TnT)
41. Voltage gated Ca⁺⁺ channel
42. Voltage gated Na⁺ channel

78. Skeletal muscle cannot be in the excited state for long, which will result in muscular fatigue. Therefore, relaxation is a must. The relaxation of the skeletal muscle involves the process of _____ (hint: regeneration of membrane potential). It is known that the muscle cell is completely depolarized at _____. During this time a neurotransmitter specific enzyme called _____ present at the synaptic cleft begins to break down the neurotransmitter called _____ thereby eliminating the stimulus from the receptor protein also known as _____. Once the neurotransmitter is removed, the protein units of the ionotropic receptor called _____ will return to its original conformation, whereby closing the pore. At this stage the _____ cations has reached its

equilibrium therefore no net movement of ions will flow across the membrane. Because there is no net movement of cations, there will be no electrical impulse generated. Lack of electrical impulse will cause the _____ channels to close while _____ channels are triggered to open whereby _____ cations begin to flow down its concentration gradient and leave the sarcoplasm to the extracellular matrix of the skeletal muscle fiber. As more cations leave the cell, the more _____ (give a specific charge) the membrane potential become. Eventually, the voltage will drops to _____ mV which is also known as _____. At the previously mentioned voltage, the _____ channel will close and active transport protein called _____ will activate. This active transport protein will send _____ (how many) _____ cations out into the extracellular matrix while returning _____ (how many) _____ cations into the sarcoplasm. Additionally, through the actions of calcium pumps the Ca⁺⁺ ions will also be sent out of the sarcoplasm. Through the actions of these active transport proteins, the membrane potential will be reset back to _____ mV.

1. +20 mV
2. Acetylcholine
3. Acetylcholinesterase
4. Ca⁺⁺
5. Depolarization
6. Hyperpolarization
7. Hypopolarization
8. K⁺
9. Na⁺
10. Na⁺/K⁺ pump
11. Negative
12. Polarization
13. Positive
14. Power stroke
15. Repolarization
16. Rigor state
17. Ryanodine receptor
18. S1 region
19. Sarcoplasmic reticulum
20. Transverse tubules
21. Triad
22. Tropomyosin
23. Troponin C (TnC)
24. Troponin I (TnI)
25. Troponin T (TnT)

26. 1
27. -103 mV
28. 2
29. 3
30. 4
31. 5
32. -55 mV
33. -58 mV
34. -85 mV
35. ADP
36. ATP
37. GDP
38. GTP
39. Ligand gated Ca⁺⁺ channel
40. Ligand gated K⁺ channel
41. Ligand gated Na⁺ channel
42. mAChR
43. Multimer
44. nAChR
45. Voltage gated Ca⁺⁺ channel
46. Voltage gated K⁺ channel
47. Voltage gated Na⁺ channel
48. α
49. β
50. γ

79. The elimination of the neurotransmitter and the initiation of repolarization causes the enlarged ends of the modified sER called _____ to collect, concentrate and reabsorb _____ cations. Please note that the reabsorption of these cations is an energy dependent step where the energy molecule called _____ is needed. If these energy molecules are not available, these cations will not be reabsorbed and a medical condition known as _____ may result. As these cations are removed from the sarcoplasm, they are also removed from the _____ and _____ receptors located at the triad. These cations are also reabsorbed from the subunit of troponin called _____ thereby returning the regulatory molecule troponin back to its original conformation. Once troponin is returned back to its original conformation it will move _____ back to its original position where it blocks the active site of _____. With the active site blocked, the cross bridge formed between the myosin and actin myofilament is broken. At this time, the heavily compressed spring-like elastic protein called _____ will release its potential energy and push myosin and actin myofilaments back to its original shape.

1. Acetylcholine
2. Actin
3. Adenine
4. ADP
5. ATP
6. Ca⁺⁺
7. CDP
8. CTP
9. Depolarization
10. DHP receptors
11. GDP
12. GTP
13. Hyperpolarization
14. Hypopolarization
15. K⁺
16. Myosin
17. Na⁺
18. Na⁺/K⁺ pump

19. Nebulin
20. Negative
21. Acetylcholinesterase
22. Polarization
23. Positive
24. Ryanodine receptor
25. TDP
26. Terminal cisternae
27. Tetany
28. Titin
29. TnC
30. TnI
31. TnT
32. Tropomyosin
33. TTP
34. UDP
35. UTP

80. Based upon the calculations, the ideal length of a sarcomere that produces maximal tension occurs at _____ of its resting length.
a. 80 to 120 percent
b. Greater than 120 percent
c. Less than 80%
d. Less than 0.80%

81. Experiments has demonstrated that when an isolated muscle fiber is stretched to the point of minimal overlap of actin and myosin myofilaments (greater than 120 percent) and then stimulated by an electrode to contract, the muscle tension (force) generated is measured to be _____.
a. Adequate
b. Maximum
c. Minimal
d. Moderate

82. When an isolated muscle fiber is stimulated to contract at maximum overlap of

the myofilaments (less than 80 percent), the muscle tension (force) produced is _____.
a. Adequate
b. Maximum
c. Minimal
d. Moderate

83. Which of the following is defined as the contraction of a muscle in response to a single motor neuron stimulus that causes an action potential in one or more muscle of its motor unit?
a. Muscle twitch
b. Muscle tension
c. Neuromuscular coupling
d. Motor end plate
e. Endomysium

84. What is the terminology given to the period between the application of a stimulus and the beginning of contraction?
a. Contraction phase
b. Lag phase
c. Post-contraction phase
d. Pre-contraction phase
e. Relaxation phase

85. What is the terminology given to the period when the energy stored from the previous hydrolysis of ATP causes the heavy meromyosin to bend at the S1 region thus resulting in the formation of a power stroke and the subsequent shortening of the sarcomere?
a. Contraction phase
b. Lag phase
c. Post-contraction phase
d. Pre-contraction phase
e. Relaxation phase

86. What is the terminology given to the period when ACh is eliminated from the receptor protein (nAChR) and the initiation of repolarization process?
a. Contraction phase
b. Lag phase
c. Post-contraction phase
d. Pre-contraction phase
e. Relaxation phase

87. Which of the following terminology is defined as the progressive activation of a muscle through the incremental recruitment of motor units?
a. Excitation-contraction coupling
b. Frequency of stimuli
c. Motor end plate
d. Multiple motor unit summation
e. Multiple wave summation
f. Neuromuscular stimulation
g. Treppe

88. Which of the following terminology is defined as the manner by which the force generated by a muscle could be intensified through the increased in the frequency of stimuli from motor nerves and the subsequent increase of tension in skeletal muscle?
a. Excitation-contraction coupling
b. Frequency of stimuli
c. Motor end plate
d. Multiple motor unit summation
e. Multiple wave summation
f. Neuromuscular stimulation
g. Treppe

89. Which of the following terminology is defined as the staircase effect where a skeletal muscle has been at rest for a prolonged period and then activated to contract?
a. Excitation-contraction coupling
b. Frequency of stimuli
c. Motor end plate
d. Multiple motor unit summation
e. Multiple wave summation
f. Neuromuscular stimulation
g. Treppe

90. Which of the following occurs when the length of the muscle remains the same but the amount of tension increase?
a. Anticoncentric contrcations
b. Concentric contractions
c. Eccentric contractions
d. Isometric contractions
e. Isotonic contractions
f. Peritonic contrcations

91. Which of the following occurs when the tension increases briefly during the initial phase just to overcome the load but remains the same for the remainder of the contraction because additional force is no longer needed?
a. Anticoncentric contrcations
b. Concentric contractions
c. Eccentric contractions
d. Isometric contractions
e. Isotonic contractions
f. Peritonic contrcations

92. Which of the following occurs when an increase in tension that also result in the shortening of the muscle?
a. Anticoncentric contrcations
b. Concentric contractions
c. Eccentric contractions
d. Isometric contractions
e. Isotonic contractions
f. Peritonic contrcations

93. Which of the following occurs when the muscle lengthen while the tension generated remains the same?
a. Anticoncentric contrcations
b. Concentric contractions
c. Eccentric contractions
d. Isometric contractions
e. Isotonic contractions
f. Peritonic contrcations

94. The following is a table describing the input and output of glycolysis. Use the alphabetized terms provided and place the correct alphabet associated with the correct terminology into the appropriate column.

Input		Output	
1.	Glucose	5.	
2.		6.	
3.		7.	
4.			

a.	1 acetyl CoA	n.	2 ATP
b.	1 CO_2	o.	30-32 ATP
c.	1 Co-enzyme A (CoA)	p.	32-34 ATP
d.	10 NADH	q.	36-38 ATP
e.	2 acetyl CoA	r.	4 ATP
f.	2 CO_2	s.	4 CO_2
g.	2 Co-enzyme A (CoA)	t.	4 NAD+
h.	2 FAD	u.	4 NADH
i.	2 $FADH_2$	v.	4 pyruvate
j.	2 NAD+	w.	6 ATP
k.	2 NADH	x.	6 NAD+
l.	2 pyruvate	y.	6 NADH
m.	2 ADP	z.	8 NADH

95. The following is a table describing the input and output of transitional reaction. Use the alphabetized terms provided and place the correct alphabet associated with the correct terminology into the appropriate column.

Input		Output	
1.		4.	
2.		5.	
3.		6.	

a.	1 acetyl CoA	n.	2 ATP
b.	1 CO_2	o.	30-32 ATP
c.	1 Co-enzyme A (CoA)	p.	32-34 ATP
d.	10 NADH	q.	36-38 ATP
e.	2 acetyl CoA	r.	4 ATP
f.	2 CO_2	s.	4 CO_2
g.	2 Co-enzyme A (CoA)	t.	4 NAD+
h.	2 FAD	u.	4 NADH
i.	2 $FADH_2$	v.	4 pyruvate
j.	2 NAD+	w.	6 ATP]
k.	2 NADH	x.	6 NAD+
l.	2 pyruvate	y.	6 NADH
m.	2 ADP	z.	8 NADH

96. The following is a table describing the input and output of the Krebs cycle. Use the alphabetized terms provided and place the correct alphabet associated with the correct terminology into the appropriate column.

Input		Output	
1.		5.	
2.		6.	
3.		7.	
4.		8.	

a.	1 acetyl CoA	n.	2 ATP
b.	1 CO_2	o.	30-32 ATP
c.	1 Co-enzyme A (CoA)	p.	32-34 ATP
d.	10 NADH	q.	36-38 ATP
e.	2 acetyl CoA	r.	4 ATP
f.	2 CO_2	s.	4 CO_2
g.	2 Co-enzyme A (CoA)	t.	4 NAD+
h.	2 FAD	u.	4 NADH
i.	2 $FADH_2$	v.	4 pyruvate
j.	2 NAD+	w.	6 ATP
k.	2 NADH	x.	6 NAD+
l.	2 pyruvate	y.	6 NADH
m.	2 ADP	z.	8 NADH

97. How many ATPs could be generated through the electron transport chain?

a. 30-32
b. 33-34
c. 34-35
d. 35-36
e. 36-38
f. 38-40

98. The human skeletal muscle is designed to possess numerous backup energy sources from which spent energy molecule such as ADP could be recharged and used during an emergency.

_____ is the enzyme designed to break down energy molecules such as TTP, CTP, GTP and UTP to recharge an ADP molecule

_____ is the enzyme designed to breakdown a molecule of ADP to recharge another molecule of ADP

_____ is the enzyme designed to breakdown creatine phosphate to recharge an ADP molecule

_____ is the enzyme designed to breakdown an GTP molecule and thereby releasing energy and a phosphate functional group

a. Adenosine triphosphatase (ATPase)
b. Adenylate cyclase
c. Adenylate kinase
d. Alpha (α) amylase
e. Creatine kinase
f. Cyclic monophosphate
g. Cytosine triphosphatase (CTPase)
h. Guanosine triphosphatase (GTPase)
i. Lipase
j. Nucleoside diphosphokinase
k. Peptidase
l. Phosphodiesterase
m. Protein kinase
n. Sucrase
o. Thyrosine triphosphatase (TTPase)
p. Urosine triphosphatase (UTPase)

99. During anaerobic respiration, how many ATPs (gross ATPs) could be generated?
a. 1
b. 2
c. 3
d. 4
e. 5
f. 6
g. 7

100. _____ is the waste product formed during the initial phase of anaerobic respiration. Often, this initial waste product will break-down to form _____.
a. Acetyl-COA
b. FADH$_2$
c. Lactate
d. Lactic acid
e. NADH
f. Pyruvate

101. Lactic acid can only be excreted form the skeletal muscle cells via which of the following transport process?
a. Facilitated diffusion
b. Antiport
c. Symport
d. Gated channels
e. Osmosis
f. Diffusion

102. The combination of the demand of ATP, lack of oxygen and the buildup of lactic acid and lactate within the skeletal muscle during heavy exercise is called _____.
a. Anaerobic respiration
b. Electron transport chain
c. Glycolysis
d. Krebs cycle
e. Oxygen debt
f. Respiration
g. Transitional reaction

103. In the liver, lactic acid proceeds through a process known as gluconeogenesis. Which of the following product is generated through this process?
a. FADH$_2$
b. Glucose
c. Lactate
d. Maltose
e. NADH
f. Pyruvate

104. Slow twitch oxidative fiber utilizes which of the following component of cellular respiration to obtain its ATP energy source?
a. Calvin cycle
b. Dihydropyridine
c. Electron transport system
d. Excitation contraction coupling
e. Glycolysis
f. Kreb cycle
g. Transitional reaction
h. Answers c, e and f

105. Fast twitch glycolytic fiber utilizes which of the following component of cellular respiration to obtain its ATP energy source?
a. Calvin cycle
b. Dihydropyridine
c. Electron transport system
d. Excitation contraction coupling
e. Glycolysis
f. Kreb cycle
g. Transitional reaction
h. Answers c, e and f

106. Fast twitch oxidative fiber utilizes which of the following component of cellular respiration to obtain its ATP energy source?
a. Calvin cycle
b. Dihydropyridine
c. Electron transport system
d. Excitation contraction coupling
e. Glycolysis
f. Kreb cycle
g. Transitional reaction
h. Answers c, e, and f

107. Which of the following type of skeletal muscle contains the highest amount of myoglobin and are the smallest in muscle fiber diameter?
a. Fast twitch glycolytic fibers
b. Fast twitch oxidative fibers
c. Slow twitch glycolytic fibers
d. Slow twitch oxidative fibers

108. Which of the following type of skeletal muscle contains the lowest amount of myoglobin and are possesses the largest muscle fiber diameter?
a. Fast twitch glycolytic fibers
b. Fast twitch oxidative fibers
c. Slow twitch glycolytic fibers
d. Slow twitch oxidative fibers

109. Tetany is a condition of hyper-contracted muscle that sometimes follows a prolonged period of repetitive, summed muscle stimulation. Knowing the above statement is true, which of the following is (are) NOT (a) manner (s) in which the human body develops this condition?
a. Failure of parathyroid glands to release parathyroid hormone, the substance responsible for the regulation of calcium concentration in the body
b. A deficiency in vitamin D, which facilitates calcium ion absorption from the gastrointestinal tract
c. Alkalosis, an excessively alkaline (basic) state of body fluids resulting from persistent vomiting, rapid breathing
d. Excess activity of the hormone aldosterone (a mineralcorticoid) which causes the excessive secretion of calcium ions form the kidneys
e. Pathogenic bacterium Clostridium tetani releasing toxins within the human body
f. Hyperextension of skeletal muscles resulting from excessive exercise in anaerobic conditions
g. Both a and c
h. Both b and d

110. Which of the following is an iron/protein complex that is designed to store oxygen in muscle fibers for emergency use? This oxygen carrying pigment is what gives the "Dark meat" the reddish / burgundy shade of color.
a. Acetylcholinesterase
b. Creatine kinase
c. ATPase
d. Hemoglobin
e. Excitation-contraction coupling
f. Myoglobin
g. Creatine phosphate
h. Adenylate kinase

ANSWERS

1.	g	2.	b	3.	b
4.	c	5.	b	6.	c
7.	a	8.	f	9.	h
10.	a	11.	b	12.	d
13.	b	14.	f	15.	c
16.	a	17.	a	18.	e
19.	b	20.	f	21.	e
22.	b	23.	c	24.	a
25.	c	26.	e	27.	d
28.	h	29.	b, a	30.	c
31.	a, b, c, h, f, d	32.	f	33.	g
34.	f	35.	e	36.	d
37.	k	38.	d	39.	h
40.	c	41.	d	42.	h
43.	c	44.	a	45.	b
46.	g, r, l, p, e, g, k, w, h	47.	h, b	48.	g
49.	f	50.	a	51.	g

52.	b	53.	b	54.	c
55.	d	56.	c	57.	e
58.	g	59.	a	60.	a
61.	c	62.	d, c, a	63.	e, m, g, f, j, d, n. h, b, a, b
64.	a, d	65.	c	66.	f
67.	e	68.	e	69.	d
70.	f	71.	b	72.	34, 38, 39, 40, 38, 40, 35, 39, 33, 41, 28, 3
73.	37, 11, 25, 13, 18, 8, 9, 12	74.	30, 5, 42, 28, 26, 27, 39, 1, 43	75.	35, 36, 13, 34, 42, 11, 32, 21, 26, 1
76.	38, 40, 39, 37, 6, 19, 31	77.	9, 33, 30, 19, 10, 9, 28	78.	15, 1, 3, 2, 44, 43, 9, 47, 46, 8, 11, 27, 6, 46, 10, 29, 9, 28, 8, 34
79.	26, 6, 5, 27, 10, 24, 29, 32, 2, 28	80.	a	81.	c
82.	c	83.	a	84.	b
85.	a	86.	e	87.	d
88.	e	89.	g	90.	d
91.	e	92.	b	93.	c
94.	j, m, n, l, k, r	95.	l, g, j, e, f, k	96.	e, m, x, h, s, n, y, i
97.	a	98.	j, c, e, h	99.	d
100.	d, c	101.	a	102.	e
103.	b	104.	h	105.	e
106.	h	107.	d	108.	a
109.	f	110.	f		

REFERENCES

Anderson JL. ST segment elevation acute myocardial infarction and complications of myocardial infarction. In: Goldman L, Schafer AI, eds. Cecil Medicine. 24th edition. Philadelphia, PA: Saunders Elsevier. 2011: Chapter 73

Dulhunty AF, Haarmann CS, Green D, Laver DR, Board PG, Casarotto MG. Interactions between dihydropyridine receptors and ryanodine receptors in striated muscle. Prog Biophys Mol Biol., 2002. 79: 45-75

Gurgel-Giannetti J., Reed U., Bang M.L., Pelin K., Donner K., Marie S.K., Carvalho M., Fireman M.A., Zanoteli E., Oliveira A.S., Zatz M., Wallgren-Pettersson C., Labeit S., Vainzof M. Nebulin expression in patients with nemaline myopathy. Nueromuscular Disorders, 2001. 11: pp. 154-162

Hoyle, G. Comparative aspects of muscle. Annu Rev Physiol., 1969. 31: pp. 43–82

Huxley, A. F. & Niedergerke, R. Structural changes in muscle during contraction: Interference microscopy of living muscle fibres. Nature, 1954. 173: pp. 971–973

Huxley, H. E. & Hanson, J. Changes in the cross-striations of muscle during contraction and stretch and their structural interpretation. Nature, 1954. 173: pp. 973–976

Hynes T.R., Block S.M., White B.T., Spudich J.A. Movement of myosin fragments in vitro: Domains involved in force production. Cell, 1987. 48: pp. 953–963

Krans JL. The Sliding Filament Theory of Muscle Contraction. Nature Education, 2010. 3: pp. 66

Kushner FG, Hand M, Smith Jr SC, King SB. Focused Updates: ACC/AHA Guidelines for the Management of Patients With ST-Elevation Myocardial Infarction (Updating the 2004 Guideline and 2007 Focused Update) and ACC/AHA/SCAI Guidelines on Percutaneous Coronary Intervention (Updating the 2005 Guideline and 2007 Focused Update). Circulation, 2009. 120: pp. 2271-2306

Labeit S., Ottenheijm C.A., Granzier H. Nebulin, a major player in muscle health and disease. Federation of American Societies for Experimental Biology Journal, 2011. 25: pp. 822-829

Ottenheijm C.A., Witt C.C., Stienen G.J., Labeit S., Beggs A.H., Granzier H. Thin filament length dysregulation contributes to muscle weakness in nemaline myopathy patients with nebulin deficiency. Human Molecular Genetics, 2009. 18: pp. 2359-2369

Patil H, Vaidya O, Bogart D. A review of causes and systemic approach to cardiac troponin elevation. Clin. Cardiol., 2011. 34: pp. 723-728

Sabatine MS, Cannon CC. Approach to the patient with chest pain. In: Bonow RO, Mann DL, Zipes DP, Libby P, eds. Bonow: Braunwald's Heart Disease: A Textbook of Cardiovascular Medicine. 9th ed. Saunders; 2011: Chap 53

Saladin KS. Anatomy & Physiology. The Unity of Form and Function 6th Edition. McGraw-Hill, New York, New York. 2010

Seeley RR, Stephens TD, Tate P. Anatomy and Physiology 6th Edition. McGraw-Hill, New York, New York. 2003

Sheriff DF. Medical Biochemistry. New Delhi, India. Jaypee Brothers Medical Publishers Ltd. 2004. pp. 418

Spudich, J. A. The myosin swinging cross-bridge model. Nature Rev Mol Cell Biol, 2001. 2: pp. 387–392

Tate P. Seeley's Principles of Anatomy & Physiology 1st Edition. McGraw-Hill, New York, New York. 2009

PHOTO AND GRAPHIC BIBLIOGRAPHY

1. Figure 1: Open Stax College, Connexions, reprint permission granted under the Creative Commons Attribution 3.0 Unported license.

2. Figure 2: Open Stax College, Connexions, reprint permission granted under the Creative Commons Attribution 3.0 Unported license.

3. Figure 3: Open Stax College, Connexions, reprint permission granted under the Creative Commons Attribution 3.0 Unported license.

4. Figure 4: Open Stax College, Connexions, reprint permission granted under the Creative Commons Attribution 3.0 Unported license.

5. Figure 5: National Institute of Health, 2013. Public domain graphics

6. Figure 6: Graphic designed by P.Y.P. Jen. Copyright ©

7. Figure 7: Open Stax College, Connexions, reprint permission granted under the Creative Commons Attribution 3.0 Unported license.

8. Figure 8: Open Stax College, Connexions, reprint permission granted under the Creative Commons Attribution 3.0 Unported license.

9. Figure 9: Open Stax College, Connexions, reprint permission granted under the Creative Commons Attribution 3.0 Unported license.

10. Figure 10: Open Stax College, Connexions, reprint permission granted under the Creative Commons Attribution 3.0 Unported license.

11. Figure 11: Graphic designed by Patrick J. Lynch, medical illustrator. This file is licensed under the Creative Commons Attribution 2.5 generic license

12. Figure 12: Open Stax College, Connexions, reprint permission granted under the Creative Commons Attribution 3.0 Unported license

13. Figure 13: Open Stax College, Connexions, reprint permission granted under the Creative Commons Attribution 3.0 Unported license.

14. Figure 14: Open Stax College, Connexions, reprint permission granted under the Creative Commons Attribution 3.0 Unported license.

15. Figure 15: Graphic designed by P.Y.P. Jen. Copyright ©

16. Figure 16: Graphic designed by P.Y.P. Jen. Copyright ©

17. Figure 17: Graphic designed by P.Y.P. Jen. Copyright ©

18. Figure 18: Graphic designed by P.Y.P. Jen. Copyright ©

19. Figure 19: Open Stax College, Connexions, reprint permission granted under the Creative Commons Attribution 3.0 Unported license

20. Figure 20: Open Stax College, Connexions, reprint permission granted under the Creative Commons Attribution 3.0 Unported license

21. Figure 21: Open Stax College, Connexions, reprint permission granted under the Creative Commons Attribution 3.0 Unported license

22. Figure 22: Open Stax College, Connexions, reprint permission granted under the Creative Commons Attribution 3.0 Unported license

23. Figure 23: Open Stax College, Connexions, reprint permission granted under the Creative Commons Attribution 3.0 Unported license

24. Figure 24: Open Stax College, Connexions, reprint permission granted under the Creative Commons Attribution 3.0 Unported license

25. Figure 25: Open Stax College, Connexions, reprint permission granted under the Creative Commons Attribution 3.0 Unported license

26. Figure 26: Open Stax College, Connexions, reprint permission granted under the Creative Commons Attribution 3.0 Unported license

27. Figure 27: Open Stax College, Connexions, reprint permission granted under the Creative Commons Attribution 3.0 Unported license

28. Figure 28: Graphic designed by P.Y.P. Jen. Copyright ©

29. Figure 29: Graphic designed by P.Y.P. Jen. Copyright ©

30. Figure 30: Graphic designed by P.Y.P. Jen. Copyright ©

31. Figure 31: Open Stax College, Connexions, reprint permission granted under the Creative Commons Attribution 3.0 Unported license

32. Figure 32: Open Stax College, Connexions, reprint permission granted under the Creative Commons Attribution 3.0 Unported license

33. Figure 33: Graphic designed by P.Y.P. Jen. Copyright ©

34. Figure 34: Graphic designed by P.Y.P. Jen. Copyright ©

35. Figure 35: Graphic designed by P.Y.P. Jen. Copyright ©

36. Figure 36: Graphic designed by Fvasconcellos, 2007. Public domain graphics

37. Figure 37: Graphic designed by P.Y.P. Jen. Copyright ©

38. Figure 38: Open Stax College, Connexions, reprint permission granted under the Creative Commons Attribution 3.0 Unported license

39. Figure 39: Graphic designed by Thomas Splettstoesser and released on July 10, 2006. Reprint permission granted under the Creative Commons Attribution-Share Alike 3.0 Unported license

CHAPTER SEVEN

NERVOUS SYSTEM

All functions that occur within the human body are controlled directly or indirectly by the nervous system. There are two major divisions of the nervous system: the central nervous system (CNS) and the peripheral nervous system (PNS). The brain and its extension, the spinal cord, comprise the CNS, while the remaining nerves outside the CNS are called the PNS. Put simply, the PNS consist of nerves that connect the CNS to all areas of the body (Figure 1).

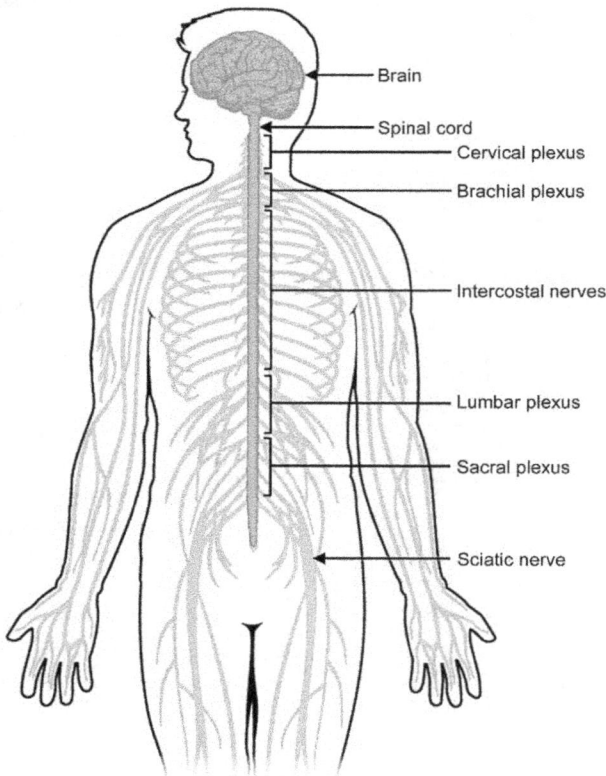

Brain
Spinal cord
Cervical plexus
Brachial plexus
Intercostal nerves
Lumbar plexus
Sacral plexus
Sciatic nerve

Figure 1: Illustration of the central nervous system (brain and spinal cord) and the peripheral nervous system

The PNS is subdivided into two divisions: sensory and motor. The sensory division, also known as afferents, provides peripheral awareness of the functions of our internal organs, as well as, the surrounding environment. The motor division, also known as the efferents, transmit action potentials from the CNS to the muscles, organs, and glands. The motor division can be further divided into the somatic motor nervous system, which is involved in the control of the voluntary movements of skeletal muscles, and the autonomic nervous system that regulates involuntary functions of cardiac muscles of the heart, smooth muscles of the organs and various glands of the human body.

The autonomic nervous system is segregated into three divisions: the sympathetic division, the parasympathetic divisions, and the enteric nervous system. In most organs, tissues and glands, the sympathetic and parasympathetic divisions of the autonomic nervous system function as antagonists. For example, the arrector pili muscles of the dermis are excited to contract by norepinephrine, also known as noradrenaline, released by the sympathetic nerves, while acetylcholine released by the parasympathetic nerves cause these muscles to relax. The enteric nervous system, on the other hand, is located in the gastrointestinal tract and functions independently of the CNS to control mechanical and chemical digestion.

The sensory division of the PNS is responsible for detecting stimuli from the internal organs, muscles and the external environments before transmitting the information to the CNS, where the information is subsequently integrated. The integration process of the brain involves inputs from the sensory nerves associated with various sensory modalities. These sensory modalities include: proprioception (position of the body and body parts), equilibrium (sense of balance), interoception (sense of physiological conditions associated with autonomic nerves), vision (photoreception), auditory sensations (sound), tactile (pressure reception), and chemosensation (sense of smell and taste). Once the various information inputs are integrated, they are analyzed and placed into memory, while a decision of a reaction is selected and sent from the brain via the motor division.

Regardless of their complexity in form and function, nervous tissues are composed of electrically excitable cells, called neurons, with various supporting cells called neuroglial cells. Neuroglial cells do not share a developmental lineage with neurons, therefore are not neurological in origin. It is understood that most neuroglial cells, with the exception of microglia, are derived from embryonic ectodermal tissue, while microglial cells are derived from hemopoietic stem cells.

Nerve Cell Structure

The neurons are designed to receive information via the dendrites. The dendrites are plasma membrane extensions of the soma (neuronal cell body). The dendrites are thin, highly branched, threadlike nervous structures that receive information transmitted from other nerve cells, muscles, glands, accessory cells, etc. Depending on the type of nerve cell, the dendrites may be highly branched (e.g. dendritic spines) to increase surface area for intercepting incoming stimuli or are associated with specific accessory cells that assist in sensory reception. For example, the rod and cone cells of the retina are non-neurological accessory cells that are designed to convert photonic energy (e.g. energy from light) into action potentials before transmitting them to the connecting bipolar neurons. Another example is the dermal pressure sensors, Pacinian corpuscles. A Pacinian corpuscle is constructed of 20 to 60 layers of concentric lamellae surrounding a single dendrite of sensory nerve. These concentric lamellae are composed of layers of accessory cells, such as modified Schwann cells (a type of neuroglial cell), fibroblasts, and fibrous connective tissue. It has been demonstrated that a deformation of the concentric lamellae will result in the subsequent bend of the single dendrite of the sensory nerve. This forceful distortion of the dendrite will result in the generation of an action potential.

There are numerous ways in which information is received by dendrites. Dendritic reception is dependent on types of gated ion channels found in the dendritic membrane, which in turn, will determine the particular function of the nerve. For example, there are three types of gated ion channels commonly found in dendrites: mechanically gated, voltage gated and ligand gated.

Mechanically gated ion channels, also known as stretch-gated ion channels, respond to any mechanical deformation of the plasma membrane. Such mechanically induced deformation will cause ion channels to open, allowing an electrical impulse to be generated, which results in depolarization. The voltage gated ion channels are designed to open and close depending on the availability of an electrical impulse. Ligand gated ion channels respond to a ligand, or a signal molecule, that triggers the opening of the ion channels when it binds to membrane bounded receptors.

Once the electrical impulse is generated, it travels down the plasma membrane of the dendrites to the soma. Please note that this initial electrical impulse is called a local current and is not an action potential.

The soma contains a centrally located nucleus and a relatively pronounced nucleolus. Within the neuroplasm (e.g. the cytoplasm of the nerve cell) there are large amounts of endoplasmic reticulum (ER), Golgi apparatus and mitochondria. The aggregates of rough endoplasmic reticulum (rER) and ribosomes are referred to as Nissl bodies and are the primary location for protein synthesis in nerve cells.

This local current will continue to spread through the soma until it reaches the initial segment also known as the trigger zone. The initial segment is located within the axonal hillock, an area that marks the end of the soma and the beginning of the axon. The initial segment is where the threshold of the neuron exists.

The threshold potential, or simply known as the threshold, is the set level of depolarization that must be reached in order to trigger an action potential. Put simply, the threshold is where the summation of excitatory postsynaptic potential (EPSP) must be greater than the summation of inhibitory postsynaptic potential (IPSP). If the EPSP is greater than IPSP, an action potential will be generated and this electrical impulse will continue down the length of the axon. However, if the EPSP is less than the IPSP, the local current will either dissipate in this area of the nerve cell, or be stored within milliseconds of another impulse, thereby allowing the local currents to be totaled via the process of summation. If the summed local potential (e.g. EPSP) is greater than the IPSP, an action potential will be generated.

Nerve fibers, also known as axons, are responsible for the transmission of electrical impulses or action potentials to all locations within the human body. The axon is a lengthening of the cell body where it forms a long thread-like extension. Like electricity traveling along an extension cord, the axon is basically designed to allow the action potential to travel down its entire length. The axon may remain as a single structure or branch to form collaterals. Put simply, a single nerve with numerous collaterals may innervate numerous postsynaptic nerve cells, muscle fibers, or glands (Figure 2).

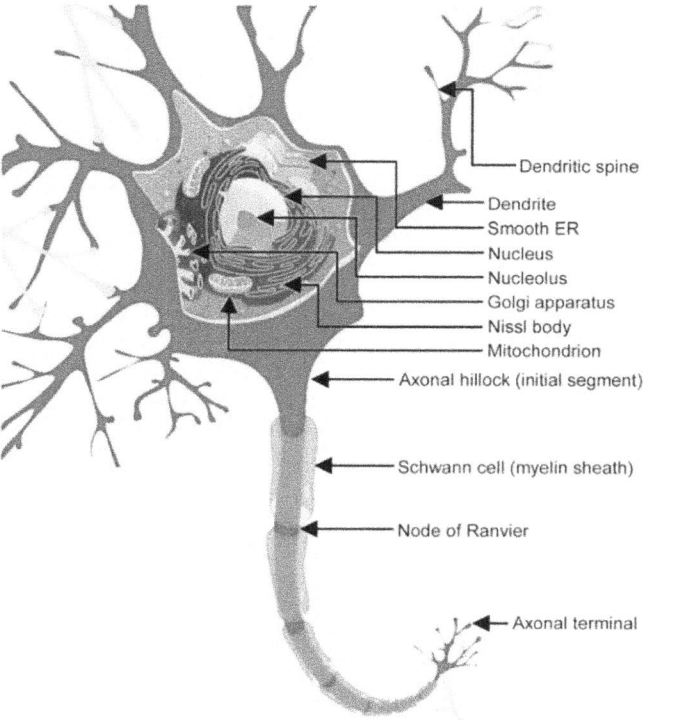

Dendritic spine
Dendrite
Smooth ER
Nucleus
Nucleolus
Golgi apparatus
Nissl body
Mitochondrion
Axonal hillock (initial segment)
Schwann cell (myelin sheath)
Node of Ranvier
Axonal terminal

Figure 2: Illustration of a neuron

At the terminal end of the axon are various extensions called axonal terminals. The axon terminals, also known as synaptic knobs, are the communication end of the nerve fiber, which allows action potentials to be transmitted to target cells. Within the axon terminal there are numerous mitochondria, as well as, vesicles containing signal molecules or neurotransmitters. For example, once the impulse reaches the synaptic knob, neurotransmitters from the axon terminals are released via exocytosis, transmitting the information from the nerve to the target cell (Figure 2).

Axonal Transport

Since the bulk of the organelles are located in the soma, an intracellular transport system is needed to move the manufactured substances in an anterograde direction down the length of the axon to the axonal terminal and to bring substances back in a retrograde direction from the axonal terminal. This bidirectional transport process is called axonal transport, and this system is divided into two portions, the slow and the fast axonal transport systems.

The slow axonal transport system is needed to transport cytoskeletal proteins and soluble enzymes, not associated with membranous organelles, to and from the axonal terminal. Overall, the slow axonal transport system is calculated to be at least two times slower, when compared to the fast axonal transport system. However, recent observations have shown that the speed by which slow axonal transport moves substances are just as rapid as fast axonal transport, but their movements are limited by rapid and brief burst of speed in the anterograde direction, frequent stops, and occasional burst of movement in the retrograde direction. These intermittent movements average at millimeters or tenths of millimeters per day or several nanometers or tens of nanometers per second. Put simply, the terms fast and slow axonal transport refer to the average rate of movement on a timescale of days or weeks and do not reflect the actual velocity of the transport.

Experiments have demonstrated that slow axonal transport consists of two components: the Slow Component a (SCa), which is the slower form of slow axonal transport, and the Slow Component b (SCb), the faster form of the slower axonal transport. SCa consists mainly of cytoskeletal proteins that form neurofilaments (i.e. intermediate filaments) and microtubules (i.e. tubulin). The SCb is composed of a complex protein structure that includes hundreds of distinct proteins ranging from cytoskeletal proteins, such as actin, small portion of tubulin, clathrin, calmodulin, and glycolytic enzymes.

Although much research has been done with respect to the components of slow axonal transport, information regarding the exact mechanism of this system is limited. One hypothesis states that the cargoes that are transported are attached to cytoskeletal polymers that are constantly organizing and reorganizing down the length of the axon. It is known that the cytoskeletal systems are dynamic and adaptable. They are capable of rearranging themselves rather rapidly depending on the needs of the cell. For example, the relatively simple protein composition of SCa suggests that the cargo may be attached to constructing neurofilaments and microtubules which could explain its stop and go characteristics (Figure 3). In regards to SCb, a similar theory has been proposed. The availability of actin in SCb suggests that its hundreds of distinct protein cargo are attached to supramolecular complexes, which move with the construction of actin filaments (moves as the length of the filaments are being lengthened) (Figure 4)

Alternatively, it has also been suggested that the supramolecular complexes of the SCb may be associated with motor proteins like kinesin or dynein, which are generally associated with fast axonal transport. It is postulated that the motor proteins will move the cargo to the end of one actin filament where it stops and waits until the attaching segment is constructed. This too would explain the frequent stop-and-go characteristics of SCb and slow axonal transport in general (Figure 4)

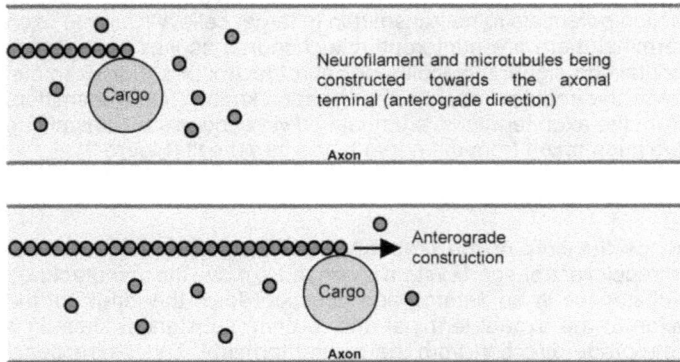

Figure 3: SCa slow axonal transport. Please note that the cargo is anterograde transported as the neurofilaments and microtubules are being constructed

Hypothesis I

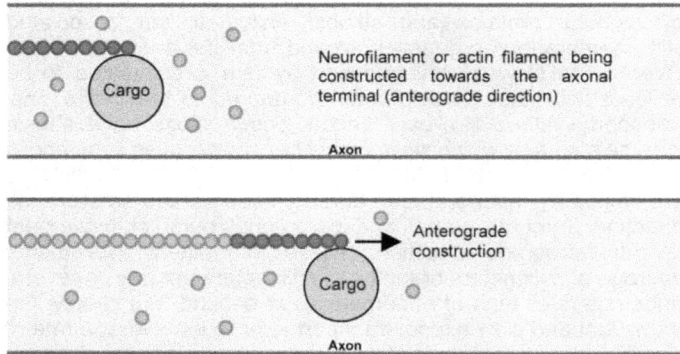

Hypothesis II

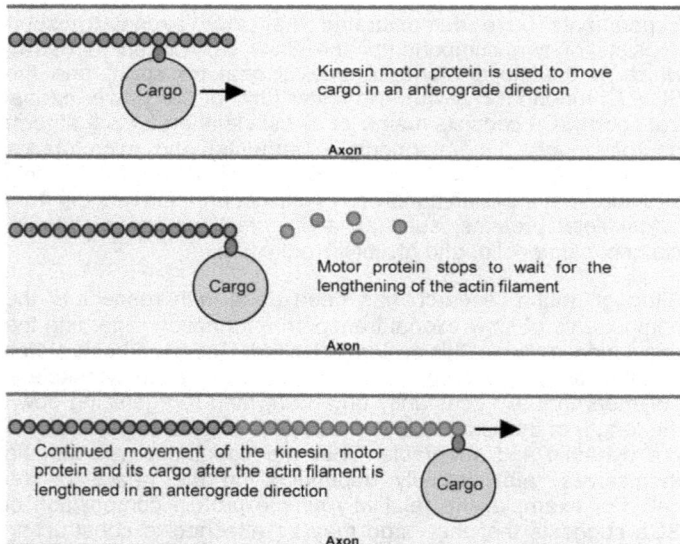

Figure 4: Illustrations of the two hypothesis of SCb slow axonal transport. Hypothesis I show that the cargo is being transported by attaching to cytoskeletal polymers that are constantly organizing and reorganizing down the length of the axon. Hypothesis II indicate that in addition to the lengthening of the cytoskeletal polymers, the transport of the cargo is being aided by motor proteins such as kinesin

Fast axonal transport represents the movement of membranous organelles, such as secretory vesicles, endocytic organelles (i.e. vesicles formed via endocytosis), phagosomes, peroxisomes, lysosomes, and mitochondria to and from the axonal terminal. It has been calculated that the movement of these cellular components occurs at rates of 200 to 400 mm per day or 2 to 10 μm per second.

In regards to the anterograde transport of vesicles from the soma to the axonal terminal, it is understood that a variety of materials move in fast anterograde transport including membrane bounded enzymes, proteins neurotransmitters, and neuropeptides etc. These materials must first be synthesized, sorted and packaged into vesicles. For example, secretory and membrane bounded proteins must be first synthesized on polyribosomes associated with the rough endoplasmic reticulum (rER). Once the secretory protein has been assembled on the polyribosome, it enters the lumen of the rER at the traslocon, which is lined with a specific receptor called docking protein, also known as the SRP receptor. Within the lumen, the protein will be folded and form a multi-subunit complex under the scrutiny of a rER lumen chaperone molecule. The chaperone molecule will then escort the polypeptide to interact with various enzymes where it could be modified by adding carbohydrate chains, forming it into a glycoprotein, or edited via deletion or rearrangement of amino acid sequences. Once the polypeptide is edited and deemed functional, the nearly completed product is sent to the Golgi apparatus via transfer vesicles. The Golgi apparatus is responsible for post-translational modification of the proteins delivered from rER. The process of post-translational modification includes glycosylation, sulfation, and proteolytic cleavage. Once assembled and packaged, a molecular motor protein, kinesin, will be added to the vesicle.

Kinesin is polarity dependent, rod-shaped protein, approximately 80 nm long, possessing two globular heads connected to a fan-shaped tail via a long stalk. Similar to the heavy meromyosin of the myosin myofilament, the globular heads of the kinesin motor protein possesses two binding sites, a microtubule binding sites and an ATP binding site (also known as ATPase). The microtubule binding site of kinesin is designed to bind to the tubulin of microtubules, while the ATPase is used to hydrolyze energy molecule, providing the energy needed for movement (Figure 5).

Unlike other cytoskeletal proteins, which are composed of various combinations of proteins, microtubules are simple in construction and are composed of tubulin. Tubulin is composed of three structurally similar polypeptide monomers, α-, β-and γ-tubulin. The γ-tubulin serves as a template or foundation for the correct assembly of microtubules, while α- and β-tubulin assemble on top of the γ-tubulin through hydrophobic interactions to form tubulin dimers. The tubulin dimers bond together (polymerize) through intermolecular interactions to form linear protofilaments.

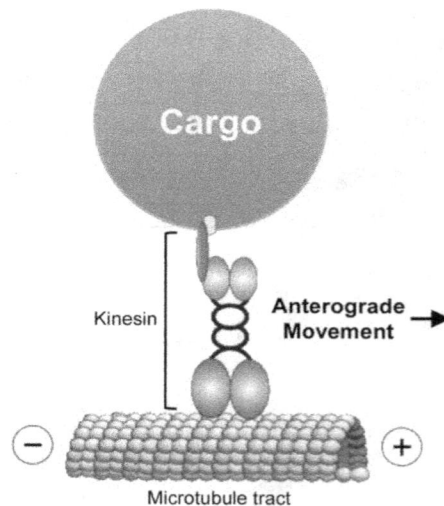

Figure 5: Illustration of the motor protein kinesin transporting cargo towards the axonal terminal

It has been calculated that a single microtubule consists of 13 protofilaments that interact with one another to form a 24 nm wide, hollow cylinder. The head-to-tail bonding of α- and β-tubulin creates a polar structure. In each protofilament, the α-tubulin is always located at the minus ⊖ end (towards the soma or the γ-tubulin), while the β-tubulin is oriented towards the growing end, also known as the plus ⊕ end (Figure 6).

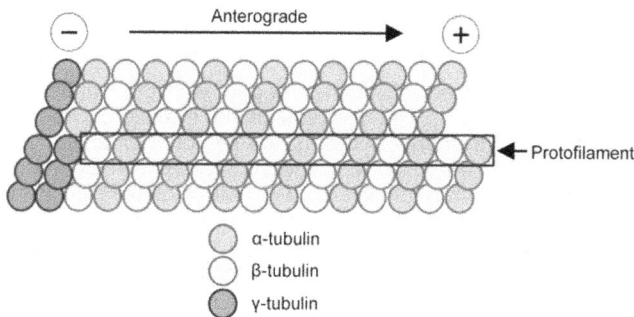

Figure 6: Illustration of microtubule formation with α- and β-tubulins assembles on top of the γ-tubulin which serves as the foundation. Please note that the 13 protofilaments interact with one another to form a 24 nm wide hollow cylinder

In neurons, microtubule tracts extend the length of both the axons and dendrites. In the axon, microtubule tracts are always organized, where the minus ⊖ end is located in the soma and the plus ⊕ end is located towards the axonal terminal. In the dendrite, the minus ⊖ end is always located in the terminal end of the dendrite, while the positive ⊕ end is located in the soma. This directional polarity is important, since the motor protein kinesin travels in a minus ⊖ to plus ⊕ direction on the microtubule tract, which would, explains the anterograde movement of vesicles from the soma to the axonal terminal (or from the dendritic terminals to the soma). Similarly, other organelles, such as mitochondria and peroxisomes, are all coated with kinesin motor proteins on their membranes, which allow them to travel from the soma to the axonal terminal (Figures 5 and 6).

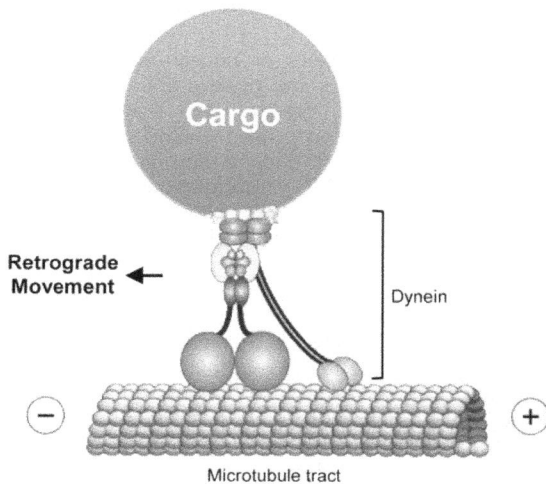

Figure 7: Illustration of the motor protein dynein transporting cargo towards the soma

Vesicles, such as phagosomes, that are moving in retrograde transport are believed to contain materials that are to be delivered to lysosomes in the soma for degradation. On average, the sizes of the membrane bounded organelles and phagosomes, are larger than those moving in the anterograde direction, which increases their drag and slows their pace. These phagosomes are associated with another motor protein, cytoplasmic dynein. Like kinesin, the head of dynein (dynein heavy chain) possesses two binding sites: an ATP binding site (ATPase) and a microtubule binding site. The microtubule binding site of cytoplasmic dynein is designed to bind to the tubulin of microtubules and the ATPase is made to hydrolyze energy molecule, providing the energy needed for movement. In contrast to kinesin, cytoplasmic dynein

travels in a plus ⊕ to minus ⊖ directionality on the microtubule tract, which would explain the retrograde movement of vesicles from the axonal terminal to the soma (Figure 7).

Neurotransmitters

Neurotransmitters are signal molecules or ligands that allow the transmission of signals from the presynaptic neuron to the postsynaptic neuron across the synapse. Nerve cells are responsible for the synthesis and storage of neurotransmitters. There are many neurotransmitters found within the human body, that are divided into two broad categories: small-molecule neurotransmitters and neuropeptide neurotransmitters (Table 1).

The small-molecule neurotransmitters are either synthesized in the axonal terminal or within the soma. The components necessary for small-molecule neurotransmitter synthesis are acquired using transport proteins located within the axolemma to move available materials from the extracellular spaces and/or the use of products from neuronal metabolic processes. Enzymes that are needed for the synthesis of small-molecule neurotransmitters are coded within the DNA. Through the processes of transcription mRNA is formed and exits the nucleus via the nuclear pores as exportins. Polyribosomes are responsible for the translation of the mRNA into numerous polypeptide sequences before inserting them through the membrane of the rER via the docking protein, also known as the SRP receptor, and into the lumen of the organelle. Within the rER, the protein will be folded, forming a multi-subunit complex. They are then modified and edited before they are transferred to the Golgi apparatus. At the Golgi apparatus, the polypeptide is further modified and placed within an intramembranous organelle or a secretory vesicle, before being transported to the axonal terminal through either SCa or SCb modes of slow axonal transport.

Table 1: List of common neurotransmitters found within the human body. Please note that the small-molecule neuro-transmitters are highlighted in blue while neuropeptides are highlighted in orange. GABA: Gamma-Aminobutyric acid; VIP: Vasoactive Intestinal Peptide; CGRP: Calcitonin Gene Related Peptide

Neurotransmitters	Function	Location
Acetylcholine	Muscle control, memory, sensory response	PNS[1], CNS[2]
Serotonin	Intestinal movement, mood, appetite, sleep, muscle control	CNS[2], ENS[3]
Dopamine	Reward pathway, role in cognition	Hypothalamus, endocrine
Norepinephrine	Adrenal medulla, sympathetic nerves	Endocrine, ANS[4]
Epinephrine	Adrenal medulla, sympathetic nerves	Endocrine, ANS[4]
GABA	Inhibitory neurotransmitter, regulation of muscle tone	CNS[2]
Glycine	Inhibitory neurotransmitter found in the brain	CNS[2]
Aspartate	Excitatory neurotransmitter found in the brain	CNS[2]
Nitric oxide	Inhibitory neurotransmitter, vasodilator	PNS[1], CNS[2], ANS[4]
Glutamate	Excitatory neurotransmitter memory formation	CNS[1]
Neuropeptide Y	Vasoconstrictor, lipogenesis	Hypothalamus, ANS[4]
Somatostatin	Inhibitory actions, control hormone secretion	CNS[2], ENS[3]
CGRP	Vasodilator, transmission of nociception	PNS[1], CNS[2]
Met-enkephalin	Modulating the impact of pain, opioid	PNS[1], CNS2
Leu-enkephalin	Modulating the impact of pain, opioid	CNS[2]
Beta-endorphin	Modulating the impact of pain, opioid	CNS[2]
VIP	Smooth muscle relaxation, regulate prolactin secretion	PNS[1], CNS[2]
Substance P	Inducing pain, sensory nerve neurotransmitter	CNS[1]

[1] Peripheral nervous system (PNS), [2] Central nervous system (CNS), [3] Enteric nervous system (ENS), [4] Autonomic nervous system (ANS)

Acetylcholine (ACh) is an example of a small-molecule neurotransmitter that is formed at the axonal terminal. ACh is constructed from choline and acetate. Choline molecules are present in the plasma or are recycled after the breakdown of

ACh by acetylcholinesterase (AChE) in the synaptic cleft and subsequently absorbed by the cholinergic nerve cells (acetylcholine secreting nerves) via membrane bounded choline transport proteins. Acetate is formed from acetyl-CoA, a product acquired from the transition reaction of cellular respiration (Figure 8).

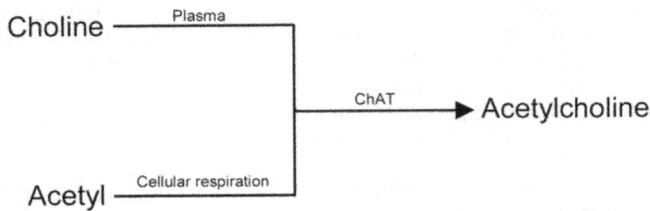

Figure 8: Acetylcholine formation. Please note that choline is transported into the axon while the acetyl is formed intracellularlly during cellular respiration

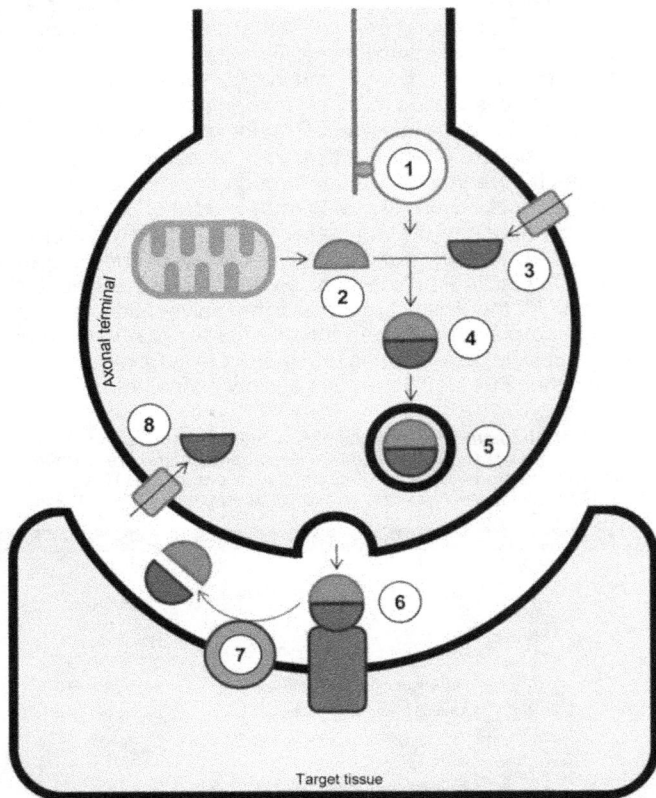

Figure 9: Illustration of ACh formation and release from the axonal terminal. ① The enzyme ChAT is transported down the length of the axon via SCb slow axonal transport. ② Acetyl formed from Acetyl-CoA, a process in cellular respiration. ③ Choline is absorbed from blood plasma. ④ Enzymatic actions of ChAT forms ACh. ⑤ ACh is packaged into a secretory vesicle and ⑥ released via exocytosis into the synaptic cleft and binds to a receptor protein. ⑦ AChE breaks down ACh back into acetyl and choline and finally, ⑧ choline is absorbed back into the axonal terminal

Choline acetyltransferase (ChAT) is formed in the soma and anterograde transported to the axonal terminal via slow axonal transport. ChAT is responsible for combining choline and acetate to form ACh. Once ACh is formed, it is transported from the axolemma into a secretory vesicle by the vesicular membrane bounded proteoglycan, vesiculin. After ACh neurotransmitters are packaged into a vesicle, they are released via exocytosis through the arrival of action potential (Figures 8 and 9).

Catecholamines (CA) are a series of related monoamine neurotransmitters, which are released by both the endocrine and the nervous systems. For example, epinephrine, also known as adrenaline, is released into the blood stream by the chromaffin cells of the adrenal medulla as a neurohormone. Post ganglionic sympathetic fibers of the autonomic nervous system release norepinephrine (NE), also known as noradrenaline, as a neurotransmitter when triggered by an action potential. Dopamine (DA) is released by the dopaminergic neurons located in the substantia nigra, found in the basal ganglia, the ventral tegmental area (VTA) of mesencephalon (mid brain), and the arcuate and periventricular nuclei of the hypothalamus as a neurotransmitter.

The production of catecholamines begins with the transport of the amino acid tyrosine from the plasma into the axon. Within the axoplasm, tyrosine is converted into L-3,4-dihydroxyphenylalanine (L-DOPA) through the enzymatic actions of tyrosine hydroxylase. Soon after the formation of L-DOPA, this molecule is further processed into the neurotransmitter DA by the enzyme DOPA decarboxylase. Once DA is formed, it is transported/packaged into secretory vesicles by vesicular monoamine transporter 2 (VMAT2). Within the vesicle, DA may be further converted into NE by the enzyme dopamine β-hydroxylase (DβH). Furthermore, NE may be further processed into epinephrine through the enzymatic actions of phenylethanolamine-N-methyltransferase (PMNT) (Figure 10). Please note that all the enzymes involved in catecholamine synthesis are produced in the soma and are anterograde transported down to the axon and axonal terminal via slow axonal transport.

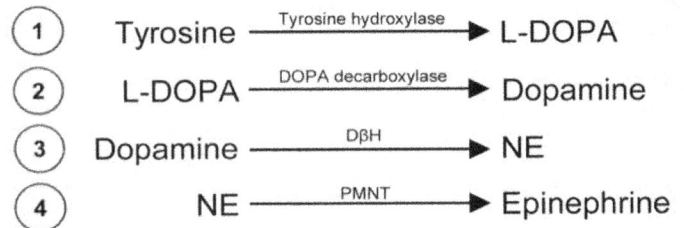

Figure 10: The catecholamine pathway. ① Tyrosine is transported into the axon and is converted into L-DOPA. ② L-DOPA is conversted into dopamine. ③ Dopamine is converted into NE and ④ NE is converted to epinephrine

As with any neurotransmitters or neurohormones, the ligand has to be removed to allow the postsynaptic nerve or tissues to repolarize or return to pre-activated state. For example, the enzyme acetylcholinesterase (AChE), found in the neuromuscular junctions and cholinergic neuron synapses, are used to hydrolyze ACh. AChE will breakdown ACh into choline and acetate, thereby effectively removing the ligand from the membrane bounded receptor protein (Figure 9).

There are several different manners by which NE and epinephrine are removed from receptor proteins. It has been estimated that approximately 90% of the NE is actively transported back into the nerve terminal by a neuronal reuptake transport system to be used again after the arrival of the subsequent action potential(s). Small amounts (approximately 5%) of NE are metabolized by catechol-O-methytransferase (COMT) and monoamine oxidase (MAO) or diffused into the circulatory system. For example, the metabolic cycle of NE involves the breakdown of this neurotransmitter by COMT into normetanephrine. This intermediate product is further processed into vanillylmandelic acid (VMA) through the enzymatic actions of MAO. NE may also be hydrolyzed into dihydroxymandelic acid and further processed into VMA by COMT. VMA is released into the circulatory system and removed by the kidneys. Epinephrine, on the other hand, can also be removed from the receptor protein on target tissues by both COMT and MAO. For instance, epinephrine may be hydrolyzed by COMT into metanephrine and further processed by MAO into VMA. Epinephrine can also be processed by MAO into dihydroxymandelic acid and further processed by COMT into VMA. Like the NE removal cycle, VMA is released into the circulatory system and eventually removed by the kidneys (Figure 11).

The second category of neurotransmitter are the neuropeptides.

On average, neuropeptides are approximately 3 to 36 amino acids in length, and are created in the soma of the neuron. For example, the genetic codes needed to manufacture neuropeptides are found within the DNA, which are transcribed into mRNA before leaving the nucleus via the nuclear pores as exportins. Polyribosomes are responsible for the translation of the mRNA into polypeptide sequences. Once the translation is complete, the polypeptide sequences will be inserted through the membrane of the rER via the docking protein, also known as the SRP receptor, into the lumen of the organelle. Within the rER, the protein will be folded, forming a multi-subunit complex, modified and edited before they are transferred to the Golgi apparatus. In the lumen of the Golgi apparatus, the polypeptides will be further modified before being placed in an intramembranous organelle or in a secretory vesicle and transported to the axonal terminal through fast axonal transport. Like the small-molecule neurotransmitters, the secretory vesicles containing neuropeptides will be released via exocytosis upon the arrival of action potentials.

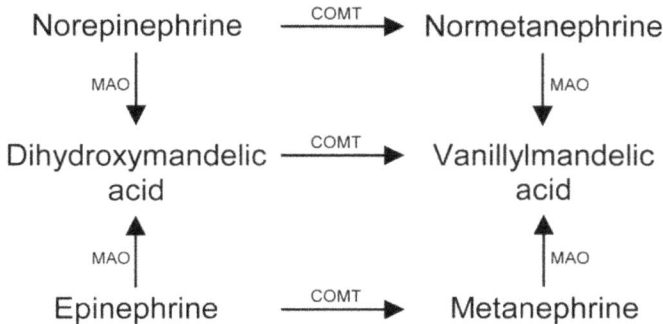

Figure 11: The degradation pathway of norepinephrine and epinephrine. Please note that catechol-O-methytransferase is abbreviated COMT, while monoamine oxidase is abbreviated MAO

Neuron Types

There are several classification schemes that may be used to categorize neurons. For example, in mammals, most neurons travel in a unidirectional manner, with the minor exception of anaxonic neurons. For most of the neurons, a local potential will form at the dendrites and be transmitted through the soma. At the initial segment of the axonal hillock, the local potential will become an action potential if the threshold is met. The action potential will continue along the axon, before finally arriving at the axonal terminal. With most nervous impulses travelling in an unidirectional manner, neurons can be classified based on the direction that the impulses are being conducted. For example, afferent neurons are sensory neurons that conduct their action potential from the periphery to the CNS. Efferent neurons are motor neurons that conduct their action potential from the CNS to the peripheral tissues. Finally, interneurons conduct action potentials from one neuron to another within the CNS.

Neurons may also be classified based on their structure. For example, a multipolar neuron has a distinctly large soma and numerous dendrites protruding from its cellular memrbane. It has a single axon that emerges at the axonal hillock as a long tail extending away from the soma. Multipolar neurons are commonly seen in the CNS and serve as motor neurons (Figure 12).

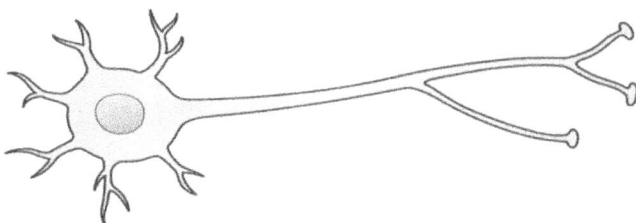

Figure 12: Illustration of a multipolar neuron

Bipolar neurons have two processes extending from the polar ends of the soma. Bipolar neurons are commonly found in sensory organs, such as the retina and the olfactory nerves of the nose (Figure 13).

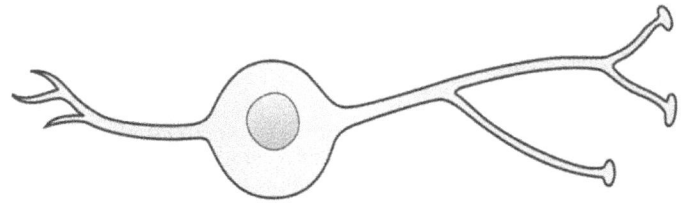

Figure 13: Illustration of a bipolar neuron

Unipolar neurons have a single process extending a short distance from the soma that divide into two branches, the dendritic and axonal branch. The dendritic branch extends to the peripheral tissue and has a sensory receptor, while the axonal branch extends into the CNS. For example, the dorsal root ganglia are composed of unipolar soma, with branches extending to the peripheral tissue and to the CNS (Figure 14).

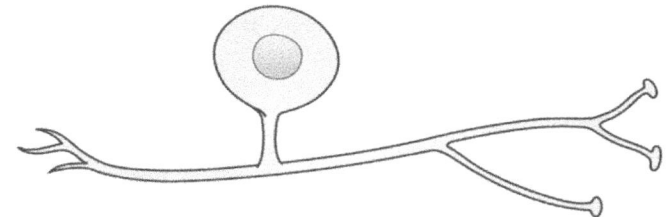

Figure 14: Illustration of a unipolar neuron

Anaxonic neurons are atypical neurons found in the CNS of mammals. Anaxonic neurons do not have axons and are components of complex interneuronal circuits, where the nervous impulses can travel in any direction and not simply the dendrite to axon format. These types of nerves are commonly found in areas of the brain that control activities involved in complex learning and formation of memory (Figure 15).

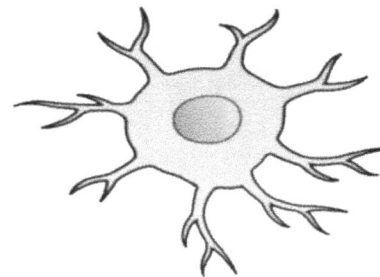

Figure 15: Illustration of a anaxonic neuron

Synapses

Synapses are the locations where nerve impulses are passed from nerve cells to target tissues. This target tissue may be another nerve, a gland or a muscle fiber. There are two types of synapses: chemical and electrical.

A chemical synapse is where an action potential triggers the release of ligands (e.g. neurotransmitters or neuropeptides) via exocytosis into the synaptic cleft. These ligands will diffuse through the fluid filled space of the synaptic cleft and bind with membrane bounded receptors located on the membrane of the target tissue. As discussed in Chapter Three, there are two types of membrane bound receptor proteins: ionotropic and metabotropic receptors. The ionotropic receptors, also known as ligand gated receptors, are transmembrane proteins that possess a ligand binding site directly linked to the ion channels to form a single molecular entity. Ionotropic receptors are composed of four to five transmembrane protein subunits, also known as multimers, organized around a pore. These ionotropic receptors are fast acting and are commonly associated with neurotransmitters (signal molecules, ligands) such as γ-Aminobutyric acid (GABA), glycine, serotonin, acetylcholine, etc. and are specific in the type of ions that they allow to flow through the pore (Na^+, K^+, Ca^{++} or Cl^-) (Figure 9).

The metabotropic receptors are not regulated as part of channel proteins. Instead, the opening and closing of the channel protein is dependent upon a guanine nucleotide-binding protein (G-protein) complex. A G-protein consists of alpha (α), beta (β) and gamma (γ) subunits on the inner surface of the cell membrane. The G-protein is activated when a ligand attaches to its binding site. Once the ligand binds to the G-protein complex, the receptor changes its shape and the three subunits (α, β and γ) separate. The β- and γ-subunits form a stable dimer (the joining of two molecules through intermolecular bonding) and is responsible for activating the α-subunit. As a result, guanosine diphosphate (GDP; spent energy molecule) bounded α-subunit is replaced with guanosine triphosphate (GTP; charged energy molecule). Subsequently, the α-subunit-GTP complex causes the channel protein to change its structural conformation and allows specific ions to flow down their concentration gradient(s).

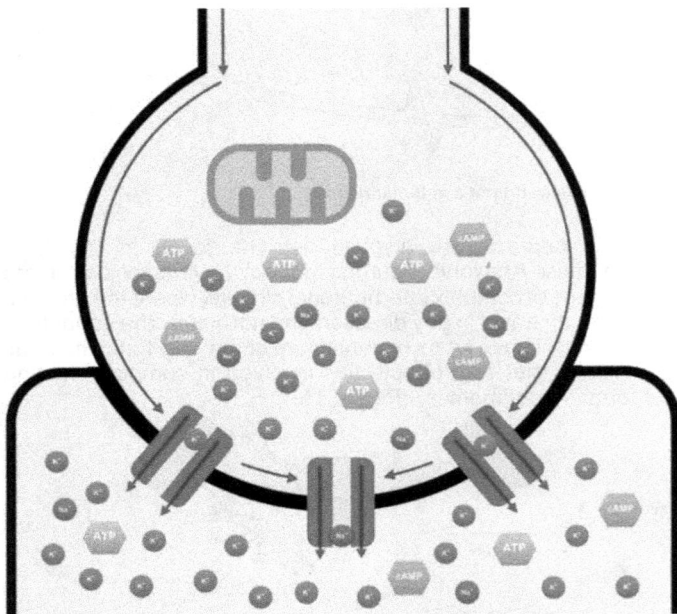

Figure 16: Illustration of an electrical synapse. Please note that the red arrow indicate the spread of action potential from the presynaptic neuron, through the gap junctions and into the postsynaptic neuron. In addition, other substances such as Na^+ and K^+ cations as well as ATP and intracellular messengers like cAMP also flows through the gap junctions

In nervous tissue, electrical synapses are a minority in comparison to chemical synapses. Nonetheless, this type of synapse is found in both the CNS and the PNS. Electrical synapses are formed when the membrane of the axonal terminal of the presynaptic neuron are interconnected through gap junctions with the plasma membrane of the postsynaptic neuron. These gap junctions allow the flow of ATP and various intracellular metabolites, such as secondary messengers, like cAMP, to be transferred between neurons. In addition, gap junctions permit electrical currents (i.e. action potential) to flow passively through the intercellular channels. It has been observed that the transmission of action potentials through the gap junctions is extremely fast; in fact it is nearly instantaneous. One of the purposes of electrical synapses is to coordinate and synchronize nervous impulses among populations of neurons. For example, the arcuate nuclei of the hypothalamus produces a peptide hormone called growth hormone releasing hormone (GHRH). The arcuate nuclei are

interconnected by electrical synapses and a single action potential can trigger the simultaneous release of GHRH into the hypothalamohypophyseal portal system, located within the infundibulum or the pituitary stalk (Figure 16).

NEUROGLIAL CELLS

Neuroglial cells, also known as glial cells, do not directly participate in generating and conducting action potentials, nor do they manufacture and secrete neurotransmitters. However, neuroglial cells serve various supportive functions in both the CNS and PNS (Figure 17).

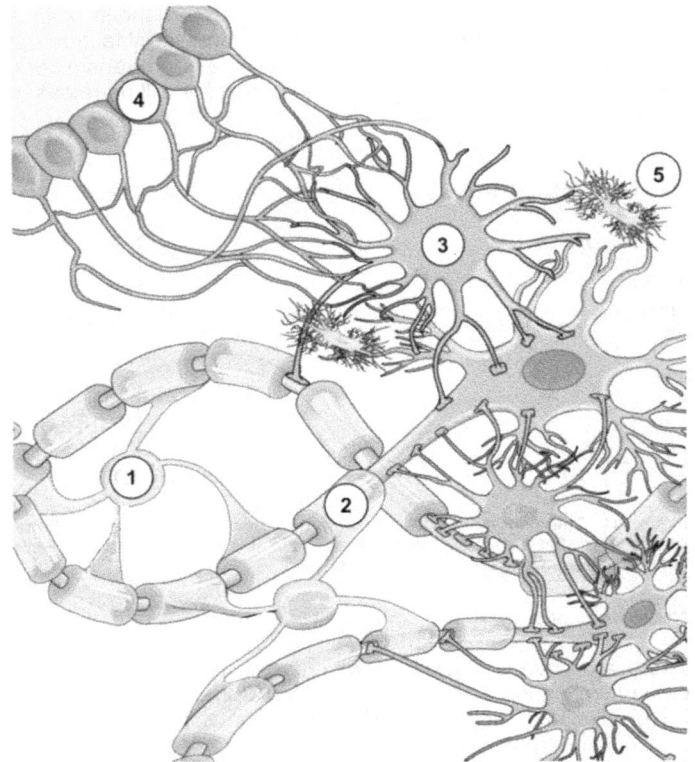

Figure 17: Illustration of neuroglial cells found in the CNS. ① Oligodendrocyte with numerous cellular extensions forming ② myelin sheath. ③ Astrocyte forming connections with a multipolar neuron. ④ Ependymal cells forming the lining of the ventricles and ⑤ the microglial cell

Astrocytes

Astrocytes are major supportive cells in the CNS and systematically overlay the entire CNS in a connecting and non-overlapping manner, as well as, forming extensive contact with neighboring astrocytes via gap junctions. Astrocytes are divided into two main subtypes: protoplasmic astrocytes, found in the gray matter, and fibrous astrocytes, found in the white matter (Figures 18 and 19).

Both astrocyte subtypes makes extensive contacts with blood vessels and help regulate blood flow in the CNS. For example, astrocytes release paracrine secretions, such as prostaglandins and nitric oxide, which causes smooth muscle contraction and dilations, respectively. In addition, astrocytes have been shown to actively transport glucose molecules from the blood vessels and distribute them to the axons at nodes of Ranvier. This glucose transport role performed by the astrocytes allows the replenishment of these energy molecules for nerve cells.

Astrocytes play an important role in the regulation of the composition of cerebral spinal fluid (CSF). Astrocytes chemically regulate the formation of tight junctions between the endothelial cells (i.e. simple squamous epithelial cells) of the capillaries, and in turn, contribute to the formation of blood brain barrier

(BBB) (Figure 19). The BBB is created as a diffusion barrier that regulates the flow of molecules entering the CNS based on the polarity and size of the molecules. In addition, BBB protects the brain from pathogens, as well as, potentially toxic substances and chemicals from entering the CNS, while allowing the exchange of nutrients and wastes between the cells of the brain and spinal cord with the circulatory system (Figure 18).

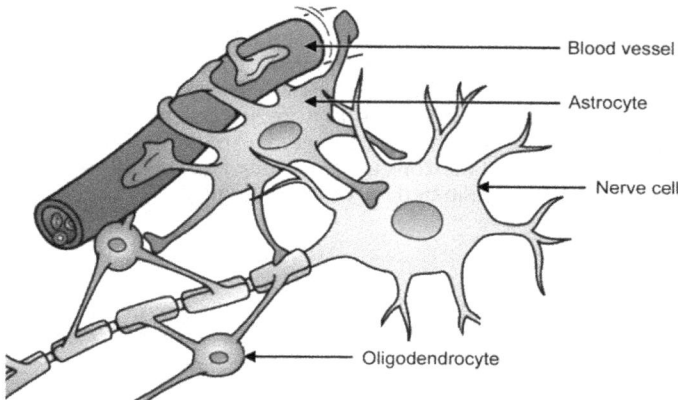

Figure 18: Illustration of an astrocyte and its interaction with a blood vessel and a nerve cell

The cellular processes of astrocytes envelope nearly all synapses in the CNS. It is understood that these neuroglial cells function to maintain the fluid, ion, pH, and neurotransmitter levels. For example, astrocyte processes found at the synapses contain transporters for glutamate, GABA, and glycine, which free the synaptic cleft of any residual neurotransmitters. Once these neuro-transmitters are actively transported into the cytoplasm of astrocytes, these ligands are rendered inert through the catabolic actions by enzymes.

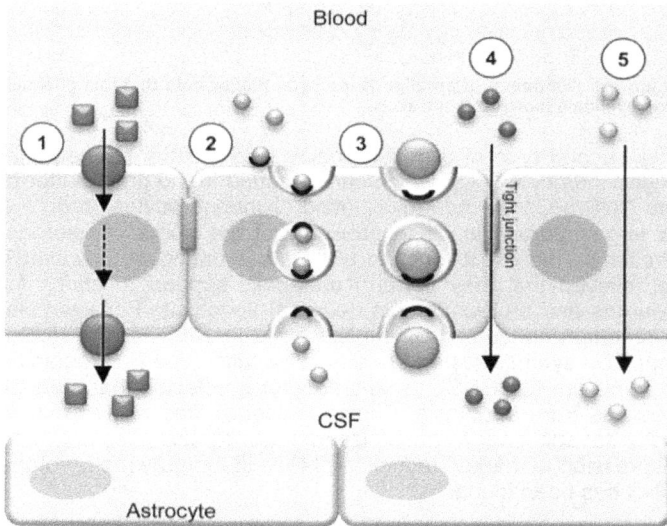

Figure 19: Illustration of the BBB. ① Glucose, amino acid and nucleosides are actively transported into the CSF. ② Insulin and other ligands are transported via receptor mediated pinocytosis. ③ Various plasma proteins are transported via phagocytosis. ④ Water soluble agents are allowed to diffuse through the tight junctions and ⑤ lipid soluble agents directly diffuse through the membrane of endothelial cells

For example, after gluatmate is absorbed by the astrocytes, it is processed by an enzyme called glutamine synthetase which converts glutamate into glutamine. Glutamine is then released back into the synapses, making it available for the reformation of glutamate by the nerve cell (Figure 20).

Once absorbed by the astrocyte, GABA is processed by the enzyme GABA transaminase and succinate-semialdehyde dehydrogenase into succinate before being sent into the Krebs

cycle for energy production (Figure 21).

Glycine, after its absorption by astrocytes, is catabolized by P-enzyme (glycine dehydrogenase), T-enzyme (aminomethyltransferase) and L-enzyme (dihydrolipoyl dehydrogenase) via a carrier protein called H-protein, in a process known as the glycine cleavage system. The glycine cleavage system converts this amino acid into 5,10-methylenetetrahydrofolate which can be used as a substrate or as a coenzyme (a "helper" compound that is assist in biochemical transformations in various metabolic cycles within the human body) in various celllar processes (Figure 22).

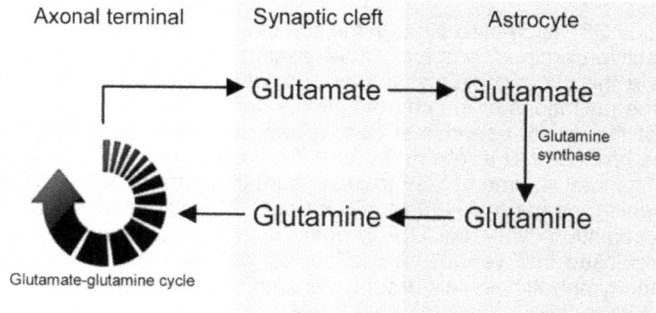

Figure 20: Illustration of the absorption of glutamate by astrocyte and its process into glutamine before being released into the synaptic cleft and reabsorbed by the axonal terminal to enter the glutamate-glutamine cycle

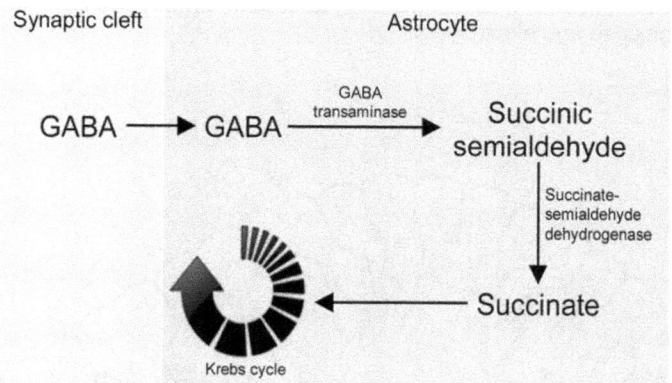

Figure 21: Illustration of the absorption of GABA by astrocyte and its process into succinate before transferring the molecule to the Krebs cycle

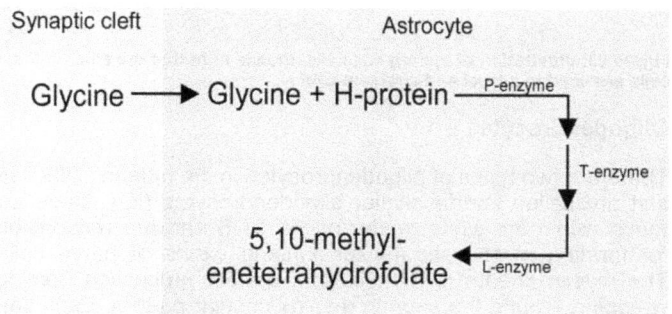

Figure 22: Illustration of the absorption of glycine by astrocyte and its process into 5,10-methyl-enetetrahydrofolate

Astrocytes are also involved in a response known as reactive astrogliosis, which may be viewed as beneficial or detrimental to the repair or regenerative process of the CNS. It is known that soon after tissue damage in the CNS has taken place, astrocytes will immediately form scar tissue to wall off the injured site in an effort to limit the spread of inflammation to other healthy nerve cells. However, in doing so, the scar tissue will hinder any possible regeneration or healing process of the damaged tissue. By impeding the healing process, astrocyte formed scar tissue will cause permanent, and at present, irreversible damage to the CNS.

Ependymal Cells

Ependymal cells are cuboidal-shaped ciliated cells found lining the ventricles of the brain, as well as, the central canal of the spinal cord. Although these cells resemble simple cuboidal epithelial cells, they do not possess a basement membrane. Instead, they have root-like structures that penetrate deep into the tissues. Ependymal cells along with capillaries form the choroid plexuses within the lateral and fourth ventricles and are responsible for creating cerebral spinal fluid (CSF).

The CSF is created through the combined processes of diffusion, active transport, and membrane assisted transport (pinocytosis) via the choroid plexus and its circulation. This is prompted by the pulsations of the choroid plexus and through the movement of the apical ependymal cilia. It has been estimated that CSF is produced at a rate of 0.2 to 0.7 ml/min or 600 to 700 ml/day. The total volume of CSF found in human adults is 140 to 270 ml, which means that 430 to 460 ml of CSF is returned to general circulation every day. CSF is absorbed by the arachnoid villi and released into venous circulation, as well as, being drained by the lymphatic vessels around the cranial cavity and spinal canal (Figure 23).

CSF serves as a fluid barrier to protect the brain from shocks. In addition, the CSF is involved in the metabolic process of the CNS because it supplies the nerve cells and neuroglial cells with glucose and electrolytes.

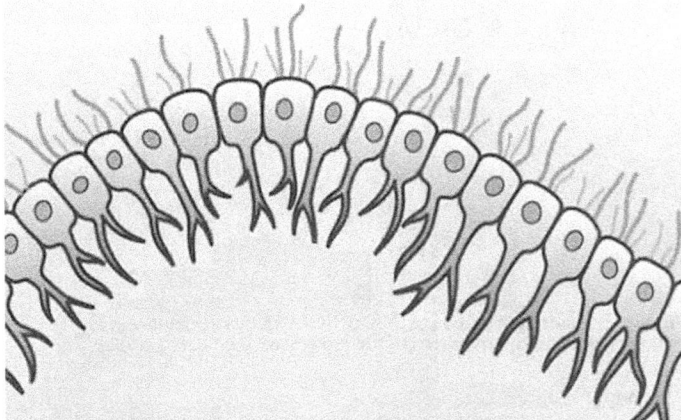

Figure 23: Illustration of ependymal cells. Please note that the cilia on these cells are used to propel and circulate CSF

Oligodendrocytes

There are two types of oligodendrocytes in the human CNS. The first are called interfascicular oligodendrocytes (IO), these are found within the white matter of the CNS and are responsible for forming myelin sheath surrounding axons of nerve cells. The myelin sheath is formed from cellular processes from IO wrapping around the axon in a spiral/circular pattern. Each time the cellular process circles, the axon cytoplasm is squeezed out of that area and pushed towards the surface of the wrapping. Put simply, myelin sheath is made out of the cellular membrane of IO minus the cytoplasm (Figure 24). Unlike the Schwann cells, which are myelinating cells of the PNS, oligodendrocytes can form myelin sheath around many axons instead of only a single axon.

Myelin sheaths form an electrical insulator along segments of the axon leaving only small exposed spaces between myelin sheaths called node of Ranvier. It is at the nodes of Ranvier that the myelinated axolemma depolarizes. For example, when the membrane at the node is excited (depolarized), the action

potential will not be allowed to flow through the insulating sheath. Nonetheless, the action potentials can "jump" over (flow out and around) the insulation and depolarize the next node. The ability of allowing the electrical current to "jump" from one node to the next node limits the length of the membrane that is needed to depolarize and increases the speed by which the action potential spreads down the length of the axon. Comparatively, in non-myelinated axons, found in gray matter, action potential is propagated by the influx of Na^+ ions through each and every voltage-gated channel along the entire length of the axon. The opening of one voltage-gated channel causes one local circuit to depolarize, where in turn, results in the depolarization of the adjacent piece of membrane in a sequential and continuous fashion. This methodical way of movement of the action potential along non-myelinated axons is equally effective, but substantially slower than the myelinated nerve fibers (Figure 27).

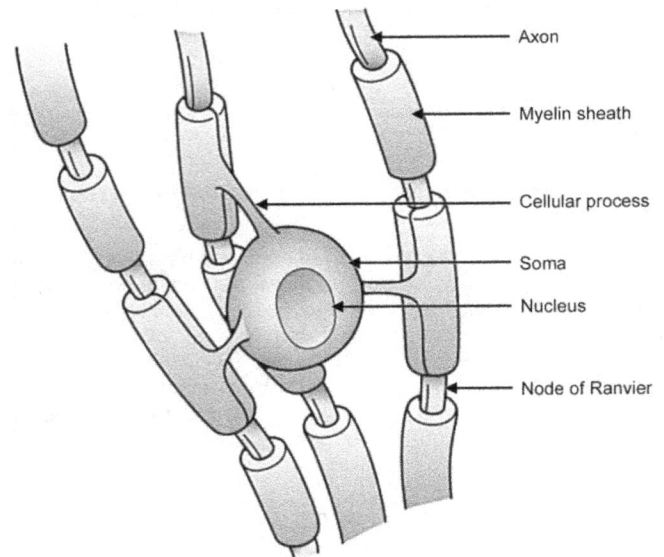

Figure 24: Illustration of an oligodendrocyte. Please note that this glial cell can myelinate more than one axon

The second type of oligodendrocyte is called the perineuronal oligodendrocyte (PO), which can be found in the gray matter of the CNS. POs are non-myelinating oligodendrocytes, attaching to large neurons in close proximity of the soma of neurons. Presently, the exact function of POs are unknown, however it is believed that they perform metabolic support functions for neurons and protect against neuronal apoptosis. For example, a research conducted by Taniike et. al in 2002, demonstrates that POs synthesizes and release Lipocalin-Type Prostaglandin D Synthase (L-PGDS), an anti-apoptotic molecule that protects neurons from apoptosis. Clinical evidence has shown that in patients suffering from schizophrenia, bipolar disorder, and depression, a marked reduction of POs in the gray matter of the CNS has been found.

Microglial Cell

Microglial cells (MC), also known as, microglia, are the phagocytic cells of the CNS. MCs are in many ways similar to macrophages and are derived from a similar embryonic origin, hematopoietic stem cells. MCs are highly mobile and they constantly survey the CNS for foreign materials or damaged cells. If there are no foreign materials detected or damage done to CNS tissues, the MCs remain in a resting stage, and are referred to as "resting MCs." However, if the neurons of the CNS suffer any damage, the cytokines released by the damaged tissue immediately activate and attract MCs and cause them to undergo a multistage activation process that transforms these cells into the "activated MCs." Once activated, these cells begin to proliferate and phagocytize damaged cells. In addition, MCs have the capacity to release various substances one of which is a toxin called reactive oxygen intermediates (ROIs), oxygen free radicals which are capable of killing neurons healthy or otherwise (Figure 16).

Schwann Cells

Schwann cells, also known as neurolemmocytes, are the neuroglial cells of the PNS. Schwann cells are subdivided into two groups: myelinating Schwann cells and nonmyelinating Schwann cells (Figure 25).

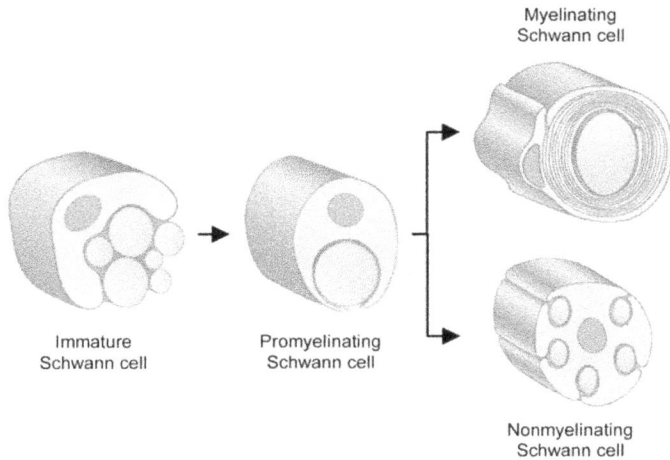

Figure 25: The Schwann cell maturation process to form myelinating or nonmyelinating Schwann cells

Myelinating Schwann cells form the myelin sheath by wrapping itself, approximately 100 times around a small segment of an axon. Each time it wraps around the axon, cytoplasm is pushed away from the inner layer until all the organelles, including the nucleus, exist only on the outer most layer of the Schwann cell. The location where all the organelles exist is called neurilemma (Figure 26).

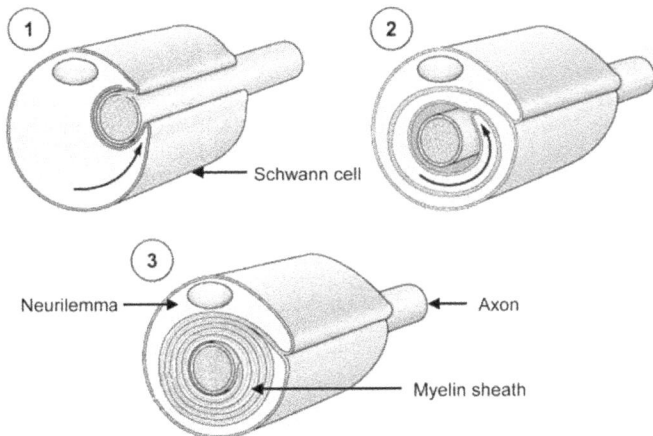

Figure 26: Illustration of myelinating Schwann cells which at first ① envelopes the axon before ② wrapping itself around the axon. ③ After several hundred wraps around the axon, myelin sheath is formed

The myelin sheath forms an electrical insulator covering a small segment of the axon. Like the myelin sheaths formed by oligodendrocytes, the segment of axon covered by myelin sheaths (wrapped by Schwann cells) will not depolarize. Instead, the only location where action potential is propagated is at the nodes of Ranvier, which are small exposed segments of axon existing in between adjacent Schwann cells. Therefore, the action potential will "jump" over (flow out and around) the insulation and depolarize one node of Ranvier at a time until it quickly transverses the entire length of the axon. The ability of allowing the electrical current to "jump" from one node to the

next node limits the length of the membrane that is needed to depolarize and increases the speed by which the action potential is spread down the length of the axon (Figure 27).

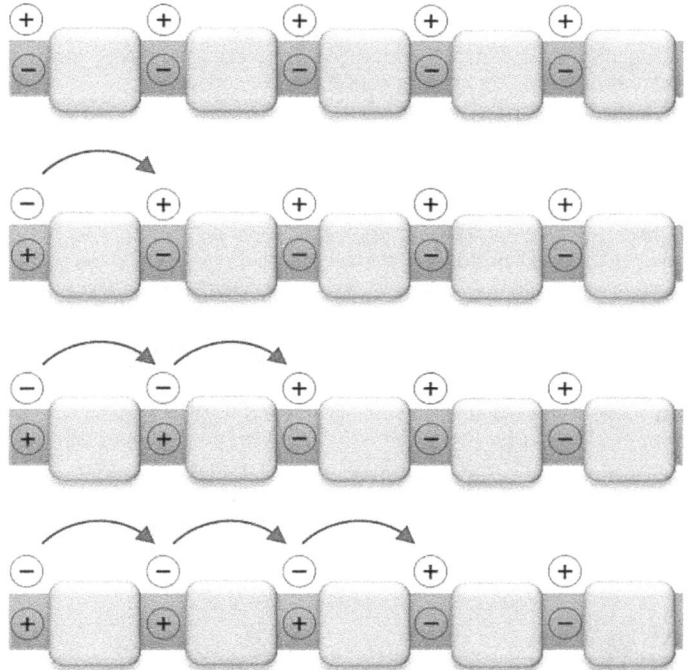

Figure 27: Illustration of how action potentials perpetuate along the length of myelinated nerve fiber. Please note that the depolarization only occur at the node of Ranvier and not within the insulated (myelinated) segments of the axon

Nonmyelinating Schwann cells exist within the non-myelinated fibers of the PNS. Instead of wrapping around the segments of the axon to form myelin sheath, nonmyelinating Schwann cells embed 1 to 12 small axons within grooves on its cellular surface. Please note that embedding is not the same as wrapping the entire cell around a segment of the axon. Embedding only allows the nerve fiber (axon) to rest upon the membrane of the Schwann cells where no myelin sheaths are formed (Figure 25).

Unlike the nerves of the CNS, which are well protected by the skull and the vertebra, the nerves of the PNS are highly vulnerable to cuts and other damage. Due to the lack of physical protection for the peripheral nerves, evolutionary contingencies were made to allow peripheral nerves to heal. As long as the soma of the damaged peripheral nerve remains intact, repair and regeneration will take place. This is in contrast to the limited abilities of central nerves to make similar repairs and regenerations due to the rapidity of scar tissue formation by astrocytes, as well as, autodigestion by microglial cell.

For example, if a somatic motor neuron is severed, the axon that is distal to the soma will not survive since it is incapable of protein synthesis. This deterioration of distal axon is also referred to as Wallerian degeneration. As the distal axon deteriorates, the myelinating Schwann cells, which depend upon the axons for maintenance, also deteriorate. A drastic infusion of macrophages will be observed as these phagocytes proceed in autodigestion of cellular debris. During this time the skeletal muscle fibers, now severed from their innervating nerve fiber, begin to demonstrate signs of denervation atrophy (Figure 28).

The soma of the injured somatic motor nerve will also show signs of abnormalities since it is disconnected from nerve growth factors supplies (via retrograde axonal transport) that had been provided by the skeletal muscle fiber. Due to the lack of growth factors, the soma will exhibit signs of hypertrophy, and the Nissl bodies will begin to breakup in a process called Nissl bodies disperse. The nucleus of the somatic motor nerve will move off center as it will either proceed into apoptosis or begin the regeneration process.

Please note that not all somatic motor nerves will survive being damaged; the ability to repair and regenerate will depend on extensiveness of the damage. Nonetheless, if the somatic motor nerve proceeds down the path of regeneration, newly formed axon stumps will begin to sprout from the soma.

The newly infused Schwann cells, along with the existing basal lamina, will form a regeneration tube (also known as Büngner) and begin secreting cell adhesion molecules and nerve growth factors. One of the newly grown stumps will find its way into the regeneration tube. The regeneration tube will guide this new-growth (one axon stump) to the correct destination while other axon stomps will deteriorate. It has been estimated that regrowth occurs rapidly, at 3 to 5 mm/day. Finally, the regeneration tube will guide the new growth back to the neuromuscular junction and reestablish synaptic contact. Eventually, myelinating Schwann cells will reform myelin sheath on the regenerated portion of the axon. Nonetheless, the newly formed myelin sheath will be thinner and shorter in comparison to the original (Figure 28).

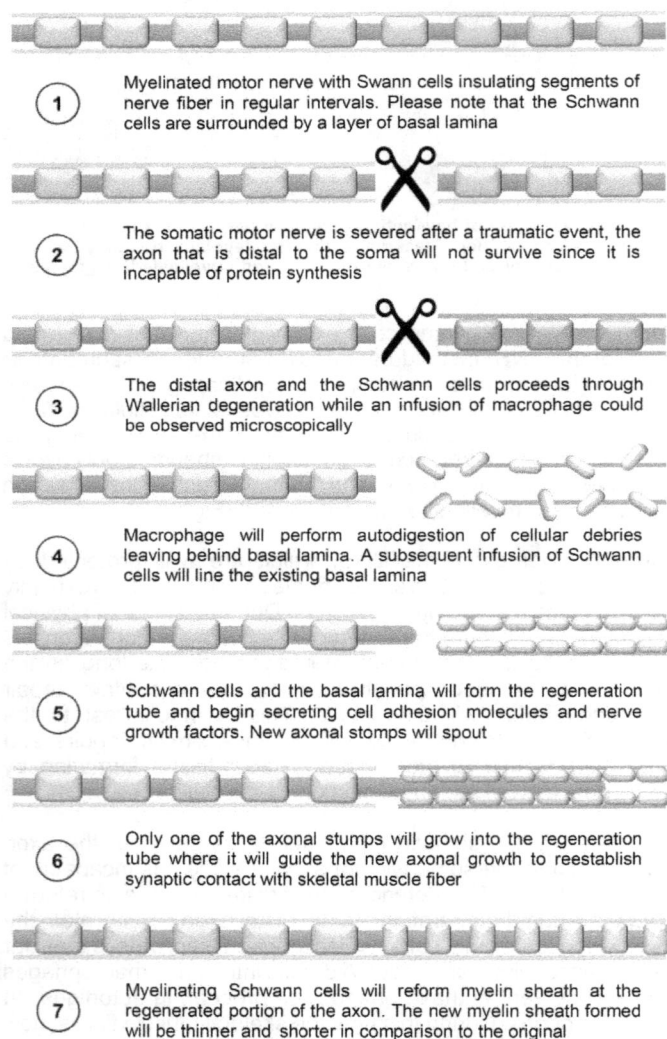

1. Myelinated motor nerve with Swann cells insulating segments of nerve fiber in regular intervals. Please note that the Schwann cells are surrounded by a layer of basal lamina

2. The somatic motor nerve is severed after a traumatic event, the axon that is distal to the soma will not survive since it is incapable of protein synthesis

3. The distal axon and the Schwann cells proceeds through Wallerian degeneration while an infusion of macrophage could be observed microscopically

4. Macrophage will perform autodigestion of cellular debries leaving behind basal lamina. A subsequent infusion of Schwann cells will line the existing basal lamina

5. Schwann cells and the basal lamina will form the regeneration tube and begin secreting cell adhesion molecules and nerve growth factors. New axonal stomps will spout

6. Only one of the axonal stumps will grow into the regeneration tube where it will guide the new axonal growth to reestablish synaptic contact with skeletal muscle fiber

7. Myelinating Schwann cells will reform myelin sheath at the regenerated portion of the axon. The new myelin sheath formed will be thinner and shorter in comparison to the original

Figure 28: Illustration of somatic motor nerve regeneration. Please note that the reformed myelin sheath are smaller and more numerous when compared with the originals

Satellite Glial Cells

Satellite glial cells (SGCs) are found surrounding the cell bodies (somas) in the sensory ganglia of the PNS. Sensory ganglia, such as the dorsal root ganglia, contain the cell bodies of sensory

neurons and serve to establish the connection between the PNS and CNS. Unlike the neurons of the CNS, sensory ganglia lack the blood-nerve barrier to isolate and protect the cell bodies. Nonetheless, these cell bodies are surrounded by great numbers of SGCs, which form an enveloping layer to provide physical support, as well as a protective barrier. SGCs possess receptors for various neuroactive agents (i.e. neurotransmitters and modulators), therefore, they are capable of receiving signals from surrounding cells in response to changes in their environment. It has been postulated that SCGs have the capabilities to influence neighboring neurons and participate in signal processing, as well as, transmission in sensory ganglia (Figure 29).

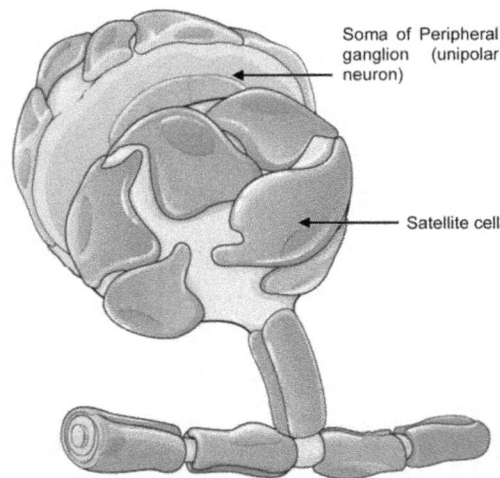

Soma of Peripheral ganglion (unipolar neuron)

Satellite cell

Figure 29: Illustration of satellite cells found within peripheral ganglia

ELECTRICAL PHYSIOLOGY

Both muscular and neuronal excitations are based on a cellular mechanism known as electrophysiology. Electrophysiology is defined as the manipulation of ionic concentrations to create an electrical potential or membrane potential. An electrical potential is a form of potential energy that can be converted into an electrical current, or kinetic energy, under the right conditions. For example, if an artificially manipulated ionic concentration is allowed to flow down its diffusion gradient, an electrical current will be generated. This mechanism generates an electrical current and because it is similar to the mechanism of batteries, which are used to power thousands of electronic gadgets that intimately occupy all aspects of our lives.

The plasma membrane of the neuron, like the sarcolemma, is electrically charged and polarized. The polarization of the plasma membrane is mainly caused by the unequal distribution of Na^+ and K^+ cations in the extracellular fluid and within the neuroplasm respectively. To a minor extent, chloride (Cl^-) anions also contribute to the polarization of the plasma membrane. This uneven distribution generates a charge difference also known as resting membrane potential (RMP). The RMP is maintained through the actions of an active transport integral protein called sodium-potassium exchange pump (SPEP) and chloride pumps that actively transport Cl^- out of the neuroplasm (Figure 30).

SPEP is an antiport (i.e. active transporter), which sends three Na^+ cations into the extracellular fluids and brings two K^+ cations into the neuroplasm. Because of the SPEP, the concentration of K^+ cations is approximately 40 times higher in the neuroplasm than the extracellular fluid. In contrast, the concentration of Na^+ cations is approximately 12 times higher in the extracellular fluid than in the neuroplasm. It has been calculated that the SPEP accounts for approximately 70% of the ATP consumption in the nerve cell (Figures 30 and 31).

Even with the high concentrations of K^+ cations, the neuroplasm is negatively charged in comparison to the extracellular fluid. This negative charge is mainly the result from negatively charged functional groups from proteins, such as phosphates (PO_4^{-3}), sulfates (SO_4^{-2}), ATP, RNA, etc. that are present within the neuroplasm (Figures 30 and 31).

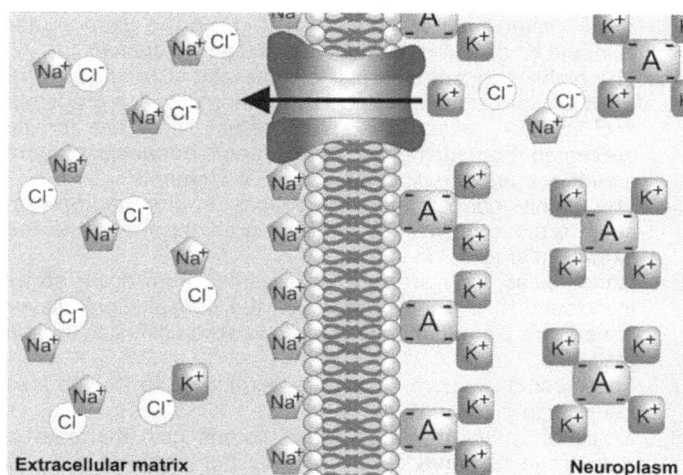

Figure 30: Illustration of the ionic concentrations of a nerve cell which results in creating the resting membrane potential. Please note that "A" represent the anions (PO_4^{-3}, SO_4^{-2}) that are present within the neuroplasm

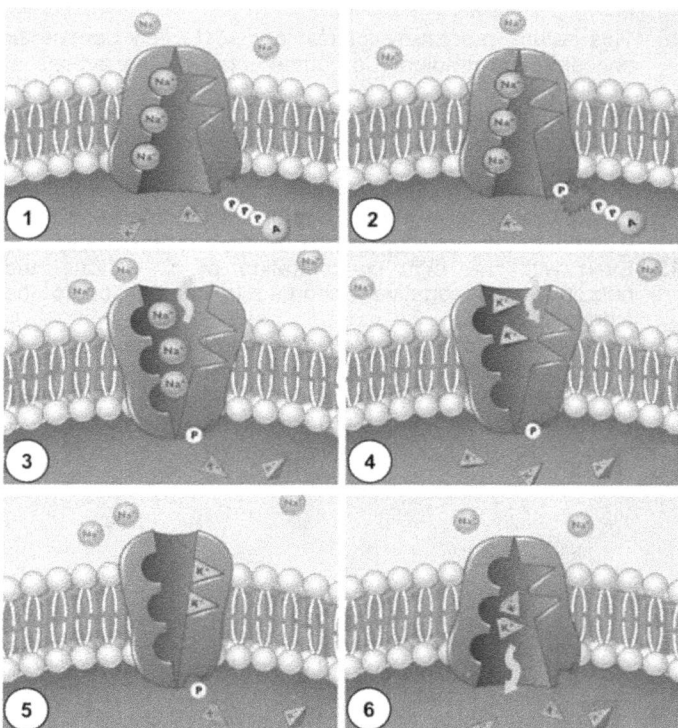

Figure 31: Illustration of sodium-potassium exchange. ① Three Na⁺ are collected by the antiport from the neuroplasm. ② Energy and phosphate is released by hydrolyzing the high energy bond of ATP. ③ The release energy and phosphate causes a conformational change in the active transport protein which sends three Na⁺ out of the cell. ④ Two K⁺ is collected from the extracellular fluid prior to the removal of the phosphate group from the transport protein. ⑤ The removal of the phosphate result in another conformational change which ⑥ brings two K⁺ into the neuroplasm

Experiments have shown that the plasma membrane of neurons is more permeable to K⁺ cations in comparison to other ions. Because of this permeability, K⁺ tends to leak out of the cell and into the extracellular fluids because of the concentration gradient, causing the neuroplasm to become even more negative. Nonetheless, the outflow of K⁺ will eventually reach equilibrium, as the increasing negative charge within the neuroplasm will prevent excessive amounts of positively charged K⁺ from leaving

the neuroplasm. Additionally, the SPEP will actively transport K+ cations back into the neuroplasm.

Even though the plasma membrane is less permeable to Na⁺, small amounts of these cations will flow down their concentration gradient and into the neuroplasm. Nonetheless, this minute amount of Na⁺ will also be actively removed by the SPEP into the extracellular fluid. By maintaining the imbalance of ionic concentrations, a charge difference between the extra-cellular fluid and neuroplasm is observed. Using a voltmeter, the average RMP of nerve cells is measured at approximately -70 millivolts (mV) (Figure 32).

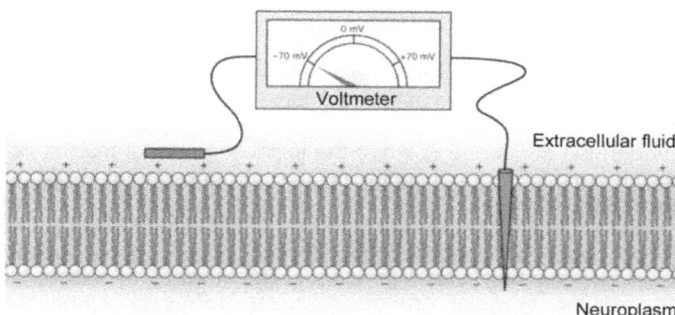

Figure 32: Illustration of the resting membrane potential which is measured at -70 mV

Local Potential

Local potential is defined as a limited depolarization or electrical discharge that may or may not result in an action potential. Local currents are generally formed only at the dendrites, and at times, the soma, due to the number of gated channels present within the plasma membrane of the cell body. For example, it has been calculated that there are approximately 50 to 75 voltage gated Na⁺ channels per square micrometer (μm) found in the membranes of both the dendrites and the soma. In contrast, beginning at the initial segment of the axonal hillock, the number of voltage gated channels increases drastically to 350 to 500 voltage gated Na⁺ channels per square micrometer (μm). Simply put, the amount of voltage gated channels found in the dendrites and soma is inadequate for the creation of an action potential.

Local potentials are graded, which signifies that the voltage generated may vary in magnitude depending upon the intensity of the stimulus. For example, a weak stimulus will only cause the opening of a few Na⁺ channels and the generation of a minor current (i.e. slight depolarization). In contrast, a stronger stimulus will result in the opening of more Na⁺ channels and the generation of a more powerful current (i.e. stronger depolarization).

Local potentials are reversible, which means that the current generated will dissipate and the RMP is reinstated, as soon as, the stimulation stops. Additionally, local potentials have a tendency to become weaker as they get further away from the point of stimulation. This decremental spread of electrical current is because Na⁺ cations have the propensity to leak out of the neuroplasm after the initial influx of these cations soon after the opening of the sodium channels. The decremental spread of generated currents prevents weaker local potentials from having long distance effects.

Local potentials can be either excitatory or inhibitory and it often depends upon the type of neurotransmitter that is binding to the receptor protein. Excitatory local potential are at times triggered by excitatory neurotransmitters, such as acetylcholine (ACh), which may result in the generation of an action potential if the strength of the stimulus reaches the threshold. Inhibitory local potential are at times caused by the binding of inhibitory neurotransmitters, such as glycine or GABA, to receptor proteins. These inhibitory neurotransmitters cause the opening of chloride

(Cl⁻) channels which prompts the influx of Cl⁻ anions into the neuroplasm. The flow of Cl⁻ into the neuroplasm causes the membrane potential to become more negative or hyperpolarized. A hyperpolarized membrane makes a nerve less sensitive and less prone to generate an electrical discharge.

In most cases, local potentials are formed at the dendritic end of the nerve cell and spread through the soma before they are summarized and weighed against the threshold at the initial segment of the axonal hillock. Summation is defined as the accumulation of stimuli from various stimulations in a short period of time, which in turn will generate a stronger local potential (Figure 33).

Action Potential

Action potentials will only be generated if the threshold is reached. As mentioned previously, local potentials are formed in the dendrite, where the stimulus is first received, then they flow through the soma and down to the initial segment of the axonal hillock where the local potential is summarized and weighed against the threshold.

It has been calculated that the threshold of a nerve cell is approximately -55 mV. Put simply, the current of the local potential has to reach -55 mV in order for an action potential to be formed or an all-or-none response is generated. If the local potential is not strong enough to reach the threshold, an action potential will not be generated and the local potential will reverse itself and dissipate (Figure 33).

Action potentials are nondecremental, which indicates that this electrical current generated will not become weaker as it travels down the length of the axon. Action potentials of all nerves will remain the same strength from the axonal hillock to the axonal terminal. Once generated, action potential is irreversible. As indicated previously, the action potential is an all-or-none response. Once the threshold is reached and action potential is generated, the nerve will generate a voltage of approximately +35 mV until it proceeds through a period of repolarization.

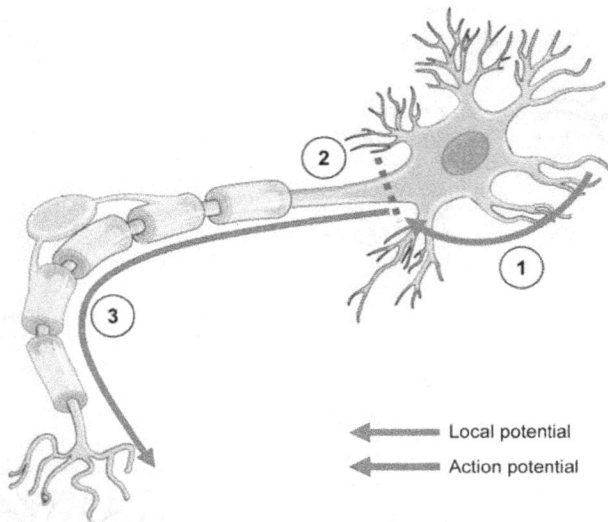

Figure 33: Illustration of the formation of an action potential. ① Local potentials are formed after dendritic stimulation and spread down the dendrite and through the soma. ② The local potentials are summarized at the initial segment located at the axonal hillock. ③ If the threshold is reached, an action potential will be formed

SUMMARY OF ACTION POTENTIAL FORMATION

Polarization

1. The active transport transmembrane protein called, the sodium-potassium exchange pump (SPEP), sends three Na⁺ out and brings in two K⁺ into the neuroplasm (Figures 30 and 31)

2. The active transport transmembrane protein, called the chloride (Cl⁻) pump, is responsible for sending Cl⁻ anions out of the neuroplasm

3. SPEP creates a Na⁺ diffusion gradient in the extracellular fluid and K⁺ diffusion gradient within the neuroplasm

4. The highly concentrated Na⁺ cations and Cl⁻ anions in the extracellular fluid seek to flow down their diffusion gradients and into the neuroplasm. Nonetheless, they are mostly prevented from doing so by the semipermeable plasma membrane and the closed Na⁺ and Cl⁻ channels

5. The highly concentrated K⁺ cations in the neuroplasm seek to move down their diffusion gradient and out to the extracellular fluid.

6. Nonetheless, they are mostly prevented from doing so by the closed K⁺ channels. However, the phospholipid bilayer of neuron's plasma membrane allows some diffusion to take place

7. The plasma membrane is more permeable to K⁺ and less permeable to Na⁺ cations.

8. Negligible amounts of Cl⁻ leak into the cell; the plasma membrane of nerve cells is mostly non-permeable to Cl⁻ anions

9. Due to the diffusion gradient and the Laws of Electromagnetic Attraction, K⁺ have the tendency to leak out of the into the extracellular fluids while Na⁺ have the affinity to diffuse into the neuroplasm

10. This natural propensity for Na⁺ and K⁺ to flow down their concentration gradients is counteracted by the actions of SPEP

11. Because of the constant actions of the SPEP, the nerve cell expends nearly 70% of its energy molecules (ATPs) in maintaining the Na⁺ and K⁺ diffusion gradient

12. Due to the high concentration of Na⁺ and the comparably less Cl⁻ concentration in the extracellular fluids, the net charge external to the neuronal plasma membrane is positive

13. Even with the high concentration of K⁺ cations, the neuroplasm is negatively charged in comparison to the extracellular fluid.

14. This negative charge is mainly caused by the negatively charged functional groups of proteins such as phosphates (PO_4^{-3}), sulfates (SO_4^{-2}), ATP, RNA, etc. that are present within the neuroplasm (Figure 34)

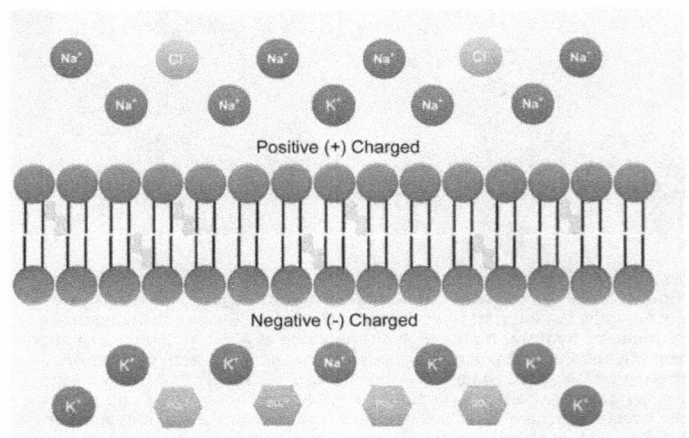

Figure 34: Illustration of Na⁺, Cl⁻ and K⁺ diffusion gradient. Please note that due to the anionic concentration of PO_4^{-3}, and SO_4^{-2}, etc. the neuroplasm is negatively charged

15. By maintaining the imbalance of Na⁺, Cl⁻, and K⁺ ionic concentrations, a notable charge difference between the extracellular fluid and neuroplasm is observed

16. Using a voltmeter, the average resting membrane potential (RMP) of nerve cells is measured at -70 millivolts (mV) (Figure 32)

Depolarization: Local Potential

17. Different types of neurons respond to different kind of stimulus
18. Nerves may respond to chemicals, light, pain, heat, or mechanical forces such as pressure
19. The type of stimulus that the nerve cell responds to is wholly dependent on the type of receptor proteins that are located in the dendrites, and at times in the soma
20. For example, in a chemical synapse, the stimulus triggers the release of signal molecules (ligands or neurotransmitters) from the presynaptic neuron. As mentioned previously, a chemical synapse is where a presynaptic nerve releases ligands (neurotransmitters or neuropeptides) via exocytosis into the synaptic cleft and triggers a local/action potential in the postsynaptic nerve.
21. These ligands will be allowed to diffuse through the fluid filled space of the synaptic cleft and bind with membrane bounded receptors located on the postsynaptic neuron or target tissue (glands or muscle fibers) (Figure 35)

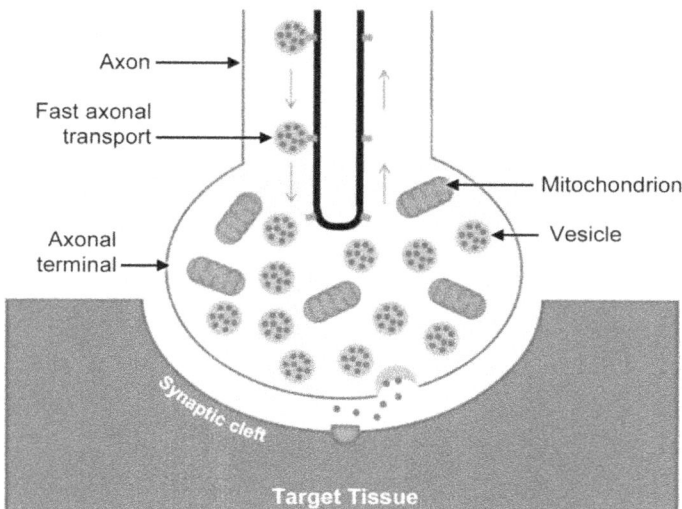

Figure 35: Illustration of a synapse. Please note that the neuropeptides contained within the vesicle are released via exocytosis into the synaptic cleft

22. If the ligand binds with an ionotropic Na^+ receptor it will cause the multimers (the 5 to 6 protein subunits) to change its shape and open the center pore. The opening of the pore will allow Na^+ cations to flow down its concentration gradient and into the neuroplasm (Figure 36)
23. If the ligand binds with metabotropic receptors, it would cause the activation of a G-protein complex. The activation of the G-protein will result in the separation of α, β and γ subunits.
24. The β- and γ-subunits form a stable dimer (joining of two molecules through intermolecular bonding) and are responsible for activating the α-subunit
25. As a result GDP (a spent energy molecule) bounded α-subunit is replaced with GTP (a charged energy molecule)
26. Subsequently, the α-subunit-GTP complex causes the Na^+ channel protein to change its conformation and allow Na^+ cations (which are kept in high concentrations in the extracellular fluids) to flow down their concentration gradient and into the neuroplasm (Figure 37)
27. The influx of Na^+ cations causes the membrane to depolarize and creates an electrical current, which is also known as, a local potential. In addition, with the influx of the Na^+ cations the membrane polarity is reversed at that area
28. The electrical current (local potential) generated will cause the adjacent voltage gated Na^+ channel to open, allowing more Na^+ flow into the neuroplasm
29. Allowing more Na^+ to enter the nerve cell, additional electrical currents will be generated
30. The additional electrical current will cause the next voltage

gated Na^+ channel to open and so forth

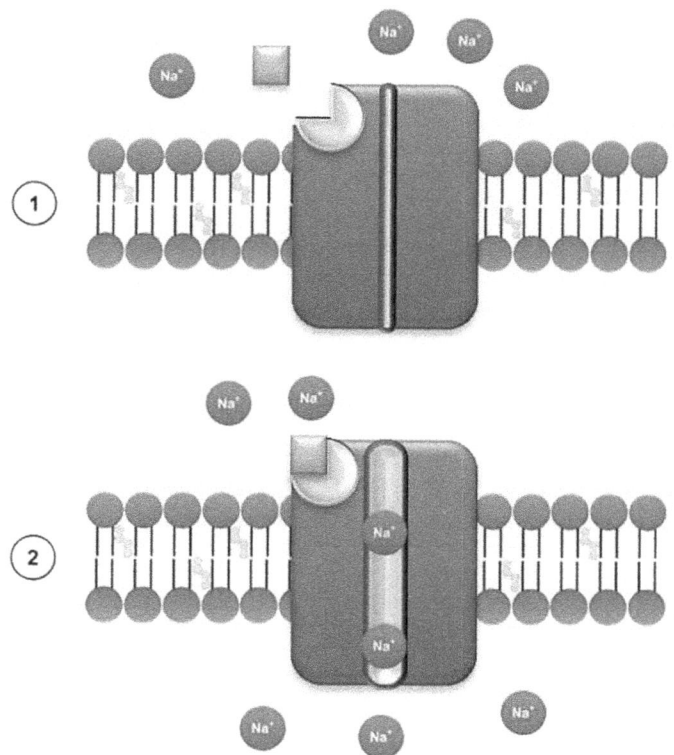

Figure 36: Please note that the ionotropic receptor is a transmembrane protein that possesses a ligand binding site directly linked to the ion channels to form a single molecular entity. ① The pore is in the closed position prior to the binding of the ligand. ② Once the signal molecule binds to the ligand binding site, the intermolecular interaction will result in the opening of the pore thereby allowing Na^+ to flow down its diffusion gradient

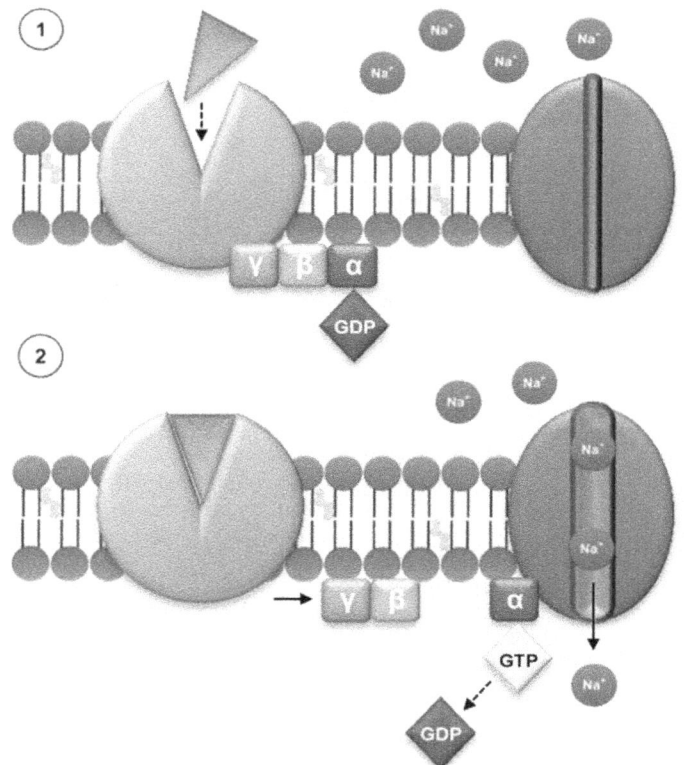

Figure 37: ① Prior to the binding of ACh the gated channel is closed. ② Binding of ACh causes the α, β and γ subunits to separate and GDP on the α subunit is exchanged with GTP. The α subunit-GTP complex causes a conformational change in the gated channel causing it to open, allowing Na^+ to flow into the cell via its diffusion gradient

31. Like a series of dominos in a cascading effect, the opening of one voltage gated Na^+ channel will cause the next voltage gated Na^+ channel to open and generate additional electrical current (Figure 38)

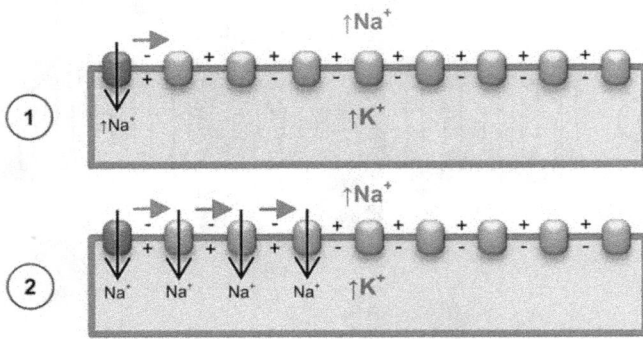

Figure 38: The perpetuation of a local potential. ① The pore of the ligand gated Na^+ channel (red) opens after the binding of ligand which allows the influx of Na^+ and the generation of an electric current (red arrow). ② Subsequently, the electric impulse will spread down the membrane and open voltage gated Na^+ channels, thereby allowing more Na^+ to flow into the cell

32. The opening of voltage gated Na^+ channels will perpetuate the local potential down the length of the dendrite and flow across the plasma membrane of the soma

33. As more Na^+ cations flow down their concentration gradient, the membrane potential will become more and more positive (Figure 39)

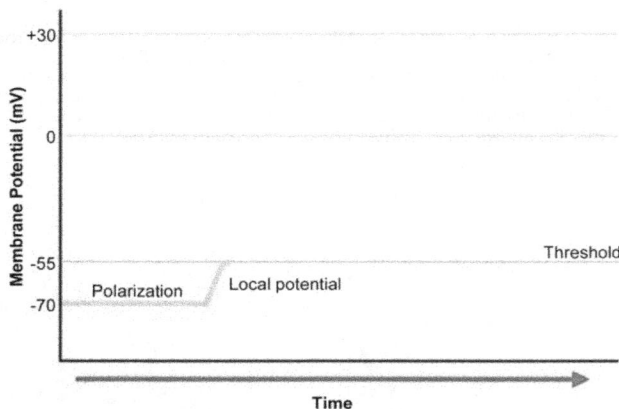

Figure 39: Illustration of the depolarization of a nerve cell and the formation of a local potential. The graph plots membrane potential (mV) versus time. Please note that the membrane potential becomes more positive (proceeding from -70 mV towards -55 mV, which is also known as the threshold) as increased amounts of Na^+ enters the neuroplasm

34. Please note, it has been calculated that there are only 50 to 75 voltage gated Na^+ channels per square micrometer (μm) in the membrane of the dendrites and the soma. Simply put, the amount of voltage gated channels found in the dendrite and soma are inadequate for the creation of an action potential. From the dendrite through the soma only local potential can be created

35. At the initial segment of the axonal hillock, the generated local potentials are summarized and weighed against the threshold (-55 mV)

36. Only when local potentials reach -55 mV, will an action potential (all-or-none response) be created

Depolarization: Action Potential

37. At the initial segment of the axonal hillock, the number of voltage gated Na^+ channels increases to 350 to 500 channels per square micrometer (μm). The quantity of voltage gated Na^+ channels available in the axon are adequate for the formation of action potential

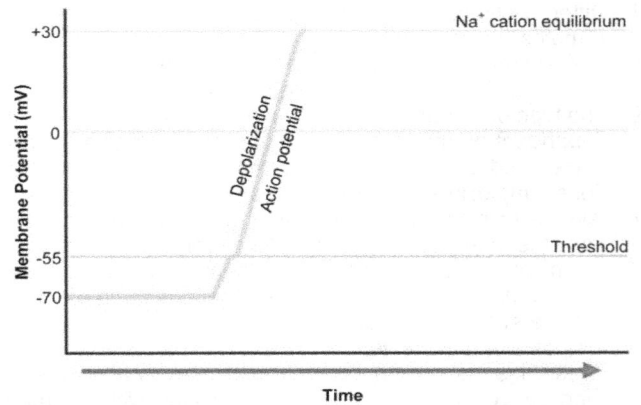

Figure 40: Plotting membrane potential (mV) versus time during an action potential. Once the threshold is reached an action potential is formed. Please note that the concentration of Na^+ reaches equilibrium at +30 mV

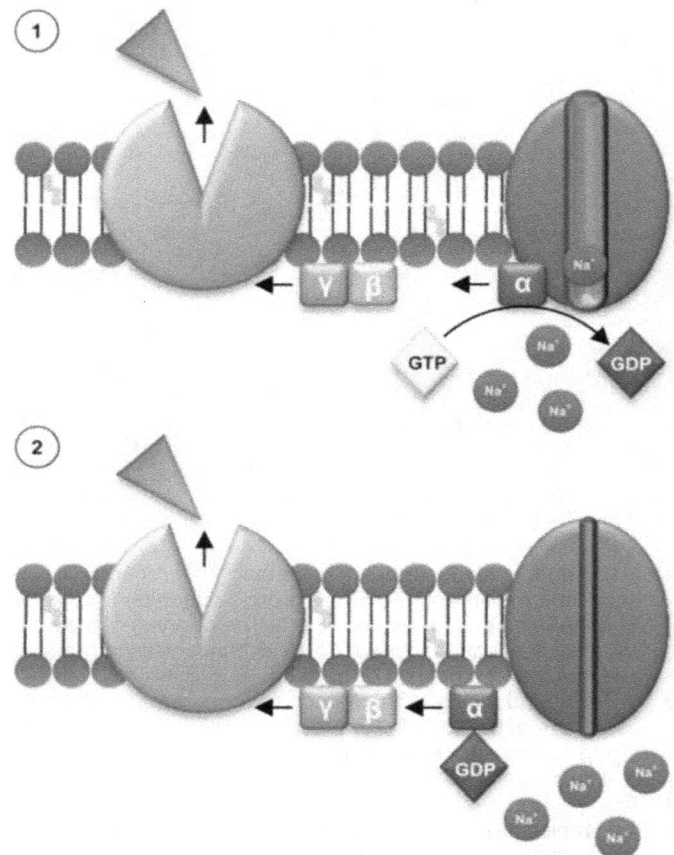

Figure 41: Illustration of a G-protein complex. ① Ligand is removed from the G-protein binding site thereby triggering α subunit to hydrolyze the energy molecule GTP releasing a phosphate molecule (Pi) and energy. ② Once the energy molecule GTP is hydrolyzed, the interaction between the α subunit and the gated channel is severed thereby returning the gated channel back to its original conformation (closed gate)

38. Once the local current reaches the threshold at the initial segment of the axonal hillock, an action potential will be formed

39. The formation of the action potential initiates the absolute refractory period, which means that the nerve cannot be stimulated to generate another local or action potential (Figure 43)

40. Just like the local potential, the formation of the action potential (electric current) will trigger more voltage gated Na^+ channels to open, which will in turn, perpetuate the

depolarization of the axolemma (plasma membrane of the axon)

41. As more voltage gated Na$^+$ channels open, more Na$^+$ cations flows into the nerve cell (neuroplasm), which causes the membrane potential to become more positive until the membrane potential reaches approximately +30 mV

42. At +30 mV the amount of Na$^+$ in the extracellular fluids is equal to the amount of Na$^+$ in the neuroplasm. In other words, the Na$^+$ cations has reached their equilibrium (Figure 40)

43. Between 0 to +30 mV, the ligand that initiated the local potential (which eventually became action potential) will be enzymatically removed

44. If the ligand is removed from an ionotropic Na$^+$ receptor it will cause the multimers (the five to six protein subunits) to reverse its conformational change and close the center pore

45. If the ligand is removed from a metabotropic receptor, the α-subunit will hydrolyze the energy molecule GTP releasing a phosphate molecule and energy, thereby breaking the interaction with the gated channel

46. Once the connection between the α-subunit-GDP complex is severed, the gated channel will reconstitute its original conformation and prevent Na$^+$ cations from entering the cytoplasm (Figure 41)

47. Also between 0 to +30 mV, the voltage gated Na$^+$ channels close

Repolarization

48. The absolute refractory period exists through the repolarization period of a nerve (Figure 42 and 43)

49. At approximately +30 mV the voltage gated K$^+$ channels open, which allows the K$^+$ cations (kept at high concentrations in the neuroplasm) to diffuse down their concentration gradient into the extracellular matrix

50. The efflux of K$^+$ cations reverses the membrane potential from positive towards the negative

51. The voltage gated K$^+$ channels stay open longer than Na$^+$ channels, therefore more K$^+$ cations leave the neuroplasm

52. Because more K$^+$ cations leave the neuroplasm, the membrane potential drops below the RMP and proceeds to settle at approximately -80 mV

53. At -80 mV, also known as hyperpolarization, the amount of K$^+$ cations in the extracellular fluid is equal to the amount of K$^+$ cations in the neuroplasm

54. This hyperpolarization of the membrane potential lasts approximately three to four milliseconds

55. At hyperpolarization, the sodium-potassium exchange pump (SPEP) activates and begins sending three Na$^+$ out into the extracellular fluid and two K$^+$ into the neuroplasm (Figure 42)

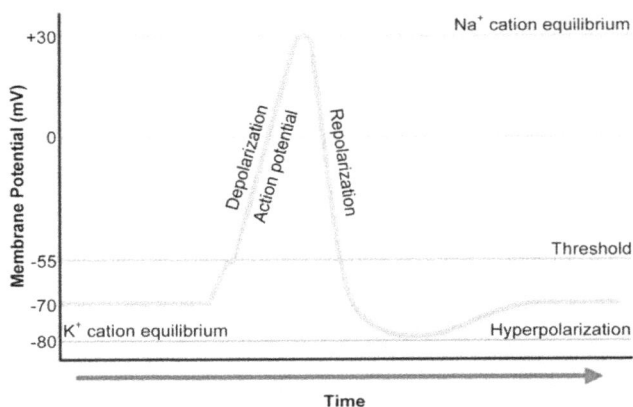

Figure 42: Illustration of repolarization and hyperpolarization of a nerve cell. Please note that at -80 mV the amount of K$^+$ within the neuroplasm is equal to the amount of K$^+$ in the extracellular matrix. This equilibrium causes the membrane potential to become hyperpolarized. It is at hyperpolarization when the SPEP is turned on and is responsible for restoring the membrane potential at -70 mV

56. Over the next 1.5 milliseconds (ms), during which time the resting membrane potential is being re-established, the nerve is in its relative refractory period, during which it cannot be re-stimulated

57. Once SPEP restores the membrane potential of the nerve cell, the relative refractory period ends and the nerve can generate local potentials and possibly another action potential (Figure 43)

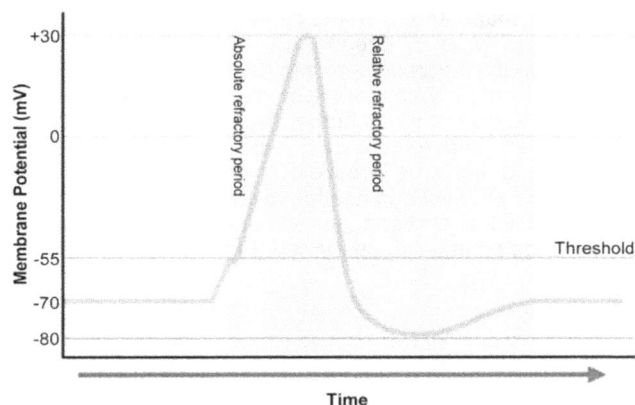

Figure 43: Illustration of absolute and relative refractory period. Please note that during both refractory periods, the nerve cell cannot be stimulated to generate an action potential

Inhibiting Action Potential Formation

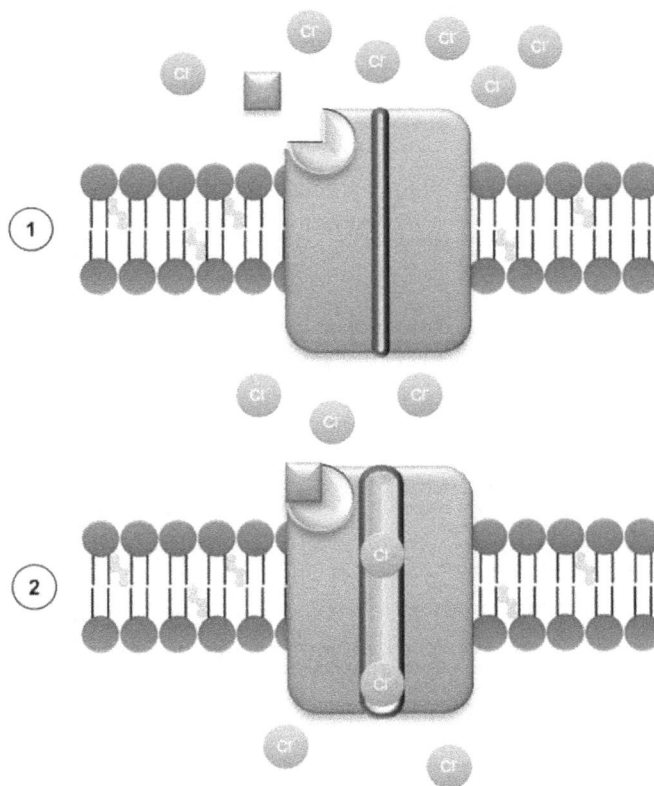

Figure 44: Illustration of a chloride (Cl$^-$) ionotropic receptor. Please note that this receptor is a transmembrane protein that possesses a ligand (glycine, glutamate or GABA) binding site directly linked to the ion channels to form a single molecule entity. ① The pore is in the closed position prior to the binding of the ligand. ② Once the signal molecule binds to the ligand binding site, the intermolecular interaction will result in the opening of the pore, thereby allowing Na$^+$ ions to flow down its diffusion gradient

58. Neurotransmitters, such as glycine and GABA, are responsible for inhibiting action potential formation

59. Glycine and GABA are released into the synaptic cleft via

exocytosis and allowed to diffuse to the target tissue and bind with their respected receptor proteins, which are ionotropic receptors

60. Ionotropic receptor proteins are composed of four to five multimers organized around a pore
61. These ionotropic receptor proteins form a chloride (Cl^-) channel
62. Once glycine or GABA binds with an ionotropic Cl^- receptor it will cause the multimers (the five to six protein subunits) to change shape and open the center pore
63. Once the pore is opened, Cl^- begins to flow down its concentration gradient and into the neuroplasm (Figure 44)
64. The influx of Cl^- will cause the membrane potential to become more negative, thereby further hyperpolarize the nerve cell
65. Increased hyperpolarization will increases the voltage needed to reach the threshold
66. Increasing the voltage needed to reach the threshold means that additional and stronger stimulus is needed in order for an action potential to be formed (Figure 45)

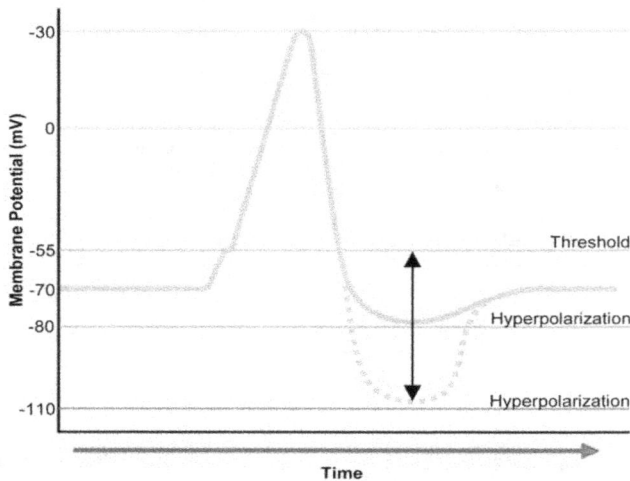

Figure 45: Influx of Cl^- anions will make the membrane potential more negative and further away from the threshold

QUESTIONS

1. Which of the following system is consists of the brain and the spinal cord?
   a. Central nervous system
   b. Circulatory system
   c. Digestive system
   d. Endocrine system
   e. Integument system
   f. Peripheral nervous system

2. The autonomic nerves are part of which of the following system?
   a. Central nervous system
   b. Circulatory system
   c. Digestive system
   d. Endocrine system
   e. Integument system
   f. Peripheral nervous system

3. Which of the following provides peripheral awareness, monitors the functions of internal organs, as well as, the surrounding environments?
   a. Autonomic nervous system
   b. Basal ganglia
   c. Brain
   d. Hypothalamus
   e. Sensory division
   f. Somatic motor division
   g. Spinal cord

4. Which of the following is involved in the control of the voluntary movements of skeletal muscles?
   a. Autonomic nervous system
   b. Basal ganglia
   c. Brain
   d. Hypothalamus
   e. Sensory division
   f. Somatic motor division
   g. Spinal cord

5. The autonomic nervous system is subdivided into _____, _____ and _____ divisions?

   a. Auditory
   b. Enteric
   c. Ophthalmic
   d. Parasympathetic
   e. Sensory
   f. Somatic motor
   g. Sympathetic

6. Which of the following function as an antagonist towards one another?
   a. Afferent and autonomic nervous systems
   b. Autonomic and peripheral nervous systems
   c. Central and peripheral nervous systems
   d. Enteric and parasympathetic nervous systems
   e. Enteric and sympathetic nervous systems
   f. Parasympathetic and sympathetic nervous systems

7. Which of the following post-synaptic nerves releases noradrenaline?
   a. Auditory
   b. Enteric
   c. Ophthalmic
   d. Parasympathetic
   e. Sensory
   f. Somatic motor
   g. Sympathetic

8. Which of the following post-synaptic nerves release acetylcholine?
   a. Auditory
   b. Enteric
   c. Ophthalmic
   d. Parasympathetic
   e. Sensory
   f. Somatic motor
   g. Sympathetic

9. Which of the following is in direct control of the gastrointestinal tract?
   a. Auditory
   b. Enteric
   c. Ophthalmic
   d. Parasympathetic
   e. Sensory
   f. Somatic motor
   g. Sympathetic

10. Which of the following is a sensory modality that involves the position of the body and body parts?
    a. Auditory
    b. Interoception
    c. Olfactory
    d. Proprioception
    e. Tactile
    f. Taste
    g. Vestibular system
    h. Vision
    i. Equilibrium

11. Which of the following is a sensory modality that involves balance?
    a. Auditory
    b. Interoception
    c. Olfactory
    d. Proprioception
    e. Tactile
    f. Taste
    g. Vestibular system
    h. Vision
    i. Equilibrium

12. Which of the following is a sensory modality that involves the sense of physiological conditions?
    a. Auditory
    b. Interoception
    c. Olfactory
    d. Proprioception
    e. Tactile
    f. Taste
    g. Vestibular system
    h. Vision
    i. Equilibrium

13. Which of the following is a sensory modality that involves the photoreception?
    a. Auditory
    b. Interoception
    c. Olfactory
    d. Proprioception
    e. Tactile
    f. Taste
    g. Vestibular system
    h. Vision
    i. Equilibrium

14. Which of the following is a sensory modality that involves the detection of sound?
    a. Auditory
    b. Interoception
    c. Olfactory

d. Proprioception
e. Tactile
f. Taste
g. Vestibular system
h. Vision
i. Equilibrium

15. Which of the following is a sensory modality that involves the detection of external pressure?
a. Auditory
b. Interoception
c. Olfactory
d. Proprioception
e. Tactile
f. Taste
g. Vestibular system
h. Vision
i. Equilibrium

16. Which of the following is a sensory modality that involves the sense of smell?
a. Auditory
b. Interoception
c. Olfactory
d. Proprioception
e. Tactile
f. Taste
g. Vestibular system
h. Vision
i. Equilibrium

17. Which of the following is ectodermal in origin?
a. Bipolar neurons
b. Neuroglial cells
c. Microglial cells
d. Unipolar neurons
e. Multipolar neurons

18. Which of the following is hemopoietic in origin?
a. Bipolar neurons
b. Neuroglial cells
c. Microglial cells
d. Unipolar neurons
e. Multipolar neurons

19. Dendritic spines are formed to perform which of the following tasks?
a. Cellular transport
b. Decrease communication with organs
c. Increase neurotransmitter production
d. Increase surface area
e. Increase transmission
f. Provide movement

20. Dendrites are designed to do which of the following tasks?
a. Fast axonal transport
b. Formation of vesicles
c. Production of neuropeptides
d. Receive information
e. Slow axonal transport
f. Transmit information
g. Both a and e
h. Both b and c

21. Axons are designed to do which of the following tasks?
a. Fast axonal transport
b. Formation of vesicles
c. Production of neurotransmitters
d. Receive information
e. Slow axonal transport
f. Transmit information

22. Soma is the location where which of the following functions/tasks take place?
a. Fast axonal transport
b. Formation of vesicles
c. Production of neuropeptides
d. Receive information
e. Slow axonal transport
f. Transmit information
g. Terminates action potential
h. Both a and f
i. Both b and c

23. Axon branches to increase the surface area. What is the name given to these branches?
a. Chromatophilic substance
b. Collaterals
c. Dendritic spines
d. Fast axonal transport
e. Golgi apparatus
f. Slow axonal transport

24. Which of the following protein channels are formed to respond to the deformation of dendritic membranes?
a. Antiports

b. Ligand gated channels
c. Mechanically gated channel
d. Symports
e. Uniports
f. Voltage gated channel

25. Which of the following protein channels are formed to respond to electrical impulses?
a. Antiports
b. Ligand gated channels
c. Mechanically gated channel
d. Symports
e. Uniports
f. Voltage gated channel

26. Which of the following protein channels are formed to respond to signal molecules?
a. Antiports
b. Ligand gated channels
c. Mechanically gated channel
d. Symports
e. Uniports
f. Voltage gated channel

27. In nerves, the aggregates of rough ER and ribosomes are referred to as _____.
a. Golgi apparatus
b. Nissl bodies
c. Nucleus
d. Ribosome
e. Rough endoplasmic reticulum
f. Smooth endoplasmic reticulum

28. Which of the following is also known as a trigger zone?
a. Axonal hillock
b. Collaterals
c. Dendritic spines
d. Initial segment
e. Kinesin
f. Nissl bodies

29. The trigger zone is located within which of the following structure of the nerve?
a. Axonal hillock
b. Collaterals
c. Dendritic spines
d. Initial segment .
e. Kinesin
f. Nissl bodies

30. From the dendrite through the soma, a type of current called _____ is generated.
a. Action potential
b. Axonal hillock
c. Initial segment
d. Local current
e. Postsynaptic potential
f. Presynaptic potential

31. The _____ is the set level of depolarization that must be reached in order to trigger an action potential.
a. Excitatory postsynaptic potential
b. Inhibitory postsynaptic potential
c. Initial segment
d. Local potential
e. Summation
f. Threshold

32. What is triggered when the summation of excitatory postsynaptic potential (EPSP) is greater than the summation of inhibitory postsynaptic potential (IPSP)?
a. Action potential
b. Dissipation of the local current
c. Ligand gated channels
d. Local potential
e. Summation
f. Threshold
g. Voltage gated channels

33. What is triggered when the summation of excitatory postsynaptic potential (EPSP) is less than the summation of inhibitory postsynaptic potential (IPSP)?
a. Action potential
b. Dissipation of the local current
c. Ligand gated channels
d. Local potential
e. Summation
f. Threshold
g. Voltage gated channels

34. Substances moving from the soma to the axonal terminal are proceeding in a _____ direction.
a. Anterograde
b. Declining
c. Forward
d. Regressive

e.   Retrograde
f.   Reverse

35. Substances moving from the axonal terminal to the soma are proceeding in a _____ direction.
a.   Anterograde
b.   Declining
c.   Forward
d.   Regressive
e.   Retrograde
f.   Reverse

36. Slow Component a (SCa) is the slower form of slow axonal transport. It is responsible for transporting which of the following substances?
a.   Clathrin
b.   Neuropeptides
c.   Neurotransmitters
d.   Proteins that forms neurofilaments
e.   Tubulin
f.   Both and e
g.   Both d and e

37. Slow Component b (SCb) is the faster form of the slower axonal transport. It is responsible for transporting which of the following substances?
a.   Clathrin
b.   Neuropeptides
c.   Neurotransmitters
d.   Proteins that forms neurofilaments
e.   Tubulin
f.   Both a and e
g.   Both d and e

38. Which of the following is the manner by which a cargo is transported by attaching to constructing neurofilaments and microtubules?
a.   Slow component a
b.   Slow component b
c.   Fast axonal transport
d.   Formation of cilia
e.   Formation of flagella
f.   Axoplasmic flow

39. Which of the following is the manner of neuronal transport where the supramolecular complexes are connected to constructing actin filaments? This is one of two possible hypotheses.
a.   Slow component a
b.   Slow component b
c.   Fast axonal transport
d.   Formation of cilia
e.   Formation of flagella
f.   Axoplasmic flow

40. Which of the following is the manner of neuronal transport where the supramolecular complexes are associated with motor proteins such as kinesin or dynein, as well as constructing actin filaments? This is one of two possible hypotheses.
a.   Slow component a
b.   Slow component b
c.   Fast axonal transport
d.   Formation of cilia
e.   Formation of flagella
f.   Axoplasmic flow

41. Which of the following is a polarity dependent, rod-shaped motor protein?
a.   Kinesin
b.   Dynein
c.   α-tubulin
d.   β-tubulin
e.   γ-tubulin
f.   myosin
g.   actin

42. The globular heads of the kinesin possesses which of the following binding site (s)?
a.   Actin binding site
b.   ATPase
c.   Intermediate filament binding site
d.   Microtubule binding site
e.   Both a and b
f.   Both b and d

43. The movement of kinesin is polarity specific. Which of the following is the direction which kinesin moves?
a.   + to – anterograde direction
b.   – to + anterograde direction
c.   + to – retrograde direction
d.   – to + retrograde direction

44. Which of the following structural proteins serves as the template for assembling microtubules?
a.   Kinesin
b.   Dynein
c.   α-tubulin
d.   β-tubulin

e.   γ-tubulin
f.   myosin
g.   actin

45. Which of the following microtubule protein components once bounded together create a polar structure?
a.   Kinesin
b.   Dynein
c.   α-tubulin
d.   β-tubulin
e.   γ-tubulin
f.   both c and e
g.   both c and d

46. The head-to-tail binding of _____ and _____ creates a polar structure. In each protofilament, the _____ is always located at the minus (-) end. The minus (-) end is located towards the _____ or the _____ microtubule protein component. The _____ is oriented towards the growing end, also known as the plus (+) end which is directed at the _____.
a.   Axonal terminal
b.   Dynein
c.   Kinesin
d.   Soma or the dendrites
e.   α-tubulin
f.   β-tubulin
g.   γ-tubulin

47. The movement of dynein is polarity specific. Which of the following is the direction which kinesin moves?
e.   + to – anterograde direction
f.   – to + anterograde direction
g.   + to – retrograde direction
h.   – to + retrograde direction

48. Acetylcholine (ACh) is an example of a _____ that is formed at the _____. ACh is constructed out of two components known as _____ and _____.
a.   Acetate
b.   Axonal terminal
c.   Axons
d.   Choline
e.   Dendrite
f.   Neuropeptides
g.   Small-molecule neurotransmitter
h.   Soma

49. The choline components of ACh are found in which of the following location (s) or formed from which of the following reaction (s)?
a.   From glycolysis
b.   From the electron transport system
c.   From the Kreb's cycle
d.   From the transition reaction
e.   In the blood plasma
f.   In the synaptic cleft after the breakdown of ACh
g.   Both c and d
h.   Both e and f

50. The acetate component of ACh is found in which of the following location (s) or formed from which of the following reaction (s)?
a.   From glycolysis
b.   From the electron transport system
c.   From the Kreb's cycle
d.   From the transition reaction
e.   In the blood plasma
f.   In the synaptic cleft after the breakdown of ACh
g.   Both c and d
h.   Both e and f

51. Which of the following enzymes is responsible for the formation of ACh?
a.   Acetylcholinesterase
b.   Catechol-O-methyltransferase
c.   Choline acetyltransferase (ChAT)
d.   DOPA decarboxylase
e.   Dopamine β-hydroxylase
f.   Monoamine oxidase
g.   Phenylethanolamine-N-methyltransferase
h.   Tyrosine hydroxylase
i.   Vesiculin
j.   Tyrosine
k.   Vesicular monoamine transporter 2 (VMAT2)

52. Once ACh is formed, it is transported from the axolemma into an empty secretory vesicle by the vesicular membrane bounded proteoglycan called _____.
a.   Acetylcholinesterase
b.   Catechol-O-methyltransferase
c.   Choline acetyltransferase (ChAT)
d.   DOPA decarboxylase
e.   Dopamine β-hydroxylase
f.   Monoamine oxidase
g.   Phenylethanolamine-N-methyltransferase
h.   Tyrosine hydroxylase
i.   Vesiculin
j.   Tyrosine

k. Vesicular monoamine transporter 2 (VMAT2)

53. Which of the following is the precursor molecule for all catecholamines?
   a. Acetylcholinesterase
   b. Catechol-O-methytransferase
   c. Choline acetyltransferase (ChAT)
   d. DOPA decarboxylase
   e. Dopamine β-hydroxylase
   f. Monoamine oxidase
   g. Phenylethanolamine-N-methyltransferase
   h. Tyrosine hydroxylase
   i. Vesiculin
   j. Tyrosine
   k. Vesicular monoamine transporter 2 (VMAT2)

54. Within the axoplasm the previously mentioned precursor molecule is converted into L-3,4-dihydroxyphenylalanine (L-DOPA) through the enzymatic actions of _____.
   a. Acetylcholinesterase
   b. Catechol-O-methytransferase
   c. Choline acetyltransferase (ChAT)
   d. DOPA decarboxylase
   e. Dopamine β-hydroxylase
   f. Monoamine oxidase
   g. Phenylethanolamine-N-methyltransferase
   h. Tyrosine hydroxylase
   i. Vesiculin
   j. Tyrosine
   k. Vesicular monoamine transporter 2 (VMAT2)

55. Soon after the formation of L-DOPA, this molecule is further processed into the neurotransmitter DA by the enzyme _____.
   a. Acetylcholinesterase
   b. Catechol-O-methytransferase
   c. Choline acetyltransferase (ChAT)
   d. DOPA decarboxylase
   e. Dopamine β-hydroxylase
   f. Monoamine oxidase
   g. Phenylethanolamine-N-methyltransferase
   h. Tyrosine hydroxylase
   i. Vesiculin
   j. Tyrosine
   k. Vesicular monoamine transporter 2 (VMAT2)

56. Once DA is formed, it is transported/packaged into secretory vesicles by _____.
   a. Acetylcholinesterase
   b. Catechol-O-methytransferase
   c. Choline acetyltransferase (ChAT)
   d. DOPA decarboxylase
   e. Dopamine β-hydroxylase
   f. Monoamine oxidase
   g. Phenylethanolamine-N-methyltransferase
   h. Tyrosine hydroxylase
   i. Vesiculin
   j. Tyrosine
   k. Vesicular monoamine transporter 2 (VMAT2)

57. Within the vesicle, DA could be further converted into NE by the enzyme _____.
   a. Acetylcholinesterase
   b. Catechol-O-methytransferase
   c. Choline acetyltransferase (ChAT)
   d. DOPA decarboxylase
   e. Dopamine β-hydroxylase
   f. Monoamine oxidase
   g. Phenylethanolamine-N-methyltransferase
   h. Tyrosine hydroxylase
   i. Vesiculin
   j. Tyrosine
   k. Vesicular monoamine transporter 2 (VMAT2)

58. Within the vesicle, NE could be further converted into epinephrine through the enzymatic actions of _____.
   a. Acetylcholinesterase
   b. Catechol-O-methytransferase
   c. Choline acetyltransferase (ChAT)
   d. DOPA decarboxylase
   e. Dopamine β-hydroxylase
   f. Monoamine oxidase
   g. Phenylethanolamine-N-methyltransferase
   h. Tyrosine hydroxylase
   i. Vesiculin
   j. Tyrosine
   k. Vesicular monoamine transporter 2 (VMAT2)

59. Which of the following enzymes/proteins are responsible for breaking down ACh?
   a. Acetylcholinesterase
   b. Catechol-O-methytransferase
   c. Choline acetyltransferase (ChAT)
   d. DOPA decarboxylase
   e. Dopamine β-hydroxylase

f. Monoamine oxidase
g. Phenylethanolamine-N-methyltransferase
h. Tyrosine hydroxylase
i. Vesiculin
j. Tyrosine
k. Vesicular monoamine transporter 2 (VMAT2)

60. It is known that the majority of NE and epinephrine are removed from receptor protein via which of the following processes?
   a. Active transport back into the neuron
   b. Diffused into the blood stream
   c. Enzymatic breakdown
   d. Processed through the Kreb's cycle
   e. Processed through the transition reaction
   f. None of the above

61. Approximately 5% of NE are metabolized by the enzymes _____ and _____.
   a. Acetylcholinesterase
   b. Catechol-O-methytransferase
   c. Choline acetyltransferase (ChAT)
   d. DOPA decarboxylase
   e. Dopamine β-hydroxylase
   f. Monoamine oxidase
   g. Phenylethanolamine-N-methyltransferase
   h. Tyrosine hydroxylase
   i. Vesiculin
   j. Tyrosine
   k. Vesicular monoamine transporter 2 (VMAT2)

62. The metabolic cycle of NE involves the breakdown of NE by the enzyme _____ into normetanephrine and this intermediate product is further processed into vanillylmandelic acid (VMA).
   a. Acetylcholinesterase
   b. Catechol-O-methytransferase
   c. Choline acetyltransferase (ChAT)
   d. DOPA decarboxylase
   e. Dopamine β-hydroxylase
   f. Monoamine oxidase
   g. Phenylethanolamine-N-methyltransferase
   h. Tyrosine hydroxylase
   i. Vesiculin
   j. Tyrosine
   k. Vesicular monoamine transporter 2 (VMAT2)

63. The metabolic cycle of NE involves the breakdown of NE by the enzyme _____ into dihydroxymandelic acid before being further processed into VMA.
   a. Acetylcholinesterase
   b. Catechol-O-methytransferase
   c. Choline acetyltransferase (ChAT)
   d. DOPA decarboxylase
   e. Dopamine β-hydroxylase
   f. Monoamine oxidase
   g. Phenylethanolamine-N-methyltransferase
   h. Tyrosine hydroxylase
   i. Vesiculin
   j. Tyrosine
   k. Vesicular monoamine transporter 2 (VMAT2)

64. Epinephrine could be hydrolyzed by _____ into metanephrine and further processed by _____ into VMA.
   a. Acetylcholinesterase
   b. Catechol-O-methytransferase
   c. Choline acetyltransferase (ChAT)
   d. DOPA decarboxylase
   e. Dopamine β-hydroxylase
   f. Monoamine oxidase
   g. Phenylethanolamine-N-methyltransferase
   h. Tyrosine hydroxylase
   i. Vesiculin
   j. Tyrosine
   k. Vesicular monoamine transporter 2 (VMAT2)

65. Epinephrine could also be processed by _____ into dihydroxymandelic acid and further processed by _____ into VMA.
   a. Acetylcholinesterase
   b. Catechol-O-methytransferase
   c. Choline acetyltransferase (ChAT)
   d. DOPA decarboxylase
   e. Dopamine β-hydroxylase
   f. Monoamine oxidase
   g. Phenylethanolamine-N-methyltransferase
   h. Tyrosine hydroxylase
   i. Vesiculin
   j. Tyrosine
   k. Vesicular monoamine transporter 2 (VMAT2)

66. Which of the following neurotransmitter is formed in the soma of the nerve cell before they are fast axonal transported to the axonal terminal?
   a. Acetate
   b. Axonal terminal
   c. Axons
   d. Choline
   e. Dendrite

f. Neuropeptides
g. Small-molecule neurotransmitter
h. Soma

67. Which of the following conduct its action potential from the periphery to the CNS?
a. Afferents
b. Autonomic
c. Efferents
d. Enteric
e. Interneurons
f. Motor
g. Sensory
h. Both a and g
i. Both c and f

68. Which of the following conduct its action potential from the CNS to the periphery?
a. Afferents
b. Autonomic
c. Efferents
d. Enteric
e. Interneurons
f. Motor
g. Sensory
h. Both a and g
i. Both c and f

69. Which of the following conduct its action potential from one neuron to another within the CNS?
a. Afferents
b. Autonomic
c. Efferents
d. Enteric
e. Interneurons
f. Motor
g. Sensory
h. Both a and g
i. Both c and f

70. Which of the following neuron possesses a distinctly large soma and numerous dendrites protruding from its cellular membrane? It has a single axon that emerges at the axonal hillock as a long tail extending away from the soma. This neuron is commonly found in the CNS.
a. Anaxonic neuron
b. Bipolar neurons
c. Interneurons
d. Multipolar neurons
e. Unipolar (pseudounipolar) neurons

71. Which of the following neuron possesses two processes extending from the polar ends of the soma? This type of neuron is commonly found in sensory organs such as the retina and the olfactory nerves.
a. Anaxonic neuron
b. Bipolar neurons
c. Interneurons
d. Multipolar neurons
e. Unipolar (pseudounipolar) neurons

72. Which of the following neuron has a single process extending a short distance from the soma and divides into two branches, the dendritic and axonal branch? The dendritic branch extends to the peripheral tissuess and has sensory receptors while the axonal branch extends into the CNS. For example, the dorsal root ganglia are composed of this type of neuron with branches extending to the peripheral tissue and to the CNS.
a. Anaxonic neuron
b. Bipolar neurons
c. Interneurons
d. Multipolar neurons
e. Unipolar (pseudounipolar) neurons

73. Which of the following is an atypical neurons found in the CNS of mammals? This type of neuron does not have axons and are components of complex interneuronal circuits where the nervous impulses could travel in any direction and not simply the dendrite to axon format. These types of nerves are commonly found in areas of the brain that controls activities involved in complex learning and formation of memory.
a. Anaxonic neuron
b. Bipolar neurons
c. Interneurons
d. Multipolar neurons
e. Unipolar (pseudounipolar) neurons

74. Which of the following neuroglial cell makes extensive contacts with blood vessels and helps in regulating blood flow in the CNS?
a. Astrocytes
b. Ependymal cells
c. Interfascicular oligodendrocytes
d. Microglial cells
e. Myelinating Schwann cells
f. Nonmyelinating Schwann cells
g. Perineuronal oligodendrocyte
h. Satellite cells

75. Which of the following neuroglial cell lines the ventricles of the brain as well as the central canal of the spinal cord?
a. Astrocytes
b. Ependymal cells
c. Interfascicular oligodendrocytes
d. Microglial cells
e. Myelinating Schwann cells
f. Nonmyelinating Schwann cells
g. Perineuronal oligodendrocyte
h. Satellite cells

76. Which of the following neuroglial cell are found within the white matter of the CNS and are responsible for forming myelin sheath surrounding axons of nerve cells?
a. Astrocytes
b. Ependymal cells
c. Interfascicular oligodendrocytes
d. Microglial cells
e. Myelinating Schwann cells
f. Nonmyelinating Schwann cells
g. Perineuronal oligodendrocyte
h. Satellite cells

77. Which of the following neuroglial cells are found in the gray matter of the CNS? These neuroglial cells are non-myelinating and attaches to large neurons in close proximity of the soma of neurons.
a. Astrocytes
b. Ependymal cells
c. Interfascicular oligodendrocytes
d. Microglial cells
e. Myelinating Schwann cells
f. Nonmyelinating Schwann cells
g. Perineuronal oligodendrocyte
h. Satellite cells

78. Which of the following neuroglial cells are phagocytic cells of the CNS? These cells are in many ways similar to macrophage and are derived from similar embryonic origins hematopoietic stem cells.
a. Astrocytes
b. Ependymal cells
c. Interfascicular oligodendrocytes
d. Microglial cells
e. Myelinating Schwann cells
f. Nonmyelinating Schwann cells
g. Perineuronal oligodendrocyte
h. Satellite cells

79. Which of the following neuroglial cell are found in the PNS and are responsible for forming myelin sheath by wrapping itself approximately 100 times around a small segment of an axon?
a. Astrocytes
b. Ependymal cells
c. Interfascicular oligodendrocytes
d. Microglial cells
e. Myelinating Schwann cells
f. Nonmyelinating Schwann cells
g. Perineuronal oligodendrocyte
h. Satellite cells

80. Which of the following neuroglial cells exist among non-myelinated fibers of the PNS? Instead of wrapping around the segments of the axon to form myelin sheath, these cells are responsible for embedding 1-12 small axons within grooves upon its cellular surface.
a. Astrocytes
b. Ependymal cells
c. Interfascicular oligodendrocytes
d. Microglial cells
e. Myelinating Schwann cells
f. Nonmyelinating Schwann cells
g. Perineuronal oligodendrocyte
h. Satellite cells

81. Which of the following neuroglial cells are found surrounding the cell bodies (somas) in sensory ganglia of the PNS?
a. Astrocytes
b. Ependymal cells
c. Interfascicular oligodendrocytes
d. Microglial cells
e. Myelinating Schwann cells
f. Nonmyelinating Schwann cells
g. Perineuronal oligodendrocyte
h. Satellite cells

82. Astrocytes are known to secrete _____ where this signal molecule causes vasoconstriction (e.g. smooth muscle contraction)
a. Acetylcholine
b. Blood brain barrier
c. GABA
d. Glucose
e. Glutamate
f. Glycine
g. Nitric oxide
h. Prostaglandins

83. Astrocytes are known to secrete _____ where this signal molecule causes vasodilation (e.g. smooth muscle relaxation)
    a. Acetylcholine
    b. Blood brain barrier
    c. GABA
    d. Glucose
    e. Glutamate
    f. Glycine
    g. Nitric oxide
    h. Prostaglandins

84. Astrocytes transport which of the following molecule which in turn allows the replenishment of the much needed energy molecules for nerve cells?
    a. Acetylcholine
    b. Blood brain barrier
    c. GABA
    d. Glucose
    e. Glutamate
    f. Glycine
    g. Nitric oxide
    h. Prostaglandins

85. Astrocytes along with the endothelial cells of the blood vessels forms tight junctions thereby creating a diffusion barrier that regulates the flow of molecules entering the CNS based upon their polarity and size of the molecules. What is the name given to this barrier?
    a. Acetylcholine
    b. Blood brain barrier
    c. GABA
    d. Glucose
    e. Glutamate
    f. Glycine
    g. Nitric oxide
    h. Prostaglandins

86. By creating a barrier that isolates the CNS from the rest of the body, the astrocytes are responsible for keeping harmful chemicals, bacteria and viruses from the brain and the spinal cord. Nonetheless, there are substances that are allowed to pass the barrier. For example, glucose, amino acid and nucleosides are _____ into the CSF.
    a. Active transport
    b. Diffusion
    c. Facilitated diffusion
    d. Phagocytized
    e. Pinocytized

87. By creating a barrier that isolates the CNS from the rest of the body, the astrocytes are responsible for keeping harmful chemicals, bacteria and viruses from the brain and the spinal cord. Nonetheless, there are substances that are allowed to pass the barrier. For example, Insulin and other ligands are transported via receptor mediated _____.
    a. Active transport
    b. Diffusion
    c. Facilitated diffusion
    d. Phagocytized
    e. Pinocytized

88. By creating a barrier that isolates the CNS from the rest of the body, the astrocytes are responsible for keeping harmful chemicals, bacteria and viruses from the brain and the spinal cord. Nonetheless, there are substances that are allowed to pass the barrier. For example, various plasma proteins are transported via _____.
    a. Active transport
    b. Diffusion
    c. Facilitated diffusion
    d. Phagocytized
    e. Pinocytized

89. By creating a barrier that isolates the CNS from the rest of the body, the astrocytes are responsible for keeping harmful chemicals, bacteria and viruses from the brain and the spinal cord. Nonetheless, there are substances that are allowed to pass the barrier. For example, water soluble agents are allowed to _____ through the tight junctions.
    a. Active transport
    b. Diffusion
    c. Facilitated diffusion
    d. Phagocytized
    e. Pinocytized

90. By creating a barrier that isolates the CNS from the rest of the body, the astrocytes are responsible for keeping harmful chemicals, bacteria and viruses from the brain and the spinal cord. Nonetheless, there are substances that are allowed to pass the barrier. For example, lipid soluble agents are able to directly _____ through the membrane.
    a. Active transport
    b. Diffusion
    c. Facilitated diffusion
    d. Phagocytized
    e. Pinocytized

91. Gluatmate is absorbed by the astrocytes and is processed by an enzyme called _____ which converts glutamate into glutamine. Glutamine is then released back to the synapses thereby making them available for reformation

of glutamate by the nerve cell
    a. 5,10-methyl-enetetrahydrofolate
    b. GABA transaminase
    c. Glutamine synthetase
    d. H-protein
    e. L-enzyme (dihydrolipoyl dehydrogenase)
    f. P-enzyme (glycine dehydrogenase)
    g. Succinate-semialdehyde dehydrogenase
    h. T-enzyme (aminomethyltransferase)
    i. Both a and d
    j. Both b and g

92. GABA is absorbed by the astrocytes and is processed by enzymes called _____ and _____ into succinate before being sent to the Krebs cycle for energy production.
    a. 5,10-methyl-enetetrahydrofolate
    b. GABA transaminase
    c. Glutamine synthetase
    d. H-protein
    e. L-enzyme (dihydrolipoyl dehydrogenase)
    f. P-enzyme (glycine dehydrogenase)
    g. Succinate-semialdehyde dehydrogenase
    h. T-enzyme (aminomethyltransferase)
    i. Both a and d
    j. Both b and g

93. Glycine, after its absorption by astrocytes, is catabolized by _____, _____ and _____ via a carrier protein called _____ via a process known as glycine cleavage system. The glycine cleavage system converts this amino acid into _____ which can be used as a substrate or as a coenzyme in various celluar processes.
    a. 5,10-methyl-enetetrahydrofolate
    b. GABA transaminase
    c. Glutamine synthetase
    d. H-protein
    e. L-enzyme (dihydrolipoyl dehydrogenase)
    f. P-enzyme (glycine dehydrogenase)
    g. Succinate-semialdehyde dehydrogenase
    h. T-enzyme (aminomethyltransferase)
    i. Lipocalin-Type Prostaglandin D Synthase (L-PGDS)
    j. Both a and d
    k. Both b and g

94. It is known that soon after tissue damage in the CNS has taken place, astrocytes will immediately form scar tissue to wall off the injured site in an effort to limit the spread of inflammation to other healthy nerve cells. What is the name given to this process?
    a. Active transport
    b. Axoplasmic flow
    c. Diffusion
    d. Facilitated diffusion
    e. Fast axonal transport
    f. Reactive astrogliosis
    g. Slow axonal transport

95. Ependymal cells along with capillaries form the _____ within the _____ and _____. These cells are responsible for creating cerebral spinal fluid (CSF).
    a. Central canal
    b. Cerebral aqueduct
    c. Choroid plexuses
    d. Fourth ventricle
    e. Interventricular foramen
    f. Lateral ventricle
    g. Third ventricle

96. Myelin sheath forms an electrical insulator along segments of the axon leaving only small exposed spaces between myelin sheaths called _____. This exposed location is the only region of a myelinated axon that will depolarize.
    a. Axonal hillock
    b. Collaterals
    c. Dendritic spine
    d. Initial segment
    e. Node of Ranvier

97. The perineuronal oligodendrocyte (PO) is believed to synthesize and release _____, an anti-apoptotic molecule that protects neurons form apoptosis.
    a. 5,10-methyl-enetetrahydrofolate
    b. GABA transaminase
    c. Glutamine synthetase
    d. H-protein
    e. L-enzyme (dihydrolipoyl dehydrogenase)
    f. P-enzyme (glycine dehydrogenase)
    g. Succinate-semialdehyde dehydrogenase
    h. T-enzyme (aminomethyltransferase)
    i. Lipocalin-Type Prostaglandin D Synthase (L-PGDS)
    j. Both a and d
    k. Both b and g

98. When a somatic motor neuron is severed, the axon that is distal to the soma will not survive since it is incapable of protein synthesis. This deterioration of distal axon is also referred to as _____.
    a. Axonal degeneration
    b. Denervation atrophy

c. Multiple sclerosis
d. Muscular dystrophy
e. Nissl bodies disperse
f. Reactive astrogliosis
g. Wallerian degeneration

99. After the severing of a somatic motor neuron, the skeletal muscle fiber will demonstrate which of the following signs?
a. Axonal degeneration
b. Denervation atrophy
c. Multiple sclerosis
d. Muscular dystrophy
e. Nissl bodies disperse
f. Reactive astrogliosis
g. Wallerian degeneration

100. After the severing of a somatic motor neuron, the soma of the injured somatic motor nerve will also show signs of abnormalities since it is disconnected from the supplies of _____ (**Answer A**) which is released by the skeletal muscle fiber and transported to the soma via _____. Due to the lack of **Answer A**, the soma will exhibit signs of hypertrophy and the ribosome and rER of the nerve, also known as _____, begins to breakup in a process called _____. The nucleus of the somatic motor nerve will move off center as it will either proceed into _____ (programmed cell death) or begin the regeneration process.

The newly infused Schwann cells, along with the existing basal lamina, will form a regeneration tube also known as _____ and begin secreting cell adhesion molecules and nerve growth factors. One of the newly grown stomps will find its way into the regeneration tube. The regeneration tube will guide this new-growth (one axon stomp) to the correct destination while other axon stomps will deteriorate. It has been estimated that the regrowth occurs rapidly at 3-5 mm/day. Finally, the regeneration tube will guide the new growth back to the _____ and reestablishes synaptic contact. Eventually, myelinating Schwann cells will reform myelin sheath at the regenerated portion of the axon.
a. Apoptosis
b. Anterograde axonal transport
c. Büngner
d. Denervation atrophy
e. Nerve growth factors
f. Neuromuscular junction
g. Nissl bodies
h. Nissl bodies disperse
i. Reactive astrogliosis
j. Retrograde axonal transport
k. Wallerian degeneration

101. Which of the following is defined as the manipulation of ionic concentration to create an electrical potential or membrane potential?
a. Büngner
b. Denervation atrophy
c. Electrophysiology
d. Nissl bodies disperse
e. Reactive astrogliosis
f. Wallerian degeneration

102. Plasma membrane of the neuron, like the sarcolemma, is electrically charged or polarized. The polarization of the plasma membrane is mainly created by the unequal distribution of _____ and _____ cations and to a minor extent, _____ anions also contribute to the polarization of the plasma membrane. This uneven distribution generates a charge difference also known as _____. This charge difference is maintained through the actions of an active transport integral protein (an antiport) called _____, as well as _____ that actively transport Cl⁻ out of the neuroplasm.
a. Calcium
b. Chloride
c. Chloride pumps
d. Depolarized
e. Hyperpolarized
f. Hypopolarized
g. Iodide
h. Polarized
i. Potassium
j. Repolarized
k. Resting membrane potential
l. Sodium
m. Sodium-potassium exchange pump

103. The resting membrane potential of a polarized nerve cell is measured at _____.
a. -55 mV
b. -60 mV
c. -65 mV
d. -70 mV
e. -85 mV
f. -103 mV

104. Which of the following is defined as a limited depolarization or electrical discharge that may or may not result in an action potential?
a. Actin potential
b. Membrane potential
c. Local potential
d. Polarized

e. Hyperpolarized
f. Depolarized
g. Depolarized

105. It has been calculated that there are approximately _____ per square micrometer (μm) found in the membrane of the dendrites and the soma. In contrast, beginning at the initial segment of the axonal hillock, the number of voltage gated channels increases drastically to _____ per square micrometer (μm). Simply put, the amount of voltage gated channels found in the dendrites and soma are _____ (Hint: adequate or inadequate) for the creation of an action potential.
a. 100-200 voltage gated Na⁺ channels
b. 200-300 voltage gated Na⁺ channels
c. 30-45 voltage gated Na⁺ channels
d. 350-500 voltage gated Na⁺ channels
e. 40-65 voltage gated Na⁺ channels
f. 50-75 voltage gated Na⁺ channels
g. Adequate
h. Graded
i. Inadequate
j. Reversible

106. Local potentials are _____ which signifies that the voltage generated vary in magnitude depending upon the intensity of the stimulus. For example, a weak stimulus will only cause the opening of a few Na⁺ channels and the generation of a minor current (i.e. slight depolarization). In contrast, a stronger stimulus will result in the opening of more Na⁺ channels and the generation of a more powerful current (i.e. stronger depolarization).

Local potential are _____ (Hint: reversible or irreversible) which means that the current generated will dissipate and the RMP reinstated as soon as the stimulation stops.
a. 100-200 voltage gated Na⁺ channels
b. 200-300 voltage gated Na⁺ channels
c. 30-45 voltage gated Na⁺ channels
d. 350-500 voltage gated Na⁺ channels
e. 40-65 voltage gated Na⁺ channels
f. 50-75 voltage gated Na⁺ channels
g. Adequate
h. Graded
i. Inadequate
j. Reversible

107. Local potential could be either excitatory or inhibitory and often depends upon the type of neurotransmitter that is binding to the receptor protein. Excitatory local potential are at times triggered by excitatory neurotransmitters such as _____ which may result in the generation of an action potential if the strength of the stimulus reaches the _____. Inhibitory local potential are at times caused by the binding of inhibitory neurotransmitters such as _____ or _____ to receptor proteins. These inhibitory neurotransmitters cause the opening of _____ which prompts the influx of _____ anions into the neuroplasm. The flow of this anion into the neuroplasm causes the membrane potential to become more _____ (positive or negative) or _____.
a. Acetylcholine (ACh)
b. Calcium
c. Chloride
d. Depolarize
e. GABA
f. Glycine
g. Hyperpolarized
h. Hypopolarized
i. Ligand gated calcium (Ca⁺) channels
j. Ligand gated chloride (Cl⁻) channels
k. Ligand gated potassium (K⁺) channels
l. Ligand gated sodium (Na⁺) channels
m. Negative
n. Polarize
o. Positive
p. Potassium
q. Repolarize
r. Resting membrane potential
s. Sodium
t. Threshold

108. It has been calculated that the threshold of a nerve cell is approximately _____. Put simply, the current of the local potential has to reach threshold in order for an action potential to be formed or an all-or-none response is generated.
a. -55 mV
b. -60 mV
c. -65 mV
d. -70 mV
e. -85 mV
f. -103 mV

109. The active transport transmembrane protein (an antiport) called _____ sends _____ (how many) _____ (name the cation) out of the neuroplasm and brings in _____ (how many) _____ (name the cation) into the neuroplasm. The active transport transmembrane protein called _____ is responsible for sending _____ anions out of the neuroplasm
a. 1
b. 2
c. 3
d. 4
e. 5
f. Calcium
g. Calcium potassium exchange pump
h. Calcium-sodium exchange pump
i. Chloride
j. Chloride pump
k. Iodide

l. Iodide pump
m. Potassium
n. Sodium
o. Sodium-potassium exchange pump

110. Due to the high concentration of _____ cation and comparably less _____ antion concentrations in the extracellular fluids, the net charge external to the neuronal plasma membrane is positive. Even with the high concentration of _____ cations, the neuroplasm is negatively charged in comparison to the extracellular fluid. This negative charge is mainly caused by the negatively charged functional groups of proteins such as, _____, _____, ATP and RNA etc. that are present within the neuroplasm.

a. Calcium
b. Chloride
c. Iodide
d. Phosphates (PO_4^{-3})
e. Potassium
f. Sodium
g. Sulfates (SO_4^{-2})

111. In a chemical synapses, the ligand released by a presynaptic nerve binds with an _____ which will cause the _____ (5-6 protein subunits) to change its shape and open the center pore. The opening of the pore will allow _____ cations to flow down its concentration gradient and into the neuroplasm.

1. Action potential
2. Action potential
3. ADP
4. ATP
5. Ca^{++} channel protein
6. Calcium
7. Chloride
8. Cl^- channel protein
9. Depolarize
10. Dimer
11. GDP
12. G-protein complex
13. GTP
14. Hyperpolarize
15. Initial segment
16. Iodide
17. Ionotropic Ca^{++} receptor
18. Ionotropic K^+ receptor
19. Ionotropic Na^+ receptor
20. Ionotropic Cl^- receptor
21. K^+ channel protein
22. Local current
23. Metabotropic receptors
24. Multimers
25. Na^+ channel protein
26. Negative
27. Polarize
28. Positive
29. Potassium
30. Repolarize
31. Sodium
32. Threshold
33. Voltage gated Ca^{++} channel
34. Voltage gated Cl^- channel
35. Voltage gated K^+ channel
36. Voltage gated Na^+ channel

112. In a chemical synapses, the ligand released by a presynaptic nerve, binds with a (an) _____, which would cause the activation of _____ (a cellular switch). The activation of this cellular switch will result in the separation of α, β and γ subunits. The β- and γ-subunits form a stable _____ (joining of two molecules through intermolecular bonding) and are responsible for activating the α-subunit. As a result _____ (a spent energy molecule) bounded α-subunit is replaced with _____ (a charged energy molecule). Subsequently, the α-subunit-energy molecule complex will cause the _____ to change its structural conformation and allow _____ cations (which are kept in high concentrations in the extracellular fluids) to flow down its concentration gradient and into the neuroplasm. The influx of these cations will cause the membrane to _____ and create an electrical current which is also known as a _____. This electrical current generated will cause the adjacent _____ to open thereby allowing more cations to flow into the neuroplasm. By allowing more cations to enter the nerve cell will generate additional electrical currents. Like a series of dominos in a cascading effect, the additional electrical current will cause the next _____ to open and so forth, thereby spreading the electrical current down the length of the dendrite and flow across the plasma membrane of the soma. At this time, the membrane potential will become more and more _____ (positive or negative). At the _____ of the axonal hillock, the generated electrical current is summarized and weighed against the _____ (-55 mV). Only when electrical current reaches -55 mV will a (an) _____ (all-or-none response) be created

1. Action potential
2. Action potential
3. ADP
4. ATP
5. Ca^{++} channel protein
6. Calcium
7. Chloride
8. Cl^- channel protein
9. Depolarize
10. Dimer
11. GDP
12. G-protein complex
13. GTP
14. Hyperpolarize
15. Initial segment
16. Iodide
17. Ionotropic Ca^{++} receptor
18. Ionotropic K^+ receptor
19. Ionotropic Na^+ receptor
20. Ionotropic Cl^- receptor
21. K^+ channel protein
22. Local current
23. Metabotropic receptors
24. Multimers
25. Na^+ channel protein
26. Negative
27. Polarize
28. Positive
29. Potassium
30. Repolarize
31. Sodium
32. Threshold
33. Voltage gated Ca^{++} channel
34. Voltage gated Cl^- channel
35. Voltage gated K^+ channel
36. Voltage gated Na^+ channel

113. Just like the local potential, the formation of the action potential (electric current) will trigger more _____ to open where in turn perpetuate the depolarization of the axolemma (plasma membrane of the axon). As more channels open, more _____ cations flow into the nerve cell (neuroplasm) which causes the membrane potential to become more positive until the membrane potential reaches approximately _____ mV. At the above mentioned mV, it is understood that the amount of _____ cation in the extracellular fluids is equal

to the amount of _____ cation in the neuroplasm. In another words, these particular cations have reached their equilibrium.

a. +20 mV
b. +30 mV
c. -103 mV
d. -55 mV
e. -60 mV
f. -65 mV
g. -70 mV
h. -80 mV
i. -85 mV
j. Calcium
k. Chloride
l. Depolarization
m. Hyperpolarization
n. Iodide
o. Polarization
p. Potassium
q. Sodium
r. Voltage gated Ca^{++} channel
s. Voltage gated Cl^- channel
t. Voltage gated K^+ channel
u. Voltage gated Na^+ channel

114. At approximately _____ mV the _____ channels open which allows the _____ cations (kept at high concentrations in the neuroplasm) to diffuse down its concentration gradient and into the extracellular matrix. The efflux of these cations reverses the membrane potential from positive towards the negative. Eventually the membrane potential drops below the RMP and proceeds to settle at approximately _____ mV which is also known as _____ (**Answer A**). At **Answer A** the amount of _____ cations in the extracellular fluid is equal to the amount of _____ cations in the neuroplasm (reached equilibrium). At **Answer A** the sodium-potassium exchange pump (SPEP) activates and begins sending _____ (how many) _____ (name the cation) into the extracellular fluid and _____ (how many) _____ (name the cation) into the neuroplasm

a. +20 mV
b. +30 mV
c. -103 mV
d. -55 mV
e. -60 mV
f. -65 mV
g. -70 mV
h. -80 mV
i. -85 mV
j. Calcium
k. Chloride
l. Depolarization
m. Hyperpolarization
n. Iodide
o. Polarization
p. Potassium
q. Sodium
r. Voltage gated Ca^{++} channel
s. Voltage gated Cl^- channel
t. Voltage gated K^+ channel
u. Voltage gated Na^+ channel
v. 1
w. 2
x. 3
y. 4
z. 5

ANSWERS

1.	a		2.	f		3.	e	
4.	f		5.	b, d, g		6.	f	
7.	g		8.	d		9.	b	
10.	d		11.	i		12.	b	
13.	h		14.	a		15.	e	
16.	c		17.	b		18.	c	
19.	d		20.	d		21.	f	
22.	h		23.	b		24.	c	
25.	f		26.	b		27.	b	
28.	d		29.	a		30.	d	
31.	f		32.	a		33.	b	
34.	a		35.	e		36.	g	
37.	f		38.	a		39.	b	
40.	b		41.	a		42.	f	
43.	b		44.	e		45.	g	
46.	e, f, e, d, g, f, a		47.	g		48.	g, b, a, d	
49.	h		50.	g		51.	c	
52.	i		53.	j		54.	h	
55.	d		56.	k		57.	e	
58.	g		59.	a		60.	a	
61.	c, f		62.	b		63.	f	
64.	b, f		65.	f, b		66.	f	
67.	a		68.	c		69.	e	
70.	d		71.	b		72.	e	
73.	a		74.	a		75.	b	
76.	c		77.	g		78.	d	
79.	e		80.	f		81.	h	
82.	h		83.	g		84.	d	
85.	b		86.	a		87.	e	
88.	d		89.	b		90.	b	
91.	c		92.	b, g		93.	f, h, e, d, a	
94.	f		95.	c, d, f		96.	a	
97.	i		98.	g		99.	b	
100.	e, j, g, h, a, c, f		101.	c		102.	l, i, b, k, m, c	
103.	d		104.	c		105.	f, d, i	
106.	h, j		107.	a, t, e, f, j, c, m, g		108.	a	

109. o, c, n, b, m, j, i	110.	f, b, e, d, g	111. 19, 24, 31

112. 23, 12, 10, 11, 13, 25, 31, 9, 22, 36, 36, 28, 15, 32, 1	113. o, q, b, q, q	114. b, t, p, h, m, p, p, x, q, w, p

44. Figure 44: Graphic created and release by P.Y.P. Jen. Copyright ©

REFERENCES

Baas PW, Buster DW. Slow axonal transport and the genesis of neuronal morphology. Journal or Neurobiology, 2003. 58: pp. 3-17

Shah JV, Cleveland DW. Slow axonal transport: fast motors in the slow lane. Curr. Opin. Cell Biol., 2002. 14: pp. 58–62

Stenoien DL, Brady ST. Slow axonal transport. Basic Neurochemistry: Molecular, Cellular and Medical Aspects. 6th ed. Chapter 28. Philadelphia: Lippincott-Raven. 1999

T. Galli, V. Haucke, Calcium-triggered exocytosis and clathrin-mediated endocytosis of synaptic vesicles. *Sci. STKE* 2005, tr1

Masako Taniike M, Mohri I, Eguchi N, Beuckmann CT, Suzuki K, Urade Y. Perineuronal Oligodendrocytes Protect against Neuronal Apoptosis through the Production of Lipocalin-Type Prostaglandin D Synthase in a Genetic Demyelinating Model. J. Neurosci., 2002. 22: pp. 4885–4896

Verkhratsky A, Butt AM. Glial Physiology and Pathophysiology. Chapter 5.1.3. West Sussex, Wiley-Blackwell. 2013

Purves D, Augustine GJ, Fitzpatrick D, Katz LC, LaMantia A-S, McNamara JO, Williams SM. Neuroscience. 2nd edition. Chapter 5. Sunderland, Sinauer Associates. 2001

Seeley RR, Stephens TD, Tate P. Anatomy and Physiology 6th Edition. McGraw-Hill, New York, New York. 2003

Tate P. Seeley's Principles of Anatomy & Physiology 1st Edition. McGraw-Hill, New York, New York. 2009

Saladin KS. Anatomy & Physiology. The Unity of Form and Function 6th Edition. McGraw-Hill, New York, New York. 2010

PHOTO AND GRAPHIC BIBLIOGRAPHY

1. Figure 1: Graphic created and release by Open Stax College, Connexions, reprint permission granted under the Creative Commons Attribution 3.0 Unported license
2. Figure 2: Graphic created and release by LadyofHats, 2007. Public domain graphics
3. Figure 3: Graphic created and release by by P.Y.P. Jen. Copyright ©
4. Figure 4: Graphic created and release by by P.Y.P. Jen. Copyright ©
5. Figure 5: Graphic created and release by Duncan JE, Goldstein LS. The Genetics of Axonal Transport and Axonal Transport Disorders. PLoS Genet., 2006. Public domain graphics
6. Figure 6: Graphic created and release by P.Y.P. Jen. Copyright ©
7. Figure 7: Graphic created and release by Duncan JE, Goldstein LS. The Genetics of Axonal Transport and Axonal Transport Disorders. PLoS Genet., 2006. Public domain graphics
8. Figure 8: Graphic created and release by P.Y.P. Jen. Copyright ©
9. Figure 9: Graphic created and release by P.Y.P. Jen. Copyright ©
10. Figure 10: Graphic created and release by P.Y.P. Jen. Copyright ©
11. Figure 11: Graphic created and release by P.Y.P. Jen. Copyright ©
12. Figure 12: Graphic created and release by Holly Fischer. University of Michigan Open learning. Public domain graphics
13. Figure 13: Graphic created and release by Holly Fischer. University of Michigan Open learning. Public domain graphics
14. Figure 14: Graphic created and release by Holly Fischer. University of Michigan Open learning. Public domain graphics
15. Figure 15: Graphic created and release by Holly Fischer. University of Michigan Open learning. Public domain graphics
16. Figure 16: Graphic created and release by P.Y.P. Jen. Copyright ©
17. Figure 17: Graphic created and release by Open Stax College, Connexions, reprint permission granted under the Creative Commons Attribution 3.0 Unported license
18. Figure 18: Graphic created and release by Holly Fischer. University of Michigan Open learning. Public domain graphics
19. Figure 19: Graphic created and release by P.Y.P. Jen. Copyright ©
20. Figure 20: Graphic created and release by P.Y.P. Jen. Copyright ©
21. Figure 21: Graphic created and release by P.Y.P. Jen. Copyright ©
22. Figure 22: Graphic created and release by P.Y.P. Jen. Copyright ©
23. Figure 23: Graphic created and release by Holly Fischer. University of Michigan Open learning. Public domain graphics
24. Figure 24: Graphic created and release by Holly Fischer. University of Michigan Open learning. Public domain graphics
25. Figure 25: Graphic created and release by Salzer JL. Switching myelination on and off. J. Cell Biol., 2008. Open I beta, United States National Library. Public domain graphics
26. Figure 26: Graphic created and release by Open Stax College, Connexions, reprint permission granted under the Creative Commons Attribution 3.0 Unported license
27. Figure 27: Graphic created and release by P.Y.P. Jen. Copyright ©
28. Figure 28: Graphic created and release by P.Y.P. Jen. Copyright ©
29. Figure 29: Graphic created and release by Open Stax College, Connexions, reprint permission granted under the Creative Commons Attribution 3.0 Unported license
30. Figure 30: Graphic created and release by Synaptidude, 2011. Public domain graphics
31. Figure 31: Graphic created and release by Crepalde MA, Faria-Campos AC, Campos SV. Modeling and analysis of cell membrane systems with probabilistic model checking. BMC Genomics, 2011. Open I beta, United States National Library. Public domain graphics
32. Figure 32: Graphic created and release by Open Stax College, Connexions, reprint permission granted under the Creative Commons Attribution 3.0 Unported license
33. Figure 33: Graphic created and release by Open Stax College, Connexions, reprint permission granted under the Creative Commons Attribution 3.0 Unported license
34. Figure 34: Graphic created and release by P.Y.P. Jen. Copyright ©
35. Figure 35: Graphic created and release by P.Y.P. Jen. Copyright ©
36. Figure 36: Graphic created and release by P.Y.P. Jen. Copyright ©
37. Figure 37: Graphic created and release by P.Y.P. Jen. Copyright ©
38. Figure 38: Graphic created and release by P.Y.P. Jen. Copyright ©
39. Figure 39: Graphic created and release by P.Y.P. Jen. Copyright ©
40. Figure 40: Graphic created and release by P.Y.P. Jen. Copyright ©
41. Figure 41: Graphic created and release by P.Y.P. Jen. Copyright ©
42. Figure 42: Graphic created and release by P.Y.P. Jen. Copyright ©
43. Figure 43: Graphic created and release by P.Y.P. Jen. Copyright ©

CHAPTER EIGHT

CENTRAL NERVOUS SYSTEM

The central nervous system (CNS) is comprised of the brain and the spinal cord and is the center of information integration and analysis within the human body. The brain is composed of three major areas: the cerebrum, cerebellum and the brain stem (Figure 1).

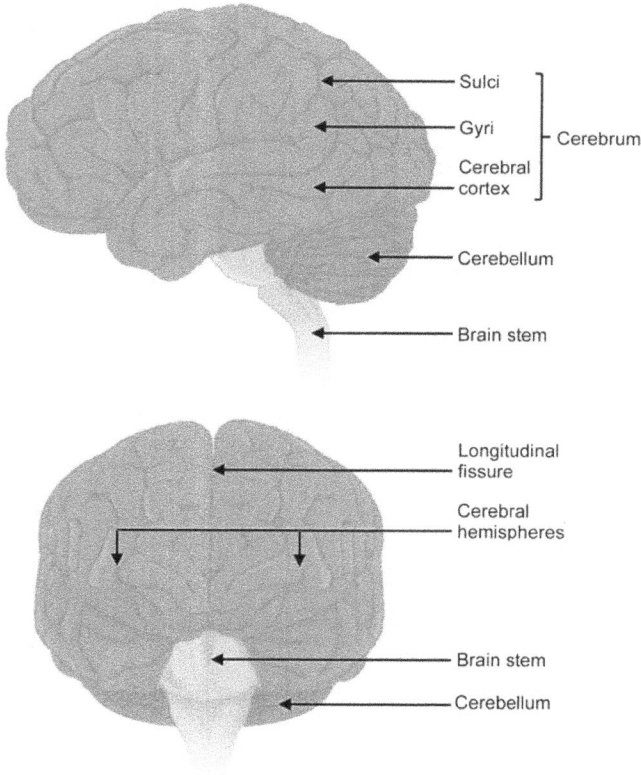

Figure 1: Illustration of the cerebrum, cerebellum, and the brain stem. Please note that the cerebral cortex comprise the most external part of the cerebrum

Cerebrum

The cerebrum is the largest portion of the human brain comprising of approximately 83% of the total brain volume. Embryonically, the cerebrum is developed from the telencephalon at approximately fifth week gestation. The cerebrum consists of two cerebral hemispheres separated by the longitudinal fissure and are connected to one another via a network of interconnecting nervous tissue called the corpus callosum. The corpus callosum consists of nerve tracts, which allow the hemispheres of the cerebrum to communicate, thereby integrating all motor and sensory information (Figure 2).

The surface of the cerebrum is composed of the cerebral cortex, which also consists of two hemispheres. The surface of the cerebral cortex is lined with thick folds (wrinkles) called gyri, which are separated from other folds by shallow grooves called sulci. Gyri and sulci allow more cerebral cortex to be contained within the cranial cavity, increasing the surface area for information processing (Figure 1).

Depending on their location, some gyri are consistent and predictable in their distribution, while others differ from hemisphere to hemisphere or person to person. Using the gyri that are uniform from one individual to another, five distinct lobes are delineated.

The frontal lobe lies immediately behind the frontal bone of the skull and is positioned superior to the eyes. The frontal lobe extends from the anterior border of the brain (rostral: towards the nose) to the central sulcus (Figure 3).

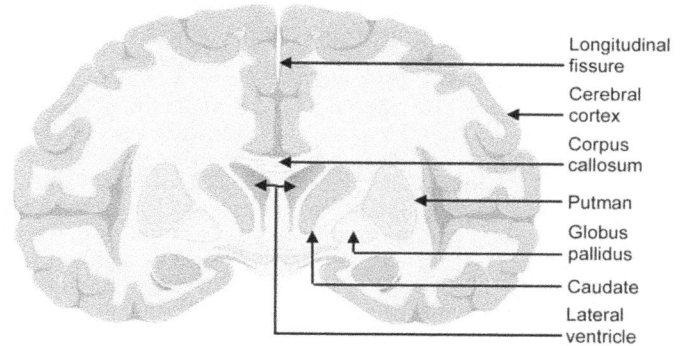

Figure 2: Frontal section of a human brain

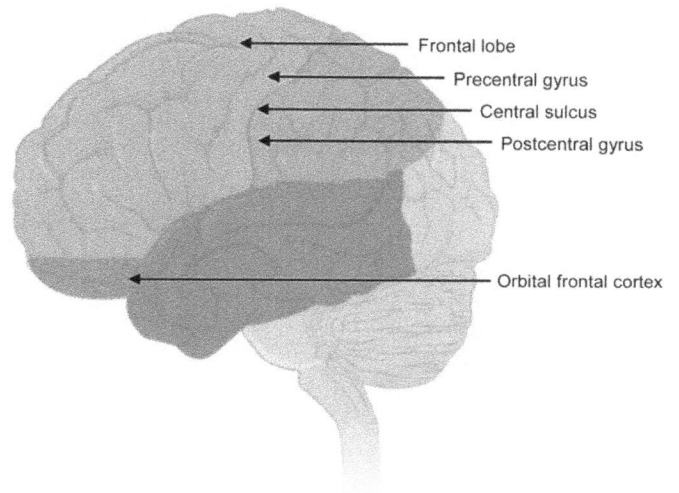

Figure 3: Illustration of the frontal lobe. Please note that the orbital frontal cortex is part of the frontal lobe

The frontal lobe consists of a right and left hemisphere and is generally acknowledged as the center of an individual's personality and emotions. The left hemisphere of the frontal lobe is involved with language capabilities, while the right hemisphere of the frontal lobe involves mainly non-verbal aspects of communication, such as the awareness of emotions in others, such as sensing the moods of others through awareness of their facial expressions, tone of their voice, etc. In general, both hemispheres of the frontal lobe are involved with voluntary motor functions, motivation, impulse control, judgment and memory. In addition, the frontal lobe is also involved with a cognitive process called executive function. Executive functions include planning, cognitive flexibility, abstract thinking, rule acquisition, inhibiting inappropriate actions, as well as, inhibiting irrelevant sensory information. People who suffer damage to the left hemisphere frontal lobe may experience problems with language capabilities, verbal skills, and positive emotions. Damaged right hemisphere frontal lobe will result in a deficiency in non-verbal communication and negative emotions.

The orbital frontal cortex (OFC) is a part of the frontal lobe that is involved in the modulation of behaviors, as well as, cognitive processing, and influencing decision making. For example, OFC is involved in the modulation of antisocial behavior, such as aggression. Based on psychiatric data, it is understood that there are two forms of aggression. First is reactive aggression, caused by frustration or threat. The second is called instrumental aggression, which is a means of securing reward(s) or accomplishing goal(s). Dysfunction in the OFC and the amygdala (discussed later in this chapter) are linked to the development of Borderline Personality Disorder (BPD), which encompasses individuals that possess fundamental failures in socialization.

The OFC also contains the secondary taste cortex, as well as, the secondary and tertiary olfactory cortical areas. In these areas of the OFC, tastes and smells are assigned reward values. Indirectly, the OFC also receives visual information from the temporal lobe and the visual cortex which are also assigned reward values. Together, the OFC integrates vision, taste, and smell sensory information through a process known as stimulus-reinforcement association, learning, and assign reward values. For example, the shape, coloration, smell, and taste of a particular food are assigned a particular value (i.e. such as great to extremely bad) by the OFC and stored within our memory. Because of this capability, we possess discriminating tastes in regards to the types of foods we enjoy consuming. In addition, this capability also provides a basic survival skill, where certain items are not edible based upon their shape, coloration, smell and taste.

The parietal lobe forms the upper most part of the brain and is immediately beneath the parietal bone of the skull. The parietal lobe extends from the central sulcus caudally to the parieto-occipital sulcus. Like all the lobes of the brain, it consists of two hemispheres (right and left parietal hemispheres). The parietal lobe is responsible for integrating sensory information such as general senses (e.g. pain, touch sensation etc.), tastes (i.e. a special sense), and is involved in visual perception (Figure 4).

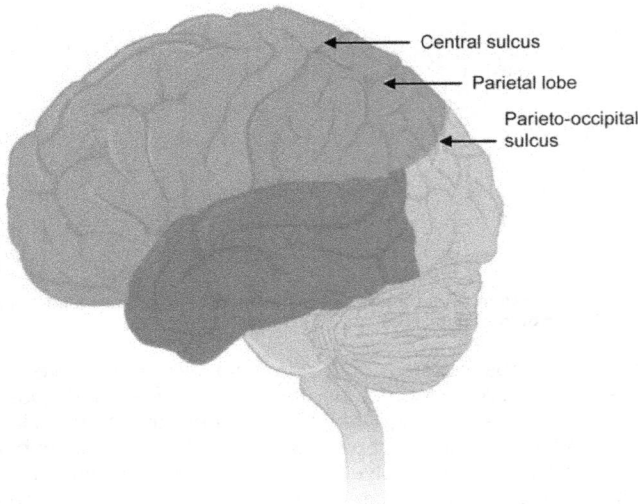

Figure 4: Illustration of the parietal lobe

The temporal lobe (right and left hemispheres) forms the lateral aspects of the human brain. The temporal lobe is situated immediately beneath the temporal bone of the skull. It is separated from the frontal and parietal lobe via the lateral sulcus. The temporal lobe is involved in sensory information processing such as hearing, smells, learning, memory, as well as, some aspects of vision and emotions (Figure 5).

The parahippocampal gyrus (PG) is a part of the temporal lobe and is involved in the recognition of places and scenes. In conjunction with other regions of the brain (e.g. right inferior parietal cortex and left frontal operculum), the PG is involved with the recognition of real and imagined events. Lesions in the PG are often due to stroke, and may lead to difficulties in distinguishing real and imagined events or the individual may suffer from the inability to visually recognize scenes (Figures 5 and 9).

The occipital lobe (e.g. right and left hemisphere occipital lobes) is located towards the caudal area (posterior side) of the brain and lies immediately beneath the occipital bone of the skull. The occipital lobe is separated from the parietal lobe via the parieto-occipital sulcus (Figure 5). The occipital lobe is the principle center for visual information processing, as well as, integrating visual information.

The insula, the last of the lobes of the cerebrum, can only be visualized by removing the overlying layers. Unlike the other lobes, the insula is less well understood, but it is believed to play a role in language skills, taste, and is involved in the integration of visual information (Figure 6).

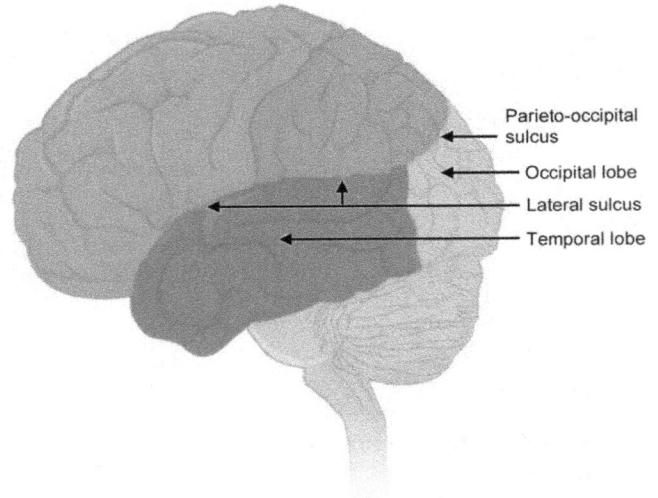

Figure 5: Illustration of the temporal and occipital lobe

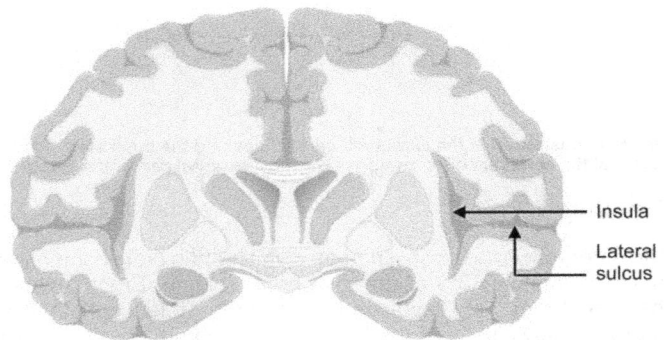

Figure 6: Illustration of the insula which is only seen via frontal section or peeling away the parietal and temporal lobes

Cerebral Cortex

The cerebral cortex is a large part of the cerebrum consisting approximately 40% of the entire brain mass (i.e. consisting or 14 to 16 billion nerve cells). The cerebral cortex covers the surfaces of the hemispheres (i.e. including the surfaces of all the lobed previously mentioned) and is composed of gray matter. The cerebral cortex is one of the three locations involved in neuronal integration (the others are the basal ganglia and the limbic system).

The cerebral cortex is constructed from two principle types of nerve cells: stellate and pyramidal. These two principle nerve cells are used to form three, five, or six layers of nervous tissue depending upon the location within the cerebral cortex.

It is estimated that 90% of the cerebral cortex is made up of six layers of tissue, called the neocortex. The neocortex is thus named because it is believed that this area of the brain is relatively new in our evolutionary development.

Depending on its location, the neocortex may have thicker or thinner layers. For example, in sensory regions of the brain, layer IV is the thickest of all the layers, while in the motor regions, layer V is the thickest. Nonetheless, all axons that leave the neocortex and enter the white matter of the brain arise from layers III, V, and VI.

The five layered area of the cerebral cortex is referred to as the paleocortex. It is believed that paleocortex is the earliest form of cortex that appeared in vertebrates. The paleocortex is found in the insula and parts of the temporal lobe and is generally involved with the sense of smell. The archicortex, or the three layer areas of the cerebral cortex, is believed to be the least developed location of the cerebral cortex. The archicortex is found in the hippocampus of the temporal lobe. The hippocampus, which will be discussed in more details in later sections, is the principle area of memory formation in humans.

Basal Ganglia

The basal ganglia consist of a small portion of the cerebrum that is associated with motor and cognitive functions. In regards to the motor functions, it is necessary to point out that the basal ganglia do not initiate motor commands; the motor commands are initiated in the motor cortex (the motor cortex will be discussed in later sections). The purpose of the basal ganglia is to modulate and subsequently relay appropriate motor commands to the lower motor hierarchy. For example, the basal ganglia are involved in the integration of previous motor movements with the motor actions that are needed to be taken presently to ensure proper movement.

On the other hand, it is necessary to point out that the basal ganglia are not directly involved in the formation of cognitive functions, but are concerned with the modulation of cognitive signals. For example, the close association of the basal ganglia with the prefrontal cortex and the limbic system allows it to select and enable various cognitive, executive functions such as planning, cognitive flexibility, abstract thinking, rule acquisition, inhibiting inappropriate actions, as well as, inhibiting irrelevant sensory information. In addition, the basal ganglia can also select and enable emotional programs that are stored in these cortical areas. Just like their motor functions, the basal ganglia are involved in the integration of the memories of previous cognitive, executive, or emotional programs with the actions needed to be taken presently to ensure an appropriate motor response.

The basal ganglia are composed of six groups of cerebral gray matter located adjacent to the thalamus and are buried deep within the white mater of the brain. The first four groups are the caudate nucleus, putamen, nucleus accumbens, and globus pallidus. Together, these four closely intertwining structures are named the corpus striatum. The remaining two groups are the subthalamic nucleus and substantia nigra, which are comprised of the pars compacta and the pars reticulate (Figure 7).

The caudate nucleus is closely associated with the lateral wall of the lateral ventricle and is involved in learning and storing memories. More specifically, research has demonstrated that the left caudate nucleus is heavily involved in the development and use of language(s), as well as, communication skills.

The putamen is involved in a very complex feedback-loop that modulates the movement of upper and lower limbs. For example, signals are transmitted through the putamen to various motor areas of the brain (e.g. motor areas of the cerebral cortex, brain stem etc.) which aid in the formation of memories of movements, which in turn, aids in deciding which movements will be made in the future depending upon the circumstances. Research has shown that lesions of the putamen caused by Parkinson's

disease or Huntington's disease will cause involuntary muscle movements such as jerking.

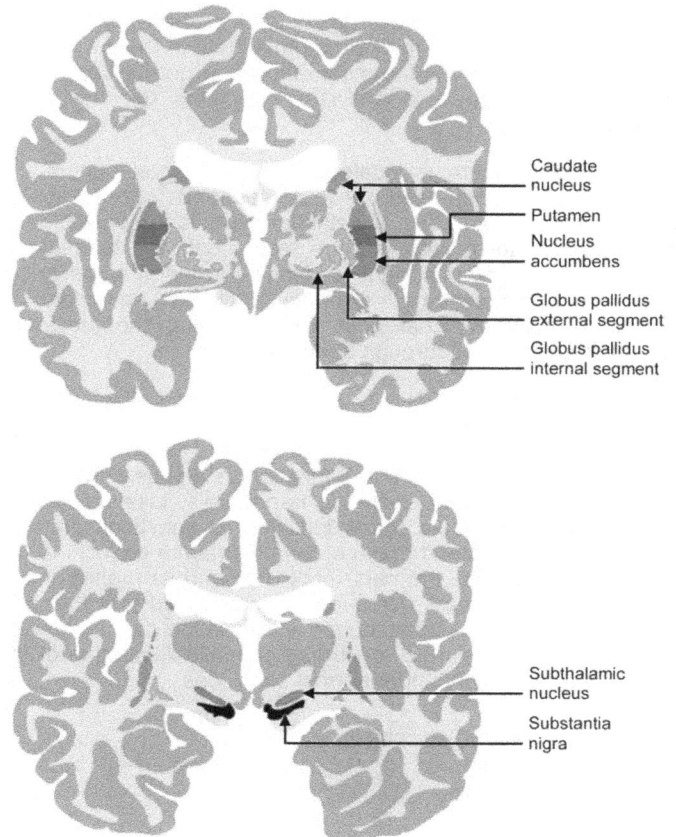

Figure 7: Illustration the basal ganglia. Please note that two different frontal sections are needed to illustrate the various parts of the structure

The nucleus accumbens is composed of dopaminergic neurons which are involved in reinforcement learning (i.e. perceives its current state and takes actions) and play a role in pleasures such as sex, laughter, and rewards etc. In addition, the nucleus accumbens is believed to modulate signals regarding addiction, fear, aggression and impulsivity. For example, research has found that the nucleus accumbens serves as a crucial link in the reward system, which involves reinforcing the effects stimuli such as food or drugs, which may lead to addiction. Addictive drugs such as cocaine, heroin, nicotine, or alcohol increase the amount of dopamine secreted by nucleus accumbens. This increase in dopamine is believed to be responsible for the reinforcement learning that triggers the desire for illicit drugs and may promote substance dependency.

The globus pallidus is composed of inhibitory, GABA secreting neurons that are used to modulate motor signals from the motor cortex. In addition, research has shown that the globus pallidus is responsible for reinforcing the desires for stimuli such as food or drugs, which may lead to addiction. The globus pallidus possesses two major components: external (GPe) and the internal segment (GPi). The GPe receives excitatory signals from the thalamus and sends output signals to other structures of the corpus striatum. In turn, GPe provides GABAergic inhibitory efferent connections to all the basal ganglia's input and output nuclei. GPi receives signals from all other structures of the corpus striatum and sends inhibitory (GABA) transmissions to the thalamus and the brainstem, etc.

The subthalamic nucleus (STN) is the only structure within the basal ganglia that secretes excitatory neurotransmitter called glutamate. Glutamate is responsible for exerting a profound excitatory influence on the associated structures of the basal ganglia. For this reason, the subthalamic nucleus is referred to as the "driving force" of the basal ganglia. In addition, the STN plays an important role in motor control. For example, research has shown that lesions within the STN cause involuntary muscle movements of the extremities. In contrast, increased activities of the STN will result in a marked improvement in motor function. For this reason, STN is the most efficient target for deep brain stimulations designed to alleviate symptoms of Parkinson's disease.

The substantia nigra is located near the ventral part of the midbrain and is composed of two components: pars compacta and pars reticulate. The substantia nigra is responsible for modulating motor movement, eye movement, and learning. In addition, the substantia nigra is involved in learning behaviors that will lead to a reward (for example food or sex) and addiction. Research has shown that most of the effects of the substantia nigra are coordinated through the corpus striatum.

The pars compacta (SNc) is a dopaminergic pathway to the corpus striatum. It is understood that the loss of neurons in this area is the cause of Parkinson's disease. The pars reticulata (SNr) is similar in structure and function as globus pallidus and is composed of GABA secreting nerve cells. Like the globus pallidus, pars reticulata forms inhibitory connections on their targets. It is understood that the corpus striatum projects to the pars reticulate, while the pars reticulate send its projections to the thalamus and pars compacta. It is believed that the pars reticulata modulates dopaminergic activity in the pars compacta. Please examine Figure 8 for a diagrammatic illustration of the basal ganglia circuitry under normal conditions.

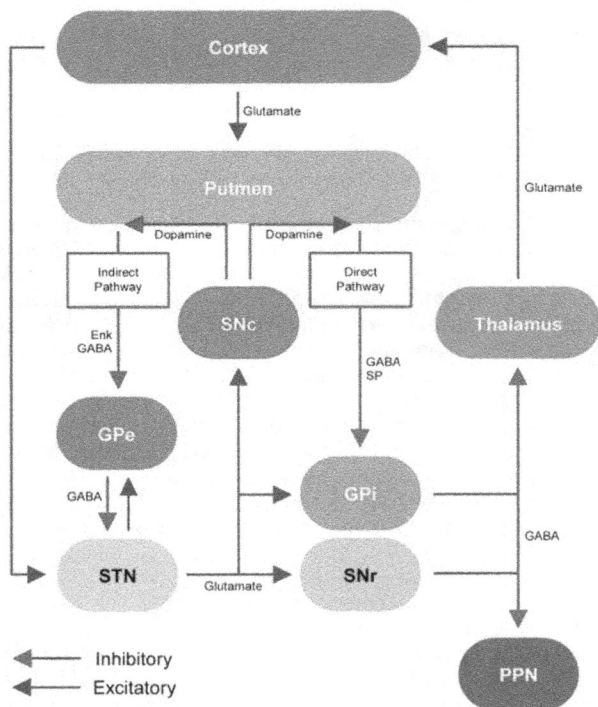

Figure 8: Diagrammatic illustration of the basal ganglia circuitry under normal condition. Please note the following abbreviations: substantia nigra pars compacta (SNc), substantia nigra pars reticulata (SNr), globus pallidus possesses two major components: external (GPe) and internal segment (GPi), and subthalamic nucleus (STN). Pedunculopontine nucleus (PPN) is located in the brain stem and helps influence sleep and waking

LIMBIC SYSTEM

The limbic system is a group of interconnecting (i.e. functionally and anatomically) cortical structures and nuclei derived from the cerebrum and portions of the diencephalon. The main cortical structures and nuclei included within the limbic system are the hypothalamus, mammillary bodies, amygdala, hippocampus,

and the prefrontal cortex. The limbic system is involved in the regulation of endocrine and autonomic functions. In addition, the system is also involved in memory formation, as well as, reinforced behaviors and motivation (Figure 9).

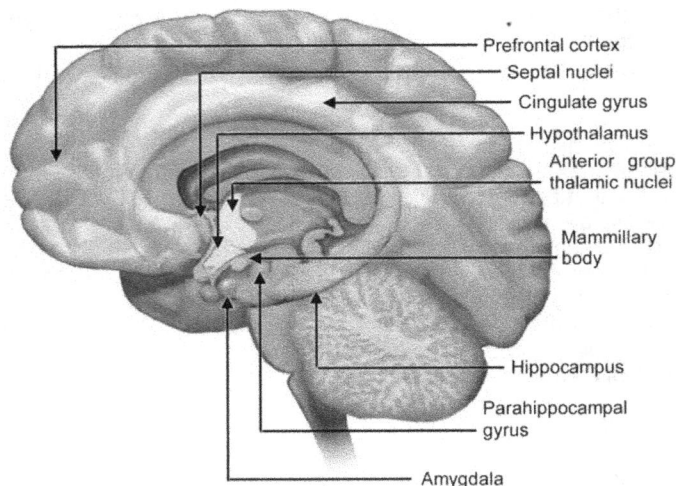

Figure 9: Illustration of the limbic system

Considered to be the command center of the endocrine system, the hypothalamus has many other functions that influence sexual and behavioral functions, as well as, autonomic functions. Known as the primary output node for the limbic system, the hypothalamus interconnects with various locations within the brain. For example, the hypothalamus is interconnected with the frontal lobes, septal nuclei, and the reticular formation (located in the brain stem). It also receives information from the hippocampus and the amygdala. For the hypothalamus to perform properly, it must receive information from the olfactory lobe, visual cortex, as well as, visceral (bodily) sensory information relayed by the thalamus. In addition, the hypothalamus possesses various sensors that are able to monitor internal temperature, osmolality, glucose levels, sodium concentrations, as well as, the production and secretion of various intercellular signals (i.e. hormones) (Figures 9 and 10).

The endocrine functions of the hypothalamus involves the direct secretion of hormones via the posterior pituitary (e.g. vasopressin and oxytocin) or the secretion of releasing or inhibiting hormones via the hypothalamo-hypophyseal portal system to control the hormone secretion of the anterior pituitary.

The hypothalamus influences autonomic functions via projections to the brain stem and spinal cord. Depending on the needs of the body, the hypothalamus may cause the increase in sympathetic activities, while decreasing parasympathetic actions or vice-versa. The hypothalamus sends its projections to the reticular formation to influence emotional reactions, generally after a traumatic event. Additionally, the hypothalamus influences complex behaviors such as the control of thirst, sleep, and sexual behavior through the actions of the preoptic area of the hypothalamus. For example, the medial arcuate nuclei of the preoptic area of the hypothalamus are responsible for producing and releasing gonadotropin-releasing hormone (GnRH). GnRH stimulates the synthesis and secretion of two anterior pituitary hormones called luteinizing hormone (LH) and follicle-stimulating hormone (FSH). In addition, this preoptic area also possesses a densely packed cluster of ovoid-shaped cells called sexually dimorphic nucleus (SDN). SDN is crucially involved in the regulation of sexually differentiated behaviors and functions in animals, as well as, humans. Interestingly, experiments have shown that the male SDN is more numerous in cell numbers and larger in cell size in comparison to female SDN, which may be one of the significant factors in the dimorphic behavioral differences between the sexes (Figure 9, 10 and 12).

The mammillary bodies are two small neurological structures located behind the infundibulum and are considered to be a part of the hypothalamus. The mammillary bodies are responsible for the feeding reflexes such as licking and swallowing. In addition, the mammillary bodies are involved in emotional responses to odors, olfactory reflexes and memory (Figures 9 and 10).

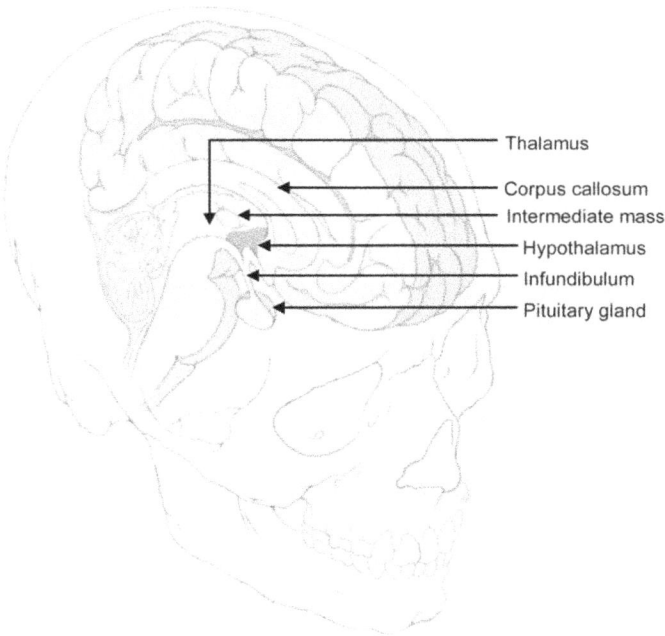

Figure 10: Sagittal section of the hypothalamus

- Thalamus
- Corpus callosum
- Intermediate mass
- Hypothalamus
- Infundibulum
- Pituitary gland

The amygdala is a structure located in the anterior temporal lobe that forms reciprocal connections with the thalamus, hypothalamus, hippocampus, the brain stem, orbital frontal cortex (which is a part of the frontal lobe), and the parahippocampal gyrus (which is a part of the temporal lobes). In addition, the amygdala also forms reciprocal connections with the septal nuclei (SN), and the cingulate gyrus (CG). The SN is subdivided into two partitions, medial and lateral SN, and is continuous with the gray matter located at the medial surface of the cerebral hemisphere. The SN forms reciprocal connections with the hippocampus and the hypothalamus and is involved in the modulation of rage and pleasure. For example, individuals with lesions in the septal nuclei will have developed uncontrollable rage behavior. In contrast, electrical stimulation of the SN will induce the sensation of pleasure. The CG, on the other hand, lies above the corpus callosum, and helps regulate emotions in association with pain. It is theorized that the CG drives the body's conscious response to unpleasant experiences and is involved in fear, as well as, the prediction and avoidance of negative consequences. It is known that the olfactory bulb is the only area that has input to the amygdala and does not receive reciprocal projections from the amygdala (Figure 9).

The amygdala is known as a center for managing behavioral responses, it is also a critical location for managing emotions and formation of long term memories. The amygdala influences the autonomic and endocrine responses to external stimuli, particularly those that elicit an emotional reaction. It is well established that the amygdala is essential for emotional learning, where visual, olfactory, taste and tactile signs are triggers for memories of either rewarding or aversive events. In addition, the amygdala is also involved in managing and coordinating responses to stress, fear and anxiety. Individuals suffering from lesions to the amygdala demonstrate reduced emotional responses to stress. Furthermore, research has shown that psychopathic individuals demonstrating significant aggression,

as well as, impairment in aversive conditioning and passive avoidance learning, will generally show evidence of amygdala dysfunction (Figure 9).

The hippocampus is located in the medial aspect of the temporal lobe and forms reciprocal connections with the cerebral cortex (i.e. neocortex), prefrontal cortex (discussed below) and the amygdala. On the other hand, the outputs of the hippocampus project to the mammillary bodies, the preoptic nuclei of the hypothalamus, the nucleus accumbens of the basal ganglia and the frontal lobe. The hippocampus is involved in retrieving memories. It is understood that there are many different types of memories. For example, explicit memory are memories of facts and events, which can be expressed in words. Implicit memory, on the other hand, is associated with memories dealing with emotional responses. The hippocampus is involved with explicit memories and plays an important role in short-term memory formation. It is necessary to note that the formation of short-term memory is necessary for the eventual formation of long term memory. Put simply, long-term memories are short-term memories that are transferred and stored within the cerebral cortex. Studies have shown that patients suffering from lesions in the hippocampus have an inability to form short-term memories but does not affect stored long-term memories (Figure 9).

The prefrontal cortex is located in the outer layers of the frontal lobe and is essential to the formation of judgment, insight, motivation and mood. The prefrontal cortex forms reciprocal connection with various areas of the limbic system (e.g. amygdala, septal nuclei, etc.), as well as, forming connections with the anterior group nuclei of the thalamus. Patients suffering from lesions to the prefrontal cortex display difficulties with abstract reasoning, judgment, and resolving problems. Individuals suffering from dysfunction in the prefrontal cortex will display depression and other mood disorders (Figure 9).

DIENCEPHALON

The diencephalon is located at the midline of the brain and consists of the thalamus, the subthalamus, the epithalamus, the hypothalamus, and the posterior pituitary gland.

The thalamus is the largest structure in the diencephalon and is bilaterally shaped with two lateral portions connected by the intermediate mass, also known as the interthalamic adhesion. It is known that all sensory information that reaches the cerebrum (with the exception of the olfactory signals) must first be transmitted through the thalamus. For this reason, the thalamus is considered to be the central relay station for sensory impulses from the ascending fibers (from the body) and most sensory input from the cranium. Once the information is received by the thalamus, it modulates these signals and assigns priority to the information. Subsequently, the thalamus transfers the signals to the appropriate region of the cerebral cortex for analysis and interpretation. For example, sensory information received by the retina are transmitted in sequence through the optic nerve, optic chiasma, optic tract and then connects with the lateral geniculate nucleus of the thalamus, before being sent to the visual cortex, located within the occipital lobe (Figure 10).

In addition to serving as the sensory information relay station in the brain, the thalamus is also involved in the modulation and control of skeletal muscle movements. For example, the thalamus is responsible for sending and receiving neurotransmissions to and from the cerebellum, basal ganglia and the motor areas of the cerebral cortex. The thalamus is also directly linked to the cerebellum through the ventral anterior (VA), thalamic nuclei and the reticular nucleus of the thalamus, while the link to the basal ganglia are made through the midline and intralaminar thalamic nuclei, the ventral anterior (VA) and the ventral lateral (VL) thalamic nuclei. The VA and VL are also involved in receiving outputs from the basal ganglia before relaying the information to the primary motor cortex. In addition, it is known that the ventral tier thalamic nuclei have direct connections with the premotor, motor, and cingulate motor cortices, which show their involvement in the control of motor functions. Clinical reports indicate that damage to the VA and VL regions of the thalamus result in ataxia (i.e. uncoordinated movement due to an inability to coordinate movements) and intention tremor (i.e. low frequency tremor), which is reminiscent of the symptoms caused by cerebellar or basal ganglia damage (Figure 11).

As mentioned in the previous section, elements of the thalamus are associated with the limbic system. The thalamus is interconnected with other structures of the limbic system to influence moods, and actions associated with emotions such as rage and fear. Please examine the section entitled Limbic System for more information.

The subthalamus is located inferior to the thalamus and comprises the subthalamic nuclei (STN) and several ascending and descending nerve tracts. It is understood that subthalamus is involved in controlling motor functions. The STN receives its afferents from the globus pallidus of the basal ganglia, cerebral cortex, thalamus and the brainstem while projecting glutaminergic efferents to both segments of the globus pallidus, substantia nigra, as well as, the striatum of the brain stem. Clinical research has shown that lesions of the STN induce abnormal muscular movements and ballism (i.e. frequent violent movements of the shoulder and arm). Depending on the side that the lesion occurred, the abnormal muscular movements and ballism will involve on the contralateral or opposite side of the body. This type of lesion is frequently observed in individuals suffering from Parkinson's disease.

The epithalamus is located superior and posterior to the thalamus and is composed of two groups of nuclei (i.e. one on each side of the cerebral hemisphere) called the habenula and the pineal body (Figure 11).

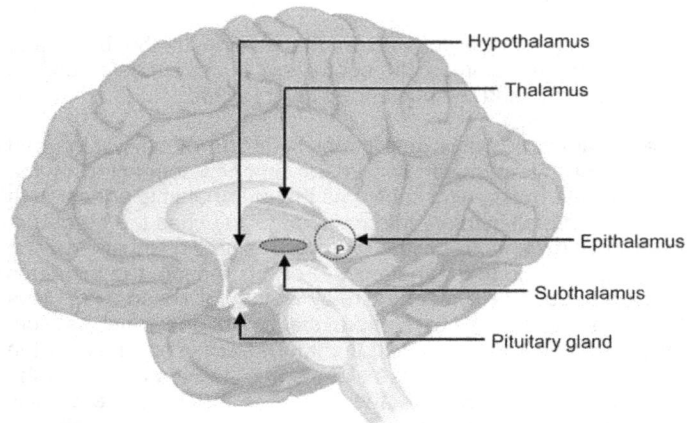

Figure 11: Illustration of the thalamus, subthalamus, epithalamus, hypothalamus and the pituitary gland. Please note that the pineal gland (P) is a part of the epithalamus

The habenula influences the brain's response to stress, anxiety, pain, and reward. Dysfunction of the habenula has been associated with individuals suffering from depression, schizophrenia, as well as, the effects of drugs abuse. The habenula is divided into the medial habenula (MHb) and the lateral habenula (LHb). The main input to the MHb is provided by the septum (a pleasure area of the brain) while the output from the MHb is mainly projected at the LHb although there are no reciprocal connections from the LHb back to the MHb. Research has shown that the MHb modulates acetylcholine release and is involved in nicotine and morphine dependencies. The LHb receives afferents primarily from limbic brain regions and the basal ganglia. The efferents of the LHb sends its projections to the forebrain and the neurons of the brainstem. For example,

the LHb sends inhibitory signals (e.g. GABA secreting nerves) to the dopaminergic neurons of the brain stem which are crucial for reward-based motor control. These dopaminergic neurons of the brain stem are responsible for sending signals that are essential for the instigation and learning of body movements. These neurons predict errors and act to promote motor behavior that may maximize rewards (Figure 11).

The pineal body, also known as the pineal gland, is a part of the endocrine system that is responsible for the secretion of a hormone called melatonin. In vertebrates, the pineal body is in control of the circadian rhythm, where the secretion of melatonin is increased during the darkness, while its secretion is suppressed during periods of light. In animals, the duration of the nocturnal melatonin secretion provides an internal calendar (e.g. photoperiodic calendar) that regulates seasonal reproductive cycles. Humans, on the other hand, are not believed to be photoperiodic, therefore the effects of the pineal gland are believed to be diminished. Nonetheless, the incidences of a condition known as seasonal affective disorder (SAD) and its treatment through the exposure to light imply that we have retained some photoperiodic responsiveness (Figure 11).

Inferior to the thalamus is the hypothalamus and it is involved with the autonomic nervous system, the the endocrine system, as well as, the limbic system as mentioned previously. The hypothalamus is subdivided into various subsections which are dedicated to various functions. For example, the lateral hypothalamic area and ventromedial nucleus are known as the feeding center of the brain. Experiments have shown that the neurons in these areas respond to systemic glucose, free fatty acid and insulin levels. It has been postulated the lateral hypothalamic area and ventromedial nucleus provides a set-point for a person's body weight. If a person goes below this particular set-pont, the lateral hypothalamic area will be activated to increase appitite and increase feeding. Experiments have shown that bilateral lesions in the lateral hypothalamic area result in anorexia. In contrast, bilateral lesions of the ventromedial nucleus will result in hyperphagia (overeating), extreme obesity, as well as, displaying chronically irritable mood and aggressive behavior, also known as hypothalamic rage (Figure 12).

The dorsal medial nucleus is responsible for the regulation of feeding, drinking, maintenance of body weight, as well as, contributing to the maintenance of the circadian rhythm. Experiments have shown that lesions in the dorsal medial nucleus result in hypophagia (i.e. reduced food intake) and hypodipsia (i.e. reduced fluid intake) (Figure 12).

The hypothalamic regulation of the temperature set-point is monitored and maintained by the anterior hypothalamic area and the preoptic nucleus, also known as preoptic-anterior hypothalamus (POAH), with a subservient role provided by the dorsal and posterior hypothalamic areas. Regulation of core body temperature is essential since most enzymes that maintain metabolic processes are highly temperature sensitive.

It is known that the POAH possesses three types of neurons that are involved in determining the temperature set-point: warm-sensitive neurons, cold-sensitive neurons, and temperature-insensitive neurons. For example, if the temperature increases above 37°C the depolarization rate of these warm-sensitive neurons increases, which in turn, activate the neurons in the paraventricular nucleus (PVN) and the lateral hypothalamus. The activation of the PVN and the lateral hypothalamus results in the increase in parasympathetic innervation, thereby increasing the dissipation of heat (e.g. sweating and relaxation of arrector pili muscles etc.). In contrast to the warm-sensitive neurons, the cold-sensitive neurons depolarize below 37°C and in turn activate neurons in the PVN and the posterior hypothalamus. The activation of the PVN and the posterior hypothalamus results in the increase in sympathetic innervation, thereby promoting the generation and conservation of heat (e.g. shivering, contraction of the arrector pili muscles, etc.).

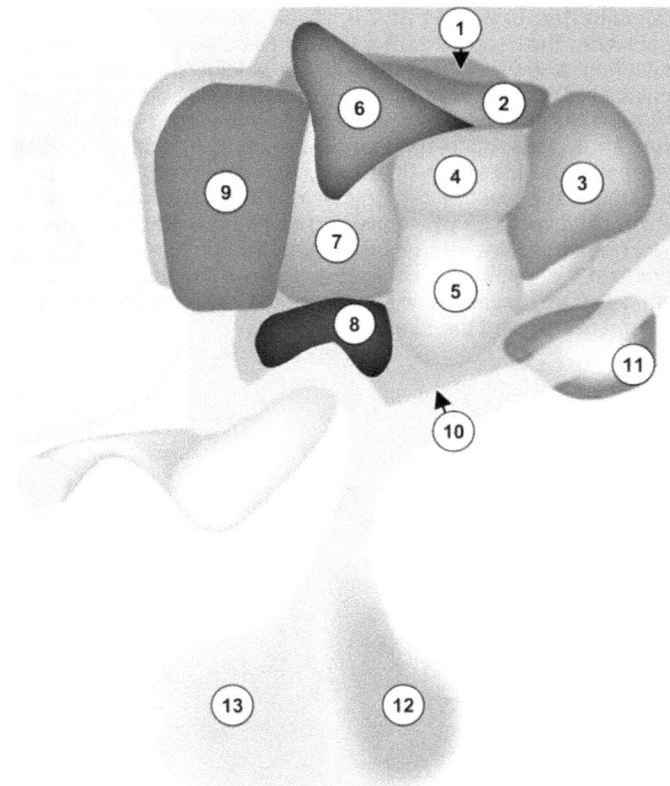

Figure 12: Illustration of the hypothalamus: ① Lateral hypothalamic area, ② dorsal hypothalamic area, ③ posterior hypothalamic area, ④ dorsal medial nucleus, ⑤ ventromedial nucleus, ⑥ paraventricular nucleus, ⑦ anterior hypothalamic area, ⑧ supraoptic nucleus, ⑨ preoptic nucleus, ⑩ arcuate nuclei. Areas not part of the hypothalamus: ⑪ mammillary body, ⑫ posterior pituitary and ⑬ anterior pituitary

The magnocellular neurons found within the supraoptic and paraventricular nuclei are responsible for producing and secreting antidiureic hormone (ADH), also known as vasopressin, as well as, a reproductive hormone called oxytocin. Vasopressin is a hormone responsible for maintaining osmolality (i.e. the concentration of dissolved particles) and fluid balance. Oxytocin, on the other hand, is a hormone responsible for uterine smooth muscle contraction during labor and the secretion of breast milk after delivery. Once formed, ADH and oxytocin are transported via the fast axonal transport down the length of the axon which is located in the hypothalamo-hypophyseal tract of the infundibulum (e.g. pituitary stalk). Once the neurohormones reach the axonal terminal, which is located at the posterior pituitary, they are released via exocytosis into circulation (Figure 13).

In addition, the neurons (e.g. parvocellular neurons) found within the paraventricular nuclei produce various hormones that control the production and release of the anterior pituitary hormones. For example, the parvocellular neurons produce corticotrophin releasing hormone, which in turn controls the production and release of anterior pituitary hormones, such as lipotropin (LPH), beta-endorphins and adrenocorticotropic hormone (ACTH). In addition, the parvocellular neurons also produce thyrotropin releasing hormones (TRH), which stimulate the production of thyroid stimulating hormone (TSH) in the anterior pituitary. TSH controls the release of thyroxin (T4) and triiodothyronine (T3) from the thyroid gland.

The arcuate nucleus, also known as the infundibular nucleus or the periventricular nucleus, is composed of neuroendocrine neurons and is a part of the endocrine system. Just like the parvocellular neurons of the paraventricular nuclei, these neuroendocrine neurons produce and secrete releasing or inhibiting hormones.

For example, the neuroendocrine neurons secretes growth hormone releasing hormone (GHRH), gonadotropin releasing hormone (GnRH), and growth hormone inhibiting hormone (GHIH), also known as somatostatin (Figure 12).

The posterior pituitary is composed of axonal projections extending from the supraoptic and paraventricular nuclei of the hypothalamus. In other words, the posterior pituitary is an extension of the hypothalamus via a thin layer of tissue composed mostly of axons and axonal terminals. As mentioned previously, this diencephalon structure is involved with the secretion of neurohormones, such as ADH (e.g. vasopressin) and oxytocin via exocytosis into circulation (Figure 13).

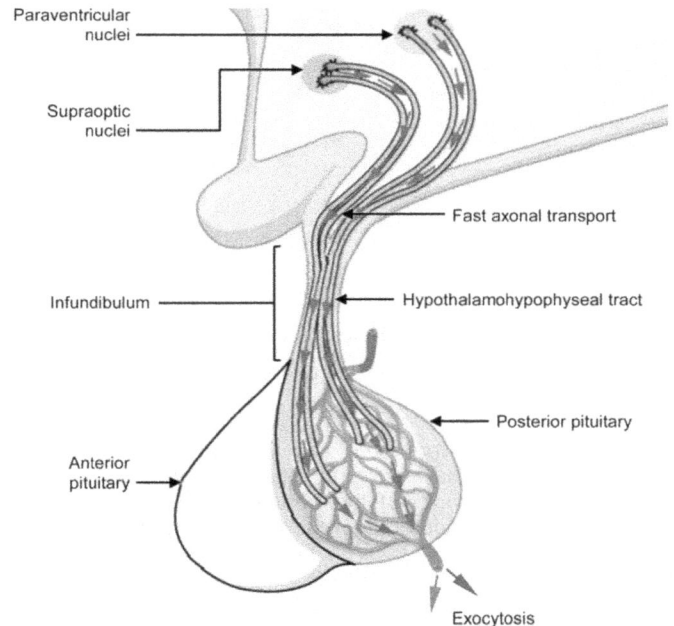

Figure 13: Illustration of the production, transport and release of ADH and oxytocin. Please note that the magnocellular neurons located in both supraoptic and paraventricular nuclei are responsible for producing these hormones

BRAIN STEM

The brain stem is located inferiorly, towards the diencephalon and consists of three areas: the medulla oblongata, pons, and the midbrain (Figure 14).

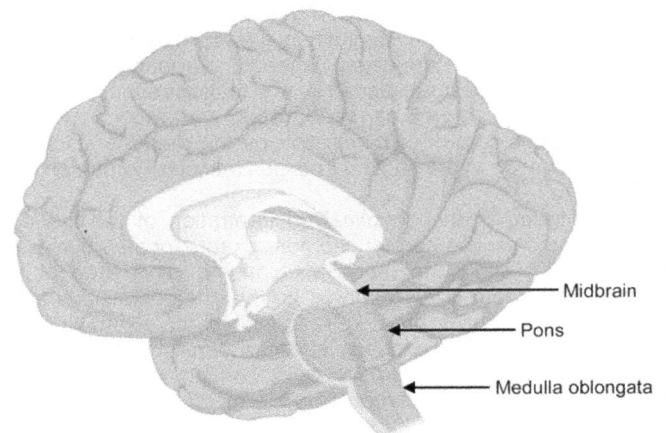

Figure 14: Illustration of the brain stem

The medulla oblongata extends from the foramen magnum to the pons, is continuous with the spinal cord and lies anteriorly from the cerebellum. Four pairs of cranial nerves arise from the

179

medulla oblongata: glossopharyngeal nerve IX, vagus nerve X, accessory nerve XI, hypoglossal nerve XII (Figure 15).

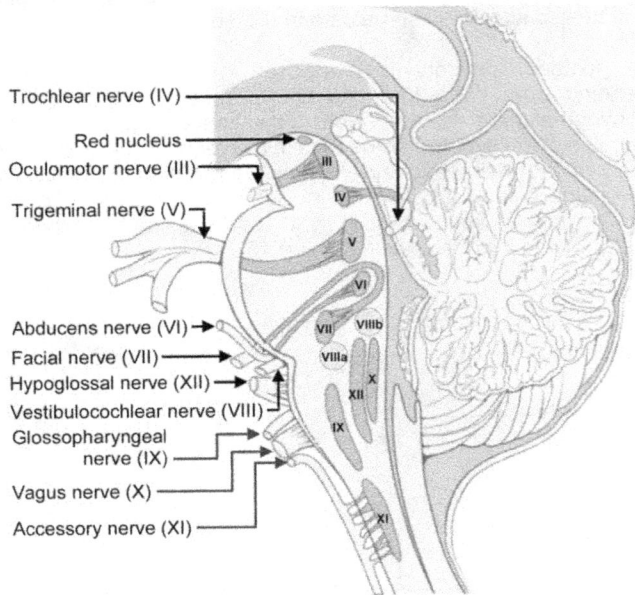

Trochlear nerve (IV)
Red nucleus
Oculomotor nerve (III)
Trigeminal nerve (V)
Abducens nerve (VI)
Facial nerve (VII)
Hypoglossal nerve (XII)
Vestibulocochlear nerve (VIII)
Glossopharyngeal nerve (IX)
Vagus nerve (X)
Accessory nerve (XI)

Figure 15: Illustration of the cranial nerves and its nucleus found within the brain stem. Please note that the nucleus forming the glossopharyngeal nerve (IX) is called nucleus ambiguus. Nucleus ambiguus also forms the motor branches of the vagus nerve (X)

Two prominent enlargements are located on the anterior surface of the medulla oblongata. These enlargements, which extend the entire length of the nervous structure, are called pyramids. The pyramids contain the descending tracts that are involved in the conscious control of skeletal muscles, as well as, maintaining balance and coordination. Towards the inferior ends of the medulla oblongata these nerve fibers of the descending tract cross or decussate to the opposite side. This process of decussation is partly accountable to each side of the brain controlling the opposite side of the body (Figure 16).

Posteriorly, the medulla oblongata contains the ascending tracts from the cerebellum, spinal cord, and cranial nerves. This cranial structure is responsible for relaying nervous impulses through the numerous ascending nerve tracts to the thalamus. In addition, the medulla oblongata contains various neural networks that are involved in both sensory and motor functions. For example, the medulla oblongata contains the neural networks that are involved in the senses of touch (i.e. pressure), temperature, and pain. The motor neuronal networks include the regulation of mastication (i.e. chewing), salivation, deglutition (i.e. swallowing), gagging, vomiting, coughing, sneezing, speech, sweating, digestion, controlling autonomic functions relative to respiration, heart rate and contractility, as well as, vasodilation and vasoconstriction. The autonomic control of respiration by the medulla oblongata will be discussed in the section entitled Reticular Formation.

The monitoring of the chemical concentration of the blood is accomplished via the central chemoreceptors located bilaterally and centrally within the medulla oblongata. This area of the medulla oblongata is also known as the chemosensitive area. The chemo-sensitive area is responsible for examining the concentrations of blood gases such as carbon dioxide (CO_2), as well as, the levels of carbonic acid. The information collected by the chemosensitive area is supplemented and integrated with the sensory information received and transmitted by the peripheral chemoreceptors located within the carotid arteries and the aorta. These chemoreceptors are also referred to as the carotid and aortic bodies. For example, the chemosensitive area of the medulla oblongata, in addition to the carotid and aortic bodies, monitor the level of carbonic acid in the blood stream. As mentioned in Chapter 2, carbonic acid is a weak acid and a

human blood buffer that is synthesized by the enzyme carbonic anhydrase, located within the membranes of erythrocytes, and maintains the blood pH at 7.4. If the level of CO_2 is high in circulation, the level of carbonic acid will also be high, thereby indicating a low pH level (low pH = acidic). The low pH will activate the chemosensitive area of the medulla oblongata, which will stimulate the respiratory center to increase the respiratory rate. An increase in respiration will expel CO_2 from the blood, increasing the blood pH. Inversely, if the level of CO_2 is low in circulation, the level of carbonic acid will also be low, indicating a high pH level (high pH = basic). The high pH will also activate the chemosensitive area of the medulla oblongata, which will stimulate the respiratory center to decrease the respiratory rate. A decrease in respiration will retain more CO_2 in the blood stream thereby decreasing the blood pH.

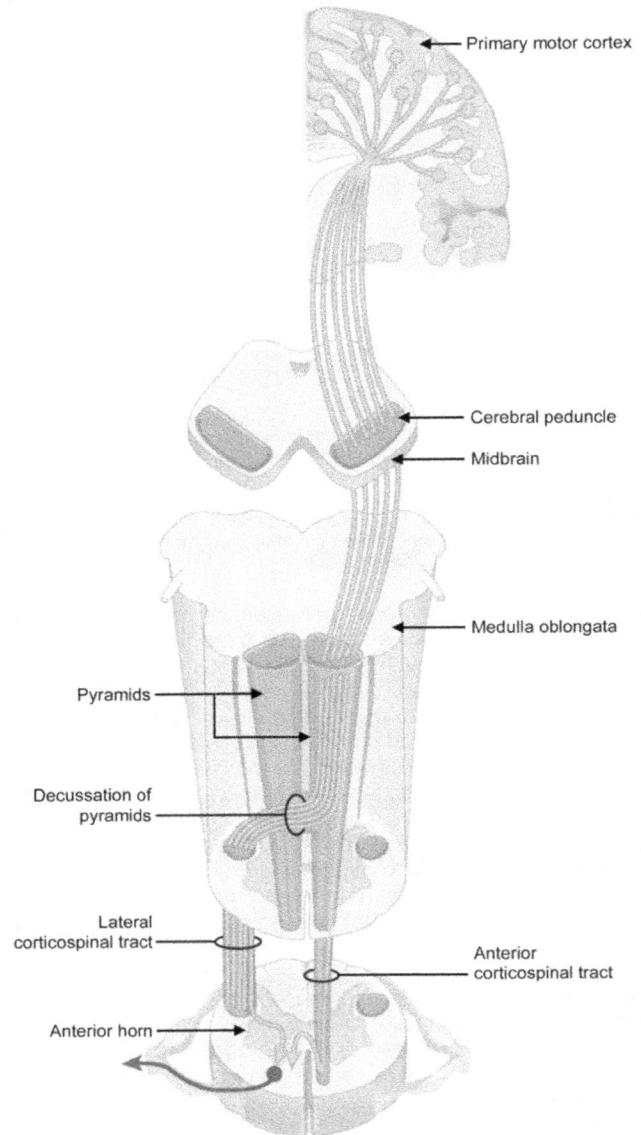

Primary motor cortex
Cerebral peduncle
Midbrain
Medulla oblongata
Pyramids
Decussation of pyramids
Lateral corticospinal tract
Anterior corticospinal tract
Anterior horn

Figure 16: Illustration of primary motor cortex's direct control over skeletal muscles. Please note that towards the inferior ends of the medulla oblongata these nerve fibers of the descending tract decussate to the opposite side. This process of decussation is partly accountable each side of the brain controlling the opposite side of the body

The baroreceptor reflex is responsible for detecting changes of the blood pressure. The information regarding blood pressure is received by the baroreceptors, also known as stretch receptors, located in the large arteries (i.e. internal carotid, aorta, etc.), before being transmitted to the cardioregulatory center of the medulla oblongata where the sensory information is integrated.

The cardioregulatory center is responsible for controlling the action potential frequencies of the autonomic nervous system. The autonomic nervous system is composed of two subdivisions: the sympathetic nerves, which release norepinephrine as postsynaptic neurotransmitter, and parasympathetic nerves, which release acetylcholine as postsynaptic neurotransmitter. These autonomic nerves project from the medulla oblongata to the heart. When the blood pressure increases, the walls of the large arteries will stretch; this in turn, activates the baroreceptors. The activated baroreceptors will send an action potential directly to the cardioregulatory center of the medulla oblongata. This action potential will prompt the cardioregulatory center to decrease sympathetic innervation and increase para-sympathetic innervation, which in turn, will decrease heart rate, thereby reducing the blood pressure. Inversely, if the blood pressure decreases, the baroreceptors will also respond by sending an action potential to the cardioregulatory center of the medulla oblongata. This action potential will prompt the cardioregulatory center to increase sympathetic innervation and decrease parasympathetic innervation, which in turn, will increase heart rate and thereby elevating the blood pressure.

The medulla oblongata is also responsible for controlling the release of neurohormones, such as, norepinephrine and epinephrine from the adrenal medulla. This influence over the adrenal medulla is called the adrenal medullary mechanism and is also part of the body's control over hypotension. For example, if a substantial decrease in the blood pressure is detected, the medulla oblongata will trigger the adrenal medulla to release norepinephrine and epinephrine. The release of these neurohormones will result in increased heart rate, stroke volume and vasoconstriction. In contrast, if there is a substantial increase in blood pressure detected, the medulla oblongata will decrease the release of the neurohormones from the adrenal medulla.

Superior to the medulla oblongata is the nervous structure called pons. Anteriorly, the pons possesses a large bulge which is dominated by ascending and descending tracts. Posteriorly, the pons contains two thick stalks called the cerebellar peduncles which serve to connect the cerebellum with the brain stem. Four pairs of cranial nerves arise from the pons: trigeminal nerves V, abducens nerve VI, facial nerve VII, and vestibulocochlear nerves VIII. It is necessary to note that although the trigeminal nerves (V) arise from the pons, they have segments that extend into the medulla oblongata (Figures 14 and 15).

In addition to being a major relay station for ascending and descending tracts, the pons also assists in controlling autonomic functions and respiration. The control of respiration by the pons is achieved through the pontine respiration group, which will be discussed in the section titled Reticular Formation.

The midbrain is the most superior part of the brainstem and contains various ascending and descending tracts. The midbrain is considered to be one of the major thoroughfares, along with medulla oblongata, and the pons, for nerve fibers which connect the brain with the rest of the body. In addition, there are two pairs of cranial nerves that arise from this neuronal location: oculomotor nerves III, trochlear nerves IV (Figures 14 and 15).

The midbrain is divided into anterior and posterior sections by the aqueduct of Sylvius. The aqueduct of Sylvius is the structure within the brainstem that facilitates the flow of cerebral spinal fluid between the third and fourth ventricles.

Located on the posterior surface of the midbrain is the corpora quadrigemina, which is composed of two superior colliculi and two inferior colliculi. The two superior colliculi are responsible for relaying input from the optic tract to the lateral geniculate bodies of the thalamus. In addition, the superior colliculi are also involved in visual attention and visual tracking (i.e. tracking a moving object), as well as reflexes such as focusing, blinking, pupillary dilation and constriction. The two inferior colliculi are responsible

for relaying auditory information to the medial geniculate bodies of the thalamus. In addition, the inferior colliculi are responsible for visual and auditory reflexes such as turning towards the origin of a particular sound.

Anterior to the aqueduct of Sylvius, two stalk-like structures called cerebral peduncles exist. The cerebral peduncles are responsible for anchoring the cerebrum to the brain stem and are subdivided into three sections: the tegmentum, the substantia nigra and the cerebral crus.

The most significant structure of the tegmentum is the red nucleus which is responsible for connecting the midbrain to the cerebellum, as well as, to the vestibular area (i.e. controlling balance) of the inner ear. The red nucleus is involved in motor coordination, in addition to collaborating with the cerebellum in fine motor control. The cerebral crus is composed of bundles of nerve fibers that connect the cerebrum to the pons. Part of this nerve fiber bundle is the descending nerve tract called the corticospinal tract. As mentioned previously, substantia nigra pars compacta and pars reticulate belong to the limbic system. Please examine the Limbic System section for more information.

Reticular Formation

The reticular formation is loosely organized gray matter composed of more than a hundred small clusters of interconnecting nuclei that run vertically through the brain stem and forms connections with various regions within the diencephalon (particularly the thalamus and the hypothalamus) and the cerebrum. Within the brain stem, the reticular formation is subdivided into four anatomical components: ① the dorsal tegmental nuclei is located in the midbrain, ② the central tegmental nuclei is found in the pons while ③ the central and ④ inferior nuclei are situated in the medulla oblongata. Functionally, the reticular formation is subdivided into two parts: the ascending and the descending formation. Together, the ascending and descending formations take part in performing regulatory functions within the brain, in addition to regulating cyclical and repetitive motor functions (Figure 17).

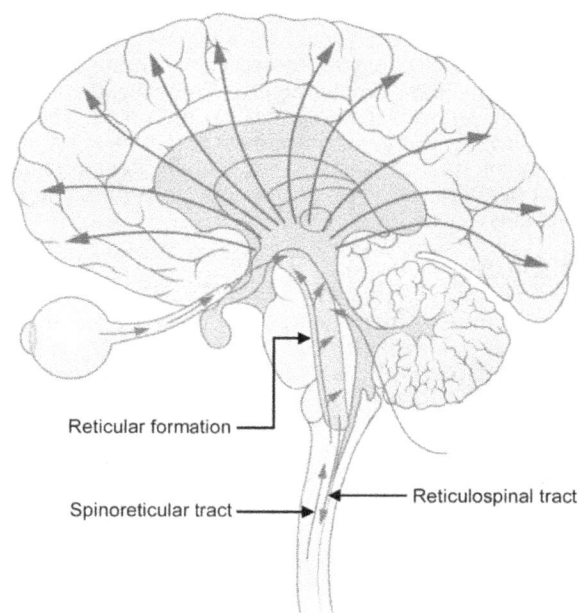

Figure 17: The reticular formation is composed of over 100 nuclei within the brain stem. The blue arrow indicates sensory input from the spinoreticular tract to the ascending reticular formation. The red arrow indicates the transmission from the descending reticular formation via the reticulospinal tract. The purple arrow indicates radiating signals from the thalamus to the cerebral cortex

The ascending reticular formation, also known as the reticular activating system, is responsible for levels of alertness and the sleep-wake cycle. The ascending reticular formation receives sensory input from sense organs, such as the eyes and ears, as well as information regarding proprioception, touch, vibration, pain, and temperature from the body via the spinoreticular fibers. Once the information is received, the ascending reticular formation will project them to the anterior and dorso-medial nuclei of the thalamus where some influence and modification of the sensory information occurs. Once the sensory information has been modified, it is then transmitted to the cerebral cortex where it comes to an individual's conscious attention. For example, the reticular formation is one of the many routes where pain signals from the lower body can reach the cerebral cortex. This particular nerve tract is called the spinoreticular tract and it transmits the pain signals resulting from tissue damage via the first order neurons to the spinal cord. At the posterior horn, the first order neurons synapse with the second order neurons. The second order neurons decussate to the opposite side the spinal cord before ascending to the reticular formation located in the medulla oblongata and the pons. At the reticular formation, this pain signal is sent to the thalamus before being relayed to the cerebral cortex, hypothalamus and the limbic system via third order neurons. The pain signals transmitted by the reticular formation activate visceral, emotional and behavioral response, such as increased alertness, nervousness, excitement, nausea, and fear.

The descending reticular formation, along with the motor areas of the cerebral cortex, is involved in maintaining muscle tone, posture and equilibrium especially during bodily movements. Together with the hypothalamus, the descending reticular formation aids in maintaining the activities of the autonomic nervous system in the control of the urinary bladder, gastrointestinal peristalsis, glandular secretions, as well as maintaining its influence over the respiratory and cardiovascular systems. Additionally, the control over swallowing, mastication, and vomiting reflexes are all equally regulated by the descending reticular formation. It is understood that the descending reticular formation maintains its influence over various systems via the reticulospinal fibers.

Although the reticular formation is subdivided into ascending and descending formations, a high level of integration between the two structures takes place in order to maintain proper bodily functions. For example, the reticular formation plays an essential part in somatic motor control. The ascending reticular formation relays sensory information regarding vestibular, auditory, and visual signals to the cerebellum. At the cerebellum, the sensory information is integrated with the transmissions by the descending reticular formation before an appropriate motor command is given.

CEREBELLUM

The cerebellum is located towards the posterior (caudal) side of the brain and is separated into the right and left cerebellar hemispheres connected by a thin layer of nervous tissue called vermis. On the surface of each cerebellar hemisphere are numerous folds called folia. Each of the folia is separated by a shallow indention called the sulci. Similar to the cerebrum, the cerebellum possesses a surface layer composed of gray matter where all inputs to the cerebellum are received, while an inner layer of white matter is where all outputs form the cerebellum originates. The white matter is distributed within the cerebellum as a branch-like pattern called arbor vitae (Figure 18).

The cerebellum is composed of ~100 billion nerve cells, which constitutes nearly 50% of the total neuronal populations of the brain. The majority of the cells found within the cerebellum are the small granule cells and the large Purkinje cells.

The cerebellum is anchored to the brain stem via three pairs of cerebral peduncles. Two inferior peduncles connect the

cerebellum to the medulla oblongata. The second pair, the middle peduncles connects it to the pons. Finally, a pair of superior peduncles connects the cerebellum to the midbrain. Within the peduncles, thick bundles of nerve fibers exist, and they are responsible for transmitting signals to and from this nervous structure. In general, most spinal inputs to the cerebellum travel via the inferior peduncles, while the inputs from the rest of the brain enter the cerebellum via the middle peduncles. It is understood that most of the outputs from the cerebellum exit via the superior peduncles (Figure 18).

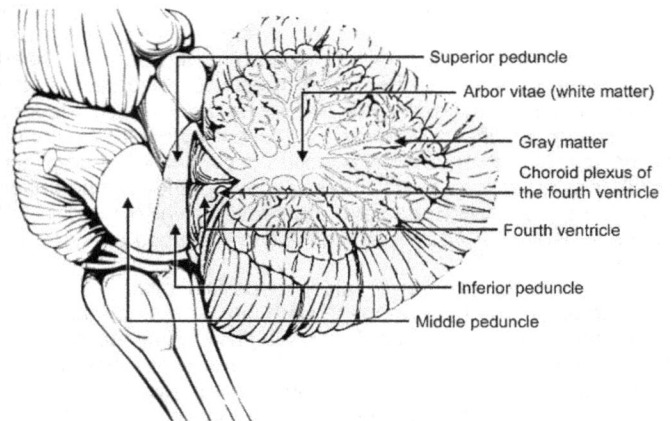

Figure 18: Illustration of the cerebellum. Please note that the branch-like patterns is the arbor vitae which is also the white matter while the surrounding tissue is the gray matter

Traditionally, the cerebellum was thought to be involved in the control of motor functions. This theory was reinforced by the fact that any damage done to the cerebellum led to impairments in motor control and posture. However, research within the last three decades has shown that the cerebellum is the location for monitoring and modifying motor commands that originate in the motor cortex. The modifications performed by this structure make the movements more accurate and adaptive. For example, the cerebellum is essential for making constant adjustments to an individual's posture in order to maintain balance. The cerebellum receives sensory inputs from various proprioceptors within the joints, as well as, vestibular receptors from the inner ear, in order to modulate commands to the descending tracts (i.e. motor nerves) to counteract shifts in body position or changes in the weight placed upon muscles.

The cerebellum is also responsible for motor learning which includes the ability of timekeeping. For example, accurate and correct movements are formed through trial-and-error are stored and can be reclaimed and used on a later date. This motor learning includes timekeeping, where an individual may predict where a moving object will be a split second later. This form of motor learning is essential when a predator is attempting to catch a mobile prey or when a linebacker is trying to tackle a zigzagging running back.

SPINAL CORD

The spinal cord is surrounded by 33 boney vertebrae in a human adult that extends from the cervical region to the coccyx region. The vertebrae are responsible for protecting the fragile spinal cord.

The spinal cord arises from the brain stem at the foramen magnum and is an extension of the brain. In adults, this thin tubular structure is approximately 45 cm long and 1.8 cm thick and extends from the foramen magnum to the first lumbar segment of the vertebrae. Slightly inferior to the first lumbar segment (L_1), the spinal cord tapers to a point called the conus medullaris and separates into bundles of nerve roots. These bundles of nerve roots occupy the vertebral canal from L_1-S_5 (5th

sacral segment). The separation of the spinal cord into individual nerves is called the terminal filum. Due to this separation into nerve bundles, and the minor resemblance of this region of the spinal cord to a horse's tail, the L_1 to S_5 region is referred to as the cauda equine. In all, the entire spinal cord, along with the terminal filum, is divided into 31 segments and serves as the information receiving and transmitting region of the CNS (Figure 19).

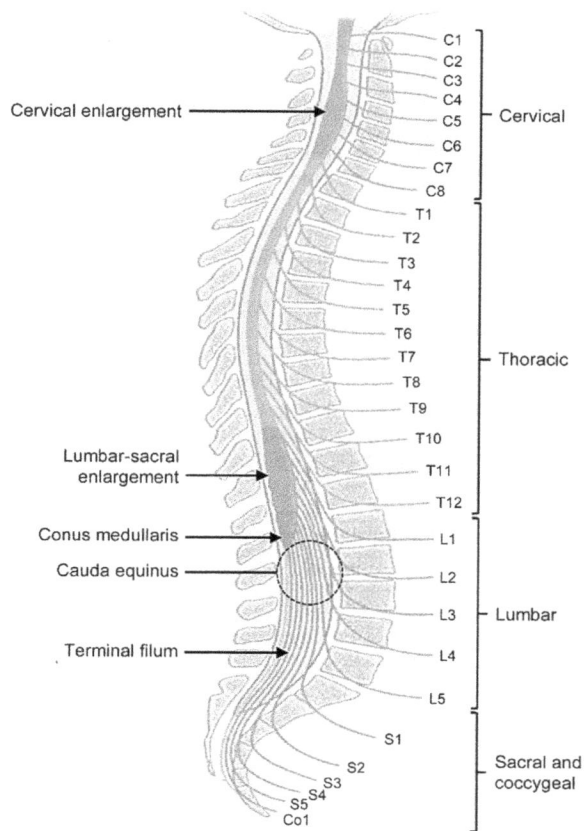

Figure 19: Illustration of the spinal cord

Anatomically the spinal cord has two distinct thickenings: the cervical and the lumbar enlargements. The cervical enlargement is an anatomical thickening, starting from the third cervical segment (C_3) and extends to the second thoracic segment (T_2) of the spinal cord. The cervical enlargement transmits to and receives impulses from the upper limbs. The lumbar-sacral enlargement is an anatomical thickening between the L_1 to the third sacral segment (S_3) of the spinal cord. The lumbar-sacral enlargement is responsible for transmitting and receiving impulses from the lower limbs.

Like the brain, the spinal cord is composed of gray matter and white matter. This segregation of the unmyelinated and myelinated tissue is visually apparent when examining a cross section of this tissue. For example, the spinal cord has a central core of gray matter that is somewhat butterfly shaped. This gray matter, composed mainly of dorsal horns, extends towards the posterior-lateral surfaces, while the ventral horns extend towards the anterior-lateral surfaces. The dorsal and ventral horns taper and connects at a point called the gray commissure and it surrounds an ependymal layer (ependymal cell lined) central canal, which is an extension of the fourth ventricle and contains cerebrospinal fluid (CSF) (Figure 20).

The dorsal and ventral horns can be further divided into posterior horns, anterior horns, a lateral column, and the lateral horns, which appear only in the thoracic regions. The posterior horns are primarily made up of sensory nuclei that receive information from sensory nerves from the peripheries of the body and whose

cell bodies are found in the dorsal root ganglia. The anterior horns consist of the somas (e.g. cell bodies) that give rise to motor nerve fibers. Finally, the lateral horn is composed mainly of interneurons, and in certain segments, the cell bodies of both parasympathetic and sympathetic preganglionic fibers (Figures 20 and 21).

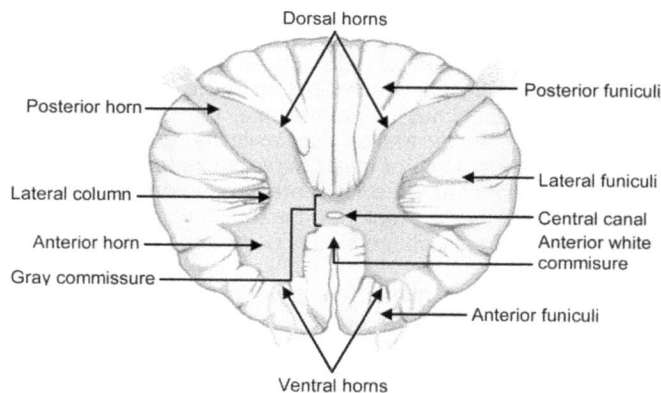

Figure 20: Illustration of a cervical spinal cord cross section

The white matter of the spinal cord consists of both sensory and motor neurons, and is divided into the anterior, lateral, and posterior funiculi. The funiculi consists of longitudinally arranged nerve pathways, called nerve tracks. These nerve tracks are known as ascending or descending tracts. The ascending tracts are found in all funiculi, whereas, the descending tracts are found only in the lateral and the anterior funiculi. The anterior white commissure lies in the center of the spinal cord. This thin structure contains crossing nerve fibers that belong to the spinothalamic tracts, the spinocerebellar tracts, and the anterior corticospinal tracts (Figure 20).

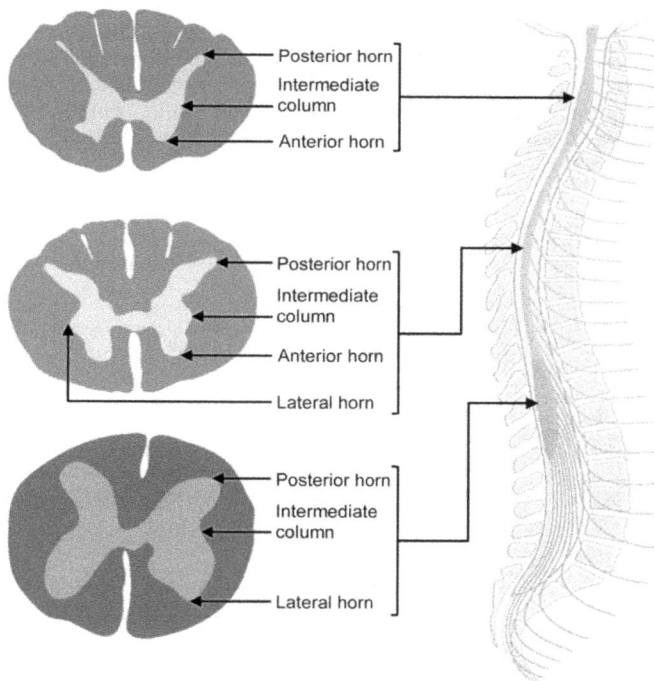

Figure 21: The changing ratio of the white to gray matter along the length of the spinal cord. Please note that the cervical cross section (in green) demonstrates more white matter than gray matter. At the lumbar cross section (in purple), the ratio of the gray matter is slightly more than the white matter

Depending on the location in the spinal cord, a variation of the shape and size of the gray matter can be observed. At higher levels (e.g. towards the brain stem) the ratio of the white matter is higher when compared with the gray matter because these higher levels possess more ascending and descending spinal tracts. Conversely, at the lower levels (e.g. towards the coccyx), the amount of spinal nerve tracts decrease, as does the ratio of white matter to gray matter (Figure 21).

Spinal Tracts

The ascending tract nerve fibers arise from the first order neurons located in the dorsal root ganglia. These long ascending nerve fibers synapse with various brainstem nuclei, the cerebellum and the dorsal thalamus. For example, the posterior funiculi possesses two nerve tracts: the fasiculus gracilis and the fasiculus cuneatus. Both of these nerve tracts transmit sensory information related to tactile, vibration, proprioception, and movement to the brain. Together, these two spinal nerve tracts are called the Dorsal Column Medial Lemniscus System.

The lateral funiculi contains two ascending tracts: the lateral spinothalamic tract and the dorsal and ventral spinocerebellar tracts. The lateral spinothalamic tract, carries sensory information regarding pain, temperature and touch from both somatic and visceral structures. This spinal tract is a part of the anterolateral system. The posterior and anterior spinocerebellar tracts, which are also located within the lateral funiculi, convey unconscious proprioception information from muscles and joints of the legs to the cerebellum. These two nerve tracts are a part of the spinocerebellar tract.

The anterior funiculi possess four ascending tracts: the anterior spinothalamic tract, the spinoolivary tract, spinoreticular tract and the spinotectal tract. The anterior spinothalamic tract transmits information regarding pain, temperature, and touch to the brain stem and the diencephalon. This nerve tract is part of the anterolateral system. The spinoolivary tract is responsible for carrying information from Golgi tendon organs, which detect changes in muscle tension, to the cerebellum. This spinal tract is also a part of the anterolateral system. The spinoreticular tract sends sensory information to the reticular formation. As mentioned in the section entitled Reticular Formation, this particular tract participates in various reflex reactions to pain, which can trigger behavioral reactions to pain. Finally, the spinotectal tract influences reflexive orientation of the head and eyes toward a somatic stimulus (Figure 22).

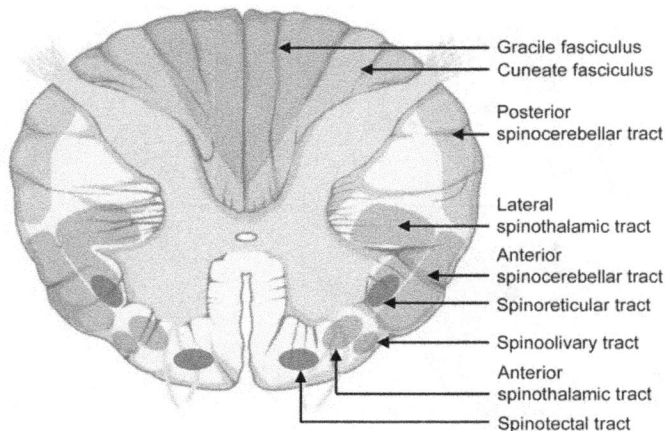

Figure 22: Illustration of ascending tracts

The descending tracts originate from the cerebral cortex and the brainstem, which later synapse with various nuclei found within the spinal cord gray matter. The descending tracts transmit information associated with maintenance of motor functions, including posture, balance, muscle tone, and both visceral and somatic reflexes. For example, there are three descending tracts located in the lateral funiculi: the lateral corticospinal tract, the rubrospinal tracts and the lateral reticulospinal tract.

Lateral corticospinal tract and the rubrospinal tracts convey motor impulses associated with voluntary movement. The lateral corticospinal tract belongs to the pyramidal tracts while the rubrospinal tracts belong to the Extrapyramidal Tracts. The lateral reticulospinal tract convey motor impulses associated with posture and muscle tone. The lateral reticulospinal tract is a part of the extrapyramidal tracts.

The anterior funiculi possesses four descending nerve tracts: the posterior reticulospinal tract, the olivospinal tract, the vestibulospinal tract, and the anterior corticospinal tract. Like the lateral reticulospinal tract, the posterior reticulospinal tract transmits motor impulses associated with posture and muscle tone. The posterior reticulospinal tract belongs to the extrapyramidal tracts. The anterior corticospinal tract mediates balance and postural movements. Anterior corticospinal tract belongs to the pyramidal tracts. The olivospinal tract transmits motor impulse related to proprioception and is a part of the extrapyramidal tracts. The vestibulospinal tract is linked with the vestibular nuclei, which are located in the pons and medulla oblongata. The vestibular nuclei receives sensory information from the vestibular branch of the vestibulocochlear nerve regarding the orientation of the cranium before transmitting motor commands through the vestibulo-spinal tract, which in turn controls muscle tone, extension, and changes the position of the limbs to maintain posture and balance in relation to the cranium. The vestibulospinal tract belongs to the extrapyramidal tracts (Figure 23).

Located at the dorsolateral funiculi, wedged between the edge of the dorsal horn, and the surface of the spinal cord, is a nerve tract called Lissauer's tract. The Lissauer's tract cannot be grouped with either ascending or descending tracts and is responsible for regulating incoming pain sensations at the spinal level (Figure 23).

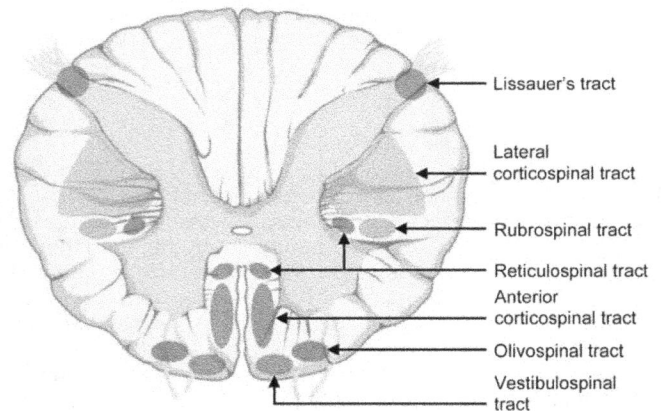

Figure 23: Illustration of descending tracts

MENINGES

Lying just beneath the bones of the skull and the vertebra, the central nervous system (CNS: brain and the spinal cord) is surrounded by a connective tissue covering called the meninges. The meninges are responsible for isolating and protecting the CNS, as well as, forming a structural framework for the blood vessels. The meninges are composed of three distinct layers: dura mater, arachnoid mater, and pia mater (Figure 24).

The superior layer or the outer most layer of the meninges is called the dura mater, which rests just beneath the flat bones of the cranium and the vertebrae. In the cranium, the dura mater is composed of two distinct layers: the outer periosteal layer and an inner meningeal layer. The outer periosteal layer is the equivalent to the periosteum of the cranial bones. In contrast, the inner meningeal layer is composed of tough fibrous connective tissue that is highly vascularized and innervated (many blood vessels and nerves). At the foramen magnum, the periosteal layer ends, while the meningeal layer continues down the vertebral canal where it forms the dural sac.

Within the cranium, the dura mater (both an outer periosteal layer and an inner meningeal layer) is tightly pressed against the cranial bones and creates no epidural spaces. Nonetheless, it is necessary to indicate that even though the dura mater is pushed against the cranial bones, they do not form a continuous bone-meninges attachment, except at a few select locations such as the sella turcica, crista galli and the foramen magnum.

The dura mater of the spinal cord, located within the vertebral canal, is composed of the inner meningeal layer. Unlike the dura mater of the cranium, this inner meningeal layer is not tightly pressed against the bones of the vertebra, but forms an epidural space. The epidural space is filled with lymphatic vessels, adipose tissue, roots of spinal nerve, arteries and the epidural venous plexus.

In the cranium, the two layers of the dura mater (i.e. periosteal and meningeal layers) are generally pressed tightly against one another with the exception of two locations. The first is called the superior sagittal sinus, which is located beneath the sagittal suture, and the second is called transverse sinus, which extends horizontally from the caudal end of the cranium towards each ear. These sinuses are formed to collect blood that has circulated through the brain before emptying into the jugular veins of the neck (Figure 24).

In certain locations, the dura mater of the cranium forms folds that segregate certain parts of the brain. For example, the falx cerebri extends into the longitudinal fissure to separate the hemispheres of the cerebrum (Figure 24). The tentorium cerebelli separates the cerebrum from the cerebellum, while the falx cerebelli separates the lobes of the cerebellum.

Beneath the dura mater and right above the arachnoid mater is a tiny opening called the subdural space. It is at this location where

blood may collect after a head injury. Depending on the amount of blood collected, the pressure could build up to a point where brain damage may result.

The arachnoid mater is a delicate fibrous membrane forming the second or middle layer of the meninges. The arachnoid mater is named because of its spider web-like appearance due to the numerous blood vessels and cranial nerves that enter and exit the brain. The arachnoid mater is continuous with the dura mater and covers both the brain and spinal cord. In some areas, the arachnoid mater projects into the sinuses formed by the dura mater. These projections are called arachnoid granulations or arachnoid villi. The arachnoid villi are responsible for transferring cerebrospinal fluid from the ventricles back into the bloodstream (Figure 24).

Beneath the arachnoid mater is a tiny opening called subarachnoid space. The subarachnoid space is connected with the cerebral ventricles via three foramina. These foramina, also known as interventricular foramen or foramina of Monro, allow the cerebral spinal fluid (CSF) to circulate continuously from the ventricles into the subarachnoid space.

The pia mater is the innermost layer of the meninges and it is the thinnest and most delicate connective tissue layer. The pia mater is continuous with the other layers of the meninges (e.g. dura mater and arachnoid mater) and covers both the brain and the spinal cord. Like the arachnoid mater, the pia mater is composed of delicate collagen fibers and fine elastic fibers with the occasional fibrocyte. Beneath the pia mater is a basement membrane layer which is situated immediately above the astrocytes of the CNS. The pia mater adheres closely to the brain, lining both the gyri and the sulci of the cerebral cortex. The pia mater fuses with the ependyma, the membranous lining of the ventricles, to create the choroid plexus. The choroid plexus is responsible for producing CSF (Figure 24).

CEREBRAL SPINAL FLUID

The choroid plexuses of the lateral ventricles, third ventricle,

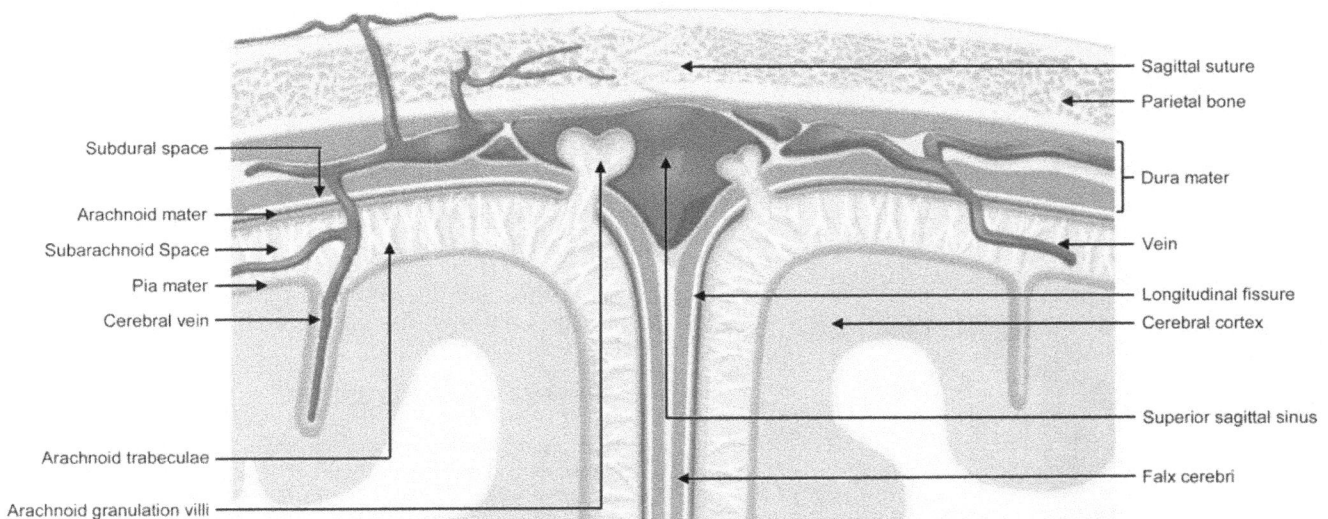

Figure 24: Graphic illustration of the meninges which consists of dura mater, arachnoid mater and the pia mater. Please note that the two layers of the dura mater (periosteal layer and the meningeal layer) are not shown

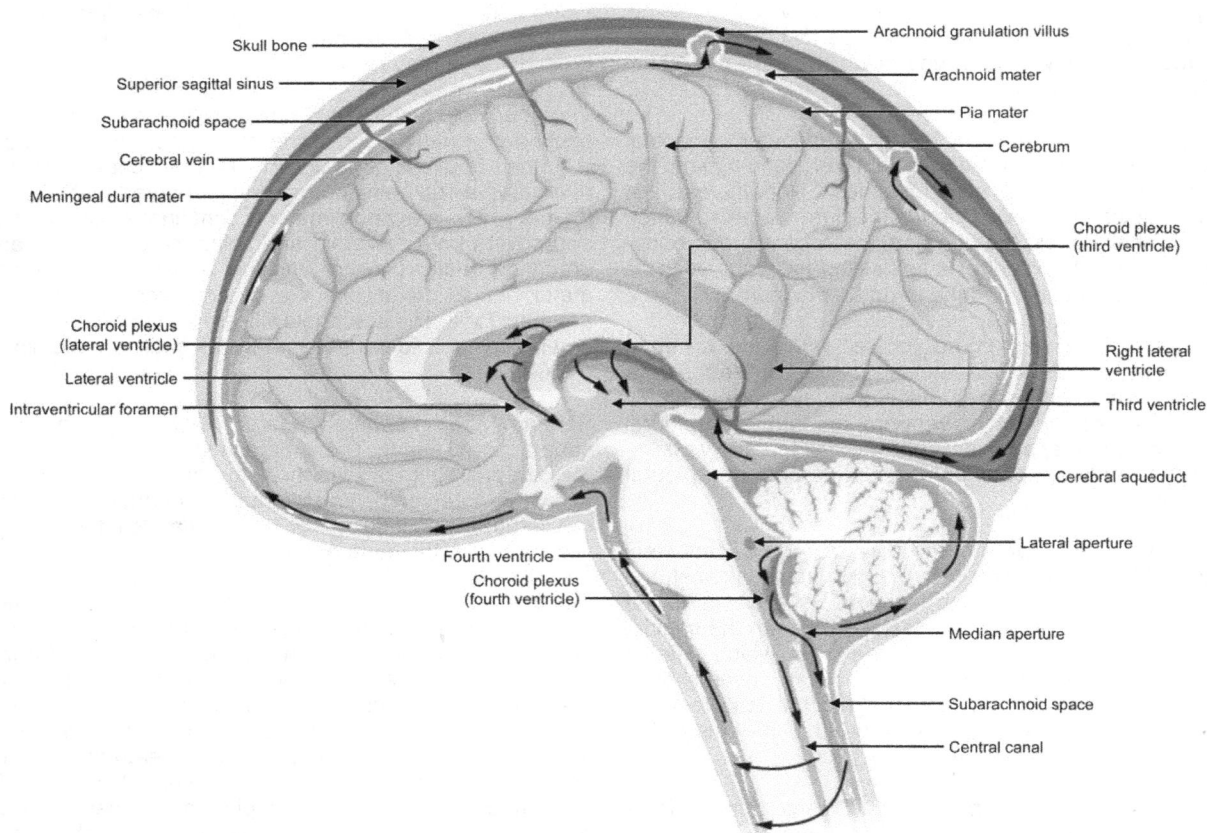

Figure 25: Illustration of the flow of CSF within the cranium and along the spinal cord. Please note that only one of the lateral aperture is shown

fourth ventricles, and the ependymal cells produce cerebrospinal fluid (CSF) from the arterial blood via a combined process of diffusion, pinocytosis and active transport. Additionally, astrocytes play an important role in the regulating the composition of the CSF. Astrocytes make extensive contact with blood vessels, as well as, formation of the blood brain barrier, help in regulating blood flow in the CNS. For example, astrocytes are responsible for releasing prostaglandins and nitric oxide, which cause smooth muscle contraction and dilations, respectively. In addition, astrocytes have been shown to actively transport glucose molecules from the blood vessels and distribute them to the axons at the nodes of Ranvier. This glucose transport role performed by the astrocytes allows for the replenishment of the much needed energy molecules in the nerve cells (Table 1).

Table 1: Comparison between the average composition of CSF and blood plasma. Please note that meq is an abbreviation for milliequivalents and mosm is an abbreviation for milliosmols

Substance	CSF	Plasma
Na^+ meq/kg H_2O	147.0	150.0
K^+ meq/kg H_2O	2.0	4.6
Ca^{+2} meq/kg H_2O	2.3	4.7
Cl^- meq/kg H_2O	114.0	99.0
HCO^{-3} meq/kg H_2O	25.1	24.8
PCO^2 mmHg	50.2	39.5
pH	7.33	7.40
Osmolarity mosm/kg H_2O	289.0	289.0
Protein mg/dL	20.0	6000.0
Glucose mg/dL	64.0	100.0
Cholesterol mg/dL	0.2	175.0

On average, the total volume of CSF in the adult is approximately 140 mL, and is produced at a rate of 600 to 700 mL per day. The circulation of CSF is aided by the pulsations of the choroid plexus and by the motion of the cilia of the ependymal cells. Subsequent

to CNS circulation, the CSF is absorbed across the arachnoid villi into systemic venous circulation, thereby keeping a constant volume.

The circulation patterns of the CSF is shown on Figure 25. The CSF flows from the lateral ventricles via the interventricular foramen, also known as foramina of Monro, into the third ventricle. From the third ventricle the fluid flows to the fourth ventricle via the cerebral aqueduct in the brainstem. Subsequently, the CSF can flow either into the central canal of the spinal cord or into the cisterns of the subarachnoid space via three small foramina, the median aperture, also known as central foramen of Magendie, and the two lateral aperture of the fourth ventricle, also known as lateral foramina of Luschka. The CSF within the spinal cord can flow all the way down to the lumbar cistern at the end of the cord, around the cauda equine.

Within the cranium, the CSF flows around the superior sagittal sinus to be reabsorbed through the arachnoid villi into the venous system.

PERIPHERAL NERVOUS SYSTEM

The peripheral nervous system is composed of two subdivisions, the sensory and motor divisions. The motor division, also known as the efferents, are responsible for transmitting signals from the CNS to various glands, organs and muscles of the body. The motor division is further divided into two subdivisions. The somatic motor division is responsible for transmitting signals from the CNS to the skeletal muscles of the body. This type of innervation results in voluntary skeletal muscle control, which occurs under conscious control, or it may be in the form of involuntary contraction, which are the somatic reflexes. The visceral motor division is also referred to as the autonomic nervous system and it is responsible for carrying nervous impulses to the glands, smooth muscles and internal organs of the body. Autonomic nervous system is completely involuntary. The visceral motor division could be further divided into two subdivisions: the sympathetic and parasympathetic division.

The sensory division, also known as afferents, are responsible for transmitting signals from various types of receptors from the peripheral tissues to the CNS. Please note that these are the receptors that generate the initial action potential and can be either simple sensory receptors or initiated from sense organs. The sensory division is further divided into two divisions. The somatic sensory division is responsible for transmitting signals from the receptors located in the skin, muscle joints and bones to the CNS. The visceral sensory division is responsible for conveying signals from the various organs of the thoracic and abdominal cavities.

The motor and sensory divisions are composed of 12 pairs of cranial nerves and 31 pairs of spinal nerves.

Cranial Nerves

With the exception of the two pairs of cranial nerves that arise from the cerebrum, the remaining ten pairs originate from the brain stem.

The first pair of cranial nerves, also known as the olfactory nerves (cranial nerve I), originate within the olfactory mucosa. From the olfactory mucosa, the axons of these bipolar neurons are bundled together as fascicles before exiting the nasal cavity via the olfactory foramina located in the cribiform plate of the ethmoid bone. On the superior side of the cribiform plate, the fascicles merge with the olfactory bulb. As their name implies, these olfactory nerves are sensory in origin and are associated with the sense of smell (Figure 26).

The second pair of cranial nerves are the optic nerves (cranial

nerve II) and they originate as the axons from the ganglion cells of the retina before leaving the orbit via the optic canal. The optic nerves are sensory in origin and are associated with transmitting visual information to the brain (Figure 26).

The oculomotor nerves (cranial nerve III) enter the orbit via the superior orbital fissure and control most of the voluntary and involuntary movement of the eyes by innervating the inferior oblique, as well as, the superior, medial and inferior rectus, (moving the eyeballs up, down and medially). In addition, this cranial nerve also controls the iris, lens, as well as, the eyelid through its innervation of the levator palpebrae superioris muscle (Figures 26 and 27).

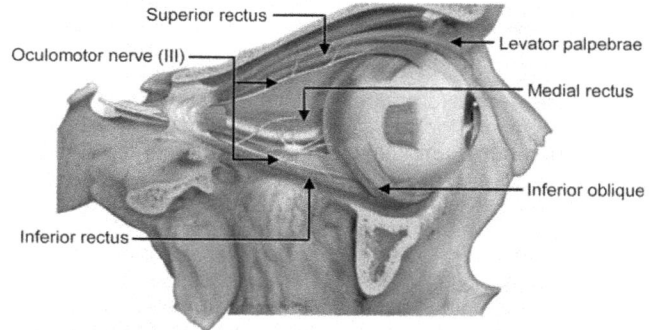

Figure 27: Illustration of the muscles of the eye innervated by the oculomotor nerve (III)

The fourth pair of cranial nerves are the trochlear nerves (cranial nerve IV). The trochlear nerves are motor in function and innervate superior oblique muscles of the eyes. The superior oblique muscle originates in the upper, medial side of the orbit and is responsible for abducting, depressing (moving the eyes downward), and rotating the eye (Figures 26 and 28).

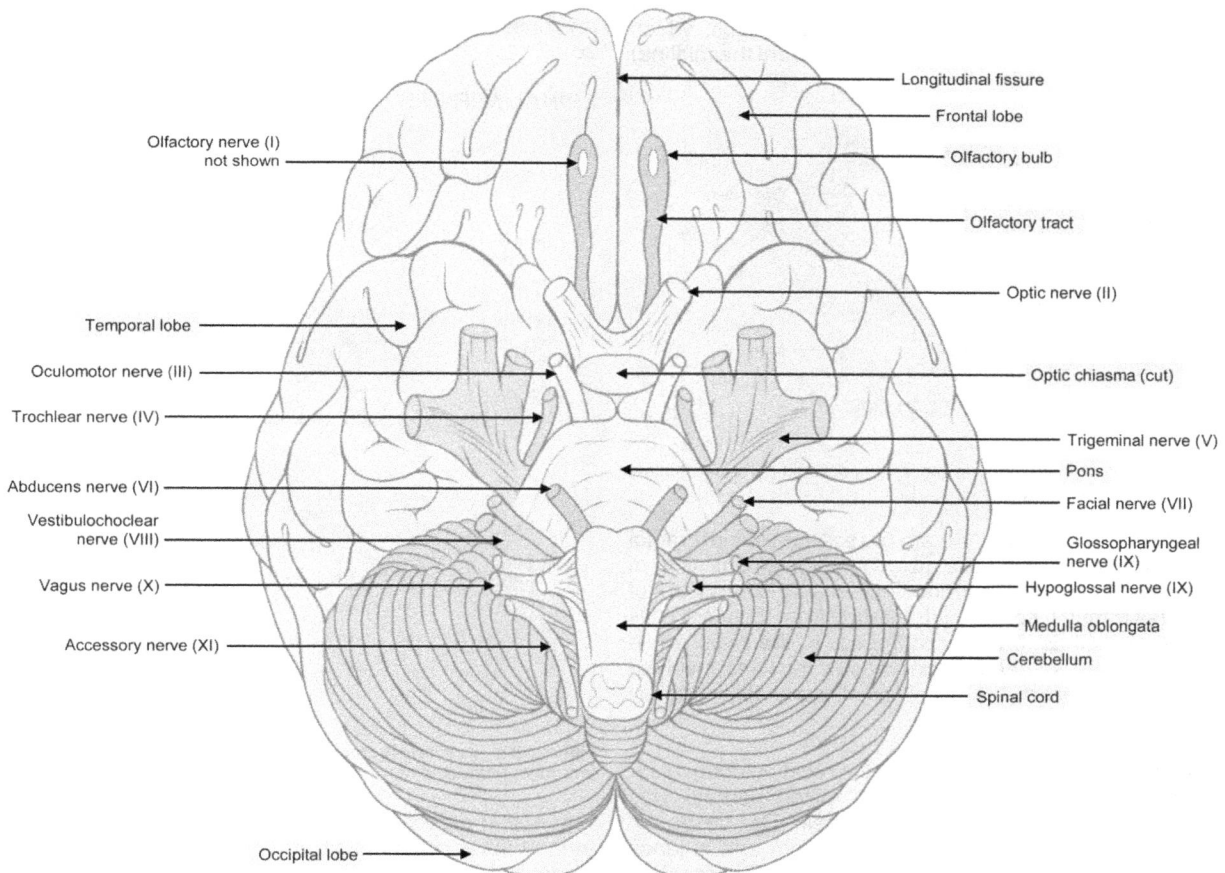

Figure 26: Illustration of the twelve pairs of cranial nerves. Two of the twelve pairs originate in the cerebrum while the remaining ten pairs

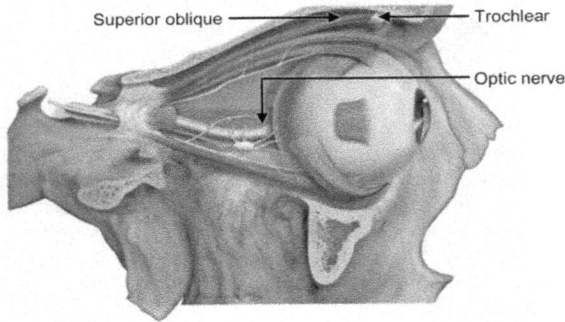

Figure 28: Illustration of the muscles of the eye innervated by the trochlear nerve (IV). Please note that the trochlea is a band of connective tissue that allows the superior oblique to rotate inferiorly towards the eye

The trigeminal nerves (cranial nerve V) are mixed nerves with both motor and sensory functions. The trigeminal nerve is subdivided into three divisions: ophthalmic, maxillary, and mandibular.

The ophthalmic division transmits pressure, temperature, and pain sensations to the brain from the surfaces of the eyes, tear glands, superior nasal mucosa, as well as, the frontal and ethmoid sinuses. The maxillary division transmits pressure, temperature, and pain sensations from the upper lip, teeth, gum, as well as the nasal mucosa, maxillary sinus and the palate to the brain. The mandibular division transmits pressure, temperature and pain sensations to the brain from the lower lip, teeth, gums, as well as, the floor of the mouth and the anterior two-thirds of the tongue. In addition, the mandibular division also possesses proprioceptors that transmit the sense of position in relation to balance from the temporomandibular joint (e.g. the joint of the jaw). The mandibular division's motor function controls the muscles of mastication (Figure 26).

The sixth pair of cranial nerves are the abducens nerves (cranial nerve VI). These nerves innervate the lateral rectus, which results in abduction (turning the line of site away from the midline) of the eye (Figures 26 and 29).

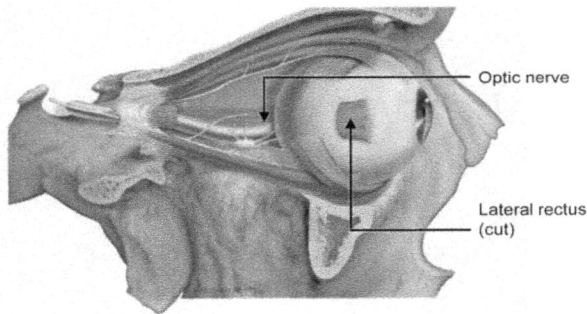

Figure 29: Illustration of the muscles of the eye innervated by the adbucens nerve (VI)

The facial nerves (cranial nerve VII) are the seventh pair of cranial nerves. These cranial nerves have mixed functions, consisting of both sensory and motor components. The sensory component of the facial nerves transmit sensory information from the taste cells of the tongue, as well as, the skin of the ear to the brain, while the motor component innervates the various muscles involved in facial expression, causes the secretion of tears, saliva, nasal, and oral mucus (Figure 26).

The eighth pair of cranial nerves are the vestibulochochlear nerves (cranial nerve VIII) and they are mixed nerves. This cranial nerve is subdivided into two branches. The vestibular branch transmits information involving equilibrium and balance from the vestibule of the inner ear to the brain. The cochlear branch transmits sensory information regarding hearing from the cochlea, as well as, sending motor inputs to the cochlea to fine tune our ability to hear (Figure 26).

The ninth pair of cranial nerves are the glossopharyngeal nerves (cranial nerve IX). This pair of cranial nerves are mixed nerves with both sensory and motor fibers. The motor aspect of the glossopharyngeal nerve is responsible for controlling salivation, deglutition, and gagging. The sensory portion of this cranial nerve is responsible for bringing the sensation of taste, touch (pressure), pain, temperature from the tongue, and the outer ear. In addition, the sensory component is responsible for monitoring blood pressure, as well as, respiration (Figure 26).

The tenth pair of cranial nerves are the vagus nerves (cranial nerve X). The vagus nerves are comprised of both motor (somatic and autonomic) and sensory divisions (somatic and visceral). They innervate the cervical, thoracic and abdominal regions of the human body. For example, the somatic motor component of the vagus nerve is involved in the control of speech and deglutition. The somatic and visceral sensory component of these nerves are involved in transmitting afferent signals from the back of ear, external auditory canal and dura matter, as well as, taste sensation and the feeling of hunger, fullness and abdominal discomfort to the brain. The autonomic component is associated with the control of cardiac contractions, gastrointestinal secretions, as well as, the control of smooth muscles found in both the thoracic and abdominal regions (Figure 26).

The accessory nerves (cranial nerve XI) are the eleventh pair of cranial nerves. The accessory nerve consists of two branches both somatic motor in function. For example, the cranial branch is responsible for transmitting impulses to the muscles of the soft palate, pharynx, and larynx, while the spinal branch innervates the trapezious and sternocleidomastoid muscles (Figure 26).

The last pair of cranial nerves are the hypoglossal nerves (cranial nerve XII). These nerves are somatic motor in function and are responsible for controlling the movements of the tongue (Figure 26).

SPINAL NERVES

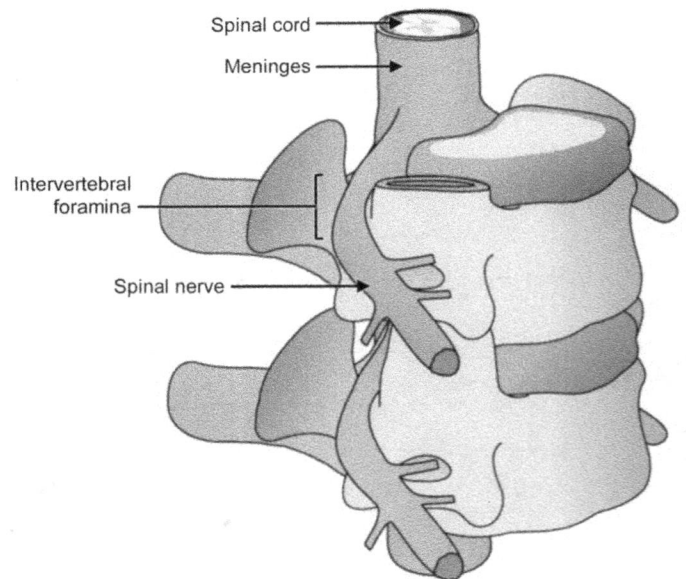

Figure 30: Illustration of spinal nerves exiting the vertebra via the intervertebral foramina, oblique view

The spinal nerves of the peripheral nervous system (PNS) are formed via a combination of the ventral and dorsal roots of the spinal cord. The ventral roots consist mainly of motor nerves, while the dorsal roots consist mainly of sensory nerves. The spinal nerves are paired (one on each side of the spinal cord) and exit the vertebrae via the intervertebral foramina (Figure 30). There are a total of 31 pairs of spinal nerves. The cervical region possesses eight pairs, the thoracic region possesses 12 pairs, the lumbar and sacral regions possess five pairs of spinal nerves each, while the coccyx region possesses only one pair of coccygeal spinal nerves.

Spinal nerves, with the exception of nearly all of the thoracic nerves, combine to form plexuses. A nerve plexus is where the spinal nerve fibers are grouped, sorted, and tracked to a peripheral organ or muscle in one combined nerve. With the exception of T_{12} and branches from T_1, T_2, most of the thoracic spinal nerves do not form plexus. Many of the thoracic nerves remain as intercostal nerves (T_1-T_{11}). The remaining nerves are grouped into the cervical, brachial, lumbosacral and coccygeal plexuses (Figure 31).

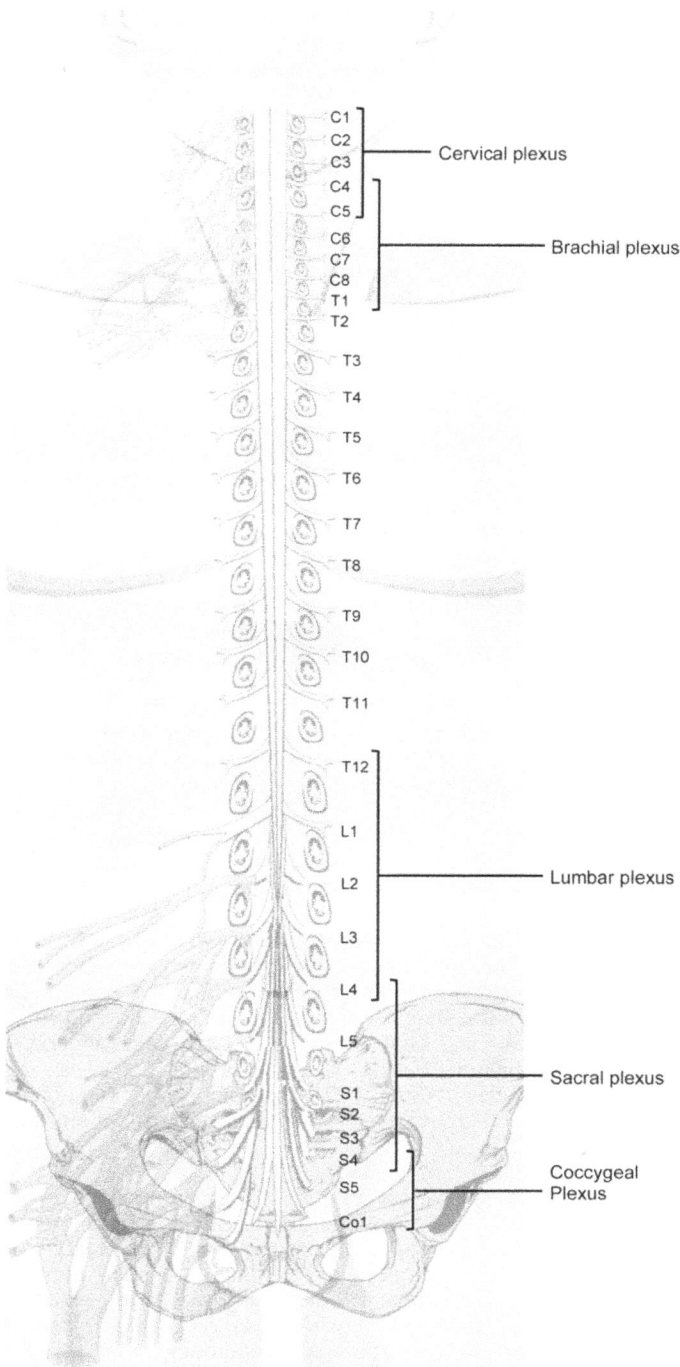

Figure 31: Illustration of spinal nerve plexus, posterior view

Cervical Plexus

The cervical plexus consists of the first five cervical nerve pairs (C_1- to C_5) and innervates the muscles and skin of the neck (Figure 31). The cervical plexus are subdivided into the following branches:

The lesser occipital branch is created from nerves extending from C_2, with minor contributions from C_3, and is a somatosensory nerve. This spinal nerve is responsible for innervating the skin on the postero-lateral aspect of the neck as well as the lateral aspect of the scalp (mainly behind the ears).

The greater auricular branch is created from nerves extending from C_2 and C_3, and is a somatosensory nerve. It is responsible for innervating the skin of the outer ear (e.g. auricles) and the skin covering the parotid gland (largest of the salivary glands).

The transverse cervical branch, also known as transverse cutaneous branch, is created from nerves extending from C_2 and C_3, and is a somatosensory nerve. It is responsible for innervating the skin on the anterior and lateral aspects of the neck.

The supraclavicular branch is subdivided into the anterior, middle, and posterior branches and both branches are somatosensory in function. These branches are created from nerves extending from C_3 and C_4. The supraclavicular nerve is responsible for innervating the skin of the shoulders and the anterior aspect of the chest.

The ansa cervicalis is subdivided into the superior and inferior roots, and is created from nerves extending from C_1 to C_3. This nerve is motor in function and is responsible for innervating infrahyoid muscle of the neck.

The phrenic nerve is formed from nerves extending from C_3 to C_4 with minor fibers contributed from C_5. This nerve is motor in function and is responsible for innervating the diaphragm and is essential in the control of breathing.

Brachial Plexus

Brachial plexus consists of the remaining cervical nerve pairs (with minor contribution from C_4) and two of the thoracic pairs of nerves (C_5 to T_1). It supplies the muscles of the skin, arms, forearms and hands (Figure 31). The brachial plexus is subdivided into the following branches:

The axillary nerve is formed from nerves extending from C_5 and C_6. It is a mixed nerve composed of both motor and somatosensory divisions. The motor branches of this nerve are responsible for innervating the deltoid and teres minor muscles. The somatosensory branches are responsible for innervating the skin around the shoulder region.

The musculocutaneous nerve is formed from nerves extending from C_5 to C_7. It is a mixed nerve composed of both motor and somatosensory divisions. The motor branches of this nerve are responsible for innervating the flexor muscles of the anterior arm (biceps brachii, brachialis, and coacobrachialis). The somatosensory branches are responsible for innervating the skin on the anterolateral forearm.

The median nerve consists of two branches – medial and lateral cords. The medial cord is formed from nerves extending from C_8 and T_1, while the lateral cord is formed from nerves extending from C_5 to C_7. These nerves are mixed nerves, composed of both motor and somatosensory divisions. The motor branches of the nerves are responsible for innervating the flexor muscles of the anterior arm (palmaris longus, flexor carpi, flexor digitorum superficialis, flexor polloicis longus, lateral half of the flexor digitorum profundus and the pronator muscle). The somatosensory branches are responsible for innervating the lateral two-thirds of the hand.

The ulnar nerve is formed from nerves extending from C_6 and T_1. It is a mixed nerve composed of both motor and somatosensory divisions. The motor branches innervate the flexor muscles in the anterior forearm (flexor carpi ulnaris, and the medial half of the flexor digitorum profundus). The somatosensory branches are responsible for innervating both the posterior and anterior skin of the medial third of the hand.

The radial nerve is formed from nerves extending from C_5 to C_8 and T_1. It is a mixed nerve composed of both motor and somatosensory divisions. The motor branch of this nerve is responsible for innervating the posterior muscle of the arm, forearm and hands (e.g. triceps brachii, anconeus, supinator, brachioradialis, extensors, carpi radialis longus and brevis, extensor carpi ulnaris, etc.) The somatosensory branches innervate the skin of the posterolateral surface of the entire limb.

The dorsal scapular branch is formed from nerves extending from C_5. This nerve is motor in function and innervates the rhomboid muscles, which pulls the scapula towards the spine, and levator scapulae muscle, which elevates the scapula.

The long thoracic branch is formed from nerves extending from C_5 to C_7. This nerve is motor in function and innervates the serratus anterior muscle, which is responsible for stabilizing and rotating the scapula.

The pectoral branch consists of two branches, the lateral and medial pectoral branches, and is formed from nerves extending from C_5 to T_1. Both the lateral and medial branches are motor in function and both branches are responsible for innervating the pectoralis minor and pectoralis major (Figure 6).

Lumbar Plexus

The lumbar plexus consists of lumbar segments L_1 to L_4 and at times involves thoracic segment T_{12}. This plexus innervates the abdominal and pelvic regions of the human body, as well as, the anterior thigh. The lumbar plexus is subdivided into the following branches (Figure 31):

The femoral branch is formed from extension of nerves from L_2 to L_4 and is a mixed nerve composed of both motor and somatosensory divisions. The motor branches innervate the anterior muscles of the thigh (e.g. quadriceps and sartorius), as well as, pectineus, and iliacus. The somatosensory branches innervate the skin of the anterior and medial thigh, the skin of the medial leg, foot, hip and knee joints.

The obturator branch is formed from extensions of nerves from L_2 to L_4 and possesses two subdivisions: motor and somatosensory. The motor division innervates the adductor magnus, adductor longus, and adductor brevis muscles, as well as, gracilis and obturator externus muscles. The somatosensory division innervates the skin of the medial thigh and the hip and knee joints.

The lateral femoral cutaneous branches are formed from nerve extensions from L_2 and L_3. This is a somatosensory nerve that innervates the skin of the lateral thigh, as well as, the connective tissues of the peritoneum.

The iliohypogastric branch is formed from nerve extending from L_1 of the spinal cord. It is a mixed nerve and serves both motor and somatosensory functions. The motor aspects innervate the internal and external obliques, as well as, the transverses abdominis. The sensory aspect innervates the skin of the lower abdomen, lower back and hip.

The ilioinguinal branch is formed from nerve extending from L_1

section of the spinal cord and is a mixed nerve containing both motor and somatosensory divisions. The motor component joins the iliohypogastric branch, innervates the internal and external obliques, and innervates the transverses abdominis. The sensory component innervates the skin of the upper thigh, scrotum, root of the penis, and/or labia majora.

The saphenous nerve is formed from extensions of nerves from L_3 and L_4. It is a somatosensory nerve which innervates the skin of the medial aspect of the leg and foot and knee joint.

The genitofemoral branch is formed from nerve extensions from L_1 and L_2. It is a somatosensory nerve which innervates the skin of the middle anterior of the thigh, the scrotum, cremaster muscles, and labia majora.

Sacral Plexus

The sacral plexus is formed from the lumbar and sacral regions of the spinal cord (L_4, L_5 and S_1 to S_4). The sacral plexus is associated with the skin, muscles of the lower abdominal wall, genitalia, buttocks, and lower limbs. The sacral plexus is subdivided into the following braches (Figure 31):

Superior gluteal nerve is formed from extensions of nerves from L_4, L_5, and S_1. It is a motor nerve which innervates gluteus minimus, gluteus medius, and tensor fasciae latae.

The inferior gluteal nerve is formed from extensions of nerves from L_5 to S_2. It is a motor nerve and innervates gluteus maximus.

The nerve to the piriformis is formed from extensions of nerves from S_1 and S_2. It is a motor nerve and it innervates the piriformis muscle.

The nerve to the quadratus femoris is formed from extensions of nerves from L_4, L_5, and S_1. It is a mixed nerve composed of both motor and somatosensory divisions. The motor component innervates quadratus femoris and gemellus inferior, while the somato-sensory component innervates the hip joint.

The nerve to the internal obturator arises from extensions of nerves from L_3 and L_4 with some contribution from L_2. It is a motor nerve and innervates the internal obturator and gemellus superior muscles.

The perforating cutaneous nerve arises from extensions of nerves from S_3 to S_5. It is a somatosensory nerve and innervates the skin of the posterior aspect of the buttocks

The posterior cutaneous nerve arises from extensions of nerves from S_1 to S_3. It is a somatosensory nerve and innervates the skin of the lower buttocks, anal region, etc.

The tibial nerve arises from extensions of nerves from L_4 to S_3 and is a part of the sciatic nerve. It is a mixed nerve composed of both motor and somatosensory divisions. The motor component innervates the semitendinosus, semimembranosus, long head of the biceps femoris, gastrocnemius, soleus, flexor digitorum longus, flexor hallucis longus, tibialis posterior, popliteus, and the intrinsic muscles of the foot. The somatosensory component innervates the skin of the posterior leg, sole of the foot and foot joints.

The common peroneal (Fibular) nerve arises from extensions of nerves from L_4 to L_5 and S_1 to S_2 and is a part of the sciatic nerve. It is a mixed nerve composed of both motor and somatosensory divisions. The motor component innervates the short head of the biceps femoris, peroneus tertius, tibialis anterior, extensor hallucis longus, extensor digitorum longus, and extensor digitorum brevis. The somatosensory component innervates the skin of the anterior distal one-third of the leg, dorsal aspects of the foot, toes and the knee joint.

The pudendal nerve arises from extensions of nerves from S_2 to S_4. It is a mixed nerve composed of both motor and somatosensory divisions. The motor component innervates the muscle of the perineum, while the somatosensory component innervates the penis, clitoris, scrotum, labia (majora and minora), and lower vagina

Coccygeal Plexus

Coccygeal plexus is formed from nerves extending from S_4, S_5 and Co_1. This plexus innervates the pelvic floor (Figure 31). The coccygeal nerve is formed form the coccygeal plexus and it is a mixed nerve (motor and a somatosensory component). The motor component innervates the muscle of the pelvic floor (e.g. levator ani and coccygeus), while the somatosensory component innervates the skin over the coccyx.

AUTONOMIC NERVOUS SYSTEM

The autonomic nervous system (ANS) is a multifaceted, highly complex neural network that is partly responsible for maintaining homeostasis. The functions of the ANS include cardiovascular regulation, thermoregulatory, gastrointestinal and genitourinary regulation, as well as, managing the ophthalmologic (pupillary) systems (Table 2).

Table 2: Comparison between somatic motor and the autonomic nervous system

Characteristics	Somatic	Autonomic
Target tissue	Skeletal muscle	Smooth muscle, cardiac muscle & glands
Regulation	Movement of skeletal muscle	Movement of smooth muscle, cardiac muscle & the inhibition or secretion of glands
Control	Voluntary & involuntary	Involuntary
Nervous Arrangements	CNS to skeletal muscle	CNS to autonomic ganglia while the post-ganglionic extensions innervates the target
Soma	Ventral horn of the spinal cord	Preganglionic neurons are located in the lateral aspect of the spinal cord
Myelination	Myelinated	Preganglionic axons are myelinated while post-ganglionic fibers are unmyelinated
Neurotransmitter	Acetylcholine (AChE)	Preganglionic neurons releases AChE while postganglionic neurons could release either AChE or noradrenaline
Receptor	Nicotinic acetylcholine receptor (nAChR)	nAChR in the autonomic ganglia while in the target tissue the receptors are muscarinic acetylcholine receptor, α or β adrenergic receptor

The ANS is divided into the sympathetic division, the parasympathetic division, and the enteric nervous system which directly innervates the digestive tract. The distinction between the sympathetic and the parasympathetic subdivisions is based upon the location of their preganglionic neuronal cell bodies within the spinal cord, the location of their autonomic ganglia and their postganglionic neurotransmitters.

Sympathetic Division

Sympathetic preganglionic neurons are located in the lateral horn (gray matter) of the spinal cord. These preganglionic neurons, also known as the thoracolumbar division, are found between the first thoracic (T_1) and the second lumbar (L_2). The axons of these preganglionic neurons pass through the ventral roots

of these segments and synapse with the paravertebral ganglia, which lie parallel on both sides of these spinal cord segments, as well as, the unpaired collateral ganglia, which are found within the abdominal cavity. The neurotransmitter of choice used by the preganglionic fibers is acetylcholine, while the receptor proteins located in the paravertebral ganglia are nicotinic acetylcholine receptors (nAChR). Subsequently, the sympathetic axons exit the various ganglia and project to most tissues of the body. It is understood that the postganglionic fibers of the sympathetic division use noradrenaline as their neurotransmitter, while the receptor proteins located at the target tissues are muscarinic acetylcholine receptors (mAChR) (Figure 32).

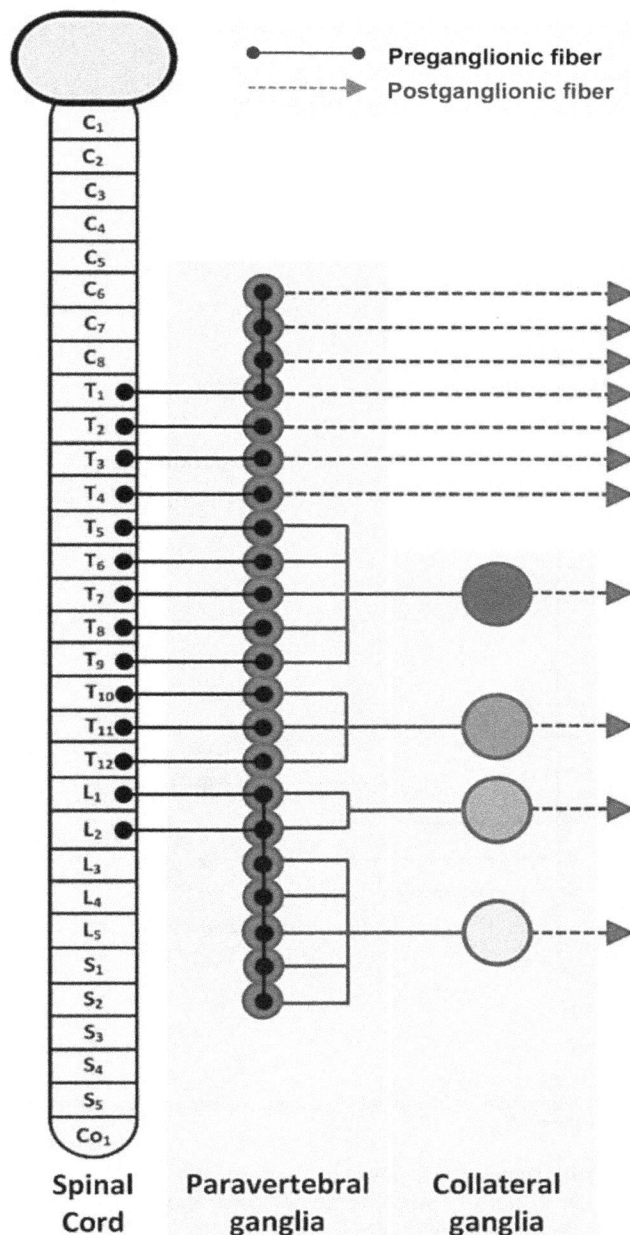

Figure 32: Illustration of the sympathetic division of the autonomic nervous system

Parasympathetic Division

The preganglionic neurons of the parasympathetic division are located within the cranial nerve nuclei found in the brain stem or within the gray matter located between the second sacral segment (S_2) to the fourth sacral segment (S_4). The axons from the preganglionic neurons are located in the brain stem track with the oculomotor (III), facial (VII), glossopharyngeal (IX), and the vagus (X) cranial nerves, while the preganglionic neurons located in the sacral region track with the pelvic splanchnic nerves before coursing off towards the terminal ganglia where they synapse with their postganglionic neurons (Figure 33).

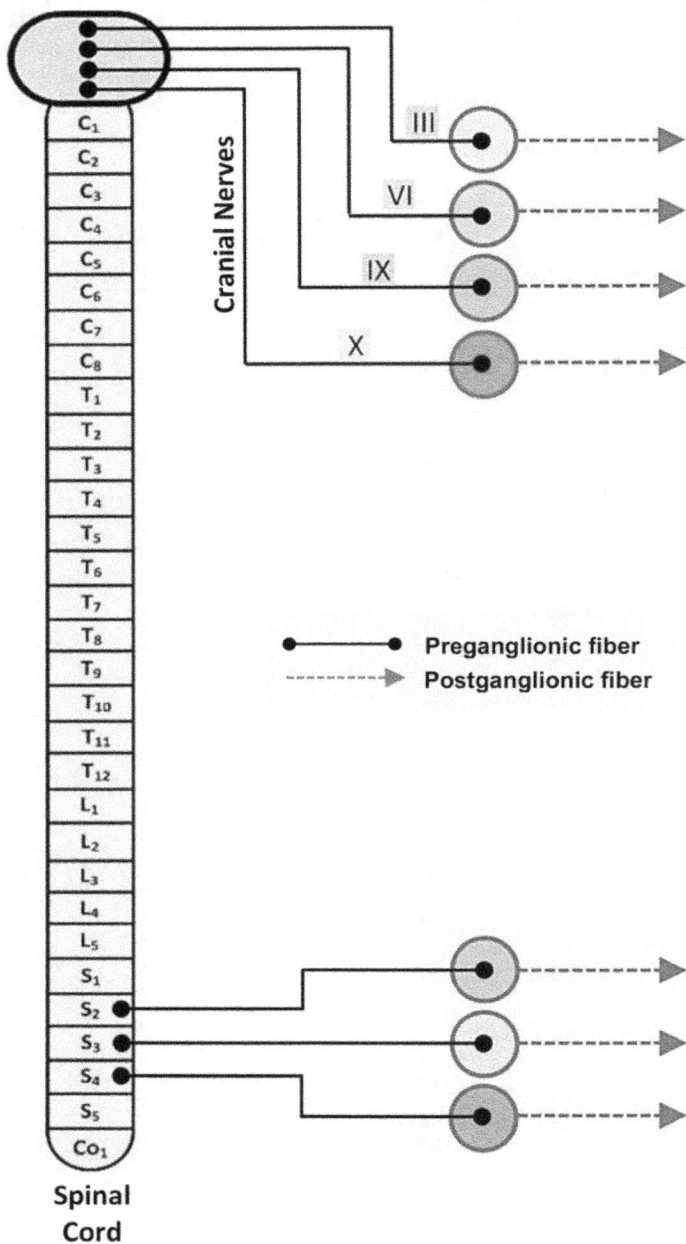

Figure 33: Illustration of the parasympathetic division of the autonomic nervous system

Unlike the sympathetic division, where the postganglionic neurons lie in parallel with the spinal cord, the postganglionic neurons for the parasympathetic division are located within a short distance or found within the targeted organs or tissues.

Reflex

A reflex is defined as an involuntary reaction towards a stimulus or excitation without the necessary intervention of consciousness. In fact, our awareness of reflexes generally occurs after the event has taken place. These reflexes are generally stereotypic, which connotes that the responses are generally the same from one stimulus to another. There are two main types of reflexes. The somatic reflexes are controlled by the motor nerves, which are

designed to prevent excessive damages that can be inflicted if the response time is delayed. The autonomic reflexes (autonomic reflex arc) are controlled by the ANS where we have no conscious awareness of its occurrence. The autonomic reflexes involve the smooth musculature of various organ systems, the cardiac musculature and exocrine glands of the body. Put simply, the autonomic reflex involves the unconscious regulation to maintain internal variables, such as blood pressure, carbon dioxide (CO_2) concentration in circulation, systemic response to stimuli of digestion (including the secretion of enzymes and digestive juices), contracting and dilating pupils, salivation, and sweating, to name a few.

The reflex arcs are the means by which various reflexes are controlled by the body. Reflex arcs are not a part of the brain and therefore they lie external to any conscious control. They consist of a small portions of the entire nervous system that is capable of receiving and responding to stimuli. It is understood that reflex arcs are made up of three basic components. For example, stimuli are sensed by sensory nerves, which transmit their action potentials to the interneurons. The interneurons are responsible for creating a junction between the sensory nerves and motor nerves. The motor nerves, in turn, innervate an effector organ (i.e. muscle) without conscious control (Figure 34).

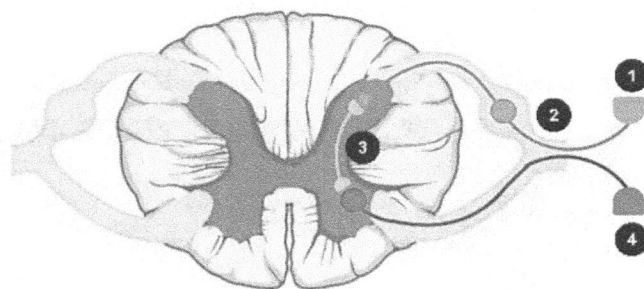

Figure 34: Illustration of a reflex arc. ① A stimulus is sensed by sensory nerves and transmits their action potentials ② through the sensory nerve's soma which are located in the dorsal root ganglia before connecting with the ③ interneuron. The interneurons are responsible for creating a junction between the sensory nerves and ④ motor nerves. The motor nerves, in turn, innervate an effector organ without conscious control

The most straightforward type of reflex is the stretch reflex. The stretch reflex is demonstrated when a muscle contracts in response to being over-stretched or when a muscle is exhibiting unnatural movements. The sensory component of this reflex consists of muscle spindles which are 3 to 10 specialized muscle fibers that are innervated by sensory nerves. The muscle spindle will detect the stretch of the muscle and the sensory nerves will transmit a signal to the posterior horn of the spinal cord. At the anterior horn the sensory nerves will directly synapse with the motor neurons and in turn activate them. These motor nerves will then cause an immediate contraction of the muscle(s).

The Golgi tendon reflex is responsible for preventing muscles from exerting excessive tension on the tendons. The sensory nerves involved in detecting mechanical forces applied on the tendon are referred to as the Golgi tendon organs. These structures are composed of encapsulated receptors such as Ruffini corpuscles. For example, intense stretch of the skeletal muscle will result in the Golgi tendon organs sending action potentials to the spinal cord. These action potentials will proceed up to the brain via the spinothalamic and reticular tract, as well as, synapse with interneurons located in the posterior horn. These interneurons will then connect with the motor neurons, which will in turn send an inhibitory action potential to the muscles. As a result, the muscle will relax, thereby relieving the tension applied to the tendon.

The withdraw reflex is responsible for moving the body away from painful stimulus. For example, after the pain receptors detect a painful sensation, it sends an action potential to the spinal cord. At the posterior horn of the spinal cord, the pain sensor will synapse with an excitatory interneuron, causing it to depolarize. Subsequently, the interneuron will cause the depolarization of the motor neuron (at the anterior horn) resulting in the contraction of the effector organ.

QUESTIONS

1. The central nervous system (CNS) is comprised of which of the following organs or structures?
   a. Brain
   b. Cranial nerves
   c. Skeletal muscles
   d. Spinal cord
   e. Spinal nerves
   f. Stomach
   g. Both a and b
   h. Both a and d

2. The brain is composed of three major areas: _____, _____, and _____
   a. Brain stem
   b. Cerebellum
   c. Cerebrum
   d. Corpus callosum
   e. Hypothalamus
   f. Medulla oblongata
   g. Pituitary gland
   h. Thalamus

3. Embryonically, the cerebrum is developed from the _____ at approximately fifth week gestation
   a. Diencephalon
   b. Hippocampus
   c. Mesencephalon
   d. Prosencephaon
   e. Rhinencephalon
   f. Telencephalon

4. What is the name of the interconnecting nervous tissue that connects the cerebral hemispheres?
   a. Brain stem
   b. Cerebellum
   c. Cerebrum
   d. Cerebral cortex
   e. Corpus callosum
   f. Hypothalamus
   g. Medulla oblongata
   h. Pituitary gland
   i. Thalamus

5. Which of the following comprise the surface of the cerebrum?
   a. Brain stem
   b. Cerebellum
   c. Cerebrum
   d. Cerebral cortex
   e. Corpus callosum
   f. Hypothalamus
   g. Medulla oblongata
   h. Pituitary gland
   i. Thalamus

6. What is the name given to the wrinkles of the brain?
   a. Corpus callosum
   b. Gyri
   c. Hypothalamus
   d. Medulla oblongata
   e. Sulci
   f. Thalamus

7. What is the name given to the shallow grooves of the brain?
   a. Corpus callosum
   b. Gyri
   c. Hypothalamus
   d. Medulla oblongata
   e. Sulci
   f. Thalamus

8. What is the purpose of having "wrinkles" upon the surface of the brain?
   a. Decrease in surface area
   b. Increase in surface area
   c. Insulation
   d. Mobility
   e. Prevent mobility
   f. Protection
   g. Structural deformity

9. The left hemisphere frontal lobe is involved with which of the following capabilities?
   a. Language capabilities
   b. Sense of smell

c. Sensing moods and emotions
   d. Sight
   e. Sound

10. The right hemisphere frontal lobe is involved with which of the following capabilities?
    a. Language capabilities
    b. Sense of smell
    c. Sensing moods and emotions
    d. Sight
    e. Sound

11. Which of the following is not a function of the frontal lobe?
    a. Executive function
    b. Impulse control
    c. Judgment
    d. Memories
    e. Memory
    f. Motivation
    g. Voluntary motor functions
    h. Vision

12. Which of the following is not a part of executive functions?
    a. Abstract thinking
    b. Cognitive flexibility
    c. Formation of memories
    d. Inhibiting inappropriate actions
    e. Inhibiting irrelevant sensory information
    f. Planning
    g. Rule acquisition

13. After a serious concussion, a patient developed problems with language capabilities, verbal skills and positive emotions. Which of the following structure of the brain is damaged?
    a. Left frontal lobe
    b. Left parietal lobe
    c. Left temporal lobe
    d. Right frontal lobe
    e. Right parietal lobe
    f. Right temporal lobe
    g. Orbital frontal cortex

14. After a serious concussion, a patient developed a deficiency in non-verbal communication and negative emotions. Which of the following structure of the brain is damaged?
    a. Left frontal lobe
    b. Left parietal lobe
    c. Left temporal lobe
    d. Right frontal lobe
    e. Right parietal lobe
    f. Right temporal lobe
    g. Orbital frontal cortex

15. Which of the following areas of the brain is involved in the modulation of behaviors, as well as, cognitive processing influencing decision making?
    a. Left frontal lobe
    b. Left parietal lobe
    c. Left temporal lobe
    d. Right frontal lobe
    e. Right parietal lobe
    f. Right temporal lobe
    g. Orbital frontal cortex

16. Which of the following is a type of aggression that is reacting towards frustration and threat?
    a. Hostile aggression
    b. Instrumental aggression
    c. Passive aggression
    d. Physical aggression
    e. Reactive aggression
    f. Relational aggression

17. Which of the following is a type of aggression that acts to securing rewards or accomplishing goals?
    a. Hostile aggression
    b. Instrumental aggression
    c. Passive aggression
    d. Physical aggression
    e. Reactive aggression
    f. Relational aggression

18. Dysfunction in the _____ and the amygdala are linked to the development of Borderline Personality Disorder which encompasses individuals that possess fundamental failure in socialization.
    a. Left frontal lobe
    b. Left parietal lobe
    c. Left temporal lobe
    d. Right frontal lobe
    e. Right parietal lobe
    f. Right temporal lobe
    g. Orbital frontal cortex

19. The OFC also contains secondary cortex(es). Which of the following is (are)

the name(s) of the (these) secondary cortex(es)?
a. Corpus callosum
b. Gyri
c. Hypothalamus
d. Medulla oblongata
e. Olfactory cortical areas
f. Taste cortex
g. Both a and d
h. Both e and f

20. The OFC assigns reward values towards the sense of smell, taste and sight before integrating this information through a process known as _____.
a. Cognitive processing
b. Decision making
c. Modulation of behaviors
d. Non-verbal communication
e. Stimulus-reinforcement association learning
f. Verbal communication

21. The _____ provides us with a basic survival skill where we can recognize that certain items are not edible based upon their shape, coloration, smell and taste.
a. Left frontal lobe
b. Left parietal lobe
c. Left temporal lobe
d. Right frontal lobe
e. Right parietal lobe
f. Right temporal lobe
g. Orbital frontal cortex

22. Which of the following area of the brain is responsible for integrating sensory information such as general senses (e.g. pain, touch sensation etc.), tastes (i.e. a special sense) and are involved in visual perception?
a. Frontal lobe
b. Insula
c. Occipital lobe
d. Parahippocampal gyrus
e. Parietal lobe
f. Temporal lobe

23. Which of the following area of the brain is involved in sensory information processing such as hearing, smell, learning, memory, as well as some aspects of vision and emotions?
a. Frontal lobe
b. Insula
c. Occipital lobe
d. Parahippocampal gyrus
e. Parietal lobe
f. Temporal lobe

24. Which of the following is involved in the recognition of places and scenes? In conjunction with other regions of the brain (e.g. right inferior parietal cortex and left frontal operculum), this area of the brain is involved with the recognition of real and imagined events.
a. Frontal lobe
b. Insula
c. Occipital lobe
d. Parahippocampal gyrus
e. Parietal lobe
f. Temporal lobe

25. An individual with a lesion in the parahippocampal gyrus may develop which of the following disorder?
a. Difficulties in distinguishing real and imagined events
b. Inability to visually recognize scenes
c. Unable to detect smells
d. Unable to distinguish sounds
e. Unable to process images
f. Both a and b
g. Both c and d

26. Which of the following area of the brain is the principle center for visual information processing as well as integrating visual information?
a. Frontal lobe
b. Insula
c. Occipital lobe
d. Parahippocampal gyrus
e. Parietal lobe
f. Temporal lobe

27. Which of the following area of the brain is believe to play a role in language skills, taste and is involved in the integration of visual information?
a. Frontal lobe
b. Insula
c. Occipital lobe
d. Parahippocampal gyrus
e. Parietal lobe
f. Temporal lobe

28. The cerebral cortex is constructed out of two principle nerve cells:_____ and _____. These two principle nerve cells are used to form three, five or six layers of nervous tissues depending upon the location of the cerebral cortex.
a. Caudate nucleus

b. **Globus pallidus**
c. Nucleus accumbens
d. Pukinji cells
e. Pyramidal cells
f. Sexually dimorphic nucleus
g. Stellate cells

29. It is estimated that 90% of the cerebral cortex is made out of six layers of tissue called the _____
a. Archicortex
b. Hippocampus
c. Motor cortex
d. Neocortex
e. Paleocortex
f. Parahippocampal gyrus

30. Which of the following layers are the thickest in the sensory regions of the brain?
a. Layer I
b. Layer II
c. Layer III
d. Layer IV
e. Layer V
f. Layer VI

31. Which of the following layers are the thickest in the motor regions of the brain?
a. Layer I
b. Layer II
c. Layer III
d. Layer IV
e. Layer V
f. Layer VI

32. It is understood that all axons that leave the neocortex cortex and enter the white matter of the brain arise from layers _____, _____ and _____
a. Layer I
b. Layer II
c. Layer III
d. Layer IV
e. Layer V
f. Layer VI

33. Which of the following area of the cerebral cortex is believed to be the earliest form of cortex that appeared in vertebrates? This cortex is found in the insula and parts of the temporal lobe and is generally involved with the sense of smell.
a. Archicortex
b. Hippocampus
c. Motor cortex
d. Neocortex
e. Paleocortex
f. Parahippocampal gyrus

34. Which of the following area of the cerebral cortex is believed to be the least developed location of the cerebral cortex? This area is found in the hippocampus of the temporal lobe and is the principle area of memory formations in humans.
a. Archicortex
b. Hippocampus
c. Motor cortex
d. Neocortex
e. Paleocortex
f. Parahippocampal gyrus

35. Which of the following area of the basal ganglia is involved in learning and storing memories, as well as being heavily involved in the development and the use of language (s) and communication skill?
a. Caudate nucleus
b. Globus pallidus
c. Nucleus accumbens
d. Putamen
e. Substantia nigra
f. Subthalamic nucleus

36. Which of the following area of the basal ganglia is involved in a very complex feedback-loop that modulates the movement of upper and lower limbs? Research has shown that lesions in this area of the basal ganglia caused by Parkinson's disease or Huntington's disease will cause involuntary muscle movements such as jerking.
a. Caudate nucleus
b. Globus pallidus
c. Nucleus accumbens
d. Putamen
e. Substantia nigra
f. Subthalamic nucleus

37. Which of the following area of the basal ganglia is involved in reinforcement learning and playing a role in pleasures such as sex, laughter and rewards etc.? In addition, this area is believed to modulate signals regarding addiction, fear, aggression and impulsivity.
a. Caudate nucleus
b. Globus pallidus
c. Nucleus accumbens

d. Putamen
e. Substantia nigra
f. Subthalamic nucleus

38. The nucleus accumbens is composed of _____ type of neurons?
    a. Adrenergic neurons
    b. Cholinergic neurons
    c. Dopaminergic neurons
    d. GABAergic neurons
    e. Gylcine neurons
    f. Nitric oxide neurons
    g. Somatostatin neurons

39. Which of the following best describes reinforcement learning?
    a. Perceives its current state and takes actions
    b. Integration of visual signals and assign rewards
    c. Integration of olfactory signals and assign rewards
    d. Integration of auditory signals and assign rewards
    e. Modulation of motor movements
    f. Development of language skills

40. Which of the following area of the basal ganglia is used to modulate motor signals from the motor cortex? In addition, research has shown that this area of the basal ganglia is responsible for reinforcing the desires for stimuli such as food or drugs which may lead to addiction.
    a. Caudate nucleus
    b. Globus pallidus
    c. Nucleus accumbens
    d. Putamen
    e. Substantia nigra
    f. Subthalamic nucleus

41. The globus pallidus is composed of _____ type of neurons?
    a. Adrenergic neurons
    b. Cholinergic neurons
    c. Dopaminergic neurons
    d. GABAergic neurons
    e. Gylcine neurons
    f. Nitric oxide neurons
    g. Somatostatin neurons

42. Which of the following component of the globus pallidus receives excitatory signals form the thalamus and sends output signals to other structures of the corpus striatum?
    a. External segment
    b. Internal segment
    c. Pars compacta
    d. Pars reticulata
    e. Subthalamic nucleus
    f. Putmen

43. Which of the following component of the globus pallidus receive signals from all other structures of the corpus striatum and send inhibitory (GABA) transmissions to the thalamus and the brainstem etc.?
    a. External segment
    b. Internal segment
    c. Pars compacta
    d. Pars reticulata
    e. Putmen
    f. Subthalamic nucleus

44. Which of the following component of the globus pallidus provides GABAergic inhibitory efferent connections to all the basal ganglia's input and output nuclei?
    a. External segment
    b. Internal segment
    c. Pars compacta
    d. Pars reticulata
    e. Putmen
    f. Subthalamic nucleus

45. Which of the following area of the basal ganglia is the only structure within basal ganglia that secrete excitatory neurotransmitter?
    a. Caudate nucleus
    b. Globus pallidus
    c. Nucleus accumbens
    d. Putamen
    e. Substantia nigra
    f. Subthalamic nucleus

46. The Subthalamic nucleus (STN) secrete neurotransmitters called _____
    a. Glutamate
    b. Nitric oxide
    c. Acetylcholine
    d. Noradrenaline
    e. Adrenaline
    f. GABA
    g. Glycine

47. Which of the following area of the basal ganglia plays an important role in motor control? Research has shown that lesions within this area of the basal ganlgia will cause involuntary muscle movements of the extremities.
    a. Caudate nucleus

b. Globus pallidus
c. Nucleus accumbens
d. Putamen
e. Substantia nigra
f. Subthalamic nucleus

48. Which of the following area of the basal ganglia is the most efficient target for deep brain stimulations designed to alleviate symptoms of Parkinson's disease?
    a. Caudate nucleus
    b. Globus pallidus
    c. Nucleus accumbens
    d. Putamen
    e. Substantia nigra
    f. Subthalamic nucleus

49. Which of the following area of the basal ganglia is responsible for modulating motor movement, eye movement, and learning? In addition, this region of the basal ganglia is involved in learning behaviors that will lead to a reward (i.e. for example food or sex) and addiction.
    a. Caudate nucleus
    b. Globus pallidus
    c. Nucleus accumbens
    d. Putamen
    e. Substantia nigra
    f. Subthalamic nucleus

50. Which of the following region of the substantia nigra is a source of dopaminergic pathway to the corpus striatum and loss of neurons in this area is the cause of Parkinson's disease?
    a. External segment
    b. Internal segment
    c. Pars compacta
    d. Pars reticulate
    e. Putamen
    f. Subthalamic nucleus

51. Which of the following region of the substantia nigra is composed of GABA secreting nerve cells? This region forms inhibitory connections on their targets.
    a. External segment
    b. Internal segment
    c. Pars compacta
    d. Pars reticulata
    e. Putamen
    f. Subthalamic nucleus

52. Which of the following region of the brain is known as the command and control center of the endocrine system?
    a. Corpus callosum
    b. Frontal lobe
    c. Hypothalamus
    d. Medulla oblongata
    e. Putmen
    f. Thalamus

53. The hypothalamus interconnects with various locations within the brain. For example, the hypothalamus transmits information to the _____, _____ and _____ while it receives information from the _____ and _____.
    a. Amygdala
    b. Frontal lobes
    c. Hippocampus
    d. Medulla oblongata
    e. Pons
    f. Reticular formation
    g. Septal nuclei

54. Which of the following structure of the limbic system is responsible for monitoring internal temperature, osmolality, glucose levels, sodium concentrations and the production and secretion of various intercellular signals?
    a. Amygdala
    b. Hippocampus
    c. Hypothalamus
    d. Mammillary bodies
    e. Medulla oblongata
    f. Prefrontal cortex

55. The hypothalamus produces hormones such as vasopressin and oxytocin and transport them via the hypothalamohypophyseal tract before releasing them from the _____.
    a. Anterior pituitary
    b. Frontal lobe
    c. Mammillary bodies
    d. Medulla oblongata
    e. Parietal lobe
    f. Pons
    g. Posterior pituitary

56. The hypothalamus produces and secretes releasing or inhibiting hormones via the _____ to control the hormone secretion of the anterior pituitary.
    a. Amygdala

b. Frontal lobes
c. Hippocampus
d. Hypothalamohypophyseal portal system
e. Hypothalamohypophyseal tract
f. Infundibulum
g. Posterior pituitary

57. The hypothalamus influences autonomic functions via projections to the _____ and _____. Depending on the need of the body, the hypothalamus may cause the increase in sympathetic activities while decreasing parasympathetic actions or vice-versa.
a. Amygdala
b. Brain stem
c. Frontal lobes
d. Hippocampus
e. Infundibulum
f. Reticular formation
g. Posterior pituitary
h. Spinal cord

58. The hypothalamus sends its projections to the _____ to influence emotional reactions generally after a traumatic event.
a. Amygdala
b. Brain stem
c. Frontal lobes
d. Hippocampus
e. Infundibulum
f. Reticular formation
g. Posterior pituitary
h. Spinal cord

59. The preopitc areas of the hypothalamus are responsible for influencing which of the following behavior(s)?
a. Heart and respiration rates
b. Hunger
c. Sexual behavior
d. Sleep
e. Thirst
f. Time management
g. Answers are c, d and e
h. Answers are a, b and e

60. Which of the following cells of the preoptic areas of the hypothalamus produces and releases gonadotropin-releasing hormone (GnRH)?
a. Dorsal medial nucleus
b. Lateral arcuate nuclei
c. Medial arcuate nuclei
d. Paraventricular nucleus
e. Preoptic nucleus
f. Sexually dimorphic nucleus
g. Supraoptic nucleus
h. Ventromedial nucleus

61. Which of the following regions of the preoptic area is involved in the regulation of sexually differentiated behaviors and functions in animals as well as humans?
a. Dorsal medial nucleus
b. Lateral arcuate nuclei
c. Medial arcuate nuclei
d. Paraventricular nucleus
e. Preoptic nucleus
f. Sexually dimorphic nucleus
g. Supraoptic nucleus
h. Ventromedial nucleus

62. Which of the following is responsible for the feeding reflexes such as licking and swallowing? Additionally this region is also involved in emotional responses to odors, olfactory reflexes and memory.
a. Amygdala
b. Hippocampus
c. Hypothalamus
d. Mammillary bodies
e. Medulla oblongata
f. Prefrontal cortex

63. Which of the following structures of the brain is located within the anterior temporal lobe and forms reciprocal connections with the thalamus, hypothalamus, hippocampus, the brain stem, orbital frontal cortex, parahippocampal gyrus, septal nuclei and the cingulate gyrus?
a. Amygdala
b. Hippocampus
c. Hypothalamus
d. Mammillary bodies
e. Medulla oblongata
f. Prefrontal cortex

64. Which of the following is continuous with the gray matter located at the medial surface of the cerebral hemisphere? This area forms reciprocal connections with hippocampus and the hypothalamus and are involved in modulation of rage and pleasure? It is known that individuals with lesions in this area will produce uncontrollable rage behavior.
a. Cingulate gyrus
b. Dorsal medial nucleus

c. Lateral arcuate nuclei
d. Medial arcuate nuclei
e. Paraventricular nucleus
f. Preoptic nucleus
g. Septal nuclei

65. Which of the following lies above the corpus callosum, and helps regulate emotions in association with pain?
a. Cingulate gyrus
b. Dorsal medial nucleus
c. Lateral arcuate nuclei
d. Medial arcuate nuclei
e. Paraventricular nucleus
f. Preoptic nucleus
g. Septal nuclei

66. Which of the following is known as a center for managing behavioral responses, as well as, a critical location for managing emotions and formation of long term memories?
a. Amygdala
b. Hippocampus
c. Hypothalamus
d. Mammillary bodies
e. Medulla oblongata
f. Prefrontal cortex

67. Which of the following area is responsible for emotional learning where visual, olfactory, taste and tactile signs are triggers for memories of either rewarding or aversive events?
a. Amygdala
b. Hippocampus
c. Hypothalamus
d. Mammillary bodies
e. Medulla oblongata
f. Prefrontal cortex

68. Which of the following region is involved in managing and coordinating responses to stress, fear and anxiety?
a. Amygdala
b. Hippocampus
c. Hypothalamus
d. Mammillary bodies
e. Medulla oblongata
f. Prefrontal cortex

69. Which of the following is located in the medial aspect of the temporal lobe and forms reciprocal connections with the cerebral cortex, prefrontal cortex and the amygdala?
a. Amygdala
b. Hippocampus
c. Hypothalamus
d. Mammillary bodies
e. Medulla oblongata
f. Prefrontal cortex

70. The outputs of the hippocampus project to the _____, _____, _____ and the _____.
a. Amygdala
b. Frontal lobe
c. Hypothalamus
d. Mammillary bodies
e. Medulla oblongata
f. Neonatal cortex
g. Nucleus accumbens
h. Preoptic nuclei

71. Which of the following is the type of memory of facts and events which could be expressed in words?
a. Explicit memory
b. Implicit memory
c. Long-term memory
d. Loss of memory
e. Punishment memory
f. Reward memory
g. Short-term memory

72. Which of the following is the type of memory of that are formed to dealing with emotional responses?
a. Explicit memory
b. Implicit memory
c. Long-term memory
d. Loss of memory
e. Punishment memory
f. Reward memory
g. Short-term memory

73. Hippocampus plays an important role in short-term memory formation. It is necessary to note that the formation of short-term memory is necessary for the eventual formation of long term memory. Put simply, long-term memories are short-term memories that are transferred and stored within the _____.
a. Amygdala
b. Cerebral cortex
c. Frontal lobe

d. Hypothalamus
e. Mammillary bodies
f. Medulla oblongata
g. Nucleus accumbens
h. Preoptic nuclei

74. Which of the following is located in the outer layers of the frontal lobe and is essential to the formation of judgment, insight, motivation and mood?
    a. Amygdala
    b. Hippocampus
    c. Hypothalamus
    d. Mammillary bodies
    e. Medulla oblongata
    f. Prefrontal cortex

75. Which of the following is located at the midline of the brain and consists of thalamus, subthalamus, epithalamus, hypothalamus and the posterior pituitary gland?
    a. Diencephalon
    b. Hippocampus
    c. Mesencephalon
    d. Prosencephaon
    e. Rhinencephalon
    f. Telencephalon

76. The two lateral portion of the thalamus is interconnected via _____.
    a. Amygdala
    b. Corpus callosum
    c. Hippocampus
    d. Interthalamic adhesion
    e. Mammillary bodies
    f. Medulla oblongata
    g. Prefrontal cortex

77. With the exception of _____ all sensory information must first be transmitted through the thalamus?
    a. Auditory information
    b. Information regarding balance
    c. Olfactory
    d. Optical information
    e. pH information
    f. Temperature information
    g. Visceral sensation

78. Which of the following is considered to be the central relay station for sensory impulses from ascending fibers (from the body) and most sensory input from the cranium?
    a. Brain stem
    b. Cerebellum
    c. Cerebrum
    d. Corpus callosum
    e. Hypothalamus
    f. Medulla oblongata
    g. Pituitary gland
    h. Thalamus

79. In addition to serving as the sensory information relay station in the brain, this region of the brain is also involved in the modulation and control of skeletal muscle movements. Clinical reports have indicated that damaged to the _____ and _____ will result in ataxia.
    a. Dorsal anterior thalamic nuclei
    b. Dorsal lateral thalamic nuclei
    c. Dorsal posterior thalamic nuclei
    d. Lateral arcuate nuclei
    e. Medial arcuate nuclei
    f. Ventral anterior thalamic nuclei
    g. Ventral lateral thalamic nuclei
    h. Ventral posterior thalamic nuclei

80. The subthalamus is located inferior to the thalamus and comprises of the subthalamic nuclei (STN) and several ascending and descending nerve tracts. It is understood that subthalamus are involved in controlling motor functions. Clinical research has shown that lesions of the STN induce abnormal muscular movements and _____ (i.e. frequent violent movements of the shoulder and arm).
    a. Ataxia
    b. Atrophy
    c. Ballism
    d. Dystonia
    e. Myoclonus
    f. Spasticity
    g. Tremor

81. The epithalamus is located superior and posterior to the thalamus and it is composed of two groups of nuclei called _____ and _____.
    a. Cingulate gyrus
    b. Dorsal medial nucleus
    c. Habenula
    d. Lateral arcuate nuclei
    e. Medial arcuate nuclei
    f. Paraventricular nucleus
    g. Pineal body
    h. Substantia nigra pars compacta

82. Which of the following influences the brain's response to stress, anxiety, pain, and reward? Dysfunction of this region has been associated with individuals suffering from depression, schizophrenia, as well as the effects of drugs of abuse.
    a. Cingulate gyrus
    b. Dorsal medial nucleus
    c. Habenula
    d. Lateral arcuate nuclei
    e. Medial arcuate nuclei
    f. Paraventricular nucleus
    g. Pineal body

83. The habenula is divided into the _____ and _____.
    a. Anterior habenula
    b. Lateral habenula
    c. Medial habenula
    d. Posterior habenula
    e. Preoptic nucleus
    f. Putmen
    g. Septal nuclei
    h. Septum

84. The pineal body, also known as the pineal gland, is a part of the endocrine system that is involved in the secretion of a hormone called _____.
    a. Acetylcholine
    b. Adrenaline
    c. GABA
    d. Melatonin
    e. Noradrenaline
    f. Oxytocin
    g. Vasopressin

85. In vertebrates, the pineal body is in control of the _____ (e.g. day and night cycles). It is known that the secretion of _____ is increased by the pineal body during night time while the same secretion is suppressed during daylight.
    a. Circadian rhythms
    b. Melatonin
    c. Noradrenaline
    d. Oxytocin
    e. Photoperiodic
    f. Seasonal affective disorder
    g. Vasopressin

86. In animals, the duration of the nocturnal melatonin secretion provides an internal calendar. What is the scientific terminology used to describe internal calendar?
    a. Circadian rhythms
    b. Melatonin
    c. Noradrenaline
    d. Oxytocin
    e. Photoperiodic
    f. Seasonal affective disorder
    g. Vasopressin

87. Which of the following area of the hypothalamus is (are) known as the feeding center of the brain? Experiments have shown that the neurons in this (these) area(s) respond(s) to systemic glucose, free fatty acid and insulin levels.
    a. Cingulate gyrus
    b. Dorsal medial nucleus
    c. Lateral arcuate nuclei
    d. Lateral hypo-thalamic area
    e. Medial arcuate nuclei
    f. Paraventricular nucleus
    g. Ventromedial nucleus
    h. Both a and b
    i. Both d and g

88. The _____ and _____ of the hypothalamus is believed to provides a set-point for for a person's body weight? Therefore, if a person goes below this particular set-pont, this area will be activated to increase appitite and increase feeding
    a. Cingulate gyrus
    b. Dorsal medial nucleus
    c. Hyperphagia
    d. Hypothalamic rage
    e. Lateral arcuate nuclei
    f. Lateral hypothalamic area
    g. Medial arcuate nuclei
    h. Paraventricular nucleus
    i. Ventromedial nucleus

89. Experiments have shown that bilateral lesions in the _____ will result in anorexia. In contrast, bilateral lesions of the _____ will result in _____ (overeating), extreme obesity, as well as, displaying chronically irritable mood and aggressive behavior also known as _____.
    a. Cingulate gyrus
    b. Dorsal medial nucleus
    c. Hyperphagia
    d. Hypophagia
    e. Hypodipsia
    f. Hyperdipsia
    g. Hypothalamic rage
    h. Lateral arcuate nuclei

i.   Lateral hypothalamic area
j.   Medial arcuate nuclei
k.   Paraventricular nucleus
l.   Ventromedial nucleus

h.   Sympathetic
i.   Temperate-sensitive neurons
j.   Temperature-insensitive neurons
k.   Warm-sensitive neurons

90. Which of the following area of the hypothalamus is responsible for the regulation of feeding, drinking, maintaining body weight, as well as, contributing to the maintenance of circadian rhythm? Experiment has shown that lesions in this area result in reduced food intake and reduced fluid intake.
   a. Cingulate gyrus
   b. Dorsal medial nucleus
   c. Lateral arcuate nuclei
   d. Lateral hypothalamic area
   e. Medial arcuate nuclei
   f. Paraventricular nucleus
   g. Ventromedial nucleus

91. What is the medical term use to describe reduce food intake?
   a. Hyperdipsia
   b. Hyperphagia
   c. Hyperthyroid
   d. Hypodipsia
   e. Hypophagia
   f. Hypothalamic rage
   g. Hypothyroid

92. What is the medical term used to describe reduced fluid intake?
   a. Hyperdipsia
   b. Hyperphagia
   c. Hyperthyroid
   d. Hypodipsia
   e. Hypophagia
   f. Hypothalamic rage
   g. Hypothyroid

93. The hypothalamic regulation of temperature set-point is monitored and maintained by the _____ and _____ with a subservient role provided by the _____ and _____.
   a. Anterior hypothalamic area
   b. Dorsal hypothalamic area
   c. Dorsal medial nucleus
   d. Lateral arcuate nuclei
   e. Lateral hypothalamic area
   f. Medial arcuate nuclei
   g. Paraventricular nucleus
   h. Posterior hypothalamic areas
   i. Preoptic nucleus
   j. Ventromedial nucleus

94. It is known that POAH possesses three types of neurons that are involved in determining the temperature set-point: _____, _____ and _____.
   a. Cold-sensitive neurons
   b. Freezing-point sensitive neurons
   c. Hot-sensitive neurons
   d. Lateral hypothalamus
   e. Parasympathetic
   f. Paraventricular nucleus
   g. Posterior hypothalamus
   h. Sympathetic
   i. Temperate-sensitive neurons
   j. Temperature-insensitive neurons
   k. Warm-sensitive neurons

95. If the temperature increases above 37 C° the depolarization rate of _____ (Hint: type of neurons) increases which in turn activates the neurons in the _____ and the _____, which in turn, results in the increase in the activities of _____ (Hint: subtype of autonomic nervous system) thereby increasing the dissipation of heat (e.g. sweating and relaxation of arrector pili muscles etc.).
   a. Cold-sensitive neurons
   b. Freezing-point sensitive neurons
   c. Hot-sensitive neurons
   d. Lateral hypothalamus
   e. Parasympathetic
   f. Paraventricular nucleus
   g. Posterior hypothalamus
   h. Sympathetic
   i. Temperate-sensitive neurons
   j. Temperature-insensitive neurons
   k. Warm-sensitive neurons

96. If the temperature decreases below 37 C° the depolarization rate of _____ (Hint: type of neurons) increases which in turn activates the neurons in the _____ and _____, which in turn, results in the increase in the activities of _____ (Hint: subtype of autonomic nervous system) thereby promoting the generation and conservation of heat (e.g. shivering, contraction of the arrector pili muscles etc.).
   a. Cold-sensitive neurons
   b. Freezing-point sensitive neurons
   c. Hot-sensitive neurons
   d. Lateral hypothalamus
   e. Parasympathetic
   f. Paraventricular nucleus
   g. Posterior hypothalamus

97. Which of the following area(s) of the hypothalamus is (are) responsible for producing vasopressin and oxytocin?
   a. Lateral arcuate nuclei
   b. Lateral hypothalamic area
   c. Medial arcuate nuclei
   d. Paraventricular nueclei
   e. Preoptic nucleus
   f. Supraoptic nuclei
   g. Ventromedial nucleus
   h. Both a and c
   i. Both d and f

98. Which of the following is the type of neuron that produces vasopressin and oxytocin?
   a. Dorsal medial nucleus
   b. Lateral arcuate nuclei
   c. Lateral hypothalamic area
   d. Magnocellular neurons
   e. Medial arcuate nuclei
   f. Paraventricular nucleus
   g. Ventromedial nucleus

99. Once formed, ADH and oxytocin are transported via the _____ (Hint: Fast or slow axonal transport) down the length of the axons from the hypothalamus to the _____. These axons are located within the _____ which is located in the _____ (e.g. pituitary stalk). Once the neurohormones reach the axonal terminal, they are released via _____ into circulation.
   a. Anterior pituitary
   b. Endocytosis
   c. Exocytosis
   d. Fast axonal transport
   e. Hypothalamohypophyseal portal system
   f. Hypothalamohypophyseal tract
   g. Infundibulum
   h. Phagocytosis
   i. Pinocytosis
   j. Posterior pituitary
   k. Slow axonal transport

100. Which of the following neuron(s) produces various hormones (Hint: releasing and inhibiting hormones) that controls the production and release of anterior pituitary hormones?
   a. Dorsal medial nucleus
   b. Lateral hypothalamic area
   c. Magnocellular neurons
   d. Medial arcuate nuclei
   e. Neuroendocrine neurons
   f. Paraventricular nucleus
   g. Parvocellular neurons
   h. Ventromedial nucleus
   i. Both a and d
   j. Both e and g

101. The cells that produces releasing and inhibiting hormone are located within _____ and _____ of the hypothalamus.
   a. Anterior hypothalamic area
   b. Arcuate nucleus
   c. Dorsal hypothalamic area
   d. Dorsal medial nucleus
   e. Lateral arcuate nuclei
   f. Lateral hypothalamic area
   g. Medial arcuate nuclei
   h. Paraventricular nucleus
   i. Posterior hypothalamic areas
   j. Preoptic nucleus
   k. Ventromedial nucleus

102. The brain stem is composed of _____, _____ and _____.
   a. Medulla oblongata
   b. Pons
   c. Midbrain
   d. Thalamus
   e. Hypothalamus
   f. Amygdala
   g. Posterior pituitary
   h. Anterior pituitary

103. Which of the following cranial nerves arises from the medulla oblongata? You may circle more than one answer.
   a. CN I – Olfactory
   b. CN II – Optic
   c. CN III – Oculomotor
   d. CN IV – Trochlear
   e. CN V – Trigeminal
   f. CN VI – Abducens
   g. CN VII – Facial
   h. CN VIII – Vestibulocochlear
   i. CN IX – Glossopharyngeal

j. CN X – Vagus
k. CN XI – Accessory
l. CN XII – Hypoglossal

104. Two prominent enlargements could be located on the anterior surface of the medulla oblongata. These enlargements, which extend the entire length of the nervous structure, are called _____. These enlargements contain the _____ (Hint: descending or ascending tracts) that are involved in the conscious control of skeletal muscles, as well as, maintaining balance and coordination. Towards the inferior ends of the medulla oblongata these nerve fibers of the descending tract crosses or _____ to the opposite side.
   a. Arcuate nucleus
   b. Ascending tracts
   c. Decussate
   d. Descending tracts
   e. Dorsal hypothalamic area
   f. Dorsal medial nucleus
   g. Lateral arcuate nuclei
   h. Lateral hypothalamic area
   i. Pyramids
   j. Thalamus

105. Posteriorly, the medulla oblongata contains the _____ (Hint: descending or ascending tracts) from the cerebellum, spinal cord and cranial nerves. This cranial structure is responsible for relying nervous impulses through the numerous ascending nerve tracts to the _____.
   a. Arcuate nucleus
   b. Ascending tracts
   c. Decussate
   d. Descending tracts
   e. Dorsal hypothalamic area
   f. Dorsal medial nucleus
   g. Lateral arcuate nuclei
   h. Lateral hypothalamic area
   i. Pyramids
   j. Thalamus

106. Medulla oblongata contains various neural networks that are involved in both sensory and motor functions. For example, the medulla oblongata contains the neural networks that are involved in the sense of touch (i.e. pressure), temperature and pain. The motor neuronal networks include the regulation of _____ (i.e. chewing), salivation, _____ (i.e. swallowing), gagging, vomiting, coughing, sneezing, speech, sweating, digestion, controlling autonomic functions relative to respiration, heart rate and contractility, as well as _____ (Hint: increasing diameter of the blood vessels) and _____ (Hint: decreasing diameter of the blood vessels)
   a. Deglutition
   b. Hyperdipsia
   c. Hyperphagia
   d. Hyperthyroid
   e. Hypodipsia
   f. Hypophagia
   g. Mastication
   h. Vasoconstrictions
   i. Vasodilation

107. The monitoring of the chemical concentration of the blood is accomplished via the central chemoreceptors located bilaterally and centrally within the medulla oblongata. This area of the medulla oblongata is also known as the _____.
   a. Baroreceptor reflex
   b. Cardioregulatory center
   c. Chemosensitive area
   d. Decussate
   e. Pyramids
   f. Respiratory center
   g. Thermosensitive area

108. The chemosensitive area of the medulla oblongata is responsible for examining the concentrations of blood gases such as _____, as well as blood buffer _____
   a. Calcium bicarbonate
   b. Carbon dioxide
   c. Carbon monoxide
   d. Carbonic acid
   e. Nitric oxide
   f. Nitrogen
   g. Nitrous oxide
   h. Oxygen

109. The peripheral chemoreceptors located within the carotid arteries and the aorta is also known as _____ and _____.
   a. Carotid bodies
   b. Aortic bodies
   c. Aortic chemosensitive areas
   d. Carotid chemosensitive areas
   e. Carotid pyramids
   f. Aortic pyramids

110. If the levels of CO_2 is high in circulation, the levels of carbonic acid will be _____ (hint: high or low) thereby indicating a _____ (Hint: high or low) pH level. At this pH level, the circulating fluids would be _____ (Hint: acidic or basic).
   a. High

b. Low
c. Acidic
d. Basic

111. If the levels of CO_2 is low in circulation, the levels of carbonic acid will be _____ (hint: high or low) thereby indicating a _____ (Hint: high or low) pH level. . At this pH level, the circulating fluids would be _____ (Hint: acidic or basic).
   a. High
   b. Low
   c. Acidic
   d. Basic

112. A low pH level will trigger the chemosensitive area of the medulla oblongata to _____ (Hint: increase or decrease) respiration rate.
   a. Increase
   b. Decrease

113. A high pH level will trigger the chemosensitive area of the medulla oblongata to _____ (Hint: increase or decrease) respiration rate.
   a. Increase
   b. Decrease

114. Which of the following is responsible for detecting changes of the blood pressure?
   a. Baroreceptor reflex
   b. Cardioregulatory center
   c. Chemosensitive area
   d. Decussate
   e. Pyramids
   f. Respiratory center
   g. Thermosensitive area

115. When the blood pressure increases, the walls of the large arteries will stretch which in turn activates the _____. The activated baroreceptors will send an action potential to the cardioregulatory center of the medulla oblongata. This action potential will prompt the cardioregulatory center to _____ (Hint: increase or decrease) sympathetic innervation and _____ (Hint: increase or decrease) para-sympathetic innervation, which in turn, will _____ (Hint: increase or decrease) heart rate and thereby _____ (Hint: increase or decrease) the blood pressure.
   a. Baroreceptors
   b. Chemoreceptors
   c. Decrease
   d. Increase
   e. pH receptors
   f. Thermoreceptors

116. When the blood pressure decreases, the walls of the large arteries will relax which in turn activates the _____. The activated baroreceptors will send an action potential to the cardioregulatory center of the medulla oblongata. This action potential will prompt the cardioregulatory center to _____ (Hint: increase or decrease) sympathetic innervation and _____ (Hint: increase or decrease) para-sympathetic innervation, which in turn, will _____ (Hint: increase or decrease) heart rate and thereby _____ (Hint: increase or decrease) the blood pressure.
   a. Baroreceptors
   b. Chemoreceptors
   c. Decrease
   d. Increase
   e. pH receptors
   f. Thermoreceptors

117. If a substantial decrease in the blood pressure is detected, the medulla oblongata will trigger the adrenal medulla to release _____ and _____. The release of these neurohormones will result in increased heart rate, stroke volume and _____.
   a. Glutamate
   b. Nitric oxide
   c. Acetylcholine
   d. Noradrenaline
   e. Adrenaline
   f. GABA
   g. Glycine
   h. Vasodilation
   i. Vasoconstriction
   j. Both a and c
   k. Both d and e

118. If a substantial increase in the blood pressure is detected, the medulla oblongata will prohibit the adrenal medulla to release _____ and _____. The decrease in the release of these neurohormones will result in decreased heart rate, stroke volume and _____.
   a. Glutamate
   b. Nitric oxide
   c. Acetylcholine
   d. Noradrenaline
   e. Adrenaline
   f. GABA
   g. Glycine
   h. Vasodilation
   i. Vasoconstriction
   j. Both a and c

k. Both d and e

119. Four pairs of cranial nerves arise from the pons: _____, _____, _____, and
_____.
a. CN I – Olfactory
b. CN II – Optic
c. CN III – Oculomotor
d. CN IV – Trochlear
e. CN V – Trigeminal
f. CN VI – Abducens
g. CN VII – Facial
h. CN VIII – Vestibulocochlear
i. CN IX – Glossopharyngeal
j. CN X – Vagus
k. CN XI – Accessory
l. CN XII – Hypoglossal

120. In addition to being a major relay station for ascending and descending tracts,
the pons also assists in controlling autonomic functions and respiration. The
control of respiration by the pons is achieved through the _____
a. Baroreceptor reflex
b. Cardioregulatory center
c. Chemosensitive area
d. Pontine respiration group
e. Pyramids
f. Respiratory center
g. Thermosensitive area

121. _____ is a cranial nerve that is associated with a single muscle - superior
oblique – which causes the eye to look downward

_____ is a cranial nerve with three subdivisions: the _____ is a sensory nerve
that brings signals to the brain from the surfaces of the eyes, tear glands
etc. The _____ transmit sensory information of the upper teeth, gum and
lip etc. and _____ is a motor and sensory component whereby the sensory
component transmits signals to the brain from the lower teeth, gums, and
lips etc. while the motor component controls the muscles involve in chewing

_____ is a mixed (autonomic and motor) cranial nerve that consists of both
somatic and autonomic nerves that spread to the chest and the abdomen.
The motor division involve in the motor control of speech and swallowing while
the autonomic division associates with the heart and smooth musculature of
the thorax and abdomen

_____ is a mixed (motor and sensory) cranial nerve that is involved in sending
impulse from the tongue, pharynx, tonsils etc. to the brain while the motor
component is associated with the act of swallowing

_____ is a motor and sensory nerve (mixed nerve) that transmits sensory
information associated with chemoreceptors (located in the taste buds) that
is found within the anterior 2/3 of the tongue while the motor component
associates with muscles involved in facial expressions

1. Abducens nerves	22. Maxillary division Medial
2. Accessory nerves	cord
3. Axillary nerve	23. Medial pectoral branch
4. Cervical plexus	24. Median nerve
5. Cochlear branch	25. Musculocutaneous nerve
6. Cranial branch	26. Obturator branch
7. Dorsal scapular branch	27. Oculomotor nerves
8. Extrapyamidal tracts	28. Olfactory nerve
9. Facial nerves	29. Osmodium nerve
10. Femoral branch	30. Ophthalmic division
11. Glossopharyngeal nerves	31. Optic nerves
12. Hypoglossal nerves	32. Pectoral branch
13. Iliohypogastric branch	33. Pyramidal tracts
14. Ilioinguinal branch	34. Radial nerve
15. Infundibulum	35. Saphenous nerve
16. Intercostal nerves	36. Spinal branch
17. Lateral femoral cutaneous	37. Spinothalamic tract
branches	38. Trigeninal nerves
18. Lateral pectoral branch	39. Trochlear nerves
19. Long thoracic branch	40. Ulnar nerve
20. Lumbar plexus	41. Vagus nerves
21. Mandibular division	42. Vestbulocochlear nerves
	43. Vestibular branch

122. _____ is a motor cranial nerve that consist of two divisions: the _____ branch
carries impulses to the muscles of the soft palate, pharynx and larynx while
the _____ branch supplies the impulses needed to control trapezious and
sternocleidomastoid muscles (muscles of the neck)

_____ is a motor cranial nerve that controls the movement of the tongue

_____ is a motor nerve that supplies the lateral rectus which results in
abduction (turning the line of site away from the midline) of the eye

_____ is a sensory cranial nerve that is associated with sense of smell

_____ is a sensory cranial nerve that is associated with vision

_____ is a sensory cranial nerve with two divisions: the _____ branch
associates with the inner ear and are involved in controlling equilibrium while

the _____ branch transmits auditory information (involved in hearing)

1. Abducens nerves	22. Maxillary division
2. Accessory nerves	23. Medial cord
3. Axillary nerve	24. Medial pectoral branch
4. Cervical plexus	25. Median nerve
5. Cochlear branch	26. Musculocutaneous nerve
6. Cranial branch	27. Obturator branch
7. Dorsal scapular branch	28. Oculomotor nerves
8. Extrapyamidal tracts	29. Olfactory nerve
9. Facial nerves	30. Osmodium nerve
10. Femoral branch	31. Ophthalmic division
11. Glossopharyngeal nerves	32. Optic nerves
12. Hypoglossal nerves	33. Pectoral branch
13. Iliohypogastric branch	34. Pyramidal tracts
14. Ilioinguinal branch	35. Radial nerve
15. Infundibulum	36. Saphenous nerve
16. Intercostal nerves	37. Spinal branch
17. Lateral femoral cutaneous	38. Spinothalamic tract
branches	39. Trigeninal nerves
18. Lateral pectoral branch	40. Trochlear nerves
19. Long thoracic branch	41. Ulnar nerve
20. Lumbar plexus	42. Vagus nerves
21. Mandibular division	43. Vestbulocochlear nerves
	44. Vestibular branch

123. Two pairs of cranial nerves arise from midbrain: _____ and _____.
a. CN I – Olfactory
b. CN II – Optic
c. CN III – Oculomotor
d. CN IV – Trochlear
e. CN V – Trigeminal
f. CN VI – Abducens
g. CN VII – Facial
h. CN VIII – Vestibulocochlear
i. CN IX – Glossopharyngeal
j. CN X – Vagus
k. CN XI – Accessory
l. CN XII – Hypoglossal

124. The midbrain is divided into anterior and posterior section by the aqueduct
of Sylvius. The aqueduct of Sylvius is the structure within the brainstem that
facilitates the flow of cerebral spinal fluids between the _____ and _____
a. Central canal
b. Cerebral aquaduct
c. Fourth ventricle
d. Interventricular foramen
e. Lateral aperture
f. Lateral ventricle
g. Third ventricle

125. Located on the posterior surface of the midbrain is the corpora quadrigemina
which is composed of four subsections. The two _____ are responsible for
relaying input from the optic tract to the _____ of the thalamus. The _____
are responsible for relaying auditory information to the _____ of the thalamus.
a. Anterior colliculi
b. Anterior geniculate bodies
c. Inferior colliculi
d. Lateral geniculate bodies
e. Medial geniculate bodies
f. Posterior colliculi
g. Superior colliculi
h. Superior geniculate bodies

126. The _____, also known as the _____, is responsible for the levels of alertness
and the sleep-wake cycle. This formation receives sensory input from
sense organs such as the eyes and ears, as well as information regarding
proprioception, touch, vibration, pain, temperature from the body via the
_____. Once the information is received, this formation will project them to
the _____ and _____ of the thalamus where some influence and modification
of the sensory information are performed. Once the sensory information
has been modified, it is then transmitted to the _____ where it comes to an
individual's conscious attention.
a. Anterior nuclei
b. Ascending reticular formation
c. Autonomic nervous system
d. Cerebral cortex
e. Corticospinal tract
f. Descending reticular formation
g. Dorso-medial nuclei
h. Reticular activating system
i. Spinoreticular fibers
j. Spinothalamic tract

127. The _____ along with the motor areas of the cerebral cortex is involved
in maintaining tone, posture and equilibrium especially during bodily
movements. Together with the hypothalamus, this area of the brain also aids
in maintaining the activities of the _____ in the control of the urinary bladder,
gastrointestinal peristalsis, glandular secretions, as well as maintaining its
influence over respiratory and cardiovascular systems. Additionally, the
control over swallowing, mastication, and vomiting reflexes are all equally
regulated by this formation
a. Anterior nuclei

b. Ascending reticular formation
c. Autonomic nervous system
d. Cerebral cortex
e. Corticospinal tract
f. Descending reticular formation
g. Dorso-medial nuclei
h. Reticular activating system
i. Spinoreticular fibers
j. Spinothalamic tract

128. The cerebellum is located towards the posterior (caudal) side of the brain and is separated into the right and left cerebellar hemispheres connected by a thin layer of nervous tissue called_____.
a. Arbor vitae
b. Dorsal medial nucleus
c. Folia
d. Gray matter
e. Gyri
f. Habenula
g. Lateral arcuate nuclei
h. Paraventricular nucleus
i. Pyramids
j. Sulci
k. Vermis
l. White matter
m. Dark matter

129. On the surface of each cerebellar hemisphere are numerous folds called _____. Each of the folds is separated by a shallow indention called the _____.
a. Arbor vitae
b. Dorsal medial nucleus
c. Folia
d. Gray matter
e. Gyri
f. Habenula
g. Lateral arcuate nuclei
h. Paraventricular nucleus
i. Pyramids
j. Sulci
k. Vermis
l. White matter

130. Similar to the cerebrum, the cerebellum possesses a surface layer composed of _____ where all inputs to the cerebellum are received while an inner layer of _____ is where all outputs form the cerebellum originates. The white matter is distributed within the cerebellum as a branch-like pattern called _____
a. Arbor vitae
b. Dorsal medial nucleus
c. Folia
d. Gray matter
e. Gyri
f. Habenula
g. Lateral arcuate nuclei
h. Paraventricular nucleus
i. Pyramids
j. Sulci
k. Vermis
l. White matter

131. Please match the number (1-33) associated with the correct terminology with the specific definitions provided for the specific spinal nerve/branch

_____ This nerve consists of two branches: Medial and Lateral cord. The Medial cord is formed from nerves extending from C8 and T1 while the Lateral cord is formed from nerves extending from C5 to C7. These nerves are mixed nerves. The motor branches of these nerves are responsible for innervating the flexor muscles of the anterior arm (palmaris longus, flexor carpi, flexor digitorum superficialis, flexor pollicis longus, lateral half of the flexor figitorum profundus and the pronator muscle) while the somatosensory branches are responsible for innervating the kind of the lateral two-thirds of the hand

_____ This nerve is created from C2 and C3 and is a somatosensory nerve. It is responsible for innervating the skin of the outer ear (e.g. auricles) and the skin covering the parotid gland (largest of the salivary glands)

_____ This nerve is created from C2 and C3 and is a somatosensory nerve. It is responsible for innervating the skin on the anterior and lateral aspects of the neck

_____ This nerve is created from C2 with minor contributions from C3 and is a somatosensory (sensory) nerve. It is responsible for innervating the skin on the posterolateral aspect of the neck as well as the lateral aspect of the scalp (mainly behind the ears)

_____ This nerve is formed from C3 to C4 with minor contributions from C5. It is motor in function and is responsible for innervating the diaphragm and is essential in the control of breathing

_____ This nerve is formed from C5 and C6 and it is a mixed nerve. The motor branches of this nerve is responsible for innervating the deltoid and teres minor muscles while the somatosensory branches are responsible for innervating the skin around the shoulder region

_____ This nerve is formed from C5 and it is a motor nerve which is responsible for innervating the rhomboid muscles which pulls the scapula towards the spine and levator scapulae muscle which elevates the scapula

_____ This nerve is formed from C5 to C7 and it is a mixed nerve. The motor branch of this nerve is responsible for innervating the flexor muscles of the anterior arm (biceps brachii, brachialis, and coacobrachialis) while the somatosensory branches are responsible for innervating the skin on the anterolateral forearm

_____ This nerve is subdivided into the anterior, middle and posterior branches and they are all somatosensory in function. These branches are created from nerves extending from C3 and C4 and are responsible for innervating the skin of the shoulders and the anterior aspect of the chest

_____ This nerve is subdivided into the superior and inferior root and is created from nerves extending from C1 to C3. This nerve is motor in function and are responsible for innervating infrahyoid muscle of the neck

1. Ansa cervicalis
2. Axillary nerve
3. Coccygeal nerve
4. Common peroneal (Fibular) nerve
5. Dorsal scapular branch
6. Femoral branch
7. Genitofemoral branch
8. Greater auricular branch
9. Iliohypogastric branch
10. Ilioinguinal branch
11. Inferior gluteal nerve
12. Intercostal nerves
13. Lateral femoral cutaneous branches
14. Lesser occipital branch
15. Long thoracic branch
16. Median nerve
17. Musculocutaneous nerve
18. Nerve to internal obturator
19. Nerve to piriformis
20. Nerve to quadratus femoris
21. Obturator branch
22. Pectoral branch
23. Perforating cutaneous nerve
24. Phrenic nerve
25. Posterior cutaneous nerve
26. Prudendal nerve
27. Radial nerve
28. Saphenous nerve
29. Superior gluteal nerve
30. Supraclavicular branch
31. Tibial nerve
32. Transverse cutaneous branch
33. Ulnar nerve

132. Please match the correct number (1-33) with the specific definitions provided for the specific spinal nerve/branch

_____ These nerves are formed by the remaining branches of the thoracic nerves that are not involves with a plexus (T1-T12). These nerves are responsible for supplying motor impulses to the intercostals muscles of the ribs and the upper abdominal wall. In addition, these nerves also receive sensory impulses from the skin of the thorax and abdomen

_____ This nerve is consists of two branches – Lateral and Medial branch and are formed from nerves extending from C5 to T1. These branches are both motor in function and both the medial and the lateral branch are responsible for innervating both pectoralis minor and major

_____ This nerve is formed from C5 to C8 and T1. It is a mixed nerve. The motor branch of this nerve is responsible for innervating the posterior muscle of the arm forearm and hands (e.g. triceps brachii, brachioradialis, extnsors carpi radialis longus and brevis, extensor carpi ulnaris etc.) while the somatosensory branches innervate the skin of the posterolateral surface of the entire limb

_____ This nerve is formed from C6 and T1. It is a mixed nerve. The motor branches innervate the flexor muscles in the anterior forearm (flexor carpi ulnaris, and the medial half of the flexor digitorum profundus) while the somatosensory branches are responsible for innervating both the posterior and anterior skin of the medial third of the hand

_____ This nerve is formed from extensions from L2 and L3 and is a somatosensory nerve that innervates the skin of the lateral thigh as well as the connective tissues of the peritoneum (lining the abdominal cavity)

_____ This nerve is formed from L2 to L4 and is a mixed nerve. The motor branches innervate the anterior muscles of the thigh (e.g. quadriceps and sartorius) as well as pectineus and iliacus while the somatosensory branch innervates the skin of the anterior and medial thigh, the skin of the medial leg, foot, hip and knee joints

_____ This nerve is formed from L2 to L4 and is a mixed nerve. The motor division innervates the adductor magnus, adductor longus and adductor brevis muscles as well as gracilis, obturator externus muscles while the somatosensory division innervates the skin of the medial thigh and the hip and knee joints

_____ This nerve is formed from nerves extending from C5 to C7 and is a motor nerve. It is responsible for innervating the serratus anterior muscle which is responsible for stabilizing and rotating the scapula

1.	Ansa cervicalis	18.	Nerve to internal obturator
2.	Axillary nerve	19.	Nerve to piriformis
3.	Coccygeal nerve	20.	Nerve to quadratus femoris
4.	Common peroneal (Fibular) nerve	21.	Obturator branch
		22.	Pectoral branch
5.	Dorsal scapular branch	23.	Perforating cutaneous nerve
6.	Femoral branch	24.	Phrenic nerve
7.	Genitofemoral branch	25.	Posterior cutaneous nerve
8.	Greater auricular branch	26.	Prudendal nerve
9.	Iliohypogastric branch	27.	Radial nerve
10.	Ilioinguinal branch	28.	Saphenous nerve
11.	Inferior gluteal nerve	29.	Superior gluteal nerve
12.	Intercostal nerves	30.	Supraclavicular branch
13.	Lateral femoral cutaneous branches	31.	Tibial nerve
		32.	Transverse cutaneous branch
14.	Lesser occipital branch	33.	Ulnar nerve
15.	Long thoracic branch		
16.	Median nerve		
17.	Musculocutaneous nerve		

133. Please match the correct number (1-33) with the specific definitions provided for the specific spinal nerve/branch

_____ This nerve arises from Co1 and it is a mixed nerve which possesses a motor and a somatosensory component. The Motor component innervates the muscle of the pelvic floor (e.g. levator ani and coccygeus) while the somatosensory component innervates the skin over the coccyx

_____ This nerve arises from the extension of the L3 and L4 with some contribution from L2. It is a motor nerve and innervates a muscle that shares the same name as the nerve as well as gemellus superior muscles.

_____ This nerve arises from the extension of the S1 – S3. It is a somatosensory nerve and innervates the skin of the lower buttocks, anal region etc.

_____ This nerve arises from the extension of the S3 – S5. It is a somatosensory nerve and innervates the skin of the posterior aspect of the buttocks

_____ This nerve is formed form the extension of the L5 to S2. It is a motor nerve and innervates gluteus maximus

_____ This nerve is formed from L1 of the spinal cord and is a mixed nerve. The sensory aspect innervates the skin of the lower abdomen, lower back and hip while the motor aspects innervate the internal and external obliques as well as the transverses abdominis

_____ This nerve is formed from L1 section of the spinal cord and is a mixed nerve. The sensory component innervates the skin of the upper thigh, scrotum, root of the penis and labia majora while the motor component joins with another spinal branch to innervate the internal and external obliques as well as the transverses abdominis

_____ This nerve is formed from the extension of nerves from L3 and L4. It is a somatosensory nerve which innervates the skin of the medial aspect of the leg and foot and knee joint

_____ This nerve is formed from the extension of the L4, L5 and S1. It is a motor nerve which innervates gluteus minimus, gluteus medius and tensor fasciae latae

_____ This nerve is formed from the extensions from L1 and L2. It is a somatosensory nerve which innervates the skin of the middle anterior of the thigh, scrotum, cremaster muscles and labia majora

1.	Ansa cervicalis	18.	Nerve to internal obturator
2.	Axillary nerve	19.	Nerve to piriformis
3.	Coccygeal nerve	20.	Nerve to quadratus femoris
5.	Common peroneal (Fibular) nerve	21.	Obturator branch
		22.	Pectoral branch
6.	Dorsal scapular branch	23.	Perforating cutaneous nerve
7.	Femoral branch	24.	Phrenic nerve
8.	Genitofemoral branch	25.	Posterior cutaneous nerve
9.	Greater auricular branch	26.	Prudendal nerve
10.	Iliohypogastric branch	27.	Radial nerve
11.	Ilioinguinal branch	28.	Saphenous nerve
12.	Inferior gluteal nerve	29.	Superior gluteal nerve

13.	Intercostal nerves	30.	Supraclavicular branch
14.	Lateral femoral cutaneous branches	31.	Tibial nerve
15.	Lesser occipital branch	32.	Transverse cutaneous branch
16.	Long thoracic branch	33.	Ulnar nerve
17.	Median nerve		
18.	Musculocutaneous nerve		

134. Please match the correct number (1-33) with the specific definitions provided for the specific spinal nerve/branch

_____ This nerve arises from the extension of the L4 to S3 and is a part of the sciatic nerve and is a mixed nerve. The motor component innervates the semitendinosus, semimembranosus, long head of the biceps femoris, gastrocnemius, Soleus, flexor digitorum longus, flexor hallucis longus, tibialis posterior, popliteus and the intrinsic muscles of the foot. The somatosensory component innervates the skin of the posterior leg, sole of the foot and foot joints

_____ This nerve arises from the extension of the L4-L5 and S1-S2 and is a part of the sciatic nerve. It is a mixed nerve where the Motor component innervates the short head of the biceps femoris, peroneus tertius, tibialis anterior, extensor hallucis longus, extensor digitorum longus and extensor digitorum brevis. The somatosensory component innervates the skin of the anterior distal one-third of the leg, dorsal aspects of the foot, toes and the knee joint

_____ This nerve arises from the extension of the S2-S4 and is a mixed nerve. The Motor component innervates the muscle of the perineum. The somatosensory component innervates the penis, clitoris, scrotum, labias (majora, minora) and lower vagina

_____ This nerve is formed from the extension of the L4, L5 and S1 and is a mixed nerve. The Motor component innervates a muscle that shares the same name as the nerve and gemellus inferior and the somatosensory component innervates the hip joint

1.	Ansa cervicalis	18.	Nerve to internal obturator
2.	Axillary nerve	19.	Nerve to piriformis
3.	Coccygeal nerve	20.	Nerve to quadratus femoris
4.	Common peroneal (Fibular) nerve	21.	Obturator branch
		22.	Pectoral branch
5.	Dorsal scapular branch	23.	Perforating cutaneous nerve
6.	Femoral branch	24.	Phrenic nerve
7.	Genitofemoral branch	25.	Posterior cutaneous nerve
8.	Greater auricular branch	26.	Prudendal nerve
9.	Iliohypogastric branch	27.	Radial nerve
10.	Ilioinguinal branch	28.	Saphenous nerve
11.	Inferior gluteal nerve	29.	Superior gluteal nerve
12.	Intercostal nerves	30.	Supraclavicular branch
13.	Lateral femoral cutaneous branches	31.	Tibial nerve
		32.	Transverse cutaneous branch
14.	Lesser occipital branch	33.	Ulnar nerve
15.	Long thoracic branch		
16.	Median nerve		
17.	Musculocutaneous nerve		

135. Please match the correct number (1-33) with the specific definitions provided for the specific spinal nerve/branch

_____ This nerve arises from the extension of the L4 to S3 and is a part of the sciatic nerve and is a mixed nerve. The Motor component innervates the semitendinosus, semimembranosus, long head of the biceps femoris, gastrocnemius, Soleus, flexor digitorum longus, flexor hallucis longus, tibialis posterior, popliteus and the intrinsic muscles of the foot. The somatosensory component innervates the skin of the posterior leg, sole of the foot and foot joints

_____ This nerve arises from the extension of the L4-L5 and S1-S2 and is also a part of the sciatic nerve. It is a mixed nerve where the Motor component innervates the short head of the biceps femoris, peroneus tertius, tibialis anterior, extensor hallucis longus, extensor digitorum longus and extensor digitorum brevis. The somatosensory component innervates the skin of the anterior distal one-third of the leg, dorsal aspects of the foot, toes and the knee joint

_____ This nerve arises from the extension of the S2-S4 and is a mixed nerve. The Motor component innervates the muscle of the perineum. The somatosensory component innervates the penis, clitoris, scrotum, labias (majora, minora) and lower vagina

_____ This nerve is formed from the extension of the L4, L5 and S1 and is a mixed nerve. The Motor component innervates a muscle that shares the same name as the nerve and gemellus inferior and the somatosensory component innervates the hip joint

1.	Ansa cervicalis	18.	Nerve to internal obturator
2.	Axillary nerve	19.	Nerve to piriformis
3.	Coccygeal nerve	20.	Nerve to quadratus femoris
4.	Common peroneal (Fibular) nerve	21.	Obturator branch
5.	Dorsal scapular branch	22.	Pectoral branch
6.	Femoral branch	23.	Perforating cutaneous nerve
7.	Genitofemoral branch	24.	Phrenic nerve
8.	Greater auricular branch	25.	Posterior cutaneous nerve
9.	Iliohypogastric branch	26.	Prudendal nerve
10.	Ilioinguinal branch	27.	Radial nerve
11.	Inferior gluteal nerve	28.	Saphenous nerve
12.	Intercostal nerves	29.	Superior gluteal nerve
13.	Lateral femoral cutaneous branches	30.	Supraclavicular branch
14.	Lesser occipital branch	31.	Tibial nerve
15.	Long thoracic branch	32.	Transverse cutaneous branch
16.	Median nerve	33.	Ulnar nerve
17.	Musculocutaneous nerve		

136. Slightly inferior to the first lumbar segment (L1), the spinal cord tapers to a point called the _____ and separates into bundles of nerve roots. These bundles of nerve roots occupy the vertebral canal from L1 to S5. The separation of the spinal cord into individual nerves is called _____. Due to this separation into nerve bundles and the minor resemblance of this region of the spinal cord to a horse's tail, the L1 to S5 region is referred to as the.
   a. Ansa cervicalis
   b. Cauda equine
   c. Conus medullaris
   d. Folia
   e. Gyri
   f. Habenula
   g. Pyramids
   h. Sulci
   i. Terminal filum
   j. Vermis

137. The spinal cord is divided into how many segments?
   a. 25
   b. 26
   c. 34
   d. 31
   e. 35
   f. 28
   g. 29

138. Cervical plexuses are formed from which of the following pairs of spinal nerves?
   a. C1-C5
   b. C2-C6
   c. C3-C7
   d. C4-C8
   e. C1-C8
   f. C5-C8
   g. C4-C7
   h. C2-C8

139. Brachial plexuses are formed from which of the following pairs of spinal nerves?
   a. C1-C8 and T1
   b. C2-C7 and T2
   c. C5-C8 and T4
   d. C5-C8 and T1
   e. C2-C4 and T1-T3
   f. C4-C7 and T2-T12
   g. C5-C8 and T2
   h. C6-C8 and T3

140. Phrenic nerves are formed from which of the following pairs of spinal nerves?
   a. C1-C3
   b. T1-T3
   c. L2-L5
   d. S1-S5 and Co1
   e. C3-C5
   f. T1-T12
   g. T5-T12
   h. L1-L5

141. Lumbosacral plexuses are formed from which of the following pairs of spinal nerves?
   a. S1-S5 and Co-1
   b. L4, L5 and S1-S4
   c. T1-T12 and L2-L5
   d. L1-L6 and S1-S5
   e. C2-C5 and T1-T6
   f. C5, T12 and L3-L5
   g. L1-L4 and T12
   h. T1-T12
   i. Both a and c

j. Both b and g

142. Intercostal nerves are formed from which of the following pairs of spinal nerves?
   a. S1-S5 and Co-1
   b. L1-L6 and S1-S5
   c. C2-C5 and T1-T6
   d. C5, T12 and L3-L5
   e. L1-L2 and S1-S5
   f. T1-T11
   g. L4, L5 and S1-S4
   h. T1-T12 and L2-L5

143. _____ is an anatomical thickening of the spinal cord that supply nerves and receive impulses from the upper limb
   a. Cauda equine
   b. Cervical enlargement
   c. Lumbar enlargement
   d. Conus medullaris
   e. Folia
   f. Gyri
   g. Habenula
   h. Pyramids
   i. Sulci
   j. Terminal filum
   k. Vermis

144. _____ is an anatomical thickening of the spinal cord that supply nerves and receive impulses from the lower limbs
   a. Cauda equine
   b. Cervical enlargement
   c. Lumbar enlargement
   d. Conus medullaris
   e. Folia
   f. Gyri
   g. Habenula
   h. Pyramids
   i. Sulci
   j. Terminal filum
   k. Vermis

145. The ascending tract nerve fibers arise from the first order neurons located in the_____
   a. Anterior funiculi
   b. Central canal
   c. Dorsal root ganglion
   d. Lateral funiculi
   e. Lateral horn
   f. Posterior funiculi
   g. Ventral horn
   h. Ventral root ganglia

146. The posterior funiculi possess two nerve tracts: _____ and _____. Both of these nerve tracts transmit sensory information related to tactile, vibration, proprioception and movement to the brain. Together, these two spinal nerve tracts are called the _____
   a. Anterior spinothalamic tract
   b. Anterolateral System
   c. Dorsal and ventral spinocerebellar tracts
   d. Dorsal column medial lemniscus system
   e. Fasiculus cuneatus
   f. Fasiculus gracilis
   g. Lateral spinothalamic tract
   h. Spinocerebellar Tracts
   i. Spinoolivary tract
   j. Spinoreticular tract
   k. Spinotectal tract

147. The lateral funiculi contain two ascending tracts: _____ and _____. These two nerve tracts are a part of the _____.
   a. Anterior spinothalamic tract
   b. Anterolateral System
   c. Dorsal and ventral spinocerebellar tracts
   d. Dorsal column medial lemniscus system
   e. Fasiculus cuneatus
   f. Fasiculus gracilis
   g. Lateral spinothalamic tract
   h. Spinocerebellar Tracts
   i. Spinoolivary tract
   j. Spinoreticular tract
   k. Spinotectal tract

148. The _____ carries sensory information regarding pain, temperature and touch from both somatic and visceral structures. This spinal tract is a part of the _____.
   a. Anterior spinothalamic tract
   b. Anterolateral System
   c. Dorsal and ventral spinocerebellar tracts
   d. Dorsal column medial lemniscus system
   e. Fasiculus cuneatus
   f. Fasiculus gracilis
   g. Lateral spinothalamic tract
   h. Spinocerebellar Tracts

i. Spinoolivary tract
j. Spinoreticular tract
k. Spinotectal tract

149. The _____ convey unconscious proprioception information from muscles and joints of the legs to the cerebellum.
a. Anterior spinothalamic tract
b. Anterolateral System
c. Dorsal and ventral spinocerebellar tracts
d. Dorsal column medial lemniscus system
e. Fasiculus cuneatus
f. Fasiculus gracilis
g. Lateral spinothalamictract
h. Posterior and anterior spinocerebellar tracts
i. Spinoolivary tract
j. Spinoreticular tract
k. Spinotectal tract

150. The anterior funiculi possess four ascending tracts: _____, _____, _____, and _____.
a. Anterior spinothalamic tract
b. Anterolateral System
c. Dorsal and ventral spinocerebellar tracts
d. Dorsal column medial lemniscus system
e. Fasiculus cuneatus
f. Fasiculus gracilis
g. Lateral spinothalamic tract
h. Spinocerebellar Tracts
i. Spinoolivary tract
j. Spinoreticular tract
k. Spinotectal tract

151. The _____ transmits information regarding pain, temperature, and touch to the brain stem and the diencephalon. This nerve tract is part of the _____.
a. Anterior spinothalamic tract
b. Anterolateral System
c. Dorsal and ventral spinocerebellar tracts
d. Dorsal column medial lemniscus system
e. Fasiculus cuneatus
f. Fasiculus gracilis
g. Lateral spinothalamic tract
h. Spinocerebellar Tracts
i. Spinoolivary tract
j. Spinoreticular tract
k. Spinotectal tract

152. The _____ is responsible for carrying information from Golgi tendon organs, which detects changes in muscle tension, to the cerebellum. This spinal tract is a part of the _____.
a. Anterior spinothalamic tract
b. Anterolateral System
c. Dorsal and ventral spinocerebellar tracts
d. Dorsal column medial lemniscus system
e. Fasiculus cuneatus
f. Fasiculus gracilis
g. Lateral spinothalamic tract
h. Spinocerebellar Tracts
i. Spinoolivary tract
j. Spinoreticular tract
k. Spinotectal tract

153. The _____ sends sensory information to the reticular formation and participates in various reflex reactions to pain which can trigger behavioral reactions to pain.
a. Anterior spinothalamic tract
b. Anterolateral System
c. Dorsal and ventral spinocerebellar tracts
d. Dorsal column medial lemniscus system
e. Fasiculus cuneatus
f. Fasiculus gracilis
g. Lateral spinothalamic tract
h. Spinocerebellar Tracts
i. Spinoolivary tract
j. Spinoreticular tract
k. Spinotectal tract

154. The _____ influences to reflex orientation of the head and eyes toward a somatic stimulus.
a. Anterior spinothalamic tract
b. Anterolateral System
c. Dorsal and ventral spinocerebellar tracts
d. Dorsal column medial lemniscus system
e. Fasiculus cuneatus
f. Fasiculus gracilis
g. Lateral spinothalamic tract
h. Spinocerebellar Tracts
i. Spinoolivary tract
j. Spinoreticular tract
k. Spinotectal tract

155. The descending tracts originated from the _____ and _____ which later synapse with various nuclei found within the spinal cord _____ (Hint: unmyelinated area of the CNS). The descending tracts transmit information

associated with maintenance of motor functions which includes posture, balance, muscle tone, and visceral and somatic reflexes.
a. Anterior corticospinal tract
b. Brainstem
c. Cerebral cortex
d. Extrapyramidal Tracts
e. Gray matter
f. Lateral corticospinal tract
g. Lateral reticulospinal tract
h. Lissauer's tract
i. Medulla oblongata
j. Olivospinal tract
k. Pons
l. Posterior reticulospinal tract
m. Pyramidal tract
n. Rubrospinal tracts
o. Vestibulocochlear nerve
p. Vestibulospinal tract
q. White matter

156. There are three descending tracts located in the lateral funiculi: _____, _____, and _____.
a. Anterior corticospinal tract
b. Brainstem
c. Cerebral cortex
d. Extrapyramidal Tracts
e. Gray matter
f. Lateral corticospinal tract
g. Lateral reticulospinal tract
h. Lissauer's tract
i. Medulla oblongata
j. Olivospinal tract
k. Pons
l. Posterior reticulospinal tract
m. Pyramidal tract
n. Rubrospinal tracts
o. Vestibulocochlear nerve
p. Vestibulospinal tract
q. White matter

157. The _____ (**Answer A**) and the _____ (**Answer B**) convey motor impulses associated with voluntary movement. **Answer A** belongs to the _____, while the **Answer B** belongs to the _____.
a. Anterior corticospinal tract
b. Brainstem
c. Cerebral cortex
d. Extrapyramidal Tracts
e. Gray matter
f. Lateral corticospinal tract
g. Lateral reticulospinal tract
h. Lissauer's tract
i. Medulla oblongata
j. Olivospinal tract
k. Pons
l. Posterior reticulospinal tract
m. Pyramidal tract
n. Rubrospinal tracts
o. Vestibulocochlear nerve
p. Vestibulospinal tract
q. White matter

158. The _____ conveys motor impulses associated with posture and muscle tone. This nerve tract is a part of the _____.
a. Anterior corticospinal tract
b. Brainstem
c. Cerebral cortex
d. Extrapyramidal Tracts
e. Gray matter
f. Lateral corticospinal tract
g. Lateral reticulospinal tract
h. Lissauer's tract
i. Medulla oblongata
j. Olivospinal tract
k. Pons
l. Posterior reticulospinal tract
m. Pyramidal tract
n. Rubrospinal tracts
o. Vestibulocochlear nerve
p. Vestibulospinal tract
q. White matter

159. The anterior funiculi possess four descending nerve tracts: _____, _____, _____, and _____.
a. Anterior corticospinal tract
b. Brainstem
c. Cerebral cortex
d. Extrapyramidal Tracts
e. Gray matter
f. Lateral corticospinal tract
g. Lateral reticulospinal tract
h. Lissauer's tract
i. Medulla oblongata
j. Olivospinal tract

k. Pons
l. Posterior reticulospinal tract
m. Pyramidal tract
n. Rubrospinal tracts
o. Vestibulocochlear nerve
p. Vestibulospinal tract
q. White matter

160. The _____ transmits motor impulses associated with posture and muscle tone. This nerve tract belongs to the _____.
a. Anterior corticospinal tract
b. Brainstem
c. Cerebral cortex
d. Extrapyramidal Tracts
e. Gray matter
f. Lateral corticospinal tract
g. Lateral reticulospinal tract
h. Lissauer's tract
i. Medulla oblongata
j. Olivospinal tract
k. Pons
l. Posterior reticulospinal tract
m. Pyramidal tract
n. Rubrospinal tracts
o. Vestibulocochlear nerve
p. Vestibulospinal tract
q. White matter

161. The _____ mediates balance and postural movements. This nerve tract belongs to the _____.
a. Anterior corticospinal tract
b. Brainstem
c. Cerebral cortex
d. Extrapyramidal Tracts
e. Gray matter
f. Lateral corticospinal tract
g. Lateral reticulospinal tract
h. Lissauer's tract
i. Medulla oblongata
j. Olivospinal tract
k. Pons
l. Posterior reticulospinal tract
m. Pyramidal tract
n. Rubrospinal tracts
o. Vestibulocochlear nerve
p. Vestibulospinal tract
q. White matter

162. The _____ transmits motor impulse related to proprioception and is a part of the _____.
a. Anterior corticospinal tract
b. Brainstem
c. Cerebral cortex
d. Extrapyramidal Tracts
e. Gray matter
f. Lateral corticospinal tract
g. Lateral reticulospinal tract
h. Lissauer's tract
i. Medulla oblongata
j. Olivospinal tract
k. Pons
l. Posterior reticulospinal tract
m. Pyramidal tract
n. Rubrospinal tracts
o. Vestibulocochlear nerve
p. Vestibulospinal tract
q. White matter

163. The _____ (**Answer A**) is linked with the vestibular nuclei, which are located in the _____ and _____. The vestibular nuclei receives sensory information from the vestibular branch of the _____ in regards to the orientation of the cranium before transmitting motor command through the **Answer A** which in turn controls the muscle tone, extend, and change the position of the limbs to maintain posture and balance in relation to the cranium. This nerve tract belongs to the _____.
a. Anterior corticospinal tract
b. Brainstem
c. Cerebral cortex
d. Extrapyramidal Tracts
e. Gray matter
f. Lateral corticospinal tract
g. Lateral reticulospinal tract
h. Lissauer's tract
i. Medulla oblongata
j. Olivospinal tract
k. Pons
l. Posterior reticulospinal tract
m. Pyramidal tract
n. Rubrospinal tracts
o. Vestibulocochlear nerve
p. Vestibulospinal tract
q. White matter

164. Located at the dorsolateral funiculi, wedged between the edge of the dorsal horn and the surface of the spinal cord, is a nerve tract called _____. The nerve tract is responsible for regulate incoming pain sensation at the spinal level.
a. Anterior corticospinal tract
b. Brainstem
c. Cerebral cortex
d. Extrapyramidal Tracts
e. Gray matter
f. Lateral corticospinal tract
g. Lateral reticulospinal tract
h. Lissauer's tract
i. Medulla oblongata
j. Olivospinal tract
k. Pons
l. Posterior reticulospinal tract
m. Pyramidal tract
n. Rubrospinal tracts
o. Vestibulocochlear nerve
p. Vestibulospinal tract
q. White matter

165. What is the name given to the connective tissue surrounding the brain and the spinal cord?
a. Astrocytes
b. Dense regular connective tissue
c. Loose connective tissue
d. Meninges
e. Myelin
f. Oligodendrocytes
g. Reticular tissue

166. The meninges composed of three distinct layers: _____, _____ and _____.
a. Arachnoid mater
b. Crista galli
c. Dense regular connective tissue
d. Dura mater
e. Foramen magnum.
f. Inner meningeal layer
g. Loose connective tissue
h. Outer periosteal layer
i. Pia mater
j. Reticular tissue
k. Sella turcica

167. The outer most layer (superior layer) layer of the connective tissue covering of the CNS is called the _____ which rests just beneath the flat bones of the cranium and the vertebra.
a. Arachnoid mater
b. Crista galli
c. Dense regular connective tissue
d. Dura mater
e. Foramen magnum.
f. Inner meningeal layer
g. Loose connective tissue
h. Outer periosteal layer
i. Pia mater
j. Reticular tissue
k. Sella turcica

168. In the cranium, the dura mater is composed of two distinct layers: _____ and _____.
a. Arachnoid mater
b. Crista galli
c. Dense regular connective tissue
d. Dura mater
e. Foramen magnum.
f. Inner meningeal layer
g. Loose connective tissue
h. Outer periosteal layer
i. Pia mater
j. Reticular tissue
k. Sella turcica

169. Which of the following layer of the dura mater (found only in the cranium) is the equivalent to the periosteum of the cranial bones?
a. Arachnoid mater
b. Crista galli
c. Dense regular connective tissue
d. Dura mater
e. Foramen magnum.
f. Inner meningeal layer
g. Loose connective tissue
h. Outer periosteal layer
i. Pia mater
j. Reticular tissue
k. Sella turcica

170. Which of the following layer of the dura mater is composed of tough fibrous connective tissue that is highly vascularized and innervated (many blood vessels and nerves)?
a. Arachnoid mater

b. Crista galli
c. Dense regular connective tissue
d. Dura mater
e. Foramen magnum.
f. Inner meningeal layer
g. Loose connective tissue
h. Outer periosteal layer
i. Pia mater
j. Reticular tissue
k. Sella turcica

171. Which of the following layer of the dura mater continues down the vertebral canal where it forms the dural sac?
a. Arachnoid mater
b. Crista galli
c. Dense regular connective tissue
d. Dura mater
e. Foramen magnum.
f. Inner meningeal layer
g. Loose connective tissue
h. Outer periosteal layer
i. Pia mater
j. Reticular tissue
k. Sella turcica

172. Within the cranium, the dura mater is attached to the cranial bones at _____, _____ and _____.
a. Arachnoid mater
b. Crista galli
c. Dense regular connective tissue
d. Dura mater
e. Foramen magnum.
f. Inner meningeal layer
g. Loose connective tissue
h. Outer periosteal layer
i. Pia mater
j. Reticular tissue
k. Sella turcica

173. The _____ is a delicate fibrous membrane forming the second or middle layer of the meninges.
a. Arachnoid mater
b. Arachnoid villi
c. Crista galli
d. Dense regular connective tissue
e. Dura mater
f. Foramen magnum.
g. Inner meningeal layer
h. Loose connective tissue
i. Outer periosteal layer
j. Pia mater
k. Reticular tissue
l. Sella turcica

174. Which of the following is formed form projections of the arachnoid mater?
a. Arachnoid mater
b. Arachnoid villi
c. Crista galli
d. Dura mater
e. Foramina of Monro
f. Pia mater
g. Reticular tissue
h. Sella turcica
i. Subarachnoid space

175. Beneath the arachnoid mater is a tiny opening called _____. This area is connected with the cerebral ventricles via three foramina. These foramina, also known as _____, allow the cerebral spinal fluid (CSF) to circulate continuously from the ventricles.
a. Arachnoid mater
b. Arachnoid villi
c. Crista galli
d. Dura mater
e. Foramina of Monro
f. Pia mater
g. Reticular tissue
h. Sella turcica
i. Subarachnoid space

176. The _____ is the innermost layer of the meninges and is a thinner and even more delicate connective tissue layer.
a. Arachnoid mater
b. Arachnoid villi
c. Crista galli
d. Dura mater
e. Foramina of Monro
f. Pia mater
g. Reticular tissue
h. Sella turcica
i. Subarachnoid space

177. Which of the following is (are) responsible for producing CSF?
a. Arachnoid mater

b. Arachnoid villi
c. Choroid plexuses
d. Crista galli
e. Dura mater
f. Ependymal cells
g. Foramina of Monro
h. Both a and g
i. Both c and f

178. The choroid plexuses exist in which of the following areas of the brain. Please circle all that applies.
a. Arachnoid mater
b. Arachnoid villi
c. Crista galli
d. Ependymal cells
e. Fourth ventricles
f. Lateral ventricles
g. Third ventricle

179. The peripheral nervous system is composed of two subdivisions, the _____ and _____
a. Afferents
b. Autonomic nervous system
c. Efferents
d. Motor
e. Parasympathetic
f. Sensory
g. Somatic motor
h. Somatic reflexes
i. Somatic sensory
j. Sympathetic
k. Visceral motor
l. Visceral sensory

180. The motor division is also known as the _____ and are responsible for transmitting signals from the CNS to various glands, organs and muscles of the body.
a. Afferents
b. Autonomic nervous system
c. Efferents
d. Motor
e. Parasympathetic
f. Sensory
g. Somatic motor
h. Somatic reflexes
i. Somatic sensory
j. Sympathetic
k. Visceral motor
l. Visceral sensory

181. The motor division could be further subdivided into two subdivisions: _____ and _____.
a. Afferents
b. Autonomic nervous system
c. Efferents
d. Motor
e. Parasympathetic
f. Sensory
g. Somatic motor
h. Somatic reflexes
i. Somatic sensory
j. Sympathetic
k. Visceral motor
l. Visceral sensory

182. The _____ is responsible for transmitting signals from the CNS to the skeletal muscles of the body. This type of innervation could result in voluntary skeletal muscle control, which occurs under conscious control, or it could be a form of involuntary contraction which is also termed _____.
a. Afferents
b. Autonomic nervous system
c. Efferents
d. Motor
e. Parasympathetic
f. Sensory
g. Somatic motor
h. Somatic reflexes
i. Somatic sensory
j. Sympathetic
k. Visceral motor
l. Visceral sensory

183. The _____ is also referring to as the _____ and it is responsible for carrying nervous impulses to the glands, smooth muscles and internal organs of the body. This motor subdivision is completely involuntary.
a. Afferents
b. Autonomic nervous system
c. Efferents
d. Motor
e. Parasympathetic
f. Sensory
g. Somatic motor
h. Somatic reflexes

i.　Somatic sensory
j.　Sympathetic
k.　Visceral motor
l.　Visceral sensory

184.　The visceral motor division could be further divided into two subdivisions: _____ and _____.
　　a.　Afferents
　　b.　Autonomic nervous system
　　c.　Efferents
　　d.　Motor
　　e.　Parasympathetic
　　f.　Sensory
　　g.　Somatic motor
　　h.　Somatic reflexes
　　i.　Somatic sensory
　　j.　Sympathetic
　　k.　Visceral motor
　　l.　Visceral sensory

185.　The sensory division is also known as _____ are responsible for transmitting signals from various types of receptors from the peripheral tissues to the CNS.
　　a.　Afferents
　　b.　Autonomic nervous system
　　c.　Efferents
　　d.　Motor
　　e.　Parasympathetic
　　f.　Sensory
　　g.　Somatic motor
　　h.　Somatic reflexes
　　i.　Somatic sensory
　　j.　Sympathetic
　　k.　Visceral motor
　　l.　Visceral sensory

186.　The sensory division could be further divided into two subdivisions: the _____ and _____.
　　a.　Afferents
　　b.　Autonomic nervous system
　　c.　Efferents
　　d.　Motor
　　e.　Parasympathetic
　　f.　Sensory
　　g.　Somatic motor
　　h.　Somatic reflexes
　　i.　Somatic sensory
　　j.　Sympathetic
　　k.　Visceral motor
　　l.　Visceral sensory

187.　The _____ is responsible for transmitting signals form the receptors located in the skin, muscle joints and bones to the CNS.
　　a.　Afferents
　　b.　Autonomic nervous system
　　c.　Efferents
　　d.　Motor
　　e.　Parasympathetic
　　f.　Sensory
　　g.　Somatic motor
　　h.　Somatic reflexes
　　i.　Somatic sensory
　　j.　Sympathetic
　　k.　Visceral motor
　　l.　Visceral sensory

188.　The _____ is responsible for conveying signals form the various organs of the thoracic and abdominal cavities. Afferents
　　a.　Autonomic nervous system
　　b.　Efferents
　　c.　Motor
　　d.　Parasympathetic
　　e.　Sensory
　　f.　Somatic motor
　　g.　Somatic reflexes
　　h.　Somatic sensory
　　i.　Sympathetic
　　j.　Visceral motor
　　k.　Visceral sensory

189.　_____ is a motor cranial nerve that consist of two divisions: the _____ branch carries impulses to the muscles of the soft palate, pharynx and larynx while the _____ branch supplies the impulses needed to control trapezious and sternocleidomastoid muscles (muscles of the neck)

　　_____ is a motor cranial nerve that controls the movement of the tongue

　　_____ is a motor nerve that supplies the lateral rectus which results in abduction (turning the line of site away from the midline) of the eye

　　_____ is a sensory cranial nerve that is associated with sense of smell

　　_____ is a sensory cranial nerve that is associated with vision

_____ is a sensory cranial nerve with two divisions: the _____ branch associates with the inner ear and are involved in controlling equilibrium while the _____ branch transmits auditory information (involved in hearing)

_____ is a cranial nerve that is associated with a single muscle, superior oblique, which causes the eye to look downward

_____ is a cranial nerve with three subdivisions: the _____ is a sensory nerve that brings signals to the brain from the surfaces of the eyes, tear glands etc. The _____ transmit sensory information of the upper teeth, gum and lip etc. and _____ is a motor and sensory component whereby the sensory component transmits signals to the brain from the lower teeth, gums, and lips etc. while the motor component controls the muscles involve in chewing

_____ is a mixed (autonomic and motor) cranial nerve that consists of both somatic and autonomic nerves that spread to the chest and the abdomen. The motor division involve in the motor control of speech and swallowing while the autonomic division associates with the heart and smooth musculature of the thorax and abdomen

_____ is a mixed (motor and sensory) cranial nerve that is involved in sending impulse of taste, touch (pressure), pain, temperature from the tongue and the outer ear to the brain while the motor component is associated with controlling salivation, deglutition and gagging

_____ is a motor and sensory nerve (mixed nerve) that transmits sensory information associated with chemoreceptors (located in the taste buds) that is found within the anterior 2/3 of the tongue while the motor component associates with muscles involved in facial expressions

1.	Abducens nerves	23. Medial cord
2.	Accessory nerves	24. Medial pectoral branch
3.	Axillary nerve	25. Median nerve
4.	Cervical plexus	26. Musculocutaneous nerve
5.	Cochlear branch	27. Obturator branch
6.	Cranial branch	28. Oculomotor nerves
7.	Dorsal scapular branch	29. Olfactory nerve
8.	Extrapyamidal tracts	30. Osmodium nerve
9.	Facial nerves	31. Ophthalmic division
10.	Femoral branch	32. Optic nerves
11.	Glossopharyngeal nerves	33. Pectoral branch
12.	Hypoglossal nerves	34. Pyramidal tracts
13.	Iliohypogastric branch	35. Radial nerve
14.	Ilioinguinal branch	36. Saphenous nerve
15.	Infundibulum	37. Spinal branch
16.	Intercostal nerves	38. Spinothalamic tract
17.	Lateral femoral cutaneous branches	39. Trigeninal nerves
18.	Lateral pectoral branch	40. Trochlear nerves
19.	Long thoracic branch	41. Ulnar nerve
20.	Lumbar plexus	42. Vagus nerves
21.	Mandibular division	43. Vestbulocochlear nerves
22.	Maxillary division	44. Vestibular branch

190.　Sympathetic preganglionic neurons are located in the _____ of the spinal cord. These preganglionic neurons, also known as the _____, are found between the _____ (Hint: name the spinal segments). The axons of these preganglionic neurons passes through the ventral roots of these segments and synapse with the _____ which lies parallel on both sides of these spinal cord segments, as well as the unpaired collateral ganglia which are found within the abdominal cavity. The neurotransmitter of choice utilized by the preganglionic fibers is _____ while the postganglionic fibers of the sympathetic division utilize _____ as their neurotransmitter
　　a.　Acetylcholine
　　b.　Brain stem
　　c.　C3-S1
　　d.　Hypothalamus
　　e.　Lateral horn
　　f.　Muscarinic acetylcholine receptors (mAChR)
　　g.　Nicotinic acetylcholine receptors (nAChR)
　　h.　Noradrenaline
　　i.　Paravertebral ganglia
　　j.　S2-S4
　　k.　T1-L2
　　l.　T2-S5
　　m.　Thalamus
　　n.　Thoracolumbar division

191.　The receptor proteins located in the paravertebral ganglia are _____.
　　a.　Acetylcholine
　　b.　Brain stem
　　c.　C3-S1
　　d.　Hypothalamus
　　e.　Lateral horn
　　f.　Muscarinic acetylcholine receptors (mAChR)
　　g.　Nicotinic acetylcholine receptors (nAChR)
　　h.　Noradrenaline
　　i.　Paravertebral ganglia
　　j.　S2-S4
　　k.　T1-L2
　　l.　T2-S5
　　m.　Thalamus

192. The receptor proteins located at the target tissues are _____.
    a. Acetylcholine
    b. Brain stem
    c. C3-S1
    d. Hypothalamus
    e. Lateral horn
    f. Muscarinic acetylcholine receptors (mAChR)
    g. Nicotinic acetylcholine receptors (nAChR)
    h. Noradrenaline
    i. Paravertebral ganglia
    j. S2-S4
    k. T1-L2
    l. T2-S5
    m. Thalamus
    n. Thoracolumbar division

193. The preganglionic neurons of the parasympathetic division are located within the cranial nerve nuclei found in the _____ or within the gray matter located between the _____.
    a. Acetylcholine
    b. Brain stem
    c. C3-S1
    d. Hypothalamus
    e. Lateral horn
    f. Muscarinic acetylcholine receptors (mAChR)
    g. Nicotinic acetylcholine receptors (nAChR)
    h. Noradrenaline
    i. Paravertebral ganglia
    j. S2-S4
    k. T1-L2
    l. T2-S5
    m. Thalamus
    n. Thoracolumbar division

ANSWERS

1. h	2. a, b, c	3. f
4. e	5. d	6. b
7. e	8. b	9. a
10. c	11. h	12. c
13. a	14. d	15. g
16. e	17. b	18. g
19. h	20. e	21. g
22. e	23. f	24. d
25. f	26. c	27. b
28. e, g	29. d	30. d
31. e	32. c, e, f	33. e
34. a	35. a	36. d
37. c	38. c	39. a
40. b	41. d	42. a
43. b	44. a	45. f
46. a	47. f	48. f
49. e	50. c	51. d
52. c	53. b, f, g, a, c	54. c
55. g	56. d	57. b, h
58. f	59. g	60. c
61. f	62. d	63. a
64. g	65. a	66. a
67. a	68. a	69. b
70. b, d, g, h	71. a	72. b
73. b	74. f	75. a
76. d	77. c	78. h
79. f, g	80. c	81. c, g
82. c	83. b, c	84. d
85. a, b	86. e	87. i
88. f, i	89. i, l, c, g	90. h
91. e	92. d	93. a, i, b, h
94. a, j, k	95. k, f, d, e	96. a, f, g, h
97. i	98. d	99. d, j, f, g, c
100. f	101. b, h	102. a, b, c
103. i, j, k, l	104. i, d, c	105. b, j
106. g, a, i, h	107. c	108. b, d
109. a, b	110. a, b, c	111. b, a, d
112. a	113. b	114. a
115. a, c, d, c, c	116. a, d, c, d, d	117. d, e, i
118. d, e, i	119. e, f, g, h	120. d

121. 39, 38, 30, 22, 21, 41, 11, 9	122. 2, 6, 37, 12, 1, 29, 32, 43, 44, 5	123. c, d
124. c, g	125. g, d, c, e	126. b, h, i, a, g, d
127. f, c	128. k	129. c, j
130. d, l, a	131. 16, 8, 32, 14, 24, 2, 5, 17, 30, 1	132. 12, 22, 27, 33, 13, 6, 21, 15
133. 3, 18, 25, 23, 11, 9, 10, 28, 29, 7	134. 31, 4, 26, 20	135. c, i
136. d	137. a	138. d
139. e	140. j	141. f
142. b	143. c	144. c
145. e, f, d	146. c, g, h	147. g, b
148. h	149. a, i, j, k	150. a, b
151. i, b	152. j	153. k
154. b, c, e	155. f, g, n	156. f, n, m, d
157. g, d	158. a, j, l, p	159. l, d
160. a, m	161. j, d	162. p, i, k, o, d
163. h	164. d	165. a, d, i
166. d	167. f, h	168. h
169. f	170. f	171. b, e, k
172. a	173. b	174. i, e
175. f	176. i	177. d, e, f, g
178. d, f	179. c	180. g, k
181. g, h	182. k, b	183. e, j
184. a	185. i, l	186. i
187. k	188. 2, 6, 37, 12, 1, 29, 32, 43, 44, 5, 40, 39, 31, 22, 21, 42, 11, 9	189. e, n, k, i, a, h
190. g	191. f	192. b, j

REFERENCES

Blair RJ. The roles of orbital frontal cortex in the modulation of antisocial behavior. Brain Cogn., 2004. 55: pp. 198-208

Rolls ET. The functions of the orbitofrontal cortex. Brain Cogn., 2004. 55: pp. 11-29

HOFMAN MA, SWAAB DF. The sexually dimorphic nucleus of the preoptic area in the human brain: a comparative morphometric study. J. Anat., 1989. 164, pp. 55-72 Nikolaus R. McFarland NR, Haber SN. Convergent Inputs from Thalamic Motor Nuclei and Frontal Cortical Areas to the Dorsal Striatum in the Primate. J. Neurosci., 2000, 20, pp. 3798–3813

Clement Hamani C, Saint-Cyr JA, Fraser J, Kaplitt M, Lozano AM. The subthalamic nucleus in the context of movement disorders. Brain, 2004, 127, pp. 4-20

Velasquez KM, Molfese DL, Salas R. The role of the habenula in drug addiction. Front. Hum. Neurosci., 2014. 8, pp. 174

Hikosaka O, Sesack SR, Lecourtier L, Shepard PD. Habenula: Crossroad between the Basal Ganglia and the Limbic System. J Neuosci., 2008. 28, pp. 11825-11829

Macchi MM, Bruce JN. Human pineal physiology and functional significance of melatonin. Front Neuroendocrinol., 2004, 25, pp. 177-95

Seeley RR, Stephens TD, Tate P. Anatomy and Physiology 6th Edition. McGraw-Hill, New York, New York. 2003

Tate P. Seeley's Principles of Anatomy & Physiology 1st Edition. McGraw-Hill, New York, New York. 2009

Saladin KS. Anatomy & Physiology. The Unity of Form and Function 6th Edition. McGraw-Hill, New York, New York. 2010

PHOTO AND GRAPHIC BIBLIOGRAPHY

1. Figure 1: Graphic designed and released by Open Stax College, Connexions, reprint permission granted under the Creative Commons Attribution 3.0 Unported license
2. Figure 2: Graphic designed and released by Open Stax College, Connexions, reprint permission granted under the Creative Commons Attribution 3.0 Unported license
3. Figure 3: Graphic designed and released by Open Stax College, Connexions, reprint permission granted under the Creative Commons Attribution 3.0 Unported license
4. Figure 4: Graphic designed and released by Open Stax College, Connexions, reprint permission granted under the Creative Commons Attribution 3.0 Unported license
5. Figure 5: Graphic designed and released by Open Stax College, Connexions, reprint permission granted under the Creative Commons Attribution 3.0 Unported license
6. Figure 6: Graphic designed and released by Open Stax College, Connexions, reprint permission granted under the Creative Commons Attribution 3.0 Unported license
7. Figure 7: Graphic designed by Mikael Häggström, reprint permission granted under the Creative Commons Attribution 3.0 Unported license
8. Figure 8: Graphic designed by P.Y.P. Jen. Copyright ©
9. Figure 9: Graphic designed by BruceBlaus, reprint permission granted under the Creative Commons Attribution 3.0 Unported license
10. Figure 10: Graphic designed by Patrick J. Lynch, medical illustrator in 2006. reprint permission granted under the Creative Commons Attribution 2.5 Unported license
11. Figure 11: Graphic designed and released by Open Stax College, Connexions, reprint permission granted under the Creative Commons Attribution 3.0 Unported license
12. Figure 12: Graphic designed and released by Open Stax College, Connexions, reprint permission granted under the Creative Commons Attribution 3.0 Unported license
13. Figure 14: Graphic designed and released by Open Stax College, Connexions, reprint permission granted under the Creative Commons Attribution 3.0 Unported license
14. Figure 15: Graphic designed by Patrick J. Lynch, medical illustrator, 2006. Creative

Commons Attribution 2.5 License. Public domain graphics

15. Figure 16: Graphic designed and released by Open Stax College, Connexions, reprint permission granted under the Creative Commons Attribution 3.0 Unported license

16. Figure 16: Graphics designed by Zlir'a reprint permission granted under Creative Commons Attribution 3.0 Unported license

17. Figure 17: Graphic designed and released by Open Stax College, Connexions, reprint permission granted under the Creative Commons Attribution 3.0 Unported license

18. Figure 18: Graphic designed and released by National Institute of Biomedical Imaging and Bioengineering, the National Institute of Neurological Disorders and Stroke, and the Christopher and Dana Reeve Foundation. Public domain graphics

19. Figure 19: Graphic designed and released by Open Stax College, Connexions, reprint permission granted under the Creative Commons Attribution 3.0 Unported license

20. Figure 20: Graphic designed and released by Kevin Dufendach reprint permission granted under the Creative Commons Attribution 3.0 Unported license. Spinal cord graphic designed by the National Institute of Biomedical Imaging and Bioengineering, the National Institute of Neurological Disorders and Stroke, and the Christopher and Dana Reeve Foundation. Public domain graphics

21. Figure 21: Graphic designed and released by Open Stax College, Connexions, reprint permission granted under the Creative Commons Attribution 3.0 Unported license

22. Figure 22: Graphic designed and released by Open Stax College, Connexions, reprint permission granted under the Creative Commons Attribution 3.0 Unported license

23. Figure 23: Graphic designed and released by Open Stax College, Connexions, reprint permission granted under the Creative Commons Attribution 3.0 Unported license

24. Figure 24: Graphic designed and released by Open Stax College, Connexions, reprint permission granted under the Creative Commons Attribution 3.0 Unported license

25. Figure 25: Graphic designed and released by Open Stax College, Connexions, reprint permission granted under the Creative Commons Attribution 3.0 Unported license

26. Figure 26: Graphic designed by Hic et nunc, reprint permission granted under the Creative Commons Attribution 3.0 Unported license

27. Figure 27: Graphic designed by Hic et nunc, reprint permission granted under the Creative Commons Attribution 3.0 Unported license

28. Figure 28: Graphic designed by Hic et nunc, reprint permission granted under the Creative Commons Attribution 3.0 Unported license

29. Figure 29: Graphic designed by Debivort, reprint permission granted under the Creative Commons Attribution 3.0 Unported license

CHAPTER NINE

SENSES

Senses are physiological methods to perceive the world that surrounds us and sensory perception is the process of acquiring, interpreting, selecting, and organizing sensory information. If by chance we were to lose our ability to sense our surroundings, it would be difficult, if not impossible, for us to realize that we even exist.

All sensory information falls into three major categories: electromagnetic (including thermal energy), mechanical force, and chemical agents. Sensory information is received by specialized sensory receptors and sense organs. These sensory receptors and sense organs are transducers that convert sensory information into action potentials before transmitting these nervous impulses to specific locations within the CNS for analysis and integration.

Sensory Perception

In general, sensory receptors transmit four types of information: modality, location, intensity, and duration.

Modality is defined as the type of stimulus or sensation that a particular type of sensory receptor produces. The most common types of modality are vision, hearing, smell, balance, and taste. Since, all the nerves transmit their information the same way, via action potentials, the brain distinguishes one modality from another via the type of sensory receptor that is transmitting the information (e.g. retinas transmit visual information, while olfactory receptors transmit information regarding smell), as well as, the specific location within the brain where the particular sensory information terminates.

The second type of information transmitted is location. Location is defined as the place from where the sensory information is being sent, as well as, the specific nerve track(s) by which the information is being transmitted. For example, all sensory neurons can receive stimuli within their assigned area known as a receptive field. On average, a pain or temperature sensory nerve in the skin of the back has the receptive field of a diameter of ~7 cm or an area ~38.5 cm^2 (area = πr^2). If an individual is jabbed in any location within this receptor field, the sensory nerve will depolarize and a pain signal will be transmitted to the CNS. However, since this particular receptor field is so large, the individual will only realize that they have been jabbed somewhere within the sensory field, but the exact location remains unknown. In contrast, the pain or temperature sensors located in the skin of the fingertips will have a receptor field of ~0.1 cm in diameter or an area of ~0.008 cm^2. With a receptor field so small, an individual will know the exact location where they have been jabbed.

The third type of sensory information transmitted is intensity, which is defined as the strength of the signal being transmitted. The strength of the sensory signal is dependent on the strength of the stimulus detected. For example, a weak and short duration sound will only cause a few sensory nerves within the cochlea of the inner ear to depolarize. In contrast, if the sound is extremely loud and persistent, the number of sensory nerves depolarizing (firing) will increase dramatically.

The last type of sensory information is duration, which is the frequency of a nerve cell firing over a period of time. For example, if the stimulus is continuous and intense, the sensory nerve will continue to fire until they eventually adapt to the stimulus.

Sensory Receptors

Receptors are classified based on how quickly they adapt to stimuli. In humans, the speed by which these receptors adapt is generally classified into two groups: phasic and tonic.

Phasic receptors are commonly found within the skin. They are hair, pain, temperature, and pressure receptors. Additionally, phasic receptors are also found among the receptors for olfactory senses (i.e. the sense of smell), the functional element of a sense organ. These phasic receptors will generate a quick burst of action potentials when they are initially stimulated, but will quickly adapt to the stimuli. In order for these phasic receptors to fire again, the stimuli will have to be more intense. For example, when a person first gets into the bath tub, the initial contact of the hot water will send a quick burst of action potentials from the temperature receptors to the CNS. However, after a brief period of exposure, these temperature receptors will adapt to the heat of the water and the action potentials transmitted will gradually subside, until such time that the individual feels as though the water is no longer hot.

Tonic receptors include the receptors of the internal organs, as well as, proprioceptors found within the muscles and joints. Tonic receptors adapt slowly to stimuli, while generating a steady and continuous stream of actions potentials to the CNS. For example, in the case of proprioceptors, these pressure receptors will continuously generate action potentials to indicate the body's position, muscle tension and movement of the joints, while adapting slowly to the continuous stimuli.

Sensory Receptor Classification

All sensory receptors are transducers, which are a device that converts one form of energy into another (stimulus → action potential). Therefore, it is rational to classify sensory receptors based on their specific stimulus modality. For example, photoreceptors are activated by photonic energy. Thermoreceptors are activated by temperature changes. Nociceptors, also known as pain receptors, are activated by intense mechanical and/or thermal stimuli. Chemoreceptors are activated by different types of ligands, which include molecules that induce the senses of taste and smell, as well as, various compounds that are essential to the body's internal fluid composition. Lastly, mechanoreceptors are activated by physical forces such as vibrations, pressure, stretch and tension.

Sensory receptors can also be classified based on the origin of the stimulus. For example, exteroceptors respond to stimuli that originate outside of the body. These exteroceptors include those that respond to sound, vision, taste, smell, pressure, temperature, and pain. Interoceptors respond to stimuli that originate inside of the body. These interoceptors are generally found within internal organs and produce sensations of stretch, pressure, nausea, and visceral pain (visceral pain is described in the following section). Proprioceptors, a type of interoceptor, are also found within the body, but they are responsible for sensing the position and movement of the body. Proprioceptors are generally found in the muscles, tendons, and joint capsules.

Sensory receptors are also classified based on their distribution within the body. For example, general somesthetic (somatosensory) senses use receptors that are commonly found within the skin, viscera (internal organs), muscles, tendons, and joint capsules. These somesthetic receptors include thermoreceptors, nociceptors, chemoreceptors, and mechanoreceptors. Special senses use receptors that are located in the head and are innervated by cranial nerves. These special sense receptors include photoreceptors that are found in the retinas, mechanoreceptors that are located in the inner ear which enable an individual to maintain equilibrium and hearing, and chemoreceptors that are found in the tongue (taste) and mucosa of the nose (smell).

SIMPLE RECEPTORS

The human body contains millions of sensory receptors that are designed to respond to two general sensory categories: special and general somesthetic senses. The special senses can be found in the cranial region and include our ability to detect odors, tastes, and light. These special senses are grouped into sense organs, which are located within our nose, on our tongue, and in our eyes respectively. The general somesthetic senses include our ability to detect pH, temperature, pain, and pressure. The capability to perceive these senses are provided by chemoreceptors, thermoreceptors, nociceptors (pain receptors), and mechanoreceptors (pressure receptors) located in our blood vessels and skin, respectively.

Most of the receptors involved in the general somesthetic senses are categorized into two types of sensory receptors based on their structure: unencapsulated and encapsulated receptors. These receptors are involved in the sensation of temperature, pain, vibration, as well as, pressure. Both unencapsulated and encapsulated receptors generally exist within the skin throughout the human body, although they are at times concentrated in great numbers within the the skin of our hands, feet, and the genitalia. For example, sensory receptors involved in sensing temperature and pain are unencapsulated, meaning they have free dendritic endings modified to detect stimuli. Temperature and pain receptors are phasic receptors, although pain receptors adapt much more slowly when compared to temperature receptors (Figure 1).

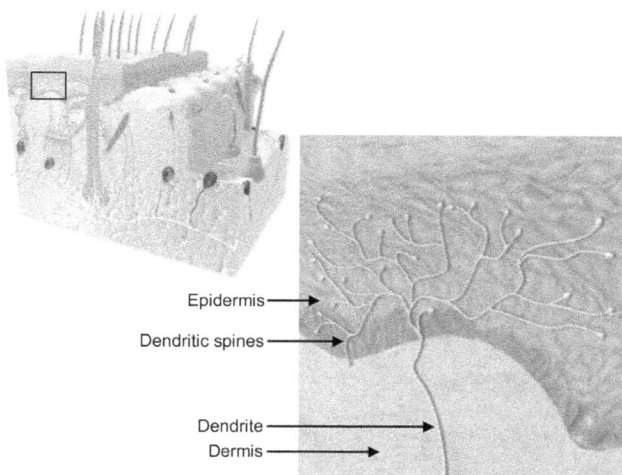

Figure 1: Illustration of an unencapsulated sensory receptor. Please note that the dendritic spines extend into the epidermis and they increase the surface area or the receptor field of this sensory nerve

Unencapsulated Thermoreceptors

One of the many sensory modalities of the skin is thermosensation. The ability to sense temperature is afforded to the body through the actions of unencapsulated phasic thermoreceptors that generate action potentials based on specific ranges in temperature (Figure 1). This capability provides the human body with the ability to maintain the body at an optimal working temperature, as well as, allowing the body to detect potentially dangerous thermal stimuli that may pose a threat to the integument and other systems.

Fluctuations in temperature cause different protein channels on the thermoreceptors to open. The opening of the different channels allow both sodium and in lesser amounts, calcium, to enter the neuroplasm, thereby producing a local potential. If the local potential manages to reach the set threshold, an action potential will be generated. Repolarization of these fibers is achieved by allowing potassium to leave the neuroplasm, while

the restoration of the membrane potential is accomplished through active transport proteins, such as, sodium, potassium, and calcium pumps.

Presently, there are 28 different temperature-sensitive channels identified and they are divided into cold-, warm-, and heat-sensitive channels. Please note that due to the limited scope of this text, only a few selected thermosensitive channels will be listed.

Cold-sensitive channels include melastatin transient receptor potential (TRPM) 8 and transient receptor potential ankyrin (TRPA) 1. TRPM8 ion channels are activated by temperatures equal or below 26°C. These thermosensitive channels are found on the dendritic ends of sensory nerves in approximately 15% of neurons in the dorsal root ganglia. In addition to their temperature specific activation, TRPM8 can also be activated by ligands that induce the sensation of cold. For example, chemicals such as menthol, eucalyptol, and the synthetic amino acid based molecule called icilin, are all capable of activating this ion channel. TRPA1 ion channels are activated by temperatures equal to or below 17°C. Like TRPM8, TRPA1 is also activated by the ligand icilin, but are insensitive to both menthol and eucalypol. The sensation of cold may also simultaneously trigger the perception of pain. If temperatures drop equal to or below 10°C, pain fibers such as Aδ (A-delta) mechanothermal nociceptors and C fibers will be activated, producing distinct sensations such as pricking, burning, and aching.

Warm-sensitive channels include vanilloid transient receptor potential (TRPV) 4 and TRPV3. TRPV4 ion channels are activated at temperatures between ~24–34°C. The TRPV3 ion channels, on the other hand, are activated at temperatures between ~31 to 39°C. In addition, TRPV3 ion channels are activated by ligands such as camphor, which causes the sensation of warmth when applied to the human skin.

Heat-sensitive receptors include TRPV1 and TRPV2. TRPV1 ion channels are activated at temperatures below or equal to 43°C. In addition to heat, TRPV1 is also activated by capsaicin the active ingredient of chili peppers and acids. TRPV2 ion channels are activated at or above 52°C, but are nonresponsive to capsaicin or acids. The activation of TRPV2 ion channels could also simultaneously trigger the perception of pain. If temperatures rise to or above 52°C, both Aδ mechanothermal nociceptors and C fibers will be activated, producing distinct sensations such as pricking, burning and aching.

Unencapsulated Nociceptors

Pain is defined as physical discomfort or distress caused by a noxious stimulation, illness, or injury. Although highly unpleasant, pain is actually one of the most important senses that allow individuals to adapt and survive the environment. Pain enables individuals to sense damages and triggers evasive responses to avoid further injuries. In addition, pain caused by certain events also allows individuals to form memories, preventing similar events from happening in the future. The value of these noxious sensations is evident when examining individuals that have lost the ability to sense pain. For example, individuals suffering from leprosy (caused by the bacterium *Mycobacterium leprae*) or life-threatening cases of diabetes mellitus will demonstrate extensive sensory nerve neuropathy (severe sensory nerve damage or loss), and therefore sense little or no pain. The absence of pain allows these individuals to be ignorant of any injuries that they may have suffered, thereby neglecting those wounds. In time, these injuries may become so dangerously infected that they could develop into a life-threatening situation.

Nociceptors are unspecialized free nerve endings that initiate the sensation of pain. In general, there are two groups of nociceptors: Aδ and C fibers. Aδ fibers are mostly myelinated phasic receptors. Because of its myelination, these Aδ fibers are also referred to as fast pain receptors. Aδ fibers are divided into two subtypes. The first is the Aδ mechanosensitive nociceptors which respond to intense mechanical (e.g. bruises or lacerations) stimuli. The second, Aδ mechanothermal nociceptors, respond to both intense mechanical and thermal stimuli. Both subtypes of Aδ fibers are associated with fast (first) pain, which produces a feeling of sharp, stabbing pain that is highly localized. Generally, but not exclusively, fast pain is associated with superficial somatic pain, which usually involves any damage(s) done to the skin.

C fibers are the second type of pain receptors that respond to both intense mechanical and thermal stimuli. These fibers are unmyelinated and are referred to as slow pain receptors. C fibers are tonic receptors that are generally associated with, but not exclusively, to deep somatic and visceral pain. Deep somatic pains are long lasting deep aching, burning, or nauseating sensations that are less localized and are generally associated with muscles and joints. Visceral pains are dull aching non-localized sensations that are associated with internal stretching, chemical irritations, and ischemia (i.e. restriction in blood supply to tissues, causing a shortage of oxygen and nutrients). Often, visceral pains are also accompanied by the sensation of nausea.

Unencapsulated Pressure Receptors

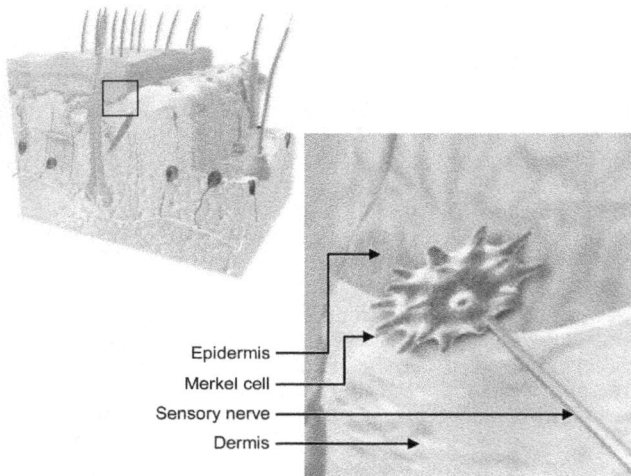

Figure 2: Illustration of Merkel receptors

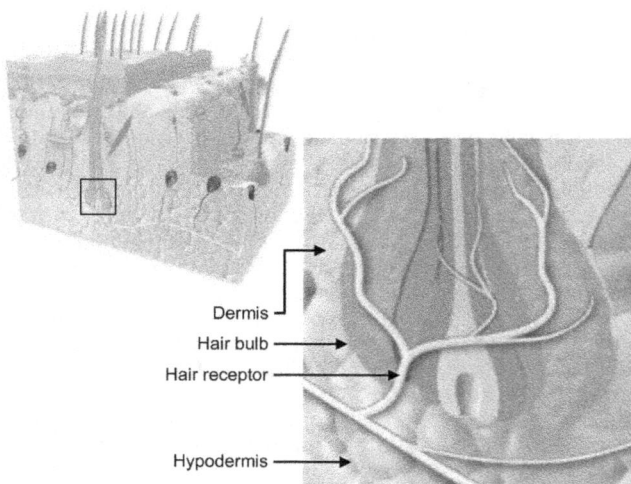

Figure 3: Illustration of a hair receptor

Tactile Merkel sensory receptors are an example of unencapsulated pressure sensors. Merkel receptors are tonic receptors that respond to light touch, texture, edges, and shapes.

These receptors are located in the base of the epidermis (basal layer) and are associated with Merkel cells, specialized sensory transducers that modify the stimulus received by the sensory nerve endings (Figure 2).

Hair receptors are found surrounding the bulb of hair follicles. These receptors are phasic mechanoreceptors and respond to any pressure that bends the hair. These receptors are activated immediately when a person places a piece of clothing over their body. However, since the hair receptors adapt quickly to the added pressure, the person will not be constantly annoyed by the constant chafing of their clothing (Figure 3).

Encapsulated Pressure Receptors

Encapsulated sensory receptors are constructed by wrapping the dendritic ends of the sensory nerve in glial cells or connective tissues. These encapsulated receptors generally consist of mechanoreceptors, which sense pressure. Most of the encapsulated sensory receptors are found in the dermal layer of the skin, scattered throughout the human body, but are especially concentrated in the hands, feet, and genitalia. Encapsulated sensory receptors includes: tactile (Meissner) corpuscles, Krause-end bulbs, lamellated (Pacinian) corpuscles and Ruffini corpuscles.

Tactile (Meissner) corpuscles are phasic mechanoreceptors that are commonly found in the dermal papillae of the skin and are concentrated in the fingertips, palms, eyelids, nipples, lips, and genital regions. They respond to texture, edges, and shapes. Tactile corpuscles are ovoid-shaped and consist of two to three nerve fibers embedded within a mass of modified Schwann cells (Figure 4).

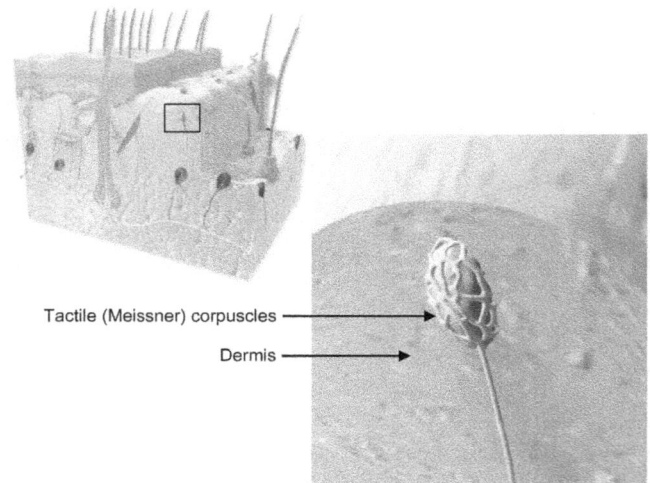

Figure 4: Illustration of a tactile (Meissner) corpuscles

Krause-end bulbs are functionally similar to the tactile (Meissner) corpuscles and are also a phasic receptor. These sensory receptors are ovoid in shape and are constructed with a single sensory nerve with numerous dendritic spines encapsulated within connective tissue. Krause-end bulbs are commonly found in the skin, lips, tongue, and conjunctiva (Figure 5).

Lamellated (Pacinian) corpuscles are phasic receptors that respond to deep pressure, stretch, tickle, and vibration. Comparatively, the Pacinian corpuscles are the largest of the encapsulated sensory receptors. On average this sensory receptor is one to two mm in diameter. Structurally, Pacinian corpuscles are constructed from a single dendrite surrounded first by modified Schwann cells in the inner layer, and outer concentric layers of fibroblasts cells. These receptors are generally found in the hands, feet, breasts, genitalia, peritoneum of bones, and joint capsules (Figure 6).

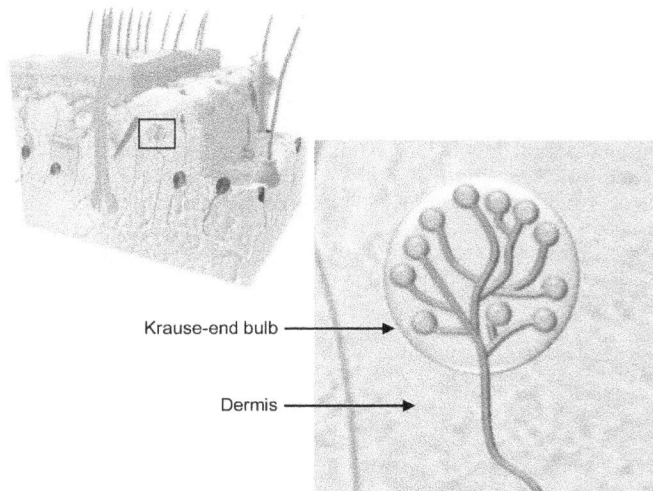

Figure 5: Illustration of a Krause-end bulb

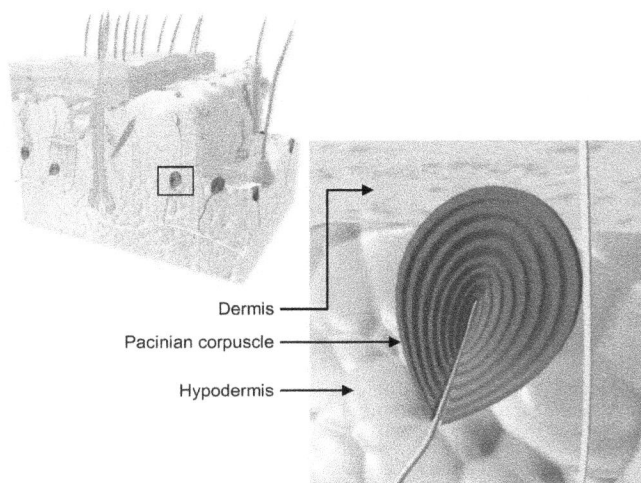

Figure 6: Illustration of a lamellated (Pacinian) corpuscle

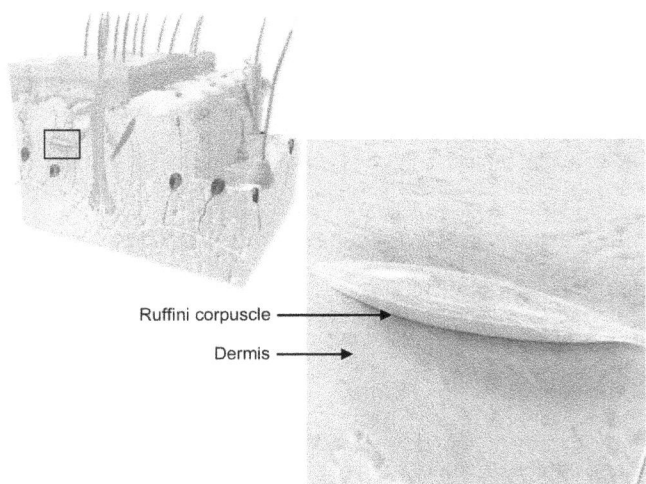

Figure 7: Illustration of a Ruffini corpuscle

Ruffini corpuscles are tonic receptors that respond to heavy touch, stretching of the skin, pressure, and joint movements. These sensory receptors are constructed with a few sensory nerve fibers surrounded by a connective tissue capsule. Ruffini corpuscles are commonly found in the dermis, subcutaneous

tissue, ligaments, tendons, and joint capsules (Figure 7).

Somatosensory Projection Pathways

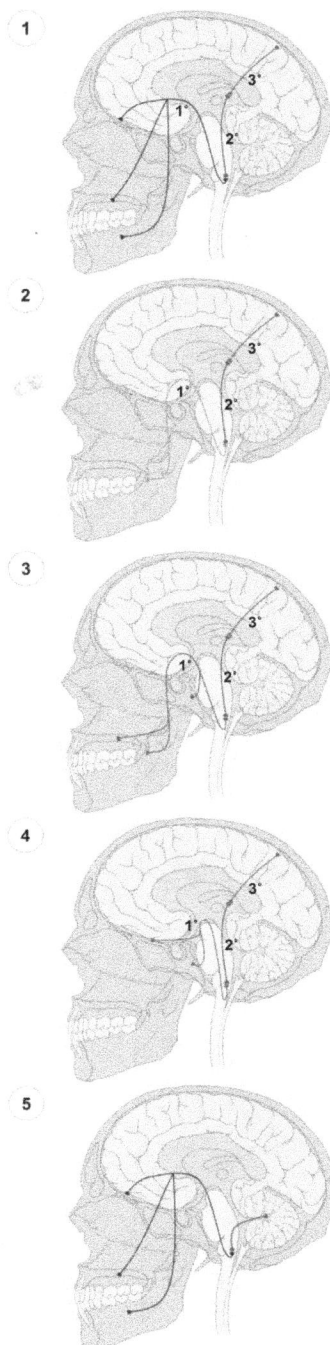

Figure 8: ① The trigeminal nerves (V) are subdivided into ophthalmic, maxillary and mandibular divisions. The sensory component of the 3 subdivisions transmits pressure, temperature, and pain sensations. Ophthalmic division sends sensory information from the surfaces of the eyes, tear glands, superior nasal mucosa, frontal, and ethmoid sinuses. Maxillary division transmits sensory information from the upper lip, teeth, gum, and the nasal mucosa, maxillary sinus and palate. Mandibular division transmits sensory information from the lower lip, teeth, gums, from the floor of the mouth and the anterior two-thirds of the tongue. ② The sensory component of the facial nerves (VII) transmits sensory information from the taste cells of the tongue as well as the skin of the outer ear. ③The sensory component of the glossopharyngeal nerves (IX) transmits the sensation of taste, touch, pressure, pain, temperature from the tongue and outer ear. In addition, its sensory component is responsible for monitoring blood pressure, as well as, respiration. ④The sensory component of the vagus nerves (X) transmits sensory signals from the back of ear, external auditory, canal, and dura matter to the brain. In addition, cranial nerve X also sends taste sensation and the feeling or hunger, fullness and abdominal discomfort. ⑤The proprioception component of the mandibular branch of the trigeminal nerve (V) transmits the sense of position in relation to balance form the muscles of mastication directly to the cerebellum

Somatosensory pathways are the way which sensory signals reach the brain. These signals arise from receptors located in the cranium, the peripheries, or the viscera. Somatosensory signals (e.g. pain, pressure, temperature, proprioception, fine, and course touch etc.) from the cranial region travels via the cranial nerves (e.g. trigeminal, facial, glossopharyngeal, and vagus nerve) to the pons and medulla oblongata. At the brain stem, these first order (1°) neurons (cranial nerves) synapse with the second order (2°) neurons. Most of these 2° neurons will decussate to the opposite side of the brain stem before proceeding to the contralateral thalamus. At the thalamus, the 2° neurons will synapse with third order (3°) neurons before being transmitted to the parietal lobe of the cerebral cortex for analysis and integration (Figure 8).

The only exception from the above transmission scheme is the proprioceptors from the mandibular branch, which transmit the sense of position in relation to balance from the muscles of mastication. These proprioceptive signals connect to the 2° neurons in the brainstem and directly carry the signals to the cerebellum for integration and analysis (Figure 8).

Somatosensory signals from the peripheries and the viscera also travel by way of three interconnected neurons called the 1°, 2°, and 3° order neurons. The 1° neurons is pseudounipolar in structure and has its cell body located within the dorsal root ganglia. The dendritic ends of the 1° neurons are stretched to the peripheral tissues, which are either encapsulated or unencapsulated sensory receptors.

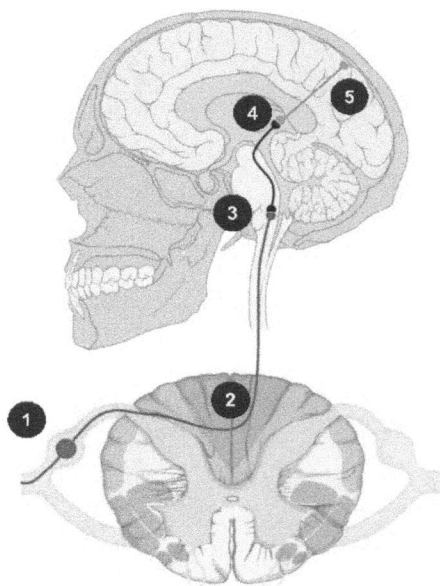

Figure 9: ① 1˚ neuron receives and transmits sensory information from the target tissue. The cell bodies of the 1° neurons are located in the dorsal root ganglia. ② Sensory signal will travel up to the brain stem via the fasiculus gracilis and the fasiculus cuneatus. ③ At the medulla oblongata, the 1° neurons will synapse with 2° neurons. The 2° neurons then decussate before proceeding to the thalamus. ④ At the thalamus the 2˚ neuron will synapse with 3° neuron which will ⑤ transmit the sensory information to the parietal lobe of the cerebral cortex

If the afferent is receiving and transmitting fine touch, proprioception or vibration sensory signals, the axon of the 1° neurons will be myelinated and fast in signal transmission. These myelinated 1° neurons will reach the brain stem via the Dorsal Column Medial Lemniscus System, which is composed of fasiculus gracilis and the fasiculus cuneatus nerve tracts. Both of these nerve tracts transmit sensory information related to tactile, vibration, proprioception, and movement to the brain. At the medulla oblongata, the 1° neurons synapse with multipolar 2° neurons. The 2° neurons then decussate across the midline to the opposite side of the medulla oblongata before proceeding

to the thalamus. At the thalamus the 2° neurons synapse with 3° multipolar neurons, which will transmit the sensory information to the parietal lobe of the cerebral cortex for analysis and integration (Figure 9).

The exception to this scheme are the signals transmitted for proprioception. Proprioception signals come from the 1° neurons that will synapse with the 2° neurons at the posterior horn. These 2° neurons will then decussate across the spinal column and travel up the spinal cord via the anterior spinocerebellar tracts, which are also located within the lateral funiculi. These nerve tracts convey unconscious proprioception signals from the muscles and joints of the legs directly to the cerebellum (Figure 10).

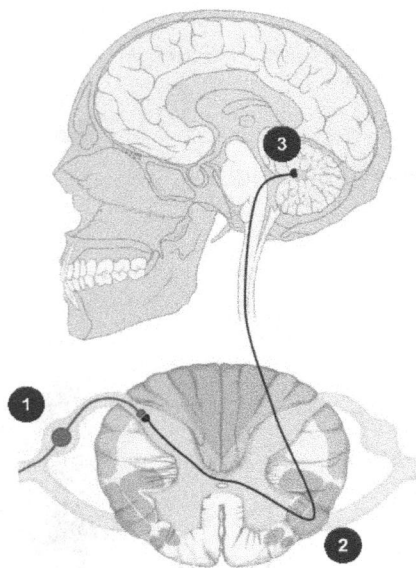

Figure 10: ① Proprioception signals from the 1° neurons could synapse with the 2° neurons at the posterior horn before ② decussate across the spinal column and travel up the spinal cord via the anterior spinocerebellar tracts. ③ These nerve tracts convey unconscious proprioception signals from muscles and joints of the legs directly to the cerebellum

If the afferent nerves are receiving and transmitting pain, temperature, and coarse touch, their 1° neurons will synapse with the 2° multipolar neurons in the posterior horn of the spinal cord (It is interesting to note that that the axons of 1° neurons for the course touch, as well as, Aδ fibers (fast pain fibers) will be myelinated, while the axons for the temperature and C-fibers (slow pain fibers) are unmyelinated). The 2° multipolar neurons will then decussate across to the opposite side of the spinal cord and join the lateral spinothalamic tract (located in the lateral funiculi), anterior spinothalamic tract (located in the anterior funiculi) and gracile fasiculus before venturing to the thalamus. At the thalamus, they will synapse with the 3° multipolar neurons, where most of the sensory information is sent to the parietal lobe of the cerebral cortex for analysis and integration. The spinothalamic tracts and the gracile fasiculus are responsible for making an injured individual aware of the damage (Figure 11).

In addition, pain signals may also travel up to the brainstem via the spinal reticular tract. Like the spinothalamic tract, 1° neurons transmitting pain signals will synapse with 2° multipolar neurons in the posterior horn of the spinal cord. The 2° multipolar neurons then decussate across to the opposite side of the spinal cord and join the spinoreticular tract before venturing to the reticular formation in the brain stem. At the reticular formation, the 2° multipolar neurons synapse with 3° multipolar neurons before the information is relayed to the limbic system and the hypothalamus. The pain signals transmitted by the spinoreticular tract activates intense arousal of the injured individuals, as well as, triggering emotional responses (e.g. nausea and fear), and behavior responses such as somatic reflexes (Figure 12).

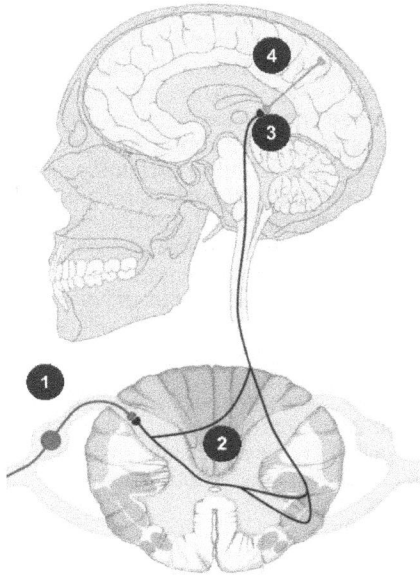

Figure 11: ① 1° neurons synapse with 2° multipolar neurons in the posterior horn of the spinal cord. ② The 2° multipolar neurons then decussate across to the opposite side of the spinal cord and join the lateral spinothalamic tract, anterior spinothalamic tract, and gracile fasciculus before venturing to the thalamus. ③ At the thalamus, they will synapse with the 3° multipolar neurons, where most of these sensory information is sent to the ④ parietal lobe of the cerebral cortex for analysis and integration

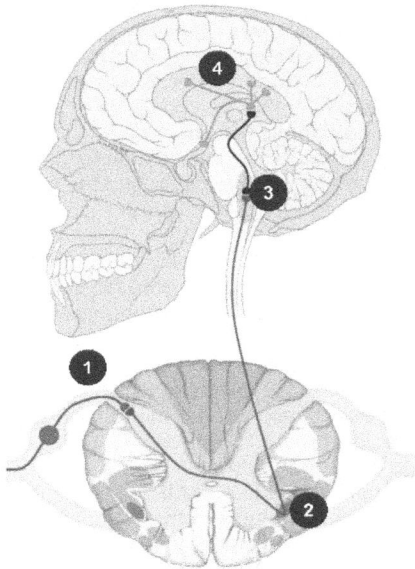

Figure 12: ① 1° neurons transmitting pain signals synapse with the 2° multipolar neurons in the posterior horn of the spinal cord. ② 2° multipolar neurons then decussate across to the opposite side of the spinal cord and join the spinoreticular tract before venturing to the reticular formation in the brain stem. ③ At the reticular formation, the 2° multipolar neurons synapse with 3° multipolar neurons before the information is relayed to the ④ limbic system and the hypothalamus

SENSE ORGANS

Taste

The interaction between chemical compounds and the taste cells of the taste buds is called gustation or taste. It has been estimated that there are approximately 10,000 taste buds found within the oral cavity of a young adult. These taste buds are mainly found on the tongue, but some are located on the interior

cheeks, soft palate, pharynx, and the epiglottis. However, as we age, these taste buds slowly deteriorate due to damage (i.e. physical or chemical) to the taste nerve fibers.

The tongue is divided into two sections via the terminal sulcus. The terminal sulcus is a V-shaped groove located on the surface of the tongue which marks the separation between the anterior oral and posterior pharyngeal regions of the oral cavity. The anterior region is comprised of approximately two-thirds of the tongue and is covered by small projections, the lingual papillae. The posterior one-third of the tongue lacks lingual papillae. This area is mainly composed of mucous secreting glands, as well as, lymphoid tissue (Figure 13).

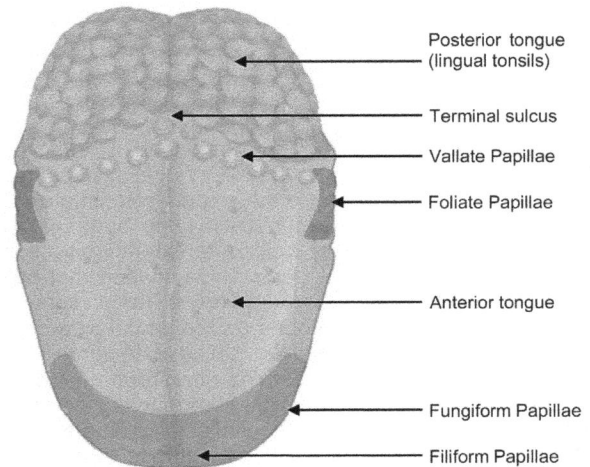

Figure 13: Illustration of a tongue. Please note that the v-shaped terminal sulcus separates the tongue into anterior and posterior portions. Please note that the shaded areas represents the location where the types of papillae are commonly found

There are many different types of lingual papillae and some of them contain taste buds. For example, the fungiform papillae, or mushroom-shaped protrusions, possesses approximately three taste buds per papilla. Fungiform papillae are widely distributed throughout the anterior portion of the tongue, but are mainly concentrated on the tip and lateral aspects. It has been reported that many of the taste buds contained within the fungiform papillae are sweet taste buds. The foliate papillae are the parallel ridges located on the side of the tongue, adjacent to the premolars and molars. These papillae are weakly developed in humans and most of the taste buds contained within these ridges deteriorate by the age of three. Foliate papillae are found on the lateral-posterior areas of the tongue (Figure 14).

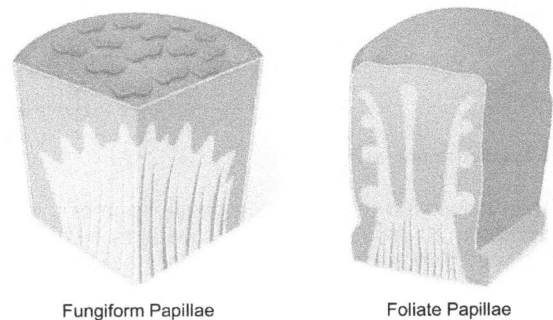

Figure 14: Illustration of fungiform and foliate papillae

Filiform papillae are tiny spike-shaped protrusions that are devoid of taste buds. These papillae are poorly developed in humans and are mainly located on the tip of the tongue. In animals, the filiform papillae are well distributed and are used for grooming purposes. The vallate (circumvallate) papillae are found towards the terminal sulcus of the tongue. These are large protrusions that contain approximately twelve taste buds per papilla. It has been reported that most of the taste buds contained within the vallate papillae are bitter taste buds (Figure 15).

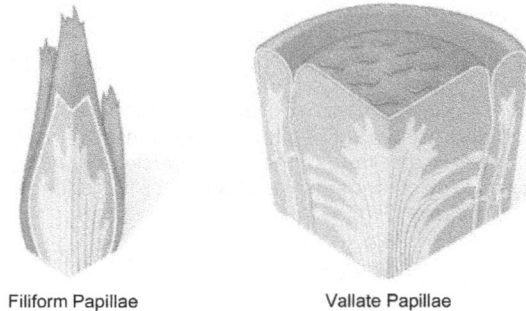

Filiform Papillae   Vallate Papillae

Figure 15: Illustration of filiform and vallate papillae

Taste buds are garlic-shaped structures composed of three types of modified epithelial cells: taste (gustatory) cells, supporting cells, and basal cells. On average, there are approximately 50 taste (gustatory) cells located within each taste bud. Each of these taste cells are connected to a dendrite extending from the taste nerve fiber. Taste cells are garlic clove-shaped that have numerous apical microvilli called taste hairs. Taste hairs are positioned in a circular pattern around on the apical side of the taste cell and protrude to an indentation, called taste pores. Taste hairs have two purposes: ① it contains numerous membrane receptors that bind to various chemical compounds from foods and ② increases the surface area available to interact with foods. Put simply, the taste cells are transducers that convert the energy generated from the binding of chemical compounds/molecules into action potentials. In turn, the generated action potentials cause the taste cells to release neurotransmitters, which will induce the depolarization of taste nerve fibers (Figure 16).

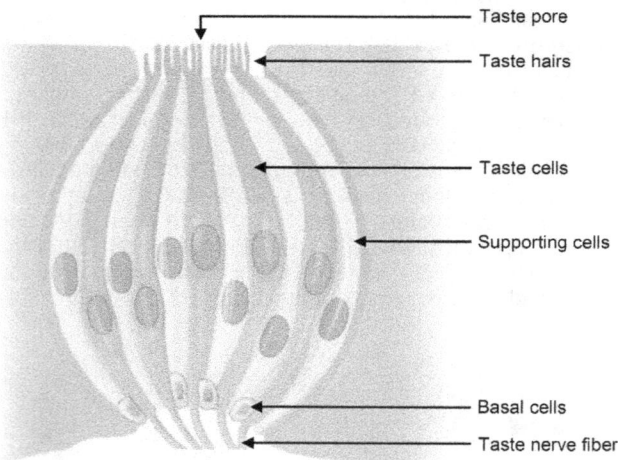

Taste pore
Taste hairs
Taste cells
Supporting cells
Basal cells
Taste nerve fiber

Figure 16: Illustration of a taste buds

It has been calculated that the average life span of a taste cell is approximately 10 days and they are constantly being replaced through mitotic division from adult stem cells called basal cells. The supporting cells are believed to aide in maintaining the ionic and nutrient concentrations, as well as, providing physical support and protection to the taste cells (Figure 16).

Taste Receptors

The receptors proteins that respond to tastes are grouped depending on the type of chemicals or compounds they can interact. An example of the transduction that takes place on the taste hairs, located on the apical domain of the taste cell, involves the activation of receptor proteins called gusducin by chemicals or compounds from foods. Once activated, gusducin, will in turn, trigger the activation of the G-protein-coupled receptors (GPCRs), also located on the taste hairs, which causes the α-, β- and γ-subunits of the GPCR to separate. The β-, and γ-dimers activate the α-subunit, causing the α-subunit to exchange GDP with GTP. The α subunit/GTP complex causes a conformational change and the subsequent activation of a membrane bounded enzyme called adenylate cyclase. Adenylate cyclase catalyzes the conversion of ATP into cyclic adenosine monophosphate (cAMP). cAMP is an intracellular (inside the cell) messenger, which sequentially activates a ligand-gated sodium (Na$^+$) ion channel and causes it to open. Na$^+$, which is kept in high concentrations in the extracellular matrix of the cell, will immediately flow down its concentration gradient and into the cytoplasm of the taste cell. This influx of Na$^+$ ions will generate a local potential. At the same time, cAMP will cause the closure of ligand-gated K$^+$ ion channels, preventing K$^+$ ions, which are kept in high concentrations in the cytoplasm, from flowing into the extracellular matrix. If the threshold is reached, an action potential will be generated and cause the release of a neurotransmitter called serotonin into the synaptic cleft located on the basal end of the taste cell and trigger a local/action potential of the connecting taste nerve fiber.

The depolarized taste nerve fibers from the taste buds will continue the transmission to the brain via three routes. The facial nerve (VII) will carry the sensory signals from the taste buds located in the anterior area of the tongue. The taste nerve fibers of the tongue connected to the taste buds that are located in the posterior area, before reaching the terminal sulcus, will send its message via the glossopharyngeal nerve (IX). Finally, the taste nerve fibers arising from the taste buds found on the cheeks (interior), soft palate, pharynx, and the epiglottis, will be transmitted by the vagus nerve (X). The taste nerve fibers, also referred to as 1° nerve fibers, found in all three cranial nerves, will converge at the solitary nucleus of the medulla oblongata and synapse with 2° neurons. The 2° neurons will transmit the sensory information to the amygdala and the hypothalamus, which activates autonomic reflexes such as salivation, gagging, and vomiting, or transmits the information to the thalamus, which will relay the information to the insula and the parietal lobe of the cerebral cortex.

The reversal of this reaction is initiated through the removal of the ligand (chemical or compound from food) binding to GPCRs. Once the ligand is removed, the α-subunit will hydrolyze GTP, releasing a phosphate molecule and energy. The release of the phosphate and energy molecule will sever the α-subunit's interaction with adenylate cyclase, which in turn, causes it to deactivate. The deactivation of the membrane bounded enzyme will discontinue the production of cAMP, which will result in the closure of the Na$^+$ channels and stop the influx of Na$^+$ ions. At the same time, the lack of cAMP will trigger the opening of the K$^+$ channels, which results in the K$^+$ ions flowing down their concentration gradient and out into the extracellular matrix. Once the taste cell reaches hyperpolarization, the sodium/potassium pump will activate and restore the taste cells' membrane potential.

Types of Taste Buds

In order to generate the sensation of taste, the molecules must first be dissolved in saliva and bind with the receptors located on the taste hairs, orientated around the taste pore. Substances that are placed on a dry tongue will generate little or no taste. Presently, there are six primary tastes sensations recognized: sour, salty, sweet, bitter, umami, and water. An individual's response to tastes is usually a combination of the sensation generated by the six primary tastes, which in turn generate thousands of distinct flavors. Each of these primary tastes is generated by distinct gustducin and GPCR complexes located within a specific population of taste buds. Nonetheless, it is interesting to note that under microscopic examination, all taste buds resemble one another, no matter the taste sensation that

216

they generate.

Sour tastes are generated by acids (substances that freely donate H^+ ions) and the intensity is proportional to the acidity of the substance. Put simply, the more acidic the substance, the more sour the sensation. Sour taste buds are generally found on the lateral margins of the tongue.

Salty tastes are usually elicited by ionized salts that dissociate in water to produce cations, such as Na^+, K^+, and Cl^-. Since these ionized salts are essential electrolytes, salty taste sensations are important for maintaining homeostasis. It is known that an electrolyte deficiency can cause cravings for salty foods. Salty taste buds are generally located on the lateral margins of the tongue.

Sweet tastes are caused by a wide variety of substances and generally, but not always, indicators of foods with high caloric values. Substance, such as carbohydrates (sugars), alcohols, aldehydes (C=O), ketones (2 CH_3-C=O), amino acids, and beryllium (Be), etc. will trigger sweet sensations. Sweet taste buds are mostly located in anterior region (particularly the tips) of the tongue, which trigger responses such as licking, salivation, and swallowing.

Comparatively, the bitter taste buds possess the lowest threshold of all the taste buds; therefore it could be easily stimulated to generate an action potential. Bitter tastes are triggered by wide variety of substances such as spoiled foods, but generally they are caused by long carbon chain organic substances containing nitrogen or alkaloids (e.g. nitrogenous ring compounds synthesized by plants). Bitter tastes trigger the autonomic reflex of gagging. Nearly all known poisons are alkaloid based compounds, therefore bitter tastes may be considered as a defense against ingesting poisonous foods. Bitter taste buds are generally located on the posterior regions (next to the terminal sulcus) of the tongue.

Umami is the Japanese word for meaty taste produced by amino acids such as aspartic and glutamic acids. Umami taste buds are scattered throughout the taste bud containing papillae (e.g. fungiform and vallate papillae in adults) of the tongue.

Water tastes are triggered by various water receptors located on the taste hairs of the taste buds. Research has shown that constant activation of the water receptors in the mouth will inhibit the production of antidiuretic hormone, also known as vasopressin, by the posterior pituitary.

Olfactory Sense

When compared to the sense of taste, the human sense of smell is highly sensitive. It has been estimated that humans can distinguish 2000 to 4000 different types of odors, although some individuals have the capability to distinguish over 10,000 types of smells. On average, women are more sensitive to smells than men. Research has found that women's sense of smell actually fluctuates from sensitive to very sensitive. For example, the sense of smell for women increases during ovulation, in comparison to other phases of the menstrual cycle, which may play a part in sexual selection.

Olfactory cells are responsible for distinguishing odors. These cells are located within the olfactory mucosa in the roof of the nasal passage, above the superior nasal concha and covering a portion of the cribriform plate, and nasal septum of each nasal fossa. The olfactory mucosa consists of approximately ten to twenty million olfactory cells, supporting cells, basal cells (the actual numbers fluctuates depending on the age of the individual) and the numerous mucous producing cells of the olfactory gland. Because of its constant exposure to the external environment, the life expectancy of an olfactory cell is approximately 60 days. Damaged or dying olfactory cells are replaced by new cells

created through the mitotic division of basal cells, also known as, adult olfactory stem cells. Supporting cells, on the other hand, are believed to aid in maintaining the ionic and nutrient concentrations as well as providing physical support and protection to the olfactory cells (Figure 17).

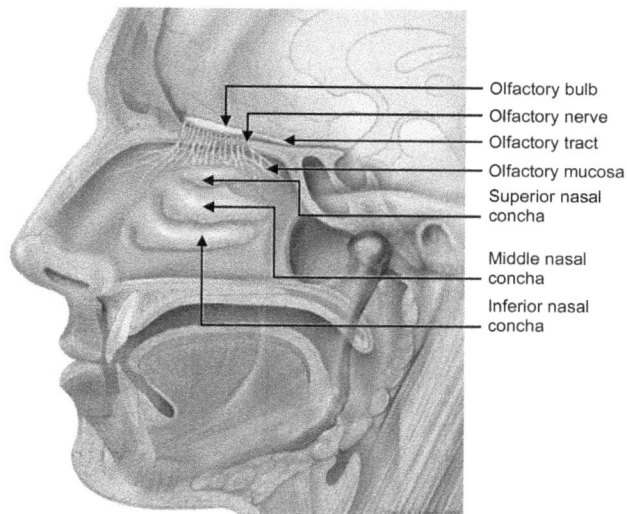

Figure 17: Illustration of the olfactory mucosa and olfactory nerve

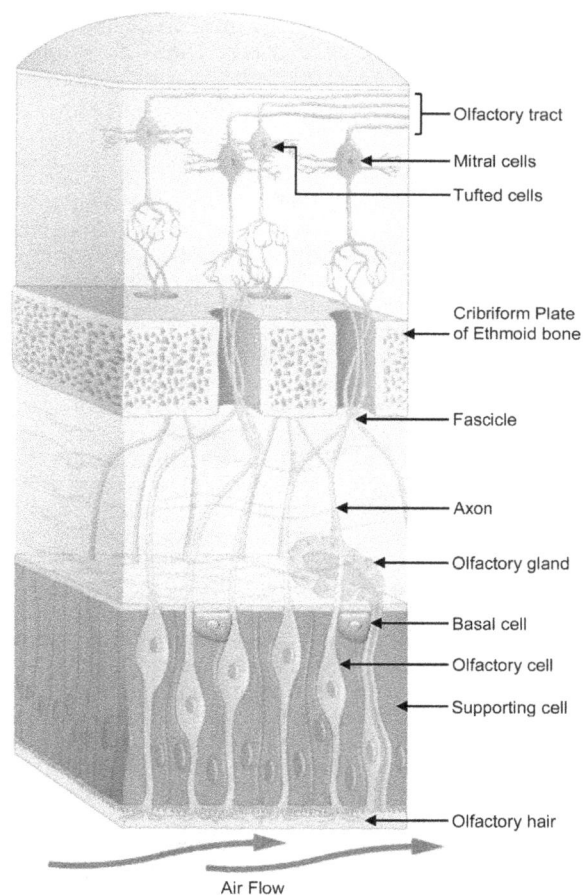

Figure 18: Illustration of the olfactory epithelium

Unlike the taste cells, which are epithelial in embryonic origin, olfactory cells are modified bipolar neurons. Olfactory cells are elongated with a large nucleus. The apical-ends of these cells are formed from modified dendrites, which form 10 to 20 olfactory hairs. The olfactory hairs increase the surface area of the olfactory cells and are the locations where the receptor proteins bind to odor molecules (also known as odorants) are found. The basal-end of the olfactory cells narrows to form an axon. Axons from groups of receptor cells will bundle together to form a fascicle. Various fascicles join together and leave the olfactory mucosa via the cribriform foramina of the ethmoid bone. As a group, the fascicles form the olfactory nerve (I), which then synapses with the dendrites of mitral and tufted cells within the pair of olfactory bulbs, located just beneath the frontal lobe of the brain. Mitral and tufted cells are commonly called projection neurons and are responsible for lateral inhibition, which aids in sharpening/refining the transmitting olfactory signals (Figure 18).

The locations where the axons of the fascicles of the olfactory nerve connect to the dendrites of the mitral and tufted cells are called glomeruli. Research shows that each glomerulus is organized to respond to a specific odor. For example, the odor for coffee is grouped into one glomerulus, while the odor for cream-filled donuts is grouped into another (Figure 18).

From the glomeruli the axons of the mitral and tufted cells form the olfactory tracts, which send their signals to the primary olfactory cortex, located in the inferior region of the temporal lobe. From the primary olfactory cortex, the signal could be transmitted to other secondary olfactory regions of the brain, such as the paleocortex of the insula (please see Chapter Eight for additional information), and the orbitofrontal cortex. The orbitofrontal cortex is located at the anterior-inferior area of the frontal lobe and receives sensory input from both taste and smell. This cerebral area is involved in the perception of flavor. In addition, other secondary areas where the primary olfactory cortex may transmit its signal include the hippocampus, amygdala, and hypothalamus (please see Chapter Eight for additional information). These areas provide the individual with visceral reactions and emotional responses to certain types of odor.

Odor Receptors

Odors are molecules that are carried by the inhaled air stream. There are two groups of odorants found in the atmosphere: hydrophilic and hydrophobic. The hydrophilic odorants are capable of diffusing through the mucous layer and bind to the receptor proteins located on the olfactory hairs. However, hydrophobic odorants require a transport molecule, called odorant-binding protein, located in the mucous to aid its passage through the water soluble layer, before binding to the receptor protein. Once an odor molecule binds to a receptor located on the olfactory hair, it activates a G-protein complex called G-protein coupled receptors (GPCR). Once activated, the α-, β-, and γ-subunits of the GPCR will separate. The β- and γ-dimer activates the α-subunit, causing the α-subunit to exchange GDP with GTP. The α subunit/GTP complex causes a conformational change and subsequent activation of a membrane bounded enzyme called adenylate cyclase. Adenylate cyclase catalyzes the conversion of ATP into cyclic adenosine monophosphate (cAMP). cAMP is an intracellular messenger and it activates ligand-gated sodium (Na^+) ion channels, causing the gates to open. Na^+, which is kept at high concentration in the extracellular matrix of the cell, will immediately flow down their concentration gradient and into the cytoplasm of the olfactory cell. The influx of Na^+ ions will generate a local potential. At the same time, cAMP will cause the closure of K^+ ion channels, preventing K^+ ions, which are kept in high concentrations in the cytoplasm, from escaping out into the extracellular matrix. If the threshold is reached, an action potential will be generated, which in turn, will cause the release of a neurotransmitter called glutamate into the synaptic cleft located on the basal end of the olfactory cells and trigger a/an local/action potential of the connecting mitral and tufted cells found in the olfactory bulbs.

The reversal of this reaction is initiated through the removal of the odorant binding to the GPCRs. Once the odorant is removed, the α-subunit will hydrolyze the energy molecule GTP, releasing a phosphate molecule and energy, thereby breaking the interaction with adenylate cyclase, and in turn causing it to deactivate. The

deactivation of the membrane bounded enzyme will discontinue the production of cAMP, which results in the closure of the Na^+ channels and stops the influx of Na^+ ions. At the same time, the lack of cAMP will trigger the opening of the K^+ channels, resulting in the flow of K^+ ions down their concentration gradient and out into the extracellular matrix. Once the olfactory cell reaches hyperpolarization, the sodium/potassium pump will activate and restore its membrane potential.

HEARING AND BALANCE

Anatomy of the Ear

The ear is separated into the outer, middle and inner ear. The outer and middle ears are areas where the sound is received and transmitted as vibrations to the inner ear. The inner ear, on the other hand, is where these vibrations are transduced into action potentials and are transmitted to the brain for integration and interpretation.

The outer (external) ear is composed of the auricle and the auditory canal. The auricle, also known as the pinna, is the fleshy visible portion of the external ear. Essentially, the auricle is a funnel that directs airborne vibrations (sound waves) into the auditory canal at optimal frequencies. The auricle is constructed of an inner layer of elastic cartilage surrounded by keratinized stratified squamous epithelium. In most animals, the auricles are controlled by muscles, which enable them to move and focus on a particular sound. However, the auricle of the human ear is immobile and relatively rudimentary by comparison (Figure 19).

The deepest depression in the auricle is called the concha, which leads to the external acoustic meatus, the opening of the auditory canal. The auditory canal is approximately three centimeters (cm) long in adults and is formed to conduct sound waves to the tympanic membrane. The auditory canal is supported by a layer of fibrocartilage at its opening and by the temporal bone of the skull for the remainder of its length (Figure 19). The auditory canal is lined by keratinized stratified squamous epithelium with ceruminous and sebaceous glands. The ceruminous glands are apocrine glands that lose part of their cell during secretion, meaning that their released products possess both cellular product and cellular fragments. Ceruminous glands produce ear wax that serves to keep the canal clean by trapping debris and pushes out dead cells that have collected within. Sebaceous glands, on the other hand are a type of holocrine gland that disintegrates completely during secretion. Sebaceous glands produce oily secretions that keep the epithelial lining of the auditory canal moist, as well as, forming a hydrophobic layer to prevent water seepage.

The middle ear is located in the tympanic cavity within the temporal bone of the skull. The tympanic cavity begins with the tympanic membrane (ear drum) and marks the end of the external ear and the beginning of the middle ear. The tympanic membrane is concave on this outer surface and is approximately one cm in diameter. The tympanic membrane is innervated with sensory branch of the vagus and the trigeminal nerves and responds to the vibration produced by sound waves. Behind the tympanic membrane is an air-filled chamber two to three cm in width. The air enters the middle ear via the auditory tube, also known as the eustachian tube, which is connected to the nasopharynx. The auditory tube is normally flattened and closed, but may be opened through yawning or swallowing. The opening of the auditory tube allows the pressure within the middle ear to equalize with the barometric pressure of the external environment (Figure 19).

Within the air-filled chamber are three smallest bones in the human body called the auditory ossicles. These auditory ossicles are the malleus (hammer), incus (anvil), and the stapes (stirrup). The malleus is attached to the inner surface of the tympanic membrane, as well as, being suspended by ligaments via the

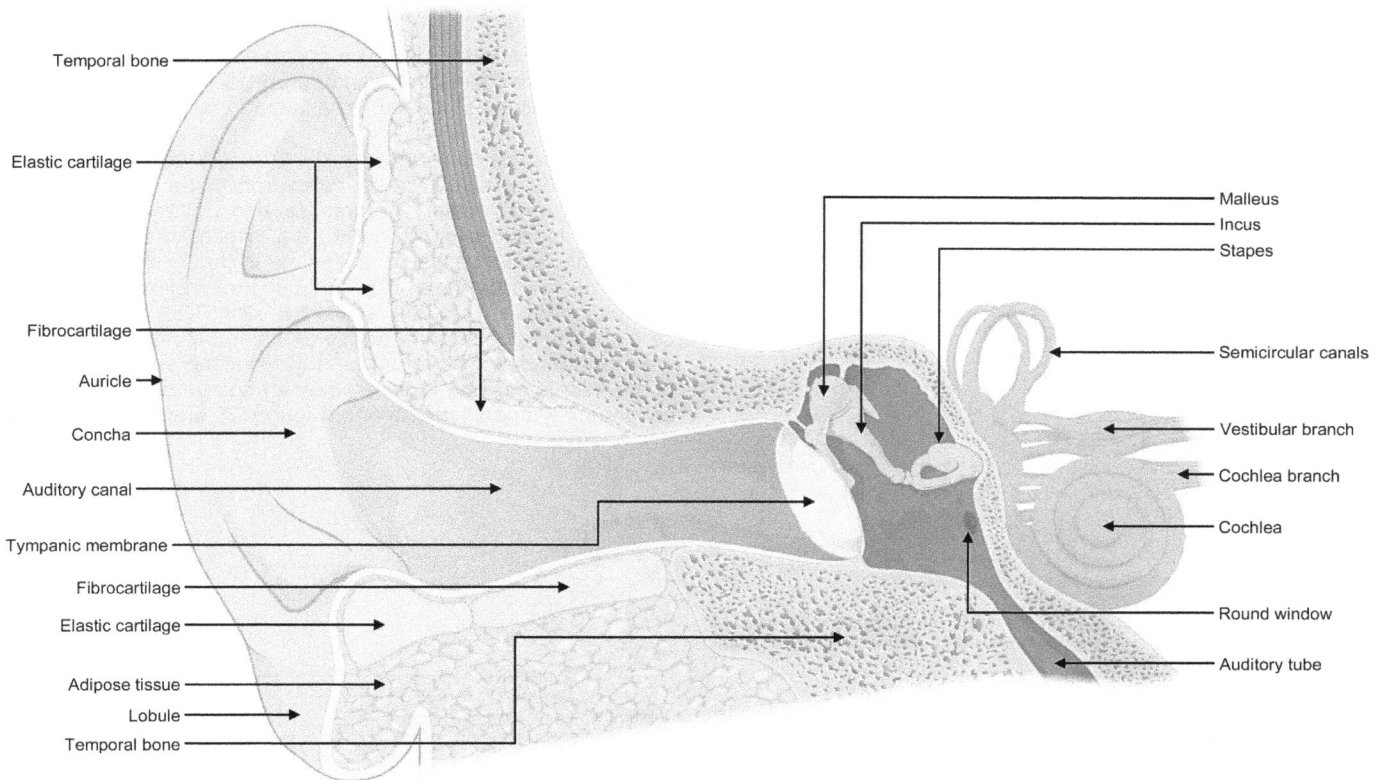

Figure 19: Illustration of the internal anatomy of the ear

walls of the cavity. The malleus articulates with the incus, which is also suspended by ligaments extending from the wall of the cavity. The incus, in turn, articulates with the stapes. The base of the stapes is attached by ligaments to an opening at the inner ear called the oval window (Figure 19). There are two involuntary striated muscles found within the cavity of the middle ear; the tensor tympani muscle, which inserts on the malleus, and the stapedius muscle, which inserts on the stapes. The contraction of these muscles are primarily activated by acoustic stimulation but they may also be triggered to contract via tactile stimulation (physical touching) of the auricle, change in the posture of the cranium, as well as, anticipation of speech or sound. These muscles act to protect the middle ear from auditory damage by lessening the transfer of energy to the inner ear through a process called tympanic reflex. For example, in anticipation of a loud noise, the tensor tympani muscle contracts and pull the tympanic membrane inward. The pull of the tensor tympani muscle causes the tympanic membrane to stiffen. At the same time, the stapedius muscle will also contact, which will reduce the movement of stapes. Together these contractions of the muscles will reduce the amount of energy reaching the inner ear.

The oval window marks the end of the middle ear and the beginning of the inner ear. The inner ear is a maze of channels called the bony labyrinth formed within the temporal bone. Lining the bony labyrinth is a fleshy tube called the membranous labyrinth. In between the bony and membranous labyrinth is a layer of fluid called the perilymph. The composition of the perilymph is similar to that of the cerebral spinal fluid. Within the membranous labyrinth is a fluid called the endolymph, which has a similar composition to filtered blood plasma.

The bony labyrinth is divided into three subsections: vestibule, semicircular canals, and the cochlea. It is understood that the vestibule, the semicircular canals, and the membranous labyrinth contained within are involved in maintaining balance, while the membranous labyrinth contained within the cochlea is involved in hearing (Figure 20).

The vestibule is a thin layer of bone that contains a pair of

endolymph-filled membranous sacs (chambers) called the saccule (anterior portion) and utricle (posterior portion). Both the saccule and utricle are a part of the membranous labyrinth. The saccule and utricle are connected by a slender passageway called the endolymphatic duct and they contain oval patches of 2 to 3 mm diameter of hair cells and supporting cells, called the maculae. The maculae in the saccule (macula sacculi) are aligned vertically on the wall of this anterior membranous sac, while the maculae in the utricle (macula utriculi) lie horizontally on the floor of this posterior membranous sac. Each of the hair cells found within the maculae possess 40 to 70 stereocilia and only a single kinocilium. The apical edges of the stereocilia and the kinocilium are embedded in a gelatinous structure called the otolithic membrane, also known as the otolith. Superior to the otolithic membrane is a layer of calcium carbonate crystals called the statoconia, also known as otoconia (Figure 21).

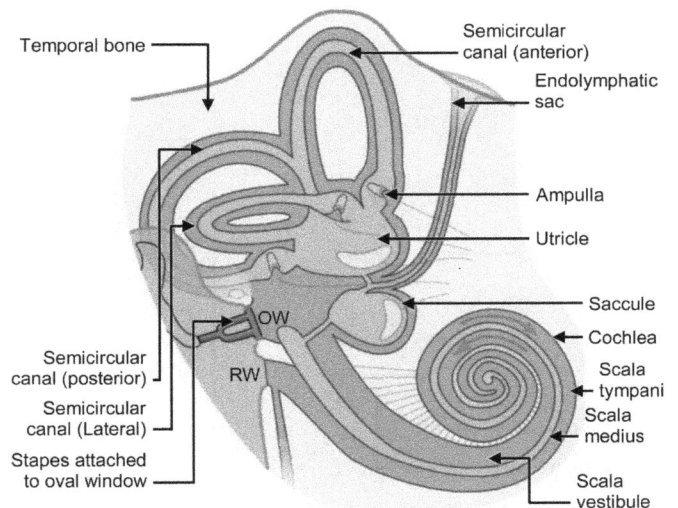

Figure 20: Anatomy of the membranous labyrinth. Please note that the abbreviation OW stands for oval window, while RW stands for round window

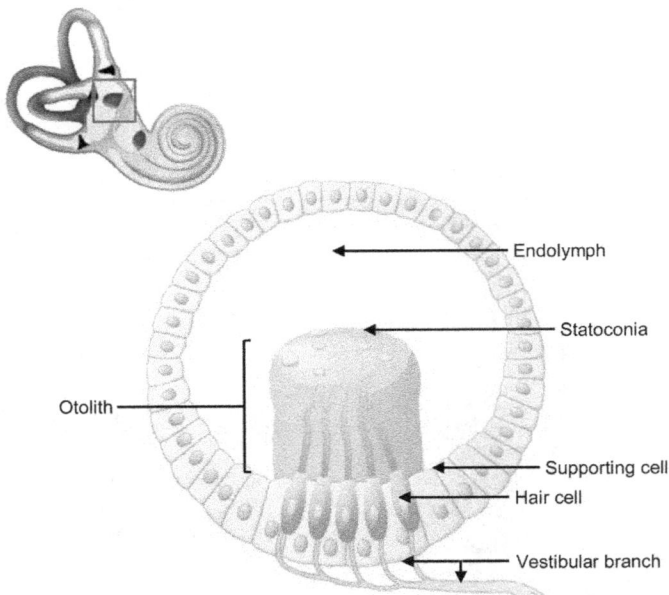

Figure 21: Cross section of the utricle and an illustration of macula. Please note that the hair cells are linked to the vestibular branch of the vestiblocochlear nerve (VIII)

The semicircular canals are a thin layer of bone that shelters and protects the semicircular ducts. The semicircular ducts, which are segments of the membranous labyrinth, are divided into anterior, posterior, and lateral semicircular ducts. The anterior and posterior semicircular ducts are positioned at right angles (90°) to one another, while the lateral semicircular duct is at an angle of approximately 30° from the horizontal/leveled plane. All three semicircular ducts are continuous with the utricle (enclosed within the vestibule) and are filled with endolymph. At the area where the semicircular ducts connect to the utricle, the semicircular ducts dilate to form a sac-like structure called the ampulla. The ampulla is an expanded region that contains a raised mass of hair cells and supporting cells called the crista ampullaris. Each of the hair cells found within the ampulla possesses 40 to 70 stereocilia and only a single kinocilium located on its apical domain (please examine the section titled Hair Cells for more information). The hair cells, as well as, their stereocilia and the kinocilium, are embedded in a gelatinous structure called the cupula. Please note that the hair cells found in the crista ampullaris are of similar conformation and function as those found in the maculae (Figure 22).

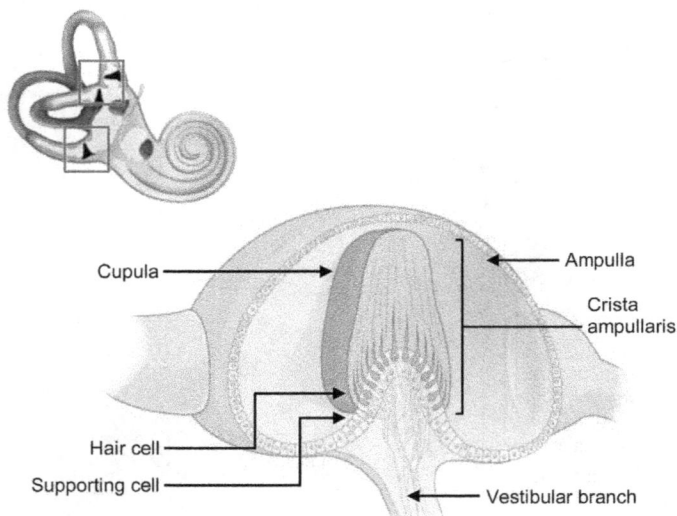

Figure 22: Illustration of an ampulla. Please note that the sensory nerves composing the vestibular branch will merge with the cochlear branch to from the vestibulocochlear nerve (VIII)

The third part of the bony labyrinth is the organ associated with hearing called the cochlea, a coiled fleshy tube that is formed anteriorly of the vestibule. The cochlea is a snail shell-shaped structure that is coiled around a spongy bone, the modiolus. The modiolus provides some structural support for the cochlea.

Within the cochlea there are three ducts separated by membranes: scala vestibule, scala media (also known as cochlear duct) and the scala tympani. The first (superior) chamber, the scala vestibule, is filled with perilymph (e.g. similar to cerebral spinal fluid). The scala vestibule arises near the oval window and spirals up towards the tip of the snail shell before transitioning (via helicotrema) into the scala tympani. The scala tympani continues to spiral in the reverse direction before terminating at the round window. The round window faces the air-filled chamber of the middle ear and is covered by a thin membrane called the secondary tympanic membrane (Figures 20 and 23).

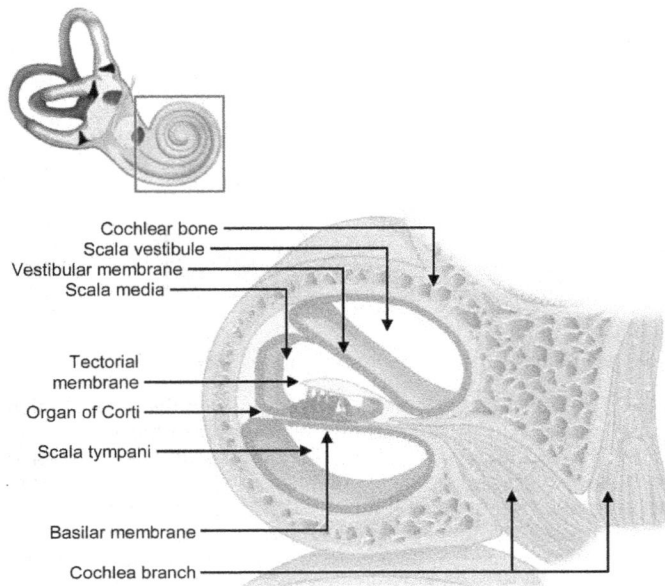

Figure 23: Illustration of a cross section of cochlea. Please note that the cochlea branch will merge with the vestibular branch to form the vestibulocochlear nerve (VIII)

The middle chamber is called the scala media, also known as the cochlear duct, which is filled with endolymph (similar to blood plasma). The scala media (cochlear duct) is a part of the membranous labyrinth and is separated from the scala vestibule by a thin membrane called the vestibular membrane. Within the scala media is a raised structure containing hair cells and supporting cells called the organ of Corti. The organ of Corti is situated on top of the basilar membrane, described below. On the apical side of the hair cells are numerous membranous protrusions called stereocilia. Situated superiorly above the stereocilia is a gelatinous layer called the tectorial membrane. The hair cells of the organ of Corti are subdivided into inner hair cells (IHC) and outer hair cells (OHC). The inner hair cells are arranged in a row along the medial side of the organ of Corti (facing the modiolus). Each of the IHC possesses a cluster of 50 to 60 stereocilia of either tall or short length. Please note that the apical tips of IHC stereocilia are not embedded in the tectorial membrane. The OHC are arranged in a row along the lateral side of the organ of Corti (away from the modiolus). Each of the OHC possesses a cluster of approximately 100 stereocilia with their apical edges embedded in the tectorial membrane.

The inferior chamber is called the scala tympani which is filled with perilymph (e.g. similar to cerebral spinal fluid). The scala tympani is separated from the scala media by a thicker membrane called the basilar membrane. As mentioned previously, the scala vestibule and scala tympani are continuous with one another through a narrow channel called the helicotrema.

Hair Cells

The hair cells are columnar in shape and are modified epithelial cells that are capable of serving as a transducer to transform vibrational energy into action potentials. Each hair cell possesses a bundle of apical cytoplasmic extensions called stereocilium or kinocilium. Depending on the location, the population of stereocilia varies from 50 to 60 in the inner hair cells of the organ of Corti (scala media of the cochlear duct), up to 100 stereocilia in the outer hair cells of the organ of Corti, 40 to 70 in the ampulla of the semicircular canal, 40 to 70 in the macula of the saccule and utricle. Nonetheless, the populations of kinocilia are rather consistent with only a single kinocilium per hair cell existing among the populations of stereocilia in almost all locations. Please note that in the cochlea of humans and other mammals, the kinocilium will develop along with stereocilia, but will recede shortly after birth.

Regardless of their names, kinocilium is the only true ciliary structure found within the inner ear. The kinocilium is constructed out of 9+2 microtubule doublet structure, similar to that of cilia and flagella. On the other hand, stereocilia are constructed from actin cytoskeletal structures, similar in construction to microvilli.

In the organ of Corti in the cochlea, the stereocilia are arranged in a graded fashion, based upon their height, from the shortest to the tallest. In the macula and the Crista ampullaris, the stereocilia are also arranged in a graded fashion based upon their height, from the shortest to the tallest apical protrusion, which is a kinocilium.

In the cochlea, the stereocilia (from the shortest to the tallest) are connected by a linear protein structure called the tip-link. The tip-links are constructed out of cadherin-23 and protocadherin-15 proteins and are anchored at the apex of each stereocilium. In the vestibular organs, the stereocilia are also interconnected by tip-links, however, the connection between the last stereocilium and the kinocilium are formed via a kinocilial link. Kinocilial links are formed between the apical-lateral side of the stereocilium and kinocilium. The kinocilial link is believed to be constructed out of the same protein structures as tip-links (Figure 24).

The availability of tip-links allows the stereocilia to bend in unison either towards the tallest stereocilia (in the cochlear) or towards the kinocilium (in the vestibular organs), when pressure is exerted. It is known that the displacement of the hair bundle toward the kinocilium or the tallest stereocilia depolarizes the hair cell. In contrast, movements parallel to this plane toward the shortest stereocilia causes hyperpolarization, while movements perpendicular to the plane of symmetry have no effect on the hair cell's membrane potential.

The resting potential (polarized) of the hair cells is approximately -45 to -60 mV and it has been calculated that the threshold of the hair cells are approximately 0.3 nm. Put simply, if the stereocilia moves 0.3 nm towards the kinocilium or the tallest stereocilia, the mechanically gated potassium (K^+) channels will open, allowing K^+ cations, which are kept at high concentrations in the extracellular matrix, to flow into the hair cell. The influx of K^+ cations depolarizes the hair cell, which in turn causes the opening of voltage gated K^+ and calcium (Ca^{2+}) channels. The opening of the voltage gated potassium channels will allow additional K^+ cations to flow into the cell, while the opening of the voltage gated calcium channels will cause Ca^{2+} cations, which are also kept in high concentration in the extracellular matrix, to flow into the hair cell, thereby generating an action potential and the subsequent release of neurotransmitter(s). The release of neurotransmitter (s) will excite the postsynaptic nerve (vestibular or cochlear), which relays the neuronal transmission to the brain. Presently, the exact neurotransmitter(s) secreted by hair cells are unknown although numerous candidates, such as acetylcholine,

nitric oxide, gamma-aminobutyric acid (GABA), dopamine, encephalin, and dynorphin have been proposed (Figure 25).

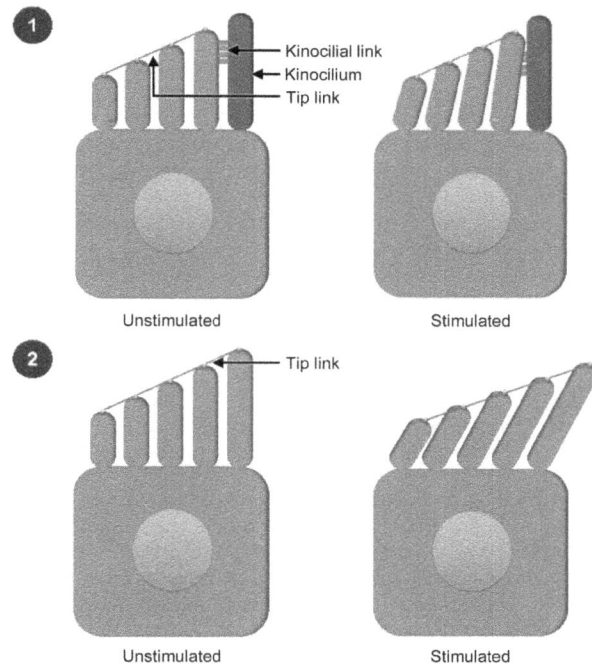

Figure 24: Illustration of ① stereocilia and kinocilium interaction in the vestibular organs and ② stereocilia and stereocilia interactions in the cochlear. Please note that the apical protrusions (stereocilia and kinocilium) are arranged in a graded fashion based upon their height

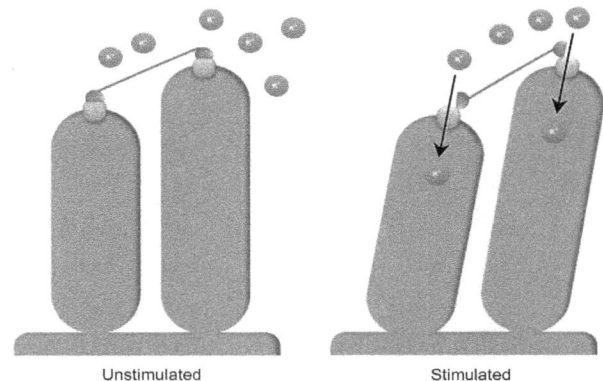

Figure 25: The opening of the K^+ mechanically gated channels via the movement of stereocilia. Please note that the initial opening of the K^+ channel will trigger the subsequent opening of the voltage gated Ca^{2+} and K^+ channels (not shown)

It is interesting to note that K^+ cations serve to depolarize and repolarize the hair cell. This unique strategy relies greatly on the epithelial origin of hair cells. Just like any other epithelial cells, the basal and apical surfaces of the hair cell are separated by tight junctions that allow for the maintenance of separate extracellular environments. For example, with the hair cells of the organ of Corti, the apical domain of the hair cell is exposed to the K^+ rich, Na^+ poor endolymph, while the basal domain is exposed to the K^+ poor and Na^+ rich perilymph. Once the threshold is reached, the K^+ cations from the endolymph will begin to flow into the apical domain of the hair cell causing depolarization. This, in turn, opens the voltage-gated K^+ and Ca^{2+} channels throughout the entire cell. The opening of the K^+ channels in the basal domain (which is exposed to the K^+ poor and Na^+ rich perilymph) favors K^+ efflux, favoring repolarization milliseconds later.

Physiology of Balance

As we have mentioned previously, the maculae found with the saccule (macula sacculi) and utricle (macula utriculi) possess hair cells. Each of these hair cells has 40 to 70 stereocilia and only a single kinocilium. The apical edges of the stereocilia and the kinocilium are embedded in a gelatinous structure called the otolithic membrane, which is weighed down by a layer of calcium carbonate crystals, called the otolith. The purpose of the otolith is to add weight and enhance the senses of motion and gravity. Please note that the only time that an action potential will be generated is when the stereocilia are bent towards the kinocilium; bending away from the kinocilium will result in hyperpolarization and will generate no action potentials. Please also recall that macula sacculi are aligned vertically on the wall of this anterior membranous sac, while the macula utriculi lie horizontally on the floor of this posterior membranous sac.

When the head is erect, the otolith membrane and the stateconia are situated directly above the hair cells of the macula utriculi. Please note that the macula utriculi are lying horizontally on the floor of this posterior membranous sac in both ears therefore no action potential will be generated. Additionally, with the head erect, the stereocilia of the macula sacculi found in both ears, which are aligned vertically on the wall of this anterior membranous sac, are minimally pulled downward, therefore only a fraction of action potentials are generated. Since only minimal action potentials are generated, the message transmitted to the brain indicates that the head is in the erect position. However, when the head is tilted it will result in a combination of movement of the otolith membrane and the stateconia, which in turn will move the stereocilia in both the macula utriculi and macula sacculi of both ears. The combination of the action potentials generated will inform the brain of the exact position of the head. Couple this information with the sensory inputs from the eyes and the proprioceptors of the neck, the brain will determine if only the head is tilting or if the entire body is at an angle (Figure 26).

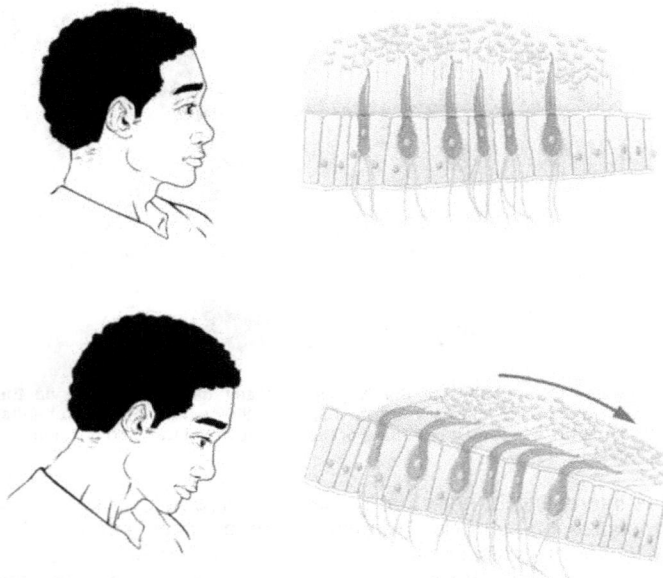

Figure 26: Illustration of the otolith membrane and the stateconia of the hair cells in the macula utriculi. Please note when the the head tilts forward, the otolith membrane and stateconia will roll forward pulling the stereocilia towards the kinocilium to generate an action potential

The macula utriculi is also important in allowing an individual to sense linear acceleration. For example, any forward movement will result in the otolithic membrane and the otoliths bending the stereocilia posteriorly, thereby generating a wave of action potential informing the brain that a movement forward is in progress. If an individual stops suddenly, the otolith membrane and the otoliths will reverse their direction and pull the stereocilia forward. This forward bending of the stereocilia will result in another wave of action potentials telling the brain that a change in linear velocity has taken place. In contrast, the macula sacculi are important for an individual's sense of vertical acceleration.

If a person is moving upward, the otolith membrane and the otoliths will bend downward in the extreme position, thereby sending waves of action potentials to the brain informing it that an upward movement is in progress. However, if the upward movement suddenly stops, the otolith membrane and the otoliths will be pulled upwards (bending the stereocilia in the opposite direction). This reaction will generate another wave of action potentials to inform the brain that the upward movement was suddenly discontinued and that a change in vertical acceleration has taken place.

The hair cells in the semicircular canal (found in the ampulla) respond to the rotational movements of the head. For example, when an individual's head rotates, the endolymph moves within the semicircular canal. This movement of the endolymph causes equal directional movements of the cupula (gel like structure) and distorts the stereocilia. Again, if the stereocilia are distorted towards the kinocilium an action potential is formed, but if the stereocilia are distorted away from the kinocilium, no action potential will be generated. Please note that when the endolymph stops moving, the stereocilia will snap back to their original state because of their elastic nature. The plane of movement is directly related to each of the three semicircular canals. A horizontal movement such as shaking your head results in the hair cells of the lateral semicircular ducts to be activated. A nod activates the hair cells of the anterior semicircular ducts. Finally, a tilt of the cranium results in the activation of the hair cells of the posterior semicircular duct (Figure 27).

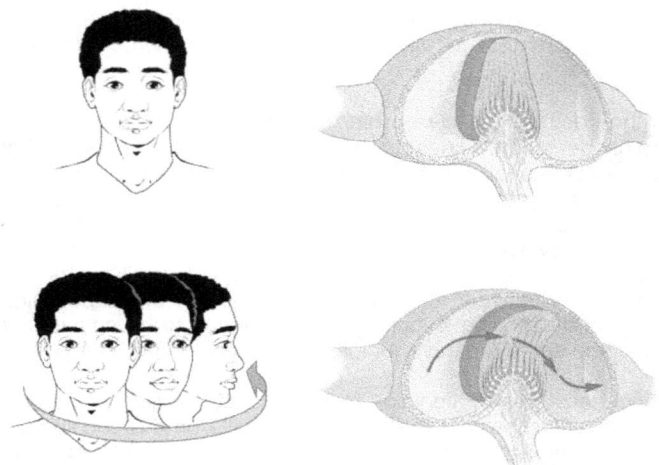

Figure 27: Illustration of the ampulla within the semicircular canal

All the action potentials generated form the hair cells in the vestibular (balance) structures are carried away by the vestibular branch. The vestibular branch will merge with the cochlear branch to form cranial nerve VIII, the vestibulochochlear nerve. Cranial nerve VIII will proceed from the inner ear to the four vestibular nuclei located on each side of the pons and medulla oblongata. Please note the vestibular nuclei are constantly communicating with one another and are thereby capable of determining the exact position and movement of the body. Subsequently, the vestibular nuclei will transmit signals to other parts of the nervous system. For example, the vestibular nuclei will provide input to the cerebellum, which in turn will properly maintain head movements, as well as, muscle tone and posture. The vestibular nuclei will also provide input to the nuclei of the oculomotor (III), trochlear (IV), and abduens nerves (VI), which will produce proper eye movements in an effort to compensate for the movement of the head. The vestibular nuclei will also provide input to the reticular formation, which will alter the respiratory rate and blood pressure due to the change in posture. Additionally, the vestibular nuclei will also provide inputs to the vestibulospinal tracts to control the extensor muscles so that quick adjustments could be made to maintain balance. Finally, the vestibular nuclei will send input to the thalamus, which will then transmit the information to the postcentral gyrus of the cerebral cortex so that the individual will be consciously aware of the position and movement of the body (Figure 28).

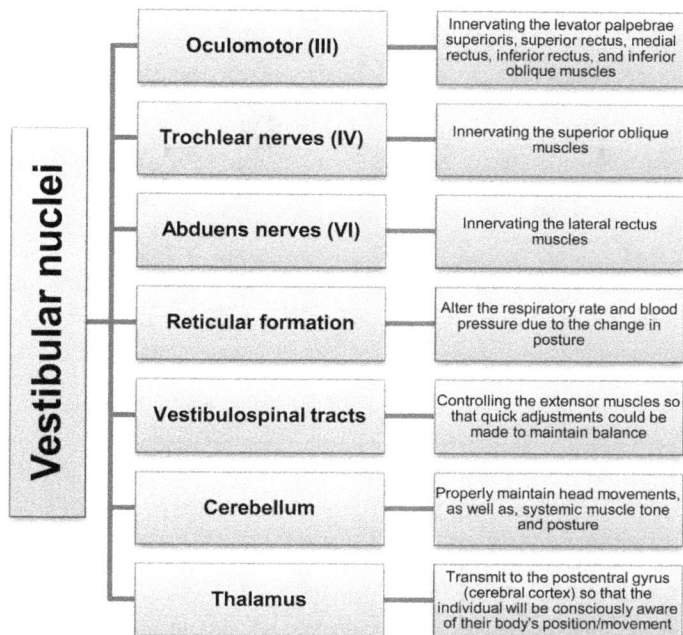

Figure 28: The multitudes of transmission/functions of the vestibular nuclei

Hearing

Sound is defined as a molecular vibration that is propagated as sound waves. These sound waves are produced by vibrations from any object, no matter if it is the vibration generated from an individual's vocal cord or an object accidentally dropped on the floor. Once initiated, the vibration will causes the surrounding air molecules to be compressed and then expanded in regular successions. Like a ripple in water, the compression and expansion of air molecules will radiate from the source in all directions. When these waves of air strike the ear drum, this energy will be transduced into action potentials, which are then translated in the brain as audible sounds. Put simply, the ability for humans to receive and translate these sound waves is called audition.

The simplest kind of sound wave is a sine wave. Sine waves are S-shaped curves that travel through the air at approximately 1235 km/h (768 mph). The wavelength of sound is calculated by measuring the distance between two adjacent wave peaks and the frequency of sound is the number of times per second that a sound wave cycles from positive to negative to positive (Figure 29).

The frequency of sound is measured in terms of cycles (waves) per second in a unit called hertz (Hz). For example, 1 Hz means that an event repeats once per second. It is understood that the frequency of sound is inversely related to wavelength. A high frequency sound (high pitch sounds) would have a short wavelength (~15,000 Hz), while a low frequency sound (low pitch sounds) would have a long wavelength (~100 Hz) (Figure 29). The intensity of sound (amplitude) is defined as the strength of a sound wave that the human ear interprets as volume (or loudness). The intensity of sound is measured in units known as decibels. For instance, human hearing is sensitive in the range of frequency of 20 Hz to 20 kHz (20,000 Hz), which is referred to as sonic. The term ultrasonic are frequencies higher than human audition (>20,000 Hz), which can be heard by animals such as dogs, cats, and bats. The term infrasonic are frequencies below human audition (<20 Hz) which are heard by animals such as

snakes, lizards, fish, whales, etc.

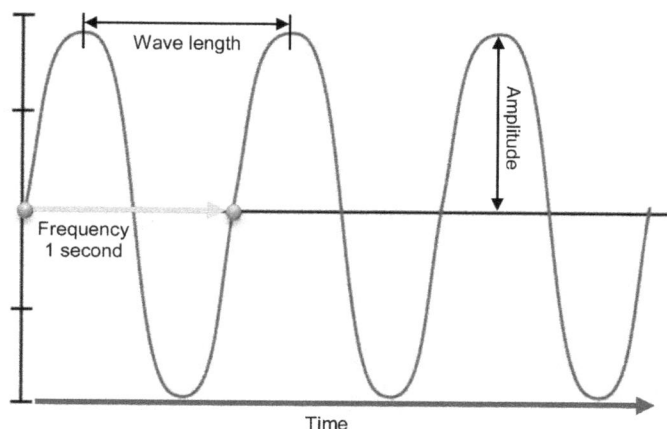

Figure 29: Illustration of a sine wave. Please note that the wave length is calculated by measuring peak to peak, amplitude is the height of each peak and the frequency is calculated by measuring the number of times per second that a sound wave cycles from positive to negative to positive

Physiology of Hearing

Sound waves are generated from an object and travel to the auricles of the ear. The auricles focus the sound into the auditory canal, which amplifies the frequencies of the sound wave in the range 3 kHz to 12 kHz. The sound waves cause the tympanic membrane to move, which in turn causes the displacement of the auditory ossicles. First the movement of the tympanic membrane pushes the malleus inward which causes the tensor tympani muscle to contract. In turn, the malleus pushes on the incus. Through their articulations, the incus pushes on the stapes which causes the stapedius muscle to contract. Please note that the contraction of the tensor tympani and the stapedius muscles are aimed to protect the middle ear from auditory damage by lessening the transfer of energy to the inner ear through a process called tympanic reflex. The movement of the stapes produces a rocking motion at the oval window. This rocking motion of the stapes generates fluid waves, also known as push-pull waves, in the perilymph of the scala vestibule. Like any fluids, the perilymph cannot be compressed therefore it respond to the pressure (push) by flowing away from the oval window. This fluid pressure will then push the vestibular membrane downward which in turn pushes the endolymph in the scala media (cochlear duct) downwards. The endolymph will push down on the basilar membrane, which sequentially pushes down on the perilymph in the scala tympani. Finally, the secondary tympanic membrane of the round window (which faces the air-filled chamber of the middle ear) is pushed outward to relieve the pressure. Inversely, the pressure generated by the pulling pressure of the stapes will result in the reverse reaction. Put simply, as the stapes rocks back and forth, the secondary tympanic membrane will also be pulled in and out while the basilar membrane moves up and down (Figures 30 and 31).

As the basilar membrane is being pushed up and down by the fluid pressure, the tectorial membrane (the gelatinous layer) remains stable and in place. Any movement upwards by the basilar membrane will result in the stereocilia being jammed into the tectorial membrane and bent forward. As the tallest of the stereocilia bends forward, the tip-link connecting it to the other stereocilia will pull them forward as well. Please note that any bend that equals or exceeds ~0.3 nm will result in the hair cell reaching threshold and the opening of the mechanically gated K^+ channels. The opening of the mechanically gated K^+ channels will result in the influx of K^+ cations. The influx of K^+ cations will cause the voltage gated K^+ and Ca^{2+} channel to open, which in turn will cause the additional influx of K^+ and Ca^{2+} cations. At this point, the hair cell is depolarized and will result in the release of neurotransmitter(s) (Figures 30 and 31).

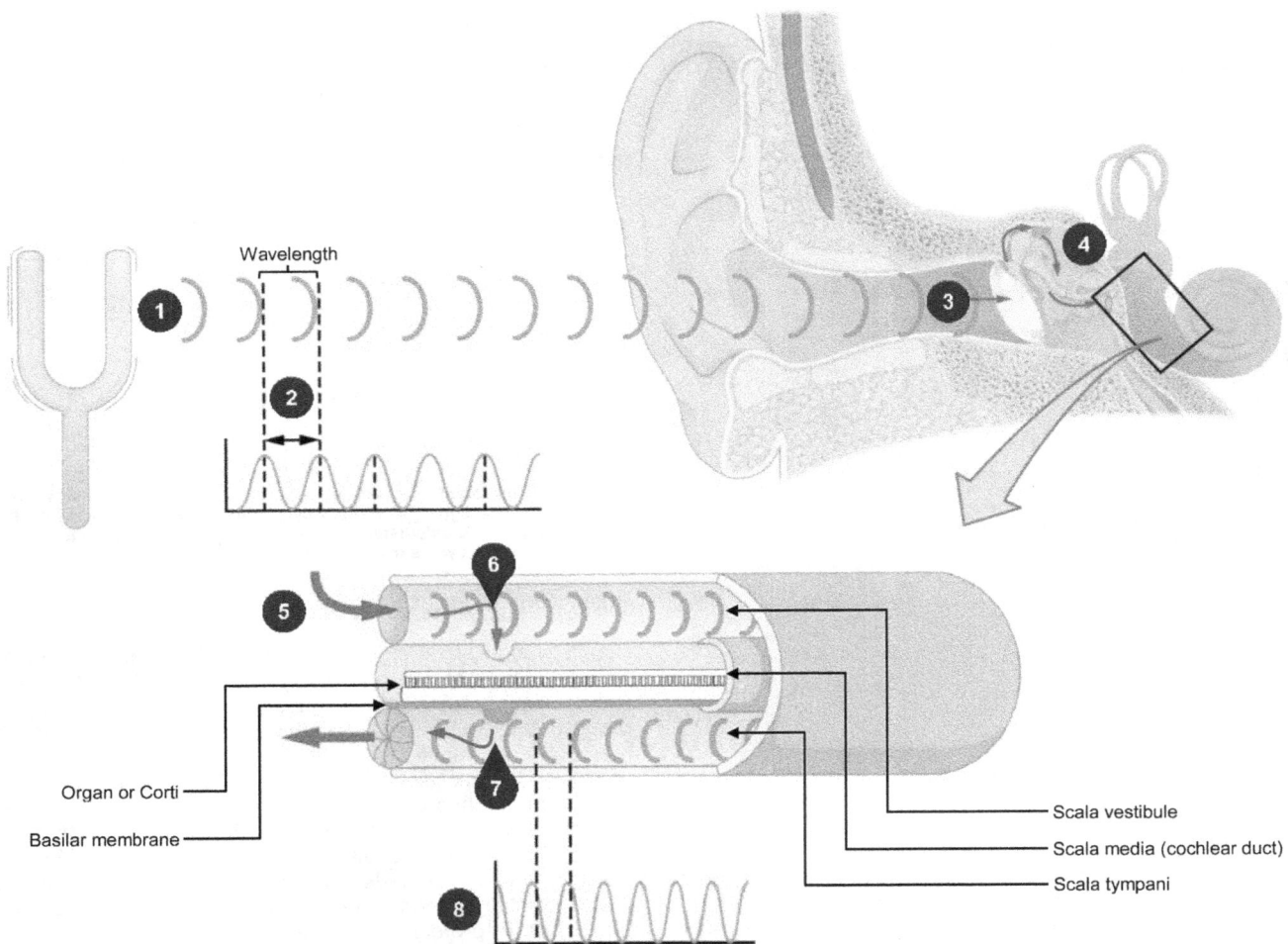

Figure 30: ① Sound wave represents alternating areas of high and low pressure and the ② frequency of sound wave measured in hertz (Hz) also known as cycles per second. ③The tympanic membrane vibrates in response to sound wave and ④ the vibrations that the waves generated are amplified across the ossicles. ⑤The vibrations against the oval window caused fluid pressure in the perilymph of the scala vestibule. ⑥The pressure bends the membrane in the cochlear duct (filled with endolymph) at a point of maximum vibration for a given frequency, ⑦ causing the hair cells of the organ of Corti and basilar membrane to bend. ⑧ Please note that the frequency of standing wave is the same as the sound wave

Variations in the intensity of sound (amplitude) will result in a vibrational variation of the basilar membrane. For example, if the amplitude of the sound is low or soft, the movement of the basilar membrane will be minimal; therefore the number of hairs cells stimulated will be limited. This limited stimulation will translate into limited action potentials being sent to the brain, which the brain will interpret the input as a soft sound. Inversely, if the amplitude of the sound is loud, the movement of the basilar membrane will be equally as drastic. Intense movement of the basilar membrane will result in greater numbers of hair cells being stimulated and equal amounts of action potentials generated. Put simply, the greater number of action potentials being sent to the brain will be interpreted as loudness.

Frequency determination is based on the structural differences of the basilar membrane. For example, at the proximal end of the cochlea, the basilar membrane is attached, making it stiff and relatively inflexible. Since the basilar membrane is attached at the proximal end of the cochlea, the organ of Corti in this area stiff and relatively inflexible. The proximal end of the cochlear duct is also where the outer hair cells (OHC) exist. On the distal end, or the apex of the cochlea duct, the basilar membrane is unattached, making it more flexible in comparison. Since the basilar membrane is unattached at the apex of the cochlea duct, the organ of Corti in this area is relatively flexible. The distal end of the cochlea duct is also where the inner hair cells (IHC) exist. It is known that the peak amplitudes for high frequency sound stimulates the proximal end, while the peak amplitudes of low frequency sounds stimulate the distal end. Therefore, if the brain receives action potential mostly form the OHC, it is interpreted

as high frequency (high pitch) sounds. If the brain receives action potentials mostly form the IHC, it will interpret them as low frequency (low pitch) sounds.

Once an action potential is generated from the hair cells it will release its neurotransmitter(s) into the synaptic cleft of the adjacent bipolar sensory neurons. The soma of these bipolar sensory neurons forms the spiral ganglion near the modiolus. The axons of the spiral ganglion form the cochlear branch, which later joins with the vestibular branch to form the vestibulochochlear nerve (VIII). Each ear sends its auditory signals via cranial nerve VIII to the cochlear nuclei in the medulla oblongata. The cochlear nuclei then send their information (via 2° neuron) to the olivary nucleus in the pons. In addition to receiving sensory information, the olivary nucleus also possesses efferent functions. For example, the olivary nucleus can send efferents back to the cochlea, which are involved in cochlear tuning, allowing individuals to better discriminate the pitch of various sounds. Additionally, the olivary nucleus can also transmit motor signals via the trigeminal nerve (V) and facial nerve (VII) to control the tensor tympani and stapedius (striated muscles) of the middle ear (Figure 32).

Cochlear tuning sharpens the abilities of the cochlea and allows individuals to better discriminate the pitch of various sounds. It is interesting to note that in addition to transmitting auditory induced sensory signals to the brain, the brain can in return adjust the OHC and IHC to fine tune the sounds. Depending on the commands by the olivary nuclei, the OHC may be commanded to shorten or lengthen. For example, if the olivary nuclei prompts the OHC to shorten by 15%, the result is a less flexible basilar membrane

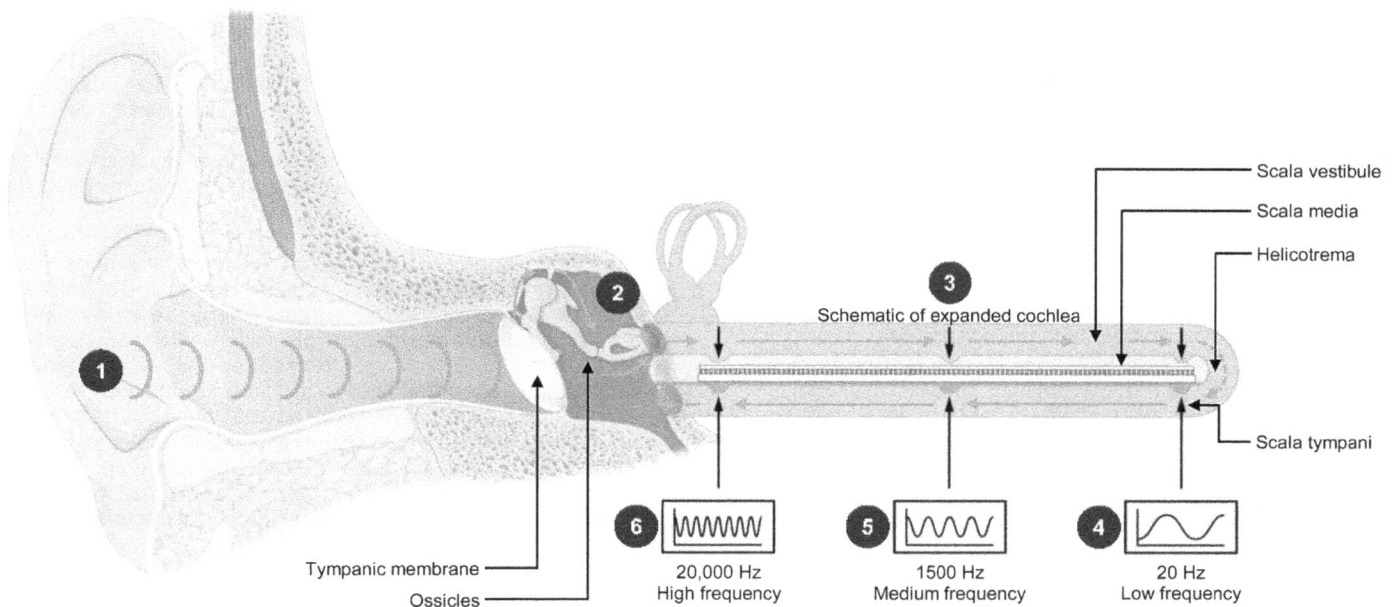

Figure 31: ① The sound waves cause the tympanic membrane to move which in turn ② causes the displacement of the auditory ossicles. ③The movement of the stapes produces a rocking motion at the oval window which generates push-pull fluid waves in the perilymph in the scala vestibule. Like any fluids, the perilymph cannot be compressed therefore it responds to the pressure (push) by flowing away from the oval window. This fluid pressure will then push the vestibular membrane downward, which in turn, pushes the endolymph in the scala media (cochlear duct) downwards as well. The endolymph will push down on the basilar membrane which sequentially will push down on the perilymph in the scala tympani. For example, ④ at low frequency (20 Hz), the deformation of the vestibular and the basilar membrane will occur at the furthest areas of the cochlea (next to the helicotrema). ⑤ At medium frequency (1500 Hz) the deformation of the vestibular and the basilar membrane will occur towards the middle of the cochlea while ⑥ at high frequency (20,000 Hz) the deformation will occur towards the oval and round windows. Finally, the secondary tympanic membrane of the round window (which faces the air-filled chamber of the middle ear) is pushed outward to relieve the pressure. Inversely, the pressure generated by the pulling pressure of the stapes will result in the reverse reaction. Any movement upwards by the basilar membrane will result in the stereocilia being jammed into the tectorial membrane and bend forward which will result in the depolarization of the hair cells

and fewer signals being transmitted by the OHC to the medulla oblongata. Having less OHC signals allows the brain to better analyze other signals being sent by adjacent hair cells (e.g. IHC), and more able to distinguish frequencies. In addition, the olivary nuclei can inhibit the sensory signals being transmitted by the IHC, which also allows the brain to better analyze other signals being sent by adjacent hair cells (e.g. OHC), and more able to distinguish frequencies.

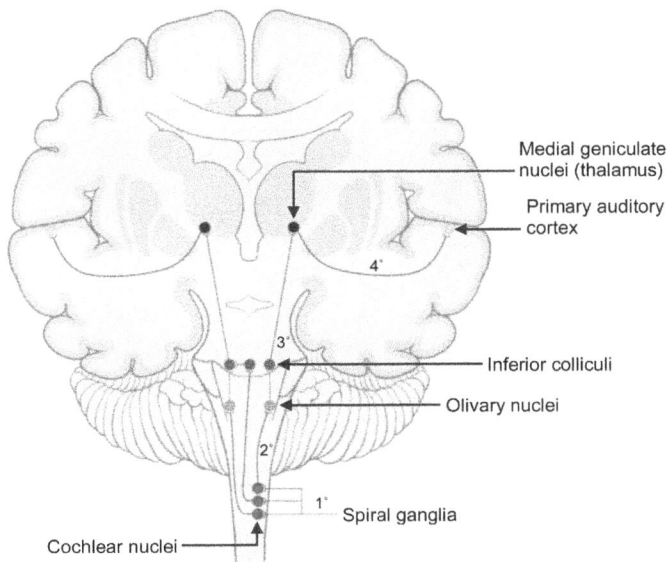

Figure 32: Illustration of the auditory pathway

The olivary nucleus is also capable of comparing signals from the right and left ear in a process called binaural hearing. The ability to compare signals from the right or left ear allows an individual to determine which direction the sounds are emanating.

From the olivary nuclei, the information is sent to the inferior colliculi of the midbrain. The inferior colliculi is responsible for helping an individual determine the origin of a particular sound, processing the fluctuating in pitch as well as, auditory reflexes as when an individual being startled by a sudden noise turning their head towards the direction of the sound. From the inferior colliculi, the sensory information is sent by the 3° neuron to the medial geniculate nuclei of the thalamus, which in turn transmits the information via the 4° neuron to the primary auditory cortex located in the temporal lobe (Figure 32).

SIGHT

The eye is the organ that is designed to detect light or visible electromagnetic radiation. Put simply, the human eye is a transducer that is designed to convert light particles, known as photons, into electrochemical signals through a process called the photochemical reaction, before transmitting the signals to the brain.

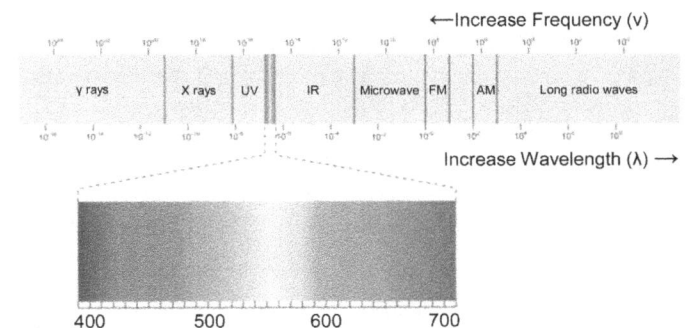

Figure 33: The electromagnetic spectrum

The human eyes are capable of detecting wavelengths of light between 400 nm to 750 nm, which is rather limited when compared to the entire electromagnetic spectrum. For example, ultraviolet (UV) radiation, which are wavelengths below 400 nm, are invisible to the human eyes. On the opposite end of the spectrum, infrared radiation (IR), which are wavelengths above 750 nm, are also invisible to the human eyes (Figure 33).

Accessory Structures of the Eyes

The eyes and their accessory structures are constructed to allow individuals to obtain visual signals from their environment, playing a part in sexual selection, and forming facial expressions to indicate mood and emotions. The accessory structures that are associated with the eyes includes eye brows, eyelids, eye lashes, conjunctiva, and the lacrimal apparatus.

The eyebrows serve to protect the eyes from glare and help to prevent perspiration from getting into the eyes. In addition, the eyebrows also play a part in enhancing facial expressions and nonverbal communications. Other follicular structures that serve to protect the eyes are the eyelashes. Eyelashes are protective structures that prevent debris from making contact with the eyes.

The eyelids, also known as palpebrae, serve to block foreign objects and debris from irritating the eyes. Like a windshield wiper, it is used to brush debris away from the surfaces of the eyes, as well as, moisturizing the conjunctiva using tears and meibum when we blink. Most importantly, the eyelids also allow an individual to have a good night's rest undisturbed from arbitrary visual stimulus. The eyelids are mainly composed of orbiculais oculi muscles (voluntary striated muscles) that are innervated by the facial nerve (VII). Each of the orbiculais oculi muscles is covered by skin and together they are responsible for closing the eyelids. Please note that the muscle responsible for opening the eyelids is called the levator palpebrae superioris, which is one of the striated muscles found in the orbit. The levator palpebrae superioris is another voluntary striated muscle, which is innervated by the oculomotor nerve (III).

The eyelids are strengthened and supported by a fibrous connective tissue plate called the tarsal plate which is mainly located at the margins of the eyelids. Within the tarsal plate there are 20 to 25 tarsal glands. Tarsal glands are modified sebaceous glands that produce and secrete meibum, an oily substance that is formed to prevent the eyelids from sticking together, and reduces the evaporation of tears from the surface of the eyes.

The upper and lower eyelids are separated from one another via the palpebral fissure (the space between the upper and lower eyelids). The palpebral fissures are joined together at the corners of the eye in a junction called the medial (towards the nose) and lateral commissures (towards the side of the face) (Figure 34).

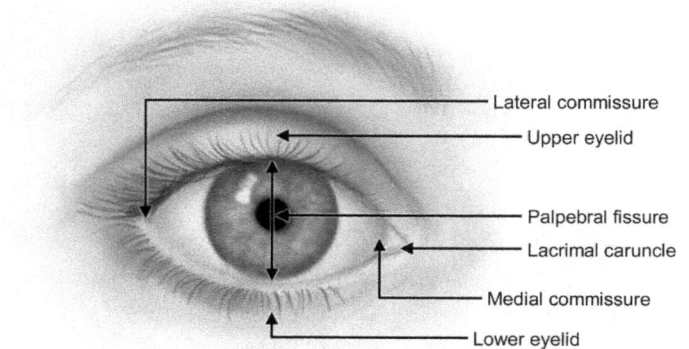

Figure 34: Illustration of palpebral fissure and the medial and lateral commissure

The conjunctiva consists of transparent mucous membranes that secrete a thin mucous film that lubricates the eyeballs. With the exception of the cornea, the conjunctiva covers the anterior aspect (e.g. sclera) of the eyes, as well as, lining the interior surfaces of the eyelids. This transparent structure is richly innervated, highly vascularized, and very sensitive to pain. If irritated, the blood vessels within the conjunctiva will dilate and give the appearance of blood shot eyes. Since the cornea is not innervated, the conjunctiva serves to inform the brain that there is debris within the eyes and triggers responses such as blinking, increasing the production of mucous and tears to remove the irritant.

The lacrimal apparatus consists of the lacrimal glands (also known as tear glands) and various ducts that are used to transport and drain their secretions. The lacrimal glands are located within the frontal bone and are responsible for producing tears. Once tears are produced, they are transported via 12 short ducts (lacrimal gland ducts) that lead to the surface of the conjunctiva of each eye. Tears are used to moisten the conjunctiva and wash away debris from the surface of the eyes. Additionally, tears also serve to act as a diffusion layer to allow gases, such as oxygen, and nutrients to be delivered to the conjunctiva of each eye. The availability of oxygen within the tears prevents anaerobic conditions, which in turn, prevents the growth of anaerobic bacteria. In addition in secreting tears, the lacrimal glands also secrete a digestive enzyme called lysozyme. Lysozyme is responsible for sterilizing the eyeball by preventing any potential infection by bacteria or viruses. After the tears are secreted, they wash over the conjunctiva and the cornea before flowing to a fleshy mass called the lacrimal caruncle, which is located at the medial aspects of the eyes. Within the lacrimal caruncles, tiny pores called the lacrimal punctums exist. The lacrimal punctum is the opening, or the drain, to the lacrimal canal that leads to the lacrimal sac, where excess tears are collected. The lacrimal sac opens to the nasolacrimal duct, which is responsible for carrying tears to the inferior meatus of the nasal cavity where it would travel up the nasal canal (on ordinary conditions), and swallowed. Nonetheless, if excess tears are produced, a running nose will result. This condition is evident when an individual is crying. Additionally, if the nasolacrimal duct is obstructed (e.g. during an allergic reaction), the tears will be prevented from drainage, resulting in a condition commonly called watery eyes (Figure 35).

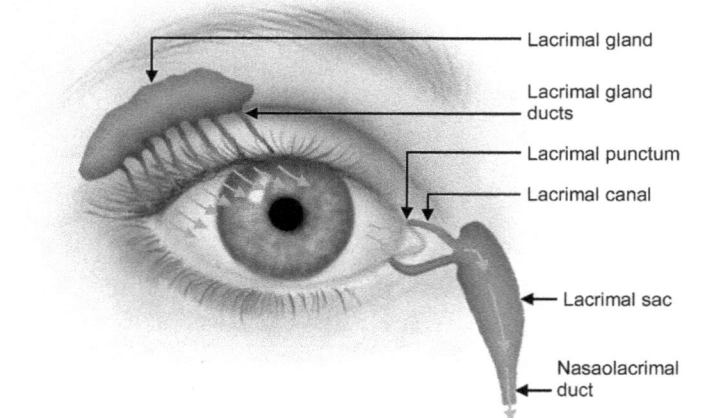

Figure 35: Illustration of the lacrimal apparatus. Please note that the nasolacrimal duct drains into the inferior meatus of the nasal cavity (not shown)

The movement of the eyes is controlled by six extrinsic muscles, which are attached to the surface of the eyes and the walls of the orbit. There are four rectus (straight) muscles and two oblique muscles (Figure 36).

The superior oblique courses along the medial walls of the orbit before its tendon inserts through a fibrocartilage ring called the trochlea, and thus redirects it inferiorly before inserting onto the anterior region of the eye. This muscle's function includes intorsion (e.g. internally rotate), abduction, and the depression of the eyes when moving from side to side. The superior oblique is innervated by the trochlear nerve (IV). The inferior oblique extends from the medial wall of the orbit to the inferior laterals side of the eyes. Inferior oblique is the direct functional antagonist to the superior oblique. The function fo this muscle includes extorsion (e.g. external rotation), adduction, and elevation of the eyes when moving from side to side. The inferior oblique is innervated by the oculomotor nerves. The superior rectus moves the eye upward, and secondarily adducts, or rotates, the top of the eye toward the nose. The superior rectus is innervated by the oculomotor nerves. Inferior rectus moves the eye downward and secondarily abducts or rotates the top of the eye away from the nose. The inferior rectus is innervated by the oculomotor nerves. Lateral rectus is responsible for abduction, or moving the eyes away from the nose. The lateral rectus is innervated by the abducens nerve. The medial rectus is responsible for adduction, or moving the eyes towards the nose. The medial rectus is innervated by the oculomotor nerves (Figure 37).

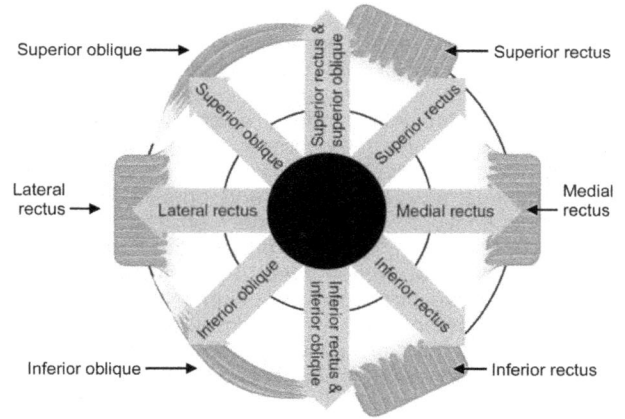

Figure 37: Illustration of coordinated muscle contraction and subsequent movement of the eye. Please note the graphic illustrates the right eye

The eyes are surrounded laterally and posteriorly by layers of adipose tissue called orbital fat. Orbital fat cushions the eyes from blunt force, but still permits freedom of movement. It is also responsible for protecting the blood vessels and nerves that are located within the orbit.

Anatomy of the Eyes

The human eyes are spherical structures that are composed of three layers of tunics, optical components, and neuronal apparatus.

The three layers of tunics of the eyes are: the tunica fibrosa, vasculosa, and interna. The tunica fibrosa is the outermost

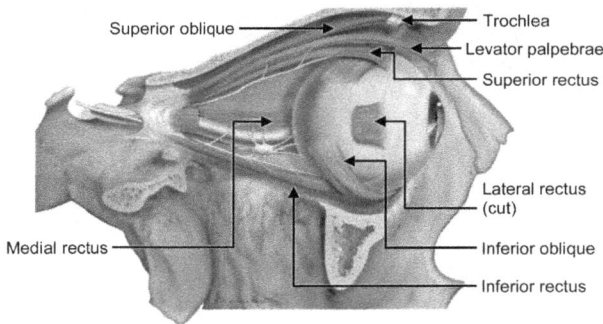

Figure 36: Illustration of extrinsic muscles of the eye. Please note that levator palpebrae muscle is responsible for elevating the eye lid

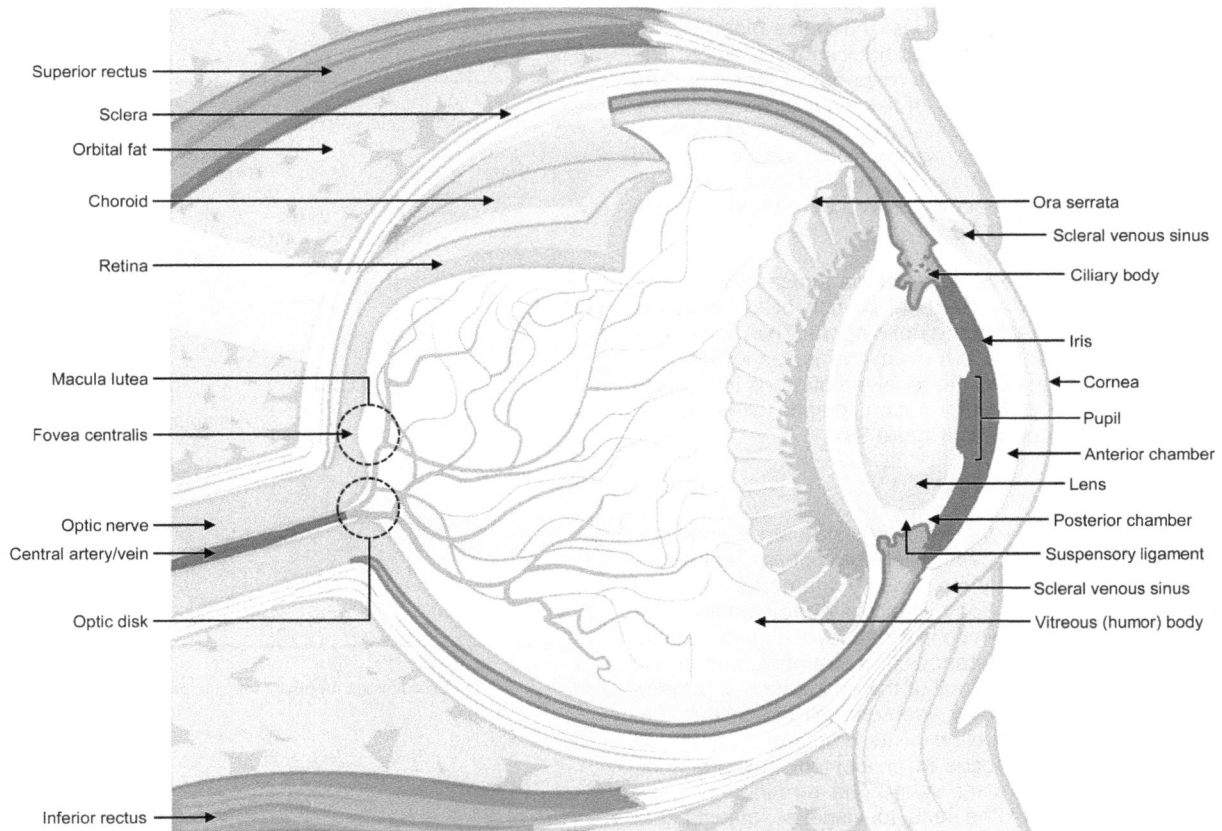

Figure 38: Sagittal section of a human eye

layer and it is subdivided into two regions: sclera and cornea. The sclera, also known as the white of the eye, covers most of the surface of the eye. The sclera is composed of dense irregular connective tissue with high collagenous content and is well innervated and vascularized. The cornea is a transparent region that is created from modified sclera, which allows light to penetrate into the eyes. The cornea is composed of collagen fibers with numerous fibroblast cells and is covered anteriorly by stratified squamous epithelium, and posteriorly by simple squamous epithelium. The anterior stratified squamous epithelium contains stem cells that provide the cornea with the ability to regenerate when damaged. Both epithelial layers are responsible for actively transporting sodium (Na^+) ions out of the cornea. It is understood that water follows the Na^+ cations via osmosis; therefore, they are actively removed from the cornea. This process is responsible for protecting the cornea from being swollen because of over hydration, and thereby prevents it from losing its transparency (Figure 38).

The middle layer of the eye is called tunica vasculosa, also known as the uvea. The tunica vasculosa consists of three regions: the choroid, the ciliary body, and the iris. The choroid is a highly pigmented and vascularized layer located immediately behind the retina. The choroid is responsible for reducing the reflected light within the eye, improving the contrast of retinal image. The ciliary body is an extension of the choroid and forms a muscular ring around the lens (it is responsible for supporting both the iris and lens). The ciliary body is a smooth muscle ring that is controlled by the parasympathetic division of the autonomic nervous system. For example, the acetylcholine secreted by the parasympathetic division causes the smooth muscle ring to contract, while the lack of this neurotransmitter causes the smooth muscle ring to relax. In addition, the ciliary body produces and secretes a plasma-like fluid, the aqueous humor. The aqueous humor is secreted into spaces between the iris and the lens, the posterior chamber, and flows into the space between cornea and iris, the anterior chamber. In the anterior chamber, the aqueous humor will be reabsorbed by the scleral venous sinus, a ring like blood vessel that is responsible for returning the fluid back into circulation. The iris is an adjustable diaphragm that is responsible for controlling the diameter of the central opening of the eyes called the pupil. The iris possesses two pigmented layers. The posterior pigment epithelium blocks stray light from reaching the retina, and the anterior border layer, which contains pigmented cells known as chromatophores (containing melanin), which determines the eye color of an individual (Figure 38).

The inner most layer of the eye is called the tunica interna. The tunica interna consists of the neuronal portion of the eye, the retina, and the initial segment of the optic nerve. The retina lines the lateral posterior interior of the eyes and is responsible for converting the light energy from the received photons into action potentials. For more detailed information, please examine the section titled Neuronal Apparatus (Figure 38).

Other Vision Apparatus

Lenses of the eyes are light focusing structures that are constructed out of transparent cells called lens fibers. Lenses are suspended behind the pupil by suspensory ligaments, which in turn are attached to the ciliary body. The ciliary body manipulates the lens through contraction, which in turn allows an individual to focus in on objects at various distances. For example, contracting the ciliary body will result in the pull of the suspensory ligament. The tightening of the suspensory ligament will cause the lens to flatten and result in a reduction of its diameter. A flattened lens allows an individual to focus in on objects that are near, while blurring things that are far away. The terminology used to describe manipulation of the lens to visualize objects that are close is termed accommodated, while focusing in on an extremely close object is characterized as fully accommodated. In contrast, relaxation of the ciliary body will relax the pull of the suspensory ligament, and in turn, relax the lens. A relaxed lens

will be allowed to reform into its rounded shape and increases its diameter. A lens with increased diameter will allow an individual to focus in on objects that are at a distance, while blurring things that are nearby. The terminology used to describe manipulation of the lens to visualize objects that are distant is termed relaxed while viewing very distant objects is characterized as totally relaxed.

The optical power of the lens in the human eye is defined as the degree to which a lens bends (converging or diverging) light. Put simply, the more that the light is bent by the lens, the greater the power. Optical power is designated with a symbol P while its SI units are the inverse meters (m^{-1}) or diopters (δ).

Optical power is defined as the reciprocal of the focal length (f) of the lens and is calculated by using the thin lens equation:

$$P = 1/f$$

It is known that the optical power of combined lenses is approximately the sum of the optical power of each lens. For example, images transmitted to the retina pass through two optical elements, the cornea and the lens. Therefore, the optic power of the human eye may be described by the following modified equation (Figure 39):

$$P = 1/d_o + 1/d_i$$

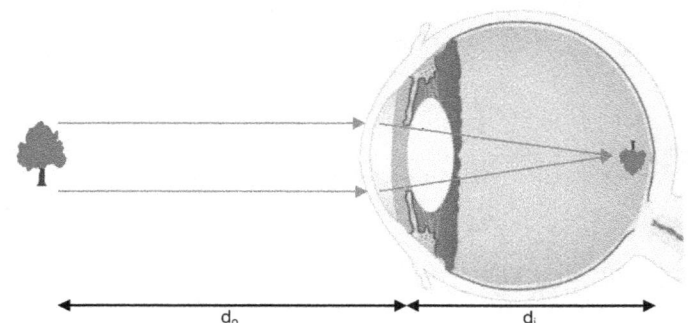

Calculations for object at a distance

d_o	=	30.0 meters
d_i	=	0.02 meters
P	=	$1/d_o + 1/d_i$
P	=	1/30.0 m + 1/.02 m
P	=	0.03 δ + 50 δ
P	=	50.03 δ

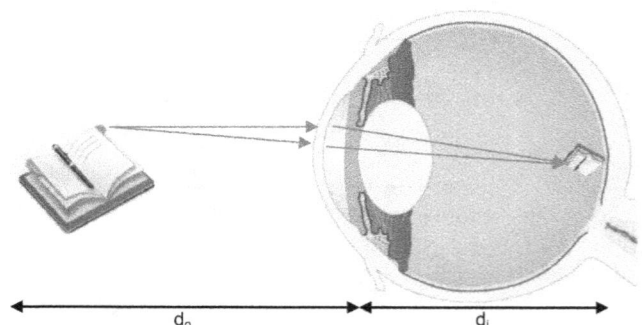

Calculations for object in close proximity

d_o	=	0.50 meters
d_i	=	0.02 meters
P	=	$1/d_o + 1/d_i$
P	=	1/0.5 m + 1/.02 m
P	=	2.0 δ + 50 δ
P	=	52.0 δ

Figure 39: Illustration and examples of calculations of optical powers

A refractive error occurs when an eye has excessive or inadequate refractive power to focus light onto the retina. For example, myopia, also known as nearsightedness is a condition of the eye where the light is focused in front of the retina when examining objects at a distance. The failure to focus the image of distant objects on the retina results in a blurred and out of focus image. Myopia is commonly corrected through the use of corrective lenses. In contrast, a hyperopic eye, also known as farsightedness, is a condition of the eye where the light is focused behind the retina when examining objects up close. Again this condition is alleviated through the use of corrective lenses (Figure 40).

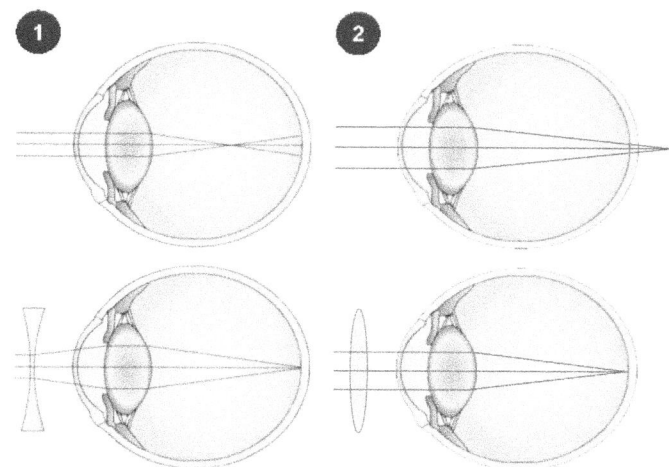

Figure 40: Illustration of the reflective errors of the human eye. ① Illustrates myopia and its improvement by the use of corrective lens. ② Illustrates hyperopic eye and its improvement by the use of corrective lens

The vitreous body is a semisolid clear gel-like secretion that fills the hollow interior of the eyeball. The vitreous body is composed of collagen fibers, vitrosin (a collagen type of protein ~100 to 150 Å in diameter) and hyaluronic acid (nonsulfated glycosaminoglycan), and water. This semisolid fluid is a stagnant substance that is not connected to blood vessels and is not constantly being regenerated or replenished.

Neuronal Apparatus

As indicated previously, the tunica intima, the innermost layer of the eye is composed of the retina and the optic nerve, which is also termed the neuronal apparatus. The retina is commonly referred to as the outermost extension of the brain. It is a thin transparent neurological membrane that is responsible for transducing light into neurological signals. The retina is attached to the eye at two points: the optic disc and the ora serrata. The optic disk is the location where the optic nerve leaves the eye, as well as, the location where blood vessels enter and exit. The optic disc possesses no optic cells, thereby producing a blind spot where no image is transmitted. The second attachment site is the ora serrata, which is the transitional junction between the retina and the ciliary body, or where the photoreceptive area transitions into a non-photoreceptive area. The remainder area of the retina is held in place by the pressure exerted by the vitreous (humor) body.

Since the retina is only partially attached to the eye, it can be separated from the choroid by a forcible blow to the head, or when the vitreous body retains insufficient pressure. This condition is known as a detached retina, and may immediately result in a blurred field of vision. Please note that the retina depends upon the choriocapillaris (capillaries of the choroid) for nourishment and waste disposal. If the detached retina is left unattended medically for an extended period of time, the retinal tissue may deteriorate, and this may result in permanent blindness.

Directly inline and posterior to the center of the lens is an area called macula lutea. At the center of the macula lutea is a tiny pit called the fovea centralis. It is known that both the macula lutea and fovea centralis contain cone cells (discussed in detail in the next section) and are responsible for providing the ability for an individual to envision detailed images (Figure 38).

Retina

The retina is composed of numerous layers of cells that contribute to visual perception. For example, the outer most (posterior) layer of the retina is the retinal pigment epithelium (RPE), which is located anterior to the choriocapillaris (capillaries of the choroid). RPE is a darkly pigmented layer that interacts with photoreceptor cells and is responsible for absorbing stray light thereby maintaining normal visual function.

Lying superior to the RPE is the photoreceptor layer composed of two types of photoreceptor cells called rods and cones. It is known that these photoreceptor cells are not nerve cells but are derived from neural glial cells, related to epidermal cells. The rod cells possess low spatial resolution, but are extremely sensitive to light. It is understood that rod cells are responsible for vision in dim light, or black and white vision. In contrast, the cone cells possess high spatial resolution, but are relatively insensitive to light. It is understood that cone cells are involved in bright-light vision and the perception of color. For example, at the low light (e.g. star light at night), only the rod cells are stimulated, which produces only low resolution visual images. This visual perception at low light levels is called scotopic vision, and provides an individual with images in shades of grey, also known as monochromatic vision. As illumination increases (e.g. twilight), cone cells are activated, increasing an individual's visual resolution. This visual perception is made through the contributions of both type of photoreceptor cells is referred to as mesopic vision. As illumination increases towards normal daylight conditions, cone cells become the dominant photoreceptor stimulated, while the contributions provided by the rod cells are nearly eliminated. This type of vision is referred to as the photopic vision.

Rod and cone cells are of similar construct and possess an outer segment that points towards the RPE and an inner segment that point towards the interior of the eye. The outer segment is a modified cilium that is designed to absorb light. The inner segment, on the other hand, contains numerous organelles, such as mitochondria, rough endoplasmic reticulum, ribosomes, etc. The two segments are separated by a narrow band that contains nine pairs of microtubules. Beneath the inner segment, towards the base of both rod and cone cells, is the cells body where the nucleus resides. At the base of the photoreceptor cell is a slight membrane extension from the cell body. This is the location where the photoreceptor cells synapse with retinal neurons.

The outer segment of the rod cells is composed of approximately 1000 membranous disks that are independent of the surrounding plasma membrane (i.e. the disks are separate entities from the plasma membrane). Each membranous disk contains visual pigments called rhodopsin (Figure 41).

Rhodopsin is a visual pigment that has a maximum sensitivity to light at a wavelength of 498 nm. Rhodopsin is composed of two subunits: rod opsin and retinal. Rod opsin is a transmembrane protein complex that is involved in mediating visual transduction in the retina (Maximum sensitivity λ_{max} at 498 nm). Retinal, on the other hand, is a yellowish photosensitive pigment that is derived from vitamin A and is covalently bound to opsin.

In complete darkness, the long hydrocarbon chain of retinal is in 11-cis structural format. However, when a photon (particle/unit of light) strikes retinal, the long hydrocarbon chain is converted from 11-cis structural format to 11-trans format via a process known as photoisomerization. The conformational change of retinal results in the activation of rod opsin, which in turn results, in the alteration of the membrane potential of the photoreceptor cell (this process will be discussed the section titled Action Potential Formation). After the light source is removed, rod opsin, and retinene reformed their covalent link while the retinene molecule returns to its original 11-cis format (Figure 42).

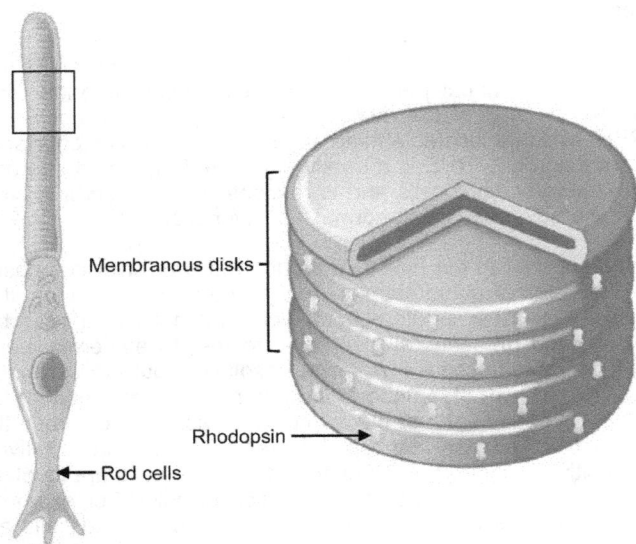

Figure 41: Illustration of membranous disk of rod cells

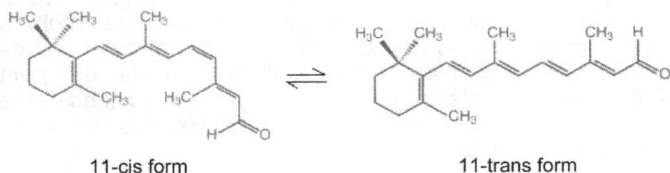

Figure 42: Photoisomerization of retinal. Please note that the maximum absorption (λ_{max}) for rhodopsin (or rod opsin) is at 498 nm

In contrast, the outer segment of the cone cell is also composed of similar numbers of membranous disks, but these are constructed out of invaginations of the plasma membrane from the apical ends of cone cells. Each of the cone disks contain the visual pigments, iodopsins.

Iodopsins are composed of retinal and the photopigment cone opsin covalently linked to one another. There are three different types of cone opsins, which are divided based upon the sensitivity of these molecules to specific wavelengths. For example, blue sensitive cone opsin (also known as short wavelength sensitive cone opsin) allows maximum absorption (λ_{max}) at 425 nm which is located in the blue region of the electromagnetic spectrum. The green sensitive cone opsin (also known as mid-wavelength sensitive cone opsin) allows maximum absorption (λ_{max}) at 530 nm which is located in the green region of the electromagnetic spectrum. Finally, the red sensitive cone opsin (also known as long wavelength sensitive cone opsin) allows maximum absorption (λ_{max}) at 560 nm, which is located towards the red region of the electromagnetic spectrum. Because there are three types of cone opsins, it is commonly stated that there are three types of cone cells that inhabit the human retina: blue sensitive cones (S-cones), Green sensitive cones (M-cones) and red sensitive cones (L-cones). Recent experimental data have indicated that there are approximately 6.5 million cone cells found within the human retina. It is estimated that 64% of the entire population of cone cells consists of L-cones (red cones), 32% M-cones (green cones) and 2% S-cones (blue cones) (Figure 43).

Human's ability to distinguish wide variations of colors is based

on the amount of color sensitive cone cells that are activated. For instance, the color cyan (~490 nm) activates blue and green cones, while yellow (~570 nm) activates green and red cones. The color orange (~590 nm) activates red cones and partially activates green cones, while magenta (~520 nm) activates red and blue cones. When all three cones are activated, our visual image indicates the shade of white.

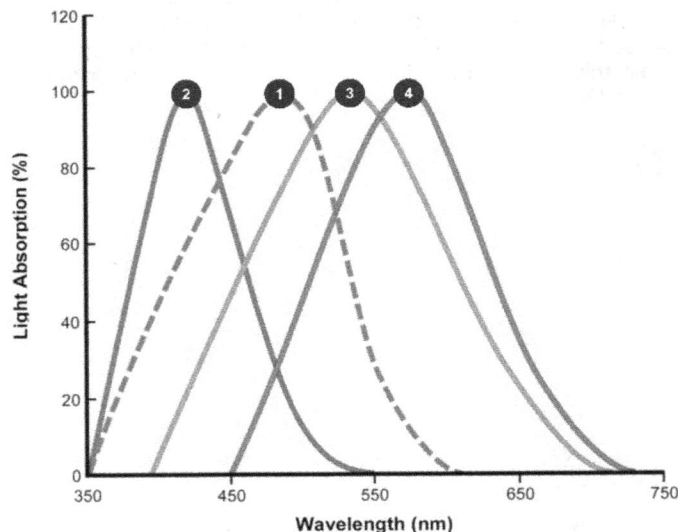

Figure 43: Illustration showing the maximum sensitivities of ① rod cells (λ_{max} 498 nm) and ② blue (λ_{max} 425 nm), ③ green (λ_{max} 530 nm), and ④ red cone cells (λ_{max} 560 nm)

Both rhodopsins and iodopsins are transmembrane glycoproteins that are synthesized in the inner segment and transported to the outer segment. Rod and cone cells continuously renew their disks by producing and introducing new disks at the base of the external segment. As each new disk is added the old disk located at the apical end of the cells is cast off and phagocytized by the pigmented cells of the RPE.

The terminals of the photoreceptor cells (rods and cones) synapse with the bipolar cells and horizontal cells at the outer plexiform layer. The bipolar cells are the most numerous cells in this layer of the retina and are the first order (1°) neurons of the visual pathway. Bipolar cells stretch from the outer plexiform layer where their dendrites exist, through the inner nuclear layer, where their cell bodies lie, and terminates in the inner plexiform layer, where the axonal terminals of these cells synapse with the ganglion cells. The horizontal cells are located in the outer plexiform layer and create lateral connections between the terminal ends of photoreceptor cells, as well as, the dendritic ends of bipolar cells. Horizontal cells play an important role in enhancing the visual systems sensitivity to luminance contrast through a broad range of light intensities, modification in the intensity of light, and the enhancement of the perception of edges.

Amacrine cells, on the other hand, are found in the inner plexiform layer and form horizontal connections between the axonal end of the bipolar cells and the dendritic end of the ganglion cells. Like the horizontal cells, amacrine cells modify and enhance visual signals that are transmitted by the bipolar cells. There are several subtypes of amacrine cells that possess a variety of neurotransmitters that make distinct contributions and provide an alternative, indirect route between bipolar and ganglion cells. One subtype of amacrine cells is responsible for transforming the persistent responses of bipolar cells to light into intermittent responses, preventing sensory overload. Other subtypes of amacrine cells are responsible for enhancing the contrasts of visual signals and changes in the intensity of light.

Like the amacrine cells, the interplexiform cells exists in the inner plexiform layer and forms feedback loops between the horizontal and amacrine cells found within the outer and inner plexiform layers respectively. These interplexiform cells are interneurons that may be either excitatory or inhibitory, depending on the cells that they form connections with and they are responsible for increasing visual signals from some photoreceptor cells, while decreasing signals from others, enhancing visual contrast (Figure 44).

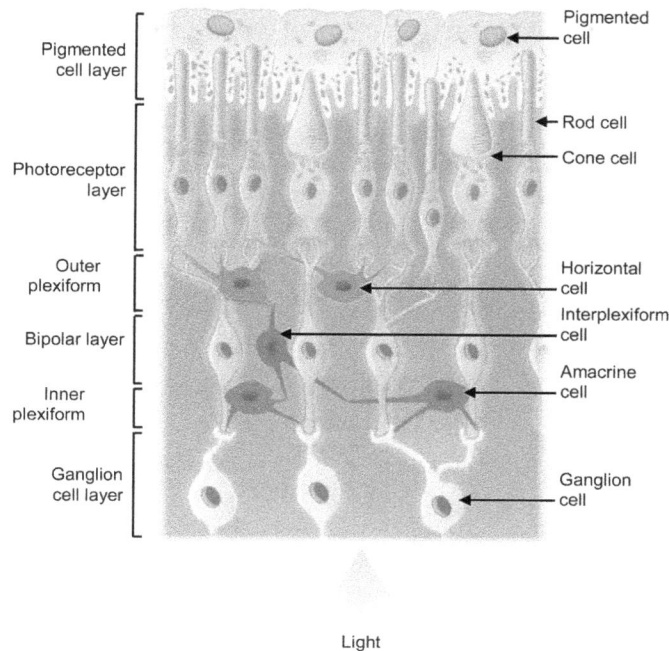

Figure 44: Histological examination of the layers of the retina

As mentioned previously, the bipolar cells synapse with ganglion cells located in the ganglion cell layer of the retina. Ganglion cells are the largest neuronal cells found in the retina and are second (2°) order neurons of the visual pathway. The ganglion cells lie immediately beside the vitreous body and their axons form the optic nerve, cranial nerve II, which crisscrosses and forms the optic chiasma. From the optic chiasma, the optic tract synapses with the lateral geniculate nucleus of the thalamus before giving rise to the third (3°) order neurons and sending the visual signal to the optic lobe of the visual cortex.

It is known that some specialized ganglion cells possess melanopsin (a type of visual pigment) which allows them to absorb light directly (bypassing the photoreceptor cells). These specialized ganglion cells then transmit the visual information to the pretectal nuclei of the brainstem. The pretectal nuclei subsequently synapse with the Edinger-Westphal nucleus, a parasympathetic cranial nerve nucleus belonging to the oculomotor nerves (cranial nerve III). The Edinger-Westphal nucleus gives rise to the preganglionic parasympathetic fibers, which synapse with the ciliary ganglion (postganglionic parasympathetic neurons). The ciliary ganglion innervates the sphincter muscle of the iris and controls diameter of the pupillary diameter.

Stargardt Disease

Stargardt disease (STGD), also known as Stargardt macular dystrophy is autosomal recessive condition that affects approximately 1 in 10,000 individuals in the United States. This disease generally demonstrates its initial symptoms around 10 to 20 years of age (the second and third decades of life) and is the most common form of inherited juvenile macular degeneration.

The mutation that causes Stargardt disease is located on chromosome 1 p13-21.19, an area of the chromosomes that encodes ABC-A4 protein. ABCA4 protein is a subgroup of 49 different ATP-binding cassette transporters (ABC transporters) that utilizes the energy of ATP to translocate an unusually diverse set of substrates, ranging from ions to lipids and peptides, across cellular membranes. ABCA4 is a ~250 kDa single-chain ABC transporter localized to the disk margins of vertebrate photoreceptor (e.g. rods and cones) outer segments. The natural substrate of ABC-A4 is retinylidene-phosphatidylethanolamine (N-retinylidene-PE), a precursor molecule of a potentially toxic compound. ABCA4 is responsible for preventing the accumulation of N-retinylidene-PE within the disks of the rod and cone cells by transporting this precursor molecule to the cytoplasmic side of the disk membrane where it can dissociate. By relocating N-retinylidene-PE, ABC-A4 allows the released all-trans-retinal to enter the visual cycle, or the conversion of photon into action potential.

STGD is divided into four stages. Stage 1 of STGD is characterized by faint and irregular pigments appearing in the retinal pigment epithelium (RPE). These pigments form a ring of flecks approximately 5 to 8 μm in diameter, which may often encircles or are found within the fovea centralis. Research has shown that these yellowish deposits are composed of A2E, a major component of N-retinylidene-PE. In general, visual acuity of the patient is generally normal at this present time (Figure 45).

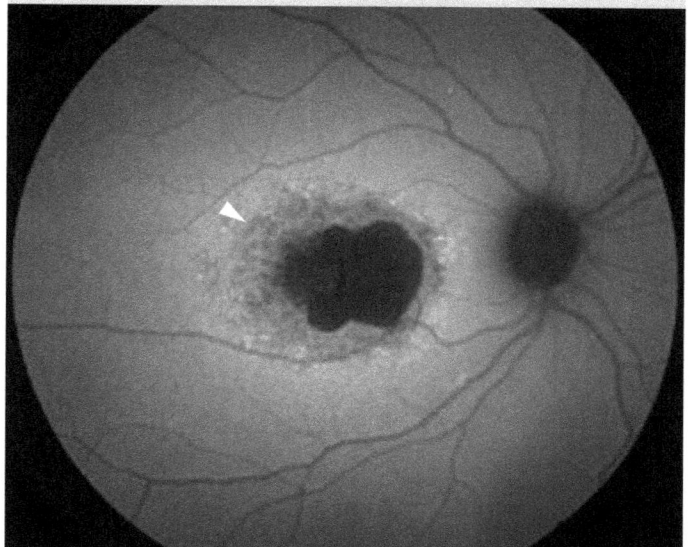

Figure 45: Autofluorescence image of a patient with Stargardt disease. Please note that flecks surrounding the macular degeneration indicated by the arrow

Stage 2 is characterized by the presence of flecks 8 μm diameter surrounding fovea centralis. Patients in this stage generally possess normal peripheral visual fields, however, subnormal cone and rod responses are observed, where these individuals display a prolonged period for dark adaptation. Dark adaptation is defined as the amount of time that it takes for rhodopsin from rod cells, or similar molecules existing in the cone cells, to reconstitute itself and adjust to the darkness. Stage 3 is characterized by the presence of diffuse yellowish flecks. At this stage the choriocapillaris of the macula lutea will demonstrate atrophy. These individuals display an even longer period of dark adaptation and an elevated threshold for the depolarization of the rod and cone cells (i.e. takes additional light to depolarize the photoreceptor cells). During this stage, individuals may experience a degree of peripheral or mid-peripheral field impairment. Stage 4 is characterized by the presence of even more diffused yellowish flecks coupled with macular atrophy, while extensive choriocapillaris as well as retinal pigment epithelial cell atrophy are also observed. Additionally, the individual may experience prolonged dark adaptation and further

Action Potential Formation

Presently, it is believed that both rod and cone photoreceptor cells initiate action potential in a similar manner. Although numerous unresolved biochemical and physiological issues remain unanswered and require further elucidation, a general overview of the process of action potential formation in the retina is provided.

- The plasma membrane located in the outer segment of the photoreceptor cells contains ligand gated sodium/calcium (Na^+/Ca^{++}) cation channels. In complete darkness, this cation channel is kept open by the presence of the ligand, cyclic guanosine mono-phosphate (cGMP), which allows the unimpeded flow of Na^+ and Ca^{++} into the cytosol of the outer segment of the photoreceptor cell.
- Also in complete darkness, the potassium channels located in the inner segment of the photoreceptor cells are kept open as potassium (K^+) cations are allowed to flow out of the cell.

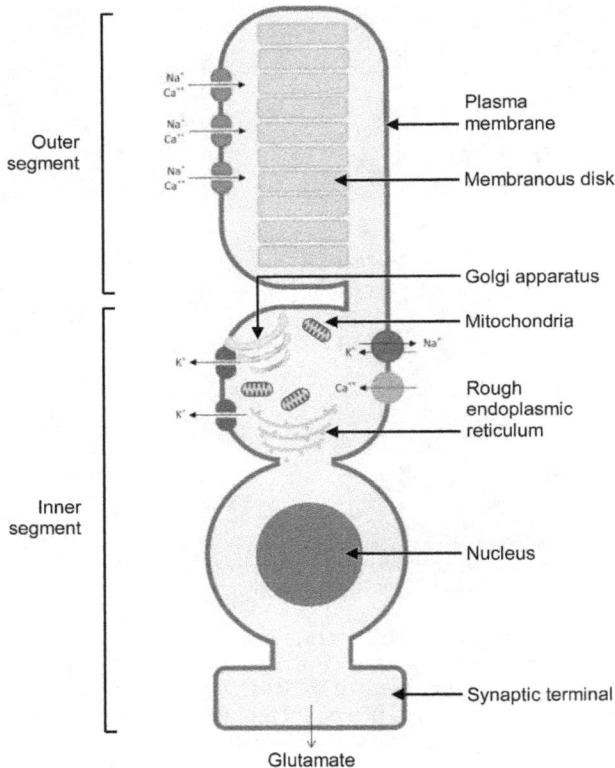

Figure 46: Illustration of the flow of cations in and out of the photoreceptor (rod) cells. Please note that the above graphic demonstrates the constant influx of Na^+ and Ca^{++}, and constant efflux of K^+ form the photoreceptor cell, which generates the dark current and causes the constant release of the neurotransmitter glutamate. Please note that due to the dark current the photoreceptor cell's membrane potential is measured at -40 mV

- At the same time, the sodium/potassium pumps and calcium pumps located in the inner segment of the photoreceptor cell continues to actively transport Na^+ and Ca^{++} out of the cell, while K^+ are actively transported into the cytosol of the photoreceptor cell. Due to the continuous influx and efflux of cations, a constant current called dark current is generated, while the membrane potential of the photoreceptor cell is measured at -40 mV rather than -60 mV to -70 mV of normal neurological tissues (Figure 46).

- This dark current causes the photoreceptor cells to continually release the neurotransmitter, glutamate.
- The constant release of glutamate by the photoreceptor cells prevents the bipolar cells from depolarizing.
- In complete darkness, the long hydrocarbon chain of retinal is in 11-cis structural format. However, when a photon (particle/unit of light) strikes retinal, the long hydrocarbon chain is converted from 11-cis structural format to 11-trans format via a process known as photoisomerization (Figure 47).

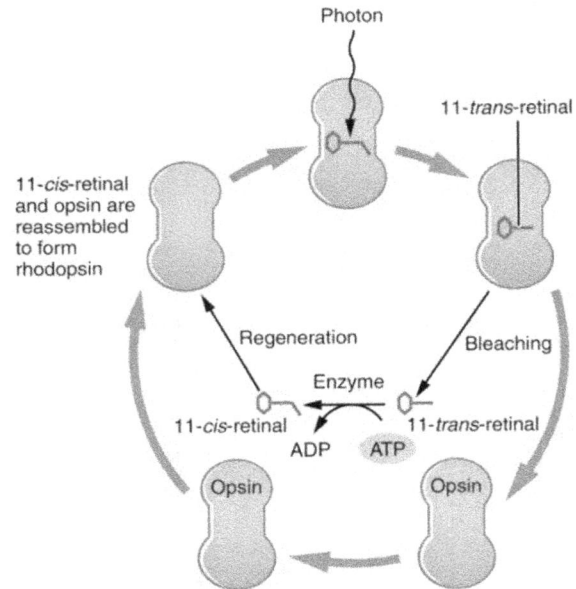

Figure 47: Illustration of photoisomerization

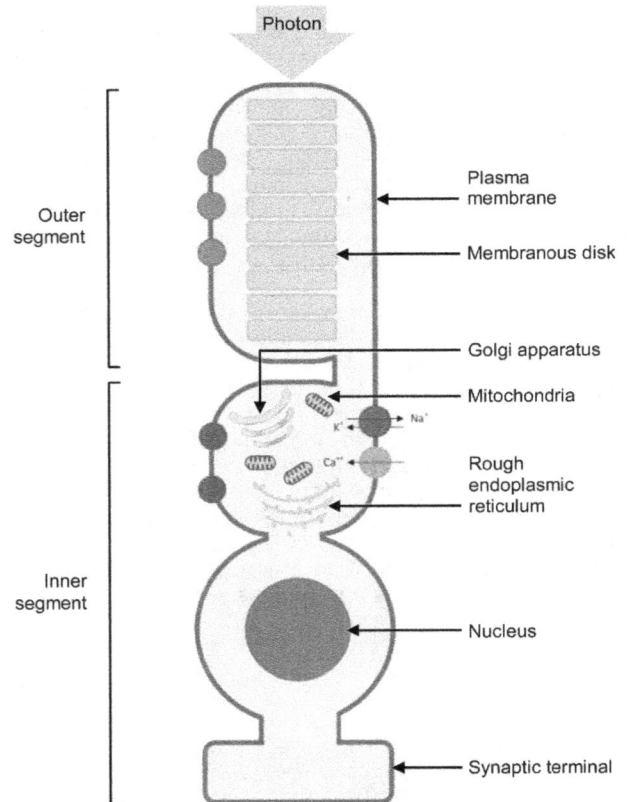

Figure 48: Illustration of a photoreceptor (rod) cells after receiving a photon. Once a photon is received by the photoreceptor cell the electrochemical reaction causes the Na^+, Ca^{++} and K^+ channels to close and the membrane potential to be restored to -60 to -70 mV. Please note that once the membrane potential is restored, the photoreceptor cell will halt the release of glutamate

- The conformational change of retinal results in the activation of opsin which in turn activates a G-protein complex called transducin.
- Once activated, transducin activates a membrane bounded enzyme called phosphodiesterase which is responsible for removing cGMP from the cation channel.
- The removal of cGMP causes an conformational change in the cation channel where it discontinues the influx of Na$^+$ and Ca^{++} into the cytosol of the outer segment of the photoreceptor cell. Nonetheless, the sodium/potassium and calcium pumps located in the inner segment of the photoreceptor cell will continue to actively transport Na$^+$ and Ca^{++} out of the cell, while bringing in K$^+$ form the extracellular matrix. Quickly, the membrane potential of the photoreceptor cell will proceed from -40 mV to -60 to -70 mV which terminates the dark current and halts the release of glutamate (Figure 48).
- Once glutamate released by the photoreceptor cells is halted, the bipolar cells become uninhibited and an action potential is generated. The bipolar cells will then stimulate the ganglion cells and in turn generate an action potential.
- This action potential generated by the ganglion cells will be carried by its axons which exit the retina at the optic disk and exits the eye as the optic nerve (II).
- The optic nerve (II) exits the orbit via the optic foramen and enters the cranial cavity. At the base of the brain just anterior to the pituitary gland, the optic nerves (II) transmitting action potential from the medial retina of both eyes will crisscrosses at the optic chiasma and projects via the optic tracts to the opposite sides of the brain. The optic nerve (II) transmitting action potential from the lateral retina will not crisscross at the optic chiasma but will be sent to the same side of the brain.

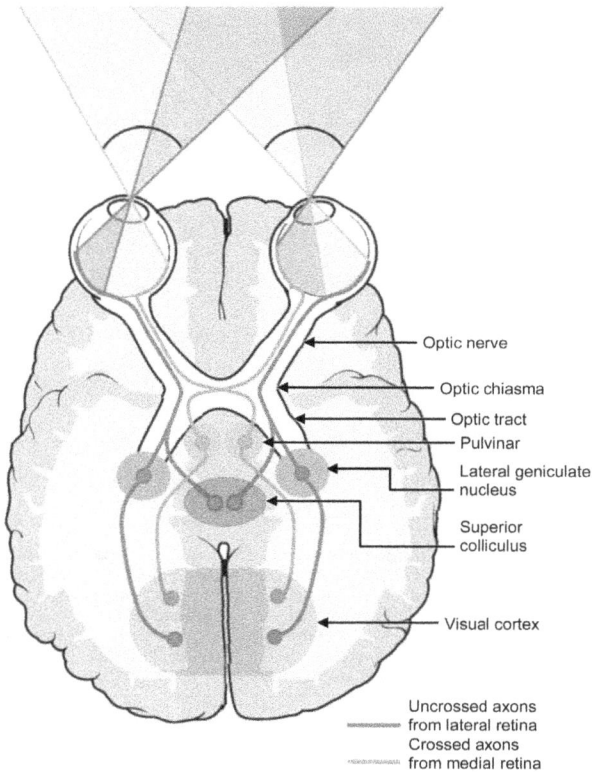

Figure 48: Illustration of the visual pathway

- Most of the axons (optic tract) from the lateral retina will synapse with the lateral geniculate nuclei of the thalamus. Some of the axons will separate from the optic tract and terminate at the superior colliculi which is the center of the visual reflexes. From the lateral geniculate nuclei of the thalamus the visual information will be transmitted via nerve

fibers of the optic radiations to the visual cortex located in the occipital lobe of the cerebrum. It is the responsibility of the visual cortex to translate the visual signals into mental images before transmitting it to various locations of the brain where it is evaluated (Figure 49).
- The axons from the medial retina will synapse with the pulvinar nucleus of the thalamus. The pulvinar nucleus is responsible for visual attention functions pertaining to perceptual temporal-order judgments and in saccadic decision. Perceptual temporal order judgments enable an individual to determine the relative timing of two spatially separate events, while saccadic decision-making is based on the accumulation of sensory evidence in favor of various alternative movement responses in a race to a final decision (decision threshold). From the pulvinar nucleus, the visual information will be transmitted via nerve fibers of the optic radiations to the visual cortex located in the occipital lobe of the cerebrum (Figure 49).

QUESTIONS

1. Which of the following is (are) the major category or categories of sensory information?
   a. Chemical agents
   b. Electromagnetic
   c. Mechanical force
   d. Metaphysical
   e. Space and time
   f. Supernatural
   g. Answers are a, b and c
   h. Answers are d, e and f

2. Which of the following term is defined as the type of stimulus or sensation that a particular type of sensory receptors produces?
   a. Duration
   b. Intensity
   c. Location
   d. Modality

3. Which of the following are the most common types of modality?
   a. Vision
   b. Hearing
   c. Smell
   d. Balance
   e. Taste
   f. All of the above
   g. None of the above

4. Since all the nerves transmit their information the same way (via action potential), the brain distinguishes one modality from another via which of the following methodology?
   a. The duration of the signal
   b. The intensity of the signal
   c. The origin of the signal
   d. The termination point of the signal
   e. The type of sensory receptor that is transmitting the signal
   f. Both a and c
   g. Both d and e

5. Which of the following term is defined as the place from where the sensory information is being sent, as well as, the specific nerve track by which the information is transmitted?
   a. Duration
   b. Intensity
   c. Location
   d. Modality

6. Which of the following term or phrase is defined as the assigned area from where all sensory neurons can receive stimuli?
   a. Area
   b. Location
   c. Receptive field
   d. Sphere of influence
   e. Termination signal

7. If the receptive field is measured at a diameter of 14 cm, what is the area of this field?
   a. 38.5 cm^2
   b. 50.25 cm^2
   c. 120 cm^2
   d. 153.9 cm^2
   e. 220.74 cm^2

8. If the receptive field is measured at a diameter of 0.2 cm, what is the area of this field?
   a. 0.031 cm^2
   b. 1.13 cm^2

c. 120 cm²
d. 153.9 cm²
e. 2.14 cm²

9. Which of the following term is defined as the strength of the signal being transmitted?
a. Duration
b. Intensity
c. Location
d. Modality

10. Which of the following term is defined as the amount or frequency in which a nerve will fire over a period of time?
a. Duration
b. Intensity
c. Location
d. Modality

11. Which of the following is a type of phasic receptor?
a. Hair receptor
b. Olfactory receptor
c. Pain receptor
d. Pressure receptors
e. Temperature receptor
f. All of the above
g. None of the above

12. Which of the following is a type of receptor that will generate a quick burst of action potentials when it is initially stimulated but will quickly adapt to the stimuli?
a. Hair receptor
b. Olfactory receptor
c. Pain receptor
d. Pressure receptors
e. Temperature receptor
f. All of the above
g. None of the above

13. If a receptor adapts to the stimuli, what could cause it to fire again?
a. Continue stimulation with the same intensity
b. Different type of stimuli
c. Stop the stimulation
d. Stronger and more intense stimuli

14. Which of the following is a type of receptor that adapt slowly and will generate a steady and continuous stream of actions potentials to the CNS?
a. Encapsulated receptor
b. Phasic receptor
c. Proprioceptor
d. Simple receptor
e. Tonic receptor
f. Both a and d
g. Both c and e

15. Which of the following is defined as a devise that is capable of convert one form of energy into another?
a. Transducer
b. Transformer
c. Transistor
d. Transmitter
e. Transmogrifier

16. Photoreceptors are activated by which of the following stimuli?
a. Intense mechanical or thermal stimuli
b. Ligands
c. Photonic energy
d. Temperature
e. Vibrations, pressure and tension

17. Chemoreceptors are activated by which of the following stimuli?
a. Intense mechanical or thermal stimuli
b. Ligands
c. Photonic energy
d. Temperature
e. Vibrations, pressure and tension

18. Nociceptors are activated by which of the following stimuli?
a. Intense mechanical or thermal stimuli
b. Ligands
c. Photonic energy
d. Temperature
e. Vibrations, pressure and tension

19. Mechanoreceptors are activated by which of the following stimuli?
a. Intense mechanical or thermal stimuli
b. Ligands
c. Photonic energy
d. Temperature
e. Vibrations, pressure and tension

20. Which of the following sensory receptor responds to stimuli originated outside of the body?
a. Exteroceptors

b. Integral receptors
c. Interoceptors
d. Multiceptors
e. Transceptors
f. Visceral receptors

21. Which of the following sensory receptor responds to stimuli originated inside of the body?
a. Exteroceptors
b. Integral receptors
c. Interoceptors
d. Multiceptors
e. Transceptors
f. Visceral receptors

22. Which of the following sensory receptor is responsible for sensing the position and movement of the body?
a. Exteroceptors
b. Integral receptors
c. Interoceptors
d. Multiceptors
e. Proprioceptors
f. Transceptors
g. Visceral receptors
h. Both a and c
i. Both c and e

23. Which of the following is (are) a part of the general somesthetic (somatosensory) senses?
a. Chemoreceptors
b. Mechanoreceptors
c. Nociceptors
d. Thermoreceptors
e. All of the above
f. None of the above

24. Which of the following is (are) a part of the special senses?
a. Photoreceptors
b. Mechanoreceptors of the inner ear
c. Chemoreceptors of the tongue
d. Chemoreceptors of the nose
e. All of the above
f. None of the above

25. Which of the following is categorized as the type of senses found in the cranial region of the body?
a. General somesthetic senses
b. Multiple senses
c. Propriosenses
d. Special senses
e. Visceral senses

26. The ability to detect pH pain, and pressure within the skin is classified as which of the following type of senses?
a. General somesthetic senses
b. Multiple senses
c. Propriosenses
d. Special senses
e. Visceral senses

27. Both temperature and pain receptors are classified as which of the following type of sensory receptor?
a. Encapsulated
b. Phasic
c. Photoreceptors
d. Proprioceptors
e. Tonic
f. Unencapsulated
g. Both a and e
h. Both b and f

28. The fluctuation in temperatures causes different _____ located on the _____ to open. The opening of the channels allows both _____ and in lesser amounts, _____ to enter the neuroplasm thereby producing a local potential. If the local potential managed to reach the pre-set all-or-none levels, also known as _____, an action potential will be generated. Repolarization of these fibers is achieved by allowing _____ to leave the neuroplasm while the restoration of the membrane potential is accomplished through active transport proteins such as _____ and _____.
a. Protein channels
b. Thermoreceptors
c. Sodium
d. Calcium
e. Potassium
f. Sodium/potassium pumps
g. Calcium pumps
h. Threshold
i. Chloride
j. Magnesium

29. At present, how many temperature-sensitive channels have been identified?
a. 5
b. 10
c. 18
d. 28
e. 32
f. 46

30. Temperature-sensitive channels are divided into _____, _____ and _____ channels.
   a. Ambient-sensitive channels
   b. Cold-sensitive channels
   c. Heat-sensitive channels
   d. Visceral-sensitive channels
   e. Warm-sensitive channels
   f. Answers are b, c and e
   g. Answers are a, d and e

31. Which of the following channel is activated by temperatures equal or below ≥26°C?
   a. Aδ
   b. C fibers
   c. TRPA1
   d. TRPM8
   e. TRPV1
   f. TRPV2
   g. TRPV3
   h. TRPV3
   i. TRPV4
   j. Answers are a, b and f

32. Which of the following channel is activated by menthol, eucalyptol and icilin?
   a. Aδ
   b. C fibers
   c. TRPA1
   d. TRPM8
   e. TRPV1
   f. TRPV2
   g. TRPV3
   h. TRPV3
   i. TRPV4

33. Which of the following channel is activated by temperatures equal or below ≥17°C?
   a. Aδ
   b. C fibers
   c. TRPA1
   d. TRPM8
   e. TRPV1
   f. TRPV2
   g. TRPV3
   h. TRPV3
   i. TRPV4
   j. Answers are a, b and f

34. Which of the following channel is activated by icilin but are insensitive towards menthol and eucalyptol?
   a. Aδ
   b. C fibers
   c. TRPA1
   d. TRPM8
   e. TRPV1
   f. TRPV2
   g. TRPV3
   h. TRPV3
   i. TRPV4
   j. Answers are a, b and f

35. It is understood that the sensation of cold could also simultaneously trigger the perception of pain. At which of the following temperature will pain receptors be activated by the decreased temperature?
   a. ≥10°C
   b. ≥12°C
   c. ≥15°C
   d. ≥18°C
   e. ≥20°C

36. Which of the following channel is (are) activated by temperatures equal or below ~24–34°C?
   a. Aδ
   b. C fibers
   c. TRPA1
   d. TRPM8
   e. TRPV1
   f. TRPV2
   g. TRPV3
   h. TRPV3
   i. TRPV4
   j. Answers are a, b and f

37. Which of the following channel is (are) activated by temperatures equal or below ~31–39°C?
   a. Aδ
   b. C fibers
   c. TRPA1
   d. TRPM8
   e. TRPV1
   f. TRPV2
   g. TRPV3
   h. TRPV3
   i. TRPV4

j. Answers are a, b and f

38. Which of the following channel is (are) activated by camphor?
   a. Aδ
   b. C fibers
   c. TRPA1
   d. TRPM8
   e. TRPV1
   f. TRPV2
   g. TRPV3
   h. TRPV3
   i. TRPV4

39. Which of the following channel is (are) activated by temperatures equal or below ≥43°C?
   a. Aδ
   b. C fibers
   c. TRPA1
   d. TRPM8
   e. TRPV1
   f. TRPV2
   g. TRPV3
   h. TRPV3
   i. TRPV4
   j. Answers are a, b and f

40. Which of the following channel is (are) activated by capsaicin and acids?
   a. Aδ
   b. C fibers
   c. TRPA1
   d. TRPM8
   e. TRPV1
   f. TRPV2
   g. TRPV3
   h. TRPV3
   i. TRPV4

41. Which of the following channel is (are) activated by temperatures equal or below ≤52°C?
   a. Aδ
   b. C fibers
   c. TRPA1
   d. TRPM8
   e. TRPV1
   f. TRPV2
   g. TRPV3
   h. TRPV3
   i. TRPV4
   j. Answers are a, b and f

42. Which of the following is the active ingredient of chili peppers?
   a. Camphor
   b. Capsaicin
   c. Eucalyptol
   d. Icilin
   e. Menthol
   f. Methanol

43. Pain is defined as a physical discomfort or distress caused by noxious stimulation, illness or injury. Although highly unpleasant, pain is actually one of the most important senses that allow individuals to between adapt and survive the environment. For example, pain enable individuals to sense damages and trigger evasive responses to avoid further injuries, as well as, formulating memories whereby preventing similar event from happening in the future. However, there are some diseases that prohibit an individual from sensing pain. Which of the following disease(s) may result in an individual from sensing pain?
   a. Amyloidosis
   b. Diabetes insipidus
   c. Diabetes mellitus
   d. Heart disease
   e. Leprosy
   f. Plague
   g. Both a and e
   h. Both c and e

44. Which of the following are comprise of mostly myelinated phasic receptors? This type of pain receptor is also referred to as fast pain receptors.
   a. Aδ
   b. C fibers
   c. TRPA1
   d. TRPM8
   e. TRPV1
   f. TRPV2
   g. TRPV3
   h. TRPV3
   i. TRPV4

45. Aδ fibers could be divided into two subtypes. Which of the following is responsible for responding to intense mechanical (bruises and lacerations etc.) stimuli?
   a. Aδ chemosensitive nociceptors
   b. Aδ ligandsensitive nociceptors

c. Aδ mechanosensitive nociceptors
d. Aδ mechanothermal nociceptors
e. C chemosensitive fibers
f. C ligandsensitive fibers
g. C mechanosensitive fibers
h. C mechanothermal fibers

46. Aδ fibers could be divided into two subtypes. Which of the following is responsible for responding to both intense mechanical and thermal stimuli?
a. Aδ chemosensitive nociceptors
b. Aδ ligandsensitive nociceptors
c. Aδ mechanosensitive nociceptors
d. Aδ mechanothermal nociceptors
e. C chemosensitive fibers
f. C ligandsensitive fibers
g. C mechanosensitive fibers
h. C mechanothermal fibers

47. Both subtypes of Aδ fibers are associated with which of the following type of pain?
a. Fast pain which produces a feeling of sharp, localized, stabbing pain that are highly localized
b. Mechanoreceptor induced pain that is generally associated with localized, stabbing pain which are generally localized
c. Slow pain that are tonic receptors that generally associated (but not exclusively) with deep somatic pain and visceral pain
d. Thermoreceptor induced pain that is generally associated with localized, stabbing pain which are generally localized

48. C fibers are associated with which of the following type of pain?
a. Fast pain which produces a feeling of sharp, localized, stabbing pain that are highly localized
b. Mechanoreceptor induced pain that is generally associated with localized, stabbing pain which are generally localized
c. Slow pain that are tonic receptors that generally associated (but not exclusively) with deep somatic pain and visceral pain
d. Thermoreceptor induced pain that is generally associated with localized, stabbing pain which are generally localized

49. Which of the following type of pain is defined as long lasting deep aching, burning or nauseating pains that are less localized and are generally associated with muscles and joints?
a. Deep somatic pain
b. Referred pain
c. Superficial somatic pain
d. Superficial visceral pains
e. Visceral pains

50. Which of the following type of pain is defined as dull aching non-localized pains that are associated with internal stretching, chemical irritations and ischemia (restriction in blood supply to tissues, causing a shortage of oxygen and nutrients)? Often, this type of pain is accompanied by the sensation of nausea.
a. Deep somatic pain
b. Referred pain
c. Superficial somatic pain
d. Superficial visceral pains
e. Visceral pains

51. Which of the following unencapsulated receptors are responsible for sensing light touch, texture, edges and shapes? These receptors are located in the base of the epidermis (basal layer) and are associated with specialized sensory transducers (amplifiers) that modify the stimulus received by the sensory nerve endings.
a. Hair receptors
b. Krause-end bulb
c. Lamellated (Pacinian) corpuscles
d. Ruffini corpuscles
e. Tactile (Meissner) corpuscles
f. Tactile Merkel sensory receptors

52. Which of the following unencapsulated receptors are found surrounding the bulb of hair follicles? These receptors are phasic mechanoreceptors respond to any pressure that bends the hair.
a. Hair receptors
b. Krause-end bulb
c. Lamellated (Pacinian) corpuscles
d. Ruffini corpuscles
e. Tactile (Meissner) corpuscles
f. Tactile Merkel sensory receptors

53. Which of the following encapsulated receptors are phasic mechanoreceptors that are commonly found in the dermal papillae of the skin? These receptors are concentrated at the fingertips, palms, eyelids, nipples, lips and genital regions, and respond to texture, edges and shapes? These encapsulated receptors are ovoid-shaped and consist of two-three nerve fibers embedded within a mass of modified Schwann cells.
a. Hair receptors
b. Krause-end bulb
c. Lamellated (Pacinian) corpuscles
d. Ruffini corpuscles
e. Tactile (Meissner) corpuscles
f. Tactile Merkel sensory receptors

54. Which of the following encapsulated receptors are phasic mechanoreceptors and respond to texture, edges and shapes? These receptors are ovoid in shape constructed with a single sensory nerve with numerous dendritic spines encapsulated by connective tissue. These receptors are commonly found in the skin, lips, tongue and conjunctiva.
a. Hair receptors
b. Krause-end bulb
c. Lamellated (Pacinian) corpuscles
d. Ruffini corpuscles
e. Tactile (Meissner) corpuscles
f. Tactile Merkel sensory receptors

55. Which of the following encapsulated receptors are phasic receptors that respond to deep pressure, stretch, tickle, and vibration? These receptors are constructed of a single dendrite surrounded first by modified Schwann cells in the inner layer and concentric layers of fibroblasts cells on the outer layers. These receptors are generally found in the hands, feet, breasts and genitalia, peritoneum of bones and joint capsules.
a. Hair receptors
b. Krause-end bulb
c. Lamellated (Pacinian) corpuscles
d. Ruffini corpuscles
e. Tactile (Meissner) corpuscles
f. Tactile Merkel sensory receptors

56. Which of the following encapsulated receptors are tonic receptors that respond to heavy touch, stretching of the skin pressure and joint movements? These sensory receptors are constructed with a few sensory nerve fibers surrounded by a connective tissue capsule. These receptors are commonly found in the dermis, subcutaneous tissue, ligaments, tendons, and joint capsules.
a. Hair receptors
b. Krause-end bulb
c. Lamellated (Pacinian) corpuscles
d. Ruffini corpuscles
e. Tactile (Meissner) corpuscles
f. Tactile Merkel sensory receptors

57. Somatosensory pathways are the manner by which sensory signals reaches the brain. Somatosensory signals from the cranial region travel via which of the following cranial nerves? Please circle all that applies.
a. Abducens
b. Facial
c. Hypoglossal
d. Oculomotor
e. Olfactory
f. Optic
g. Accessory
h. Trigeminal
i. Trochlear
j. Vagus
k. Vestibulocochlear
l. Glossopharyngeal

58. Somatosensory signals from the cranial region travel to the brain stem from the cranial nerves. Which of the following term is also used to describe these cranial nerves?
a. First order neurons
b. Second order neurons
c. Third order neurons
d. Fourth order neurons
e. Fifth order neurons
f. Sixth order neurons

59. Somatosensory signals from the cranial region travel from the cranial nerves to the _____ and _____ of the brain stem.
a. Cerebellum
b. Cerebrum
c. Hypothalamus
d. Medulla Oblongata
e. Mid brain
f. Pons
g. Telencephalon
h. Thalamus

60. Somatosensory signals from the cranial region travel to the brain stem. At the brain stem, these nerves will synapse with the _____ before decussate to the opposite side of the brain stem. From the brain stem, after decussating to the opposite side of the brain stem, these nerves will proceed to _____, synapse with _____ before being transmitted to the _____ of the cerebral cortex for analysis and integration.

The only exception from the above transmission scheme is the proprioceptors from the _____ which transmit the sense of position in relation to balance from the muscles of chewing or also known as _____. These proprioceptive signals connect to the _____ in the brainstem and directly carry the signals to the _____ for integration and analysis.

a. Cerebellum
b. Cerebrum
c. 5° order neurons
d. 1° order neurons
e. 4° order neurons
f. Frontal lobe
l. Mid brain
m. Respiration
n. Parietal lobe
o. Pons
p. 2° order neurons
q. 6° order neurons

g. Hypothalamus
h. Mastication
i. Maxilla branch
j. Mandibular branch
k. Medulla Oblongata

r. Ipsilateral
s. thalamus
t. Temporal lobe
u. Contralateral Thalamus
v. 3° order neurons

b. Goblet cell
c. Lingual papillae
d. Supporting cells
e. Taste buds
f. Taste cells
g. Taste hairs
h. None of the above

61. Somatosensory signals from the peripheries and the viscera also travels by the way of three interconnected neurons called the 1°, 2° and 3° order neurons. The 1° neurons are _____ in structure and has its cell body located within the _____. The dendritic end of the 1° neurons is stretched to the peripheral tissues which could be either encapsulated or unencapsulated.

If the afferent is receiving and transmitting fine touch, proprioception and vibration sensory signals, the axon of the 1° neurons will be _____ and fast in signal transmission. These 1° neurons will venture up to the brain stem via the _____ which is composed of _____ and the _____ nerve tracts. Both of these nerve tracts transmit sensory information related to tactile, vibration, proprioception and movement to the brain. At the _____ (**Answer A**), the 1° neurons will synapse with the 2° neurons, which are _____ in structure. The 2° neurons then decussate across the midline to the opposite side of the **Answer A** before proceeding to the _____ (**Answer B**). At the **Answer B** the 2° neuron will synapse with 3° neurons, which are _____ in structure. These 3° neurons will then transmit the sensory information to the _____ of the cerebral cortex for analysis and integration.

The exception of the above stated scheme is the transmitted signals for proprioception. Proprioception signals from the 1° neurons could synapse with the 2° neurons at the _____. This 2° neurons will then decussate across the spinal column and travel up the spinal cord via the _____ which are located within the _____. These nerve tracts convey unconscious proprioception signals from muscles and joints of the legs directly to the _____.

1. Anterior funiculi
2. Anterior horn
3. Anterior spinocerebellar tracts
4. Anterior spinothalamic tract
5. Bipolar
6. Brain stem
7. Cerebellum
8. Cerebrum
9. Dorsal root ganglia
10. Fasiculus cuneatus
11. Fasiculus gracilis
12. Frontal lobe
13. Gracile fasciculus
14. Hypothalamus
15. Lateral funiculi
16. Lateral Horn
17. Lateral spinothalamic tract
18. Limbic system
19. Dorsal column Medial Lemniscus System
20. Medulla oblongata
21. Mid brain
22. Multipolar
23. Myelinated
24. Occipital lobe
25. Parietal lobe
26. Pons
27. Posterior horn
28. Pseudounipolar
29. Spinoreticular tract
30. Temporal lobe
31. Thalamus
32. Unmyelinated
33. Ventral horn
34. Ventral Root

62. If the afferent is receiving and transmitting pain, temperature and course touch, the 1° neurons synapse with the 2° multipolar neurons in the _____ of the spinal cord. The 2° multipolar neurons will then decussate across to the opposite side of the spinal cord and join the ① _____, located in the _____, ② _____, located in the _____ and ③ _____ before venturing to the _____ (**Answer A**). At **Answer A**, the 2° multipolar neurons will synapse with the 3° multipolar neurons where most of the sensory information is sent to the _____ of the cerebral cortex for analysis and integration.

In addition, pain signals may also travel up to the brainstem via the spinal reticular tract. For example, the 1° neurons transmitting pain signals will synapse with the 2° multipolar neurons in the _____ of the spinal cord. The 2° multipolar neurons will then decussate across to the opposite side of the spinal cord and join the _____ before venturing to the reticular formation in the _____. At the reticular formation, the 2° multipolar neurons will synapse with 3° multipolar neurons before the information is relayed to the _____ and the _____ (Hint: the command and control center for the endocrine system). The pain signals transmitted by the spinoreticular tract activate intense arousal of the injured individual, as well as, triggering emotional responses such as nausea and fear and behavior responses such as somatic reflexes.

1. Anterior funiculi
2. Anterior horn
3. Anterior spinocerebellar tracts
4. Anterior spinothalamic tract
5. Bipolar
6. Brain stem
7. Cerebellum
8. Dorsal Column
9. Dorsal root ganglia
10. Fasiculus cuneatus
11. Fasiculus gracilis
12. Frontal lobe
13. Gracile fasciculus
14. Hypothalamus
15. Lateral funiculi
16. Lateral Horn
17. Lateral spinothalamic tract
18. Limbic system
19. Medial Lemniscus System
20. Medulla oblongata
21. Mid brain
22. Multipolar
23. Myelinated
24. Occipital lobe
25. Parietal lobe
26. Pons
27. Posterior horn
28. Pseudounipolar
29. Spinoreticular tract
30. Temporal lobe
31. Thalamus
32. Unmyelinated
33. Ventral horn
34. Ventral Root

63. The anterior region comprises of approximately two-thirds of the tongue and it is covered by small projections called _____.

a. Basal cells

64. What is the name given to the mushroom-shaped protrusions found on the anterior 2/3 of the tongue? It is estimated that ~3 taste buds are found located within each of these protrusion.
a. Filiform papillae
b. Foliate papillae
c. Fungiform papillae
d. Lingua villosa
e. Lingual papillae
f. Terminal sulcus
g. Vallate (circumvallate) papillae

65. What is the name given to the parallel ridges (multiple protrusions) located on the side of the tongue, adjacent to the premolars and molar of the teeth? These protrusions are weakly developed in humans and most of the taste buds contain within these ridges deteriorates by the age of three.
a. Filiform papillae
b. Foliate papillae
c. Fungiform papillae
d. Lingua villosa
e. Lingual papillae
f. Terminal sulcus
g. Vallate (circumvallate) papillae

66. What is the name given to the tiny spike-shaped protrusions that are devoid of taste buds? These protrusions are weakly developed in humans and are mainly located at the tips of the tongue. In animals, the filiform papillae are well distributed and are used for grooming purposes.
a. Filiform papillae
b. Foliate papillae
c. Fungiform papillae
d. Lingua villosa
e. Lingual papillae
f. Terminal sulcus
g. Vallate (circumvallate) papillae

67. What is the name given to the protrusions that are found towards the terminal sulcus of the tongue? These are large protrusions that contain approximately twelve taste buds per papilla. It has been reported that most of the taste buds contained within the vallate papillae are bitter taste buds.
a. Filiform papillae
b. Foliate papillae
c. Fungiform papillae
d. Lingua villosa
e. Lingual papillae
f. Terminal sulcus
g. Vallate (circumvallate) papillae

68. Taste buds are garlic-shaped structures composed of three types of modified epithelial cells: _____, _____ and _____.
a. Basal cells
b. Endothelial cells
c. Fascicles
d. Goblet cells
e. Merkel cells
f. Statoconia
g. Supporting cells
h. Taste (gustatory) cells

69. It is known that each taste cell is connected to a (an) _____ extending from the taste nerve fiber.
a. Axon
b. Basal cell
c. Chromatophilic substance
d. Collateral
e. Dendrite
f. Soma
g. Supporting cell

70. Taste hairs possess numerous membrane receptors that are intended to bind to various chemical compounds found in foods and additionally, it is designed for which of the following purpose?
a. Increase action potential intensity
b. Increase action potential rate
c. Increase surface area
d. Movement
e. Production of enzymes
f. Production of mucous
g. Production of neurotransmitters

71. It has been calculated that the average life span of a taste cell is approximately 10 days and they are constantly being replaced through _____ (Hint: cellular division) from the adult stem cells called _____.
a. Basal cells
b. Endothelial cells
c. Fascicles
d. Goblet cells

e. Meiosis
f. Merkel cells
g. Mitosis
h. Statoconia
i. Supporting cells

72. The receptors proteins responding to tastes are grouped depending upon the type of chemicals or compounds it interacts with. An example of the transduction involves the activation of receptor proteins called _____, which in turn activates _____ located on the taste hairs. Once activated, the _____, _____, and _____ of the metabotropic receptor protein will separate. The _____ and _____ will form a dimer which will activate the _____ causing the exchange of a spent energy molecule called _____ with and charged energy molecule called _____. Subsequently, the subunit/energy molecule complex, also known as _____, causes a conformation change and subsequent activation of a membrane bounded enzyme called _____ (**Answer A**). **Answer A** catalyzes the conversion of an energy molecule called _____ into an intracellular ligand called _____ which in turn activates a channel protein called _____ and causes the gate to open. _____ (Hint: ions) which are kept at high concentration in the extracellular matrix of the cell will immediately flow down its concentration gradient and into the cytoplasm of the taste cell. The influx of these ions will generate a local potential. At the same time, the intracellular ligand will cause the closure of _____ thereby preventing _____ (Hint: ions that are kept in high concentrations in the cytoplasm) to escape out into the extracellular matrix. If the threshold is reached, an action potential will be generated and cause the release of a neurotransmitter called _____ into the synaptic cleft located on the basal end of the taste cell and trigger a local/action potential of the connecting taste nerve fiber.

The depolarized taste nerve fiber from the taste buds will continue its transmission to the brain via three routes. _____ will carry the sensory signals from the taste buds located in the front area of the tongue. The taste nerve fibers connected to taste buds that are located in the back portion (before reaching the terminal sulcus) of the tongue will send its message via _____. Finally, the taste nerve fiber arising from the taste buds found on the cheeks (interior), soft palate, pharynx and the epiglottis will be transmitted by the _____.

The taste nerve fibers (also referred to as 1° nerve fibers) found in all three cranial nerve will converge at the _____ of the medulla oblongata and synapse with the 2° neurons. The 2° neurons will transmit the sensory information to the _____ and the _____, which will activate autonomic reflexes such as salivation, gagging and vomiting, or transmit the information to the _____ which will relay the information to the _____ and the _____ of the cerebral cortex.

The reversal of this reaction is initiated through the removal of the ligand (chemical or compound form food) binding to the _____. Once the ligand is removed, the _____ (Hint a subunit) will hydrolyze the energy molecule _____ releasing a phosphate molecule and energy thereby breaking the interaction with **Answer A**, which in turn, causing it to deactivate. The deactivation of the membrane bounded enzyme will discontinue the production of the intracellular ligand called _____ which will result in the closure of the _____ (Hint: an ion channel) and stop the influx of _____ ions. At the same time the lack of cAMP will trigger the opening of the _____ (Hint: an ion channel) which result in the _____ flowing down its concentration gradient and out into the extracellular matrix. Once the taste cell reaches hyperpolarization, the sodium/potassium pump will activate and restore its membrane potential.

1.	Abducens	24.	Nitric oxide
2.	Accessory	25.	Noradrenaline
3.	Acetylcholine	26.	Oculomotor nerve
4.	Adenylate cyclase	27.	Olfactory nerve
5.	Amygdala	28.	Optic nerve
6.	ADP	29.	Parietal lobe
7.	ATP	30.	Potassium
8.	Calcium	31.	Serotonin
9.	Chloride	32.	Sodium
10.	cAMP	33.	Solitary nucleus
11.	Facial nerve	34.	Thalamus
12.	GDP	35.	Trigeminal nerve
13.	G-protein-coupled receptors (GPCRs)	36.	Trochlear nerve
14.	Glossopharyngeal nerve	37.	Vagus nerve
15.	GTP	38.	Vestibulocochlear
16.	Gusducin	39.	α-subunit
17.	Hypoglossal nerve	40.	α-subunit/GTP
18.	Hypothalamus	41.	β-subunit
19.	Insula	42.	β-subunit/GTP
20.	Ligand-gated Ca++ ion channel	43.	γ-subunits
		44.	γ-subunits/GTP
		45.	δ-subunits
21.	Ligand-gated Cl- ion channel	46.	δ-subunits/GTP
		47.	ε-subunits
22.	Ligand-gated K+ ion channel	48.	ε-subunits/GTP
		49.	μ-subunits
23.	Ligand-gated Na+ ion channel	50.	μ-subunits/GTP
		51.	σ-subunits
		52.	σ-subunits/GTP

73. Which of the following taste is generated by acidic substances?
a. Sour
b. Salty
c. Sweet

d. Bitter
e. Water
f. Umami

74. The intensity of the sour taste is proportional to the _____.
a. The amount of carbohydrates (sugars), alcohols, aldehydes (C=O), ketones (2 CH₃-C=O), etc.
b. The amount of free hydrogen ions (H⁺) available
c. The amount of free hydroxide ions (OH⁻) available
d. The amount of ionized salts (Na⁺, K⁺ and Cl⁻) available
e. The availability of aspartic and glutamic acids
f. The availability of by a long carbon chain organic substances containing nitrogen or alkaloids
g. The presence of water

75. The intensity of the salty taste is proportional to the _____.
a. The amount of carbohydrates (sugars), alcohols, aldehydes (C=O), ketones (2 CH₃-C=O), etc.
b. The amount of free hydrogen ions (H⁺) available
c. The amount of free hydroxide ions (OH⁻) available
d. The amount of ionized salts (Na⁺, K⁺ and Cl⁻) available
e. The availability of aspartic and glutamic acids
f. The availability of by a long carbon chain organic substances containing nitrogen or alkaloids
g. The presence of water

76. The intensity of the sweet taste is proportional to the _____.
a. The amount of carbohydrates (sugars), alcohols, aldehydes (C=O), ketones (2 CH₃-C=O), etc.
b. The amount of free hydrogen ions (H⁺) available
c. The amount of free hydroxide ions (OH⁻) available
d. The amount of ionized salts (Na⁺, K⁺ and Cl⁻) available
e. The availability of aspartic and glutamic acids
f. The availability of by a long carbon chain organic substances containing nitrogen or alkaloids
g. The presence of water

77. The intensity of the bitter taste is proportional to the _____.
a. The amount of carbohydrates (sugars), alcohols, aldehydes (C=O), ketones (2 CH₃-C=O), etc.
b. The amount of free hydrogen ions (H⁺) available
c. The amount of free hydroxide ions (OH⁻) available
d. The amount of ionized salts (Na⁺, K⁺ and Cl⁻) available
e. The availability of aspartic and glutamic acids
f. The availability of by a long carbon chain organic substances containing nitrogen or alkaloids
g. The presence of water

78. The intensity of the meaty taste is proportional to the _____.
a. The amount of carbohydrates (sugars), alcohols, aldehydes (C=O), ketones (2 CH₃-C=O), etc.
b. The amount of free hydrogen ions (H⁺) available
c. The amount of free hydroxide ions (OH⁻) available
d. The amount of ionized salts (Na⁺, K⁺ and Cl⁻) available
e. The availability of aspartic and glutamic acids
f. The availability of by a long carbon chain organic substances containing nitrogen or alkaloids
g. The presence of water

79. Water tastes are triggered by various water receptors located on the taste hairs of the taste buds. Research has shown that constant activation of the water receptors in the mouth will inhibit the production of _____ by the posterior pituitary.
a. Antidiuretic hormone (ADH) or vasopressin
b. Follicle stimulating hormone
c. Growth hormone
d. Lipotropin
e. Luteinizing hormone
f. Melatonin
g. Oxytocin
h. Prolactin

80. During which of the following phase are women more sensitive to smell?
a. Corpus luteum
b. Follicular phase
c. Luteal Phase
d. Mittelschmerz
e. Ovulatory Phase (Ovulation)
f. Premenstrual Syndrome

81. Olfactory cells are responsible for distinguishing odors. These cells are located within the _____ (**Answer A**) in the roof of the nasal passage; above the _____ and covering a portion of the _____, and nasal septum of each nasal fossa. **Answer A** consists of approximately ten to twenty million of 3 types of cells. These cells are: _____, _____, and _____, as well as, numerous mucous producing cells of the olfactory glands. Because of its constant exposure to the external environment, the life expectancy of an olfactory cell is approximately sixty days. Damaged or dying olfactory cells are replaced by new cells created through _____ (Hint: type of division) of the _____. _____ (Hint: type of cell), on the other hand, are believed to aide in maintaining the ionic and nutrient concentrations as well as providing physical support and protection to the olfactory cells.

a. Basal cells (adult olfactory stem cells)
b. Cribriform plate
c. Fission
d. Frontal bone
e. Inferior nasal concha
f. Meiosis
g. Middle nasal concha
h. Mitosis
i. Olfactory cells
j. Olfactory epithelium
k. Olfactory meniscus
l. Olfactory mucosa
m. Parietal bone
n. Superior nasal concha
o. Supporting cells
p. Temporal bone

82. The olfactory cells are _____ in embryonic origin.
a. Modified bipolar neurons
b. Modified multipolar neurons
c. Modified unipolar neurons
d. Modified Epithelial cells
e. Modified Glandular cells
f. Modified Muscular cells
g. Modified gastrointestinal cells

83. Olfactory cells are elongate-shaped with a large nucleus. The apical-ends of the olfactory cells are formed from modified _____ which form 10-20 _____ (**Answer A**). **Answer As** are formed to _____ (Hint: identify the function) of the olfactory cells and are the locations where the receptor proteins that bind to odor molecules, also known as _____ exist. The basal-end of olfactory cells narrows to form a (an) _____ (**Answer B**). **Answer Bs** from groups of receptor cells bundles together to form a _____ (**Answer C**). Various **Answer Cs** join together and leave the olfactory mucosa via the _____ (Hint: name the structure) of the _____ (Hint: name the bone). As a group, the **Answer Cs** form the _____ (Hint: name the cranial nerve), which then synapses with the dendrites of _____ (**Answer D**) and _____ (**Answer E**) cells within the pair of olfactory bulbs located just beneath the frontal lobe of the brain. It is understood that the locations where the olfactory nerve connects to **Answer D** and **E** is called _____ (**Answer F**). **Answer D** and **E** are commonly called projection neurons and are responsible for lateral inhibition which aid in sharpening the transmitting olfactory signals.

From **Answer F** the axons of **Answer D** and **E** forms the olfactory tracts which send its signals to the _____ located in the inferior region of the_____. From this location, the signal could be transmitted to other secondary olfactory regions of the brain such as the _____ and _____.

1. Abducens nerve
2. Accessory nerve
3. Axons
4. Chromatophilic substances
5. Cribriform foramina
6. Dendrites
7. Ethmoid bone
8. Facial nerve
9. Fascicle
10. Frontal bone
11. Frontal lobe
12. Glossopharyngeal nerve
13. Glomeruli
14. Hypoglossal
15. Increases the surface area
16. Mitral cells
17. Occipital bone
18. Occipital lobe
19. Oculomotor nerve
20. Odorants
21. Olfactory nerve
22. Olfactory cilia
23. Olfactory hairs
24. Olfactory microvilli
25. Orbitofrontal cortex
26. Optic nerve
27. Parietal bone
28. Parietal lobe
29. Paleocortex of the insula
30. Primary olfactory cortex
31. Provide movement
32. Soma
33. Temporal bone
34. Temporal lobe
35. Trigeminal nerve
36. Trochlear nerve
37. Tufted cells
38. Vagus nerve
39. Vestibulocochlear nerve

84. Which of the following secondary olfactory region(s) provides the perception of flavor?
a. Amygdala
b. Cribriform foramina
c. Fascicle
d. Glomeruli
e. Glossopharyngeal
f. Hippocampus
g. Hypoglossal
h. Hypothalamus
i. Orbitofrontal cortex
j. Paleocortex of the insula
k. Answers are a, f, h
l. Answers are b d e

85. Which of the following secondary olfactory region(s) provides the visceral reactions and emotional response to certain type of odor?
a. Amygdala
b. Cribriform foramina
c. Fascicle
d. Glomeruli
e. Glossopharyngeal
f. Hippocampus
g. Hypoglossal
h. Hypothalamus
i. Orbitofrontal cortex
j. Paleocortex of the insula
k. Answers are a, f, h
l. Answers are b d e

86. Odors are molecules that are able to evaporate and be carried by the inhaled

air stream. There are two groups of odorants found in the atmosphere. The _____ are capable of diffusing through the mucous layer and binds to the receptor protein located on the olfactory hairs. _____, on the other hand, require a transport molecule called _____ located in the mucous to aid in its passage through the water soluble layer before binding to the receptor protein.
a. Albumin
b. Fibrinogen
c. Hydrophilic odorants
d. Hydrophobic odorants
e. Intercellular odorants
f. Intracellular odorants
g. Odorant-binding protein
h. Paravascular odorants
i. Perivascular odorants

87. Once the odor molecule binds a receptor located on the _____, it activates a G-protein complex called _____. Once activated, the _____, _____, and _____ subunits of the GPCR will separate. The _____ and _____ will form a dimer, which in turn, activates the _____ causing it to exchange a spent energy molecule called _____ with a charged energy molecule called _____. Subsequently, the previously mentioned subunit and energy molecule forms an complex and causes a conformation change and subsequent activation of a membrane bounded enzyme called _____. This enzyme catalyzes the conversion of a charged energy molecule called _____ into an intracellular ligand called _____. This intracellular ligand, in turn, activates a _____ (Hint: ion channel) and causes the gate to open. _____ (Hint: a type of cation), which is kept at high concentration in the extracellular matrix of the cell will immediately flow down its concentration gradient and into the cytoplasm of the olfactory cell. This influx of previousl mentioned cation will generate a local potential. At the same time, the intracellular ligand will cause the closure of _____ (Hint: a type of ion channel) thereby preventing _____ (Hint: a type of cation), which is kept in high concentrations in the cytoplasm, from escaping out into the extracellular matrix. If the threshold is reached, an action potential will be generated and causes the release of a neurotransmitter called _____ into the synaptic cleft located on the basal end of the olfactory cells and trigger a local/action potential of the connecting _____ and _____ cells found in the olfactory bulbs.

The reversal of this reaction is initiated through the removal of the odorant binding to the _____. Once the odorant is removed, the _____ will hydrolyze the energy molecule _____ releasing a phosphate molecule and energy thereby breaking the interaction with the membrane bounded enzyme called _____, which in turn, causing it to deactivate. The deactivation of the membrane bounded enzyme will discontinue the production of the intracellular ligand called _____ which will prevent the result in the closure of the _____ (Hint: a type of channel) and stop the influx of _____ (Hint: a type of cation). At the same time the lack of cAMP will trigger the opening of the _____ (Hint: a type of channel), which results in the flow of _____ (Hint: a type of cation) down its concentration gradient and out into the extracellular matrix. Once the olfactory cell reaches hyperpolarization, the _____ (Hint: an active transport) will activate and restore its membrane potential.

a. Acetylcholine
b. Adenylate cyclase
c. ADP
d. ATP
e. cAMP
f. GDP
g. Glutamate
h. G-protein coupled receptors (GPCR)
i. GTP
j. Ligand-gated potassium (K⁺) ion channel
k. Ligand-gated sodium (Ca⁺⁺) ion channel
l. Ligand-gated sodium (Na⁺) ion channel
m. Sodium
n. Potassium
o. Mitral cells
p. Olfactory hair
q. Sodium/potassium pump
r. Tufted cells
s. Voltage-gated potassium (K⁺) ion channel
t. Voltage-gated sodium (Ca⁺⁺) ion channel
u. Voltage-gated sodium (Na⁺) ion channel
v. α-subunit
w. β-subunit
x. γ-subunit
y. δ-subunit
z. μ-subunit

88. The ear is divided into 3 sections. Which of the following is (are) designed to receive and transmitted vibrations?
a. Outer ear
b. Middle ear
c. Inner ear
d. Both a and b
e. Both b and c

89. The ear is divided into 3 sections. Which of the following is (are) designed to convert vibrations into action potentials and sending the signal to the brain?
a. Outer ear
b. Middle ear
c. Inner ear
d. Both a and b
e. Both b and c

90. The outer (external) ear is composed of the _____ (also known as the pinna), and the _____ (**Answer A**). Pinna is the fleshy visible portion of the external ear and serves as a funnel that directs airborne vibrations (sound waves) into the **Answer A** at optimal frequencies. The pinna is constructed out of an inner layer of _____ surrounded by keratinized _____ epithelium. In most animals, the pinna is controlled by muscles which enable them to move and focus on a particular sound.

a. Auditory canal
b. Auricle
c. Elastic cartilage
d. Fibrocartilage
e. Hyaline cartilage
f. Pseudostratified columnar
g. Simple columnar
h. Simple cuboidal
i. Simple squamous
j. Stratified columnar
k. Stratified cuboidal
l. Stratified squamous
m. Temporal bone
n. Transitional
o. Auditory tube

91. Which of the following is the term given to the deepest depression in the auricle?
a. Auditory canal
b. Auditory ossicles
c. Auditory tube
d. Concha
e. External acoustic meatus
f. Temporal bone
g. Tympanic cavity
h. Tympanic membrane

92. Which of the following is the term given to the ~3 cm long channel of the external ear that is formed to conduct/transmit sound waves?
a. Auditory canal
b. Auditory ossicles
c. Auditory tube
d. Concha
e. External acoustic meatus
f. Temporal bone
g. Tympanic cavity
h. Tympanic membrane

93. The auditory canal is supported by a layer of _____ at its opening and by the _____ of the skull for the remainder of its length.
a. Auditory canal
b. Auricle
c. Elastic cartilage
d. Fibrocartilage
e. Hyaline cartilage
f. Pseudostratified columnar
g. Simple columnar
h. Simple cuboidal
i. Simple squamous
j. Stratified columnar
k. Stratified cuboidal
l. Stratified squamous
m. Temporal bone
n. Transitional

94. The auditory canal is lined with which of the following epithelium?
a. Pseudostratified columnar
b. Simple columnar
c. Simple cuboidal
d. Simple squamous
e. Stratified columnar
f. Stratified cuboidal
g. Stratified squamous

95. The epithelial lining of the auditory canal possesses two glands. Which of the following is responsible for producing ear wax?
a. Ceruminous gland
b. Sebaceous glands
c. Sudoriferous glands
d. Apocrine glands
e. Eccrine gland

96. The epithelial lining of the auditory canal possesses two glands. Which of the following is responsible for produce an oily secretion called sebum that keeps the epithelial lining of the auditory canal moist as well as forming a hydrophobic layer to prevent water seepage?
a. Ceruminous gland
b. Sebaceous glands
c. Sudoriferous glands
d. Apocrine glands
e. Eccrine gland

97. The middle ear is located in the _____ (Hint a cavity) found within the _____ bone of the skull.
a. Abdominal cavity
b. Auricle
c. Eustachian tube
d. Frontal bone
e. Occipital bone
f. Parietal bone
g. Pelvic cavity
h. Temporal bone
i. Thoracic cavity
j. Tympanic cavity
k. Vertebral cavity

98. Which of the following is the term given to the structure of the middle ear commonly known as the ear drum?
a. Auditory canal
b. Auditory ossicles
c. Auditory tube
d. Concha
e. External acoustic meatus
f. Temporal bone
g. Tympanic cavity
h. Tympanic membrane
i. Bony labyrinth

99. The ear drum is innervated by the sensory branches of which of the following cranial nerve(s)? You may select more than one answer.
a. CN I – Olfactory
b. CN II – Optic
c. CN III – Oculomotor
d. CN IV – Trochlear
e. CN V – Trigeminal
f. CN VI – Abducens
g. CN VII – Facial
h. CN VIII – Vestibulocochlear
i. CN IX – Glossopharyngeal
j. CN X – Vagus
k. CN XI – Accessory
l. CN XII – Hypoglossal

100. Behind the tympanic membrane is an air-filled chamber 2-3 cm in width. The air enters this middle ear via the _____ which is connected to the nasopharynx.
a. Auditory canal
b. Auditory ossicles
c. Auditory tube
d. Concha
e. External acoustic meatus
f. Temporal bone
g. Tympanic cavity
h. Tympanic membrane

101. The eustachian tube is normally flattened and closed but could be opened through yawning or swallowing. The opening of this tube allows which of the following actions to occur?
a. Allows for the movements of the auditory ossicles
b. Allows more resonance to develop within the middle ear
c. Enables the transmission of the action potential
d. Equalizing the pressure of the middle ear with the atmospheric pressure
e. Permits additional sounds to enter the middle ear
f. Protect the middle ear from auditory damage by lessening the transfer of energy to the inner ear

102. Within the air-filled chamber are the three smallest bones in the human body. What is the general name given to these bones?
a. Auditory canal
b. Auditory ossicles
c. Auditory tube
d. Concha
e. External acoustic meatus
f. Temporal bone
g. Tympanic cavity
h. Tympanic membrane

103. Which of the following bone of the middle ear is also known as the 'hammer'?
a. Bony labyrinth
b. Incus
c. Macula sacculi
d. Macula utriculi
e. Malleus
f. Stapes
g. Statoconia

104. Which of the following bone of the middle ear is also known as the 'anvil'?
a. Bony labyrinth
b. Incus
c. Macula sacculi
d. Macula utriculi
e. Malleus
f. Stapes
g. Statoconia

105. Which of the following bone of the middle ear is also known as the 'stirrup?'
a. Bony labyrinth
b. Incus
c. Macula sacculi
d. Macula utriculi
e. Malleus
f. Stapes
g. Statoconia

106. The 'hammer' is attached to the inner surface of the _____ via ligaments.
a. Auditory canal
b. Auditory ossicles
c. Auditory tube
d. Concha
e. External acoustic meatus
f. Incus
g. Malleus
h. Oval window
i. Round window
j. Stapes
k. Temporal bone
l. Tympanic cavity
m. Tympanic membrane

107. The base of the 'stirrup' is attached to the _____ via ligaments.
a. Auditory canal
b. Auditory ossicles
c. Auditory tube
d. Concha
e. External acoustic meatus
f. Incus
g. Malleus
h. Oval window
i. Round window
j. Stapes
k. Temporal bone
l. Tympanic cavity
m. Tympanic membrane

108. The 'hammer' is attached to an involuntary striated muscle called _____.
   a. Macula sacculi
   b. Macula utriculi
   c. Malleus
   d. Stapedius
   e. Statoconia
   f. Tensor tympani

109. The 'stirrup' is attached to an involuntary striated muscle called _____.
   a. Macula sacculi
   b. Macula utriculi
   c. Malleus
   d. Stapedius
   e. Statoconia
   f. Tensor tympani

110. The muscle of the middle is designed to perform which of the following function(s)?
   a. Allows for the movements of the auditory ossicles
   b. Allows more resonance to develop within the middle ear
   c. Enables the transmission of the action potential
   d. Equalizing the pressure of the middle ear with the atmospheric pressure
   e. Permits additional sounds to enter the middle ear
   f. Protect the middle ear from auditory damage by lessening the transfer of energy to the inner ear

111. Which of the following is the terminology given to the process of decreasing the transfer of energy to the inner ear through the contraction of the involuntary striated muscles of the middle ear?
   a. Accommodation reflex
   b. Ankle jerk reflex
   c. Arthrokinetic reflex
   d. Asymmetric tonic neck reflex
   e. Auditory reflec
   f. Babinski reflex
   g. Tympanic reflex

112. The _____ marks the end of the middle ear and the beginning of the inner ear. The inner ear is a maze of channels called _____ (Answer A) formed within the _____ (Hint: a bone of the skull). Lining Answer A is a fleshy tube called the _____ (Answer B). In between the Answer A and B is a layer of fluids called the _____. The composition of this fluid is similar to that of the _____. Within Answer B is a fluid called the _____ which has a similar composition to _____.
   a. Bony labyrinth
   b. Cerebral spinal fluid
   c. Ear wax
   d. Endolymph
   e. Frontal bone
   f. Membranous labyrinth
   g. Occipital bone
   h. Oval window
   i. Parietal bone
   j. Perilymph
   k. Plasma
   l. Temporal bone

113. The bony labyrinth is subdivided into three subsections: _____, _____ and _____.
   a. Auricle
   b. Cochlea
   c. Concha
   d. Incus
   e. Malleus
   f. Semicircular canals
   g. Stapes
   h. Vestibule

114. Which of the following area of the bony labyrinth is involved in maintaining balance?
   a. Auricle
   b. Cochlea
   c. Concha
   d. Incus
   e. Malleus
   f. Semicircular canals
   g. Stapes
   h. Vestibule
   i. Both a and b
   j. Both f and h

115. Which of the following area of the bony labyrinth is involved in hearing?
   a. Auricle
   b. Cochlea
   c. Concha
   d. Incus
   e. Malleus
   f. Semicircular canals
   g. Stapes
   h. Vestibule
   i. Both a and b
   j. Both f and h

116. The vestibule is a thin layer of bone which contains a pair of endolymph-filled membranous sacs (chambers) called the _____ (anterior portion) and _____ (posterior portion).
   a. Bony labyrinth
   b. Endolymphatic duct
   c. Helicotrema
   d. Incus
   e. Macula sacculi
   f. Macula utriculi
   g. Maculae
   h. Malleus
   i. Saccule
   j. Stapes
   k. Statoconia
   l. Utricle

117. These endolymph-filled membranous sacs are interconnected via a slender passageway called _____.
   a. Bony labyrinth
   b. Endolymphatic duct
   c. Helicotrema
   d. Incus
   e. Macula sacculi
   f. Macula utriculi
   g. Maculae
   h. Malleus
   i. Saccule
   j. Stapes
   k. Statoconia
   l. Utricle

118. These endolymph-filled membranous sacs possess oval patches of 2-3 mm diameter of hair cells and supporting cells called the _____. For example, the oval patches found in the saccule are called _____, while the oval patches of hair cells found in the utricle are called _____.
   a. Bony labyrinth
   b. Endolymphatic duct
   c. Helicotrema
   d. Incus
   e. Macula sacculi
   f. Macula utriculi
   g. Maculae
   h. Malleus
   i. Saccule
   j. Stapes
   k. Statoconia
   l. Utricle

119. The macula sacculi are aligned _____ on the wall of this anterior membranous sac, while the macula utriculi are located _____ on the floor of this posterior membranous sac.
   a. Horizontally
   b. Laterally
   c. Obliquely
   d. Transversely
   e. Vertically

120. Each of the hair cells found within the maculae possess 40-70 _____ (Answer A) and only a single _____ (Answer B). The apical edges of the Answer A and B are embedded in a gelatinous structure called the _____. Superior to this gelatinous structure is a layer of calcium carbonate crystals called the _____.
   a. Endolymphatic duct
   b. Helicotrema
   c. Kinocilium
   d. Macula sacculi
   e. Macula utriculi
   f. Otolithic membrane
   g. Statoconia
   h. Stereocilia

121. The semicircular canals are a thin layer of bone that shelters and protects the semicircular ducts. The semicircular ducts, which are segments of the membranous labyrinth, are divided into anterior, posterior and lateral semicircular ducts. All three semicircular ducts are continuous with the _____ (enclosed within the vestibule) and are filled with a fluid called _____.
   a. Ampulla
   b. Bony labyrinth
   c. Crista ampullaris
   d. Cupula
   e. Endolymph
   f. Endolymphatic duct
   g. Helicotrema
   h. Incus
   i. Macula sacculi
   j. Macula utriculi
   k. Maculae
   l. Malleus
   m. Perilymph
   n. Saccule
   o. Stapes
   p. Statoconia
   q. Utricle

122. At the area where the semicircular ducts connect to the utricle, the semicircular ducts dilate to form a sac-like structure called the _____ (Answer A). Answer A is an expanded region that contains a raised mass of hair cells and supporting cells called the _____. Each of the hair cells found within the ampulla possesses 40-70 _____ and only a single _____ located on its apical domain. The hair cells as well as its apical extensions are embedded in a gelatinous structure called the _____.
   a. Ampulla
   b. Bony labyrinth
   c. Crista ampullaris
   d. Cochlea
   e. Cupula
   f. Endolymph
   g. Endolymphatic duct
   h. Helicotrema
   i. Kinocilium
   j. Macula sacculi
   k. Macula utriculi
   l. Maculae
   m. Malleus
   n. Modiolus
   o. Perilymph
   p. Saccule
   q. Scala media
   r. Scala tympani
   s. Scala vestibule
   t. Stereocilia
   u. Statoconia
   v. Utricle

123. The third part of the bony labyrinth is the organ associated with hearing called the _____, a coiled fleshy tube that formed anteriorly of the vestibule. This structure is a snail shell-shaped construct that coiled around a spongy bone called the _____, which provides some structure support.
   a. Ampulla
   b. Bony labyrinth
   c. Crista ampullaris
   d. Cochlea
   e. Cupula
   f. Endolymph
   g. Endolymphatic duct
   h. Helicotrema
   i. Incus
   j. Macula sacculi
   k. Macula utriculi
   l. Maculae
   m. Malleus
   n. Modiolus
   o. Perilymph
   p. Saccule
   q. Scala media
   r. Scala tympani
   s. Scala vestibule
   t. Stapes
   u. Statoconia
   v. Utricle

124. Within the cochlea are three ducts separated by membranes. The first (superior) chamber is called _____, which is filled with _____ (e.g. similar to cerebral spinal fluid; Answer A). The middle chamber is called the _____ also known as the cochlear duct which is filled with _____ (similar to blood plasma). The inferior chamber, also known as _____ (filled with Answer A) arises near the oval window and spirals up towards the tip of the snail shell before spiraling back down and terminates at the round window.
   a. Ampulla
   b. Bony labyrinth
   c. Crista ampullaris
   l. Maculae
   m. Malleus
   n. Modiolus

d. Cochlea
e. Cupula
f. Endolymph
g. Endolymphatic duct
h. Helicotrema
i. Incus
j. Macula sacculi
k. Macula utriculi

o. Perilymph
p. Saccule
q. Scala media
r. Scala tympani
s. Scala vestibule
t. Stapes
u. Statoconia
v. Utricle

125. Which of the following is an air-filled chamber of the middle ear and is covered by a thin membrane called the secondary tympanic membrane.

a. Auditory canal
b. Auditory ossicles
c. Auditory tube
d. Concha
e. External acoustic meatus
f. Incus

g. Malleus
h. Oval window
i. Round window
j. Stapes
k. Temporal bone
l. Tympanic cavity
m. Tympanic membrane

126. The scala media (cochlear duct) is a part of the membranous labyrinth and is separated from the scala vestibule by a thin membrane called _____. Within the scala media is a raised structure containing hair cells and supporting cells called the _____ (**Answer A**). **Answer A** is situated on top of the _____. On the apical side of the hair cells are numerous membranous protrusions called _____. Situated superiorly above these membranous protrusions is a gelatinous layer called the _____.

a. Basilar membrane
b. Incus
c. Inner hair cells (IHC)
d. Malleus
e. Organ of Corti
f. Outer hair cells (OHC)
g. Oval window

h. Round window Stapes
i. Stereocilia
j. Tectorial membrane
k. Temporal bone
l. Tympanic cavity
m. Tympanic membrane
n. Vestibular membrane

127. The hair cells of the organ of Corti are subdivided into _____ (**Answer A**) and _____ (**Answer B**). **Answer A** are arranged in a row along the medial side of the organ of Corti (facing the modiolus). Each of the **Answer A** possesses a cluster of 50-60 stereocilia of either tall or short in length. Please note that the apical tips of **Answer A** stereocilia are **NOT** embedded in the _____. The **Answer B** is arranged in a row along the lateral side of the organ of Corti (away from the modiolus). Each of the **Answer B** possesses a cluster of approximately 100 stereocilia with their apical edges embedded in the _____.

a. Basilar membrane
b. Helicotrema
c. Incus
d. Inner hair cells (IHC)
e. Malleus
f. Organ of Corti
g. Outer hair cells (OHC)
h. Oval window

i. Round window
j. Stapes
k. Stereocilia
l. Tectorial membrane
m. Temporal bone
n. Tympanic cavity
o. Tympanic membrane
p. vestibular membrane

128. The scala tympani is separated from the scala media by a thicker membrane called _____. Please note that the scala vestibule and scala tympani are continuous with one another through a narrow channel called _____.

a. Basilar membrane
b. Helicotrema
c. Incus
d. Inner hair cells (IHC)
e. Malleus
f. Organ of Corti
g. Outer hair cells (OHC)
h. Oval window

i. Round window
j. Stapes
k. Stereocilia
l. Tectorial membrane
m. Temporal bone
n. Tympanic cavity
o. Tympanic membrane
p. vestibular membrane

129. Which of the following is constructed out of 9+2 microtubule doublets?
a. Cilia
b. Kinocilium
c. Microvilli
d. Stereocilia
e. Both a and b
f. Both c and d

130. Which of the following is constructed out of actin cytoskeletal structures?
a. Cilia
b. Kinocilium
c. Microvilli
d. Stereocilia
e. Both a and b
f. Both c and d

131. Which of the following apical protrusion do not possess a kinocilium?
a. Ampulla
b. Organ of Corti
c. Saccule
d. Utricle

132. In the cochlea, the stereocilia (from the shortest to the tallest) are connected by a linear protein structure called the _____ (**Answer A**). The linear protein structure is constructed out of _____ and _____ proteins and is anchored at the apex of each stereocilia. In the vestibular organs, the stereocilia are also interconnected by the previously metioned linear protein structure, however, the connection between the last stereocilium and the kinocilium are formed via _____. This link between the last stereocilium and the kinocilium are formed between the apical-lateral sides of these structures.

The availability of **Answer A** allows the stereocilia to bend in unison either towards the tallest stereocilia (in the cochlear) or towards the kinocilium (in the vestibular organs) when pressure is exerted. It is known that the displacement of the hair bundle toward the kinocilium or the tallest stereocilia _____ (Hint: depolarize, hyperpolarize, repolarize) the hair cell, movements parallel to this plane toward the shortest stereocilia causes _____ (Hint: depolarize, hyperpolarize, repolarize), while movements perpendicular to the plane of symmetry have no effect on the hair cell's membrane potential.

a. 0.07 nm
b. 0.3 nm
c. 1 nm
d. 2 nm
e. -25 to -35 mV
f. -30 to 50 mV
g. -45 to -60 mV
h. Ca^{2+} cations
i. Cadherin-23 protein
j. Cl^- anions
k. Depolarizes
l. Hyperpolarizes
m. K^+ cations
n. Kinocilial link
o. Mechanically gated Ca^{2+} channel
p. Mechanically gated Cl^- channel

q. Mechanically gated K^+ channel
r. Mechanically gated Na^+ channel
s. Na^+ cations
t. Protocadherin-15 proteins
u. Repolarizes
v. Tip-link
w. Voltage gated Ca^{2+} channels
x. Voltage gated Cl^- channels
y. Voltage gated K^+ channels
z. Voltage gated Na^+ channels

133. The resting potential (polarized) of the hair cells is approximately _____ and it has been calculated that the threshold of the hair cells are approximately _____. Put simply, if the stereocilia moves the particular distance that the threshold is set towards the kinocilium or the tallest stereocilia, a particular channel called _____ will open thereby allowing _____ which are kept at high concentrations in the extracellular matrix to flow into the hair cell. The influx of these cations depolarizes the hair cell which in turn causes the opening of _____ (Hint: **Answer A**; type of K^+ channels) and _____ (Hint: **Answer B**; type of Ca^{2+} channels). The opening of **Answer A** will allow additional _____ (Hint: type of cations) to flow into the cell while the opening of **Answer B** will cause _____ (Hint: type of cations), which are also kept in high concentration in the extracellular matrix, to flow into the hair cell thereby generating an action potential and the subsequent release of neurotransmitter(s). The release of neurotransmitter (s) will excite the postsynaptic nerve (vestibular or cochlear) which relays the neuronal transmission to the brain. Presently, the exact neurotransmitter (ls) secreted by hair cells are unclear although numerous candidates such as acetylcholine, nitric oxide, gamma-aminobutyric acid (GABA), dopamine, enkephalin and dynorphin have been proposed.

a. 0.07 nm
b. 0.3 nm
c. 1 nm
d. 2 nm
e. -25 to -35 mV
f. -30 to 50 mV
g. -45 to -60 mV
h. Ca^{2+} cations
i. Cadherin-23 protein
j. Cl^- anions
k. Depolarizes
l. Hyperpolarizes
m. K^+ cations
n. Kinocilial link
o. Mechanically gated Ca^{2+} channel
p. Mechanically gated Cl^- channel

q. Mechanically gated K^+ channel
r. Mechanically gated Na^+ channel
s. Na^+ cations
t. Protocadherin-15 proteins
u. Repolarizes
v. Tip-link
w. Voltage gated Ca^{2+} channels
x. Voltage gated Cl^- channels
y. Voltage gated K^+ channels
z. Voltage gated Na^+ channels

134. When the head is erect, the otolith membrane and the stateconia are positioned _____ (Hint: identify the location) the hair cells of the macula utriculi, which is lying horizontally on the floor of this posterior membranous sac, in both ears. At this position, _____ will be generated.
a. A single action potential
b. Directly above
c. Inverted from
d. No action potential
e. To the side of
f. Downward
g. Rotated above

135. With the head erect, the stereocilia of the macula sacculi found in both ears (aligned vertically on the wall of this anterior membranous sac) are minimally pulled _____ (Hint: identify the location), therefore only a fraction of action potentials are generated indicating the head is in the erect position.
a. A single action potential
b. Directly above
c. Inverted from
d. No action potential
e. To the side of
f. Downward
g. Rotated above

136. When the head is tilted, the otolith membrane and the stateconia are positioned _____ (Hint: identify the location) the hair cells of the macula utriculi, which is lying horizontally on the floor of this posterior membranous sac, in both ears. At this position, _____ will be generated.

a. A single action potential
b. Directly above
c. Inverted from
d. No action potential
e. To the side of
f. Pulled downward
g. Rotated above

137. Please indicate if the following statement is true or false: with the head tilted, the stereocilia of the macula sacculi found in both ears (aligned vertically on the wall of this anterior membranous sac) begins to shift towards the direction of the movement. The movements of the stereocilia of the macula sacculi will result in the generation of an action potential.
a. True
b. False

138. The hair cells in the semicircular canal (found in the ampulla) respond to the rotational movements of the head. For example, when an individual's head rotates, the _____ (Hint: type of fluid) moves within the semicircular canal. This movement of the fluids cause equal directional movements of the _____ (gel like structure) and thereby distorts the _____. Again, if the _____ are distorted towards the _____ an action potential is formed but if the _____ are distorted away from the _____, no action potential will be generated.
a. Cupula
b. Endolymph
c. Kinocilium
d. Macula sacculi
e. Macula utriculi
f. Otolith membrane
g. Perilymph
h. Stateconia
i. Stereocilia

139. The plane of movement is directly related to each of the three semicircular canals. A horizontal movement such as shaking your head results in the hair cells of the _____ to be activated. A nod activates the hair cells of the _____. Finally, a tilt of the cranium results in the activation of the hair cells of the _____.
a. Anterior semicircular ducts
b. Inferior semicircular ducts
c. Lateral semicircular ducts
d. Medial semicircular ducts
e. Posterior semicircular duct
f. Superior semicircular ducts
g. Vertical semicircular ducts

140. All the action potential generated form the hair cells in the vestibular (balance) structures are carried away by the _____ (hint a branch or division) of the _____ (Hint: cranial nerve). This cranial nerve will proceed from the inner ear to the four _____ (Answer A) located on each side of the _____ and _____. Subsequently, the Answer As will transmit signals to other parts of the nervous system. For example, the Answer A will provide input to the _____ which in turn will properly maintain head movements, as well as, muscle tone and posture. Answer A will also provide input to the nuclei of the cranial nerves _____, _____ and _____, which will produce proper eye movements in an effort to compensate for the movement of the head. It will also provide input to the _____ which will alter the respiratory rate and blood pressure due to the change in posture. It will also provide inputs to the _____ to control the extensor muscles so that quick adjustments could be made to maintain balance. Finally, Answer A will send inputs to the _____ (Hint: major switch board of the brain) which will then transmit the information to the _____ of the cerebral cortex so that the individual will be consciously aware of the position and movement of the body.

a. Cerebellum
b. CN I – Olfactory
c. CN II – Optic
d. CN III – Oculomotor
e. CN IV – Trochlear
f. CN IX – Glossopharyngeal
g. CN V – Trigeminal
h. CN VI – Abducens
i. CN VII – Facial
j. CN VIII – Vestibulocochlear
k. CN X – Vagus
l. CN XI – Accessory
m. CN XII – Hypoglossal
n. Cochlear branch
o. Mandibular division
p. Maxillary division
q. Medulla oblongata
r. Ophthalmic division
s. Pons
t. Postcentral gyrus
u. Reticular formation
v. Thalamus
w. Vestibular branch
x. Vestibular nuclei
y. Vestibulospinal tracts

141. Which of the following is the simplest kind of sound wave?
a. Alpha waves
b. Beta waves
c. Delta waves
d. Epsilon waves
e. Omega waves
f. Sine wave

142. The simplest kind of sound wave travel through the air at approximately _____ km/h.
a. 100 km/h
b. 1235 km/h

c. 200 km/h
d. 2070 km/h
e. 275 km/h
f. 500 km/h

143. The frequency of sound is measured in terms of cycles (waves) per second in a unit called _____.
a. Amperes
b. Amplitude
c. Decibels
d. Hertz
e. Kilograms
f. Ounces
g. Pounds
h. Volt

144. If a sound wave is traveling at 300 cycles per second, what is its measurement in Hz?
a. 1000 Hz
b. 120 Hz
c. 150 Hz
d. 275 Hz
e. 300 Hz
f. 600 Hz

145. A high frequency sound (high pitch sounds) would have a _____ wavelength (~15,000 Hz) while a low frequency sound (low pitch sounds) would have a _____ wavelength (~100 Hz).
a. Short
b. Long

146. The intensity of sound is defined as the strength of a sound wave where the human ear interprets as volume (or loudness). The intensity of sound is measured in units known as _____.
a. Amperes
b. Amplitude
c. Decibels
d. Hertz
e. Kilograms
f. Ounces
g. Pounds
h. Volt

147. Which of the following frequency is defined as sonic?
a. <20 Hz
b. >20 kHz
c. 20 Hz to 20 kHz

148. Which of the following frequency is defined as ultrasonic?
a. <20 Hz
b. >20 kHz
c. 20 Hz to 20 kHz

149. Which of the following frequency is defined as infrasonic?
a. <20 Hz
b. >20 kHz
c. 20 Hz to 20 kHz

150. Sound waves are generated from an object and travels to the _____ (Hint: external ear structure) of the ear. This external ear structure focuses the sound into the _____ (Hint: where the Q-tip goes) which amplifies the frequencies of the sound wave in the range _____. The sound waves cause the _____ (Hint: ear drum) to move which in turn causes the displacement of the bones of the middle ear also known as _____. First the movement of the ear drum pushes the _____ (Hint: first bone) inward which causes the _____ (Hint: the skeletal muscle attached) to contract. In turn, the 'first bone' pushes on the 'second bone' called _____. Through their articulations, the second bone pushes on the _____ (Hint: third bone) which causes the _____ (Hint: the skeletal muscle attached) to contract. The movement of the 'third bone' produces a rocking motion at the _____ (Answer A). This rocking motion of the 'third bone' generates fluid (perilymph) waves (push-pull waves) in the _____. Like any fluids, the perilymph cannot be compressed therefore it responds to the pressure (push) by flowing away from the Answer A. This fluid pressure will then push the _____ (Hint: a membrane) downward which in turn pushes the endolymph in the _____ (Hint: cochlear duct) downwards as well. Sequentially, the endolymph will push down on the _____ (Hint: a membrane; Answer B) which will in turn push down on the perilymph in the _____. Finally, the secondary tympanic membrane of the _____ (which faces the air-filled chamber of the middle ear) is pushed outward to relieve the pressure. Inversely, the pressure generated by the pulling pressure will result in the reverse reaction.

As Answer B is being pushed up and down by the fluid pressure, the _____ (the gelatinous layer) remains stable and in place. Any movement upwards by Answer B will result in the stereocilia being jammed into the tectorial membrane and bend forward. As the tallest of the stereocilia bends forward, the _____ connecting to the other stereocilia will pull them forward as well. Please note that any bend that equal or exceed ~0.3 nm will result in the hair cell reaching threshold and the opening of the _____ and results in the influx of _____. The influx of these cations will cause the _____ and _____ (Hint: ion channels) to open which in turn will result in the additional influx of _____ and _____ cations. At this point, the hair cell is depolarized and resulting in the release of neurotransmitter(s).

1.	1 kHz to 2 kHz	17.	Oval window
2.	13 kHz to 22 kHz	18.	Round window
3.	3 kHz to 12 kHz	19.	Scala media
4.	Auditory canal	20.	Scala tympani
5.	Auditory ossicles	21.	Scala vestibule
6.	Auricles	22.	Stapedius
7.	Basilar membrane	23.	Stapes
8.	Ca++ cations	24.	Tectorial membrane
9.	Incus	25.	Tensor tympani
10.	K+ cations	26.	Tip-link
11.	Malleus	27.	Tympanic membrane
12.	Mechanically gated Ca^{2+} channel	28.	Vestibular membrane
13.	Mechanically gated Cl^- channel	29.	Voltage gated Ca^{2+} channels
14.	Mechanically gated K^+ channel	30.	Voltage gated Cl^- channels
15.	Mechanically gated Na^+ channel	31.	Voltage gated K^+ channels
16.	Na+ cations	32.	Voltage gated Na^+ channels

151. Variations in the intensity of sounds, also known as _____ will result in a vibrational variation in the basilar membrane. For example, if the variations in the intensity of sounds is _____, the movement of the basilar membrane will be minimal thereby the number of hairs cells stimulated will be limited. This limited stimulation will translate into limited action potentials being sent to the brain which the brain will interpret the input as a _____. Inversely, if the amplitude of the sound is _____, the movement of the basilar membrane will be equally as drastic. Intensive movement of the basilar membrane will result in greater numbers of hair cells being stimulated and equal numbers of actions potentials generated. The greater number of action potentials being sent to the brain will be interpreted as _____.

Frequency determination is based upon the structural differences of the _____ (Hint: a membrane; **Answer A**). For example, at the proximal end of the cochlea, **Answer A** is attached thereby making it stiff and relatively inflexible therefore making the _____ in this area stiff and relatively inflexible. The proximal end of the cochlea duct is also where the _____ (Hint: type of hair cells; **Answer B**) exists. On the distal (apex of the cochlea duct), **Answer A** is unattached thereby making it more flexible by comparison therefore making the _____ in this area relatively flexible. The distal end of the cochlea duct is also where the _____ (Hint: type of hair cells; **Answer C**) exists. It is known that the peak amplitude of high frequency sound is found at the _____ (Hint: proximal or distal end) while the peak amplitude of low frequency sounds is found at the _____ (Hint: proximal or distal end). Therefore, if the brain receives action potential mostly from **Answer B**, it is interpreted as high frequency (high pitch) sounds. If the receives action potential mostly form the **Answer C**, it will interpret it as low frequency (low pitch) sounds.

a.	Amperes	j.	Loud
b.	Amplitude	k.	Loudness
c.	Anterior semicircular ducts	l.	Low or soft
d.	Basilar membrane	m.	Organ of Corti
e.	Decibels	n.	Outer hair cell (OHC)
f.	Hertz	o.	Soft sound
g.	Inferior semicircular ducts	p.	Tectorial membrane
h.	Inner hair cells (IHC)	q.	Tympanic membrane
i.	Lateral semicircular ducts	r.	Proximal
		s.	Distal

152. Once an action potential is generated form the hair cells it will release its neurotransmitter(s) into the synaptic cleft of the adjacent bipolar sensory neurons. The soma of these bipolar sensory neurons forms the _____ (**Answer A**) near the modiolus. The axons of **Answer A** form the _____ (Hint: branch or division) of the _____ (Hint: a cranial nerve). Each ear sends its auditory signals via this cranial nerve to the _____ (**Answer B**) found in the _____. The **Answer B** sends its information (2° neuron) to the _____ (**Answer C**) located in the _____. In addition to receiving sensory information, **Answer C** also possesses efferent functions. For example, **Answer C** can send efferent back to the cochlea which are involved in cochlear tuning which allow individuals to better discriminate the pitch of various sounds. Additionally, **Answer C** could also sent motor signals via _____ and _____ (Hint: cranial nerves) to control the tensor tympani and stapedius (striated muscles) of the middle ear.

1.	Cerebellum	15.	Cochlear nuclei
2.	CN I – Olfactory	16.	Mandibular division
3.	CN II – Optic	17.	Maxillary division
4.	CN III – Oculomotor	18.	Medulla oblongata
5.	CN IV – Trochlear	19.	Olivary nucleus
6.	CN IX – Glossopharyngeal	20.	Ophthalmic division
7.	CN V – Trigeminal	21.	Pons
8.	CN VI – Abducens	22.	Parietal lobe
9.	CN VII – Facial	23.	Postcentral gyrus
10.	CN VIII – Vestibulocochlear	24.	Reticular formation
11.	CN X – Vagus	25.	Spiral ganglion
12.	CN XI – Accessory	26.	Thalamus
13.	CN XII – Hypoglossal	27.	Vestibular branch
14.	Cochlear branch	28.	Vestibular nuclei
		29.	Vestibulospinal tracts

153. The skeletal muscle that is responsible for opening the eyelid is called _____, while the skeletal muscle that is responsible for closing the eyelid is called _____. These muscles are innervated by _____ (Hint: cranial nerve). The eyelids are strengthened and supported by a fibrous connective tissue

plate called the _____ which is mainly located at the margins of the eyelids. Within the tarsal plate there are approximately 20-25 _____ glands which are modified sebaceous glands that produce and secrete _____ which is an oily substance that are formed to prevent the eyelids from sticking together and reduce the evaporation of tears from the surfaces of the eyes.

1.	Abduction	16.	Extorsion
2.	Adduction	17.	Inferior oblique
3.	CN I – Olfactory	18.	Inferior rectus
4.	CN II – Optic	19.	Intorsion
5.	CN III – Oculomotor	20.	Lateral rectus
6.	CN IV – Trochlear	21.	Levator palpebrae superioris
7.	CN IX – Glossopharyngeal	22.	Medial rectus
8.	CN V – Trigeminal	23.	Meibum
9.	CN VI – Abducens	24.	Orbiculais oculi
10.	CN VII – Facial	25.	Superior oblique
11.	CN VIII – Vestibulocochlear	26.	Superior rectus
12.	CN X – Vagus	27.	Tarsal glands
13.	CN XI – Accessory	28.	Tarsal plate
14.	CN XII – Hypoglossal	29.	Trochlea
15.	Cochlear branch		

154. The _____ (**Answer A**) consist of transparent mucous membranes that secrete a thin mucous film which lubricates the eyeballs. With the exception of the _____, **Answer A** covers the anterior aspect or the _____ of the eyes, as well as, lining the interior surfaces of the eyelids. This transparent structure is richly innervated, vascularized and highly sensitive to pain. If irritated, the blood vessels within **Answer A** will dilate and provide the appearance of blood shot eyes.

a.	Alpha amylase	h.	Lacrimal glands
b.	Conjunctivas	i.	Lacrimal punctums
c.	Cornea	j.	Lacrimal sac
d.	Lacrimal apparatus	k.	Lysozyme
e.	Lacrimal canal	l.	Nasolacrimal duct
f.	Lacrimal caruncle	m.	Sclera
g.	Lacrimal gland ducts	n.	Retina

155. The _____ consists of the _____ (tear glands) and various ducts that are utilized to transport and drain their secretions. These glands are located within the frontal bone and are responsible for producing tears. Once tears are produced, it is transported via 12 short ducts also called _____ that lead to the surface of the _____ (**Answer A**). Tears are used to moisten **Answer A** and washes debris away from the surfaces of the eyes. It also serves to act as a diffusion layer to allow gases such as oxygen and nutrients to be delivered to the conjunctiva. In addition to secreting tears, these glands also secrete a digestive enzyme called _____ which is responsible for sterilizing the eyeballs by preventing any potential infection by bacteria or virus.

a.	Alpha amylase	g.	Lacrimal glands
b.	Conjunctivas	h.	Lacrimal punctums
c.	Lacrimal apparatus	i.	Lacrimal sac
d.	Lacrimal canal	j.	Lysozyme
e.	Lacrimal caruncle	k.	Nasolacrimal duct
f.	Lacrimal gland ducts	l.	Retina

156. After the tears are secreted, it washes over the conjunctiva and the cornea before flowing to a fleshy mass called the _____ which is located at the medial aspects of the eyes. Within this fleshy mass, tiny pores called the _____ exists. These pores are the opening or the drain that leads to the _____ and eventually the _____ (**Answer A**) where excess tears collect. **Answer A** opens to the _____ which is responsible for carrying tears to the inferior meatus of the nasal cavity where it would travel up the nasal canal (on ordinary conditions) and swallowed.

a.	Alpha amylase
b.	Conjunctivas
c.	Lacrimal apparatus
d.	Lacrimal canal
e.	Lacrimal caruncle
f.	Lacrimal gland ducts
g.	Lacrimal glands
h.	Lacrimal punctums
i.	Lacrimal sac
j.	Lysozyme
k.	Nasolacrimal duct

157. The movement of the eyes is controlled by six extrinsic eye muscles which are attached to the surface of the eyes and the walls of the orbit.

The _____ course along the medial walls of the orbit before its tendon inserts through a fibrocartilage ring called the _____ and thus redirects it inferiorly before inserting onto the anterior region of the eyes. This muscle's function includes _____ (internally rotate), abduction and depress the eyes when moving from side to side. This muscle is innervated by the _____ (Hint: a cranial nerve).

The _____ extends form the medial wall of the orbit to the inferior laterals side of the eyes. This muscle's functions includes _____ (external rotation), adduction and elevates the eyes when moving from side to side. This muscle is innervated by the _____ (Hint: a cranial nerve).

The _____ moves the eye upward and secondarily _____ or rotates the top of the eye toward the nose. This muscle is innervated by the _____ (Hint: a cranial nerve).

The _____ moves the eye downward and secondarily _____ or rotates the top of the eye away from the nose. This muscle is innervated by the _____

(Hint: a cranial nerve).

The _____ is responsible for _____ or moving the eyes away from the nose. This muscle is innervated by the _____ (Hint: a cranial nerve).

Medial rectus is responsible for _____ or moving the eyes towards the nose. This muscle is innervated by the _____ (Hint: a cranial nerve).

1. Abduction
2. Adduction
3. CN I – Olfactory
4. CN II – Optic
5. CN III – Oculomotor
6. CN IV – Trochlear
7. CN IX – Glossopharyngeal
8. CN V – Trigeminal
9. CN VI – Abducens
10. CN VII – Facial
11. CN VIII – Vestibulocochlear
12. CN X – Vagus
13. CN XI – Accessory
14. CN XII – Hypoglossal
15. Cochlear branch
16. Extorsion
17. Inferior oblique
18. Inferior rectus
19. Intorsion
20. Lateral rectus
21. Levator palpebrae superioris
22. Medial rectus
23. Meibum
24. Orbiculais oculi
25. Superior oblique
26. Superior rectus
27. Tarsal glands
28. Tarsal plate
29. Trochlea

158. The human eyes are spherical structures and are composed of three layers of tunics, optical components and neuronal apparatus.

The _____ is the outer most layer and it is subdivided into two regions: _____ (**Answer A**) and _____ (**Answer B**). **Answer A,** also known as a white of the eye covers most of the surface of the eye. **Answer A** is composed of _____ (Hint: type of connective tissue) with high collagenous content and is well innervated and vascularized. **Answer B** is a transparent region that is created from modified **Answer A** which allows light to penetrate the eyes. **Answer B** is composed of collagen fibers with numerous fibroblast cells and is covered anteriorly by stratified squamous epithelium and posteriorly by simple squamous epithelium. The anterior stratified squamous epithelium contains _____ that provide the cornea with the ability to regenerate when damaged.

The middle layer of the eyes is called _____ also known as the _____. This layer consists of three regions. The first region _____ is a highly pigmented and vascularized layer located immediately behind the retina. This region is responsible for reducing reflected light within the eye thereby improving the contrast of retinal image. The second region is also known as _____. This region forms a muscular ring around the lens (it is responsible for supporting the iris and lens). This region is a smooth muscle ring that is controlled by the _____ of the autonomic nervous system. For example, the neurotransmitter known as _____ is secreted by this autonomic division causes the smooth muscle ring to contract while the lack of this neurotransmitter causes the smooth muscle relaxation. In addition, this region produces and secretes a plasma-like fluid known as _____. This plasma-like fluid is secreted into spaces between the iris and the lens called the _____ and flow into the space between cornea and iris called _____. The third region is called _____ and it is an adjustable diaphragm that is responsible for controlling the diameter of the central opening of the eyes called _____. This third region possesses two pigmented layer. The _____ which blocks stray light from reaching the retina and the _____, which contains pigmented cells known as _____ (contains melanin), which determines the eye color of an individual.

The inner most layer of the eyes is called _____ and this layer is consists of the neuronal portion of the eyes called the retina and the initial segment of the optic nerve.

a. Acetylcholine
b. Anterior border layer
c. Anterior chamber
d. Aqueous humor
e. Choroid
f. Chromatophores
g. Ciliary body
h. Cornea
i. Dense irregular
j. Dense regular
k. Iris
l. Noradrenaline
m. Parasympathetic division
n. Posterior chamber
o. Posterior pigment epithelium
p. Pupil
q. Sclera
r. Stem cells
s. Sympathetic division
t. Tunica fibrosa
u. Tunica interna
v. Tunica vasculosa
w. Uvea

159. Lenses of the eyes are light focusing structure that is constructed out of transparent cells called _____. Lenses are suspended behind the pupil by _____ which in turn are attached to the _____ (**Answer A**). **Answer A** manipulates the lens through contraction which in turn allows an individual to focus in on objects at various distances.

a. Ciliary body
b. Inferior oblique
c. Inferior rectus
d. Intorsion
e. Lateral rectus
f. Lens fibers
g. Levator palpebrae superioris
h. Medial rectus
i. Meibum
j. Orbiculais oculi
k. Superior oblique
l. Superior rectus
m. Suspensory ligaments
n. Tarsal glands
o. Tarsal plate
p. Trochlea

160. The contraction of the ciliary body will result in the pull of the suspensory ligament. The tightening of the suspensory ligament will cause the lens to _____, resulting in a _____ of its diameter. This allows an individual to focus in on objects that are _____. In contrast, relaxation of the ciliary body will relax

the pull of the suspensory ligament and in turn relaxes the lens. A relaxed lens will be allowed to reform its rounded shape and thereby _____ its diameter. A lens with _____ diameter will allow an individual to focus in on objects that are _____.

a. Accommodated
b. Far away
c. Flatten
d. Fully accommodated
e. Increase
f. Near
g. Reduction/decrease
h. Relaxed
i. Totally relaxed

161. The terminology used to describe manipulation of the lens to visualize objects that are distant is _____ while viewing very distant objects is characterized as _____.
a. Accommodated
b. Far away
c. Flatten
d. Fully accommodated
e. Increase
f. Near
g. Reduction/decrease
h. Relaxed
i. Totally relaxed

162. The terminology used to describe manipulation of the lens to visualize objects that are close is _____ while focusing in on an extremely close object is characterized as _____.
a. Accommodated
b. Far away
c. Flatten
d. Fully accommodated
e. Increase
f. Near
g. Reduction/decrease
h. Relaxed
i. Totally relaxed

163. What is the optical power (P) if the object is 40 meters (d_o) away and the distance between the cornea/lens and the retina is .02 meters (d_i).
a. 0.025 δ
b. 0.03 δ
c. 50 δ
d. 50.005 δ
e. 50.025 δ
f. 50.03 δ
g. 51.25 δ
h. 55.0 δ

164. What is the optical power (P) if the object is 200 meters (d_o) away and the distance between the cornea/lens and the retina is .02 meters (d_i).
a. 0.025 δ
b. 0.03 δ
c. 50 δ
d. 50.005 δ
e. 50.025 δ
f. 50.03 δ
g. 51.25 δ
h. 55.0 δ

165. What is the optical power (P) if the object is 0.2 meters (d_o) away and the distance between the cornea/lens and the retina is .02 meters (d_i).
a. 0.025 δ
b. 0.03 δ
c. 50 δ
d. 50.005 δ
e. 50.025 δ
f. 50.03 δ
g. 51.25 δ
h. 55.0 δ

166. Which of the following is defined as nearsighted?
a. Amblyopia
b. Asthenopia
c. Emmetropia
d. Hyperopic
e. Myopia
f. Refractive error
g. Both e and f
h. Both d and f

167. Which of the following is defined as farsighted?
a. Amblyopia
b. Asthenopia
c. Emmetropia
d. Hyperopic
e. Myopia
f. Refractive error
g. Both e and f
h. Both d and f

168. The retina commonly referred to as the outermost extension of the brain, is

a thin transparent neurological membrane that is responsible for transducing light signals into neurological signals. The retina is attached to the eye at two points. The first is called _____ and it is the location where the optic nerve leaves the eye, as well as the location where blood vessels enters and exits the eye. This location possesses no optic cells thereby it produces a _____ where no images are transmitted. The second attachment site is the _____ which is defined as the transitional junction between the retina and the ciliary body or where the photoreceptive area transitioning into a non-photoreceptive area. The remainder area of the retina is held in place by the pressure exerted by the _____.

Since the retina is only partially attached to the eye, it can be separated from the choroid of the eye by a forcible blow to the head or when the vitreous body retains insufficient pressure. This condition is known as _____ and could immediately result in blurred field of vision. Please note that the retina depends upon the _____ (capillaries of the choroid) for nourishment and waste disposal. If the detached retina is left medically unattended for an extended period of time, the retinal tissue may deteriorate and result in permanent blindness.

Directly inline and posterior to the center of the lens is an area called _____ and at its center is a tiny pit called _____. It is known that these locations are responsible for providing an individual to envision detailed images.

a. Anterior border layer
b. Anterior chamber
c. Aqueous humor
d. Blind spot
e. Choriocapillaris
f. Choroid
g. Chromatophores
h. Ciliary body
i. Cones
j. Cornea
k. Detached retina

l. Fovea centralis
m. Macula lutea
n. Mesopic vision
o. Photopic vision
p. Retinal pigment epithelium (RPE)
q. Rods
r. Optic disc
s. Ora serrata
t. Vitreous humor

169. The retina is composed of numerous layers of cells that contribute to visual perceptions. For example, the outer most (posterior) layer of the retina is the _____ (**Answer A**) which is located anterior to the _____ (capillaries of the choroid). This location is a darkly pigmented layer that interacts with photoreceptor cells and is responsible for absorbing stray light thereby normal visual function could be maintained.

Lying superior to **Answer A** is the photoreceptor layer composed of two types of photoreceptor cells called _____ (**Answer B**) and _____ (**Answer C**). It is known that these photoreceptor cells are not nerve cells but are derived from neural glial cells related to epidermal cells. **Answer Bs** possess low spatial resolution but are extremely sensitive to light. It is understood that these cells are responsible for vision in dim light or _____. In contrast, **Answer Cs** possess high spatial resolution but are relatively insensitive towards light. It is understood that these cells are involved in bright-light vision or _____.

Visual perception is made through the contributions of both types of photoreceptor cells are referred to as the _____. As illumination increases towards normal daylight conditions, **Answer Cs** become the dominant photoreceptor while the contributions provided by the **Answer Bs** are nearly eliminated. This type of vision is referred to as the _____.

a. Anterior border layer
b. Anterior chamber
c. Aqueous humor
d. Blind spot
e. Choriocapillaris
f. Choroid
g. Chromatophores
h. Ciliary body
i. Cones
j. Cornea
k. Detached retina

l. Fovea centralis
m. Macula lutea
n. Mesopic vision
o. Photopic vision
p. Retinal pigment epithelium (RPE)
q. Rods
r. Optic disc
s. Ora serrata
t. Vitreous humor
u. Black and white vision
v. Color vision

170. Rhodopsin is a visual pigment that has a maximum sensitivity to light at wavelength _____. Rhodopsin is composed of two subunits. _____ (**Answer A**) is a transmembrane protein complex that is involved in mediating visual transduction in the retina. _____ (**Answer B**), on the other hand, is a yellowish photosensitive pigment that is derived from vitamin A and is covalently bounded to **Answer A**.

In complete darkness, the long hydrocarbon chain of **Answer B** is in _____ structural format. However, when a photon (particle/unit of light) strikes this molecule, the long hydrocarbon chain is converted to _____ format via a process known as _____. The conformational change of **Answer B** results in the activation of **Answer A** which in turn results in the alteration of the membrane potential of the photo-receptor cell.

In contrast, the outer segment of the cone cells is also composed of similar numbers of membranous disks but these are constructed out of invaginations of the plasma membrane from the apical ends of cone cells. Each of the cone disks contains the visual pigments called _____ (**Answer C**).

Answer Cs are composed of _____ and the photopigment _____ (**Answer D**) covalently linked to one another. There are three different types of **Answer Ds** based upon the wavelengths which they are most sensitive. For example, _____ (also known as short wavelength sensitive) allows maximum absorption (λ_{max}) at _____. The _____ (also known as mid-wavelength sensitive) allows

maximum absorption (λ_{max}) at _____. Finally, the _____ (also known as long wavelength sensitive) allows maximum absorption (λ_{max}) at _____.

a. 11-cis
b. 11-trans
c. 425 nm
d. 498 nm
e. 530 nm
f. 560 nm
g. Blue sensitive cone opsin
h. Cone opsin

i. Green sensitive cone opsin
j. Iodopsins
k. L-cone
l. M-cone
m. Photoisomerization
n. Red sensitive cone opsin
o. Retinal
p. Rod opsin
q. S-cone

171. Because there are three types of cone opsins, it is commonly stated that there are three types of cone cells that inhabit the human retina: blue sensitive cone also known as _____. Green sensitive cone, also known as _____, while red sensitive cone also known as _____.

a. 11-cis
b. 11-trans
c. 425 nm
d. 498 nm
e. 530 nm
f. 560 nm
g. Blue sensitive cone opsin
h. Cone opsin

i. Green sensitive cone opsin
j. Iodopsins
k. L-cone
l. M-cone
m. Photoisomerization
n. Red sensitive cone opsin
o. Retinal
p. Rod opsin
q. S-cone

172. The terminals of the photoreceptor cells (rods and cones) synapse with the _____ (**Answer A**) and _____ (**Answer B**). These cells are located at the _____ (**Answer C**) of the retina. **Answer As** are the most numerous cells in this layer of the retina and are the first order (1°) neurons of the visual pathway. **Answer As** stretches from the **Answer C** where its dendrites exist, through the _____, where their cell bodies lie and terminate in the _____ where the axonal terminals of these cells synapse with the _____. The **Answer Bs** located in the **Answer C** creates lateral connections between the terminal ends of photoreceptor cells, as well as, the dendritic ends of **Answer As**. **Answer Bs** play an important role in enhancing visual systems sensitivity to luminance contrast through a broad range of light intensities, modification in the intensity of light and the enhancing the perception of edges of objects.

_____, on the other hand, are found in the _____ and form horizontal connections between the axonal end of the _____ and the dendritic end of the _____. Like the **Answer Bs**, These cells modify and enhance visual signals that are transmitted.

a. Amacrine cells
b. Bipolar cells
c. Ganglion cells
d. Horizontal cells
e. Inner nuclear layer
f. Interplexiform cells
g. Lateral geniculate nucleus
h. Optic chiasma
i. Optic nerve
j. Optic tract
k. Outer plexiform layer
l. Inner plexiform layer

173. The _____ exists in the inner plexiform layer and forms feedback loops between the horizontal and amacrine cells found within the outer and inner plexiform layers respectively. These cells are interneurons that could be either excitatory or inhibitory depending on the cells that they are form connections and are responsible for increasing visual signals from some photoreceptor cells while decreasing signals from others, thereby enhancing visual contrast.

a. Amacrine cells
b. Bipolar cells
c. Ganglion cells
d. Horizontal cells
e. Inner nuclear layer
f. Interplexiform cells
g. Lateral geniculate nucleus
h. Optic chiasma
i. Optic nerve
j. Optic tract
k. Outer plexiform layer
l. Outer plexiform layer

174. Ganglion cells are the largest neuronal cells found in the retina and are second (2°) order neurons of the visual pathway. The ganglion cells lie immediately beside the vitreous body, and their axons form the _____ which crisscrosses and forms the optic chiasma. From this crisscross location, the _____ synapse with the _____ of the thalamus before giving rise to the third (3°) order neuron and send the visual signal to the optic lobe of the visual cortex.

a. Amacrine cells
b. Bipolar cells
c. Ganglion cells
d. Horizontal cells
e. Inner nuclear layer
f. Interplexiform cells
g. Lateral geniculate nucleus
h. Optic chiasma
i. Optic nerve
j. Optic tract

k. Outer plexiform layer
l. Outer plexiform layer

175. Presently, it is believed that both rod and cone photoreceptor cells initiate action potential in similar manner.

The plasma membrane located in the outer segment of the photoreceptor cells contains _____ (Hint: ion channels). In complete darkness, this cation channel is kept open by the presence of the intracellular ligand _____ which allows the unimpeded flow of _____ and _____ ions into the cytosol of the outer segment of the photoreceptor cell. Also in complete darkness the _____ located in the inner segment of the photoreceptor cells are kept open as _____ cations are allowed to flow out of the cell.

At the same time, the active transport proteins known as _____ and _____ located in the inner segment of the photoreceptor cell continue to actively transport the _____ and _____ ions out of the cell while _____ ions are actively transported into the cytosol of the photoreceptor cell. Due to the continuous influx and efflux of ions, a constant current called _____ is generated while the membrane potential of the photoreceptor cell is measured at -40 mV rather than -60 mV to -70 mV. This current cause the photoreceptor cells to continuously release neurotransmitters called _____.

In complete darkness, the long hydrocarbon chain of retinal is in _____ structural format. However, when a photon (particle/unit of light) strikes retinal, the long hydrocarbon chain is converted into _____ format via a process known as _____. The conformational change of retinal results in the activation of opsin which in turn activates a G-protein complex called _____. Once activated, this G-protein complex activates a membrane bounded enzyme called _____ which is responsible for removing the intracellular ligand called _____ from the cation channel.

The removal of the intracellular ligand causes an conformational change in the cation channel where it discontinue the influx of _____ and _____ ions into the cytosol of the outer segment of the photoreceptor cell. Nonetheless, the active transport proteins called _____ and _____ located in the inner segment of the photoreceptor cell will continue to actively transport _____ and _____ ions out of the cell. Quickly, the membrane potential of the photoreceptor cell will proceed from -40 mV to -60 to -70 mV which terminates the current called _____ and halts the release of the previously mentioned neurotransmitter.

Once neurotransmitter released by the photoreceptor cells is halted, the bipolar cells become uninhibited and an action potential is generated. The bipolar cells will then stimulate the ganglion cells and in turn generate an action potential. This action potential generated by the ganglion cells will be carried by its axons which exit the retina at the optic disk and exits the eye as the optic nerve (II).

a. 11-cis
b. 11-trans
c. Acetylcholine
d. Ca^{++}
e. Cl^-
f. Ca^{++} channels
g. Calcium pumps
h. cAMP
i. cGMP
j. cTMP
k. cUMP
l. Dark current
m. Glutamate
n. Ligand-gated K^+ channels
o. K^+

p. Ligand gated Cl^-/Ca^{++} channels
q. Ligand gated K^+/Ca^{++} channels
r. Ligand gated Na^+/Ca^{++} channels
s. Na^+
t. Na^+ channels
u. Nitric Oxide
v. Noradrenaline
w. Phosphodiesterase
x. Photoisomerization
y. Sodium/potassium pumps
z. Transducin

ANSWERS

1. g	2. d	3. f
4. g	5. c	6. c
7. d	8. a	9. b
10. a	11. f	12. f
13. d	14. g	15. a
16. c	17. b	18. a
19. e	20. a	21. c
22. i	23. e	24. e
25. d	26. a	27. h
28. a, b, c, d, h, e, f, g	29. d	30. b, e, c
31. d	32. d	33. c
34. c	35. a	36. i
37. g	38. g	39. e
40. e	41. j	42. b
43. h	44. a	45. c
46. d	47. a	48. c
49. a	50. e	51. f
52. a	53. e	54. b
55. c	56. d	57. b, h, j, l
58. a	59. d, f	60. p, t, u, n, j, h, p, a

61. 28, 9, 23, 19, 10, 11, 20, 22, 31, 22, 25, 27, 3, 15, 7	62. 27, 17, 15, 3, 1, 13, 31, 25, 27, 29, 6, 18, 14	63. C
64. c	65. b	66. a
67. g	68. a, g, h	69. e
70. c	71. g, a	72. 16, 13, 39, 41, 43, 41, 43, 39, 12, 15, 40, 4, 7, 10, 23, 32, 22, 30, 31, 11, 14, 37, 33, 5, 18, 34, 19, 39, 13, 39, 15, 10, 23, 32, 22, 30
73. a	74. b	75. d
76. a	77. f	78. e
79. a	80. e	81. l, n, b, i, o, a, h, a, o
82. a	83. 6, 23, 15, 20, 3, 9, 5, 7, 21, 16, 37, 13, 30, 34, 39, 25	84. i
85. k	86. c, d, g	87. p, h, v, w, x, w, x, v, f, i, b, d, e, l, m, j, n, g, o, r, h, v, i, b, e, l, m, j, n, q
88. d	89. c	90. a, b, c, l
91. d	92. a	93. d, m
94. g	95. a	96. b
97. j, h	98. h	99. e, j
100. c	101. d	102. b
103. e	104. b	105. f
106. m	107. h	108. f
109. d	110. f	111. g
112. h, a, l, f, j, b, d, k	113. b, f, h	114. j
115. b	116. i, l	117. b
118. g, e, f	119. e, a	120. h, c, f, g
121. q, e	122. a, c, t, i, e	123. d, n
124. s, o, q, f, r	125. i	126. o, e, a, j, k
127. d, g, l, l	128. a, b	129. e
130. f	131. b	132. v, i, t, n, k, l
133. g, b, q, m, y, w, m, h	134. b, d	135. f
136. e, a	137. a	138. b, a, i, i, c, i, c
139. c, a, e	140. w, j, x, s, q, a, d, e, h, u, y, v, t	141. f
142. b	143. d	144. e
145. a, b	146. c	147. c
148. b	149. a	150. 6, 4, 3, 27, 5, 11, 25, 9, 23, 22, 17, 21, 28, 19, 7, 20, 18, 24, 26, 14, 10, 29, 31, 10, 8
151. b, l, o, j, k, d, m, n, m, h, r, s	152. 25, 14, 10, 15, 18, 19, 21, 7, 9	153. 21, 24, 5, 28, 27, 23
154. b, c, m	155. c, g, f, b, j	156. e, h, d, i, k
157. 25, 29, 19, 6, 17, 16, 5, 26, 2, 5, 18, 1, 5, 20, 1, 9, 2, 5	158. T, q, h, i, r, v, w, e, g, m, a, d, n, c, k, p, o, b, f, u	159. f, m, a
160. c, g, f, e, e, b	161. h, i	162. a, d
163. g	164. d	165. h
166. g	167. h	168. r, d, s, t, k, e, m, l
169. p, e, q, i, n, o	170. d, p, o, a, b, m, j, o, h, g, c, i, e, n, f	171. l, q, k
172. b, d, k, e, l, c, a, l, b, c	173. f	174. i, j, g
175. r, i, s, d, n, o, y, g, s, d, o, l, m, a, b, x, z, w, i, s, d, y, g, s, d, o, l		

REFERENCES

Schepers RJ, Ringkamp M. Thermoreceptors and thermosensitive afferents. Neuroscience and Biobehavioral Reviews. 2010: 34, pp. 177–184

F. Guoping, Copray S, Huang EJ, Jones K, Yan Q, Walro J, Jaenisch R, Kucera J. Formation of a Full Complement of Cranial Proprioceptors Requires Multiple Neurotrophins. Developmental Synamics. 2000: 218, pp. 359–370

Lefkowitz RJ, Kobilka BK. Studies of G-protein-coupled receptors. The Royal Swedish Academy of Science. The Nobel Prize in Chemistry 2012

Sakaguchi H, Tokita J, Müller U, Kachar B. Tip links in hair cells: molecular composition and role in hearing loss. Curr Opin Otolaryngol Head Neck Surg. 2009: 17(5), pp. 388-93

Purves D, Augustine GJ, Fitzpatrick D, Katz LC, LaMantia A-S, McNamara JO, Williams SM. Sunderland Neuroscience, 2nd edition. 2001: pp. 283-336

PHOTO AND GRAPHIC BIBLIOGRAPHY

1. Figure 1: Graphic is designed by BruceBlaus and is reprinted under the license of Creative Commons Attribution 3.0 Unported
2. Figure 2: Graphic is designed by BruceBlaus and is reprinted under the license of Creative Commons Attribution 3.0 Unported
3. Figure 3: Graphic is designed by BruceBlaus and is reprinted under the license of Creative Commons Attribution 3.0 Unported
4. Figure 4: Graphic is designed by BruceBlaus and is reprinted under the license of Creative Commons Attribution 3.0 Unported
5. Figure 5: Graphic is designed by BruceBlaus and is reprinted under the license of Creative Commons Attribution 3.0 Unported
6. Figure 6: Graphic is designed by BruceBlaus and is reprinted under the license of Creative Commons Attribution 3.0 Unported
7. Figure 7: Graphic is designed by BruceBlaus and is reprinted under the license of Creative Commons Attribution 3.0 Unported
8. Figure 8: Graphic designed by Patrick J. Lynch, medical illustrator, 2006. Reprint permission granted under Creative Commons Attribution 2.5 License
9. Figure 9: Graphic designed by Patrick J. Lynch, medical illustrator, 2006. Reprint permission granted under Creative Commons Attribution 2.5 License. Additional graphics used is designed and released by Open Stax College, Connexions, reprint permission granted under the Creative Commons Attribution 3.0 Unported license
10. Figure 10: Graphic designed by Patrick J. Lynch, medical illustrator, 2006. Reprint permission granted under Creative Commons Attribution 2.5 License. Additional graphics used is designed and released by Open Stax College, Connexions, reprint permission granted under the Creative Commons Attribution 3.0 Unported license
11. Figure 11: Graphic designed by Patrick J. Lynch, medical illustrator, 2006. Reprint permission granted under Creative Commons Attribution 2.5 License. Additional graphics used is designed and released by Open Stax College, Connexions, reprint permission granted under the Creative Commons Attribution 3.0 Unported license
12. Figure 12: Graphic designed by Patrick J. Lynch, medical illustrator, 2006. Reprint permission granted under Creative Commons Attribution 2.5 License. Additional graphics used is designed and released by Open Stax College, Connexions, reprint permission granted under the Creative Commons Attribution 3.0 Unported license
13. Figure 13: Graphic designed and released by Open Stax College, Connexions, reprint permission granted under the Creative Commons Attribution 3.0 Unported license
14. Figure 14: Graphic designed and released by Open Stax College, Connexions, reprint permission granted under the Creative Commons Attribution 3.0 Unported license
15. Figure 15: Graphic designed and released by Open Stax College, Connexions, reprint permission granted under the Creative Commons Attribution 3.0 Unported license
16. Figure 16: Graphic designed and released by Open Stax College, Connexions, reprint permission granted under the Creative Commons Attribution 3.0 Unported license
17. Figure 17: Graphic designed and released by Open Stax College, Connexions, reprint permission granted under the Creative Commons Attribution 3.0 Unported license
18. Figure 18: Graphic designed and released by Open Stax College, Connexions, reprint permission granted under the Creative Commons Attribution 3.0 Unported license
19. Figure 19: Graphic designed and released by Open Stax College, Connexions, reprint permission granted under the Creative Commons Attribution 3.0 Unported license
20. Figure 20: Graphic designed by Jordi March i Nogué Reprint permission granted under Creative Commons Attribution 3.0 Unported license
21. Figure 21: Graphic designed and released by Open Stax College, Connexions, reprint permission granted under the Creative Commons Attribution 3.0 Unported license
22. Figure 22: Graphic designed and released by Open Stax College, Connexions, reprint permission granted under the Creative Commons Attribution 3.0 Unported license
23. Figure 23: Graphic designed and released by Open Stax College, Connexions, reprint permission granted under the Creative Commons Attribution 3.0 Unported license
24. Figure 24: Graphic designed by P.Y.P. Jen. Copyright ©
25. Figure 25: Graphic designed by P.Y.P. Jen. Copyright ©
26. Figure 26: Graphic designed and released by Open Stax College, Connexions, reprint permission granted under the Creative Commons Attribution 3.0 Unported license
27. Figure 27: Graphic designed and released by Open Stax College, Connexions, reprint permission granted under the Creative Commons Attribution 3.0 Unported license
28. Figure 28: Graphic designed by P.Y.P. Jen. Copyright ©
29. Figure 29: Graphic designed by P.Y.P. Jen. Copyright ©
30. Figure 30: Graphic designed and released by Open Stax College, Connexions, reprint permission granted under the Creative Commons Attribution 3.0 Unported license
31. Figure 31: Graphic designed and released by Open Stax College, Connexions, reprint permission granted under the Creative Commons Attribution 3.0 Unported license
32. Figure 32: Graphic designed by Patrick J. Lynch, medical illustrator. Reprint under the permission of Creative Commons Attribution 2.5 License 2006
33. Figure 33: Graphic designed by Zedh. Reprint under the permission of Creative Commons Attribution 2.5 License 2007
34. Figure 34: Graphic designed by Terese Winslow 2007. Public domain graphics
35. Figure 35: Graphic designed by Terese Winslow 2007. Public domain graphics. Additional graphics contributed by P.Y.P. Jen. Copyright ©
36. Figure 36: Graphic designed by Hic et nunc, reprint permission granted under the Creative Commons Attribution 3.0 Unported license
37. Figure 37: Graphic designed and released by Open Stax College, Connexions, reprint permission granted under the Creative Commons Attribution 3.0 Unported license
38. Figure 38: Graphic designed and released by Open Stax College, Connexions, reprint permission granted under the Creative Commons Attribution 3.0 Unported license
39. Figure 39: Graphic designed and released by Open Stax College, Connexions, reprint permission granted under the Creative Commons Attribution 3.0 Unported license
40. Figure 40: Graphic designed by Gumenyuk I.S., reprint permission granted under the Creative Commons Attribution 3.0 Unported license
41. Figure 41: Graphic designed and released by Open Stax College, Connexions, reprint permission granted under the Creative Commons Attribution 3.0 Unported license
42. Figure 42: Graphic designed and released by Open Stax College, Connexions, reprint permission granted under the Creative Commons Attribution 3.0 Unported license
43. Figure 43: Graphic designed by P.Y.P. Jen. Copyright ©
44. Figure 44: Graphic designed and released by Open Stax College, Connexions, reprint permission granted under the Creative Commons Attribution 3.0 Unported license
45. Figure 45: Photograph taken and released by Department of Health and Human Services, The National Institutes of Health. Public domain photo
46. Figure 46: Graphic designed by P.Y.P. Jen. Copyright ©
47. Figure 47: Graphic designed and released by Open Stax College, Connexions, reprint permission granted under the Creative Commons Attribution 3.0 Unported license
48. Figure 48: Graphic designed by P.Y.P. Jen. Copyright ©
49. Figure 49: Graphic designed and released by Open Stax College, Connexions, reprint permission granted under the Creative Commons Attribution 3.0 Unported license

Phillip Yuan Pei Jen received his doctorate in Human Anatomy and Physiology from The Chinese University of Hong Kong in 1995. Presently, Dr. Jen is an Associate Professor of Biology at Gordon State College located in Georgia.

Amanda Duffus received her doctorate in Biological Sciences from Queen Mary University of London in 2010. Presently, Dr. Duffus is an Assistant Professor of Biology at Gordon State College located in Georgia.